# Tire and Vehicle Dynamics

# Tire and Vehicle Dynamics

Third edition

**Hans B. Pacejka**
Delft University of Technology
The Netherlands

Contributing author
**Igo Besselink**
Eindhoven University of Technology
(Formerly TNO-Automotive)

AMSTERDAM • BOSTON • HEIDELBERG • LONDON
NEW YORK • OXFORD • PARIS • SAN DIEGO
SAN FRANCISCO • SINGAPORE • SYDNEY • TOKYO

Butterworth-Heinemann is an imprint of Elsevier

Butterworth-Heinemann is an imprint of Elsevier
The Boulevard, Langford Lane, Oxford OX5 1GB, UK
225 Wyman Street, Waltham, MA 02451, USA

First edition 2002
Second edition 2006
Third edition 2012



**British Library Cataloguing-in-Publication Data**
A catalogue record for this book is available from the British Library

**Library of Congress Cataloging-in-Publication Data**
A catalog record for this book is available from the Library of Congress

ISBN: 978-0-08-097016-5

For information on all Butterworth-Heinemann publications
visit our website at elsevierdirect.com

Printed in Great Britain

12 13 14 15 10 9 8 7 6 5 4 3 2 1



*Note from the publisher: This edition published by Elsevier and distributed in conjunction with SAE International. The SAE International logo is used with permission from SAE International.*

# Contents

# Exercises



We may also refer to the online information, cf. App. 2, containing MATLAB applications.

# Preface

The operational properties of the road vehicle are the result of the dynamic interaction of the various components of the vehicle structure, possibly including modern control elements. A major role is played by the pneumatic tire.

"The complexity of the structure and behavior of the tire are such that no complete and satisfactory theory has yet been propounded. The characteristics of the tire still presents a challenge to the natural philosopher to devise a theory which shall coordinate the vast mass of empirical data and give some guidance to the manufacturer and user. This is an inviting field for the application of mathematics to the physical world".

In this way, Temple formulated his view on the situation almost 50 years ago (Endeavor, October 1956). Since that time, in numerous institutes and laboratories, the work of the early investigators has been continued. Considerable progress in the development of the theory of tire mechanics has been made during the past decades. This has led to better understanding of tire behavior and its role as a vehicle component. Thanks to new and more refined experimental techniques and to the introduction of the electronic computer, the goal of formulating and using more realistic mathematical models of the tire in a wide range of operational conditions has been achieved.

From the point of view of the vehicle dynamicist, the mechanical behavior of the tire needs to be investigated systematically in terms of its reaction to various inputs associated with wheel motions and road conditions. It is convenient to distinguish between symmetric and anti-symmetric (in-plane and out-of-plane) modes of operation. In the first type of mode, the tire supports the load and cushions the vehicle against road irregularities while longitudinal driving or braking forces are transmitted from the road to the wheel. In the second mode of operation, the tire generates lateral, cornering, or camber forces to provide the necessary directional control of the vehicle. In more complex situations, e.g. braking in a turn, combinations of these pure modes of operation occur. Moreover, one may distinguish between steady-state performance and transient or oscillatory behavior of the rolling tire. The contents of the book have been subdivided according to these categories. The development of theoretical models has always been substantiated through experimental evidence.

Possibly one of the more difficult aspects of tire dynamic behavior to describe mathematically is the generation of forces and moments when the tire rolls over rough roads with short obstacles while being braked and steered in a time-varying fashion. In the book, tire modeling is discussed while gradually increasing its complexity, thereby allowing the modeling range of operation to become wider in terms of slip intensity, wavelength of wheel motion, and frequency. Formulas based on empirical observations and relatively simple approximate physical models have been used to describe tire mechanical behavior. Rolling over obstacles has been modeled by making use of effective road inputs. This approach forms a contrast to the derivation of complex models, which are based on more or less refined physical descriptions of the tire.

Throughout the book, the influence of tire mechanical properties on vehicle dynamic behavior has been discussed. For example, handling diagrams are introduced for both cars and motorcycles to clearly illustrate and explain the role of the tire non-linear steady-state side force characteristics in achieving certain understeer and oversteer handling characteristics of the vehicle. The wheel shimmy phenomenon is discussed in detail in connection with the non-steady-state description of the out-of-plane behavior of the tire and the deterioration of ABS braking performance when running over uneven roads is examined with the use of an in-plane tire dynamic model. The complete scope of the book may be judged best from the table of contents.

The material covered in the book represents a field of automotive engineering practice that is attractive to the student to deepen his or her experience in the application of basic mechanical engineering knowledge. For that purpose, a number of problems have been added. These exercises have been listed at the end of the table of contents.

Much of the work described in this book has been carried out at the Vehicle Research Laboratory of the Delft University of Technology, Delft, the Netherlands. This laboratory was established in the late 1950s through the efforts of professor Van Eldik Thieme. With its unique testing facilities realistic tire steady-state (over the road), transient, and obstacle traversing (on flat plank) and dynamic (on rotating drum) characteristics could be assessed. I wish to express my appreciation to the staff of this laboratory and to the Ph.D. students who have given their valuable efforts to develop further knowledge in tire mechanics and its application in vehicle dynamics. The collaboration with TNO Automotive (Delft) in the field of tire research opened the way to produce professional software and render services to the automotive and tire industry, especially for the *Delft-Tire* product range that includes the *Magic Formula* and *SWIFT* models described in Chapters 4, 9, and 10. I am indebted to the Vehicle Dynamics group for their much appreciated help in the preparation of the book.

Professors Peter Lugner (Vienna University of Technology) and Robin Sharp (Cranfield University) have carefully reviewed major parts of the book (Chapters 1–6 and Chapter 11, respectively). Igo Besselink and Sven Jansen of TNO Automotive reviewed the Chapters 5–10. I am most grateful for their valuable suggestions to correct and improve the text. Finally, I thank the editorial and production staff of Butterworth-Heinemann for their assistance and cooperation.

Hans B. Pacejka
Rotterdam,
May, 2002

## NOTE ON THE SECOND EDITION

In this new edition, many small and larger corrections and improvements have been introduced. Recent developments on tire modeling have been added. These concern mainly camber dynamics (Chapter 7) and running over three-dimensional uneven road surfaces (Chapter 10). Section 10.2 has been added to outline the structure of three advanced dynamic tire models that are important for detailed computer simulation studies of vehicle dynamic performance. In the new Chapter 12, an overview has been given of tire testing facilities that are designed to measure tire steady-state characteristics both in the laboratory and over the road, and to investigate the dynamic performance of the tire subjected to wheel vibrations and road unevennesses.

Hans B. Pacejka
Rotterdam,
September, 2005

## NOTE ON THE THIRD EDITION

In this new edition, again many improvements have been introduced. Some chapters have been reorganized, notably Chapter 10, and a new Chapter 13 has been introduced outlining three advanced dynamic tire models. We express our thanks to the two guest authors Michael Gipser and Christian Oertel (the original model developers) for their contributions in this chapter. Igo Besselink has contributed to the preparation of the new edition. He wrote his part of Chapter 13 and the extended Appendices. He prepares and maintains information online, notably on dynamic vehicle measurements, see website: http://www.elsevierdirect.com/companion.jsp?ISBN=9780080970165.

We much appreciate and are grateful for the help provided by TNO Automotive, Helmond, the Netherlands, especially for making available motorcycle tire measurement data obtained with the new test facility that can handle large camber angles. An image of the device appears on the front cover of the book. This TNO facility is shown mounted on the old Delft Tire Test Trailer but is now installed in the new TNO Tire Test Semi-Trailer, cf. Chapter 12. We also thank TNO for allowing us to use their vehicle measurement data. We are grateful to Antoine Schmeitz, who checked and made important remarks, notably on Chapter 4 and on the revised Chapter 10.

We thank Manfred Plöchl of the Vienna Technical University for carefully checking Chapters 1 and 11. Bill Milliken (now 101 years old! See his wonderful engineering autobiography (2006)) responded very competently on a question of mine regarding the origin of the *Similarity Method*. We are also grateful to readers who sent – often small – remarks that have served to improve the book.

In the book, vehicle dynamic problems have been addressed and a number of tire models have been discussed. Applications of these tire models serve to illustrate their use and the influence of relevant aspects in vehicle dynamic behavior. Two tire models are not in the first place meant to be used seriously in applications. They have been discussed for providing insight and for studying the main typical aspects of tire force and moment generation in the steady state (the *Brush Model*, Chapter 3) and in the transient state (the *String Model*, Chapter 5). In Chapter 1, the basic form of the *Magic Formula* has been introduced. This empirical model is well suited to be applied in vehicle dynamics studies. For high standard applications requiring great accuracy, the full model discussed in Chapter 4 may be employed. For relatively low frequency and larger wavelength phenomena, the transient tire model featuring the relaxation length, developed in Chapter 7 and applied in Chapter 8, is often sufficiently accurate. In case one is interested in higher frequency and shorter wavelength responses of the tire with the first natural frequencies included, the advanced dynamic tire model developed in Chapter 9 is recommended. This so-called *SWIFT* model works in conjunction with the *Magic Formula*. Rolling over uneven roads including short obstacles can be handled by using the special geometric filtering technique treated in Chapter 10.

Hans B. Pacejka
Rotterdam,
August, 2011

Chapter 1

# Tire Characteristics and Vehicle Handling and Stability

**Chapter Outline**



**Tire and Vehicle Dynamics. DOI: 10.1016/B978-0-08-097016-5.00001-2**

## 1.1. INTRODUCTION

This chapter is meant to serve as an introduction to vehicle dynamics with emphasis on the influence of tire properties. Steady-state cornering behavior of simple automobile models and the transient motion after small and large steering inputs and other disturbances will be discussed. The effects of various shape factors of tire characteristics (cf. Figure 1.1) on vehicle handling properties will be analyzed. The slope of the side force $F_y$ vs slip angle $\alpha$ near the origin (the cornering or side slip stiffness) is the determining parameter for the basic linear handling and stability behavior of automobiles. The possible offset of the tire characteristics with respect to their origins may be responsible for the occurrence of the so-called tire-pull phenomenon. The further nonlinear shape of the side (or cornering) force characteristic governs the handling and stability properties of the vehicle at higher lateral accelerations. The load dependency of the curves, notably the nonlinear relationship of cornering stiffness with tire normal load, has a considerable effect on the handling characteristic of the car. For the (quasi)-steady-state handling analysis, simple single track (two-wheel) vehicle models will be used. Front and rear axle effective side force characteristics are introduced to represent effects that result from suspension and steering system design factors such as steering compliance, roll steer, and lateral load transfer. Also, the effect of possibly applied (moderate) braking and driving forces may be incorporated in the effective characteristics. Large braking forces may result in wheel lock and possibly large deviations from the undisturbed path. The motion resulting from wheel lock will be dealt with in an application of the theory of a simple physical tire model in Chapter 3 (the brush model). The application of the handling and stability theory to the dynamics of heavy trucks will also be briefly dealt with in this chapter. Special attention will be given to the phenomenon of oscillatory instability that may show up with the car-trailer combination.

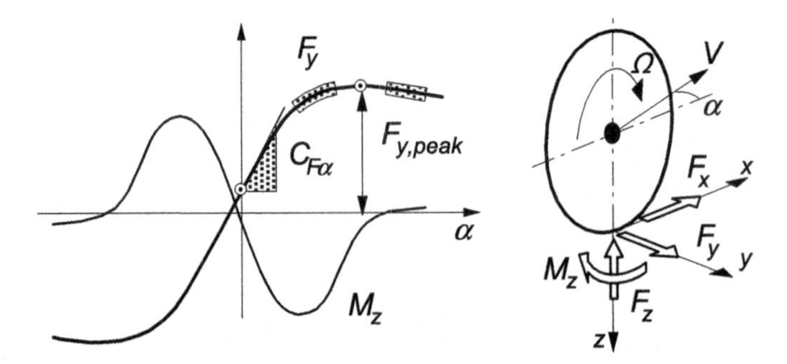


**FIGURE 1.1** Characteristic shape factors (indicated by points and shaded areas) of tire or axle characteristics that may influence vehicle handling and stability properties. Slip angle and force and moment positive directions, cf. App. 1.

When the wavelength of an oscillatory motion of the vehicle that may arise from road unevenness, brake torque fluctuations, wheel unbalance, or instability (shimmy) is smaller than say 5 m, a non-steady-state or transient description of tire response is needed to properly analyze the phenomenon. In Chapters 5–8, these matters will be addressed. Applications demonstrate the use of transient and oscillatory tire models and provide insight into the vehicle dynamics involved. Chapter 11 is especially devoted to the analysis of motorcycle cornering behavior and stability.

## 1.2. TIRE AND AXLE CHARACTERISTICS

Tire characteristics are of crucial importance for the dynamic behavior of the road vehicle. In this section, an introduction is given to the basic aspects of the force- and moment-generating properties of the pneumatic tire. Both the pure and combined slip characteristics of the tire are discussed and typical features presented. Finally, the so-called effective axle characteristics are derived from the individual tire characteristics and the relevant properties of the suspension and steering system.

### 1.2.1. Introduction to Tire Characteristics

The upright wheel rolling freely, that is without applying a driving torque, over a flat level road surface along a straight line at zero side slip, may be defined as the starting situation with all components of slip equal to zero. A relatively small pulling force is needed to overcome the tire-rolling resistance, and a side force and (self)-aligning torque may occur as a result of the not completely symmetric structure of the tire. When the wheel moves in a way that the condition of zero slip is no longer fulfilled, wheel slip occurs that is accompanied by a buildup of additional tire deformation and possibly partial sliding in the contact patch. As a result, (additional) horizontal forces and the aligning torque are generated. The mechanism responsible for this is treated in detail in the subsequent chapters. For now, we will suffice with some important experimental observations and define the various slip quantities that serve as inputs into the tire system and the moment and forces that are the output quantities (positive directions according to Figure 1.1). Several alternative definitions are in use as well. In Appendix 1, various sign conventions of slip, camber, and output forces and moments together with relevant characteristics have been presented.

For the freely rolling wheel, the forward speed $V_x$ (longitudinal component of the total velocity vector $V$ of the wheel center) and the angular speed of revolution $\Omega_o$ can be taken from measurements. By dividing these two quantities, the so-called effective rolling radius $r_e$ is obtained:

$$r_e = \frac{V_x}{\Omega_o} \tag{1.1}$$

Although the effective radius may be defined also for a braked or driven wheel, we restrict the definition to the case of free rolling. When a torque is applied about the wheel spin axis, a longitudinal slip arises that is defined as follows:

$$\kappa = -\frac{V_x - r_e\Omega}{V_x} = -\frac{\Omega_o - \Omega}{\Omega_o} \tag{1.2}$$

The sign is taken such that for a positive $\kappa$, a positive longitudinal force $F_x$ arises, that is, a driving force. In that case, the wheel angular velocity $\Omega$ is increased with respect to $\Omega_o$ and consequently $\Omega > \Omega_o = V_x/r_e$. During braking, the fore-and-aft slip becomes negative. At wheel lock, obviously, $\kappa = -1$. At driving on slippery roads, $\kappa$ may attain very large values. To limit the longitudinal slip $K$ to a maximum equal to one, in some texts the longitudinal slip is defined differently in the driving range of slip: in the denominator of (1.2), $\Omega_o$ is replaced by $\Omega$. This will not be done in the present text.

Lateral wheel slip is defined as the ratio of the lateral and the forward velocity of the wheel. This corresponds to minus the tangent of the slip angle $\alpha$ (Figure 1.1). Again, the sign of $\alpha$ has been chosen such that the side force becomes positive at positive slip angle.

$$\tan\alpha = -\frac{V_y}{V_x} \tag{1.3}$$

The third and last slip quantity is the so-called spin which is due to rotation of the wheel about an axis normal to the road. Both the yaw rate resulting in path curvature when $\alpha$ remains zero and the wheel camber or inclination angle $\gamma$ of the wheel plane about the $x$ axis contribute to the spin. The camber angle is defined to be positive when, if looking from behind the wheel, the wheel is tilted to the right. In Chapter 2, more precise definitions of the three components of wheel slip will be given. The forces $F_x$ and $F_y$ and the aligning torque $M_z$ are results of the input slip. They are functions of the slip components and the wheel load. For steady-state rectilinear motions, we have, in general,

$$F_x = F_x(\kappa,\alpha,\gamma,F_z), \quad F_y = F_y(\kappa,\alpha,\gamma,F_z), \quad M_z = M_z(\kappa,\alpha,\gamma,F_z) \tag{1.4}$$

The vertical load $F_z$ may be considered as a given quantity that results from the normal deflection of the tire. The functions can be obtained from measurements for a given speed of travel and road and environmental conditions.

Figure 1.1 shows the adopted system of axes ($x$, $y$, $z$) with associated positive directions of velocities and forces and moments. The exception is the vertical force $F_z$ acting from road to tire. For practical reasons, this force is defined to be positive in the upward direction and thus equal to the normal load of the tire. Also, $\Omega$ (not provided with a $y$ subscript) is defined positive with respect to the negative $y$-axis. Note that the axes system is in accordance with SAE standards (SAE J670e 1976). The sign of the slip angle, however, is chosen opposite with respect to the SAE definition, cf. Appendix 1.

In Figure 1.2 typical pure lateral ($\kappa = 0$) and longitudinal ($\alpha = 0$) slip characteristics have been depicted together with a number of combined slip curves. The camber angle $\gamma$ was kept equal to zero. We define pure slip to be the situation when either longitudinal or lateral slip occurs in isolation. The figure indicates that a drop in force arises when the other slip component is added. The resulting situation is designated as combined slip. The decrease in force can be simply explained by realizing that the total horizontal frictional force $F$ cannot exceed the maximum value (radius of 'friction circle') which is dictated by the current friction coefficient and normal load. Later, in Chapter 3 this becomes clear when considering the behavior of a simple physical tire model. The diagrams include the situation when the brake slip ratio has finally attained the value of 100% ($\kappa = -1$) which corresponds to wheel lock.

The slopes of the pure slip curves at vanishing slip are defined as the longitudinal and lateral slip stiffnesses, respectively. The longitudinal slip stiffness is designated as $C_{F\kappa}$. The lateral slip or cornering stiffness of the tire, denoted by $C_{F\alpha}$, is one of the most important property parameters of the tire and is crucial for the vehicle's handling and stability performance. The slope of minus the aligning torque versus slip angle curve (Figure 1.1) at zero slip angle is termed as the aligning stiffness and is denoted by $C_{M\alpha}$. The ratio of minus the aligning torque and the side force is the pneumatic trail $t$ (if we neglect the so-called residual torque to be dealt with in Chapter 4). This length is the distance behind the contact center (projection of wheel center onto the ground in wheel plane direction) to the point where the resulting lateral force acts. The linearized force and moment characteristics (valid at small levels of slip) can be

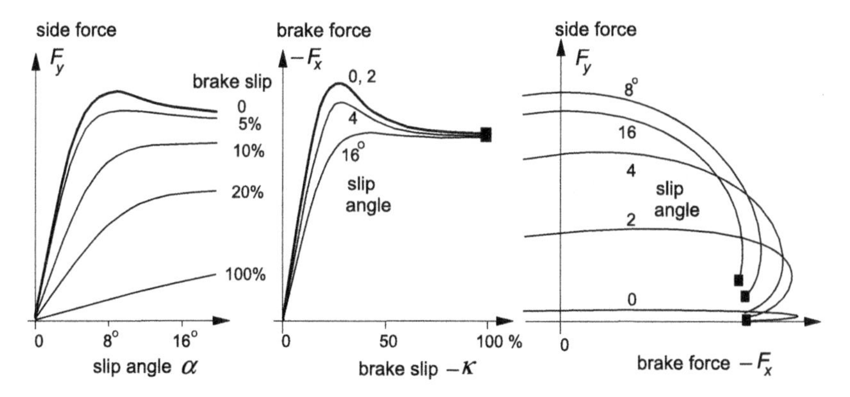


FIGURE 1.2 Combined side force and brake force characteristics.

represented by the following expressions in which the effect of camber has been included:

$$\begin{aligned} F_x &= C_{F\kappa}\,\kappa \\ F_y &= C_{F\alpha}\,\alpha + C_{F\gamma}\gamma \\ M_z &= -C_{M\alpha}\,\alpha + C_{M\gamma}\gamma \end{aligned} \tag{1.5}$$

These equations have been arranged in such a way that all the coefficients (the force, moment slip, and camber stiffnesses) become positive quantities.

It is of interest to note that the order of magnitude of the tire cornering stiffness ranges from about 6 to about 30 times the vertical wheel load when the cornering stiffness is expressed as force per radian. The lower value holds for the older bias-ply tire construction and the larger value for modern racing tires. The longitudinal slip stiffness has been typically found to be about 50% larger than the cornering stiffness. The pneumatic trail is approximately equal to a quarter of the contact patch length. The dry friction coefficient usually equals ca. 0.9 on very sharp surfaces and ca. 1.6 on clean glass; racing tires may reach 1.5–2.

For the side force which is the more important quantity in connection with automobile handling properties, a number of interesting diagrams have been presented in Figure 1.3. These characteristics are typical for truck and car tires and are based on experiments conducted at the University of Michigan Transportation Research Institute (UMTRI, formerly HSRI), cf. Ref. (Segel et al. 1981). The car tire cornering stiffness data stem from newer findings. It is observed that the cornering stiffness changes in a less than proportional fashion with the normal wheel load. The maximum normalized side force $F_{y,\text{peak}}/F_z$ appears to decrease with increasing wheel load. Marked differences in level and slope occur for the car and truck tire curves also when normalized with respect to the rated or nominal load. The cornering force vs slip angle characteristic shown at different speeds and road conditions indicates that the slope at zero

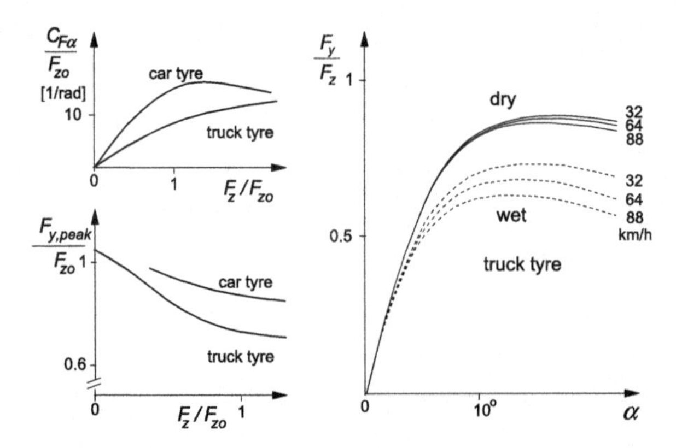


**FIGURE 1.3** Typical characteristics for the normalized cornering stiffness, peak side force, and side force vs normalized vertical load and slip angle respectively. $F_{zo}$ is the rated load.

slip angle is not or hardly affected by the level of speed and by the wet or dry condition. The peak force level shows only little variation if the road is dry. On a wet road, a more pronounced peak occurs and the peak level drops significantly with increasing speed.

Curves which exhibit a shape like the side force characteristics of Figure 1.3 can be represented by a mathematical formula called *'magic formula'*. A full treatment of the empirical tire model associated with this formula is given in Chapter 4. For now we can suffice with showing the basic expressions for the side force and the cornering stiffness:

$$\begin{aligned} &F_y = D\sin[C\arctan\{B\alpha - E(B\alpha - \arctan(B\alpha))\}] \\ &\text{with stiffness factor} \\ &B = C_{F\alpha}/(CD) \\ &\text{peak factor} \\ &D = \mu F_z (= F_{y,\text{peak}}) \\ &\text{and cornering stiffness} \\ &C_{F\alpha}(= BCD) = c_1 \sin\{2\arctan(F_z/c_2)\} \end{aligned} \tag{1.6}$$

The shape factors $C$ and $E$ as well as the parameters $c_1$ and $c_2$ and the friction coefficient $\mu$ (possibly depending on the vertical load and speed) may be estimated or determined through regression techniques.

### 1.2.2. Effective Axle Cornering Characteristics

For the basic analysis of (quasi)-steady-state turning behavior, a simple two-wheel vehicle model may be used successfully. Effects of suspension and steering system kinematics and compliances such as steer compliance, body roll, and also load transfer may be taken into account by using effective axle characteristics. The restriction to (quasi)-steady-state becomes clear when we realize that for transient or oscillatory motions, exhibiting yaw and roll accelerations and differences in phase, variables like roll angle and load transfer can no longer be written as direct algebraic functions of one of the lateral axle forces (front or rear). Consequently, we should drop the simple method of incorporating the effects of a finite center of gravity height if the frequency of input signals such as the steering wheel angle cannot be considered small relative to the body roll natural frequency. Since the natural frequency of the wheel suspension and steering systems is relatively high, the restriction to steady-state motions becomes less critical in the case of the inclusion of e.g., steering compliance in the effective characteristic. Chiesa and Rinonapoli (1967) were among the first to employ effective axle characteristics or 'working curves' as these were referred to by them. Vågstedt (1995) determined these curves experimentally.

Before assessing the complete nonlinear effective axle characteristics, we will first direct our attention to the derivation of the effective cornering stiffnesses which are used in the simple linear two-wheel model. For these to be determined, a more comprehensive vehicle model has to be defined.

Figure 1.4 depicts a vehicle model with three degrees of freedom. The forward velocity $u$ may be kept constant. As motion variables, we define the lateral velocity $v$ of reference point $A$, the yaw velocity $r$, and the roll angle $\varphi$. A moving axes system ($A$, $x$, $y$, $z$) has been introduced. The $x$-axis points forwards and lies both in the ground plane and in the plane normal to the ground that passes through the so-called roll axis. The $y$-axis points to the right and the $z$-axis points downward. This latter axis passes through the center of gravity when the roll angle is equal to zero. In this way, the location of the point of reference $A$ has been defined. The longitudinal distance to the front axle is $a$ and the distance to the rear axle is $b$. The sum of the two distances is the wheel base $l$. For convenience, we may write $a = a_1$ and $b = a_2$.

In a curve, the vehicle body rolls about the roll axis. The location and attitude of this virtual axis are defined by the heights $h_{1,2}$ of the front and rear roll centers. The roll axis is assessed by considering the body motion with respect to the four contact centers of the wheels on the ground under the action of an external lateral force that acts on the center of gravity. Due to the symmetry of the vehicle configuration and the linearization of the model, these locations can be considered as fixed. The roll center locations are governed by suspension kinematics and possibly suspension lateral compliances. The torsional springs depicted in the figure represent the front and rear roll stiffnesses $c_{\varphi 1,2}$ which result from suspension springs and antiroll bars.

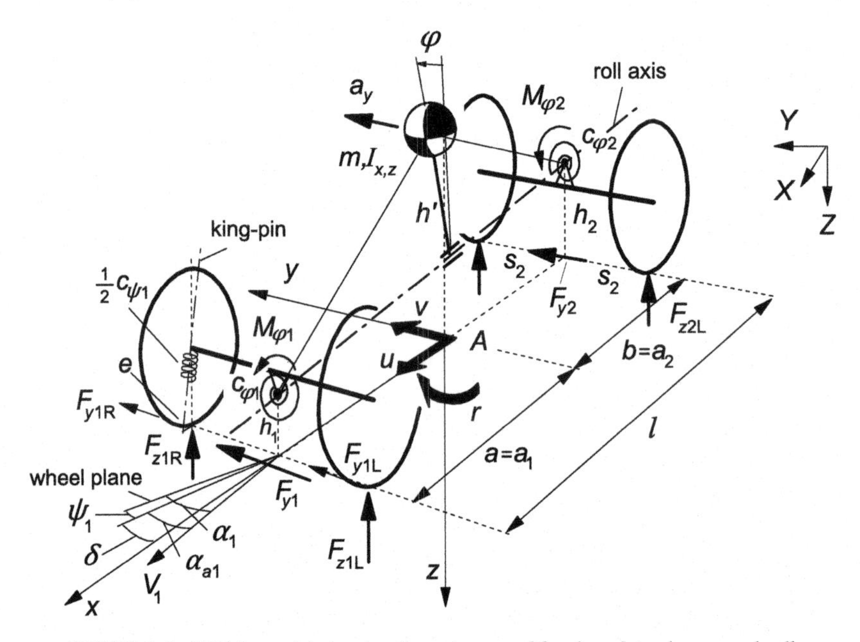


**FIGURE 1.4** Vehicle model showing three degrees of freedom: lateral, yaw, and roll.

The fore-and-aft position of the center of gravity of the body is defined by $a$ and $b$; its height follows from the distance $h'$ to the roll axis. The body mass is denoted by $m$ and the moments of inertia with respect to the center of mass and horizontal and vertical axes by $I_x$, $I_z$, and $I_{xz}$. These latter quantities will be needed in a later phase when the differential equations of motion are established. The unsprung masses will be neglected or they may be included as point masses attached to the roll axis and thus make them part of the sprung mass, that is, the vehicle body.

Furthermore, the model features torsional springs around the steering axes. The king-pin is positioned at a small caster angle that gives rise to the caster length $e$ as indicated in the drawing. The total steering torsional stiffness, left plus right, is denoted by $c_{\psi 1}$.

## *Effective Axle Cornering Stiffness*

Linear analysis, valid for relatively small levels of lateral accelerations, allows the use of approximate tire characteristics represented by just the slopes at zero slip. We will first derive the effective axle cornering stiffness that may be used under these conditions. The effects of load transfer, body roll, steer compliance, side force steer, and initial camber and toe angles will be included in the ultimate expression for the effective axle cornering stiffness.

The linear expressions for the side force and the aligning torque acting on a tire have been given by Eqns (1.5). The coefficients appearing in these expressions are functions of the vertical load. For small variations with respect to the average value (designated with subscript $o$), we write for the cornering and camber force stiffnesses the linearized expressions:

$$\begin{aligned} C_{F\alpha} &= C_{F\alpha o} + \zeta_\alpha \Delta F_z \\ C_{F\gamma} &= C_{F\gamma o} + \zeta_\gamma \Delta F_z \end{aligned} \tag{1.7}$$

where the increment of the wheel vertical load is denoted by $\Delta F_z$ and the slopes of the coefficient versus load curves at $F_z = F_{zo}$ are represented by $\zeta_{\alpha,\gamma}$.

When the vehicle moves steadily around a circular path, a centripetal acceleration $a_y$ occurs and a centrifugal force $K = ma_y$ can be said to act on the vehicle body at the center of gravity in the opposite direction. The body roll angle $\varphi$ that is assumed to be small is calculated by dividing the moment about the roll axis by the apparent roll stiffness which is reduced with the term $mgh'$ due to the additional moment $mgh'\varphi$:

$$\varphi = \frac{-ma_y h'}{c_{\varphi 1} + c_{\varphi 2} - mgh'} \tag{1.8}$$

The total moment about the roll axis is distributed over the front and rear axles in proportion to the front and rear roll stiffnesses. The load transfer $\Delta F_{zi}$

from the inner to the outer wheels that occurs at axle $i$ (= 1 or 2) in a steady-state cornering motion with centripetal acceleration $a_y$ follows from the formula:

$$\Delta F_{zi} = \sigma_i m a_y \tag{1.9}$$

with the load transfer coefficient of axle $i$:

$$\sigma_i = \frac{1}{2s_i}\left(\frac{c_{\varphi i}}{c_{\varphi 1} + c_{\varphi 2} - mgh'}h' + \frac{l - a_i}{l}h'\right) \tag{1.10}$$

The attitude angle of the roll axis with respect to the horizontal is considered small. In the formula, $s_i$ denotes half the track width, $h'$ is the distance from the center of gravity to the roll axis, and $a_1 = a$ and $a_2 = b$. The resulting vertical loads at axle $i$ for the left ($L$) and right ($R$) wheels become, after considering the left and right increments in load,

$$\begin{aligned} \Delta F_{ziL} &= \Delta F_{zi}, & \Delta F_{ziR} &= -\Delta F_{zi} \\ F_{ziL} &= \frac{1}{2}F_{zi} + \Delta F_{zi}, & F_{ziR} &= \frac{1}{2}F_{zi} - \Delta F_{zi} \end{aligned} \tag{1.11}$$

The wheels at the front axle are steered about the king-pins with the angle $\delta$. This angle relates directly to the imposed steering wheel angle $\delta_{\rm stw}$ through the steering ratio $n_{\rm st}$:

$$\delta = \frac{\delta_{\rm stw}}{n_{\rm st}} \tag{1.12}$$

In addition to this imposed steer angle, the wheels may show a steer angle and a camber angle induced by body roll through suspension kinematics. The functional relationships with the roll angle may be linearized. For axle $i$, we define

$$\begin{aligned} \psi_{ri} &= \varepsilon_i \varphi \\ \gamma_{ri} &= \tau_i \varphi \end{aligned} \tag{1.13}$$

Steer compliance gives rise to an additional steer angle due to the external torque that acts about the king-pin (steering axis). For the pair of front wheels, this torque results from the side force (and, of course, also from here not considered driving or braking forces) that exerts a moment about the king-pin through the moment arm which is composed of the caster length $e$ and the pneumatic trail $t_1$. With the total steering stiffness $c_{\psi 1}$ felt about the king-pins with the steering wheel held fixed, the additional steer angle becomes when for simplicity the influence of camber on the pneumatic trail is disregarded:

$$\psi_{c1} = -\frac{F_{y1}(e + t_1)}{c_\psi} \tag{1.14}$$

In addition, the side force (but also the fore-and-aft force) may induce a steer angle due to suspension compliance. The so-called side force steer reads

$$\psi_{sfi} = c_{sfi} F_{yi} \tag{1.15}$$

For the front axle, we should separate the influences of moment steer and side force steer. For this reason, side force steer at the front is defined to occur as a result of the side force acting in a point on the king-pin axis.

Besides the wheel angles indicated above, the wheels may have been given initial angles that already exist at straight ahead running. These are the toe angle $\psi_o$ (positive pointing outward) and the initial camber angle $\gamma_o$ (positive: leaning outward). For the left and right wheels, we have the initial angles:

$$\begin{aligned} \psi_{iLo} &= -\psi_{io}, \quad \psi_{iRo} = \psi_{io} \\ \gamma_{iLo} &= -\gamma_{io}, \quad \gamma_{iRo} = \gamma_{io} \end{aligned} \tag{1.16}$$

Adding all relevant contributions (1.12) to (1.16) together yields the total steer angle for each of the wheels.

The effective cornering stiffness of an axle $C_{\text{eff},i}$ is now defined as the ratio of the axle side force and the virtual slip angle. This angle is defined as the angle between the direction of motion of the center of the axle $i$ (actually at road level) when the vehicle velocity would be very low and approaches zero (then also $F_{yi} \to 0$) and the direction of motion at the actual speed considered. The virtual slip angle of the front axle has been indicated in Figure 1.4 and is designated as $\alpha_{a1}$. We have, in general,

$$C_{\text{eff},i} = \frac{F_{yi}}{a_{ai}} \tag{1.17}$$

The axle side forces in the steady-state turn can be derived by considering the lateral force and moment equilibrium of the vehicle:

$$F_{yi} = \frac{l - a_i}{l} m a_y \tag{1.18}$$

The axle side force is the sum of the left and right individual tire side forces. We have

$$\begin{aligned} F_{yiL} &= \left(1/_2 C_{F\alpha i} + \zeta_{\alpha i} \Delta F_{zi}\right)(\alpha_i - \psi_{io}) + \left(1/_2 C_{F\gamma i} + \zeta_{\gamma i} \Delta F_{zi}\right)(\gamma_i - \gamma_{io}) \\ F_{yiR} &= \left(1/_2 C_{F\alpha i} - \zeta_{\alpha i} \Delta F_{zi}\right)(\alpha_i + \psi_{io}) + \left(1/_2 C_{F\gamma i} - \zeta_{\gamma i} \Delta F_{zi}\right)(\gamma_i + \gamma_{io}) \end{aligned} \tag{1.19}$$

where the average wheel slip angle $\alpha_i$ indicated in the figure is

$$\alpha_i = \alpha_{ai} + \psi_i \tag{1.20}$$

and the average additional steer angle and the average camber angle are

$$\begin{aligned} \psi_i &= \psi_{ri} + \psi_{ci} + \psi_{cfi} \\ \gamma_i &= \gamma_{ri} \end{aligned} \tag{1.21}$$

The unknown quantity is the virtual slip angle $\alpha_{ai}$ which can be determined for a given lateral acceleration $a_y$. Next, we use Eqns (1.8, 1.9, 1.13, 1.18, 1.14, 1.15), substitute the resulting expressions (1.21) and (1.20) in (1.19), and add up the two equations (1.19). The result is a relationship between the axle slip angle $\alpha_{ai}$ and the axle side force $F_{yi}$. We obtain, for the slip angle of axle $i$,

$$\begin{aligned}\alpha_{ai} &= \frac{F_{yi}}{C_{\text{eff},i}} \\ &= \frac{F_{yi}}{C_{F\alpha i}}\left[\begin{array}{l} 1+\dfrac{l(\varepsilon_i C_{F\alpha i}+\tau_i C_{F\gamma i})h'}{(l-a_i)(c_{\varphi 1}+c_{\varphi 2}-mgh')}+\dfrac{C_{F\alpha i}(e_i+t_i)}{c_{\psi i}} \\ -C_{F\alpha i}c_{sfi}+\dfrac{2l\sigma_i}{l-a_i}(\zeta_{\alpha i}\psi_{io}+\zeta_{\gamma i}\gamma_{io}) \end{array}\right]\end{aligned} \tag{1.22}$$

The coefficient of $F_{yi}$ constitutes the effective axle cornering compliance, which is the inverse of the effective axle cornering stiffness (1.17). The quantitative effect of each of the suspension, steering, and tire factors included can be easily assessed. The subscript $i$ refers to the complete axle. Consequently, the cornering and camber stiffnesses appearing in this expression are the sum of the stiffnesses of the left and right tire:

$$\begin{aligned} C_{F\alpha i} &= C_{F\alpha iL}+C_{F\alpha iR} = C_{F\alpha iLo}+C_{F\alpha iRo} \\ C_{F\gamma i} &= C_{F\gamma iL}+C_{F\gamma iR} = C_{F\gamma iLo}+C_{F\gamma iRo} \end{aligned} \tag{1.23}$$

in which (1.7) and (1.11) have been taken into account. The load transfer coefficient $\sigma_i$ follows from Eqn (1.10). Expression (1.22) shows that the influence of lateral load transfer only occurs if initially, at straight ahead running, side forces are already present through the introduction of e.g., opposite steer and camber angles. If these angles are absent, the influence of load transfer is purely nonlinear and is only felt at higher levels of lateral accelerations. In the next subsection, this nonlinear effect will be incorporated in the effective axle characteristic.

### *Effective Nonlinear Axle Characteristics*

To illustrate the method of effective axle characteristics, we will first discuss the determination of the effective characteristic of a front axle showing steering compliance. The steering wheel is held fixed. Due to tire side forces and self-aligning torques (left and right), distortions will arise resulting in an incremental steer angle $\psi_{c1}$ of the front wheels ($\psi_{c1}$ will be negative in Figure 1.5 for the case of just steer compliance). Since load transfer is not considered in this example, the situation at the left and right wheels is identical (initial toe and camber angles being disregarded). The front tire slip angle is denoted by $\alpha_1$ and the 'virtual' slip angle of the axle is denoted by $\alpha_{a1}$ and equals (cf. Figure 1.5):

$$\alpha_{a1} = \alpha_1 - \psi_{c1} \tag{1.24}$$

where both $\alpha_1$ and $\psi_{c1}$ are related to $F_{y1}$ and $M_{z1}$. The subscript 1 refers to the front axle and thus to the pair of tires. Consequently, $F_{y1}$ and $M_{z1}$ denote the sum of the left and right tire side forces and moments. The objective is to find the function $F_{y1}(\alpha_{a1})$ which is the effective front axle characteristic.

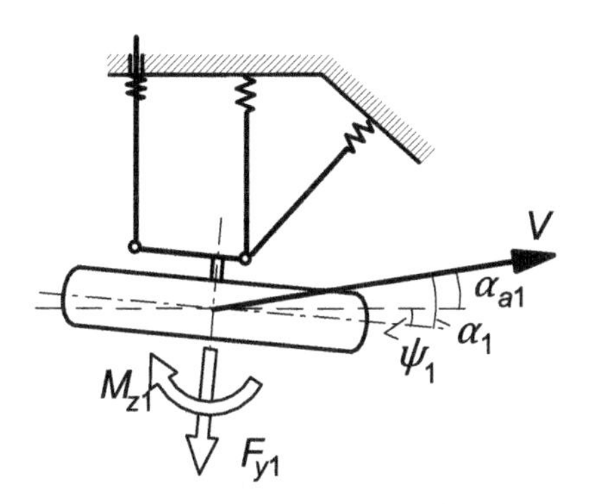


**FIGURE 1.5** Wheel suspension and steering compliance resulting in additional steer angle $\psi_1$.

Figure 1.6 shows a graphical approach. According to Eqn (1.24), the points on the $F_{y1}(\alpha_1)$ curve must be shifted horizontally over a length $\psi_{c1}$ to obtain the sought $F_{y1}(\alpha_{a1})$. The slope of the curve at the origin corresponds to the effective axle cornering stiffness found in the preceding subsection. Although the changes with respect to the original characteristic may be small, they can still be of considerable importance since it is the difference of slip angles front and rear which largely determines the vehicle's handling behavior.

The effective axle characteristic for the case of roll steer can be easily established by subtracting $\psi_{ri}$ from $\alpha_i$. Instead of using the linear relationships (1.8) and (1.13), nonlinear curves may be adopted, possibly obtained from measurements. For the case of roll camber, the situation becomes more complex. At a given axle side force, the roll angle and the associated camber

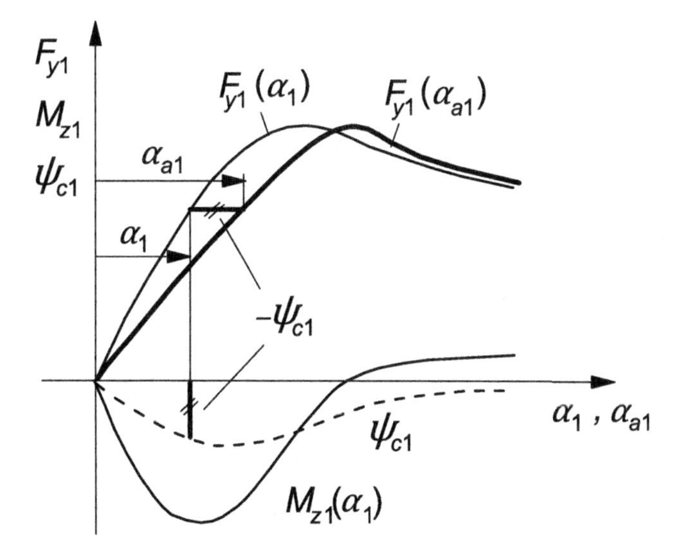


**FIGURE 1.6** Effective front axle characteristic $F_{y1}(\alpha_{a1})$ influenced by steering compliance.

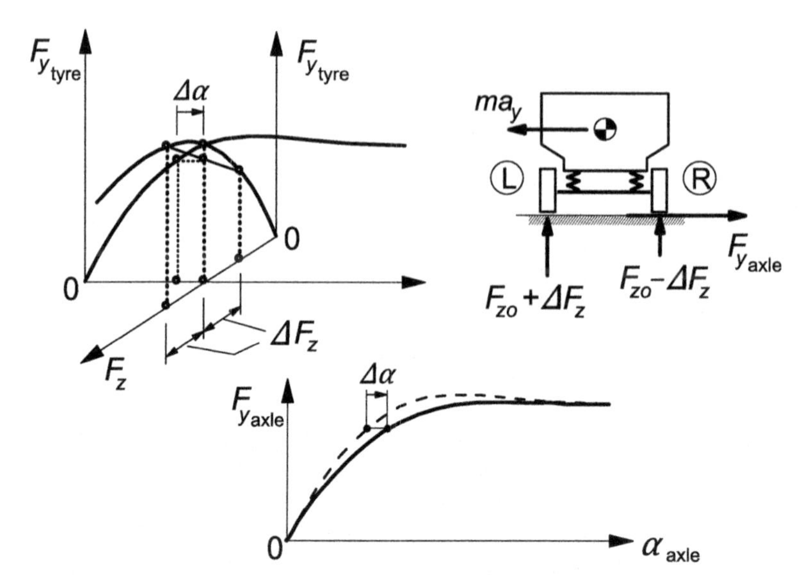


**FIGURE 1.7** The influence of load transfer on the resulting axle characteristic.

angle can be found. The cornering characteristic of the pair of tires at that camber angle is needed to find the slip angle belonging to the side force considered.

Load transfer is another example that is less easy to handle. In Figure 1.7 a three-dimensional graph is presented for the variation of the side force of an individual tire as a function of the slip angle and of the vertical load, the former at a given load and the latter at a given slip angle. The diagram illustrates that at load transfer, the outer tire exhibiting a larger load will generate a larger side force than the inner tire. Because of the nonlinear degressive $F_y$ vs $F_z$ curve, however, the average side force will be smaller than the original value it had in the absence of load transfer. The graph indicates that an increased $\Delta\alpha$ of the slip angle would be needed to compensate for the adverse effect of load transfer. The lower diagram gives a typical example of the change in characteristic as a result of load transfer. At the origin the slope is not affected, but at larger slip angles an increasingly lower derivative appears to occur. The peak diminishes and may even disappear completely. The way to determine the resulting characteristic is the subject of the next exercise.

---

**Exercise 1.1 Construction of Effective Axle Characteristics at Load Transfer**

For a series of tire vertical loads, $F_z$, the characteristics of the two tires mounted on, say, the front axle of an automobile are given. In addition, it is known how the load transfer $\Delta F_z$ at the front axle depends on the centrifugal force $K (= mg\, F_{y1}/F_{z1} = mg\, F_{y2}/F_{z2})$ acting at the center of gravity. From these data, the resulting cornering characteristic of the axle considered (at steady-state cornering) can be determined.

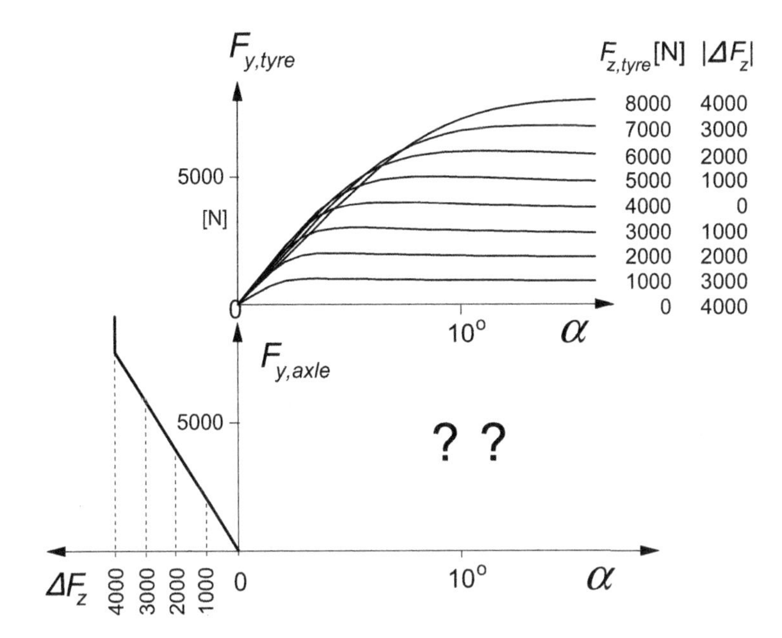


**FIGURE 1.8** The construction of the resulting axle cornering characteristics at load transfer (Exercise 1.1).

1. Find the resulting characteristic of one axle from the set of individual tire characteristics at different tire loads $F_z$ and the load transfer characteristic, both shown in Figure 1.8.
   Hint: First draw, in the lower diagram, the axle characteristics for values of $\Delta F_z = 1000, 2000, 3000$, and 4000 N and then determine which point on each of these curves is valid considering the load transfer characteristic (left-hand diagram). Draw the resulting axle characteristic.
   It may be helpful to employ the *Magic Formula* (1.6) and the parameters shown below:
   side force: $F_y = D \sin[C \arctan\{B\alpha - E(B\alpha - \arctan(B\alpha))\}]$
   with factors: $B = C_{F\alpha}/(CD)$, $C = 1.3$, $D = \mu F_z$, $E = -3$, with $\mu = 1$
   cornering stiffness: $C_{F\alpha} = c_1 \sin[2 \arctan\{F_z/c_2\}]$
   with parameters: $c_1 = 60000$ [N/rad], $c_2 = 4000$ [N]
   In addition, we have given, for the load transfer, $\Delta F_z = 0.52 F_{y,\mathrm{axle}}$ (up to lift-off of the inner tire, after which the other axle may take over to accommodate the increased total load transfer).
2. Draw the individual curves of $F_{yL}$ and $F_{yR}$ (for the left and right tire) as a function of $\alpha$ which appear to arise under the load transfer condition considered here.
3. Finally, plot these forces as a function of the vertical load $F_z$ (ranging from 0 to 8000 N). Note the variation of the lateral force of an individual (left or right) tire in this same range of vertical load which may be covered in a left- and in a right-hand turn at an increasing speed of travel until (and possibly beyond) the moment that one of the wheels (the inner wheel) lifts from the ground.

## 1.3. VEHICLE HANDLING AND STABILITY

In this section, attention is paid to the more fundamental aspects of vehicle horizontal motions. Instead of discussing results of computer simulations of complicated vehicle models, we rather take the simplest possible model of an automobile that runs at constant speed over an even horizontal road and thereby gain considerable insight into the basic aspects of vehicle handling and stability. Important early work on the linear theory of vehicle handling and stability has been published by Riekert and Schunck (1940), Whitcomb and Milliken (1956), and Segel (1956). Pevsner (1947) studied the nonlinear steady-state cornering behavior at larger lateral accelerations and introduced the handling diagram. One of the first more complete vehicle model studies has been conducted by Pacejka (1958) and by Radt and Pacejka (1963).

For more introductory or specialized study, the reader may be referred to books on the subject, cf. e.g.: Gillespie (1992), Mitschke (1990), Milliken and Milliken (1995), Kortüm and Lugner (1994), and Abe (2009).

The derivation of the equations of motion for the three degree of freedom model of Figure 1.4 will be treated first after which the simple model with two degrees of freedom is considered and analyzed. This analysis comprises the steady-state response to steering input and the stability of the resulting motion. Also, the frequency response to steering fluctuations and external disturbances will be discussed, first for the linear vehicle model and subsequently for the nonlinear model where large lateral accelerations and disturbances are introduced.

The simple model to be employed in the analysis is presented in Figure 1.9. The track width has been neglected with respect to the radius of the cornering motion which allows the use of a two-wheel vehicle model. The steer and slip angles will be restricted to relatively small values. Then, the variation of the geometry may be regarded to remain linear, that is, $\cos\alpha \approx 1$ and $\sin\alpha \approx \alpha$ and similarly for the steer angle $\delta$. Moreover, the driving force required to keep the speed constant is assumed to remain small with respect to the lateral tire force.

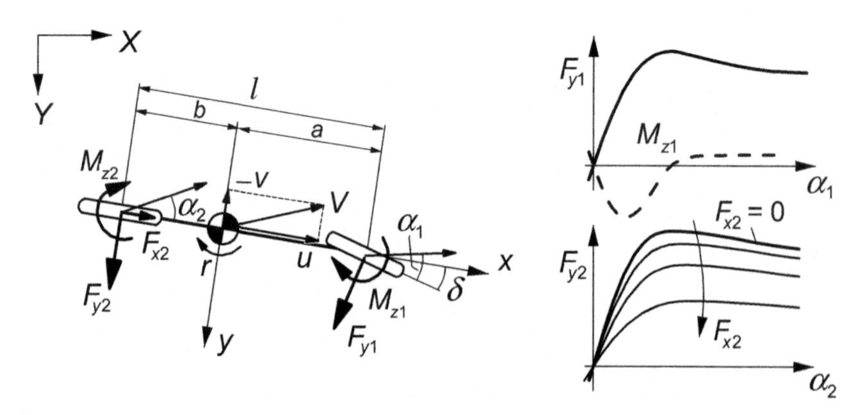


**FIGURE 1.9** Simple car model with side force characteristics for front and rear (driven) axle.

Considering combined slip curves like those shown in Figure 1.2 (right), we may draw the conclusion that the influence of $F_x$ on $F_y$ may be neglected in that case.

In principle, a model as shown in Figure 1.9 lacks body roll and load transfer. Therefore, the theory is actually limited to cases where the roll moment remains small, that is, at low friction between tire and road or a low center of gravity relative to the track width. This restriction may be overcome by using the effective axle characteristics in which the effects of body roll and load transfer have been included while still adhering to the simple (rigid) two-wheel vehicle model. As mentioned before, this is only permissible when the frequency of the imposed steer angle variations remains small with respect to the roll natural frequency. Similarly, as demonstrated in the preceding section, effects of other factors like compliance in the steering system and suspension mounts may be accounted for.

The speed of travel is considered to be constant. However, the theory may approximately hold also for quasi-steady-state situations for instance at moderate braking or driving. The influence of the fore-and-aft force $F_x$ on the tire or axle cornering force vs slip angle characteristic ($F_y$, $\alpha$) may then be regarded (cf. Figure 1.9). The forces $F_{y1}$ and $F_{x1}$ and the moment $M_{z1}$ are defined to act upon the single front wheel and similarly we define $F_{y2}$ etc. for the rear wheel.

### 1.3.1. Differential Equations for Plane Vehicle Motions

In this section, the differential equations for the three degree of freedom vehicle model of Figure 1.4 will be derived. In first instance, the fore-and-aft motion will also be left free to vary. The resulting set of equations of motion may be of interest for the reader to further study the vehicle's dynamic response at somewhat higher frequencies where the roll dynamics of the vehicle body may become of importance, cf. App. 2. From these equations, the equations for the simple two-degree-of-freedom model of Figure 1.9 used in the subsequent section can be easily assessed. In Subsection 1.3.6, the equations for the car with trailer will be established. The possible instability of the motion will be studied.

We will employ Lagrange's equations to derive the equations of motion. For a system with $n$ degrees of freedom $n$ (generalized) coordinates, $q_i$ are selected which are sufficient to completely describe the motion while possible kinematic constraints remain satisfied. The moving system possesses kinetic energy $T$ and potential energy $U$. External generalized forces $Q_i$ associated with the generalized coordinates $q_i$ may act on the system. Internal forces acting from dampers to the system structure may be regarded as external forces taking part in the total work $W$. The equation of Lagrange for coordinate $q_i$ reads

$$\frac{\mathrm{d}}{\mathrm{d}t}\frac{\partial T}{\partial \dot{q}_i} - \frac{\partial T}{\partial q_i} + \frac{\partial U}{\partial q_i} = Q_i \qquad (1.25)$$

The system depicted in Figure 1.4 and described in the preceding subsection performs a motion over a flat level road. Proper coordinates are the Cartesian coordinates $X$ and $Y$ of reference point $A$, the yaw angle $\psi$ of the moving $x$-axis with respect to the inertial $X$-axis and finally the roll angle $\varphi$ about the roll axis. For motions near the $X$-axis and thus small yaw angles, Eqn (1.25) is adequate to derive the equations of motion. For cases where $\psi$ may attain large values, e.g., when moving along a circular path, it is preferred to use modified equations where the velocities $u$, $v$, and $r$ of the moving axes system are used as generalized motion variables in addition to the coordinate $\varphi$. The relations between the two sets of variables are (the dots referring to differentiation with respect to time)

$$\begin{aligned} u &= \dot{X}\cos\psi + \dot{Y}\sin\psi \\ v &= -\dot{X}\sin\psi + \dot{Y}\cos\psi \\ r &= \dot{\psi} \end{aligned} \tag{1.26}$$

The kinetic energy can be expressed in terms of $u$, $v$, and $r$. Preparation of the first terms of Eqn (1.25) for the coordinates $X$, $Y$, and $\psi$ yields

$$\begin{aligned} \frac{\partial T}{\partial \dot{X}} &= \frac{\partial T}{\partial u}\frac{\partial u}{\partial \dot{X}} + \frac{\partial T}{\partial v}\frac{\partial v}{\partial \dot{X}} = \frac{\partial T}{\partial u}\cos\psi - \frac{\partial T}{\partial v}\sin\psi \\ \frac{\partial T}{\partial \dot{Y}} &= \frac{\partial T}{\partial u}\frac{\partial u}{\partial \dot{Y}} + \frac{\partial T}{\partial v}\frac{\partial v}{\partial \dot{Y}} = \frac{\partial T}{\partial u}\sin\psi + \frac{\partial T}{\partial v}\cos\psi \\ \frac{\partial T}{\partial \dot{\psi}} &= \frac{\partial T}{\partial r} \\ \frac{\partial T}{\partial \psi} &= \frac{\partial T}{\partial u}v - \frac{\partial T}{\partial v}u \end{aligned} \tag{1.27}$$

The yaw angle $\psi$ may now be eliminated by multiplying the final equations for $X$ and $Y$ successively with cos $\psi$ and sin $\psi$ and subsequently adding and subtracting them. The resulting equations represent the equilibrium in the $x$- and $y$- (or $u$ and $v$) direction, respectively.

We obtain the following set of modified Lagrangian equations for the first three variables $u$, $v$, and $r$ and subsequently for the remaining real coordinates (for our system only $\varphi$):

$$\begin{aligned} \frac{\mathrm{d}}{\mathrm{d}t}\frac{\partial T}{\partial u} - r\frac{\partial T}{\partial v} &= Q_u \\ \frac{\mathrm{d}}{\mathrm{d}t}\frac{\partial T}{\partial v} + r\frac{\partial T}{\partial u} &= Q_v \\ \frac{\mathrm{d}}{\mathrm{d}t}\frac{\partial T}{\partial r} - v\frac{\partial T}{\partial u} + u\frac{\partial T}{\partial v} &= Q_r \\ \frac{\mathrm{d}}{\mathrm{d}t}\frac{\partial T}{\partial \dot{\varphi}} - \frac{\partial T}{\partial \varphi} + \frac{\partial U}{\partial \varphi} &= Q_\varphi \end{aligned} \tag{1.28}$$

The generalized forces are found from the virtual work:

$$\delta W = \sum_{j-1}^{4} Q_j \delta q_i \tag{1.29}$$

with $q_j$ referring to the quasi-coordinates $x$ and $y$ and the coordinates $\psi$ and $\varphi$. Note that $x$ and $y$ cannot be found from integrating $u$ and $v$. For that reason, the term 'quasi' coordinate is used. For the vehicle model, we find for the virtual work as a result of the virtual displacements $\delta x$, $\delta y$, $\delta\psi$, and $\delta\varphi$:

$$\delta W = \sum F_x \delta x + \sum F_y \delta y + \sum M_z \delta\psi + \sum M_\varphi \delta\varphi \tag{1.30}$$

where apparently

$$\begin{aligned} Q_u &= \sum F_x = F_{x1}\cos\delta - F_{y1}\sin\delta + F_{x2} \\ Q_v &= \sum F_y = F_{x1}\sin\delta + F_{y1}\cos\delta + F_{y2} \\ Q_r &= \sum M_z = aF_{x1}\sin\delta + aF_{y1}\cos\delta + M_{z1} - bF_{y2} + M_{z2} \\ Q_\varphi &= \sum M_\varphi = -(k_{\varphi 1} + k_{\varphi 2})\dot{\varphi} \end{aligned} \tag{1.31}$$

The longitudinal forces are assumed to be the same at the left and right wheels, and the effect of additional steer angles $\psi_i$ is neglected here. Shock absorbers in the wheel suspensions are represented by the resulting linear moments about the roll axes with damping coefficients $k_{\varphi i}$ at the front and rear axles.

With the roll angle $\varphi$ and the roll axis inclination angle $\theta_r \approx (h_2 - h_1)/l$ assumed small, the kinetic energy becomes

$$\begin{aligned} T = {} & \frac{1}{2}m\{(u - h'\varphi r)^2 + (v + h'\dot{\varphi})^2\} + \frac{1}{2}I_x\dot{\varphi}^2 \\ & + \frac{1}{2}I_y(\varphi r)^2 + \frac{1}{2}I_z(r^2 - \varphi^2 r^2 + 2\theta_r r\dot{\varphi}) - I_{xz}r\dot{\varphi} \end{aligned} \tag{1.32}$$

The potential energy $U$ is built up in the suspension springs (including the radial tire compliances) and through the height of the center of gravity. We have, again for small angles,

$$U = \frac{1}{2}(c_{\varphi 1} + c_{\varphi 2})\,\varphi^2 - \frac{1}{2}mgh'\varphi^2 \tag{1.33}$$

The equations of motion are finally established by using expressions (1.31), (1.32), and (1.33) in the Eqn (1.28). The equations will be linearized in the supposedly small angles $\varphi$ and $\delta$. For the variables $u$, $v$, $r$, and $\varphi$, we obtain successively

$$m(\dot{u} - rv - h'\varphi\dot{r} - 2h'r\dot{\varphi}) = F_{x1} - F_{y1}\delta + F_{x2} \tag{1.34a}$$

$$m(\dot{v} + ru + h'\ddot{\varphi} - h'r^2\varphi) = F_{x1}\delta + F_{y1} + F_{y2} \tag{1.34b}$$

$$\begin{aligned} I_z\dot{r} + (I_z\theta_r - I_{xz})\ddot{\varphi} - mh'(\dot{u} - rv)\varphi &= aF_{x1}\delta + aF_{y1} + M_{z1} \\ &- bF_{y1} + M_{z1} - bF_{y2} + M_{z2} \end{aligned} \tag{1.34c}$$

$$\begin{aligned} &(I_X + mh'^2)\ddot{\varphi} + mh'(\dot{v} + ru) + (I_z\theta_r - I_{xz})\dot{r} \\ &-(mh'^2 + I_y - I_z)r^2\varphi + (k_{\varphi 1} + k_{\varphi 2})\dot{\varphi} \\ &+ (c_{\varphi 1} + c_{\varphi 2} - mgh')\varphi = 0 \end{aligned} \tag{1.34d}$$

Note that the small additional roll and compliance steer angles $\psi_i$ have been neglected in the assessment of the force components. The tire side forces depend on the slip and camber angles front and rear and on the tire vertical loads. We may need to take the effect of combined slip into account. The longitudinal forces are either given as a result of brake effort or imposed propulsion torque or they depend on the wheel longitudinal slip which follows from the wheel speed of revolution requiring four additional wheel rotational degrees of freedom. The first Eqn (1.34a) may be used to compute the propulsion force needed to keep the forward speed constant.

The vertical loads and more specifically the load transfer can be obtained by considering the moment equilibrium of the front and rear axle about the respective roll centers. For this, the roll moments $M_{\varphi i}$ (cf. Figure 1.4) resulting from suspension springs and dampers as appeared in Eqn (1.34d) through the terms with subscript 1 and 2 respectively, and the axle side forces appearing in Eqn (1.34b) are to be regarded. For a linear model the load transfer can be neglected if initial (left/right opposite) wheel angles are disregarded. We have at steady state (effect of damping vanishes)

$$\Delta F_{zi} = \frac{-c_{\varphi 1}\varphi + F_{yi}h_i}{2s_i} \tag{1.35}$$

The front and rear slip angles follow from the lateral velocities of the wheel axles and the wheel steer angles with respect to the moving longitudinal $x$-axis. The longitudinal velocities of the wheel axles may be regarded the same left and right and equal to the vehicle longitudinal speed $u$. This is allowed when $s_i|r| \ll u$. Then the expressions for the assumedly small slip angles read

$$\alpha_1 = \delta + \psi_1 - \frac{v + ar - e\dot{\delta}}{u}$$

$$\alpha_2 = \psi_2 - \frac{v - br}{u} \tag{1.36}$$

The additional roll and compliance steer angles $\psi_i$ and the wheel camber angles $\gamma_i$ are obtained from Eqn (1.21) with (1.13–15) or corresponding nonlinear expressions. Initial wheel angles are assumed to be equal to zero. The influence of the steer angle velocity appearing in the expression for the front slip angle is relatively small and may be disregarded. The small products of the caster length $e$ and the time rate of change of $\psi_i$ have been neglected in the above expressions.

Equations (1.34) may be further linearized by assuming that all the deviations from the rectilinear motion are small. This allows neglecting all products of variable quantities which vanish when the vehicle moves straight ahead. The side forces and moments are then written as in Eqn (1.5) with the subscripts $i = 1$ or 2 provided. If the moment due to camber is neglected and the pneumatic trail is introduced in the aligning torque, we have

$$\begin{aligned} F_{y1} &= F_{y\alpha i} + F_{y\gamma i} = C_{F\alpha i}\alpha_i + C_{F\gamma i}\gamma_i \\ M_{zi} &= M_{z\alpha i} = -C_{M\alpha i}\alpha_i = -t_i F_{y\alpha i} = -t_i C_{F\alpha i}\alpha_i \end{aligned} \tag{1.37}$$

The three linear equations of motion for the system of Figure 1.4 with the forward speed $u$ kept constant finally turn out to read, if expressed solely in terms of the three motion variables $v$, $r$, and $\varphi$:

$$\begin{aligned} m(\dot{v} + ur + h'\ddot{\varphi}) &= C_{F\alpha 1}\{(1 + c_{sc1})(u\delta + e\dot{\delta} - v - ar)/u + c_{sr1}\varphi\} \\ &+ C_{F\alpha 2}\{(1 + c_{sc2})(-v + br)/u + c_{sr2}\varphi\} + (C_{F\gamma 1}\tau_1 + C_{F\gamma 2}\tau_2)\varphi \end{aligned} \tag{1.38a}$$

$$\begin{aligned} I_z\dot{r} + (I_z\theta_r - I_{xz})\ddot{\varphi} &= (a - t_1)C_{F\alpha 1}\{(1 + c_{sc1})(u\delta + e\dot{\delta} - v - ar)/u + c_{sr1}\varphi\} \\ &- (b + t_2)C_{F\alpha 2}\{(1 + c_{sc2})(-v + br)/u + c_{sr2}\varphi\} + (aC_{F\gamma 1}\tau_1 - bC_{F\gamma 2}\tau_2)\varphi \end{aligned} \tag{1.38b}$$

$$\begin{aligned} &(I_x + mh'^2)\ddot{\varphi} + mh'(\dot{v} + ur) + (I_z\theta_r - I_{xz})\dot{r} \\ &+(k_{\varphi 1} + k_{\varphi 2})\dot{\varphi} + (c_{\varphi 1} + c_{\varphi 2} - mgh')\varphi = 0 \end{aligned} \tag{1.38c}$$

In these equations, the additional steer angles $\psi_i$ have been eliminated by using expressions (1.21) with (1.13–15). Furthermore, the resulting compliance steer and roll steer coefficients for $i = 1$ or 2 have been introduced:

$$c_{sci} = \frac{A_i C_{F\alpha i}}{1 - A_i C_{F\alpha i}}, \quad c_{sri} = \frac{\varepsilon_i + \tau_i A_i C_{F\gamma i}}{1 - A_i C_{F\alpha i}} \tag{1.39}$$

with

$$A_i = c_{sfi} - \frac{e_i + t_i}{c_{\psi i}}$$

where the steer stiffness at the rear $c_{\psi 2}$ may be taken equal to infinity. Furthermore, we have the roll axis inclination angle:

$$\theta_r = \frac{h_2 - h_1}{l} \tag{1.40}$$

In Chapters 7 and 8, the transient properties of the tire will be addressed. The relaxation length denoted by $\sigma_i$ is an important parameter that controls the lag of the response of the side force to the input slip angle. For the Laplace transformed version of Eqn (1.38) with the Laplace variable $s$ representing differentiation with respect to time, we may introduce tire lag by replacing the slip angle $\alpha_i$ with the filtered transient slip angle. This may be accomplished by replacing the cornering stiffnesses $C_{F\alpha i}$ appearing in (1.38) and (1.39) with the 'transient stiffnesses':

$$C_{F\alpha i} \rightarrow \frac{C_{F\alpha i}}{1 + s\sigma_i/u} \tag{1.41}$$

A similar procedure may be followed to include the tire transient response to wheel camber variations. The relaxation length concerned is about equal to the one used for the response to side slip variations. At nominal vertical load, the relaxation length is of the order of magnitude of the wheel radius. A more precise model of the aligning torque may be introduced by using a transient pneumatic trail with a similar replacement as indicated by (1.41) but with a much smaller relaxation length approximately equal to half the contact length of the tire. For more details, we refer to Chapter 9 that is dedicated to short wavelength force and moment response.

### 1.3.2. Linear Analysis of the Two-Degree-of-Freedom Model

From Eqns (1.34b and c), the reduced set of equations for the two-degree-of-freedom model can be derived immediately. The roll angle $\varphi$ and its derivative are set equal to zero and, furthermore, we will assume the forward speed $u$ ($\approx V$) to remain constant and neglect the influence of the lateral component of the longitudinal forces $F_{xi}$. The equations of motion of the simple model of Figure 1.9 for $v$ and $r$ now read

$$m(\dot{v} + ur) = F_{y1} + F_{y2} \tag{1.42a}$$

$$I\dot{r} = aF_{y1} - bF_{y2} \tag{1.42b}$$

with $v$ denoting the lateral velocity of the center of gravity and $r$ the yaw velocity. The symbol $m$ stands for the vehicle mass and $I$ ($= I_z$) denotes the moment of inertia about the vertical axis through the center of gravity. For the matter of simplicity, the rearward shifts of the points of application of the forces $F_{y1}$ and $F_{y2}$ over a length equal to the pneumatic trail $t_1$ and $t_2$ respectively (that is the aligning torques) have been disregarded. Later, we come back to this. The side forces are functions of the respective slip angles:

$$F_{y1} = F_{y1}(\alpha_1) \quad \text{and} \quad F_{y2} = F_{y2}(\alpha_2) \tag{1.43}$$

and the slip angles are expressed by

$$\alpha_1 = \delta - \frac{1}{u}(v + ar) \quad \text{and} \quad \alpha_2 = -\frac{1}{u}(v - br) \tag{1.44}$$

neglecting the effect of the time rate of change of the steer angle appearing in Eqn (1.36). For relatively low-frequency motions, the effective axle characteristics or effective cornering stiffnesses according to Eqns (1.17, 1.22) may be employed.

When only small deviations with respect to the undisturbed straight-ahead motion are considered, the slip angles may be assumed to remain small enough to allow linearization of the cornering characteristics. For the side force, the relationship with the slip angle reduces to the linear equation

$$F_{yi} = C_i\alpha_i = C_{F\alpha i}\alpha_i \tag{1.45}$$

where $C_i$ denotes the cornering stiffness. This can be replaced by the symbol $C_{F\alpha i}$ which may be preferred in more general cases where also camber and aligning stiffnesses play a role.

The two linear first-order differential equations now read:

$$\begin{aligned} m\dot{v} + \frac{1}{u}(C_1 + C_2)\,v + \left\{mu + \frac{1}{u}(aC_1 - bC_2)\right\} r &= C_1\delta \\ I\dot{r} + \frac{1}{u}(a^2C_1 - b^2C_2)\,r + \frac{1}{u}(aC_1 - bC_2)v &= aC_1\delta \end{aligned} \tag{1.46}$$

After elimination of the lateral velocity $v$, we obtain the second-order differential equation for the yaw rate $r$:

$$\begin{aligned} &Imu\ddot{r} + \{I(C_1 + C_2) + m(a^2C_1 + b^2C_2)\}\,\dot{r} \\ &+ \frac{1}{u}\{C_1C_2l^2 - mu^2(aC_1 - bC_2)\}r = muaC_1\dot{\delta} + C_1C_2l\delta \end{aligned} \tag{1.47}$$

Here, as before, the dots refer to differentiation with respect to time, $\delta$ is the steer angle of the front wheel, and $l$ $(= a + b)$ represents the wheel base. The equations may be simplified by introducing the following quantities:

$$\begin{aligned} C &= C_1 + C_2 \\ Cs &= aC_1 - bC_2 \\ Cq^2 &= a^2C_1 + b^2C_2 \\ mk^2 &= I \end{aligned} \tag{1.48}$$

Here $C$ denotes the total cornering stiffness of the vehicle, $s$ is the distance from the center of gravity to the so-called neutral steer point $S$ (Figure 1.11), $q$ is a length corresponding to an average moment arm, and $k$ is the radius of gyration. Equations (1.46) and (1.47) now reduce to

$$m(\dot{v} + ur) + \frac{C}{u}v + \frac{Cs}{u}r = C_1\delta, \qquad mk^2\dot{r} + \frac{Cq^2}{u}r + \frac{Cs}{u}v = C_1 a\delta \tag{1.49}$$

and, with $v$ eliminated,

$$m^2k^2u^2\ddot{r} + mC(q^2 + k^2)u\dot{r} + (C_1C_2l^2 - mu^2Cs)r = mu^2aC_1\dot{\delta} + uC_1C_2l\delta \tag{1.50}$$

The neutral steer point $S$ is defined as the point on the longitudinal axis of the vehicle where an external side force can be applied without changing the vehicle's yaw angle. If the force acts in front of the neutral steer point, the vehicle is expected to yaw in the direction of the force and, if behind, then against the force. The point is of interest when discussing the steering characteristics and stability.

### *Linear Steady-State Cornering Solutions*

We are interested in the path curvature ($1/R$) that results from a constant steer angle $\delta$ at a given constant speed of travel $V$. Since we have at steady state

$$\frac{1}{R} = \frac{r}{V} \approx \frac{r}{u} \tag{1.51}$$

the expression for the path curvature becomes by using (1.47) with $u$ replaced by $V$ and the time derivatives omitted:

$$\frac{1}{R} = \frac{C_1C_2l}{C_1C_2l^2 - mV^2(aC_1 - bC_2)}\delta \tag{1.52}$$

By taking the inverse, the expression for the steer angle required to negotiate a curve with a given radius $R$ is obtained:

$$\delta = \frac{1}{R}\left(l - mV^2\frac{aC_1 - bC_2}{lC_1C_2}\right) \tag{1.53}$$

It is convenient to introduce the so-called understeer coefficient or gradient $\eta$. For our model, this quantity is defined as

$$\eta = -\frac{mg}{l}\frac{aC_1 - bC_2}{C_1C_2} = -\frac{s}{l}\frac{mgC}{C_1C_2} \tag{1.54}$$

with $g$ denoting the acceleration due to gravity. After having defined the lateral acceleration which in the present linear analysis equals the centripetal acceleration,

$$a_y = Vr = \frac{V^2}{R} \tag{1.55}$$

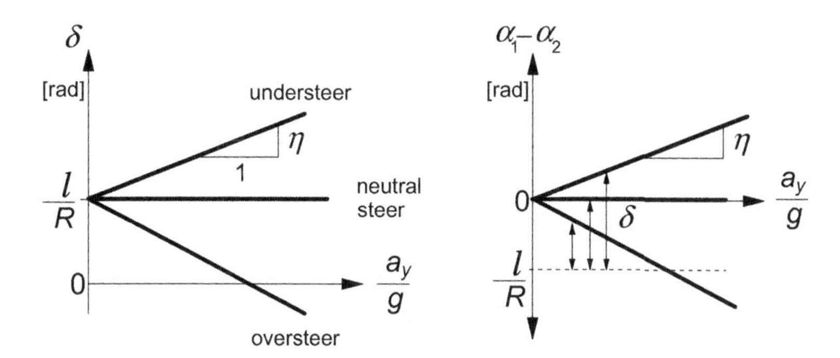


**FIGURE 1.10** The steer angle versus lateral acceleration at constant path curvature (left graph). The difference in slip angle versus lateral acceleration and the required steer angle at a given path curvature (right graph). The understeer gradient $\eta$.

Equation (1.53) can be written in the more convenient form:

$$\delta = \frac{l}{R}\left(1 + \eta \frac{V^2}{gl}\right) = \frac{l}{R} + \eta \frac{a_y}{g} \tag{1.56}$$

The meaning of understeer versus oversteer becomes clear when the steer angle is plotted against the centripetal acceleration while the radius $R$ is kept constant. In Figure 1.10 (left-hand diagram) this is done for three types of vehicles showing understeer, neutral steer, and oversteer. Apparently, for an understeered vehicle, the steer angle needs to be increased when the vehicle is going to run at a higher speed. At neutral steer the steer angle can be kept constant, while at oversteer a reduction in steer angle is needed when the speed of travel is increased and at the same time a constant turning radius is maintained.

According to Eqn (1.56), the steer angle changes sign when, for an oversteered car, the speed increases beyond the critical speed that is expressed by

$$V_{\text{crit}} = \sqrt{\frac{gl}{-\eta}} \quad (\eta < 0) \tag{1.57}$$

As will be shown later, the motion becomes unstable when the critical speed is surpassed. Apparently, this can only happen when the vehicle shows oversteer.

For an understeered car, a counterpart has been defined which is the so-called characteristic speed. It is the speed where the steer angle required to maintain the same curvature increases twice the angle needed at speeds approaching zero. We may also say that at the characteristic speed the path curvature response to steer angle has decreased to half its value at very low speed. Also interesting is the fact that at the characteristic speed, the yaw rate response to steer angle $r/\delta$ reaches a maximum (the proof of which is left to the reader). We have, for the characteristic velocity,

$$V_{\text{char}} = \sqrt{\frac{gl}{\eta}} \quad (\eta > 0) \tag{1.58}$$

Expression (1.54) for the understeer gradient $\eta$ is simplified when the following expressions for the front and rear axle loads are used:

$$F_{z1} = \frac{b}{l}mg \quad \text{and} \quad F_{z2} = \frac{a}{l}mg \tag{1.59}$$

We obtain

$$\eta = \frac{F_{z1}}{C_1} - \frac{F_{z2}}{C_2} \tag{1.60}$$

which says that a vehicle exhibits an understeer nature when the relative cornering compliance of the tires at the front is larger than at the rear. It is important to note that in (1.59) and (1.60), the quantities $F_{z1,2}$ denote the vertical axle loads that occur at stand-still and thus represent the mass distribution of the vehicle. Changes of these loads due to aerodynamic down forces and fore-and-aft load transfer at braking or driving should not be introduced in expression (1.60).

In the same diagram, the difference in slip angle front and rear may be indicated. We find, for the side forces,

$$F_{y1} = \frac{b}{l}ma_y = F_{z1}\frac{a_y}{g}, \quad F_{y2} = \frac{a}{l}ma_y = F_{z2}\frac{a_y}{g} \tag{1.61}$$

and hence, for the slip angles,

$$\alpha_1 = \frac{F_{z1}}{C_1}\frac{a_y}{g}, \quad \alpha_2 = \frac{F_{z2}}{C_2}\frac{a_y}{g} \tag{1.62}$$

The difference now reads when considering relation (1.59):

$$\alpha_1 - \alpha_2 = \eta\frac{a_y}{g} \tag{1.63}$$

Apparently, the sign of this difference is dictated by the understeer coefficient. Consequently, it may be stated that according to the linear model, an understeered vehicle ($\eta > 0$) moves in a curve with slip angles larger at the front than at the rear ($\alpha_1 > \alpha_2$). For a neutrally steered vehicle, the angles remain the same ($\alpha_1 = \alpha_2$) and with an oversteered car the rear slip angles are bigger ($\alpha_2 > \alpha_1$). As shown by the expressions (1.54), the signs of $\eta$ and $s$ are different. Consequently, as one might expect when the centrifugal force is considered as the external force, a vehicle acts oversteered when the neutral steer point lies in front of the center of gravity and understeered when $S$ lies behind the c.g. As we will see later on, the actual nonlinear vehicle may change its steering character when the lateral acceleration increases. It appears then that the difference in slip angle is no longer directly related to the understeer gradient.

Considering Eqn (1.56) reveals that on the left-hand graph of Figure 1.10, the difference in slip angle can be measured along the ordinate starting from the value $l/R$. It is of interest to convert the diagram into the graph shown on the right-hand side of Figure 1.10 with ordinate equal to the difference in slip angle. In that way, the diagram becomes more flexible because the value of the curvature $l/R$ may be selected afterward. The horizontal dotted line is then shifted vertically according to the value of the relative curvature $l/R$ considered. The distance to the handling line represents the magnitude of the steer angle.

Figure 1.11 depicts the resulting steady-state cornering motion. The vehicle side slip angle $\beta$ has been indicated. It is of interest to note that at low speed this angle is negative for right-hand turns. Beyond a certain value of speed, the tire slip angles have become sufficiently large and the vehicle slip angle changes into positive values. In Exercise 1.2 the slip angle $\beta$ will be used.

### Influence of the Pneumatic Trail

The direct influence of the pneumatic trails $t_i$ may not be negligible. In reality, the tire side forces act a small distance behind the contact centers. As a consequence, the neutral steer point should also be considered to be located at a distance approximately equal to the average value of the pneumatic trails, more to the rear, which means actually more understeer. The correct

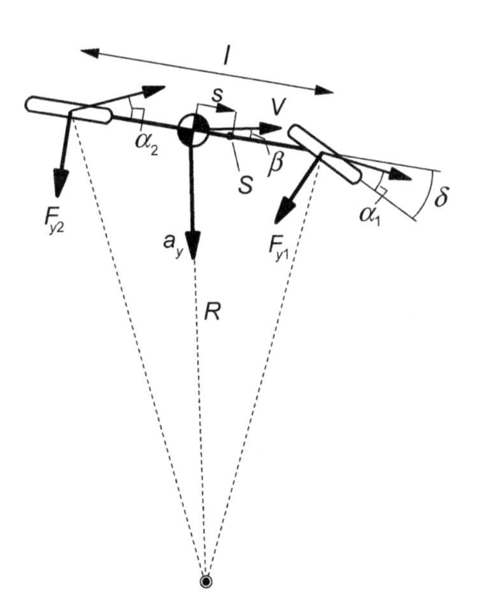


**FIGURE 1.11** Two-wheel vehicle model in a cornering maneuver.

values of the position $s$ of the neutral steer point and of the understeer coefficient $\eta$ can be found by using the effective axle distances $a' = a - t_1$, $b' = b + t_2$, and $l' = a' + b'$ in Eqns (1.48) and (1.59) instead of the original quantities $a$, $b$, and $l$.

### *Stability of the Motion*

Stability of the steady-state circular motion can be examined by considering the differential Eqn (1.47) or (1.50). The steer angle is kept constant so that the equation gets the form

$$a_0\ddot{r} + a_1\dot{r} + a_2 r = b_1\delta \tag{1.64}$$

For this second-order differential equation, stability is assured when all coefficients $a_i$ are positive. Only the last coefficient $a_2$ may become negative which corresponds to divergent instability (spin-out without oscillations). As already indicated, this will indeed occur when, for an oversteered vehicle, the critical speed (1.57) is exceeded. The condition for stability reads

$$a_2 = C_1C_2l^2\left(1 + \eta\frac{V^2}{gl}\right) = C_1C_2l^2\left(\frac{\delta}{l/R}\right)_{ss} > 0 \tag{1.65}$$

with the subscript $ss$ referring to steady-state conditions, or

$$V < V_{\text{crit}} = \sqrt{\frac{gl}{-\eta}} \quad (\eta < 0) \tag{1.66}$$

The next section will further analyze the dynamic nature of the stable and unstable motions.

It is of importance to note that when the condition of an automobile subjected to driving or braking forces is considered, the cornering stiffnesses front and rear will change due to the associated fore-and-aft axle load transfer and the resulting state of combined slip. In expression (1.60) for the understeer coefficient $\eta$, the quantities $F_{zi}$ represent the static vertical axle loads obtained through Eqn (1.59) and are to remain unchanged! In Subsection 1.3.4, the effect of longitudinal forces on vehicle stability will be further analyzed.

### *Free Linear Motions*

To study the nature of the free motion after a small disturbance in terms of natural frequency and damping, the eigenvalues, that is the roots of the characteristic equation of the linear second-order system, are to be assessed.

The characteristic equation of the system described by the Eqn (1.49) or (1.50) reads after using the relation (1.54) between $s$ and $\eta$:

$$m^2k^2V^2\lambda^2 + mC(q^2 + k^2)V\lambda + C_1C_2l^2\left(1 + \frac{\eta}{gl}V^2\right) = 0 \qquad (1.67)$$

For a single mass–damper–spring system shown in Figure 1.13 with $r$ the mass displacement, $\delta$ the forced displacement of the support, $M$ the mass, $D$ the sum of the two damping coefficients $D_1$ and $D_2$, and $K$ the sum of the two spring stiffnesses $K_1$ and $K_2$, a differential equation similar in structure to Eqn (1.50) arises:

$$M\ddot{r} + D\dot{r} + Kr = D_1\dot{\delta} + K_1\delta \qquad (1.68)$$

and the corresponding characteristic equation:

$$M\lambda^2 + D\lambda + K = 0 \qquad (1.69)$$

When an oversteered car exceeds its critical speed, the last term of (1.67) becomes negative which apparently corresponds to a negative stiffness $K$. An inverted pendulum is an example of a second-order system with negative last coefficient showing monotonous (diverging) instability.

The roots $\lambda$ of Eqn (1.67) may have loci in the complex plane as shown in Figure 1.12. For positive values of the cornering stiffnesses, only the last coefficient of the characteristic equation can become negative which is responsible for the limited types of eigenvalues that can occur. As we will see in Subsection 1.3.3, possible negative slopes beyond the peak of the nonlinear axle characteristics may give rise to other types of unstable motions connected with two positive real roots or two conjugated complex

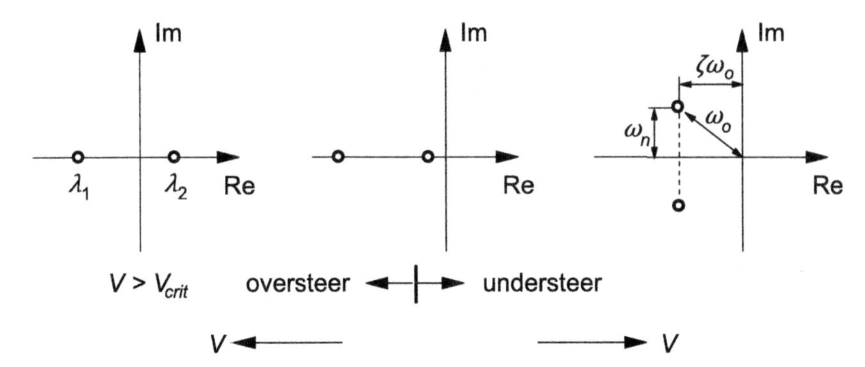


**FIGURE 1.12** Possible eigenvalues for the over- and understeered car at lower and higher speeds.

roots with a positive real part. For the linear vehicle model, we may have two real roots in the oversteer case and a pair of complex roots in the understeer case, except at low speeds where the understeered vehicle can show a pair of real negative roots.

As indicated in the figure, the complex root is characterized by the natural frequency $\omega_o$ of the undamped system ($D = 0$), the damping ratio $\zeta$, and the resulting actual natural frequency $\omega_n$. Expressions for these quantities in terms of the model parameters are rather complex. However, if we take into account that in normal cases $|s| \ll l$ and $q \approx k \approx \frac{1}{2} l$, we may simplify these expressions and find the following useful formulas:

*The natural frequency of the undamped system*:

$$\omega_o^2 = \frac{K}{M} \approx \left(\frac{C}{mV}\right)^2 \cdot \left(1 + \frac{\eta}{gl} V^2\right) \tag{1.70}$$

*The damping ratio*:

$$\zeta = \frac{D}{2M\omega_o} \approx \frac{1}{\sqrt{1 + \frac{\eta}{gl} V^2}} \tag{1.71}$$

*The natural frequency*:

$$\omega_n^2 = \omega_o^2 (1 - \zeta^2) \approx \left(\frac{C}{m}\right)^2 \cdot \frac{\eta}{gl} \tag{1.72}$$

The influence of parameters has been indicated in Figure 1.13. An arrow pointing upward represents an increase of the quantity in the same column of the matrix.

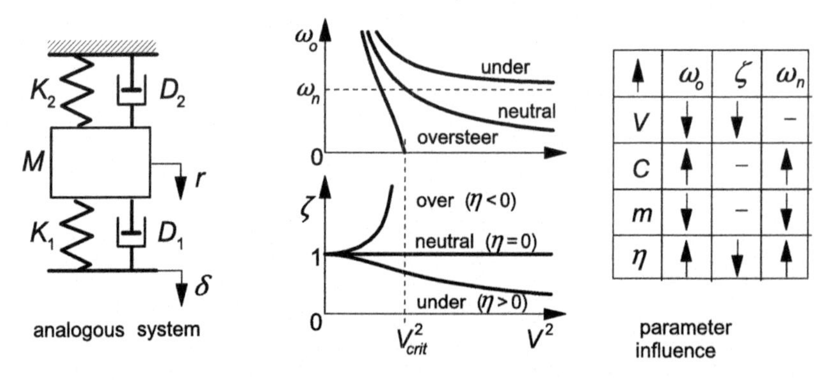


| ↑ | $\omega_o$ | $\zeta$ | $\omega_n$ |
|---|---|---|---|
| V | ↓ | ↓ | – |
| C | ↑ | – | ↑ |
| m | ↓ | – | ↓ |
| $\eta$ | ↑ | ↓ | ↑ |

**FIGURE 1.13** The influence of parameters on natural frequency and damping.

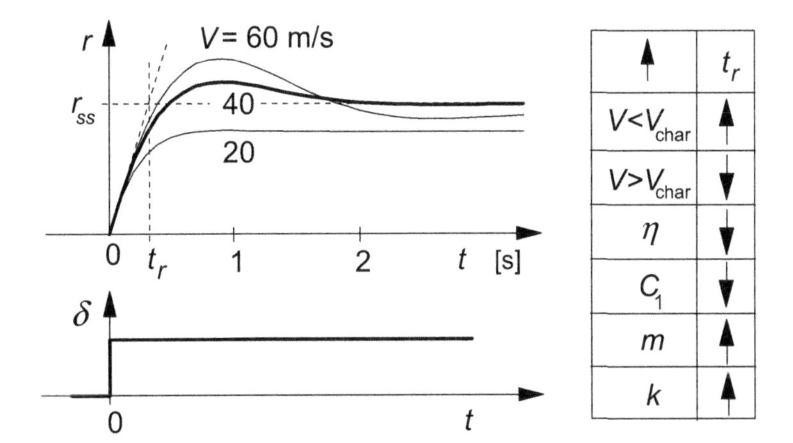


**FIGURE 1.14** Step response of yaw rate to steer angle. Parameters according to Table 1.1. Parameter influence on the rise time $t_r$.

The yaw rate response to a step change in steer angle is typified by the rise time $t_r$ indicated in Figure 1.14 and expressed in terms of the parameters as follows:

$$t_r = \frac{r_{ss}}{\left(\dfrac{\partial r}{\partial t}\right)_{t=0}} = \frac{mk^2V}{aC_1 l\left(1+\dfrac{\eta}{gl}V^2\right)} = \frac{mk^2V}{\dfrac{a}{l}\left\{C_1 l^2 + \left(b - a\dfrac{C_1}{C_2}\right)mV^2\right\}} \tag{1.73}$$

which may be readily obtained with the aid of Eqns (1.46, 1.47).

The parameter influence has been indicated in the figure. The results correspond qualitatively well with the 90% response times found in vehicle model simulation studies. A remarkable result is that for an understeered automobile, the response time is smaller than for an oversteered car.

### *Forced Linear Vibrations*

The conversion of the equations of motion (1.46) into the standard state space representation is useful when the linear system properties are the subject of investigation. The system at hand is of the second order and hence possesses two state variables for which we choose $v$ and $r$. The system is subjected to a single input signal: the steer angle $\delta$. Various variables may be of interest to analyze the vehicle's response to steering input oscillations. The following quantities are selected to illustrate the method and to study the dynamic behavior of the vehicle: the lateral acceleration $a_y$ of the center of gravity of the vehicle, the yaw rate $r$, and the vehicle slip angle $\beta$ defined at the center of gravity. In matrix notation, the equation becomes

$$\begin{aligned} \dot{\boldsymbol{x}} &= \boldsymbol{Ax} + \boldsymbol{Bu} \\ \boldsymbol{y} &= \boldsymbol{Cx} + \boldsymbol{Du} \end{aligned} \tag{1.74}$$

with

$$\dot{\boldsymbol{x}} = \begin{pmatrix} \dot{v} \\ \dot{r} \end{pmatrix}, \quad \boldsymbol{u} = \delta, \quad \boldsymbol{y} = \begin{pmatrix} a_y \\ r \\ \beta \end{pmatrix} = \begin{pmatrix} \dot{v} + Vr \\ r \\ -v/V \end{pmatrix} \tag{1.75}$$

and

$$\boldsymbol{A} = -\begin{pmatrix} \dfrac{C}{mV} & V + \dfrac{Cs}{mV} \\ \dfrac{Cs}{mk^2V} & \dfrac{Cq^2}{mk^2V} \end{pmatrix}, \quad \boldsymbol{B} = \begin{pmatrix} \dfrac{C_1}{m} \\ \dfrac{C_1 a}{mk^2} \end{pmatrix}$$

$$\boldsymbol{C} = -\begin{pmatrix} \dfrac{C}{mV} & \dfrac{Cs}{mV} \\ 0 & -1 \\ 1/V & 0 \end{pmatrix}, \quad \boldsymbol{D} = \begin{pmatrix} \dfrac{C_1}{m} \\ 0 \\ 0 \end{pmatrix}$$

The frequency response functions have been computed using Matlab software. Figure 1.15 presents the amplitude and phase response functions for each of the three output quantities and at three different values of speed of travel. The values of the chosen model parameters and a number of characteristic quantities have been listed in Table 1.1.

Explicit expressions of the frequency response functions in terms of model parameters are helpful to understand and predict the characteristic aspects of these functions which may be established by means of computations or possibly through full-scale experiments.

From the differential Eqn (1.50), the frequency response function is easily derived. Considering the quantities formulated by (1.70) and (1.71) and the steady-state response $(r/\delta)_{ss} = (V/R)/\delta$ obtained from (1.56), we find

$$\frac{r}{\delta}(j\omega) = \left(\frac{r}{\delta}\right)_{ss} \cdot \frac{1 + \dfrac{mVa}{C_2 l} j\omega}{1 - \left(\dfrac{\omega}{\omega_o}\right)^2 + 2\zeta\left(\dfrac{j\omega}{\omega_o}\right)}; \quad \left(\frac{r}{\delta}\right)_{ss} = \frac{V/l}{1 + \dfrac{\eta}{gl}V^2} \tag{1.77}$$

Similarly, the formula for the response of lateral acceleration $a_y$ can be derived:

$$\frac{a_y}{\delta}(j\omega) = \left(\frac{a_y}{\delta}\right)_{ss} \cdot \frac{1 - \dfrac{mk^2}{C_2 l}\omega^2 + \dfrac{b}{V} j\omega}{1 - \left(\dfrac{\omega}{\omega_o}\right)^2 + 2\zeta\left(\dfrac{j\omega}{\omega_o}\right)}; \quad \left(\frac{a_y}{\delta}\right)_{ss} = \frac{V^2/l}{1 + \dfrac{\eta}{gl}V^2} \tag{1.78}$$

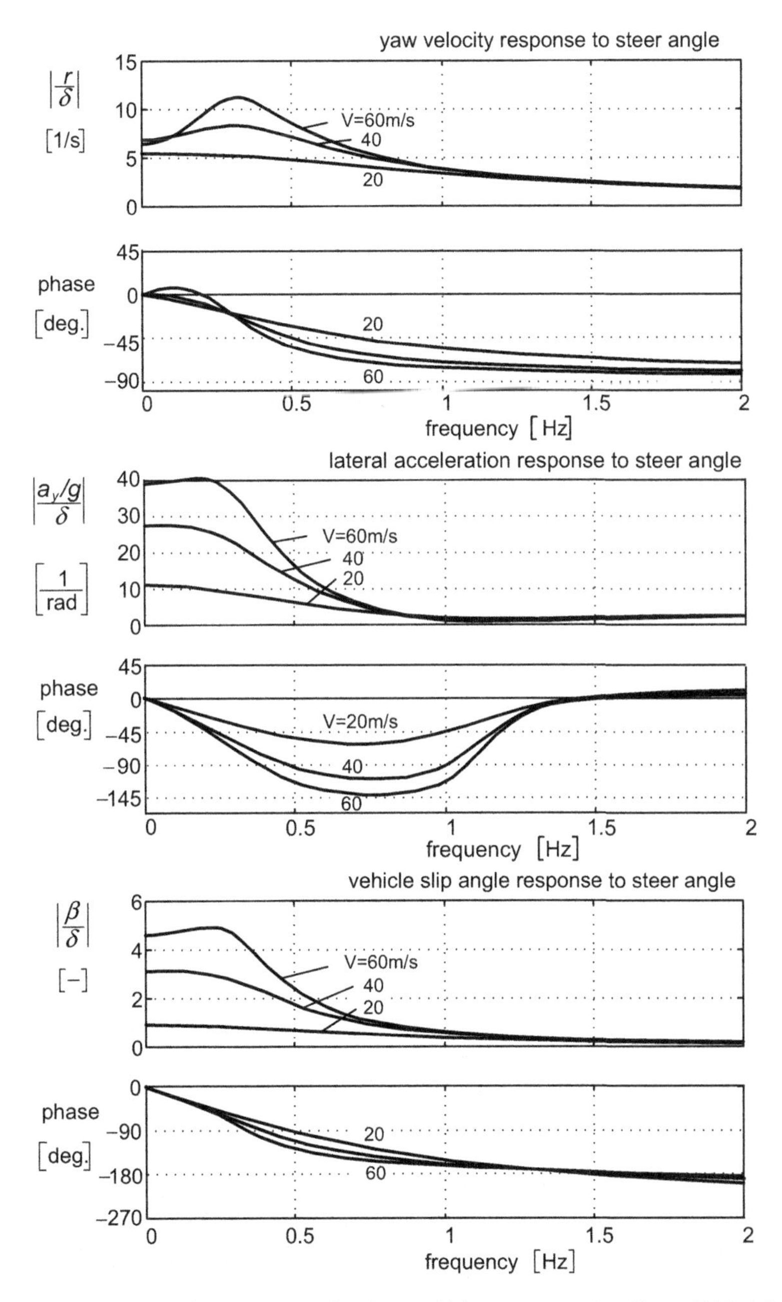


**FIGURE 1.15** Frequency response functions; vehicle parameters according to Table 1.1.

**TABLE 1.1 Parameter Values and Typifying Quantities**

| Parameters | | Derived Typifying Quantities | | | | | |
|---|---|---|---|---|---|---|---|
| $a$ | 1.4 m | $l$ | 3 m | $V$ [m/s] | 20 | 40 | 60 |
| $b$ | 1.6 m | $F_{z1}$ | 8371 N | $\omega_o$ [rad/s] | 4.17 | 2.6 | 2.21 |
| $C_1$ | 60000 N/rad | $F_{z2}$ | 7325 N | $\zeta$ [–] | 0.9 | 0.7 | 0.57 |
| $C_2$ | 60000 N/rad | $q$ | 1.503 m | $\omega_n$ [rad/s] | 1.8 | 1.8 | 1.82 |
| $m$ | 1600 kg | $s$ | −0.1 m | $t_r$ [s] | 0.23 | 0.3 | 0.27 |
| $k$ | 1.5 m | $\eta$ | 0.0174 rad (~1° extra steer/$g$ lateral acceleration) | | | | |

And, for the slip angle $\beta$,

$$\frac{\beta}{\delta}(j\omega) = \left(\frac{\beta}{\delta}\right)_{ss} \cdot \frac{1 - \dfrac{mk^2V}{amV^2 - bC_2 l} j\omega}{1 - \left(\dfrac{\omega}{\omega_o}\right)^2 + 2\zeta\left(\dfrac{j\omega}{\omega_o}\right)}; \quad \left(\frac{\beta}{\delta}\right)_{ss} = -\frac{b}{l}\frac{1 - \dfrac{a}{l}\dfrac{m}{C_2 l}V^2}{1 + \dfrac{\eta}{gl}V^2} \tag{1.79}$$

By considering Eqn (1.77), it can now be explained that for instance at higher frequencies the system exhibits features of a first-order system: because of the $j\omega$ term in the numerator, the yaw rate amplitude response tends to decay at a 6 dB per octave rate (when plotted in log–log scale) and the phase lag approaches 90 degrees. The phase increase at low frequencies and higher speeds is due to the presence of the same $jw$ term in the numerator. At speeds beyond approximately the characteristic speed, the corresponding (last) term in the denominator has less influence on the initial slope of the phase characteristic. The lateral acceleration response (1.78) shown in the center graph of Figure 1.15 gives a finite amplitude at frequencies tending to infinity because of the presence of $\omega^2$ in the numerator. For the same reason, the phase lag goes back to zero at large frequencies. The side slip phase response tends to −270 degrees (at larger speeds) which is due to the negative coefficient of $j\omega$ in the numerator of (1.79). This is in contrast to that coefficient of the yaw rate response (1.77).

It is of interest to see that the steady-state slip angle response, indicated in (1.79), changes sign at a certain speed $V$. At low speeds where the tire slip angles are still very small, the vehicle slip angle obviously is negative for positive steer angle (considering positive directions as adopted in Figure 1.11). At larger velocities, the tire slip angles increase and, as a result, $\beta$ changes to the positive direction.

**Exercise 1.2. Four-Wheel Steer at the Condition that the Vehicle Slip Angle Vanishes**

Consider the vehicle model of Figure 1.16. Both the front and the rear wheels can be steered. The objective is to make the vehicle move at a slip angle $\beta$ remaining equal to zero. In practice, this may be done to improve handling qualities of the automobile (reduces to first-order system!) and to avoid excessive side slipping motions of the rear axle in lane change maneuvers. Adapt the equations of motion (1.46) and assess the required relationship between the steer angles $\delta_1$ and $\delta_2$. Do this in terms of the transfer function between $\delta_2$ and $\delta_1$ and the associated differential equation. Find the steady-state ratio $(\delta_2/\delta_1)_{ss}$ and plot this as a function of the speed $V$. Show also the frequency response function $\delta_2/\delta_1$ $(j\omega)$ for the amplitude and phase at a speed $V = 30$ m/s. Use the vehicle parameters supplied in Table 1.1.

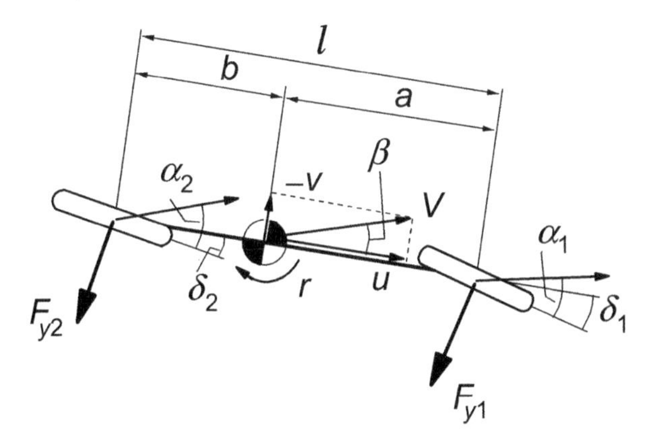


FIGURE 1.16 'Four-wheel' steering to make slip angle $\beta = 0$ (Exercise 1.2).

### 1.3.3. Nonlinear Steady-State Cornering Solutions

From Eqns (1.42) and (1.59) with the same restrictions as stated below Eqn (1.60), the following force balance equations can be derived (follows also from Eqn (1.61)). The effect of the pneumatic trails will be dealt with later on.

$$\frac{F_{y1}}{F_{z1}} = \frac{F_{y2}}{F_{z2}} = \frac{a_y}{g} \quad \left(= \frac{K}{mg}\right) \tag{1.80}$$

where $K = ma_y$ represents the centrifugal force. The kinematic relationship

$$\delta - (\alpha_1 - \alpha_2) = \frac{l}{R} \tag{1.81}$$

follows from Eqns (1.44) and (1.51). In Figure 1.11 the vehicle model has been depicted in a steady-state cornering maneuver. It can easily be observed from this diagram that relation (1.81) holds approximately when the angles are small.

The ratio of the side force and vertical load as shown in (1.80) plotted as a function of the slip angle may be termed as the normalized tire or axle characteristic. These characteristics subtracted horizontally from each other produce the 'handling curve'. Considering the equalities (1.80), the ordinate may be replaced by $a_y/g$. The resulting diagram with abscissa $\alpha_1-\alpha_2$ is the nonlinear version of the right-hand diagram of Figure 1.10 (rotated 90° anti-clockwise). The diagram may be completed by attaching the graph that shows, for a series of speeds $V$, the relationship between lateral acceleration (in $g$ units) $a_y/g$ and the relative path curvature $l/R$ according to Eqn (1.55).

Figure 1.17 shows the normalized axle characteristics and the completed handling diagram. The handling curve consists of a main branch and two side lobes. The different portions of the curves have been coded to indicate the corresponding parts of the original normalized axle characteristics they

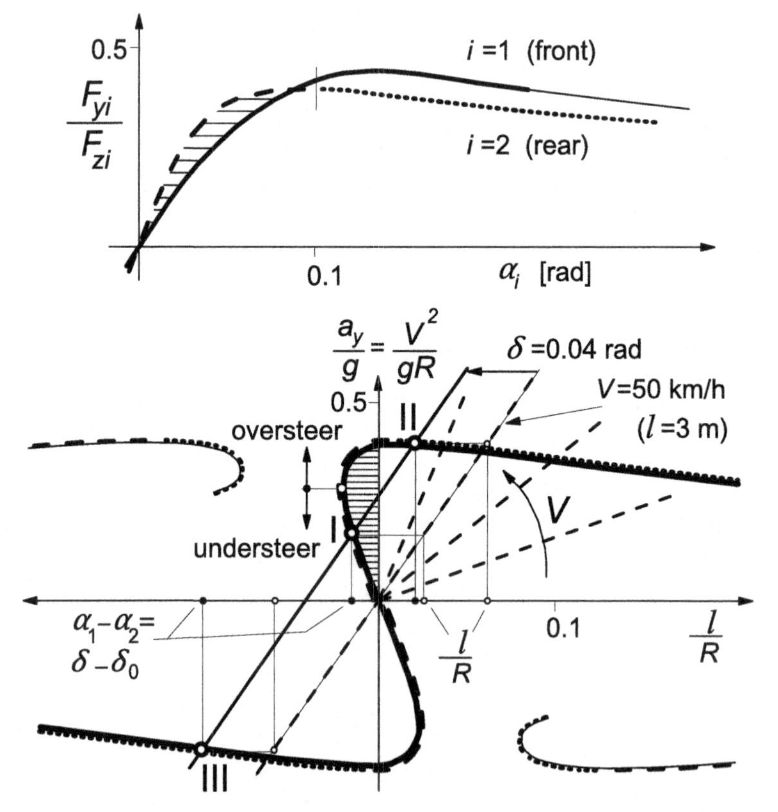


**FIGURE 1.17** Handling diagram resulting from normalized tire characteristics. Equilibrium points I, II, and III (steady turns), of which only I is stable, arise for speed $V = 50$ km/h and steer angle $\delta = 0.04$ rad. From the different line types, the manner in which the curves are obtained from the upper diagram may be retrieved.

originate from. Near the origin, the system may be approximated by a linear model. Consequently, the slope of the handling curve in the origin with respect to the vertical axis is equal to the understeer coefficient $\eta$. In contrast to the straight handling line of the linear system (Figure 1.10), the nonlinear system shows a curved line. The slope changes along the curve which means that the degree of understeer changes with increasing lateral acceleration. The diagram of Figure 1.17 shows that the vehicle considered changes from understeer to oversteer. We define

$$\begin{aligned} &\text{understeer if:} \quad \left(\frac{\partial \delta}{\partial V}\right)_R > 0 \\ &\text{oversteer if:} \quad \left(\frac{\partial \delta}{\partial V}\right)_R < 0 \end{aligned} \tag{1.82}$$

The family of straight lines represents the relationship between acceleration and curvature at different levels of speed. The speed line belonging to $V = 50$ km/h has been indicated (wheel base $l = 3$ m). This line is shifted to the left over a distance equal to the steer angle $\delta = 0.04$ rad and three points of intersection with the handling curve arise. These points I, II, and III indicate the possible equilibrium conditions at the chosen speed and steer angle. The connected values of the relative path curvature $l/R$ can be found by going back to the speed line. As will be shown further on, only point I refers to a stable cornering motion. In points II and III ($R < 0$!), the motion is unstable.

At a given speed $V$, a certain steer angle $\delta$ is needed to negotiate a circular path with given radius $R$. The steer angle required can be read directly from the handling diagram. The steer angle needed to negotiate the same curve at very low speed ($V \rightarrow 0$) tends to $l/R$. This steer angle is denoted by $\delta_o$. Consequently, the abscissa of the handling curve $\alpha_1 - \alpha_2$ may as well be replaced by $\delta - \delta_o$. This opens the possibility to determine the handling curve with the aid of simple experimental means, i.e., measuring the steering wheel input (reduced to equivalent road wheel steer angle by means of the steering ratio, which method automatically includes steering compliance effects) at various speeds running over the same circular path.

Subtracting normalized characteristics may give rise to very differently shaped handling curves only by slightly modifying the original characteristics. As Figure 1.17 shows, apart from the main branch passing through the origin, isolated branches may occur. These are associated with at least one of the decaying ends of the pair of normalized tire characteristics.

In Figure 1.18 a set of four possible combinations of axle characteristics have been depicted together with the resulting handling curves. This collection of characteristics shows that the nature of steering behavior is entirely governed by the normalized axle characteristics and in particular their relative shape with respect to each other.

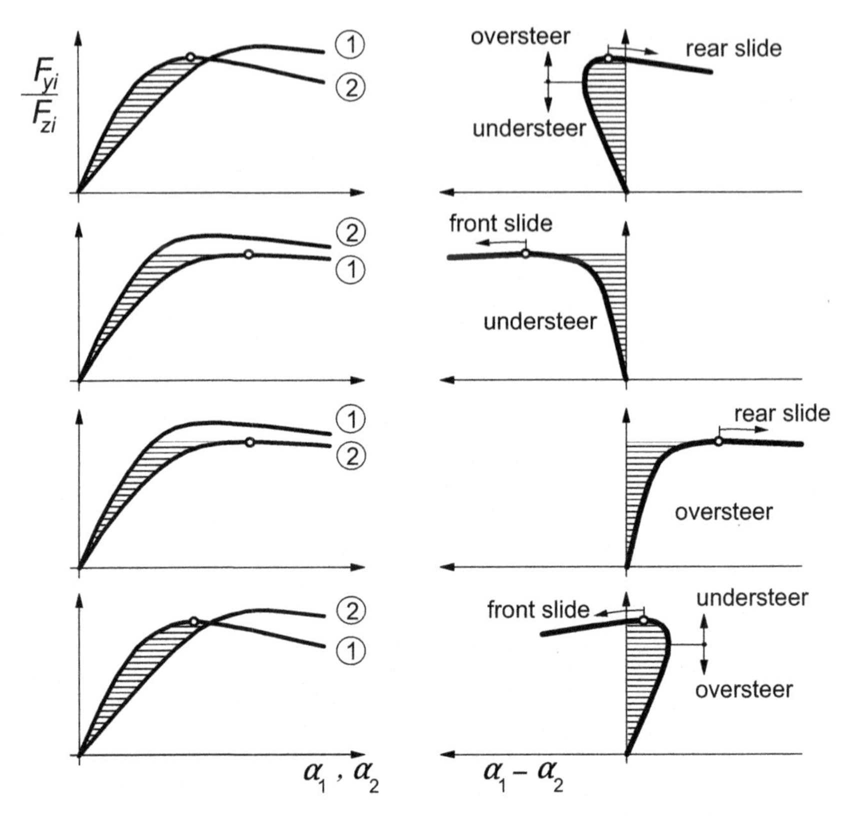


**FIGURE 1.18** A number of handling curves arising from the pairs of normalized tire characteristics shown in left. Only the main branch of the handling curve has been drawn (1: front, 2: rear).

The way in which we can use the handling diagram is presented in Figure 1.19. The speed of travel may be kept constant and the lateral acceleration is increased by running over a spiral path with decreasing radius. The required variation of the steer angle follows from the distance between the handling curve and the speed line. Similarly, we can observe what happens when the path curvature is kept constant and the speed is increased. Also, the resulting variation of the curvature at a constant steer angle and increasing speed can be found. More general cases of quasi-steady-state motions may be studied as well.

### *Stability of the Motion at Large Lateral Accelerations*

The nonlinear set of Eqns (1.42–44) may be linearized around the point of operation, which is one of the equilibrium states indicated above. The resulting second-order differential equation has a structure similar to Eqn (1.64) or (1.47)

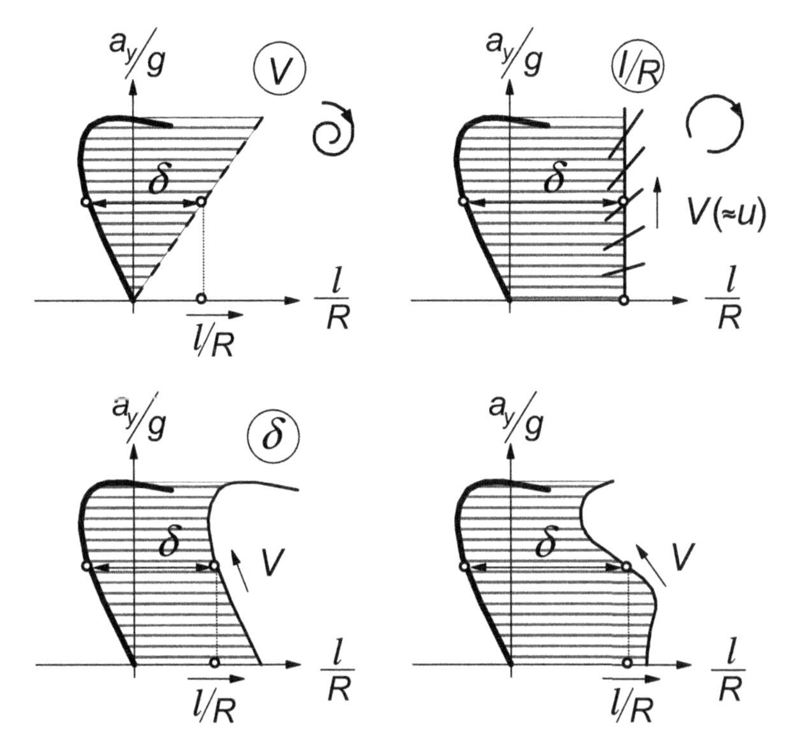


**FIGURE 1.19** Types of quasi-steady-state maneuvers.

but with the variables replaced by their small variations with respect to the steady-state condition considered. Analysis of the coefficients of the characteristic equation reveals if stability exists. Also the nature of stability (monotonous, oscillatory) follows from these coefficients. This is reflected by the type of singular points (node, spiral, saddle) representing the equilibrium solutions in the phase plane as treated in the next section.

It now turns out that not only the last coefficient can become negative but also the second coefficient $a_1$. Instead of the cornering stiffnesses $C$ defined in the origin of the tire cornering characteristics, the slope of the normalized characteristics at a given level of $a_y/g$ becomes now of importance. We define

$$\Phi_i = \frac{1}{F_{zi}} \frac{\partial F_{yi}}{\partial \alpha_i} \quad (i = 1, 2) \tag{1.83}$$

The conditions for stability, that is, second and last coefficient of equation comparable with Eqn (1.47) must be positive, read after having introduced the radius of gyration $k$ ($k^2 = I/m$):

$$b(k^2 + a^2)\Phi_1 + a(k^2 + b^2)\Phi_2 > 0 \tag{1.84}$$

$$\Phi_1 \Phi_2 \left( \frac{\partial \delta}{\partial 1/R} \right)_V > 0 \tag{1.85}$$

The subscript $V$ refers to the condition of differentiation with $V$ kept constant, that is while staying on the speed line of Figure 1.17. The first condition (1.84) may be violated when we deal with tire characteristics showing a peak in side force and a downward sloping further part of the characteristic. The second condition corresponds to condition (1.65) for the linear model. Accordingly, instability is expected to occur beyond the point where the steer angle reaches a maximum while the speed is kept constant. This, obviously, can only occur in the oversteer range of operation. In the handling diagram the stability boundary can be assessed by finding the tangent to the handling curve that runs parallel to the speed line considered.

In the upper diagram of Figure 1.20 the stability boundary that holds for the right part of the diagram ($a_y$ vs $l/R$) has been drawn for the system of Figure 1.17 that changes from initial understeer to oversteer. In the middle diagram, a number of shifted $V$-lines, each for a different steer angle $\delta$, has been indicated. In each case, the points of intersection represent possible steady-state solutions. The highest point represents an unstable solution as the corresponding point on the speed line lies in the unstable area. When the steer angle is increased, the two points of intersections move toward each other. It turns out that for this type of handling curve, a range of $\delta$ values exists without intersections with the positive half of the curve. The fact that both right-hand turn solutions may vanish has serious implications which follow from the phase plot. At increased steer angle, however, new solutions may show up. At first, these solutions appear to be unstable, but at rather large steer angles of more than about 0.2 rad we find again stable solutions. These occur on the isolated branch where $\alpha_2$ is small and $\alpha_1$ is large. Apparently, we find that the vehicle that increases its speed while running at a constant turning radius will first cross the stability boundary and may then recover its stability by turning the steering wheel to a relatively large angle. In the diagram, the left part of the isolated branch is reached where stable spirals appear to occur. This phenomenon may correspond to similar experiences in the racing practice, cf. Jenkinson (1958).

The lower diagram depicts the handling curve for a car that remains understeered throughout the lateral acceleration range. Everywhere the steady-state cornering motion remains stable. Up to the maximum of the curve, the tangents slope to the left and cannot run parallel to a speed line. Beyond the peak, however, we can find a speed line parallel to the tangent, but at the same time one of the slopes ($\Phi_1$) of the normalized axle characteristics starts to show a negative sign so that condition (1.85) is still satisfied. Similarly, the limit oversteer vehicle of the upper graph remains unstable beyond the peak. On the isolated part of the handling curve of the lower diagram, the motion remains unstable. It will be clear that the isolated branches vanish when we deal with axle characteristics that do not show a peak and decaying part of the curve.

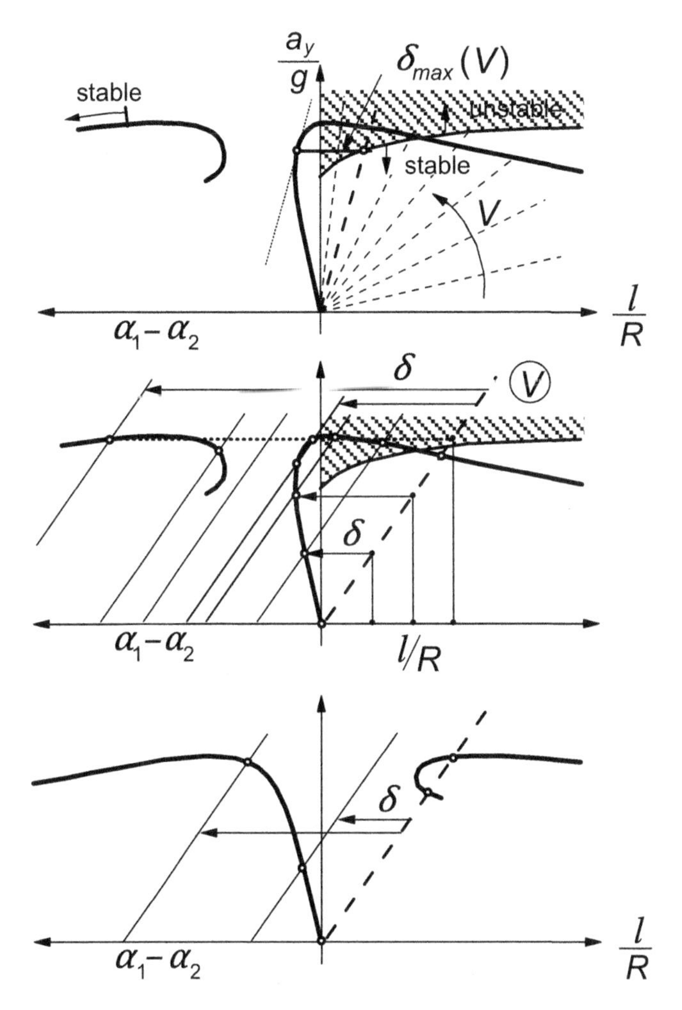


**FIGURE 1.20** Construction of stability boundary (upper diagram, from Figure 1.17). On the isolated branch a stable range may occur (large steer angle as indicated in middle diagram). The lower diagram shows the case with complete understeer featuring a stable main branch.

It may seem that the establishment of unstable solutions has no particular value. It will become clear, however, that the existence and the location of both stable and unstable singular points play an important role in shaping the trajectories in the phase plane. Also, the nature of stability or instability in the singular points is of importance.

---

**Exercise 1.3 Construction of the Complete Handling Diagram from Pairs of Axle Characteristics**

We consider three sets of hypothetical axle characteristics (a, b, and c) shown in the graph of Figure 1.21. The dimensions of the vehicle model are: $a = b = \frac{1}{2} l = 1.5$ m.

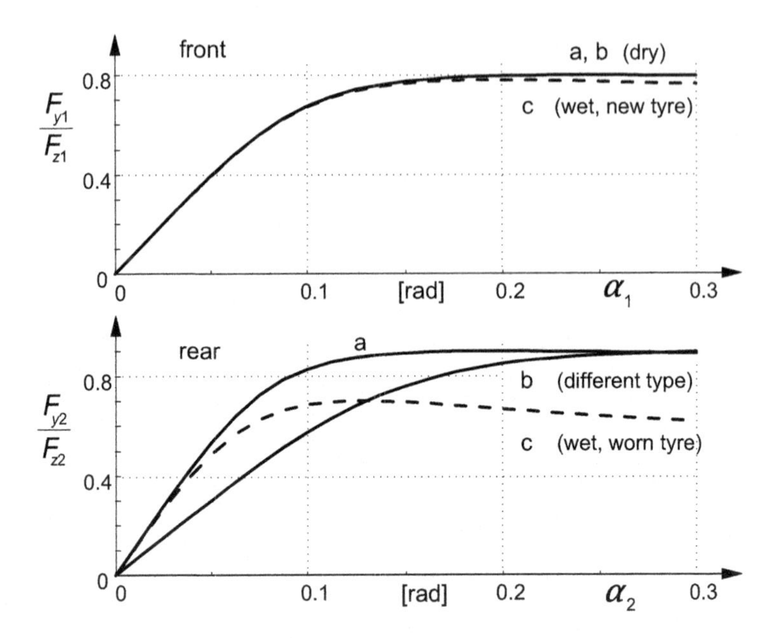


**FIGURE 1.21** Three sets of hypothetical axle cornering characteristics (Exercise 1.3).

For the tires, we may employ axle characteristics described by the *magic formula* (1.6):

$$F_y = D \sin[C \arctan\{B\alpha - E(B\alpha - \arctan(B\alpha))\}]$$

We define the peak side force $D = \mu F_z$ and the cornering stiffness $C_{F\alpha} = BCD = c_{F\alpha} F_z$ so that $B = c_{F\alpha}/(C\mu)$. For the six tire/axle configurations, the parameter values have been given in the table below.

| axle | case | $\mu$ | $c_{F\alpha}$ | $C$ | $E$ |
|---|---|---|---|---|---|
| front | a,b | 0.8 | 8 | 1.2 | −2 |
| | c | 0.78 | 8 | 1.3 | −2 |
| rear | a | 0.9 | 11 | 1.2 | −2 |
| | b | 0.9 | 6 | 1.2 | −2 |
| | c | 0.65 | 11 | 1.5 | −1 |

Determine, for each of the three combinations (two dry, one wet),

1. The handling curve (cf. Figure 1.17).
2. The complete handling diagram (cf. Figure 1.17).
3. The portion of t curves where the vehicle shows an oversteer nature.

4. The stability boundary (associated with these oversteer ranges) in the ($a_y/g$ versus $l/R$) diagram (= right-hand side of the handling diagram) (cf. Figure 1.20).
5. Indicate in the diagram (or in a separate graph):
   a. the course of the steer angle $\delta$ required to negotiate a curve with radius $R = 60$ m as a function of the speed $V$. If applicable, indicate the stability boundary, that is, the critical speed $V_{crit}$, belonging to this radius.
   b. the course of steer angle $\delta$ as a function of relative path curvature $l/R$ at a fixed speed $V = 72$ km/h. If applicable, assess the critical radius $R_{crit}$.

For the vehicle systems considered so far, a unique handling curve appears to suffice to describe the steady-state turning behavior. Cases may occur, however, where more curves are needed, one for each velocity. A simple example is the situation when the car runs over a wet surface where the tire characteristics change considerably with speed. Also, as a result of the down forces acting on e.g., the body of a racing car, the tire loads increase with speed. Consequently, the tire characteristics change accordingly which requires an adaptation of the handling curve.

A more difficult and fundamentally different situation occurs when the vehicle is equipped with a third axle. Also in this case multiple handling curves arise. A tandem rear axle configuration of a heavy truck, for example, strongly opposes movement along a curved track. The slip angles of the two rear axles are different so that a counteracting torque arises. This torque gets larger when the turning radius becomes smaller. This may for instance occur at a given level of lateral acceleration. When at this level the speed becomes lower, the curvature must become larger and the opposing torque will increase which entails an increased front steer angle to generate a larger side force needed to balance the vehicle. This increased steer angle goes on top of the steer angle which was already larger because of the increased $l/R$. Here, $l$ is the average wheel base. Consequently, in the handling diagram, the points on the handling curve belonging to the lower speed lie more to the left. For a detailed study on this special subject, we refer to Winkler (1998).

### Assessment of the Influence of Pneumatic Trail on Handling Curve

So far the direct influence of the pneumatic trails has not been taken into account. As with the linear analysis, we may do this by considering the effective axle positions

$$a' = a - t_1, \quad b' = b + t_2 \quad \text{and} \quad l' = a' + b' \tag{1.86}$$

The difficulty we have to face now is the fact that these pneumatic trails $t_i$ will vary with the respective slip angles. We have, if the residual torques are neglected,

$$t_i(\alpha_i) = -\frac{M_{zi}(\alpha_i)}{F_{yi}(\alpha_i)} \tag{1.87}$$

Introducing the effective axle loads

$$F'_{z1} = \frac{b'}{l'}mg, \quad F'_{z2} = \frac{a'}{l'}mg \tag{1.88}$$

yields for the lateral force balance instead of (1.80):

$$\frac{F_{y1}}{F'_{z1}} = \frac{F_{y2}}{F'_{z2}} = \frac{a_y}{g} \tag{1.89}$$

or after some rearrangements:

$$\frac{a'}{a}\frac{F_{y1}}{F_{z1}} = \frac{b'}{b}\frac{F_{y2}}{F_{z2}} = Q\frac{a_y}{g} \tag{1.90}$$

where

$$Q = \frac{l}{l'}\frac{a'b'}{ab} \approx 1 \tag{1.91}$$

The corrected normalized side force characteristics as indicated in (1.90) can be computed beforehand and drawn as functions of the slip angles, and the normal procedure to assess the handling curve can be followed. This can be done by taking the very good approximation $Q = 1$ or we might select a level of $Qa_y/g$ and then assess the values of the slip angles that belong to that level of the corrected normalized side forces and compute $Q$ according to (1.91) and from that the correct value of $a_y/g$.

### *Large Deviations with Respect to the Steady-State Motion*

The variables $r$ and $v$ may be considered as the two state variables of the second-order nonlinear system represented by the Eqn (1.42). Through computer numerical integration, the response to a given arbitrary variation of the steer angle can be easily obtained. For motions with constant steer angle $\delta$ (possibly after a step change), the system is autonomous and the phase-plane representation may be used to find the solution. For that, we proceed by eliminating the time from Eqn (1.42). The result is a first-order nonlinear equation (using $k^2 = I/m$):

$$\frac{\mathrm{d}v}{\mathrm{d}r} = k^2\frac{F_{y1} + F_{y2} - mVr}{aF_{y1} - bF_{y2}} \tag{1.92}$$

Since $F_{y1}$ and $F_{y2}$ are functions of $\alpha_1$ and $\alpha_2$, it may be easier to take $\alpha_1$ and $\alpha_2$ as the state variables. With (1.44), we obtain

$$\frac{\mathrm{d}\alpha_2}{\mathrm{d}\alpha_1} = \frac{\dfrac{\mathrm{d}v}{\mathrm{d}r} - b}{\dfrac{\mathrm{d}v}{\mathrm{d}r} + a} \tag{1.93}$$

which becomes, with (1.92),

$$\frac{d\alpha_2}{d\alpha_1} = \frac{\dfrac{F_{y2}(\alpha_2)}{F_{z2}} - (\delta - \alpha_1 + \alpha_2)\dfrac{V^2}{gl}}{\dfrac{F_{y1}(\alpha_1)}{F_{z1}} - (\delta - \alpha_1 + \alpha_2)\dfrac{V^2}{gl}} \tag{1.94}$$

For the sake of simplicity, we assumed $I/m = k^2 = ab$.

By using Eqn (1.94), the trajectories (solution curves) can be constructed in the $(\alpha_1, \alpha_2)$ plane. The isocline method turns out to be straightforward and simple to employ. The pattern of the trajectories is strongly influenced by the so-called singular points. In these points the motion finds an equilibrium. In the singular points, the motion is stationary and consequently the differentials of the state variables vanish.

From the handling diagram, $K/mg$ and $l/R$ are readily obtained for given combinations of $V$ and $\delta$. Used in combination with the normalized tire characteristics $F_{y1}/F_{z1}$ and $F_{y2}/F_{z2}$, the values of $\alpha_1$ and $\alpha_2$ are found, which form the coordinates of the singular points. The manner in which a stable turn is approached and from what collection of initial conditions such a motion can or cannot be attained may be studied in the phase plane. One of the more interesting results of such an investigation is the determination of the boundaries of the domain of attraction in case such a domain with finite dimensions exists. The size of the domain may give indications as to the so-called stability in the large. In other words, the question may be answered: does the vehicle return to its original steady-state condition after a disturbance and to what degree does this depend on the magnitude and point of application of the disturbance impulse?

For the construction of the trajectories, we draw isoclines in the $(\alpha_1, \alpha_2)$ plane. These isoclines are governed by Eqn (1.94) with slope $d\alpha_2/d\alpha_1$ kept constant. The following three isoclines may already provide sufficient information to draw estimated courses of the trajectories. We have, for $k^2 = ab$,

*vertical intercepts* ($d\alpha_2/d\alpha_1 \rightarrow \infty$):

$$\alpha_2 = \frac{gl}{V^2}\frac{F_{y1}(\alpha_1)}{F_{z1}} + \alpha_1 - \delta \tag{1.95}$$

*horizontal intercepts* ($d\alpha_2/d\alpha_1 \rightarrow 0$):

$$\alpha_1 = -\frac{gl}{V^2}\frac{F_{y2}(\alpha_2)}{F_{z2}} + \alpha_2 + \delta \tag{1.96}$$

*intercepts under 45°* ($d\alpha_2/d\alpha_1 = 1$):

$$\frac{F_{y1}(\alpha_1)}{F_{z1}} = \frac{F_{y2}(\alpha_2)}{F_{z2}} \tag{1.97}$$

Figure 1.22 illustrates the way these isoclines are constructed. The system of Figure 1.17 with $k = a = b$, $\delta = 0.04$ rad, and $V = 50$ km/h has been considered. Note that the normalized tire characteristics appear in the left-hand diagram for the construction of the isoclines. The three points of intersection of the isoclines are the singular points. They correspond to the points I, II, and III of Figure 1.17. The stable point is a focus (spiral) point with a complex pair of solutions of the characteristic equation with a negative real part. The two unstable points are of the saddle type corresponding to a real pair of solutions, one of which is positive. The direction in which the motion follows the trajectories is still a question to be examined. Also for this purpose, the alternative set of axes with $r$ and $v$ as coordinates (multiplied with a factor) has been introduced in the diagram after using the relations (1.44).

From the original Eqn (1.42), it can be found that the isocline (1.97) forms the boundary between areas with $\dot{r} > 0$ and $\dot{r} < 0$ (indicated in Figure 1.22). Now it is easy to ascertain the direction along the trajectories. We note that the system exhibits a bounded domain of attraction. The boundaries are called separatrices. Once outside the domain, the motion finds itself in an unstable situation. If the disturbance remains limited in magnitude, so that resulting initial conditions of the state variables stay within the boundaries, then ultimately the steady-state condition is reached again.

For systems with normalized characteristics showing everywhere a positive slope, a handling curve arises that consists of only the main branch through the origin. If the rear axle characteristic (at least in the end) is higher than the front axle characteristic, the vehicle will show (at least in the limit) an understeer

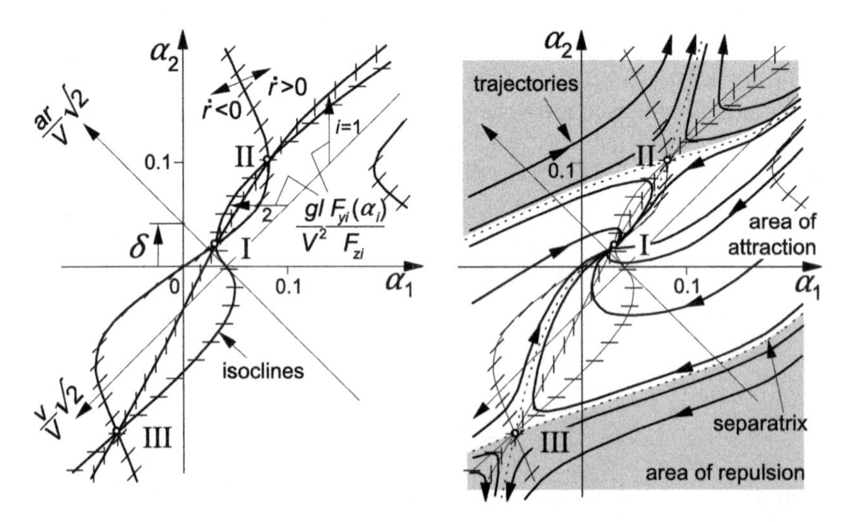


**FIGURE 1.22** Isoclines for the construction of trajectories in the phase plane. Also shown: the three singular points I, II, and III (cf. Figure 1.17) and the separatrices constituting the boundary of the domain of attraction. Point I represents the stable cornering motion at steer angle $\delta$.

nature and unstable singular points cannot occur. It will occur at least if in the case of initial oversteer the speed remains under the critical speed. In such cases, the domain of attraction is theoretically unbounded so that for all initial conditions ultimately the stable equilibrium is attained. The domain of Figure 1.22 appears to be open on two sides which means that initial conditions, in a certain range of ($r/v$) values, do not require to be limited in order to reach the stable point. Obviously, disturbance impulses acting in front of the center of gravity may give rise to such combinations of initial conditions.

In Figures 1.23 and 1.24, the influence of an increase in steer angle $\delta$ on the stability margin (distance between stable point and separatrix) has been shown for the two vehicles considered in Figure 1.20. The system of Figure 1.23 is clearly much more sensitive. An increase in $\delta$ (but also an increase in speed $V$) reduces the stability margin until it is totally vanished as soon as the two singular points merge (also the corresponding points I and II on the handling curve of Figure 1.17) and the domain breaks open. As a result, all trajectories starting above the lower separatrix tend to leave the area. This can only be

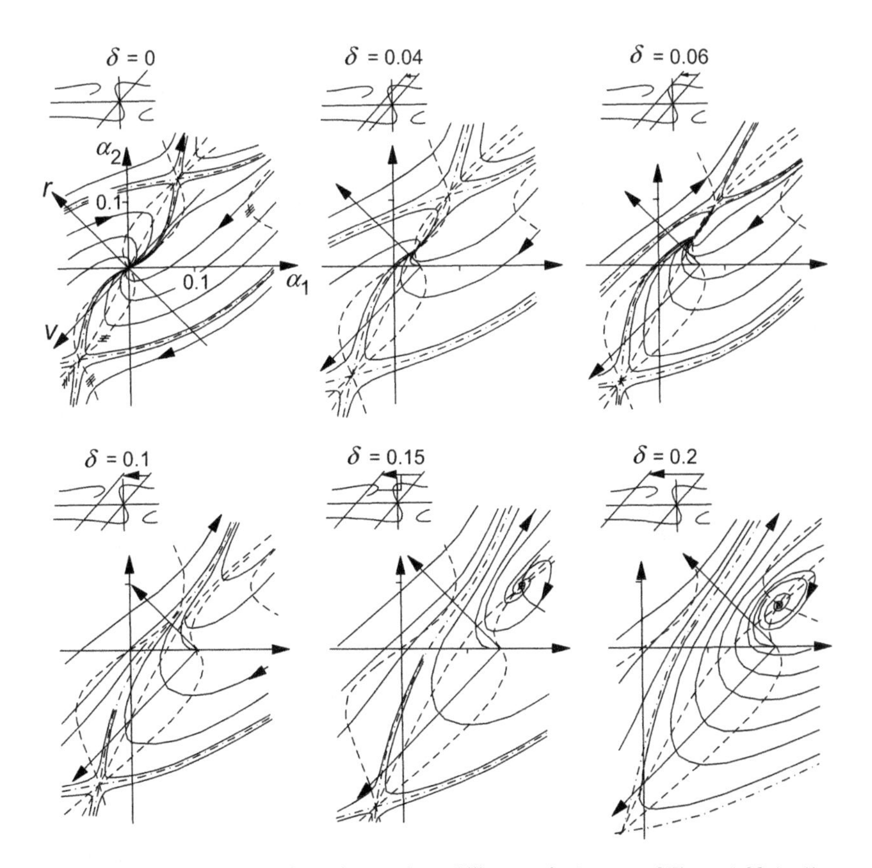


**FIGURE 1.23** Influence of steering on the stability margin (system of Figure 1.20 (top)).

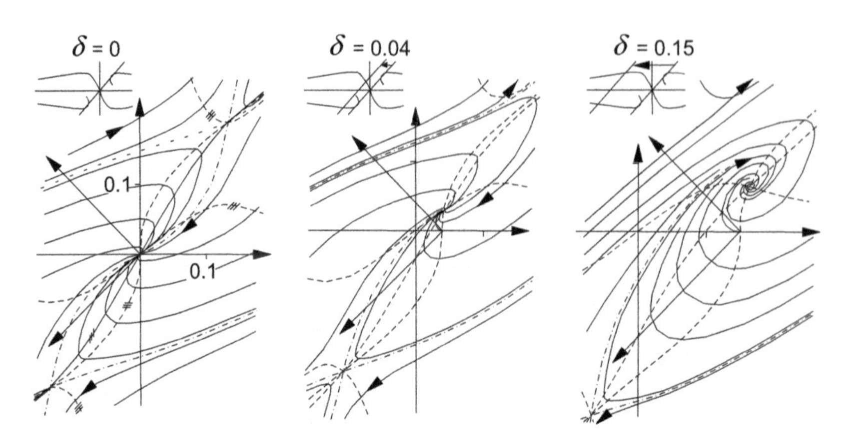


**FIGURE 1.24** Influence of steering on the stability margin (system of Figure 1.20 (bottom)).

stopped by either quickly reducing the steer angle or enlarging $\delta$ to around 0.2 rad or more. The latter situation appears to be stable again (focus) as has been stated before. For the understeered vehicle of Figure 1.24, stability is practically always ensured.

For a further appreciation of the phase diagram, it is of interest to determine the new initial state ($r_o$, $v_o$) after the action of a lateral impulse to the vehicle (cf. Figure 1.25). For an impulse $S$ acting at a distance $x$ in front of the center of gravity, the increase in $r$ and $v$ becomes

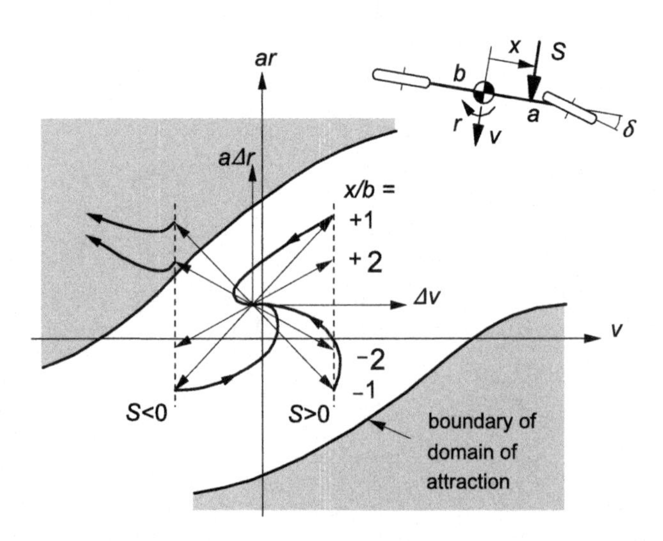


**FIGURE 1.25** Large disturbance in a curve. New initial state vector ($\Delta v$, $\Delta r$) after the action of a lateral impulse $S$. Once outside the domain of attraction, the motion becomes unstable and may get out of control.

$$\Delta r = \frac{Sx}{l}, \quad \Delta v = \frac{S}{m} \tag{1.98}$$

which results in the direction

$$\frac{a\Delta r}{\Delta v} = \frac{x}{b}\frac{ab}{k^2} \tag{1.99}$$

The figure shows the change in state vector for different points of application and direction of the impulse $S$ ($k^2 = I/m = ab$). Evidently, an impulse acting at the rear (in outward direction) constitutes the most dangerous disturbance. On the other hand, an impulse acting in front of the center of gravity about half way from the front axle does not appear to be able to get the new starting point outside of the domain of attraction irrespective of the intensity of the impulse.

When the slip angles become larger, the forward speed $u$ may no longer be considered as a constant quantity. Then, the system is described by a third-order set of equations. In the paper (Pacejka 1986), the solutions for the simple automobile model have been presented also for yaw angles $> 90°$.

## 1.3.4. The Vehicle at Braking or Driving

When the vehicle is subjected to longitudinal forces that may result from braking or driving actions possibly to compensate for longitudinal wind drag forces or down or upward slopes, fore-and-aft load transfer will arise (Figure 1.26). The resulting change in tire normal loads causes the cornering stiffnesses and the peak side forces of the front and rear axles to change. Since, as we assume here, the fore-and-aft position of the center of gravity is not affected (no relative car body motion), we may expect a change in handling behavior indicated by a rise or drop of the understeer gradient. In addition, the

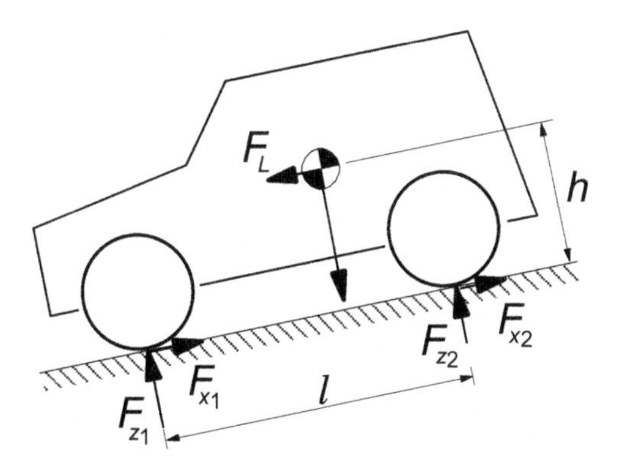


**FIGURE 1.26** The automobile subjected to longitudinal forces and the resulting load transfer.

longitudinal driving or braking forces give rise to a state of combined slip, thereby affecting the side force in a way as shown in Figure 1.2.

For moderate driving or braking forces, the influence of these forces on the side force $F_y$ is relatively small and may be neglected for this occasion. This means that, for now, the cornering stiffness may be considered to be dependent on the normal load only. The upper-left diagram of Figure 1.3 depicts typical variations of the cornering stiffness with vertical load.

The load transfer from the rear axle to the front axle that results from a forward longitudinal force $F_L$ acting at the center of gravity at a height $h$ above the road surface ($F_L$ possibly corresponding to the inertial force at braking) becomes

$$\Delta F_z = \frac{h}{l} F_L \tag{1.100}$$

The understeer gradient reads according to Eqn (1.60):

$$\eta = \frac{F_{z1o}}{C_1(F_{z1})} - \frac{F_{z2o}}{C_2(F_{z2})} \tag{1.101}$$

The static axle loads $F_{zio}$ ($i = 1$ or 2) are calculated according to Eqn (1.59), while the actual loads $F_{zi}$ front and rear become

$$F_{z1} = F_{z1o} + \Delta F_z, \quad F_{z2} = F_{z2o} - \Delta F_z \tag{1.102}$$

At moderate braking with deceleration $-a_x = F_L/m$, the load transfer remains small and we may use the linearized approximation of the variation of cornering stiffness with vertical load:

$$C_i = C_{io} + \zeta_{\alpha i} \Delta F_{zi} \text{ with } \zeta_{\alpha i} = \left( \frac{\partial C_i}{\partial F_{zi}} \right)_{F_{zio}} \tag{1.103}$$

The understeer gradient (1.101) can now be expressed in terms of the longitudinal acceleration $a_x$ (which might be minus the forward component of the acceleration due to gravity parallel to the road). We obtain

$$\eta = \eta_o + \lambda \frac{a_x}{g} \tag{1.104}$$

with the determining factor $\lambda$ approximately expressed as

$$\lambda = \zeta_{\alpha 1} \frac{h}{b} \left( \frac{F_{z1o}}{C_{1o}} \right)^2 + \zeta_{\alpha 2} \frac{h}{a} \left( \frac{F_{z2o}}{C_{2o}} \right)^2 \tag{1.105}$$

and $\eta_o$ denoting the original value not including the effect of longitudinal forces. Obviously, since $\zeta_{\alpha 1,2}$ is usually positive, negative longitudinal accelerations $a_x$, corresponding to braking, will result in a decrease of the degree of understeer.

To illustrate the magnitude of the effect, we use the parameter values given in Table 1.1 (p. 34) and add the c.g. height $h = 0.6$ m and the cornering stiffness versus load gradients $\zeta_{\alpha i} = 0.5C_{io}/F_{zio}$. The resulting factor appears to take the value $\lambda = 0.052$. This constitutes an increase of $\eta$ equal to $0.052a_x/g$. Apparently, the effect of $a_x$ on the understeer gradient is considerable when regarding the original value $\eta_o = 0.0174$.

As illustrated by Figure 1.9, the peak side force will be diminished if a longitudinal driving or braking force is transmitted by the tire. This will have an impact on the resulting handling diagram in the higher range of lateral acceleration. The resulting situation may be represented by the second and third diagrams of Figure 1.18 corresponding to braking (or driving) at the front or rear respectively. The problem becomes considerably more complex when we realize that at the front wheels, the components of the longitudinal forces perpendicular to the $x$-axis of the vehicle are to be taken into account. Obviously, we find that at braking of the front wheels, these components will counteract the cornering effect of the side forces and thus will make the car more understeer. The opposite occurs when these wheels are driven (more oversteer). For a more elaborate discussion on this item, we may refer to Pacejka (1973b).

At hard braking, possibly up to wheel lock, stability and steerability may deteriorate severely. This more complex situation will be discussed in Chapter 3 where more information on the behavior of tires at combined slip is given.

### 1.3.5. The Moment Method

Possible steady-state cornering conditions, stable or unstable, have been portrayed in the handling diagram of Figure 1.17. In Figure 1.22, motions tending to or departing from these steady-state conditions have been depicted. These motions are considered to occur after a sudden change in steer angle. The potential available to deviate from the steady turn depends on the margin of the front and rear side forces to increase in magnitude. For each point on the handling curve, it is possible to assess the degree of maneuverability in terms of the moment that can be generated by the tire side forces about the vehicle's center of gravity. Note that at the steady-state equilibrium condition, the tire side forces are balanced with the centrifugal force and the moment equals zero.

In general, the handling curve holds for a given speed of travel. That is so, when e.g., the aerodynamic down forces are essential in the analysis. In Figure 1.27 a diagram has been presented that is designated as the *MMM* diagram (the Milliken moment method diagram) and is computed for a speed of 60 mph. The force–moment concept was originally proposed by W.F. Milliken in 1952 and thereafter continuously further developed by the Cornell Aeronautical Laboratory staff and by Milliken research associates. A detailed description is given in Milliken's book (1995).

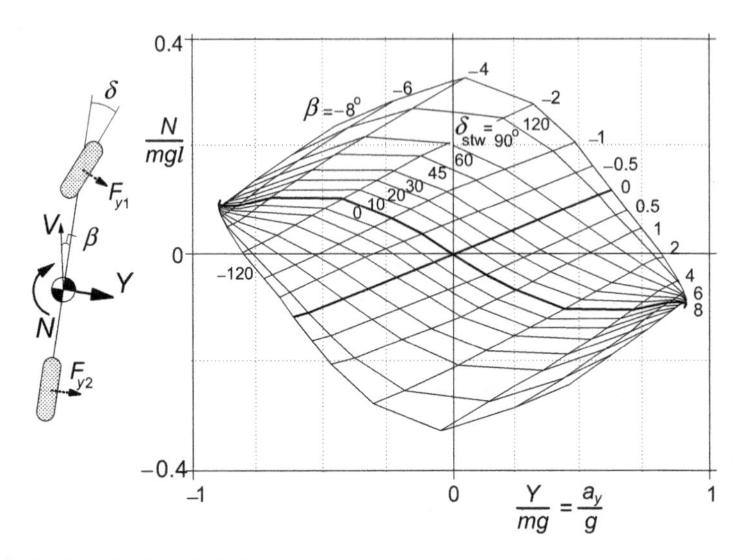


**FIGURE 1.27** The *MMM* diagram portraying the car's potential maneuvering capacity.

The graph shows curves of the resulting tire moment $N$ vs the resulting tire side force $Y$ in nondimensional form. The resulting force and moment result from the individual side forces and act from ground to vehicle. For greater accuracy, one may take the effect of the pneumatic trails into consideration. Two sets of curves have been plotted: one set for constant values of the vehicle side slip angle $\beta$ with the steering wheel angle $\delta_{stw}$ as parameter and the other set for constant steer angle and varying slip angle. Along the horizontal axis, the moment is zero and we have the steady-state equilibrium cornering situation that corresponds to the handling curve. It is observed that for the constant speed considered in the diagram, the steer angle increases when the total side force $Y$ or lateral acceleration $a_y$ is chosen larger which indicates that the motion remains stable. At the limit (near number 2), the maximum steady-state lateral acceleration is attained. At that point, the ability to generate a positive moment is exhausted. Only a negative moment may still be developed by the car that tends to straighten the curve that is being negotiated. As we have seen in Figure 1.18, second diagram, there is still some side force margin at the rear tire which can be used to increase the lateral acceleration in a transient fashion. At the same time, however, the car yaws outward because the associated moment is negative (cf. Figure 1.27, the curve near number 8). How to get at points below the equilibrium point near the number 2 is a problem. Rear wheel steering is an obvious theoretical option. In that way, the vehicle slip angle $\beta$ and front steer angle $\delta$ can remain unchanged while the rear steer angle produces the desired rear tire slip angle.

Of course, the diagram needs to be adapted in the case of rear wheel steering. Another more practical solution would be to bring the vehicle in the desired attitude ($\beta \rightarrow 8°$) by briefly inducing large brake or drive slip at the rear that lowers the cornering force and lets the car swing to the desired slip angle while at the same time the steering wheel is turned backward to even negative values.

The *MMM* diagram, which is actually a Gough plot (for a single tire, cf. Figures 3.5 and 3.29) established for the whole car at different steer angles, may be assessed experimentally through either outdoor or indoor experiments. On the proving ground, a vehicle may be attached at the side of a heavy truck or railway vehicle and set at different slip angles while the force and moment are being measured (tethered testing), cf. Milliken (1995). Figure 1.28 depicts the remarkable laboratory MMM test machine. This MTS Flat-Trac Roadway Simulator™ uses four flat belts which can be steered and driven independently. The car is constrained in its center of gravity but is free to roll and pitch.

## 1.3.6. The Car-Trailer Combination

In this section we will discuss the role of the tire in connection with the dynamic behavior of a car that tows a trailer. More specifically, we will study the possible unstable motions that may show up with such a combination. Linear differential equations are sufficient to analyze the stability of the straight-ahead motion. We will again employ Lagrange's equations to set up the equations of motion. The original Eqn (1.25) may be employed because the yaw angle is assumed to remain small. The generalized coordinates $Y$, $\psi$, and $\theta$ are used to describe the car's lateral position and the yaw angles of car and

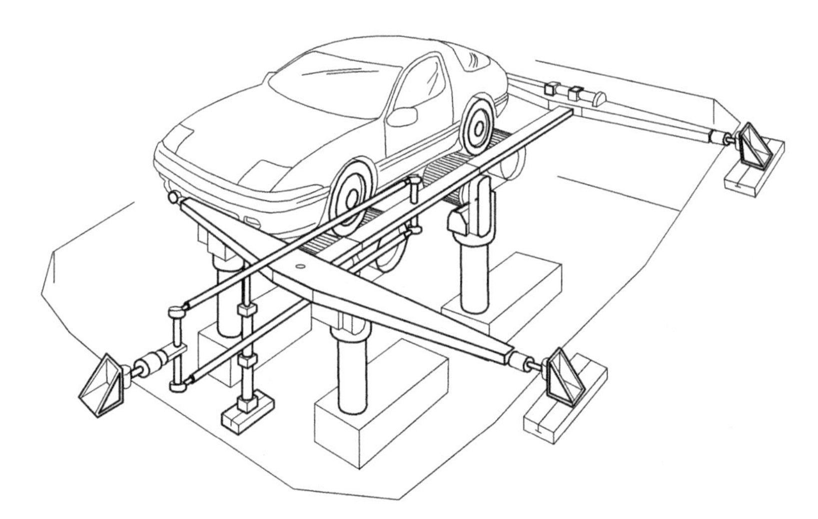

**FIGURE 1.28** The MTS Flat-Trac Roadway Simulator™, Milliken (1995).

trailer respectively. The forward speed d*X*/d*t* ($\approx V \approx u$) is considered to be constant. Figure 1.29 gives a top view of the system with three degrees of freedom. The alternative set of three variables $v$, $r$, the articulation angle $\varphi$, and the vehicle velocity $V$ (a parameter) which are not connected to the inertial axes system ($O$, $X$, $Y$) has been indicated as well and will be employed later on. The kinetic energy for this system becomes, if we neglect all the terms of the second order of magnitude (products of variables),

$$T = \frac{1}{2}m(\dot{X}^2 + \dot{Y}^2) + \frac{1}{2}I\dot{\psi}^2 + \frac{1}{2}m_c\left\{\dot{X}^2 + (\dot{Y} - h\dot{\psi} - f\dot{\theta})^2\right\} + \frac{1}{2}I_c\dot{\theta}^2 \quad (1.106)$$

The potential energy remains zero:

$$U = 0 \quad (1.107)$$

and the virtual work done by the external road contact forces acting on the three axles reads

$$\delta W = F_{y1}\delta(Y + a\psi) + F_{y2}\delta(Y - b\psi) + F_{y3}\delta(Y - h\psi - g\theta) \quad (1.108)$$

With the use of Eqns (1.25) and (1.29), the following equations of motion are established for the generalized coordinates $Y$, $\psi$, and $\theta$:

$$(m + m_c)\ddot{Y} - m_c(h\ddot{\psi} + f\ddot{\theta}) = F_{y1} + F_{y2} + F_{y3} \quad (1.109)$$

$$(I_c + m_c f^2)\ddot{\theta} - m_c f(\ddot{Y} - h\ddot{\psi}) = -gF_{y3} \quad (1.110)$$

$$(I + m_c h^2)\ddot{\psi} - m_c h(\ddot{Y} - f\ddot{\theta}) = aF_{y1} - bF_{y2} - hF_{y3} \quad (1.111)$$

This constitutes a system of the sixth order. By introducing the velocities $v$ and $r$, the order can be reduced to four. In addition, the angle of articulation $\varphi$ will be used. We have the relations

$$\dot{Y} = V\psi + v, \quad \dot{\psi} = r, \quad \theta = \psi - \varphi \quad (1.112)$$

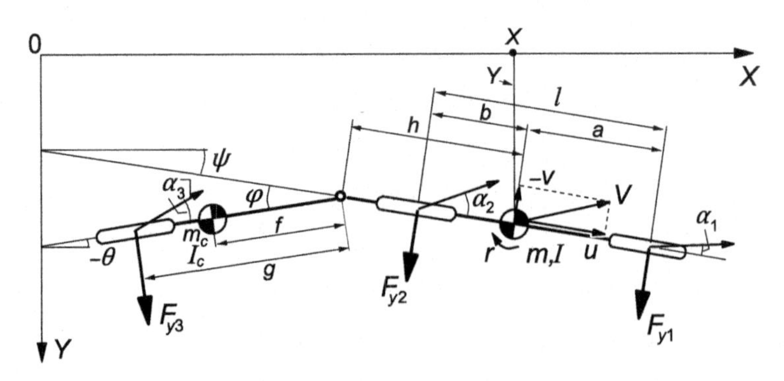


**FIGURE 1.29** Single track model of car-trailer combination.

And, with these the equations for $v$, $r$, and $\varphi$,

$$(m+m_c)(\dot{v}+Vr)-m_c\{(h+f)\dot{r}-f\ddot{\varphi}\} = F_{y1}+F_{y2}+F_{y3} \tag{1.113}$$

$$\{I+m_ch(h+f)\}\dot{r}-m_ch(\dot{v}+Vr+f\ddot{\varphi}) = aF_{y1}-bF_{y2}-hF_{y3} \tag{1.114}$$

$$(I_c+m_cf^2)(\ddot{\varphi}-\dot{r})+m_cf(\dot{v}+Vr-h\dot{r}) = gF_{y3} \tag{1.115}$$

The right-hand members are still to be expressed in terms of the motion variables. With the axle cornering stiffnesses $C_1$, $C_2$, and $C_3$, we have

$$\begin{aligned} F_{y1} &= C_1\alpha_1 = -C_1\frac{v+ar}{V} \\ F_{y2} &= C_2\alpha_2 = -C_2\frac{v-br}{V} \\ F_{y3} &= C_3\alpha_3 = -C_3\left(\frac{v-hr-g(r-\dot{\varphi})}{V}+\varphi\right) \end{aligned} \tag{1.116}$$

From the resulting set of linear differential equations, the characteristic equation may be derived which is of the fourth degree. Its general structure is

$$a_0s^4+a_1s^3+a_2s^2+a_3s+a_4 = 0 \tag{1.117}$$

The stability of the system can be investigated by considering the real parts of the roots of this equation or we might employ the criterion for stability according to Routh-Hurwitz. According to this criterion, the system of order $n$ is stable when all the coefficients $a_i$ are positive and the Hurwitz determinants $H_{n-1}$, $H_{n-3}$ etc. are positive. For our fourth-order system, the complete criterion for stability reads

$$H_3 = \begin{bmatrix} a_1 & a_0 & 0 \\ a_3 & a_2 & a_1 \\ 0 & a_4 & a_3 \end{bmatrix} = a_1a_2a_3-a_1^2a_4-a_0a_3^2>0 \tag{1.118}$$

$$a_i>0 \quad \text{for } i = 0,1,\ldots 4$$

In Figure 1.30, the boundaries of stability have been presented in the caravan axle cornering stiffness vs speed parameter plane. The three curves belong to the three different sets of parameters for the position $f$ of the caravan's center of gravity and the caravan's mass $m_c$ as indicated in the figure. An important result is that a lower cornering stiffness promotes oscillatory instability: the critical speed beyond which instability occurs decreases. Furthermore, it appears from the diagram that moving the caravan's center of gravity forward ($f$ smaller) stabilizes the system which is reflected by the larger critical speed. A heavier caravan ($m_c$ larger) appears to be bad for stability. Furthermore, it has been found that a larger draw bar length $g$ is favorable for stability.

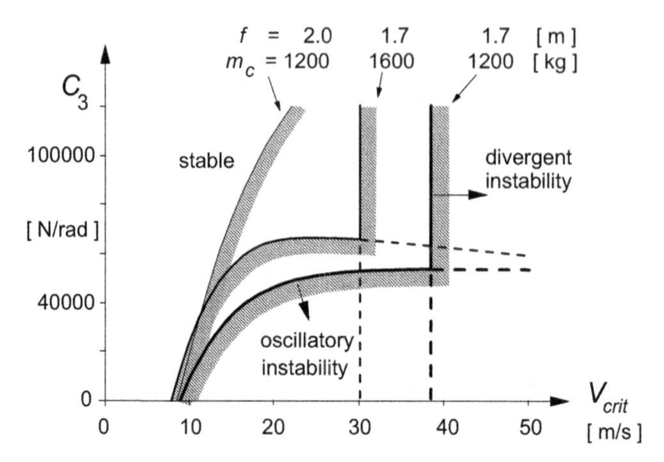


**FIGURE 1.30** Stability boundaries for the car caravan combination in the caravan cornering stiffness vs critical speed diagram. Vehicle parameters according to Table 1.1, in addition: $h = 2$ m, $g = 2$ m, $k_c = 1.5$ m ($I_c = m_c k_c^2$), cf. Figure 1.29.

It turns out that a second type of instability may show up. This occurs when the portion of the weight of the caravan supported by the coupling point becomes too large. This extra weight is felt by the towing vehicle and makes it more oversteer. The critical speed associated with this phenomenon is indicated in the diagram by the vertical lines. This divergent instability occurs when (starting out from a stable condition) the last coefficient becomes negative, that is, $a_n = a_4 < 0$.

The oscillatory instability connected with the 'snaking' phenomenon arises as soon as (from a stable condition) the second highest Hurwitz determinant becomes negative, $H_{n-1} = H_3 < 0$ (then also $H_n < 0$), cf. Klotter (1960) or Leipholz (1987). When the critical speed is surpassed, self-excited oscillations are created which shows an amplitude that, in the actual nonlinear case, does not appear to limit itself. This is in contrast to the case of the wheel shimmy phenomenon to be treated in Chapter 5 where a stable limited oscillation appears to arise. The cause of the unlimited snaking oscillation is that with increasing amplitudes also the slip angle increases which lowers the average cornering stiffness as a consequence of the degressively nonlinear cornering force characteristic. From the diagram we found that this will make the situation increasingly worse. As has been seen from full vehicle/caravan model simulations, the whole combination will finally overturn. Another effect of this reduction of the average cornering stiffness is that when the vehicle moves at a speed lower than the critical speed, the originally stable straight-ahead motion may become unstable if, through the action of an external disturbance (side wind gust), the slip angle of the caravan axle becomes too large (surpassing of the associated unstable limit cycle). This is an unfortunate, possibly dangerous situation! We refer to Troger and Zeman (1984) for further details.

**Exercise 1.4 Stability of a Trailer**

Consider the trailer of Figure 1.31 that is towed by a heavy steadily moving vehicle at a forward speed $V$ along a straight line. The trailer is connected to the vehicle by means of a hinge. The attachment point shows a lateral flexibility that is represented by the lateral spring with stiffness $c_y$. Furthermore, a yaw torsional spring and damper are provided with coefficients $c_\varphi$ and $k_\varphi$.

Derive the equations of motion of this system with generalized coordinates $y$ and $\varphi$. Assume small displacements so that the equations can be kept linear. The damping couple $k_\varphi\dot{\varphi}$ may be considered as an external moment acting on the trailer or we may use the dissipation function $D = (1/2)k_\varphi\dot{\varphi}^2$ and add $+\partial D/\partial\dot{q}_i$ to the left-hand side of Lagrange's Eqn (1.25). Obviously, the introduction of this extra term will be beneficial particularly when the system to be modeled is more complex.

Assess the condition for stability for this fourth-order system. Simplify the system by putting $g = f$ and $c_\varphi = k_\varphi = 0$. Now find the explicit conditional statement for the cornering stiffness $C$.

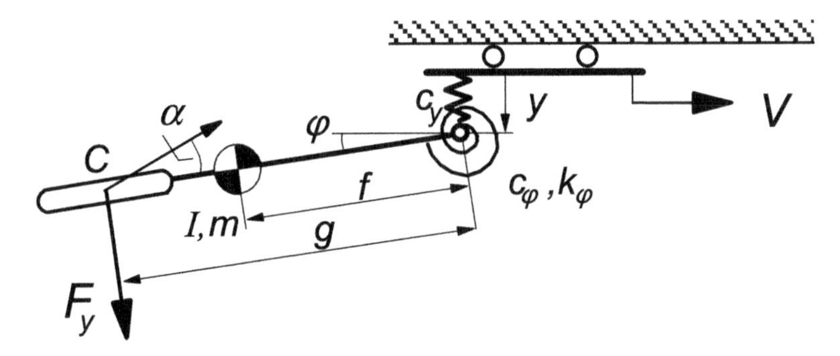


**FIGURE 1.31** On the stability of a trailer (Exercise 1.4).

## 1.3.7. Vehicle Dynamics at More Complex Tire Slip Conditions

So far, relatively simple vehicle dynamics problems have been studied in which the basic steady-state cornering force versus slip angle characteristic plays the dominant role. The situation becomes more complex when matters like combined slip at hard braking, wheel camber, tire transient and vibrational properties, and e.g., obstacle crossings are to be considered.

In the subsequent chapters, tire performance and modeling will be treated in greater detail which enables us to introduce relevant tire properties in the analysis. The following specific subjects will be studied as applications of the tire modeling theory:

- Vehicle stability at excessive braking and wheel lock (Chapter 3)
- Motorcycle steady-state cornering (Chapter 11)

- Wheel shimmy (Chapter 6)
- Steering vibrations (Chapter 8)
- Motorcycle weave and wobble (Chapter 11)
- Tire out-of-roundness (Chapter 8)
- Cornering on uneven roads (Chapters 5,8)
- ABS on uneven roads (Chapter 8)
- Traversing short obstacles (Chapter 10)
- Parking (Chapter 9)

Online, cf. App. 2, several types of car handling experiments are discussed and compared with theoretical results of Sec. 1.3.2. *MATLAB* programs are presented, also for the three degree of freedom model, for the benefit of the interested reader.

Chapter 2

# Basic Tire Modeling Considerations

Chapter Outline



## 2.1. INTRODUCTION

The performance of a tire as a force- and moment-generating structure is a result of a combination of several aspects. Factors that concern the primary tasks of the tire may be distinguished from factors that involve (often-important) secondary effects.

In Table 2.1 these factors are presented in matrix form. A further distinction is made between (quasi) steady state and vibratory behavior and, additionally, between symmetric (or in-plane) and anti-symmetric (or out-of-plane) aspects. The primary task factors appear in bold letters. The remaining factors are considered secondary factors.

The primary requirements to transmit forces in the three perpendicular directions ($F_x$, $F_y$, and $F_z$) and to cushion the vehicle against road irregularities involve secondary factors such as lateral and longitudinal distortions and slip. Although regarded as secondary phenomena, some of the quantities involved are crucial for the generation of the deformations and the associated forces and will be treated as input variables into the system.

Figure 2.1 presents the 'vectors' of input and output components. In this diagram, the tire is assumed to be uniform and to move over a flat road surface. The input vector stems from motions of the wheel relative to the road. A precise definition of these input quantities is given in the next section.

Tire and Vehicle Dynamics. DOI: 10.1016/B978-0-08-097016-5.00002-4

**TABLE 2.1 Tire Factors**

| | Primary Task Functions and Secondary Effects | | |
|---|---|---|---|
| | (Quasi) Steady-State ↔ Transient/Vibratory State | | |
| symmetric (in-plane) | **load carrying braking/ driving** rolling resistance | radial deflection tangential slip and distortion | **cushioning** dynamic coupling natural vibrations |
| anti-symmetric (out-of-plane) | **cornering** pneumatic trail overturning couple | lateral and spin/ turn slip and distortion | phase lag destabilization natural vibrations |

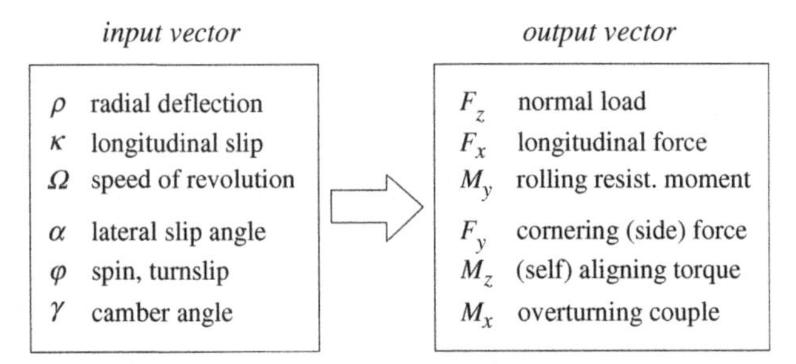


**FIGURE 2.1** Input/output quantities (road surface considered flat).

For small deviations from the straight-ahead motion a linear description of behavior may be given. Then, it is advantageous to recognize the fact that the responses to the symmetric and anti-symmetric motions of the assumedly symmetric wheel–tire system can be considered as uncoupled. Figure 2.2 shows the separate function blocks with the input and output quantities. Here we have also considered the possibility of input from variations in road surface geometry and from tire nonuniformities resulting in e.g., out-of-roundness, stiffness variations, and 'built-in' forces.

The forces and moments are considered as output quantities. It is sometimes beneficial to assume these forces to act on a rigid disk with inertial properties equal to those of the tire when considered rigid. These forces may differ from the forces acting between the road and the tire because of the dynamic forces acting on the tire when vibrating relative to the wheel rim. The motions of the wheel rim and the profile of the road, represented by its height $w$ and its forward and transverse slope at or near the contact center, are regarded as input quantities to the tire. Braking and driving torques $M_a$ are considered to act on the rotating wheel inertia $I_w$. For the freely rolling tire (then, by definition $M_a = 0$), the wheel angular motion about the spindle axis is governed by only the internal

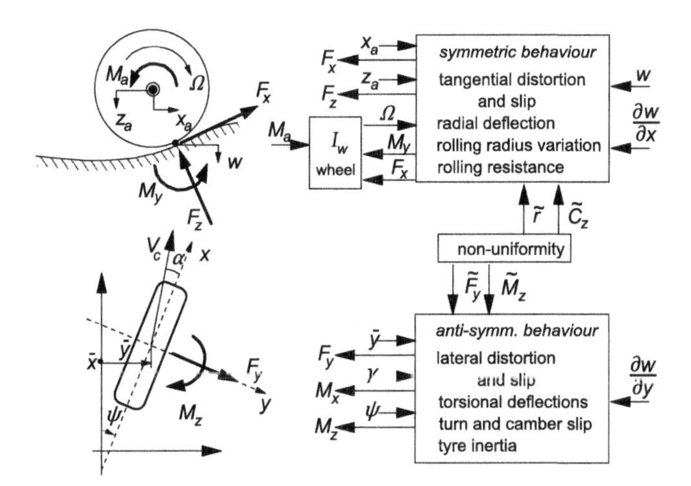


**FIGURE 2.2** Wheel axle motion and road surface coordinates and possibly uncoupled tire system blocks valid for small deviations from the steady-state straight ahead motion.

moment $M_\eta$ acting between the rim and the tire. The five motion components of the wheel-spin axis may then remain to serve as input vector.

The discussion on the force generation and dynamic properties of the tire will be conducted along the two main lines: symmetric and anti-symmetric behavior. Interaction between these main groups of input motions complicates the situation (combined slip). These interactions become important if at least one of the input motions or more precisely, one of the associated slip components becomes relatively large. Because of its relative simplicity, steady-state behavior will be treated first (Chapter 3). The discussion on tire dynamic behavior starts in Chapter 5.

## 2.2. DEFINITION OF TIRE INPUT QUANTITIES

If the problem that is going to be investigated involves road irregularities, then the location and the orientation of the stub axle (spindle axis) must be known with respect to the specific irregularity met on the road. The road surface is defined with respect to a coordinate system of axes attached to the road. If the position and orientation of the axle are known with respect to the fixed triad, then the exact position of the wheel with respect to the possibly irregular road surface can be determined. This relative position and orientation of the wheel with respect to the road are important to derive the radial tire deflection and the relative attitude (camber) and to assess the current value of the friction coefficient, which may vary due to e.g., slippery spots or nonhomogeneous surface conditions (grooves). The time rate of change of this relative position is needed not only for possible hysteresis effects but also mainly for the determination of the so-called 'slip' of the wheel with respect to the ground.

If the road surface near the contact patch can be approximated by a flat plane (that is, when the smallest considered wavelength of the decomposed surface vertical profile is large with respect to the contact length and its amplitude small), the distance of the wheel center to the road plane and the angle between wheel plane and the normal to the road surface will suffice in addition to the several slip quantities and the running speed of the wheel.

For the definition of the various motion and position input quantities listed in Figure 2.1, it is helpful to consider Figure 2.3. A number of planes have been drawn. The road plane and the wheel-center plane (with line of intersection along the unit vector $\boldsymbol{l}$) and two planes normal to the road plane, one of which contains the vector $\boldsymbol{l}$ and the other the unit vector $\boldsymbol{s}$, which is defined along the wheel-spin axis. From the figure follows the definition of the contact center $C$, also designated as the point of intersection (of the three planes). The unit vector $\boldsymbol{t}$ lies in the road plane and is directed perpendicular to $\boldsymbol{l}$. The vector $\boldsymbol{r}$ forms the connection between wheel center $A$ and the contact center $C$. Its length, $r$, is defined as the loaded radius of the tire. The position and attitude of the wheel with respect to the inertial triad are completely described by the vectors $\boldsymbol{b} + \boldsymbol{a}$ and $\boldsymbol{s}$. The road plane is defined at the contact center by the position vector of that point $\boldsymbol{c}$ and the normal to the road in that point represented by the unit vector $\boldsymbol{n}$ (positive upward). Figure 2.3 also shows two systems of axes (besides the inertial triad). First, we have introduced the road contact axes system ($C$, $x$, $y$, and $z$) of which the $x$-axis points forward along the line of intersection ($\boldsymbol{l}$), the $z$-axis points downward normal to the road plane ($-\boldsymbol{n}$), and the $y$-axis points to

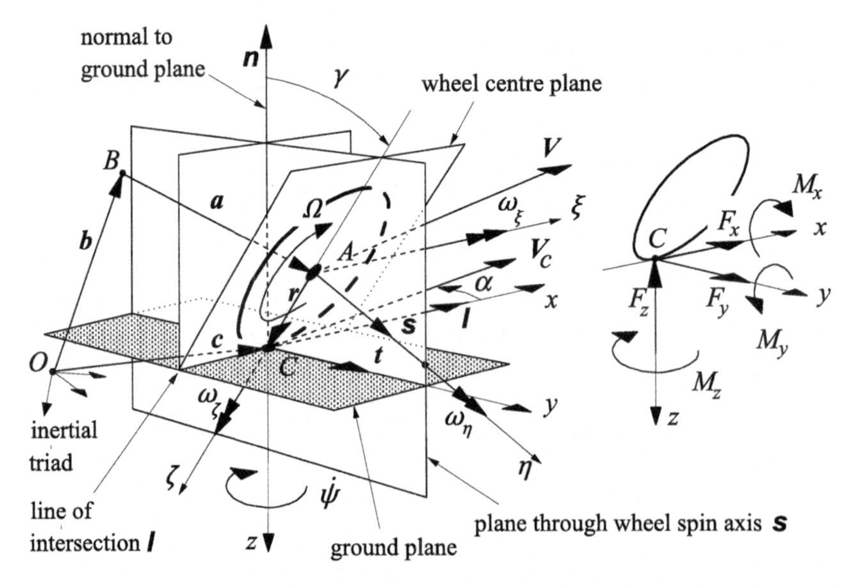


**FIGURE 2.3** Definition of position, attitude, and motion of the wheel and the forces and moments acting from the road on the wheel. Directions shown are defined as positive.

the right along the transverse unit vector $\boldsymbol{t}$. Second, the wheel axle system of axes ($A$, $\xi$, $\eta$, and $\zeta$) has been defined with the $\xi$ axis parallel to the $x$-axis, the $\eta$ axis along the wheel spindle axis ($\boldsymbol{s}$), and the $\zeta$ axis along the radius ($\boldsymbol{r}$).

Sign conventions in the literature are not uniform. For the sake of convenience and to reduce sources of making errors, we have chosen a sign convention that avoids working with negative quantities as much as possible.

The radial deflection of the tire $\rho$ is defined as the reduction of the tire radius from the unloaded situation $r_f$ to the loaded case $r$:

$$\rho = r_f - r \tag{2.1}$$

For positive $\rho$ the wheel load $F_z$ (positive upward) is positive as well.

The tangential or longitudinal slip $\kappa$ requires deeper analysis. For the sake of properly defining the longitudinal slip, the so-called slip point $S$ is introduced. This point is thought to be attached to the rim or wheel body at a radius equal to the slip radius $r_s$ and forms the center of rotation when the wheel rolls at a longitudinal slip equal to zero. The slip radius is the radius of the slip circle. At vanishing longitudinal slip, this slip circle rolls purely over an imaginary surface parallel to the road plane. The length of the slip radius depends on the definition of longitudinal slip that is adopted. A straightforward definition would be to make the slip radius equal to the loaded wheel radius. This, however, would already lead to a considerable magnitude of the longitudinal force $F_x$ that would be generated at longitudinal slip equal to zero. A more convenient and physically proper definition corresponds to the situation that $F_x = 0$ at zero longitudinal slip. Because of the occurrence of rolling resistance, measurements of tire characteristics would then require the application of a driving torque to reach the condition of slip equal to zero! This may become of importance, especially when experiments are conducted at large camber angles where the drag may become considerable (motorcycle tires). An alternative, often-used definition takes the effective rolling radius $r_e$ defined at free rolling ($M_a = 0$) as the slip radius. Under normal conditions, the resulting $F_x$ vs $\kappa$ diagrams according to the latter two definitions are very close. A small horizontal shift of the curves is sufficiently accurate to change from one definition to the other. The drawback of the last definition is that when testing on very low friction (icy) surfaces, the rolling resistance may be too large to let the wheel rotate without the application of a driving torque. Consequently, the state of free rolling cannot be realized under these conditions. Nevertheless, we will adopt the last definition where $r_s = r_e$ and consequently, point $S$ is located at a distance $r_e$ from the wheel center. Figure 2.4 depicts this configuration.

According to this definition we will have the situation that when a wheel rolls freely (that is, at $M_a = 0$) at constant speed over a flat even road surface, the longitudinal slip $\kappa$ is equal to zero. This notwithstanding the fact that at free rolling some fore and aft deformations will occur because of the presence of hysteresis in the tire that generates a rolling resistance moment $M_y$. Through this a rolling resistance force $F_r = M_y / r$ arises, which necessarily is accompanied by

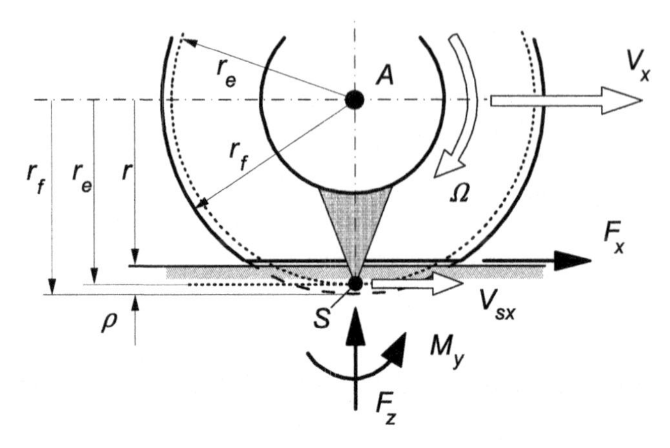


**FIGURE 2.4** Effective rolling radius and longitudinal slip velocity.

tangential deformations. We may agree that at the instant of observation, point $S$, which lies on the slip circle and is attached to the wheel rim, has reached its lowest position, that is: on the line along the radius vector $\boldsymbol{r}$. At free rolling, its velocity has then become equal to zero and point $S$ has become the center of rotation of the motion of the wheel rim. We have at free rolling on a flat road for a wheel in upright position ($\gamma = 0$) and/or without wheel yaw rate ($\dot{\psi} = 0$), cf. Figure 2.3, a velocity of the wheel center in the forward ($x$ or $\xi$) direction:

$$V_x = r_e \Omega \tag{2.2}$$

with $\Omega$ denoting the speed of revolution of the wheel body to be defined hereafter. By using this relationship, the value of the effective rolling radius can be assessed from an experiment. The forward speed and the wheel speed of revolution are both measured while the wheel axle is moved along a straight line over a flat road. Division of both quantities leads to the value of $r_e$. The effective rolling radius will be a function of the normal load and the speed of travel. We may possibly have to take into account the dependency on the camber angle and the slip angle.

If at braking or driving the longitudinal slip is no longer zero, point $S$ will move with a longitudinal slip speed $V_{sx}$ that differs from zero. We obviously obtain if again $\gamma\dot{\psi} = 0$:

$$V_{sx} = V_x - r_e \Omega \tag{2.3}$$

The longitudinal slip (sometimes called the slip ratio) is denoted by $\kappa$ and may be tentatively defined as the ratio of longitudinal slip velocity $-V_{sx}$ of point $S$ and the forward speed of the wheel center $V_x$:

$$\kappa = -\frac{V_{sx}}{V_x} \tag{2.4}$$

or with (2.3):

$$\kappa = -\frac{V_x - r_e \Omega}{V_x} \tag{2.5}$$

This again holds for a wheel on a flat road and with $\gamma\dot{\psi} = 0$. A more general and precise definition of $\kappa$ will be given later on. The sign of the longitudinal slip $\kappa$ has been chosen such that at driving, when $F_x > 0$, $\kappa$ is positive and at braking, when $F_x < 0$, $\kappa$ is negative. When the wheel is locked ($\Omega = 0$), we obviously have $\kappa = -1$. In the literature, the symbol $s$ (or $S$) is more commonly used to denote the slip ratio.

The angular speed of rolling $\Omega_r$, more precisely defined for the case of moving over undulated road surfaces, is the time rate of change of the angle between the radius connecting $S$ and $A$ (this radius is thought to be attached to the wheel) and the radius $\boldsymbol{r}$ defined in Figure 2.3 (always lying in the plane normal to the road through the wheel-spin axis). Figure 2.5 illustrates the situation.

The linear speed of rolling $V_r$ is defined as the velocity with which an imaginary point $C^*$, which is positioned on the line along the radius vector $\boldsymbol{r}$ and coincides with point $S$ at the instant of observation, moves forward (in $x$ direction) with respect to point $S$ that is fixed to the wheel rim:

$$V_r = r_e \Omega_r \tag{2.6}$$

For a tire freely rolling over a flat road we have, $\Omega_r = \Omega$ and with $\gamma\dot{\psi} = 0$ in addition: $V_r = V_x$. Note, that at wheel lock ($\Omega = 0$) the angular speed of rolling $\Omega_r$ is not equal to zero when the wheel moves over a road with a curved vertical profile (then not always the same point of the wheel is in contact with the road). For a cambered wheel showing a yaw rate $\dot{\psi}$, pure rolling can occur on a flat road even when the speed of the wheel center $V_x = 0$. In that case a linear speed of rolling arises that is equal to $V_r = r_e\, \dot{\psi} \sin \gamma$ and consequently an angular speed of rolling $\Omega_r = \dot{\psi} \sin \gamma$.

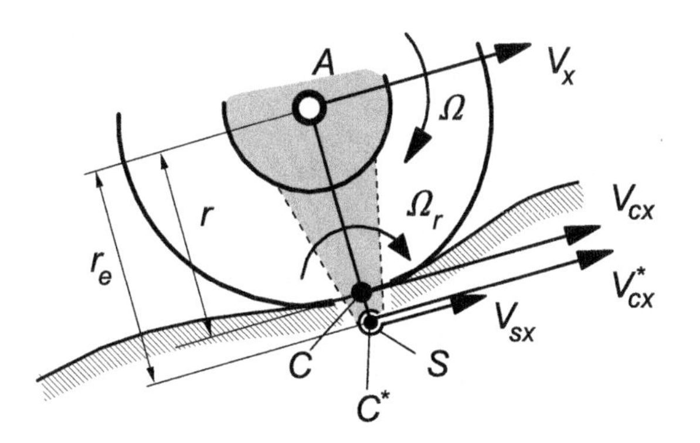


**FIGURE 2.5** Rolling and slipping of a tire over an undulated road surface.

In the normal case of an approximately horizontal road surface, the wheel speed of revolution $\Omega$ may be defined as the angular speed of the wheel body (rim) seen with respect to a vertical plane that passes through the wheel spindle axis. On a flat level road, the angular speed of rolling $\Omega_r$ and the speed of revolution of the wheel $\Omega$ are equal to each other. The absolute speed of rotation of the wheel about the spindle axis $\omega_\eta$ will be different from $-\Omega$ when the wheel is cambered and a yaw rate about the normal to the road occurs of the plane that passes through the spindle axis and is oriented normal to the road. Then (cf. Figure 2.6):

$$\omega_\eta = -\Omega + \dot{\psi} \sin \gamma \tag{2.7}$$

This equation forms a correct basis for a general definition of $\Omega$ also on nonlevel road surfaces. Its computation is straightforward if $\omega_\eta$ is available from wheel dynamics calculations.

The longitudinal running speed $V^*_{cx}$ is defined as the longitudinal component of the velocity of propagation of the imaginary point $C^*$ (on radius vector $\boldsymbol{r}$) in the direction of the $x$-axis (vector $\boldsymbol{l}$). In case the wheel is moved in such a way that the same point remains in contact with the road, we would have $V^*_{cx} = V_{sx}$. This corresponds to wheel lock when the road is flat and the vehicle pitch rate is zero. For a freely rolling tire the longitudinal running speed equals the linear speed of rolling: $V^*_{cx} = V_r$. On a flat road and at zero camber or zero yaw rate

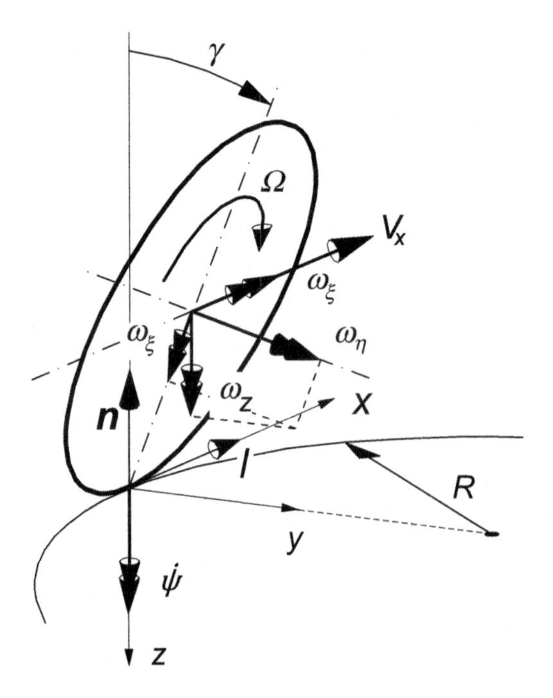


**FIGURE 2.6** Rotational slip resulting from path curvature and wheel camber (slip angle = 0).

$(\gamma\dot{\psi} = 0)$, we obtain $V^*_{cx} = V_x$. The general definition for longitudinal slip now reads:

$$\kappa = -\frac{V_{sx}}{V^*_{cx}} \tag{2.8}$$

The lateral slip is defined as the ratio of the lateral velocity $-V_{cy}$ of the contact center $C$ and the longitudinal running speed $V^*_{cx}$. We have in terms of the slip angle $\alpha$:

$$\tan\alpha = -\frac{V_{cy}}{V^*_{cx}} \tag{2.9}$$

which for a wheel, not showing camber rate $\dot{\gamma}$ nor radial deflection rate $\dot{\rho}$ and yaw rate $\dot{\psi}$ at nonzero camber angle $\gamma$, when running on a flat road reduces to the ratio of lateral and forward speed of the wheel center:

$$\tan\alpha = -\frac{V_y}{V_x} \tag{2.10}$$

In practice, points $C$ and $C^*$ lie close together and making a distinction between the longitudinal or the lateral velocities of these points is only of academic interest and may be neglected. Instead of $V^*_{cx}$ in the denominator, we may write $V_{cx}$ and if we wish, instead of $V_{cy}$ in the numerator the lateral speed of point $S$ (parallel to road plane), which is $V_{sy}$. The definitions of the slip components then reduce to

$$\kappa = -\frac{V_{sx}}{V_{cx}} \tag{2.11}$$

$$\tan\alpha = -\frac{V_{sy}}{V_{cx}} \tag{2.12}$$

The slip velocities $V_{sx}$ and $V_{sy}$ form the components of the slip speed vector $\boldsymbol{V}_s$ and $\kappa$ and $\tan\alpha$ the components of the slip vector $\boldsymbol{s}_s$. We have

$$\boldsymbol{V}_s = \begin{pmatrix} V_{sx} \\ V_{sy} \end{pmatrix} \tag{2.13}$$

and

$$\boldsymbol{s}_s = \begin{pmatrix} \kappa \\ \tan\alpha \end{pmatrix} \tag{2.14}$$

The 'spin' slip $\varphi$ is defined as the component $-\omega_z$ of the absolute speed of rotation vector $\boldsymbol{\omega}$ of the wheel body along the normal to the road plane $\boldsymbol{n}$ divided by the speed of $C$. We obtain the expression in terms of yaw rate $\dot{\psi}$ and camber angle $\gamma$ (cf. Figure 2.6):

$$\varphi = -\frac{\omega_z}{V_c} = -\frac{\dot{\psi} - \Omega\sin\gamma}{V_c} \tag{2.15}$$

The minus sign is introduced again to remain consistent with the definitions of longitudinal and lateral slip (2.11, 2.12). Then, we will have as a result of a positive $\varphi$ a positive moment $M_z$. It turns out that then also the resulting side force $F_y$ is positive. The yaw rate $\dot{\psi}$ is defined as the speed of rotation of the line of intersection (unit vector $\boldsymbol{l}$) about the $z$-axis normal to the road cf. Figure 2.3. If the side slip angle remains constant ($\dot{\alpha} \equiv 0$) and the wheel moves over a flat road, Eqn (2.15) may be written as

$$\varphi = -\frac{1}{R} + \frac{\Omega_r}{V_c}\sin\gamma = -\frac{1}{R} + \frac{1}{r_e}\frac{V_r}{V_c}\sin\gamma \tag{2.16}$$

When the tire rolls freely (then $V_{sx} = 0$, $V_c = V_r$), we obviously obtain

$$\varphi = -\frac{1}{R} + \frac{1}{r_e}\sin\gamma \tag{2.17}$$

with $1/R$ denoting the momentary curvature of the path of $C^*$ or approximately of the contact center $C$.

For a tire we shall distinguish between spin due to path curvature and spin due to wheel camber. For a homogeneous ball, the effect of both input quantities is the same. For further use, we define turn slip as

$$\varphi_t = -\frac{\dot{\psi}}{V_c} \quad \left( = -\frac{1}{R} \text{ if } \alpha \text{ is constant} \right) \tag{2.18}$$

Wheel camber or wheel inclination angle $\gamma$ is defined as the angle between the wheel-center plane and the normal to the road. With Figure 2.3 we find

$$\sin\gamma = -\boldsymbol{n}\cdot\boldsymbol{s} \tag{2.19}$$

or on flat level roads:

$$\sin\gamma = s_z \tag{2.20}$$

where $s_z$ represents the vertical component of the unit vector $\boldsymbol{s}$ along the wheel-spin axis.

## 2.3. ASSESSMENT OF TIRE INPUT MOTION COMPONENTS

The location of the contact center $C$ and the magnitude of the wheel radius $r$ result from the road geometry and the position of the wheel axle. We consider the approximate assumption that the road plane is defined by the plane touching the surface at point $Q$ located vertically below the wheel center $A$. The position of point $Q$ with respect to the inertial frame ($O^o$, $x^o$, $y^o$, $z^o$) is given by vector $\boldsymbol{q}$. The normal to the road plane is defined by unit vector $\boldsymbol{n}$. The location of a reference point $B$ of the vehicle is defined by vector $\boldsymbol{b}$ and the location of the wheel center $A$ by $\boldsymbol{b} + \boldsymbol{a}$ (cf. Figure 2.3). The orientation of the wheel-spin axis is given by unit vector $\boldsymbol{s}$ and the location of the contact center $C$ by

$$\boldsymbol{c} = \boldsymbol{b} + \boldsymbol{a} + \boldsymbol{r} \tag{2.21}$$

where $\boldsymbol{r}$ is still to be determined. The expression for $\boldsymbol{r}$ is derived from the equations:

$$\boldsymbol{r} = r\boldsymbol{l} \times \boldsymbol{s} \tag{2.22}$$

with

$$\boldsymbol{l} = \lambda \boldsymbol{n} \times \boldsymbol{s} \tag{2.23}$$

(with $\lambda$ resulting from the condition that $|\boldsymbol{l}| = 1$) and with (2.21) in order to obtain the magnitude of the loaded radius $r$:

$$\boldsymbol{c} \cdot \boldsymbol{n} = \boldsymbol{q} \cdot \boldsymbol{n} \tag{2.24}$$

which indicates that contact point $C$ and road point $Q$ lie on the same plane perpendicular to $\boldsymbol{n}$. On flat level roads, the above equations become a lot simpler since in that case $\boldsymbol{n}^{\mathrm{T}} = (0, 0, -1)$ and the $z$ components of $\boldsymbol{c}$ and $\boldsymbol{q}$ become zero.

For small camber the radial tire deflection $\rho$ is now readily obtained from (cf. Figure 4.27 and Eqn (7.46) for the deflection normal to the road):

$$\rho = r_f - r \tag{2.25}$$

with $r_f$ the free unloaded radius. For a given tire the effective rolling radius $r_e$ is a function of among other things the unloaded radius, the radial deflection, the camber angle, and the speed of travel.

The vector for the speed of propagation of the contact center $\boldsymbol{V}_c$ representing the magnitude and direction of the velocity with which point $C$ moves over the road surface is obtained by differentiation with respect to time of position vector $\boldsymbol{c}$ (2.21):

$$\boldsymbol{V}_c = \dot{\boldsymbol{c}} = \dot{\boldsymbol{b}} + \dot{\boldsymbol{a}} + \dot{\boldsymbol{r}} = \boldsymbol{V} + \dot{\boldsymbol{r}} \tag{2.26}$$

with $\boldsymbol{V}$ the velocity vector of the wheel center $A$ (Figure 2.3). The speed of propagation of point $C^*$ represented by vector $\boldsymbol{V}_c^*$ becomes (cf. Figure 2.5 and assume $r_e/r$ constant):

$$\boldsymbol{V}_c^* = \boldsymbol{V} + \frac{r_e}{r}\dot{\boldsymbol{r}} \tag{2.27}$$

The velocity vector of point $S$ that is fixed to the wheel body results from

$$\boldsymbol{V}_s = \boldsymbol{V} + \frac{r_e}{r}\boldsymbol{\omega} \times \boldsymbol{r} \tag{2.28}$$

with $\boldsymbol{\omega}$ being the angular velocity vector of the wheel body with respect to the inertial frame. On the other hand, this velocity is equal to the speed of point $C^*$ minus the linear speed of rolling:

$$\boldsymbol{V}_s = \boldsymbol{V}_c^* - V_r\boldsymbol{l} \tag{2.29}$$

from which $V_r$ follows

$$V_r = \boldsymbol{l} \cdot (\boldsymbol{V}_c^* - \boldsymbol{V}_s) \tag{2.30}$$

or

$$V_r = V_{cx}^* - V_{sx} \tag{2.31}$$

The linear speed of rolling is, according to (2.6), related to the angular speed of rolling:

$$\Omega_r = \frac{1}{r_e} V_r \tag{2.32}$$

Of course, on flat roads $\Omega_r = \Omega$, which is the wheel speed of revolution and may be directly calculated by using the relationship (2.7).

The lateral slip speed $V_{cy}$ is obtained by taking the lateral component of $\boldsymbol{V_c}$ (2.26):

$$V_{cy} = \boldsymbol{V}_c \cdot \boldsymbol{t} \tag{2.33}$$

with

$$\boldsymbol{t} = \boldsymbol{l} \times \boldsymbol{n} \tag{2.34}$$

The lateral slip $\tan\alpha$ reads

$$\tan\alpha = -\frac{V_{cy}}{V_{cx}^*} = -\frac{\boldsymbol{V}_c \cdot \boldsymbol{t}}{\boldsymbol{V}_c^* \cdot \boldsymbol{l}} \tag{2.35}$$

The longitudinal slip speed $V_{sx}$ is obtained in a similar way:

$$V_{sx} = \boldsymbol{V}_s \cdot \boldsymbol{l} \tag{2.36}$$

The longitudinal slip $\kappa$ now becomes

$$\kappa = -\frac{V_{sx}}{V_{cx}^*} = -\frac{\boldsymbol{V}_s \cdot \boldsymbol{l}}{\boldsymbol{V}_c^* \cdot \boldsymbol{l}} \tag{2.37}$$

The turn slip according to definition (2.18) is derived as follows:

$$\varphi_t = -\frac{\dot{\psi}}{V_c} = -\frac{\dot{\boldsymbol{l}} \cdot \boldsymbol{t}}{V_c} \tag{2.38}$$

with the time derivative of the unit vector $\boldsymbol{l}$ in the numerator. The wheel camber angle is obtained as indicated before:

$$\sin\gamma = -\boldsymbol{n} \cdot \boldsymbol{s} \tag{2.39}$$

**Exercise 2.1 Slip and Rolling Speed of a Wheel Steered About a Vertical Axis**

The vehicle depicted in Figure 2.7 runs over a flat level road. The rear frame moves with velocities $u$, $v$, and $r$ with respect to an inertial triad (choose ($O^o$, $x^o$, $y^o$, $z^o$) that at the instant considered is positioned parallel to the moving triad ($B$, $x$, $y$, $z$) attached to the rear frame). The front frame can be turned with a rate $\dot{\delta}$ (= $d\delta/dt$) with respect to the rear frame. At the instant considered, the front frame is steered over an angle $\delta$.

It is assumed that the effective rolling radius is equal to the loaded radius ($r_e = r$, $C^* = C$). The longitudinal slip at the front wheels is assumed to be equal to zero ($V_{sx} = 0$).

Derive expressions for the lateral slip speed $V_{cy}$, the linear speed of rolling $V_r$, and the lateral slip $\tan\alpha$ for the right front wheel.

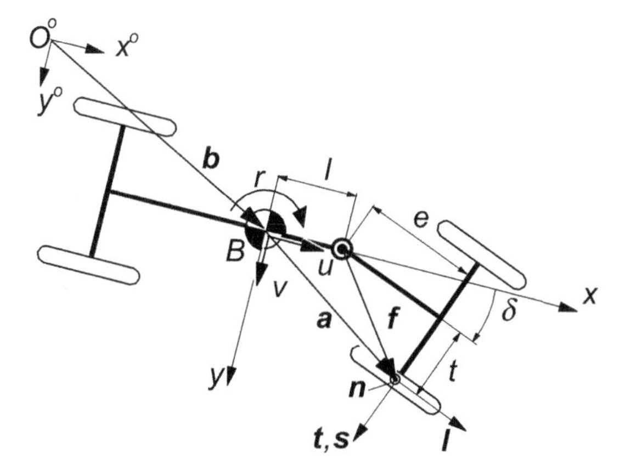


**FIGURE 2.7** Top view of vehicle (Exercise 2.1).

**Exercise 2.2 Slip and Rolling Speed of a Wheel Steered About an Inclined Axis (Motorcycle)**

The wheel shown in Figure 2.8 runs over a flat level road surface. Its center $A$ moves along a horizontal straight line at a height $H$ with a speed $u$. The rake angle $\varepsilon$ is 45°. The steer axis $BA$ (vector **a**) translates with the same speed $u$. There is no wheel slip in the longitudinal direction ($V_{sx} = 0$). Again we assume $r_e = r$. For the sake of simplifying this complex problem, it is assumed that the wheel center height $H$ is a given constant.

Derive expressions for the lateral slip speed $V_{cy}$, the linear speed of rolling $V_r$, and the turn slip speed $\dot{\psi}$ in terms of $H$, $u$, and $\dot{\delta}$ for $\delta = 0°$, 30°, and 90°. Also show the expressions for the slip angle $\alpha$, the camber angle $\gamma$, and the spin slip $\varphi$ with contributions both from turning and camber. Note that in reality height $H$ depends on $\delta$ and changes with $\dot{\delta}$.

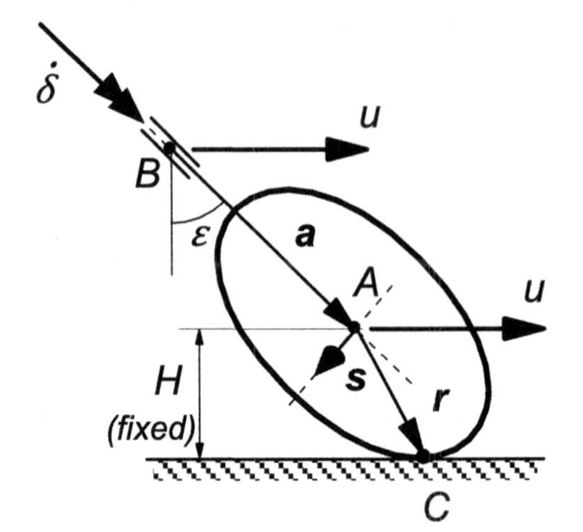


**FIGURE 2.8** Side view of the front part of vehicle (motorcycle) with wheel turned over angle $\delta$ about steer axis (***a***) (Exercise 2.2).

## 2.4. FUNDAMENTAL DIFFERENTIAL EQUATIONS FOR A ROLLING AND SLIPPING BODY

A wheel with tire that rolls over a smooth level surface and at the same time performs longitudinal and lateral slipping motions will develop horizontal deformations as a result of the presence of frictional forces that attempt to prevent the tire particles, which have entered the contact area, from sliding over the road. Besides areas of adhesion, areas of sliding may occur in the contact patch. The latter condition will arise when the deflection generated in the range of adhesion would have become too large to be maintained by the available frictional forces. In the following, a set of partial differential equations will be derived that governs the horizontal tire deflections in the contact area in connection with possibly occurring velocities of sliding of the tire particles. For a given physical structure of the tire, these equations can be used to develop the complete mathematical description of tire model behavior as will be demonstrated in subsequent chapters.

Consider a rotationally symmetric elastic body representing a wheel and tire rolling over a smooth horizontal rigid surface representing the road. As indicated in Figure 2.9 a system of axes ($O^o$, $x^o$, $y^o$, $z^o$) is assumed to be fixed to the road. The $x^o$ and $y^o$ axes lie in the road surface and the $z^o$ axis points downward. Another coordinate system ($C$, $x$, $y$, $z$) is introduced of which the axes $x$ and $y$ lie in the ($x^o$, $O^o$, $y^o$) plane and $z$ points downward. The $x$-axis is defined to lie in the wheel center plane and the $y$-axis forms the vertical projection of the wheel spindle axis. The origin $C$, which is the so-called contact center or, perhaps better: the point of intersection, travels with an assumedly constant speed $V_c$ over the ($x^o$, $O^o$, $y^o$) plane. The traveled distance $s$ is

$$s = V_c t \tag{2.40}$$

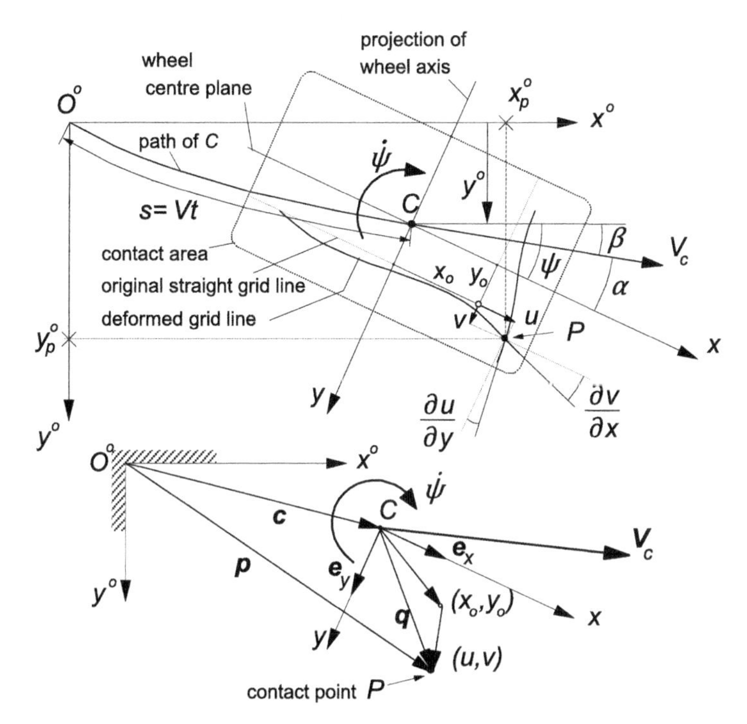


**FIGURE 2.9** Top view of tire contact area showing its position with respect to the system of axes ($O^o$, $x^o$, $y^o$, $z^o$) fixed to the road and its deformations ($u$, $v$) with respect to the moving triad ($C$, $x$, $y$, $z$).

where $t$ denotes the time. The tangent to the orbit of $C$ makes an angle $\beta$ with the fixed $x^o$-axis. With respect to this tangent, the $x$-axis is rotated over an angle $\alpha$, defined as the slip angle. The angular deviation of the $x$-axis with respect to the $x^o$-axis (that is the yaw angle) becomes

$$\psi = \beta + \alpha \tag{2.41}$$

For angle $\beta$ the following relation with $y^o$, the lateral displacement of $C$, holds

$$\sin\beta = \frac{dy^o}{ds} \tag{2.42}$$

As a result of friction, horizontal deformations may occur in the contact patch. The corresponding displacements of a contact point with respect to the position this material point would have in the horizontally undisturbed state (defined to occur when rolling on a frictionless surface), with coordinates ($x_o$, $y_o$), are indicated by $u$ and $v$ in $x$ and $y$ directions, respectively. These displacements are functions of coordinates $x$ and $y$ and of the traveled distance $s$ or the time $t$.

The position in space of a material point of the rolling and slipping body in contact with the road (cf. Figure 2.9) is indicated by the vector:

$$\boldsymbol{p} = \boldsymbol{c} + \boldsymbol{q} \tag{2.43}$$

where $\boldsymbol{c}$ indicates the position of the contact center $C$ in space and $\boldsymbol{q}$ the position of the material point with respect to the contact center. We have for the latter vector

$$\boldsymbol{q} = (x_o + u)\boldsymbol{e}_x + (y_o + v)\boldsymbol{e}_y \tag{2.44}$$

with $\boldsymbol{e}_x$ (= $\boldsymbol{l}$) and $\boldsymbol{e}_y$ (= $\boldsymbol{t}$) representing the unit vectors in $x$ and $y$ directions. The vector of the sliding velocity of the material point relative to the road obviously becomes

$$\begin{aligned} \boldsymbol{V}_g &= \dot{\boldsymbol{p}} = \dot{\boldsymbol{c}} + \dot{\boldsymbol{q}} \\ &= \boldsymbol{V}_c + (\dot{x}_o + \dot{u})\boldsymbol{e}_x + (\dot{y}_o + \dot{v})\boldsymbol{e}_y + \dot{\psi}\{(x_o + u)\boldsymbol{e}_y - (y_o + v)\boldsymbol{e}_x\} \end{aligned} \tag{2.45}$$

where $\boldsymbol{V}_c = \dot{\boldsymbol{c}}$ denotes the vector of the speed of propagation of contact center $C$.

The coordinates $x_o$ and $y_o$ of the material contact point of the horizontally undisturbed tire (zero friction) will change due to rolling. Then, the point will move through the contact area from the leading edge to the trailing edge. In the general case, e.g., of an elastic ball rolling over the ground, we may have rolling in both the forward and lateral directions. Then, both coordinates of the material point will change with time. In the present analysis of the rolling wheel, we will disregard the possibility of sideways rolling.

To assess the variation in the coordinates of the point on the zero friction surface, let us first consider an imaginary road surface that is in the same position as the actual surface but does not transmit forces to the wheel. Then, the tire penetrates the imaginary surface without deformation. When the general situation is considered of a wheel-spin axis that is inclined with respect to the imaginary road surface, i.e., rolling at a camber angle $\gamma$, the coordinates $x$ and $y$ of the material point change with time as follows:

$$\begin{aligned} \dot{x} &= \frac{\mathrm{d}x}{\mathrm{d}t} = -r_o(y) \cdot \Omega \\ \dot{y} &= \frac{\mathrm{d}y}{\mathrm{d}t} = -x \sin\gamma \cdot \Omega \end{aligned} \tag{2.46}$$

If $(1/2)\,a|\dot{\gamma}| \ll r_o|\gamma|\Omega$, the effect of the time rate of change of the camber angle (wheel plane rotation about the $x$-axis) on the partial derivative $\partial y/\partial t$ and thus on $\dot{y}$ may be neglected. This effect is directly connected with the small instantaneous or so-called nonlagging response to camber changes. Similar instantaneous responses may occur as a result of normal load changes when the tire shows conicity or ply-steer. The terms associated with $\partial y/\partial t$ have been neglected in the above equation and related neglects will be performed in subsequent formulas for $\dot{y}_o$.

The radius of the outer surface in the undeformed state $r_o$ may depend on the lateral coordinate $y$. If the rolling body shows a touching surface that is already parallel to the road before it is deformed, we would have in the neighborhood of the center of contact: $r_o(y) = r_o(0) - y \sin\gamma$. In case of a cambered car tire, a distortion of belt and carcass is needed to establish contact over a finite area with the ground. The shape of the tire cross section in the undeformed state governs the dependency of the free radius with the distance to the wheel center plane.

In Figure 2.10, an example is given of two different cases. The upper part corresponds to a rear view and the lower one to a plan view of a motorcycle tire and of a car tire pressed against an assumedly frictionless surface ($\mu = 0$). It may be noted that in case of a large camber angle like with the motorcycle tire, one might decide to redefine the position of the $x$- and $z$-axes. The contact line which is the part of the peripheral line that touches the road surface has been indicated in the figure.

When the tire is loaded against an assumedly frictionless rigid surface, deformations of the tire will occur. These will be due to: (1) lateral and longitudinal compression in the contact region, (2) a possibly not quite symmetric structure of the tire resulting in effects known as ply-steer and conicity, and (3) loading at a camber angle that will result in distortion of the carcass and belt. The deformations occurring in the contact plane will be denoted as $u_o$ and $v_o$. We will introduce the functions $\theta_{x,y}(x, y)$ representing the partial derivatives of these normal-load-induced longitudinal and lateral deformations with respect to $x$. We have

$$\theta_x(x,y) = \frac{\partial u_o}{\partial x} \quad \text{and} \quad \theta_y(x,y) = \frac{\partial v_o}{\partial x} \tag{2.47}$$

These functions depend on the vertical load and on the camber angle.

If the tire is considered to roll on a frictionless flat surface at a camber angle or with nonzero $\theta$s, lateral and longitudinal sliding of the contact points will occur even when the wheel does not exhibit lateral, longitudinal, or turn slip. Since horizontal forces do not occur in this imaginary case, $u$ and $v$ are defined to be zero. The coordinates with respect to the moving axes system $(C, x, y, z)$ of the contact point sliding over the hypothetical frictionless surface were denoted as $x_o$ and $y_o$. The time rates of change of $x_o$ and $y_o$ depend on the position in the contact patch, on the speed of rolling $\Omega$, and on the camber angle $\gamma$. We find after disregarding the effect of $\partial y_o/\partial t$:

$$\begin{aligned}
\dot{x}_o &= \dot{x} + \dot{u}_o = -\{1 + \theta_x(x,y)\} \cdot r_o(y) \cdot \Omega \\
\dot{y}_o &= \dot{y} + \dot{v}_o = -\{x_o \sin\gamma + \theta_y(x,y) \cdot r_o(y)\} \cdot \Omega
\end{aligned} \tag{2.48}$$

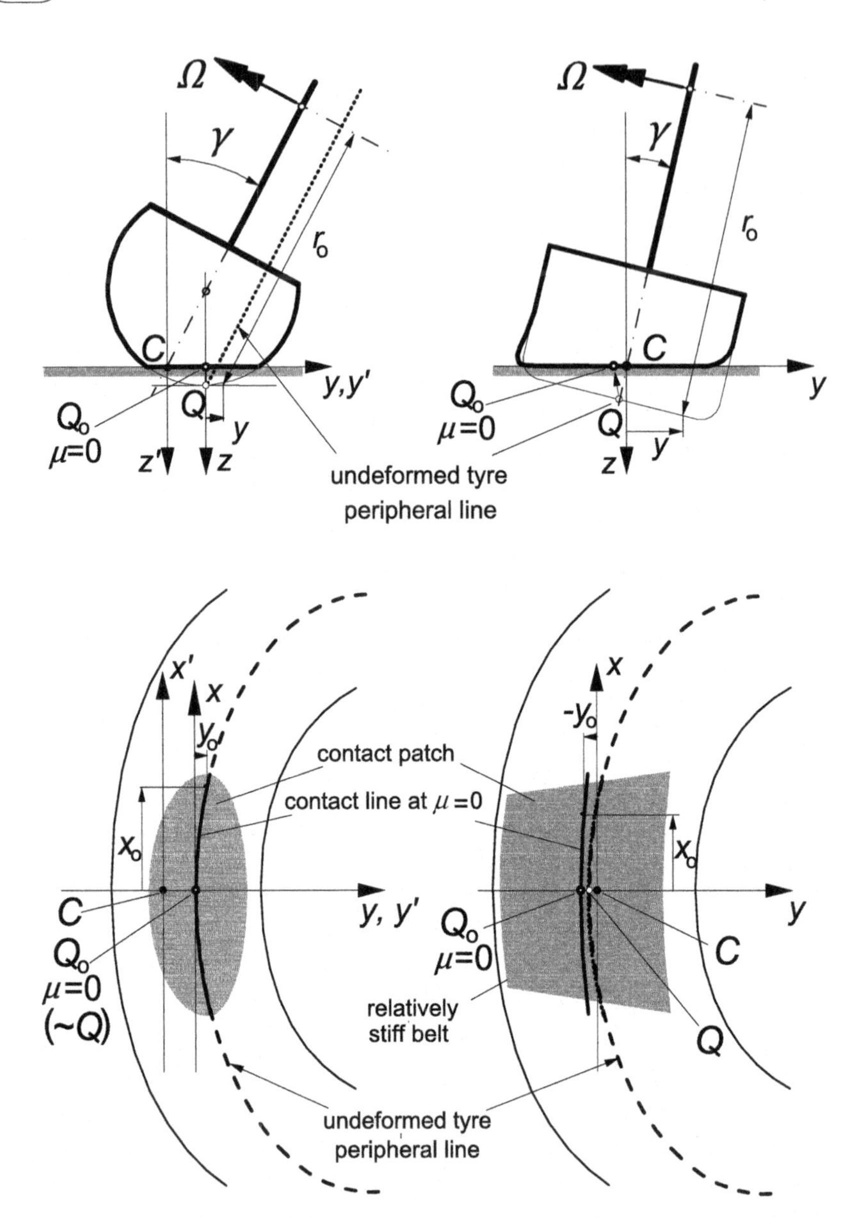


FIGURE 2.10 A motorcycle tire (left) and a car tire (right) in cambered position touching the assumedly frictionless road surface. The former without and the latter with torsion and bending of the carcass and belt. Top: rear view; bottom: top view.

For homogeneous rolling bodies (e.g., a railway wheel or a rubber ball) with counter surfaces already parallel before touching, torsion about the longitudinal axis and lateral bending do not occur, ply-steer and conicity are absent, and horizontal compression may be neglected ($\theta$'s vanish). Moreover, we then had: $r_o(y) = r_o(0) - y \sin \gamma$ so that for this special case (2.48) reduces to

$$\dot{x}_o = \dot{x} = -\{r_o(0) - y\sin\gamma\} \cdot \Omega$$
$$\dot{y}_o = \dot{y} = -x\sin\gamma \cdot \Omega \qquad (2.49)$$

For a tire the terms with $\theta$ are appropriate. Circumferential compression ($\theta_x < 0$) decreases the effective rolling radius $r_e$ that is defined at free rolling. At camber, due to the structure of a car tire with a belt that is relatively stiff in lateral bending, the compression/extension factor $\theta_x$ will not be able to compensate for the fact that a car tire shows only a relatively small variation in the free radius $r_o(y)$ across the width of the tread, while for the same reason the lateral distortion factor $\theta_y$ appears to be capable of considerably counteracting the effect of the term with $\sin\gamma$ in the second Eqn (2.48). For a motorcycle tire with a cross section approximately forming a sector of a circle, touching at camber is accomplished practically without torsion about a longitudinal axis and the associated lateral bending of the tire near the contact zone. Figure 2.10 illustrates the expected deformations and the resulting much smaller curvature of the contact line on a frictionless surface for the car tire relative to the curvature exhibited by the motorcycle tire. Since on a surface with friction the rolling tire will be deformed to acquire a straight contact line, this observation may explain the relatively low camber stiffness of the car and truck tire.

As mentioned above, circumferential compression of belt and tread resulting from the normal loading process (somewhat counteracted by the presence of hysteresis also represented by the factor $\theta_x$) gives rise to a decrease in the effective rolling radius. The compression may not be uniform along the $x$-axis. For our purposes, however, we will disregard the resulting secondary effects. The coefficient in (2.48) that relates passage velocity $-\dot{x}_o$ and speed of revolution $\Omega$ of the wheel is designated as local effective rolling radius $r_{ej}$ at e.g., row $j$ of tread elements. This radius depends on lateral position $y_j$ and will change with vertical load $F_z$ and camber angle $\gamma$. The velocity through the contact zone of the elements of row $j$ is: $-\dot{x}_o = r_{ej}\Omega$. We may write

$$r_{ej} = \{1 + \theta_x(y_j)\}r_o(y_j) = r_e + \Delta r_{ej}(y_j) \qquad (2.50)$$

in which use has been made of the overall tire effective rolling radius $r_e$ defined according to Eqns (2.2, 2.6) and $\Delta r_{ej}(y_j)$, the possibly anti-symmetric variation in the local effective rolling radius over the tread width with respect to $r_e$ due to loading at camber or conicity. The (average) effective rolling radius $r_e$ is expected to depend on the camber angle as well.

We will now try to make a distinction between the contributions to the anti-symmetric variation originating from conicity and from camber. We write for the local passage velocity:

$$\dot{x}_o = -[\{1 + \theta_{\text{conx}}(y)\}r_e - \{1 + \theta_{\gamma x}(y)\}y_o \sin\gamma] \cdot \Omega \tag{2.51}$$

In this way, we have achieved a structure closely related to the first equation of (2.49). For a rolling elastic body such as a ball or a motorcycle tire, $\theta_{\gamma x}$ will be (close to) zero, while a car, truck or racing car tire is expected to have a $\theta_{\gamma x}$ closer to $-1$. A bias-ply tire featuring a more compliant carcass and tread band and a rounder cross-section profile is expected to better 'recover' from the torsion effects, resulting in a value in-between the two extremes. It is true that in general bias-ply tires show considerably larger side forces as a result of camber than radial ply car tires.

When the product of $\theta_y$ and $r_o - r_e$ is considered to be negligible, we write for the lateral velocity of the points over a frictionless surface with respect to the $x$-axis instead of (2.48):

$$\dot{y}_o = -\{x_o \sin\gamma + \theta_y(x,y)\cdot r_e\} \cdot \Omega \tag{2.52}$$

The resulting expressions (2.51, 2.52) may now be substituted in expression (2.45) for the sliding speed components.

For a given material point of the tire outer surface that now rolls on a road surface with friction reintroduced, the associated deflections $u$ and $v$ (which occur on top of the initial load induced deflections $u_o$ and $v_o$) are functions of its location in the contact patch and of the time: $u = u(x_o, y_o, t)$ and $v = v(x_o, y_o, t)$. Hence, we have for the time rates of change:

$$\begin{aligned}\dot{u} &= \frac{\mathrm{d}u}{\mathrm{d}t} = \frac{\partial u}{\partial x_o}\frac{\mathrm{d}x_o}{\mathrm{d}t} + \frac{\partial u}{\partial y_o}\frac{\mathrm{d}y_o}{\mathrm{d}t} + \frac{\partial u}{\partial t}\\ \dot{v} &= \frac{\mathrm{d}v}{\mathrm{d}t} = \frac{\partial v}{\partial x_o}\frac{\mathrm{d}x_o}{\mathrm{d}t} + \frac{\partial v}{\partial y_o}\frac{\mathrm{d}y_o}{\mathrm{d}t} + \frac{\partial v}{\partial t}\end{aligned} \tag{2.53}$$

in which the expressions (2.51, 2.52) apply. If we disregard sideways rolling and neglect a possible small effect of the variation in $y_o$ that occurs at camber, conicity, or ply-steer, the second terms of the right-hand members disappear. The remaining expressions are substituted in (2.45).

According to Eqn (2.29), the linear speed of rolling and the velocity of the contact center are related through the slip speed. We have (when disregarding the generally very small difference between the velocities of $C^*$ and $C$) the vector relationship:

$$\boldsymbol{V}_c = \boldsymbol{V}_s + V_r\boldsymbol{e}_x \quad \text{with} \quad V_r = r_e\Omega \tag{2.54}$$

After simplifying the equations by neglecting products of assumedly small quantities (i.e., deflections and input (slip) quantities) and using the expression for $\omega_z$, which is the absolute angular velocity of the wheel body about the $z$-axis as indicated in Eqn (2.15), components of sliding velocity become

$$V_{gx} = V_{sx} - \left(\frac{\partial u}{\partial x} + \theta_{\text{conx}}(y)\right)V_r + \frac{\partial u}{\partial t} - y_o\omega_z + \theta_{\gamma x}(y) - y_o \sin\gamma \cdot \Omega \quad (2.55)$$

$$V_{gy} = V_{sy} - \left(\frac{\partial v}{\partial x} + \theta_y(x, y)\right)V_r + \frac{\partial v}{\partial t} + x_o\omega_z \quad (2.56)$$

With all $\theta$'s omitted we would return to the basic case of e.g., a ball rolling and slipping over a flat rigid surface. In Chapter 3, a physical tire model is developed using the above equations. There, the $\theta$'s are expressed in terms of the camber angle and of equivalent camber and slip angles $\gamma_{\text{con}}$ and $\alpha_{\text{ply}}$ to account for conicity and ply-steer (cf. Eqns (3.108, 3.109)).

Let us now consider the special case of a freely rolling tire subjected to only small lateral slip and spin ($\kappa = 0, |\alpha| \ll 1, |\gamma| \ll 1, r_e \ll |R|$, with $R$ the instantaneous radius of path curvature, and thus $|\varphi_t| \ll 1/r_e$) and neglect effects of initial non-parallelity of touching surfaces as well as 'built-in' load- and camber-induced deformation effects (all $\theta$'s = 0). We then have (approximately)

$$V_{sx} = 0, \;\; V_{sy} = -V_c\alpha, \;\; V_r = V_c, \;\; \omega_z(= \dot{\psi} - \Omega \sin\gamma) = -V_c\varphi \quad (2.57)$$

Using the traveled distance $s$ (2.40) instead of the time $t$ as an independent variable, we obtain the following expressions for the sliding velocities of a contact point with coordinates ($x$, $y$) of a freely rolling tire at small lateral slip and spin:

$$V_{gx} = V_c\left(y\varphi - \frac{\partial u}{\partial x} + \frac{\partial u}{\partial s}\right) \quad (2.58)$$

$$V_{gy} = V_c\left(-\alpha - x\varphi - \frac{\partial v}{\partial x} + \frac{\partial v}{\partial s}\right) \quad (2.59)$$

In the contact area, regions of adhesion may occur as well as regions where sliding takes place. In the region of adhesion, the tire particles touching the road do not move and we have $V_{gx} = V_{gy} = 0$. In this part of the contact area, the frictional shear forces acting from road to tire on a unit area (with components denoted by $q_x$ and $q_y$ not to be confused with the components of the position vector (2.44) used earlier) do not exceed the maximum available frictional force per unit area. The maximum frictional shear stress is governed by coefficient of friction $\mu$ and normal contact pressure $q_z$. In the adhesion region, Eqns (2.58, 2.59) reduce to

$$\frac{\partial u}{\partial x} - \frac{\partial u}{\partial s} = -y\varphi \tag{2.60}$$

$$\frac{\partial v}{\partial x} - \frac{\partial v}{\partial s} = -\alpha - x\varphi \tag{2.61}$$

Furthermore, we have the condition:

$$\sqrt{q_x^2 + q_y^2} < \mu q_z \tag{2.62}$$

In the region of sliding, Eqns (2.58, 2.59) hold. If the deformation gradients were known, the velocity components $V_{gx}$ and $V_{gy}$ might be obtained from these equations. For the frictional stress vector, we obtain:

$$(q_x, q_y) = -\mu q_z (V_{gx}, V_{gy})/V_g \tag{2.63}$$

with

$$V_g = \sqrt{V_{gx}^2 + V_{gy}^2} \tag{2.64}$$

For the simpler case that only lateral slip occurs ($\varphi = 0$), the equations for the lateral deformations reduce to:
*in the adhesion region*:

$$\begin{aligned} \frac{\partial v}{\partial x} - \frac{\partial v}{\partial s} &= -\alpha \\ |q_y| &< \mu q_z \end{aligned} \tag{2.65}$$

*in the sliding region*:

$$\begin{aligned} \frac{\partial v}{\partial x} - \frac{\partial v}{\partial s} &= -\alpha - \frac{V_{gx}}{V_c} \\ q_y &= -\mu q_z \operatorname{sgn} V_{gy} \end{aligned} \tag{2.66}$$

The Eqns (2.55, 2.56) apply in general. Their solutions contain constants of integration that depend on the selected physical model description of the tire. Applications of the differential equations derived above will be demonstrated in subsequent chapters both for steady-state and non-steady-state conditions. In the case of steady-state motion of the wheel, the partial derivatives with respect to time $t$ or traveled distance $s$ become equal to zero ($\partial v/\partial s = \partial u/\partial s = 0$). Then, the deformation gradients in the area of adhesion ($V_g = 0$) follow easily from the then ordinary differential equations. In the last simple case (2.66) we would obtain: $\mathrm{d}v/\mathrm{d}x = -\alpha$, which means that the contact line is straight and runs parallel to the speed vector $\boldsymbol{V}_c$.

**Exercise 2.3 Partial Differential Equations with Longitudinal Slip Included**

Establish the differential equations for the sliding velocities similar to the Eqns (2.58, 2.59) but now with longitudinal slip $\kappa$ included ($\alpha$ and $\varphi$ remain small). Note that in that case $V_r \neq V_c$ and that $V_r$ may be expressed in terms of $V_{cx}$ ($\approx V_c$) and $\kappa$. Also, find the partial differential equations governing the deflections in the adhesion zone similar to the Eqns (2.60, 2.61).

## 2.5. TIRE MODELS (INTRODUCTORY DISCUSSION)

Several types of mathematical models of the tire have been developed during the past half-century; each type for a specific purpose. Different levels of accuracy and complexity may be introduced in the various categories of utilization. This often involves entirely different ways of approach. Figure 2.11

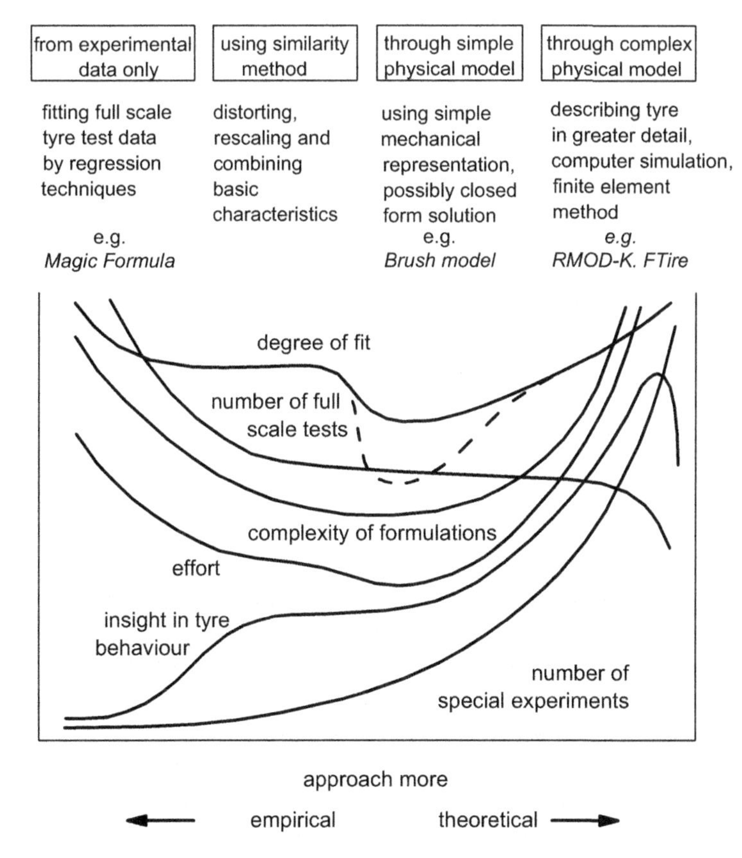


**FIGURE 2.11** Four categories of possible types of approach to develop a tire model.

roughly illustrates how the intensity of various consequences associated with different ways of attacking the problem tends to vary. From left to right the model is based less on full-scale tire experiments and more on the theory of the behavior of the physical structure of the tire. In the middle, the model will be simpler but possibly less accurate while at the far right the description becomes complex and less suitable for application in the simulation of vehicle motions and may be more appropriate for the analysis of detailed tire performance in relation to its construction.

At the left-hand category, we have mathematical tire models that describe measured tire characteristics through tables or mathematical formulas and certain interpolation schemes. These formulas have a given structure and possess parameters that are usually assessed with the aid of regression procedures to yield a best fit to the measured data. A well-known empirical model is the *Magic Formula* tire model treated in Chapter 4. This model is based on a sin(arctan) formula that not only provides an excellent fit for the $F_y$, $F_x$, and $M_z$ curves but in addition features coefficients that have clear relationships with typical shape and magnitude factors of the curves to be fitted.

The similarity approach (second category) is based on the use of a number of basic characteristics typically obtained from measurements. Through distortion, rescaling, and multiplications, new relationships are obtained to describe certain off-nominal conditions. Chapter 4 introduces this method, which is particularly useful for application in vehicle simulation models that require rapid (e.g., real-time) computations.

Depending on the type of the physical model chosen, a simple formulation may already provide sufficient accuracy for limited fields of application. The HSRI model depicted in Figure 2.12 developed by Dugoff, Fancher, and Segel (1970) and later corrected and improved by Bernard, Segel, and Wild (1977) is a good example. The figure illustrates the considerable simplification with respect to a more realistic representation of tire deformation (Figure 2.13) that is needed to keep the resulting mathematical formulation manageable for vehicle dynamics simulation purposes and still include important matters such as the representation of combined slip and a coefficient of friction that may drop with speed of sliding.

The model of Figure 2.13 exhibits carcass flexibility and shows a more realistic parabolic pressure distribution. For such a model, (approximate) analytical solutions are feasible only when pure side slip (possibly including camber) occurs and the friction coefficient is considered constant (e.g., Fiala 1954).

Relatively simple physical models of this third category such as the 'brush model' of Figure 2.12 are especially useful to get a better understanding of tire behavior. The brush model with a parabolic pressure distribution will be discussed at length in Chapter 3.

The right-most group of Figure 2.11 is aimed primarily at more detailed analysis of the tire. The complex finite element- or segment-based models

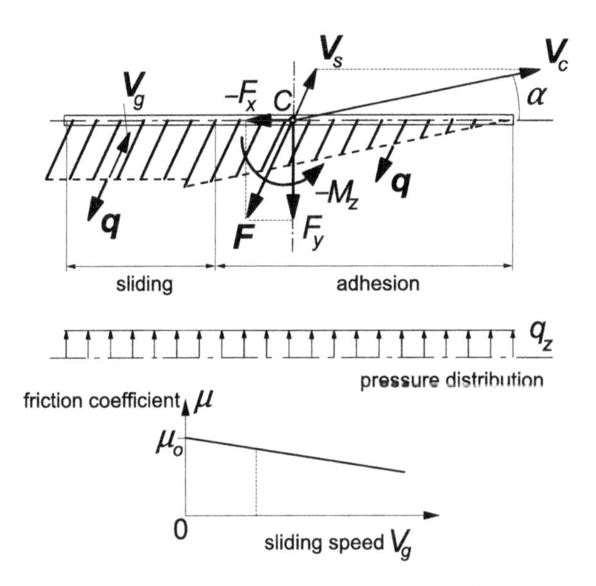


**FIGURE 2.12** Top: The brush type tire model at combined longitudinal (brake) slip and lateral slip in case of equal longitudinal and lateral stiffnesses. Bottom: The linearly decaying friction coefficient.

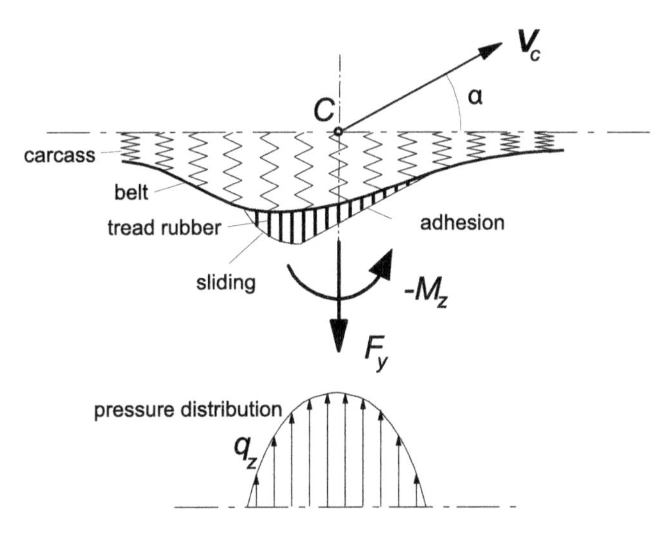


**FIGURE 2.13** Tire model with flexible carcass at steady-state rolling with slip angle $\alpha$.

belong to this category (e.g., *RMOD-K* and *FTire*, cf. Chap. 13). A simpler representation of carcass compliance that is experienced in the lower part of the tire near the contact patch considerably speeds up the computation. In addition, the way in which the tread elements are handled is crucial. The computer

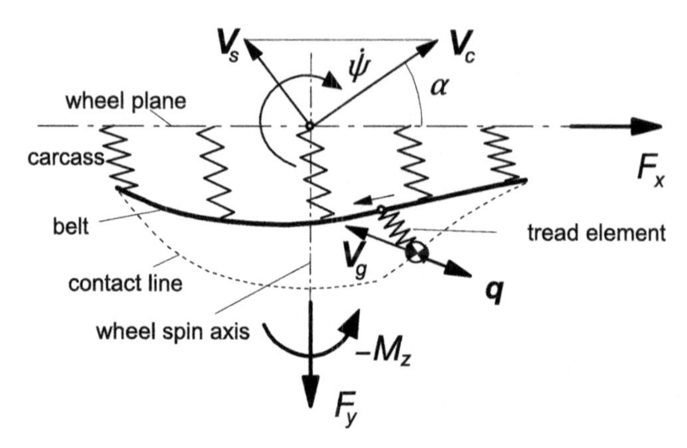


**FIGURE 2.14** Computer simulation tire model with flexible carcass, arbitrary pressure distribution, and friction coefficient functions. Forces acting on a single tread element mass during one passage through the contact length are integrated to obtain the total forces and moment $F_x$, $F_y$, and $M_z$.

simulation tread-element-following method is attractive and allows considerable freedom to choose pressure distribution and friction coefficient functions of sliding velocity and local contact pressure. The physical model that forms the basis of the latter method has been depicted in Figure 2.14.

Influence (Green) functions may be used to describe the carcass horizontal compliance in the contact zone and possibly several rows of tread elements may be considered to move through the contact patch. One element per row is followed while it travels through the length of contact (or several elements through respective sub-zones). During such a passage the carcass deflection is kept constant, the motion of the single mass-spring (tread element) system that is dragged over the ground is computed, the frictional forces are integrated, the total forces and moment determined, and the carcass deflection is updated. Instead of using the dynamic way of solving for the deflection of the tread element while it runs through the contact patch, an iteration process may be employed. The model is capable of handling non-steady-state conditions. A relatively simple application of the tread-element-following method will be shown in the subsequent chapter when dealing with the 'brush model' subjected to combined slip with camber, a condition that is too difficult to deal with analytically. In addition, the introduction of carcass compliance will be demonstrated. A method based on modal synthesis to model tire deflection has been employed by Guan et al. (1999) and by Shang et al. (2002). For further study we refer to Sec. 3.3 and to the original work of Willumeit (1969), Pacejka, and Fancher (1972a), Sharp and El-Nashar (1986), Gipser et al. (cf. Sec.13.3), Guo and Liu (1997), and Mastinu (1997) and the state-of-the-art paper of Pacejka and Sharp (1991).

Although it is possible to develop a model for non-steady-state conditions by purely empirical means, most relatively simple and more complex transient and dynamic tire models are based on the physical nature of the tire. It is of interest to note that for a proper description of tire behavior at time-varying conditions an essential property must be represented in the physical models belonging to both right-hand categories of Figure 2.11. That is the lateral and sometimes also the fore and aft compliance of the carcass. Less complex non-steady-state tire models feature only carcass compliance without the inclusion of elastic tread elements. In steady-state models, the introduction of such a flexibility is often not required. Only to represent properly the self-aligning torque in case of a braked or driven wheel, is carcass lateral compliance needed. Tire inertia becomes important at higher speeds and frequencies of the wheel motion. The problem of establishing non-steady-state tire models is addressed in Chapters 5, 7, 8, 9, and 10 in successive levels of complexity to meet conditions of increasing difficulty.

Conditions become more demanding when for example: (1) the wheel motion gives rise to larger values of slip, which no longer permits an approximate linear description of the force- and moment-generating properties; (2) combined slip occurs, possibly including wheel camber and turn slip; (3) large camber occurs, which may necessitate the consideration of the dimensions of the tire cross section; (4) the friction coefficient cannot be approximated as a constant quantity but may vary with sliding velocity and speed of travel as occurs on wet or icy surfaces; (5) the wavelength of the path of contact points at non-steady-state conditions can no longer be considered large, which may require the introduction of the lateral and longitudinal compliance of the carcass; (6) the wavelength becomes relatively short, which may necessitate the consideration of a finite contact length (retardation effect) and possibly contact width (at turn slip and camber); (7) the speed of travel is large so that tire inertia becomes of importance, in particular its gyroscopic effect; (8) the frequency of the wheel motion has reached a level that requires the inclusion of the first or even higher modes of vibration of the belt; (9) the vertical profile of the road surface contains very short wavelengths with appreciable amplitudes as would occur in the case of rolling over a short obstacle or cleat, then, among other things, the tire enveloping properties should be accounted for; and (10) motions become severe (large slip and high speed), which may necessitate modeling the effect of the warming up of the tire involving possibly the introduction of the tire temperature as a model parameter. All items mentioned (except the last one) will be accounted for in the remainder of this book.

Chapter 3

# Theory of Steady-State Slip Force and Moment Generation

## Chapter Outline



## 3.1. INTRODUCTION

This chapter is devoted to the analysis of the properties of a relatively simple theoretical tire model belonging to the third category of Figure 2.11. The mathematical modeling of the physical model shown in Figure 2.13 has been a challenge to various investigators. Four fundamental factors play a role: frictional properties in the road-tire interface, distribution of the normal contact pressure, compliance of the tread rubber, and compliance of the belt/carcass.

Models of the carcass with belt and side walls with encapsulated pressurized air that are commonly encountered in the tire modeling literature are based either on an elastic beam or on a stretched string both suspended on an elastic foundation with respect to the wheel rim. The representation of the belt by a beam instead of by a stretched string is more difficult because of the fact that the differential equation that governs the lateral deflection of the belt

**Tire and Vehicle Dynamics. DOI: 10.1016/B978-0-08-097016-5.00003-6**

under the action of a lateral force becomes of the fourth instead of the second order. For the study of steady-state tire behavior, most authors approximate the more or less exact expressions for the lateral deflection of the beam or string.

As an extension to the original 'brush' model of Fromm and of Julien (cf. Hadekel (1952) for references) who did not consider carcass compliance, Fiala (1954) and Freudenstein (1961) developed theories in which the carcass deflection is approximated by a symmetric parabola. Böhm (1963) and Borgmann (1963), the latter without the introduction of tread elements, used asymmetric approximate shapes determined by both the lateral force and the aligning torque. Pacejka (1966, 1981) established the steady-state side-slip characteristics for a stretched-string-tire model without and with the inclusion of tread elements attached to the string. The lateral stiffness distribution as measured on a slowly rolling tire in terms of influence or Green's functions (cf. Savkoor 1970) may be employed in a model for the side-slipping tire possibly in connection with the tread-element-following method that was briefly discussed in the preceding section (cf. Pacejka 1972, 1974) and will be demonstrated later on in Section 3.3 of the present chapter.

Frank (1965a) has carried out a thorough comparative investigation of the various one-dimensional models. He employed a general fourth-order differential equation with which tire models can be examined that feature a stretched string, a beam, or a stretched beam provided with elastic tread elements. Frank obtained a solution of the steady-state slip problem with the aid of a special analog computer circuit. A correlation with Fourier components of the measured deformation of real tires reveals that the stretched string-type model seems to be more suitable for the simulation of a bias ply tire, whereas the beam model is probably more appropriate for representing the radial ply tire.

Figure 3.1 (from Frank 1965b) presents the calculated characteristics of several types of carcass models provided with tread elements. The curves represent: *a*. stretched string model, *b*. beam model, *c*. approximation based on Fiala's model (symmetric parabolic carcass deflection), *d*. model of Fromm (brush model with rigid carcass). The tread element stiffness is the same for each model. The parameters in the cases *a*, *b*, and *c* have been chosen in such a way as to give a best fit to experimental data for the peak side force and the cornering force at small slip angles (that is, same cornering stiffness). It appears then that model *c* shows close correspondence with curve *a* for the side force. Curves *d* show the result when the carcass elasticity is neglected and only the flexibility of the tread elements is taken into account. When the tread element stiffness of model *d* is adapted (i.e., lowered) in such a way that the cornering stiffness becomes equal to that of the other three models, no difference between the side force characteristics according to Fromm's and Fiala's models appears to occur. Due to approximations introduced by Fiala, the coefficients in the expression for the side force versus slip angle (if parabolic pressure distribution is adopted) become equal to those obtained directly by Fromm.

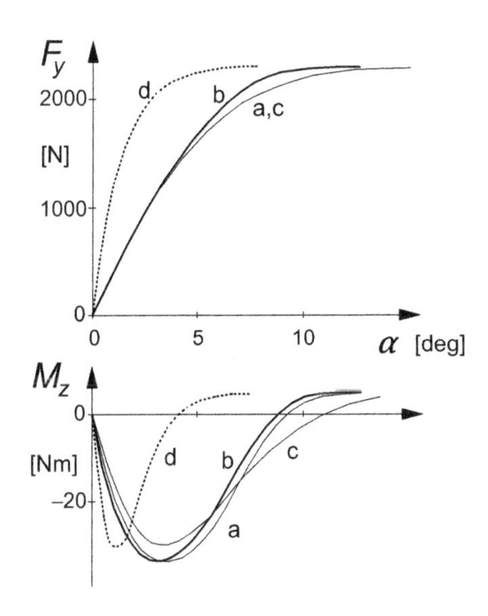


**FIGURE 3.1** Comparison of calculated characteristics for four different tire models with tread elements and nonsymmetrical pressure distribution at a given wheel vertical load (a: string, b: beam, c: Fiala, d: brush i.e., with rigid carcass) (from Frank 1965a,b).

In the calculations for Figure 3.1, Frank employed a constant coefficient of friction $\mu$ and a slightly asymmetric vertical pressure distribution $q_z(x)$, found from measurements. The positive aligning torque obtained at larger values of the slip angle $\alpha$ arises as a result of this asymmetry. The phenomenon that in practice the aligning torque indeed varies in this way is due to a combination of several effects. The main cause is probably connected with the asymmetric pressure distribution of the rolling tire (due to hysteresis of the tire compound) resulting in a small forward shift of the point of application of the normal load (giving rise to rolling resistance) and, consequently, at full sliding also of the resulting side force. Another important factor causing the moment to become positive is the fact that the coefficient of friction is not a constant but tends to decrease with sliding velocity. As may be derived from e.g., Eqn (2.59), the sliding velocity attains its largest values in the rear portion of the contact area where the slope $\partial v/\partial x$ becomes largest. Consequently, we expect to have larger side forces acting in the front half of the contact area at full sliding conditions than in the rear half. The rolling resistance force that due to the lateral distortion acts slightly beside the wheel plane may also contribute to the sign change of $M_z$. A $\mu$ that decreases with the sliding velocity (not considered by Frank) causes the creation of a peak in the $F_y(\alpha)$ curves and a further slight decay. This has often been observed to occur in practice, especially on wet and icy roads. In the longitudinal force characteristic the peak is usually more pronounced.

The influence of different but symmetric shapes for the vertical force distribution along the $x$-axis has been theoretically investigated by Borgmann (1963). He finds that, especially for tires exhibiting a relatively large carcass compliance, the influence of the pressure distribution is of importance. Nonsymmetric more general distributions were studied by Guo (1994). Many authors adopt, for the purpose of mathematical simplicity, the parabolic distribution (Fiala 1954, Freudenstein 1961, Bergman 1965, Pacejka 1958, Sakai (also n-th degree parabola, 1989), Dugoff et al. 1970 (uniform, rectangular distribution) and Bernard et al. 1977 (trapezium shape)). Models of Fiala and Freudenstein feature a flexible carcass while the remaining authors have restricted themselves to a rigid carcass or a uniformly deflected belt. That, however, enabled them to include the description of the more difficult case of combined slip. The introduction of a nonconstant friction coefficient has been treated by others: Böhm (1963), Borgmann (1963), Dugoff et al. (1970), Sakai (1981), and Bernard et al. (1977).

Figure 3.1 shows that, when the model parameters are chosen properly, the choice of the type of carcass model (beam, string, or rigid) has only a limited effect. Qualitatively, the resulting curves are identical. The rigid carcass model with elastic tread elements is often referred to as the brush tire model. Because of its simplicity and qualitative correspondence with experimental tire behavior, we will give a full treatment of its properties, mainly to provide understanding of steady-state tire slipping properties which may also be helpful in the development of more complex models. A uniform carcass deflection will be considered to improve the aligning torque representation at combined slip.

In Section 3.3, we deal with the effect of nonuniform carcass deflection, nonconstant friction coefficient, and the inclusion of camber and turning (path curvature) combined with side slip and braking/driving. For this purpose, the tread simulation model will be employed.

## 3.2. TIRE BRUSH MODEL

The brush model consists of a row of elastic bristles that touches the road plane and can deflect in a direction parallel to the road surface. These bristles may be called tread elements. Their compliance represents the elasticity of the combination of carcass, belt, and actual tread elements of the real tire. As the tire rolls, the first element that enters the contact zone is assumed to stand perpendicularly with respect to the road surface. When the tire rolls freely (that is without the action of a driving or braking torque) and without side slip, camber, or turning, the wheel moves along a straight line parallel to the road and in the direction of the wheel plane. In that situation, the tread elements remain vertical and move from the leading edge to the trailing edge without developing a horizontal deflection and consequently without generating a fore-and-aft or side force. A possible presence of rolling resistance is disregarded. When the wheel speed vector $V$ shows an angle with respect to the wheel plane, side slip occurs. When the wheel velocity of revolution $\Omega$ multiplied with the effective rolling radius $r_e$ is not equal to the forward component of the wheel speed $V_x = V\cos\alpha$, we have fore-and-aft slip. Under these conditions, depicted in Figure 3.2, horizontal deflections are developed and corresponding forces and moment arise. The tread elements move from the leading edge (on the

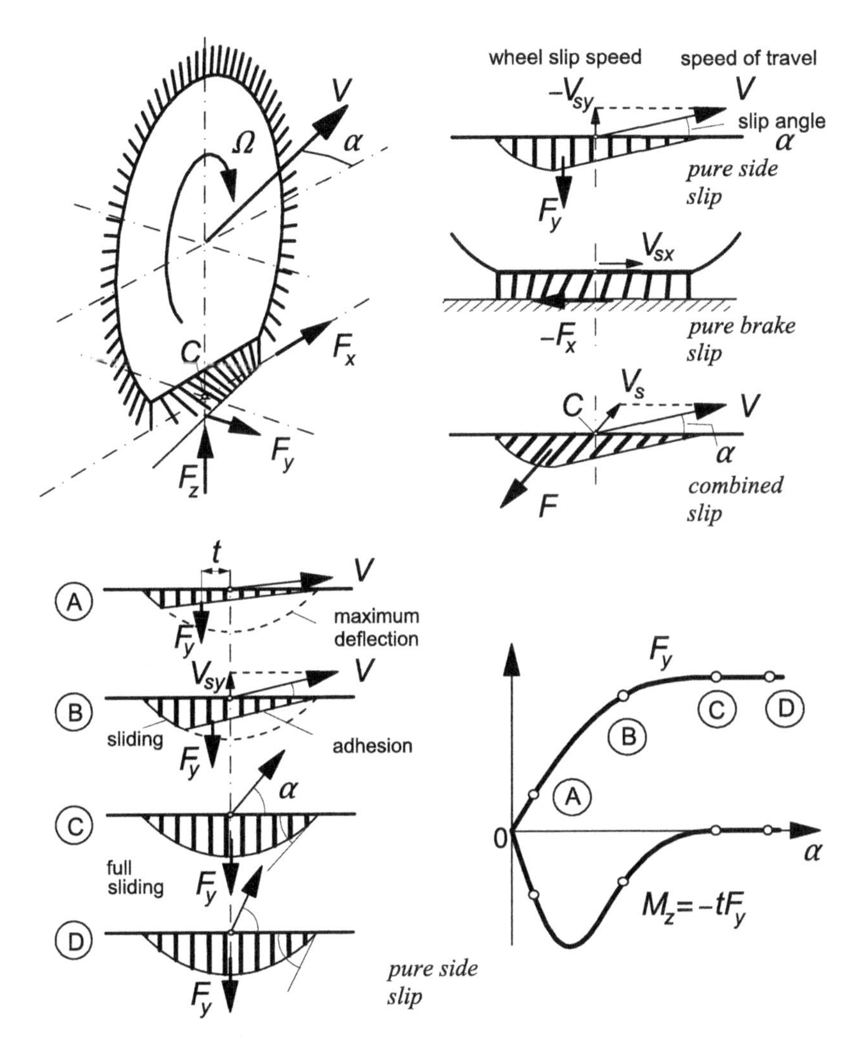


**FIGURE 3.2** The brush tire model. Top left: view of driven and side-slipping tire. Top right: the tire at different slip conditions. Bottom left: the tire at pure side slip, from small to large slip angle. Bottom right: the resulting side force and aligning torque characteristics.

right-hand side of the pictures) to the trailing edge. The tip of the element will, as long as the available friction allows, adhere to the ground (that is, it will not slide over the road surface). Simultaneously, the base point of the element remains in the wheel plane and moves backward with the linear speed of rolling $V_r$ (that is equal to $r_e\Omega$) with respect to wheel axis or better: with respect to the contact center $C$. With respect to the road, the base point of the element moves with a velocity that is designated as the slip speed $V_s$ of the wheel.

In the lower part of the figure, the model is shown at pure side slip. The slip changes from very small to relatively large. We observe that the deflection increases while the element moves further through the contact patch. The deflection rate is equal to the supposedly constant slip speed. The resulting deflection varies linearly with the distance to the leading edge and the tips form a straight contact line that lies in a direction parallel to the wheel speed vector $V$. The figure also shows the maximum possible deflection that can be reached by the element depending on its position in the contact region. This maximum is governed by the (constant) coefficient of friction $\mu$, the vertical force distribution $q_z$, and the stiffness of the element $c_{py}$. The pressure distribution and consequently also the maximum deflection $v_{max}$ have been assumed to vary according to a parabola. As soon as the straight contact line intersects the parabola, sliding will start. The remaining part of the contact line will coincide with the parabola for the maximum possible deflection. At increasing slip angle, the side force that is generated will increase. The distance of its line of action behind the contact center is termed the pneumatic trail $t$. The aligning torque arises through the nonsymmetric shape of the deflection distribution and will be found by multiplying the side force with the pneumatic trail. As the slip increases, the deformation shape becomes more symmetric and, as a result, the trail gets smaller. This is because the point of intersection moves forward, thereby increasing the sliding range and decreasing the range of adhesion. This continues until the wheel speed vector runs parallel to the tangent to the parabola at the foremost point. Then, the point of intersection has reached the leading edge and full sliding starts to occur. The shape has now become fully symmetric. The side force attains its maximum and acts in the middle so that the moment vanishes. That situation remains unchanged when the slip angle increases further. The resulting characteristics for the side force and the aligning torque have been depicted in the same figure. In the part to follow next, the mathematical expressions for these relationships will be derived, first for the case of pure side slip.

### 3.2.1. Pure Side Slip

The brush model moving at a constant slip angle has been depicted in greater detail in Figure 3.3. It shows a contact line which is straight and parallel to the velocity vector $V$ in the adhesion region and curved in the sliding region where the available frictional force becomes lower than the force which would be required for the tips of the tread elements to follow the straight line further. In the adhesion region the linear variation of the deformation is in accordance with the general equation (2.65) (where $\tan\alpha$ has been assumed small and replaced by $\alpha$) with at steady state $\partial v/\partial s = 0$. For this simple model, the deformation of the tread element at the leading edge vanishes. Consequently, the lateral deformation in the adhesion region reads

$$v = (a - x)\tan\alpha \tag{3.1}$$

where $a$ denotes half the contact length.

In the case of vanishing sliding, that will occur for $\alpha \to 0$ or for $\mu \to \infty$, expression (3.1) is valid for the entire region of contact. With the lateral

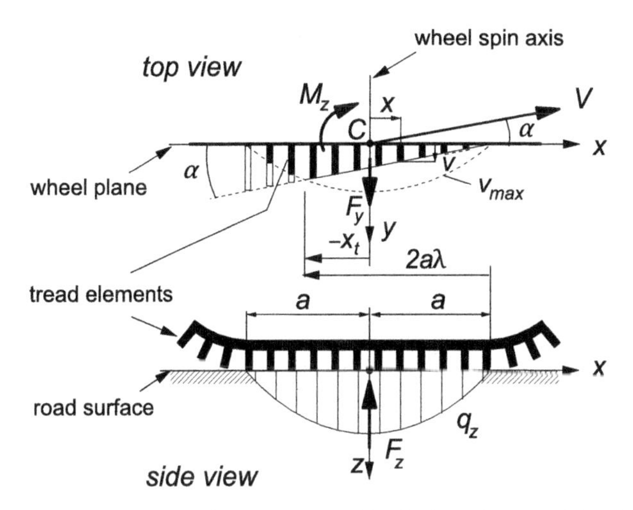


**FIGURE 3.3** Brush model moving at pure side slip shown in top and side view.

stiffness $c_{py}$ of the tread elements per unit length of the assumedly rectangular contact area, the following integrals and expressions for the cornering force $F_y$ and the aligning torque $M_z$ hold:

$$\begin{aligned} F_y &= c_{py} \int_{-a}^{a} v\mathrm{d}x = 2\,c_{py}a^2\alpha \\ M_z &= c_{py} \int_{-a}^{a} vx\mathrm{d}x = -\frac{2}{3}\,c_{py}a^3\alpha \end{aligned} \tag{3.2}$$

Consequently, the cornering stiffness and the aligning stiffness become, respectively,

$$\begin{aligned} C_{F\alpha} &= \left(\frac{\partial F_y}{\partial \alpha}\right)_{\alpha=0} = 2c_{py}a^2 \\ C_{M\alpha} &= -\left(\frac{\partial M_z}{\partial \alpha}\right)_{\alpha=0} = \frac{2}{3}\,c_{py}a^3 \end{aligned} \tag{3.3}$$

Next, we will consider the case of finite $\mu$ and a pressure distribution which gradually drops to zero at both edges. For the purpose of simplicity, we assume a parabolic distribution of the vertical force per unit length as expressed by

$$q_z = \frac{3F_z}{4a} \cdot \left\{1 - \left(\frac{x}{a}\right)^2\right\} \tag{3.4}$$

where $F_z$ represents the vertical wheel load. Hence, the largest possible side force distribution becomes

$$|q_{y,\max}| = \mu q_z = \frac{3}{4}\,\mu F_z \frac{a^2 - x^2}{a^3} \tag{3.5}$$

In Figure 3.3 the maximum possible lateral deformation $v_{\max} = q_{y,\max}/c_{py}$ has been indicated. For the sake of abbreviation, the following composite tire model parameter is introduced:

$$\theta_y = \frac{2c_{py}a^2}{3\mu F_z} \tag{3.6}$$

The distance from the leading edge to the point, where the transition from the adhesion to the sliding region occurs, is written as $2a\lambda$ and is determined by the factor $\lambda$. The value of this nondimensional quantity is found by realizing that at this point, where $x = x_t$, the deflection in the adhesion range becomes equal to that of the sliding range. Hence, with Eqns (3.1, 3.5, 3.6) the following equality holds:

$$|q_y| = c_{py}(a - x_t)|\tan\alpha| = |q_{y,\max}| = \frac{c_{py}}{2a\theta_y}(a - x_t)(a + x_t) \tag{3.7}$$

and thus for $\lambda = (a - x_t)/2a$ we obtain the relationship with the slip angle $\alpha$:

$$\lambda = 1 - \theta_y|\tan\alpha| \tag{3.8}$$

From this equation, the angle $\alpha_{sl}$, where total sliding starts ($\lambda = 0$), can be calculated:

$$\tan\alpha_{sl} = \frac{1}{\theta_y} \tag{3.9}$$

As the distribution of the deflections of the elements has now been established, the total force $F_y$ and the moment $M_z$ can be assessed by integration over the contact length (like in Eqn (3.2) but now separate for the sliding range $-a < x < x_t$ and the adhesion range $x_t < x < a$). For convenience, we introduce the notation for the slip:

$$\sigma_y = \tan\alpha \tag{3.10}$$

The resulting formula for the force reads:
if $|\alpha| \le \alpha_{sl}$

$$\begin{aligned} F_y &= \mu F_z(1 - \lambda^3)\operatorname{sgn}\alpha \\ &= 3\mu F_z\theta_y\sigma_y\left\{1 - |\theta_y\sigma_y| + \frac{1}{3}(\theta_y\sigma_y)^2\right\} \end{aligned} \tag{3.11}$$

and, if $|\alpha| \ge \alpha_{sl}$ (but $< \frac{1}{2}\pi$),

$$F_y = \mu F_z \operatorname{sgn}\alpha \tag{3.11a}$$

and for the moment:
if $|\alpha| \le \alpha_{sl}$

$$\begin{aligned} M_z &= -\mu F_z\lambda^3 a(1 - \lambda)\operatorname{sgn}\alpha \\ &= -\mu F_z\, a\,\theta_y\sigma_y\left\{1 - 3|\theta_y\sigma_y| + 3(\theta_y\sigma_y)^2 - |\theta_y\sigma_y|^3\right\} \end{aligned} \tag{3.12}$$

with peak value $27\mu F_z/256$ at $\sigma_y = 1/(4\theta_y)$;
if $|\alpha| \geq \alpha_{sl}$ (but < ½ π):

$$M_z = 0 \tag{3.12a}$$

The pneumatic trail $t$, which indicates the distance behind the contact center $C$ where the resultant side force $F_y$ is acting, becomes
if $|\alpha| \leq \alpha_{sl}$

$$t = -\frac{M_z}{F_y} = \frac{1}{3}\,a\,\frac{1 - 3|\theta_y\sigma_y| + 3(\theta_y\sigma_y)^2 - |\theta_y\sigma_y|^3}{1 - |\theta_y\sigma_y| + \frac{1}{3}(\theta_y\sigma_y)^2} \tag{3.13}$$

and if $|\alpha| \geq \alpha_{sl}$ (but < ½ π)

$$t = 0 \tag{3.13a}$$

These relationships have been shown graphically in Figure 3.4. At vanishing, slip angle expression (3.13) reduces to

$$t = t_o = -\left(\frac{M_z}{F_y}\right)_{\alpha\to 0} = \frac{1}{3}\,a \tag{3.14}$$

This value is smaller than normally encountered in practice. The introduction of an elastic carcass will improve this quantitative aspect. Then, the more realistic value of $t \approx 0.5a$ may be achieved (cf. Pacejka 1966 for a stretched string model to represent the elastic carcass and Section 3.3 for a more generally applicable way of approach to account for the carcass compliance).

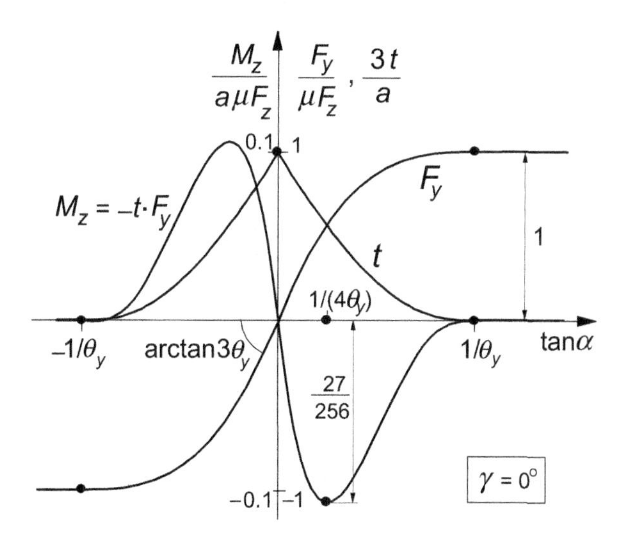


**FIGURE 3.4** Characteristics of the simple brush model: side force, aligning torque, and pneumatic trail vs slip angle.

Another point in which the simple model deviates considerably from experimental results concerns the effect of changing the vertical wheel load $F_z$. With the assumption that the contact length $2a$ changes quadratically with radial tire deflection $\rho$ and that $F_z$ depends linearly on $\rho$, so that $a^2 \sim F_z$, it can be easily shown that for the brush model $F_y$ and $M_z$ vary proportionally with $F_z$ and $F_z^{3/2}$, respectively. Experiments, however, show that $F_y$ varies less than linearly with $F_z$. In most cases, the $F_y$ vs $F_z$ characteristic, obtained at a small value of the slip angle, even shows a maximum after which the cornering force drops with increasing wheel load. Obviously, the same holds for the cornering stiffness $C_{F\alpha}$ (cf. Figure 1.3). Also in this respect the introduction of an elastic carcass (in particular when its lateral stiffness decreases with increasing normal load) improves the agreement with experiments. When considering a deflected cross section of a tire with side walls modeled as membranes under tension encapsulating pressurized air, such a decrease in lateral stiffness can be found to occur in theory (cf. Pacejka 1981, pp. 729 and 730).

An interesting diagram is the so-called Gough plot, in which $F_y$ is plotted vs $M_z$ for a series of constant values of $F_z$ (or possibly of $\mu$ at constant $F_z$) and of $\alpha$ respectively. This produces two sets of curves shown in Figures 3.5 and 3.6. In the first figure, the characteristics for the brush model have been presented. The left-hand diagram shows that when made nondimensional, a single curve results. The second figure presents the measured curves for a truck tire and in addition a diagram according to the brush model. The model has been adapted to include the nonlinear relationship of the cornering stiffness vs vertical load by making the tread element stiffness decrease linearly with vertical load. As can be seen from the left-hand plot (at large values of $\alpha$ where saturation of $F_y$ occurs), the truck tire also exhibits a decay in friction coefficient with increasing vertical load (not included in the model calculations), cf. also Figure 1.3. Further, as expected, the actual tire generates an aligning torque larger than that according to the model.

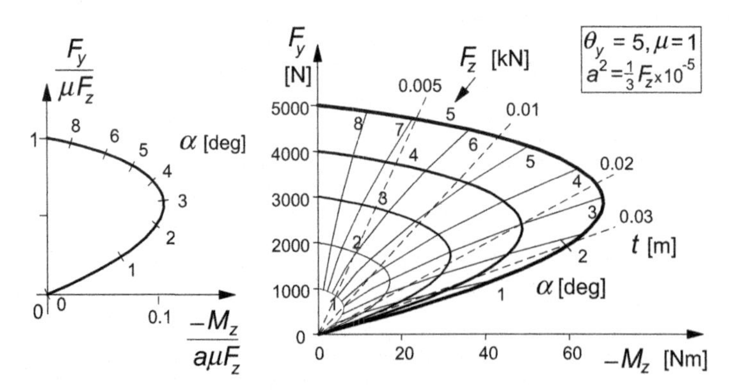


**FIGURE 3.5** The so-called Gough plot for the brush model, nondimensional, and with dimension using an assumed load vs contact length relationship.

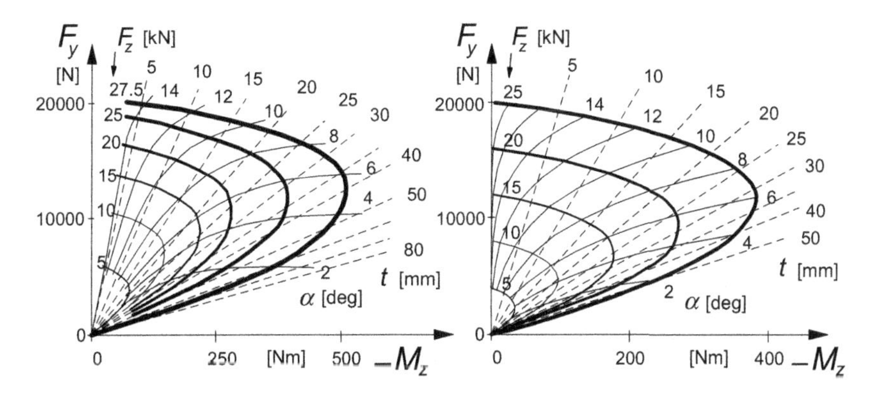


**FIGURE 3.6** Left: Gough plot of truck tire 9.00–20, measured on a dry road surface (Freudenstein 1961). Right: brush model with tread element stiffness $c_p$ decaying linearly with $F_z$.

### 3.2.2. Pure Longitudinal Slip

For the brush-type tire model with tread elements flexible in longitudinal direction, the theory for longitudinal (braking or driving) force generation develops along similar lines as those set out in Section 3.2.1 where the side force and aligning torque response to slip angle have been derived. To simplify the discussion, we restrict ourselves here to non-negative values of the forward speed $V_x$ and of the speed of revolution $\Omega$.

In Figure 3.7 a side view of the brush model has been shown. As indicated before, the so-called slip point $S$ is introduced. This is an imaginary point attached to the wheel rim and is located, at the instant considered, a distance equal to the effective rolling radius $r_e$ (defined at free rolling) below the wheel center. At free rolling, by definition, the slip point $S$ has a velocity equal to zero. Then, it forms the instantaneous center of rotation of the wheel rim. We may

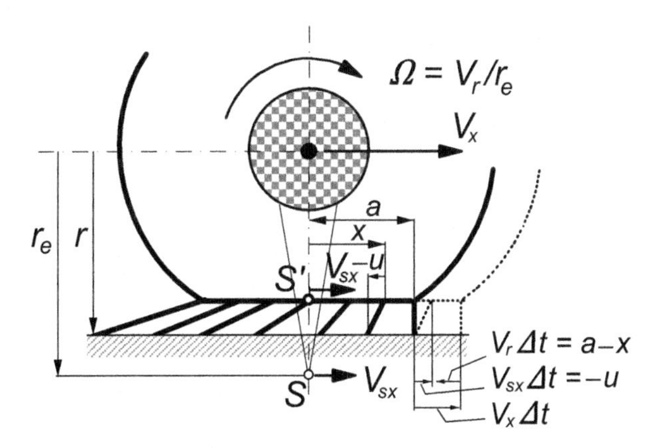


**FIGURE 3.7** Side view of the brush tire model at braking (no sliding considered).

think of a slip circle with radius $r_e$ that in the case of free rolling rolls perfectly, that is, without sliding, over an imaginary road surface that touches the slip circle in point $S$. When the wheel is being braked, point $S$ moves forward with the longitudinal slip velocity $V_{sx}$. When driven, the slip point moves backward with consequently a negative slip speed. In the model, a point $S'$ is defined that is attached to the base line at its center (that is, at the base point of the tread element below the wheel center, cf. Figure 3.7). By definition, the velocity of this point is the same as that of point $S$. That means that $S'$ also moves with the same slip speed $V_{sx}$. It is assumed that the tread elements attached at their base points to the circumferentially rigid carcass enter the contact area in vertical position. At free rolling with slip speed $V_{sx}$ (of both points $S$ and $S'$) equal to zero, the orientation of the elements remains vertical while moving from front to rear through the contact zone. Consequently, no longitudinal force is being transmitted and we have a wheel speed of revolution:

$$\Omega = \Omega_o = \frac{V_x}{r_e} \tag{3.15}$$

Here it is assumed that the longitudinal component of the speed of propagation of the contact center $C$ is equal to the longitudinal component of the speed of the wheel center ($V_{cx} = V_x$). As has been seen in the previous chapter, this will occur on a flat road surface at vanishing $\gamma\dot{\psi}$. When $\Omega$ differs from its value at free rolling $\Omega_o$, the wheel is being braked or driven and the longitudinal slip speed $V_{sx}$ becomes

$$V_{sx} = V_x - \Omega r_e \tag{3.16}$$

In the model, the base points of all the tread elements move with the same speed $V_{sx}$. A base point progresses backward through the contact zone with a speed $V_r$ called the linear speed of rolling. Apparently, we have

$$V_r = \Omega r_e = V_x - V_{sx} \tag{3.17}$$

An element the tip of which adheres to the ground and the base point is moved toward the rear over a distance $a - x$ from the leading edge (for which a time span $\Delta t = (a - x)/V_r$ is needed) has developed a deflection in longitudinal direction:

$$u = -V_{sx}\frac{a - x}{V_r} \tag{3.18}$$

The same expression may be obtained by integration of the fundamental equation (2.55) and noting that $V_{gx} = \omega_z = \partial u/\partial t = \theta$'s $= 0$ and finally using the boundary condition $u = 0$ at $x = a$.

We may write the longitudinal deflection $u$ in terms of the 'practical' longitudinal slip $\kappa = -V_{sx}/V_x$:

$$u = -(a - x)\frac{V_{sx}}{V_x - V_{sx}} = (a - x)\frac{\kappa}{1 + \kappa} \tag{3.19}$$

In terms of the alternative definition of longitudinal slip, the 'theoretical' slip is to be used in the subsequent section and defined as

$$\sigma_x = -\frac{V_{sx}}{V_r} = \frac{\kappa}{1+\kappa} \tag{3.20}$$

(note, we restrict ourselves to non-negative speeds of rolling: $V_r \geq 0$ and $\kappa \geq -1$). We obtain

$$u = (a - x)\sigma_x \tag{3.21}$$

In Section 3.2.1, we found with Eqns (3.1) and (3.10) for the lateral deflection at pure side slip ($\sigma_y = \tan\alpha$):

$$v = (a - x)\sigma_y \tag{3.22}$$

Comparison of the Eqns (3.21) and (3.22) shows that the longitudinal deformations $u$ will be equal in magnitude to the lateral deformations $v$ if $\sigma_y = \tan\alpha$ equals $\sigma_x = \kappa/(1 + \kappa)$. For equal tread element stiffnesses ($c_{px} = c_{py}$) and friction coefficients ($\mu_x = \mu_y$) in lateral and longitudinal directions, the slip force characteristics in both directions are identical when $\tan\alpha$ and $\kappa/(1 + \kappa)$ are used as abscissa (cf. Figure 3.8). Also, Eqn (3.11) holds for the longitudinal force $F_x$ if the subscripts $y$ are replaced by $x$ and $\tan\alpha$ by $\kappa$.

Obviously, total sliding will start at $\sigma_x = \kappa/(1 + \kappa) = \pm 1/\theta_x$ or in terms of the practical slip at

$$\kappa = \kappa_{sl} = \frac{-1}{1 \pm \theta_x} \tag{3.23}$$

with

$$\theta_x = \frac{2}{3}\frac{c_{px}a^2}{\mu F_z} \tag{3.24}$$

Linearization for small values of slip $\kappa$ yields a deflection at coordinate $x$:

$$u = (a - x)\kappa \tag{3.25}$$

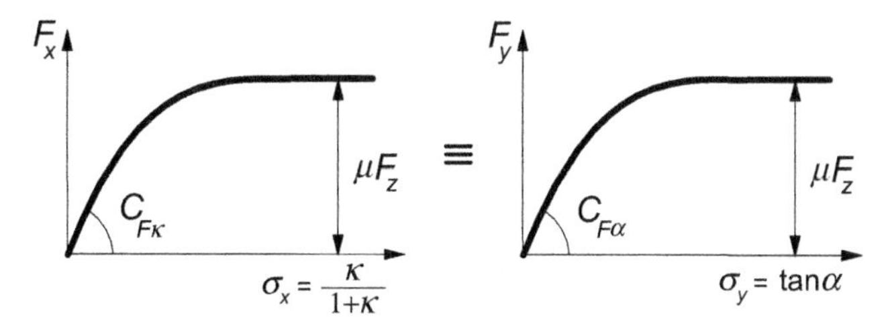


**FIGURE 3.8** Equality of the two pure slip characteristics for an isotropic tire model if plotted against the theoretical slip.

and a fore-and-aft force

$$F_x = 2c_{px}a^2\kappa \tag{3.26}$$

with $c_{px}$ the longitudinal tread element stiffness per unit length. This relation contains the longitudinal slip stiffness:

$$C_{F\kappa} = \left(\frac{\partial F_x}{\partial \kappa}\right)_{\kappa=0} = 2c_{px}a^2 \tag{3.27}$$

For equal longitudinal and lateral stiffnesses ($c_{px} = c_{py}$), we obtain equal slip stiffnesses $C_{F\kappa} = C_{F\alpha}$. In reality, however, appreciable differences between the measured values of $C_{F\kappa}$ and $C_{F\alpha}$ may occur (say $C_{F\kappa}$ about 50% larger than $C_{F\alpha}$) which is due to the lateral (torsional) compliance of the carcass of the actual tire. Still, it is expected that qualitative similarity of both pure slip characteristics remains.

### 3.2.3. Interaction between Lateral and Longitudinal Slip (Combined Slip)

For the analysis of the influence of longitudinal slip (or longitudinal force) on the lateral force and moment generation properties, we shall, for the sake of mathematical simplicity, restrict ourselves to the case of equal longitudinal and lateral stiffnesses of the tread elements (isotropic model), i.e.,

$$c_p = c_{px} = c_{py} \tag{3.28}$$

and equal and constant friction coefficients

$$\mu = \mu_x = \mu_y \tag{3.29}$$

Again a parabolic pressure distribution is considered.

Figure 3.9 depicts the deformations which arise when the tire model which runs at a given slip angle $\alpha$ is driven or braked. Due to the equal stiffness in all horizontal directions and the isotropic friction properties, the deflections are directed opposite to the slip speed vector $\boldsymbol{V}_s$, also in the sliding region. In this latter region, the tips of the elements slide over the road with sliding speed $\boldsymbol{V}_g$ directed opposite to the local friction force $\boldsymbol{q}$ (per unit contact length). The whole deformation history of a tread element, while running through the contact area, is a one-dimensional process along the direction of $\boldsymbol{V}_s$.

The velocity of progression of a base point through the contact length is the rolling speed $V_r$ (again assumed non-negative). The deflection rate of an element in the adhesion region is equal to the slip speed $V_s$. The time which elapses from the point of entrance to the point at a distance $x$ in front of the contact center equals

$$\Delta t = \frac{a - x}{V_r} \tag{3.30}$$

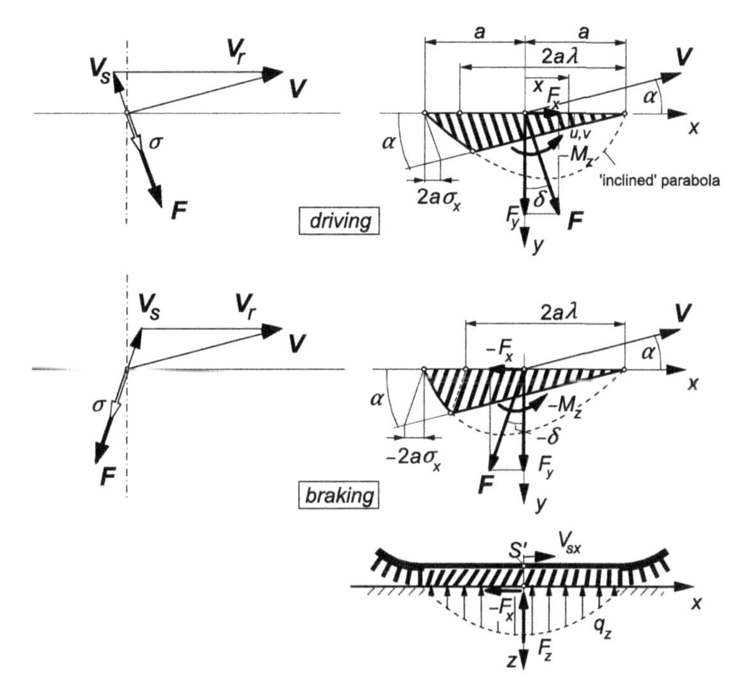


**FIGURE 3.9** Vector diagram and deformation of the brush model running at a given slip angle for the cases of driving and braking.

In this position, the deflection of an element that is still in adhesion becomes in vectorial form

$$\boldsymbol{e} = \begin{pmatrix} u \\ v \end{pmatrix} = \boldsymbol{V}_s \, \Delta t = -\frac{\boldsymbol{V}_s}{V_r}(a - x) \tag{3.31}$$

It seems natural at this stage to introduce the alternative (theoretical) slip quantity again, but now in vectorial form:

$$\boldsymbol{\sigma} = \begin{pmatrix} \sigma_x \\ \sigma_y \end{pmatrix} = -\frac{\boldsymbol{V}_s}{V_r} = -\frac{1}{V_r}\begin{pmatrix} V_{sx} \\ V_{sy} \end{pmatrix} \tag{3.32}$$

with the linear speed of rolling

$$V_r = V_x - V_{sx} \tag{3.33}$$

The relations of these theoretical slip quantities with the practical slip quantities $\kappa$ $(= -V_{sx}/V_x)$ and $\tan\alpha$ $(= -V_{sy}/V_x)$ are

$$\begin{aligned} \boldsymbol{\sigma}_x &= \frac{\kappa}{1+\kappa} \\ \boldsymbol{\sigma}_y &= \frac{\tan\alpha}{1+\kappa} \end{aligned} \tag{3.34}$$

The deflection of an element in the adhesion region now reads

$$\boldsymbol{e} = (a - x)\boldsymbol{\sigma} \tag{3.35}$$

from which it is apparent that longitudinal and lateral deflections are governed by $\sigma_x$ and $\sigma_y$ respectively and independent of each other. This would not be the case if expressed in terms of the practical slip quantities $\kappa$ and $\alpha$!

The local horizontal contact force acting on the tips of the elements (per unit contact length) reads

$$\boldsymbol{q} = c_p(a - x)\boldsymbol{\sigma} \qquad \text{(adhesion region)} \tag{3.36}$$

As soon as

$$q = |\boldsymbol{q}| = \sqrt{q_x^2 + q_y^2} > \mu q_z \tag{3.37}$$

the sliding region is entered. Then the friction force vector becomes

$$\boldsymbol{q} = -\frac{\boldsymbol{V}_s}{V_s}\mu q_z = \frac{\boldsymbol{\sigma}}{\sigma}\mu q_z \qquad \text{(sliding region)} \tag{3.38}$$

where

$$V_s = \sqrt{V_{sx}^2 + V_{sy}^2} \tag{3.39}$$

and

$$\sigma = \sqrt{\sigma_x^2 + \sigma_y^2} \tag{3.40}$$

Similarly, the magnitude of the deflection of an element becomes

$$e = |\boldsymbol{e}| = \sqrt{u^2 + v^2} \tag{3.41}$$

The point of transition from adhesion to sliding region is obtained from the condition

$$c_p e = \mu q_z \tag{3.42}$$

or

$$c_p \sigma(a - x_t) = \frac{3}{4}\mu F_z \frac{a^2 - x_t^2}{a^3} \tag{3.43}$$

which yields

$$x_t = \frac{4}{3}\frac{c_p a^3 \sigma}{\mu F_z} - a = a(2\theta\sigma - 1) \tag{3.44}$$

or, in similar terms as Eqn (3.8),

$$\lambda = 1 - \theta\sigma \tag{3.45}$$

where analogous to expressions (3.6) and (3.24) for the isotropic model parameter $\theta$ reads

$$\theta = \theta_y = \theta_x = \frac{2}{3}\frac{c_p a^2}{\mu F_z} \tag{3.46}$$

From Eqn (3.45), the slip $\sigma_{sl}$ at which total sliding starts can be calculated. We get, analogous to (3.9),

$$\sigma_{sl} = \frac{1}{\theta} \tag{3.47}$$

The magnitude of the total force $F = |\boldsymbol{F}|$ now easily follows in accordance with (3.11):

$$\begin{aligned} F &= \mu F_z(1-\lambda^3) = \mu F_z\left\{3\theta\sigma - 3(\theta\sigma)^2 + (\theta\sigma)^3\right\} && \textit{for } \sigma \le \sigma_{sl} \\ F &= \mu F_z && \textit{for } \sigma \ge \sigma_{sl} \end{aligned} \tag{3.48}$$

and obviously follows the same course as those shown in Figure 3.8. The force vector $\boldsymbol{F}$ acts in a direction opposite to $\boldsymbol{V}_s$ or $-\sigma$. Hence,

$$\boldsymbol{F} = F\frac{\boldsymbol{\sigma}}{\sigma} \tag{3.49}$$

from which the components $F_x$ and $F_y$ may be obtained.

The moment $-M_z$ is obtained by multiplication of $F_y$ with the pneumatic trail $t$. This trail is easily found when we realize that the deflection distribution over the contact length is identical with the case of pure side slip if $\tan\alpha_{eq} = \sigma$ (cf. Figure 3.10). Consequently, formula (3.13) represents the pneumatic trail at combined slip as well if $\theta_y\sigma_y$ is replaced by $\theta\sigma$. We have, with (3.13),

$$M_z = -t(\sigma)\cdot F_y \tag{3.50}$$

In Figures 3.11 and 3.12, the dramatic reduction of the pure slip forces (the side force and the longitudinal force respectively) that occurs as a result of the simultaneous introduction of the other slip component (the longitudinal slip and the side slip respectively) has been indicated. We observe an (almost)

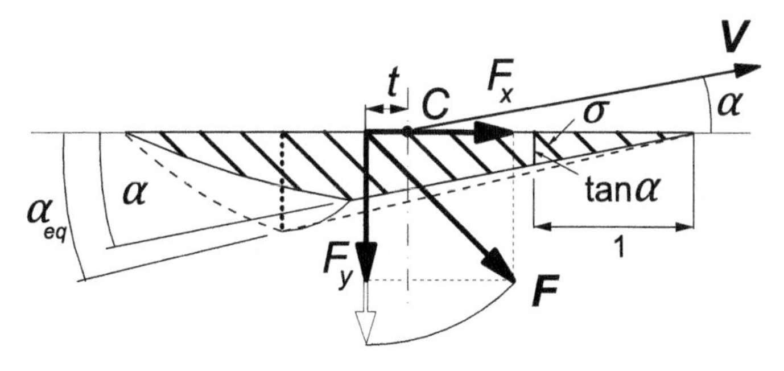


**FIGURE 3.10** Equivalent side-slip angle producing the same pneumatic trail $t$.

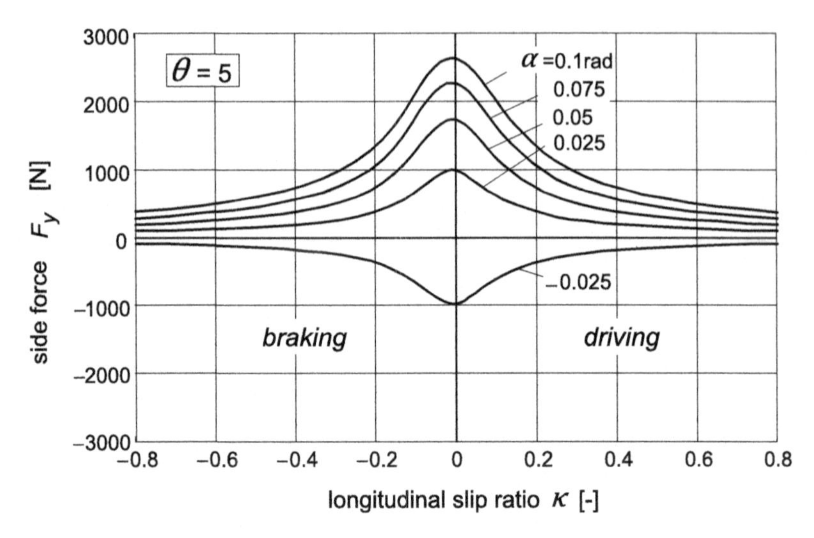


**FIGURE 3.11** Reduction of side force due to the presence of longitudinal slip.

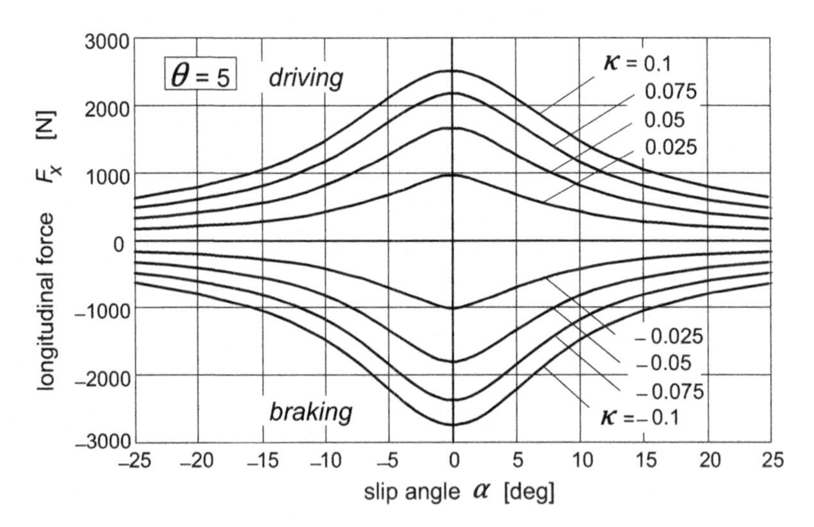


**FIGURE 3.12** Reduction of longitudinal force due to the presence of side slip.

symmetric shape of these interaction curves. The peak of the side force vs longitudinal slip curves at constant values of the slip angle appears to be slightly shifted toward the braking side. This phenomenon will be further discussed in connection with the alternative representation of the same results according to Figure 3.13. At very large longitudinal slip that is when $|V_{sx}|/V_x \rightarrow \infty$, the side force approaches zero and the same occurs for the longitudinal force when the lateral slip tan $\alpha$ goes to infinity ($\alpha \rightarrow 90°$) at

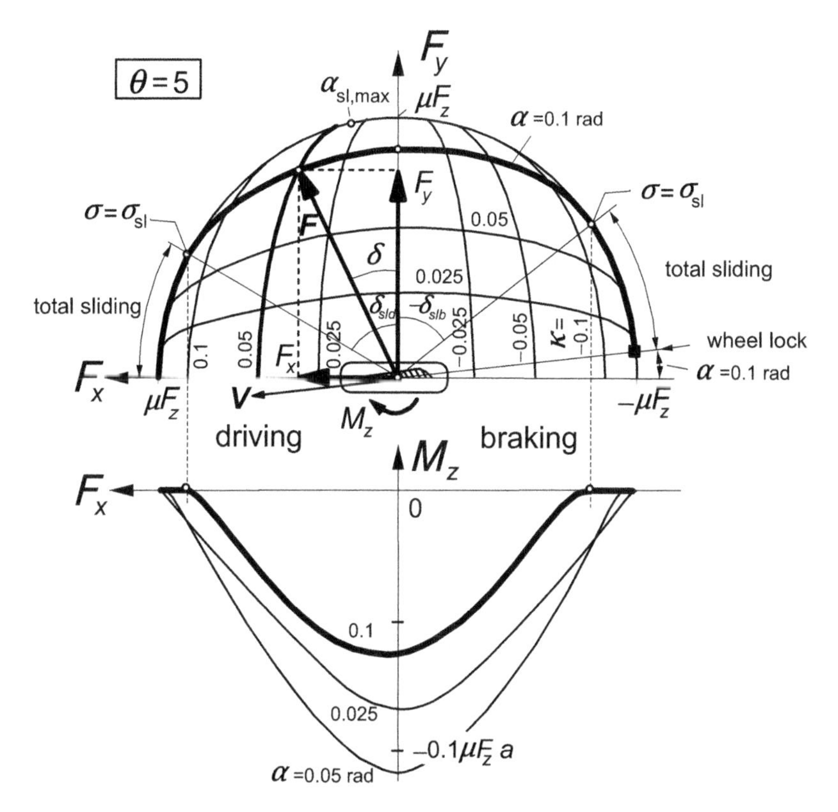


**FIGURE 3.13** Cornering force and aligning torque as functions of longitudinal force at constant slip angle $\alpha$ or longitudinal slip $\kappa$.

a given value of the longitudinal slip because obviously in that case with $V_x \rightarrow 0$ also the longitudinal slip speed must vanish. For a locked wheel with $V_{sx} = V_x$ and $\kappa = -1$, we have $F_y = \mu F_z \sin\alpha$ and $F_x = -\mu F_z \cos\alpha$.

In the diagram of Figure 3.13, the calculated variations of $F_y$ and $M_z$ with $F_x$ have been plotted for several fixed values of $\alpha$. Also, the curves for constant $\kappa$ have been depicted. For clarification of the nature of the $F_y - F_x$ diagram, the deflection of an element near the leading edge has been shown in Figure 3.14. Since the distance from the leading edge has been defined for this occasion to be equal to unity, the deflection $\boldsymbol{e}$ of the element equals the slip $\boldsymbol{\sigma}$. The radius of the circle denoting maximum possible deflections is equal to $\sigma_{sl}$. The two points on the circle where the slip angle $\alpha$ considered the sliding boundary $\sigma_{sl}$ is attained correspond to the points on the $\alpha$ curve of the force diagram of Figure 3.13. This also explains the slightly inclined nature of the $\alpha$ curves. At braking, $F_y$ appears to be a little larger than at driving. This is at least true for the two cases of Figure 3.9, one at driving and the other at braking, showing the same slip angle and the same magnitude of the deviation angle $\boldsymbol{\delta}$ of the slip

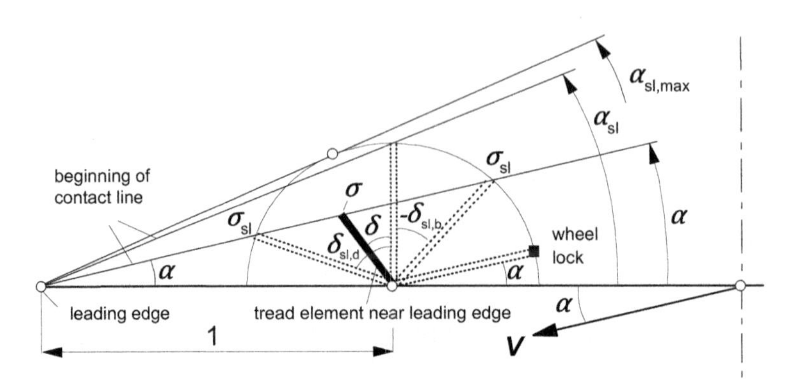


**FIGURE 3.14** The situation near the leading edge. Various deflections of an element at a distance equal to 1 from the leading edge are shown corresponding with points in the upper diagram of Figure 3.13.

velocity vector and thus of the force vector with respect to the $y$-axis. Then, the slip speeds $V_s$ are equal in magnitude, but at braking the speed of rolling $V_r$ is obviously smaller. When considering the definition of the theoretical slip $\sigma$ (3.32) and the functional relationship (3.48) with the force $F$, it becomes clear that $F$ must be larger for the case of braking because of the then larger magnitude of $\sigma$. Finally, the case of wheel lock is pointed at. Then, the force vector $\boldsymbol{F}$, which in magnitude is equal to $\mu F_z$, is directed opposite to the speed vector of the wheel $\boldsymbol{V}$ which then coincides with the slip speed vector $\boldsymbol{V}_s$.

Experimental evidence (e.g., Figure 3.17) supports the nature of the theoretical curves for the forces as shown in Figure 3.13. Often, the shape appears to be more asymmetric than predicted by the simple brush model. This may be due to a slight increase in the contact length while braking (making the tire stiffer) and by the brake force-induced slip angle of the contact patch at side slip. This is accomplished through the torsion of the carcass induced by the moment about the vertical axis that is exerted by the braking force which has shifted its line of action due to the lateral deflection connected with the side force. The more advanced model to be developed further on in this chapter will take the latter effect into account.

The moment curves presented in the lower diagram of Figure 3.13 show a more or less symmetrical bell shape. As expected, the aligning torque becomes equal to zero when total sliding occurs ($\sigma \geq \sigma_{sl}$). Later on, we will see that these computed moment characteristics may appreciably deviate from experimentally obtained curves.

At this stage, we will first apply the knowledge gained so far to the analysis of a practical situation that occurs with a wheel that is braked or driven (at constant brake pressure and throttle respectively) while its slip angle is varied.

In Figure 3.15, the force diagram is shown in combination with the corresponding velocity diagram. At a given braking force $-F_x$ and wheel speed of

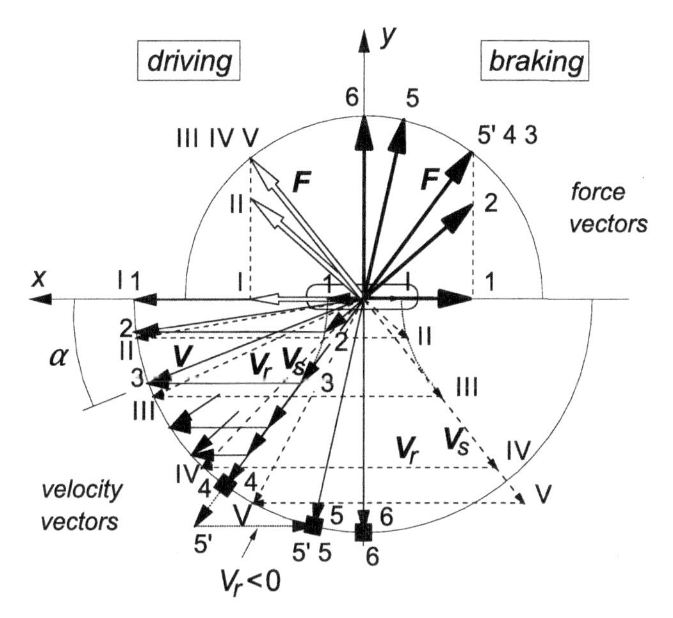


**FIGURE 3.15** The $F_x - F_y$ diagram extended with the corresponding velocity diagram. The brake (pedal) force is kept constant while the slip angle is increased. The same is done at a constant driving effort. In the case of braking, the speed of rolling $V_r$ decreases until the wheel gets locked. In the special case 5′, the wheel is rotated backward to keep the slip speed $\boldsymbol{V}_s$ and thus the force vector $\boldsymbol{F}$ in the original direction.

travel $V$, the slip angle $\alpha$ is changed from zero to 90°. The variation of slip speed $V_s$ and rolling speed $V_r$ may be followed from case 1 where $\alpha = 0$ to case 3 where total sliding starts and further to case 4 where $V_r$ and thus $\Omega$ vanish and the wheel becomes locked. A further increase of $\alpha$ (at constant brake pedal force) as in case 5 will necessarily lead to a reduction in braking force $-F_x$ unless the wheel is rotated in opposite direction ($\Omega < 0$) as represented by case 5′. In the cases of driving indicated by Roman numerals, the driving force $F_x$ can be maintained irrespective of the value of $\alpha$ (with $|\alpha| < 90°$).

The nature of the resulting $F_y$–$\alpha$ characteristics at given driving or braking effort is shown in Figure 3.16. Plotting of $F_y$ versus sin $\alpha$ is advantageous because the portion where wheel lock occurs is then represented by a straight line.

Another important advantage of putting sin $\alpha$ instead of tan $\alpha$ along the abscissa is that (after having completed the diagram for negative values of sin $\alpha$ resulting in an oddly symmetric graph) the complete range of $\alpha$ is covered: the speed vector $\boldsymbol{V}$ may swing around over the whole range of 360°. An application in vehicle dynamics will be discussed in Section 3.4 (yaw instability at locked rear wheels).

For illustration we have shown in Figures 3.17 and 3.18 experimentally assessed characteristics. The force diagrams correspond reasonably well with

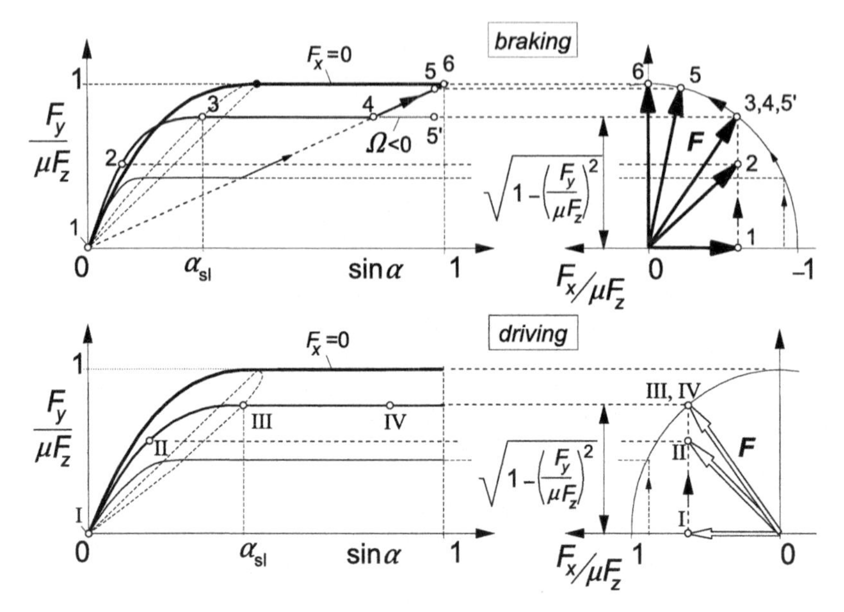


**FIGURE 3.16** Tire characteristics at constant brake (pedal) force and driving force. Numerals correspond to those of Figure 3.15.

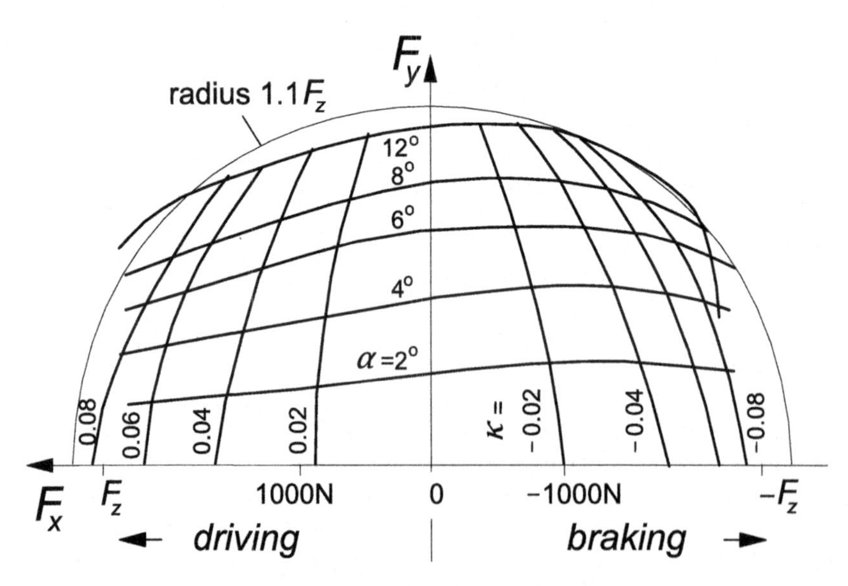


**FIGURE 3.17** $F_y - F_x$ characteristics for 6.00–13 tire measured on dry internal drum with diameter of 3.8 m (from Henker 1968).

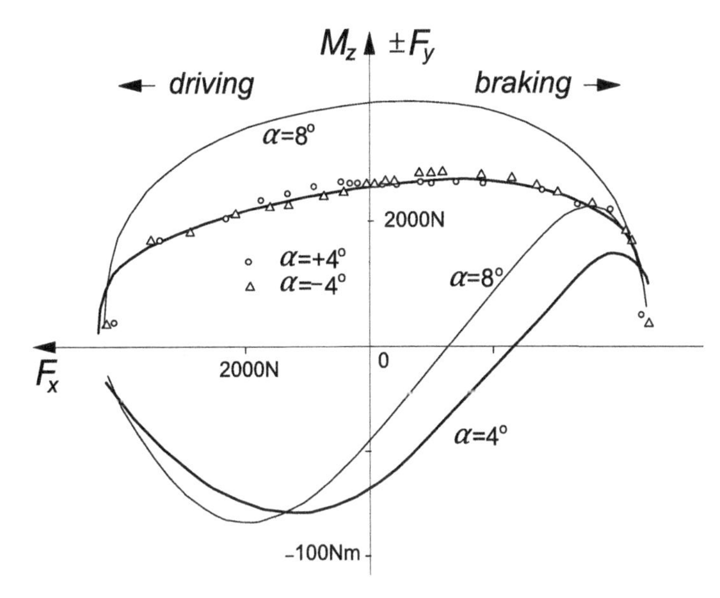


**FIGURE 3.18** Combined slip side force and moment vs longitudinal force characteristics measured at two values of the slip angle for a 7.60–15 tire on a dry flat surface (from Nordeen and Cortese 1963).

the theoretical observations. The moment curves, however, deviate considerably from the theoretical predictions (compare Figure 3.18 with Figure 3.13). It appears that according to this figure, $M_z$ changes its sign in the braking half of the diagram. This phenomenon can not be explained with the simple tire brush model that has been employed thus far.

The introduction of a laterally flexible carcass seems essential for properly modeling $M_z$ that acts on a driven or braked wheel. In Figure 3.19,

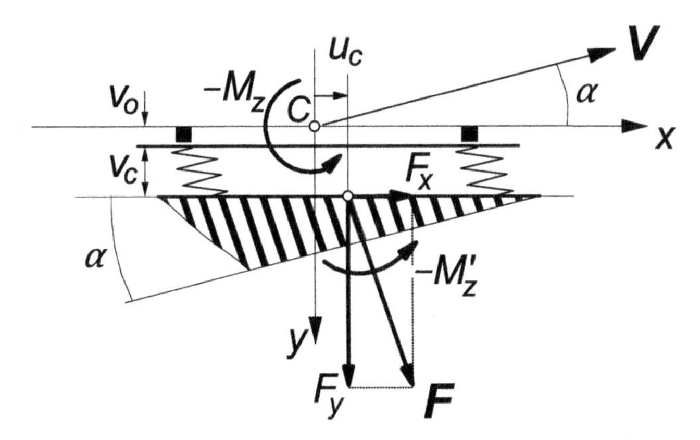


**FIGURE 3.19** Extended tire brush model showing offset and deflection of carcass line (straight and parallel to wheel plane).

a possible extension of the brush model is depicted. The carcass line is assumed to remain straight and parallel to the wheel plane in the contact region. A lateral and longitudinal compliance with respect to the wheel plane is introduced. In addition, a possible initial offset of the line of action of the longitudinal force with respect to the wheel center plane is regarded. Such an offset is caused by asymmetry of the construction of the tire or by the presence of a camber angle.

With this model, the moment $M_z$ is composed of the original contribution $M_z'$ established by the brush model and those due to the forces $F_y$ and $F_x$ which show lines of action shifted with respect to the contact center $C$ over the distances $u_c$ and $v_o + v_c$, respectively. The self-aligning torque now reads

$$\begin{aligned} M_z &= M_z' - F_x(v_o + v_c) + F_y u_c \\ &= M_z' - cF_xF_y - F_x v_o \end{aligned} \tag{3.51}$$

where the compliance coefficient $c$ has been introduced that is defined by

$$c = \frac{\varepsilon_y}{C_{cy}} - \frac{\varepsilon_x}{C_{cx}} \tag{3.52}$$

Here $C_{cx}$ and $C_{cy}$ denote the longitudinal and lateral carcass stiffnesses respectively and $\varepsilon_x$ and $\varepsilon_y$ the effective fractions of the actual displacements. These fractions must be considered because of concurrent lateral and longitudinal rolling of the tire lower section against the road surface under the action of a lateral and fore-and-aft force respectively which will change the normal force distribution in the contact patch and thereby reduce the actual displacements of the lines of action of both horizontal forces. The resulting calculations can be performed in a direct straightforward manner, because the slip angle of the extended model is the same as the one for the internal brush model. Later on, in Section 3.3, the effect of the introduction of a torsional and bending stiffness of the carcass and belt will be discussed. The resulting model, however, is a lot more complex and closed form solutions are no longer possible. In Section 3.3, the technique of the tread element following method will be employed in the tread simulation model to determine the response.

The combined slip response of the simple extended model of Figure 3.19 is given in Figures 3.20 and 3.21. It is observed that in Figure 3.20, the aligning torque changes its sign in the braking range. This is due to the term in Eqn (3.51) with the compliance coefficient $c$. The resulting qualitative shape is quite similar to the experimentally found curves of Figure 3.18. In Figure 3.21, the effect of an initial offset of $v_o = 5$ mm has been depicted.

Here carcass compliance has been disregarded and only the last term of (3.51) has been added. We see that a moment is generated already at zero slip angle. The curves found in Figure 3.13 for nonzero side slip are then simply

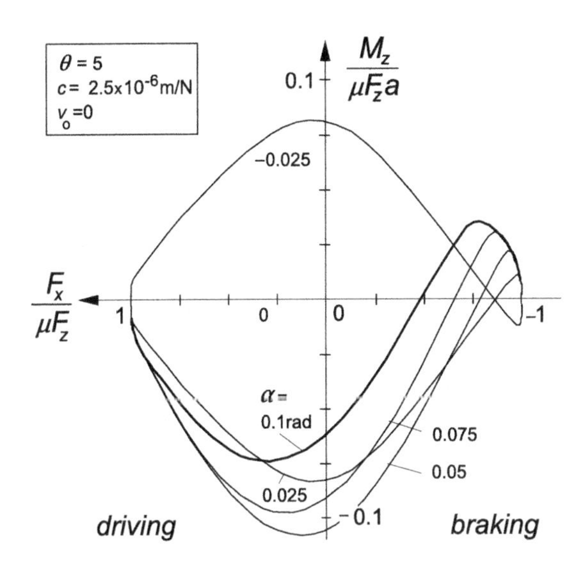


**FIGURE 3.20** The influence of lateral carcass compliance on the aligning torque for a braked or driven wheel according to the extended brush model.

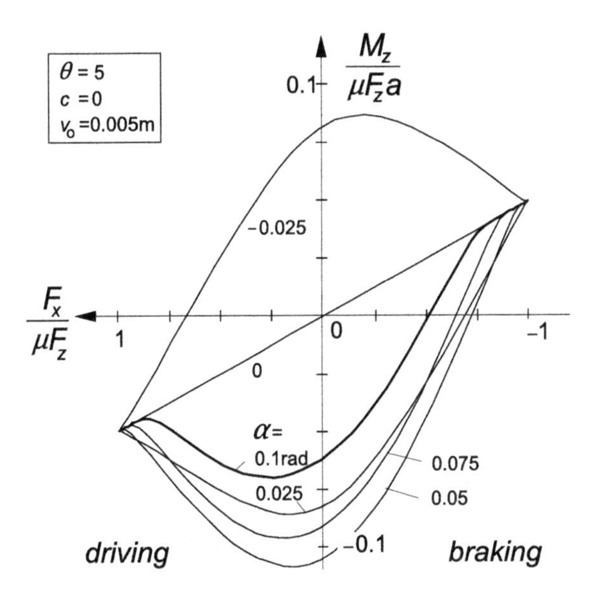


**FIGURE 3.21** The influence of an initial lateral offset $v_o$ of $F_x$ on the aligning torque for a braked or driven wheel according to the extended brush model.

added to the inclined straight line belonging to $\alpha = 0$. The type of curves that result are often found experimentally. The effect of lateral compliance may then be very small or canceled out by the effect of the fore-and-aft compliance of the carcass (second term of right-hand member of (3.52)).

**Exercise 3.1. Characteristics of the Brush Model**

Consider the brush tire model as treated in Section 3.2. Elastic and frictional properties are the same in all horizontal directions and Eqn (3.46) holds.

For $0 \le \alpha \le \frac{1}{2}\pi$, the side force characteristic is described by (according to Eqn (3.11))

$$F_y = \mu F_z(3\theta \tan\alpha - 3\theta^2 \tan^2\alpha + \theta^3 \tan^3\alpha) \qquad (\alpha \le \alpha_{sl})$$
$$F_y = \mu F_z \qquad (\alpha \ge \alpha_{sl})$$

1. Calculate the value of $\theta$ and $\tan\alpha_{sl}$ for
   $\mu F_z = 2000$ N
   $C_{F\alpha} = 18000$ N/rad
2. Sketch the $F_y(\tan\alpha)$ characteristic and also the $M_z(\tan\alpha)$ and $t(\tan\alpha)$ curves (according to Eqns (3.12, 3.13)) for
   $a = 0.1$ m
3. Replace in the $F_y(\tan\alpha)$ diagram the ordinate $F_y$ by $F$ and the abscissa $\tan\alpha$ by $\sigma$, thereby assessing the total force vs total slip diagram. Calculate the slip values $\sigma_x$, $\sigma_y$, and $\sigma$ using Eqns (3.34) and (3.40) for one value of $\tan\alpha = 0.15$ and a number of values of $\kappa$ in a suitable range (e.g., from $-1/\theta$ to $+1/\theta$). Determine the force vector $\boldsymbol{F}$ for each of the $\kappa$ values and sketch the $F_y - F_x$ curve for $\tan\alpha = 0.15$. Draw the friction circle with radius $F_{max} = \mu F_z$ in which the curve will appear. Note the two points where $\sigma = \sigma_{sl}$ ($= \tan\alpha_{sl}$) where the curve touches the circle. Indicate the point where wheel lock occurs.
4. Replace in the $t(\tan\alpha)$ diagram the abscissa by $\sigma$, thereby establishing the $t(\sigma)$ diagram. Determine the values of $M_z' = -tF_y$ for the same series of $\kappa$ values and $\tan\alpha = 0.15$.
5. Now use Eqn (3.51) and calculate the torque $M_z$ for a lateral carcass stiffness $C_{cy} = 60000$ N/m ($C_{cx} \to \infty$) and disregard the correction factor $\varepsilon_y$ (=1).
6. Draw for the cases mentioned in question 3 where the $\alpha$ curve touches the friction circle ($\sigma = \sigma_{sl}$), the force, and velocity diagram according to Figure 3.15. Do the same for the case of wheel lock.
7. Sketch the $F_y$ (sin $\alpha$) characteristics (Figure 3.16) for $F_x = 0$ and also for that constant brake pedal force corresponding to the value of $-F_x$ where the curve for $\alpha = 0.15$ touches the friction circle ($\sigma = \sigma_{sl}$).

### 3.2.4. Camber and Turning (Spin)

For the study of horizontal cornering, one should not only consider side slip but also the influence of two other effects, which in most cases (except for the motorcycle) are of much less importance than side slip. These two input variables whose introduction completes the description of the out-of-plane tire force and moment generation are firstly the wheel camber or tilt angle $\gamma$ between the wheel plane and the normal to the road (cf. Figure 2.6), and secondly the turn slip $\dot{\psi}/V_x$. Both are components of the total spin $\varphi$. For a general discussion on spin, we may refer to Pacejka (2004). First, we will analyze the situation in the absence of lateral and longitudinal wheel slip.

### Pure Spin

In the steady-state case, the turn slip equals the curvature $1/R$ of a circular path with radius $R$. For homogeneous rolling bodies (solid rubber ball, steel railway wheel), the mechanisms to produce side force and moment as a result of camber and turning are equal as they both originate from the same spin motion (cf. Eqn (2.17)). For a tire with its rather complex structure, the situation may be quantitatively different for the two components of spin.

As depicted in Figure 3.22, the wheel is considered to move tangentially to a circular horizontal path with radius $R$ while the wheel plane shows a constant camber angle $\gamma$ and apparently the slip angle is kept equal to zero. The wheel is picturized here as a part of an imaginary ball. When lifted from the ground, the intersection of wheel plane and ball outer surface forms the peripheral line of the tire. When loaded vertically, the ball and consequently the peripheral line are assumed to show no horizontal deformations, which in reality will approximately be the case for a homogeneous ball showing a relatively small contact area.

We apply the theory of a rolling and slipping body and consider Eqns (2.55, 2.56) and restrict ourselves to the case of steady-state pure spin, that is, with $\alpha = \kappa = 0$. Then with turn slip velocity $\omega_{zt} = \dot{\psi}$ in Figure 3.22, we find

$$V_{sx} = 0, \; V_{sy} = 0, \; V_r = V_c,$$
$$\omega_z = \dot{\psi} - \Omega \sin\gamma = V_c\left(\frac{1}{R} - \frac{1}{r_e}\sin\gamma\right) \tag{3.53}$$

Furthermore, the difference between $(x, y)$ and $(x_o, y_o)$ will be neglected and tire conicity and ply-steer disregarded. The correction factors $\theta$ attributed to camber may be approximated by

$$\theta_{\gamma x}(y) = -\varepsilon_{\gamma x}, \quad \theta_y(x, y) = -\varepsilon_{\gamma y}\frac{x}{r_e}\sin\gamma \tag{3.54}$$

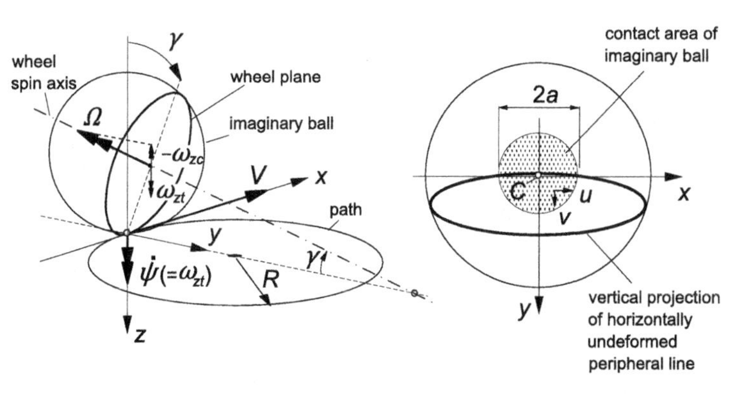


**FIGURE 3.22** Wheel rolling at a camber angle while turning along a circular path without side slip. At the right: a top view of the peripheral line of the nonrolling wheel considered as being a part of an imaginary ball pressed against a flat surface.

with both reduction factors $\varepsilon$ equal or close to zero for a railway wheel or a motorcycle tire and expected to be closer to unity for a steel belted car or truck tire, cf. discussion below Eqn (3.117). With both factors assumed equal and denoted as $\varepsilon_\gamma$, we may define a total actual spin

$$\varphi = -\frac{1}{V_c}\{\dot{\psi} - (1 - \varepsilon_\gamma)\,\Omega \sin\gamma\} = -\frac{1}{R} + \frac{1 - \varepsilon_\gamma}{r_e}\sin\gamma \tag{3.55}$$

The last term represents the curvature $-1/R_\gamma$ of the tire peripheral line touching a frictionless surface at a cambered position. When disregarding a possible uniform offset of this line, we obtain, when integrating (2.52) by using (3.54) and approximating (2.51) by taking $\dot{x}_o = \dot{x} = r_e\Omega$,

$$y_o = y_{\gamma o} = -\frac{1 - \varepsilon_\gamma}{2r_e}(a^2 - x^2)\sin\gamma \tag{3.56}$$

which reduces expressions (2.55, 2.56) for the sliding velocities with (3.53–55) to

$$\begin{aligned} V_{gx} &= -\left(\frac{\partial u}{\partial x} - y\varphi\right)V_c \\ V_{gy} &= -\left(\frac{\partial v}{\partial x} + x\varphi\right)V_c \end{aligned} \tag{3.57}$$

In the range of adhesion where the sliding velocities vanish, the deflection gradients become

$$\frac{\mathrm{d}u}{\mathrm{d}x} = y\varphi, \quad \frac{\mathrm{d}v}{\mathrm{d}x} = -x\varphi \tag{3.58}$$

Integration yields the following expressions for the horizontal deformations in the contact area:

$$\begin{aligned} u &= yx\varphi + C_1 \\ v &= -\frac{1}{2}x^2\varphi + C_2 \end{aligned} \tag{3.59}$$

The second expression is, of course, an approximation of the actual variation which is according to a circle. The approximation is due to the assumption made that the deflections $v$ are much smaller than the path radius $R$. The constants of integration follow from boundary conditions which depend on the tire model employed and on the slip level. As an example, consider the simple brush model with horizontal deformations through elastic tread elements only. The contact area is assumed to be rectangular with length $2a$ and width $2b$ and filled with an infinite number of tread elements. In Figure 3.23, three rows of tread elements have been shown in the deformed situation. For this model, the following boundary conditions apply:

$$x = a: \;\; v = u = 0 \tag{3.60}$$

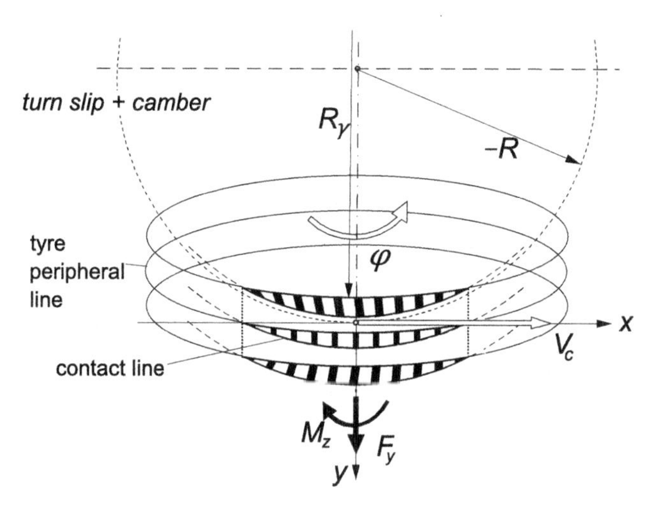


**FIGURE 3.23** Top view of cambered tire model rolling in a curve with radius $-R$.

with the use of (3.59), the formulas for the deformations in the adhesion zone starting at the leading edge read

$$
\begin{aligned}
u &= -y(a-x)\varphi \\
v &= \frac{1}{2}(a^2 - x^2)\varphi
\end{aligned}
\tag{3.61}
$$

After introducing $c'_{px}$ and $c'_{py}$ denoting the stiffness of the tread rubber per unit area in $x$- and $y$-direction respectively and assuming small spin and hence vanishing sliding, we can calculate the lateral force and the moment about the vertical axis by integration over the contact area. We obtain

$$
\begin{aligned}
F_y &= \frac{4}{3}\, c'_{py} a^3 b\, \varphi = C_{F\varphi}\varphi \\
M_z &= \frac{4}{3}\, c'_{px} a^2 b^3\, \varphi = C_{M\varphi}\varphi
\end{aligned}
\tag{3.62}
$$

or, in terms of path curvature (if $\alpha \equiv 0$ or constant) and camber angle ($\gamma$ small),

$$
\begin{aligned}
F_y &= -C_{F\varphi}\left(\frac{1}{R} - \frac{1-\varepsilon_\gamma}{r_e}\gamma\right) = -C_{F\varphi}\,\frac{1}{R} + C_{F\gamma}\gamma \\
M_z &= -C_{M\varphi}\left(\frac{1}{R} - \frac{1-\varepsilon_\gamma}{r_e}\gamma\right) = -C_{M\varphi}\,\frac{1}{R} + C_{M\gamma}\gamma
\end{aligned}
\tag{3.63}
$$

In the case of pure turning, the force acting on the tire is directed away from the path center and the moment acts opposite to the sense of turning. Consequently, both the force and the moment try to reduce the curvature $1/|R|$. In the case of pure camber, the force on the wheel is directed toward the point of intersection of the wheel axis and the road plane, while the moment tries to turn the rolling wheel toward this point of intersection. No resulting force or torque is expected to occur when $(1 - \varepsilon_\gamma) \sin \gamma = r_e/R$. For the special case that $\varepsilon_\gamma = 0$, this will occur when the point of intersection and the path center coincide. As the lateral deflection shows a symmetric distribution, the moment must be caused solely by the longitudinal forces. The generation of the moment may be explained by considering three wheels rigidly connected to each other, mounted on one axle. The wheels rotate at the same rate but in a curve the wheel centers travel different distances in a given time interval and, when cambered, these distances are equal but the effective rolling radii are different. In both situations, opposite longitudinal slip occurs, which results in a braked and a driven wheel (in Figure 3.23 the right- and the left-hand wheel respectively) and consequently in a couple $M_z$.

Up to now we have dealt with the relatively simple case of complete adhesion. When sliding is allowed by introducing a limited value of the coefficient of friction $\mu$, the calculations become quite complicated. When a finite width $2b$ is considered, complete adhesion will only occur for vanishing values of spin. We expect that sliding will start at the left and right rear corners of the contact area, since in these points the available horizontal contact forces reduce to zero and the longitudinal deformations $u$ would become maximal in the hypothetical case that $\mu \rightarrow \infty$. The zones of sliding grow with increasing spin and will thereby cause a less than proportional variation of $F_y$ and $M_z$ with $\varphi$. The case of finite contact width is too difficult to handle by a simple analysis. It will be dealt with later on in Section 3.3 when the tread element following simulation method is introduced.

For now we assume a thin tire model with $b = 0$. If, as before, a parabolic pressure distribution is assumed with a similar variation of the maximum possible lateral deflection $v_{max}$, it is obvious from Eqn (3.61) showing that the lateral deflection is also (approximately) parabolic that no sliding will occur up to a certain critical value of spin $\varphi_{sl}$, where the adhesion limit is reached throughout the contact length. Up to this point, $F_y$ varies linearly with $\varphi$ and $M_z$ remains equal to zero.

According to Eqn (3.63) with $\varepsilon_\gamma$ assumed to take a value that is minimally equal to zero, spin due to camber theoretically cannot exceed the value $1/r_e$. Consequently, at larger values of spin turn slip must be involved. Beyond the critical value $\varphi_{sl}$ the situation becomes quite complex. The discussion may be simplified by considering a turntable on top of which the wheel rolls with its spin axis fixed. The condition of adhesion is satisfied when the deflections remain within the boundaries given by the parabola's $\pm v_{max}$ as indicated in Figure 3.24a. In the same figure, the corresponding situation at camber has been indicated at the same value of spin with curvature $1/R_\gamma$ of the deflected peripheral line ($y_{\gamma o}$, Eqn (3.56)) on a $\mu = 0$ surface equal to the path curvature $1/R$.

Sliding occurs as soon as the points of the table surface can cross the adhesion boundary $\pm v_{max}$. This occurs simultaneously for all the points in the

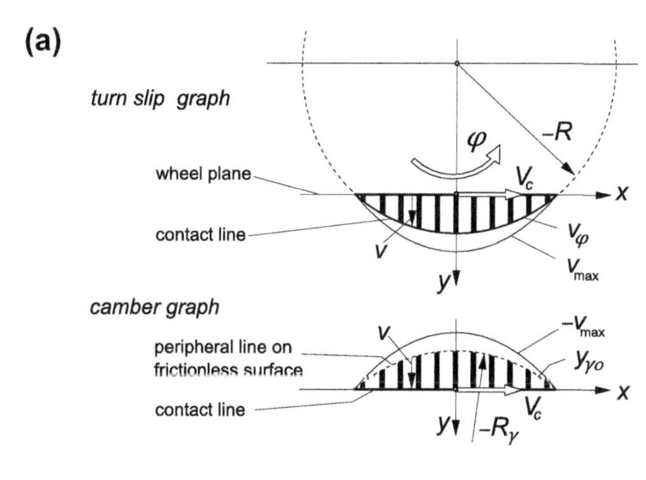


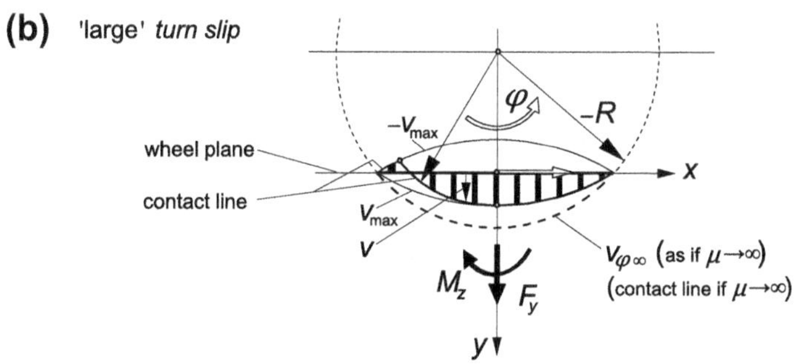


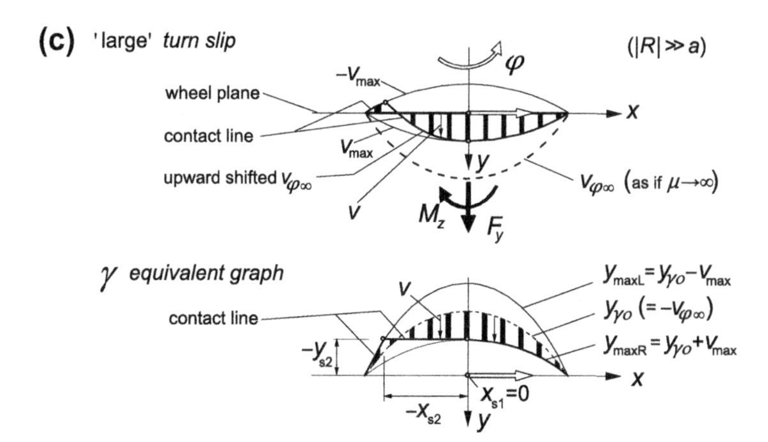


**FIGURE 3.24** The tire brush model with zero width rolling while turning or at a camber angle (a) at full adhesion. Turning at large spin showing sliding at the front half and at the rear end (b,c), with parabolic approximation of the circular path in (c).

front half of the contact line when the path curvature exceeds the curvature of the adhesion boundary. Once the contact point arrives in the rear half of the contact zone (at $x = 0$), the point can maintain adhesion because now the point on the table moves toward the inside of the adhesion boundaries. The point follows a circle until the opposite boundary is reached at $x = x_{s2}$ (cf. Figure 3.24b,c) where the deformation $v$ is opposite in sign and reaches its maximum value $v_{\max}$, after which sliding occurs again. With increasing turn slip $\varphi$ ($= -1/R$), this latter sliding zone grows. At the same time, the side force $F_y$ decreases and the torque $M_z$, that arises for $\varphi > \varphi_{sl}$, increases until the situation is reached where $R$ and $F_y$ approach zero and $M_z$ attains its maximum value (tire standing still and rotating about its vertical axis). In Figure 3.24c the radius $R$ has been considered large with respect to half the contact length $a$. Then the circle segments may be approximated by parabolas which make the analysis a lot simpler. In the lower part of the figure, a camber equivalent graph has been depicted. The curved contact line in the adhesion range is then converted to a straight (horizontal) line. The graph is obtained from the turn slip graph by subtracting from all the curves the deflection $v_{\varphi\infty}$ that would occur if full adhesion can be maintained (e.g., at $\mu \to \infty$). We use this graph as the basis for the calculation of the force and moment response to spin: first for the case of pure spin and then for the case of combined spin and side slip.

The parameter $\theta_y$ defined by Eqn (3.6) is used again and the maximum possible lateral deflection reads according to (3.5):

$$v_{\max} = \frac{a^2 - x^2}{2a\theta_y} \tag{3.64}$$

The deflection at assumed full adhesion would become

$$v_{\varphi\infty} = \frac{1}{2}\varphi(a^2 - x^2) \tag{3.65}$$

Equating the deflection at full sliding to the one at full adhesion yields the spin at the verge of sliding:

$$\varphi_{sl} = \frac{\operatorname{sgn}\varphi}{a\theta_y} \tag{3.66}$$

The locations of the transition points indicated in Figure 3.24c are found to be given by the coordinates:

$$\begin{aligned} x_{s1} &= 0, \quad x_{s2} = \frac{-a}{\sqrt{\frac{1}{2}(a\theta_y|\varphi| + 1)}} \\ y_{s1} &= y_{s2} = -\frac{a\theta_y|\varphi| - 1}{2\theta_y} a \operatorname{sgn}\varphi \end{aligned} \tag{3.67}$$

The deflections become in the first and second sliding regions and in the adhesion region respectively:

$$v_1 = -v_3 = v_{\max}\operatorname{sgn}\varphi, \qquad v_2 = y_{s1} + v_{\varphi\infty} \tag{3.68}$$

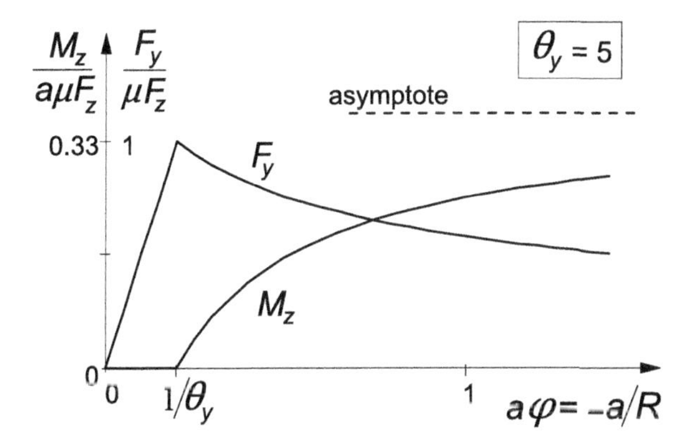


**FIGURE 3.25** Force and moment vs nondimensional spin for single row brush model.

Integration over the contact length yields, for the side force,

$$F_y = \mu F_z \sqrt{2}\,\frac{\operatorname{sgn}\varphi}{\sqrt{a\theta_y|\varphi| + 1}} \tag{3.69}$$

and, for the moment,

$$M_z = \frac{3}{8}\,\mu F_z a\,\frac{a\theta_y|\varphi| - 1}{a\theta_y|\varphi| + 1}\operatorname{sgn}\varphi \tag{3.70}$$

At $\varphi = \varphi_{sl}$, the force reduces to $\mu F_z \operatorname{sgn}\varphi$ and the moment to zero. The same can be obtained from the expressions (3.62) holding for the case of full adhesion when $2\,c'_{py}b$ is replaced by $c_{py}$ and the width $2b$ is taken equal to zero.

For spin approaching infinity, that is when the radius $R \to 0$, the force $F_y$ vanishes while the moment reaches its maximum value:

$$F_y = 0,\;\; M_z = M_{z,\max} = \frac{3}{8}\,\mu F_z a \operatorname{sgn}\varphi \tag{3.71}$$

In Figure 3.25, the pure spin characteristics have been presented according to the above expressions. It is of interest to note that although the parabola does not resemble the circular path so well at smaller radii, the resulting response seems to be acceptable at least if the deflections remain sufficiently small.

## *Spin and Side Slip*

In the analysis of combined spin and side slip, we may distinguish again between the cases of small and large spin. As before, at small spin we have only a sliding range that starts at the rear end of the contact line, whereas at large spin we have an additional sliding zone that starts at the leading edge. First we will consider the simpler case of small spin.

In Figure 3.26, an example is shown of the single row brush model, both for the cases of turning and (equivalent) camber. In the camber graph, the curves

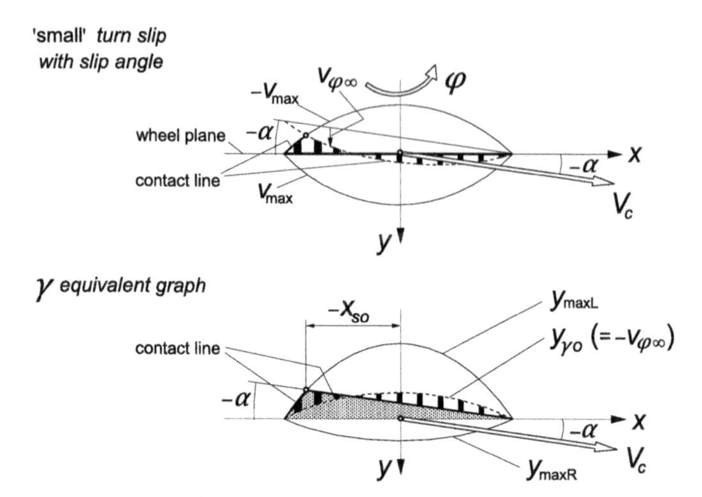


**FIGURE 3.26** The brush model turning at 'small' spin or subjected to camber at the same level of spin while running at a slip angle.

$y_{\mathrm{maxL}}$ and $y_{\mathrm{maxR}}$ have been drawn which indicate the maximum possible displacements of the tips of the elements to the left and to the right with respect to the $y_{\gamma o}$ line. The shaded area corresponds to the deformation of the brush model when subjected solely to the slip angle with the maximum possible deflection $-v_{\max}$ replaced by (in this case with negative slip angle) $y_{\mathrm{maxL}}$. This would correspond to the introduction of an adapted friction coefficient $\mu$ or parameter $\theta_y$. Apparently, the actual deflection of the elements at camber and side slip is then obtained by adding the deflection $v_{\varphi\infty}$ that would occur when the model would be subjected to spin only and full adhesion is assumed. The adapted parameter $\theta_y$ turns out to read

$$\theta_y^* = \frac{\theta_y}{1 - a\varphi\theta_y \operatorname{sgn} \alpha} \quad \text{with} \quad \theta_y = \frac{2c_{py}a^2}{3\mu F_z} \tag{3.72}$$

with the condition for the spin to be 'small':

$$|\varphi| < \varphi_l = \frac{1}{a\theta_y} \tag{3.73}$$

The side force becomes similar to Eqn (3.11) with the camber force at full adhesion added:

if $|\tan \alpha| = |\sigma_y| \leq \sigma_{y,sl} = 1/\theta_y^*$

$$F_y = 3\mu F_z\theta_y\sigma_y\left\{1 - \left|\theta_y^*\sigma_y\right| + \frac{1}{3}(\theta_y^*\sigma_y)^2\right\} + \frac{2}{3}c_{py}a^3\varphi \tag{3.74}$$

and if $|\tan \alpha| = |\sigma_y| > \sigma_{y,sl}$

$$F_y = \mu F_z \operatorname{sgn} \alpha \tag{3.74a}$$

For the moment, we obtain

if $|\tan\alpha| = |\sigma_y| \leq \sigma_{y,sl} = 1/\theta_y^*$

$$M_z = -\mu F_z a \theta_y \sigma_y \left\{ 1 - 3|\theta_y^* \sigma_y| + 3(\theta_y^* \sigma_y)^2 - |\theta_y^* \sigma_y|^3 \right\} \tag{3.75}$$

and if $|\tan\alpha| = |\sigma_y| > \sigma_{y,sl}$

$$M_z = 0 \tag{3.75a}$$

We may introduce a pneumatic trail $t_\alpha$ that multiplied with the force $F_{y\alpha}$ due to the slip angle (with the last term of (3.74) omitted) produces the moment $-M_z$:

if $|\tan\alpha| - |\sigma_y| > \sigma_{y,sl} = 1/\theta_y^*$

$$t_\alpha = -\frac{M_z}{F_{y\alpha}} = \frac{1}{3} a \frac{1 - 3|\theta_y^* \sigma_y| + 3(\theta_y^* \sigma_y)^2 - |\theta_y^* \sigma_y|^3}{1 - |\theta_y^* \sigma_y| + \frac{1}{3}(\theta_y^* \sigma_y)^2} \tag{3.76}$$

and else $t_\alpha = 0$.

The graph of Figure 3.27 clarifies the configuration of the various curves and their mutual relationship. For different values of the camber angle $\gamma$, the characteristics for the force and the moment versus the slip angle have been calculated with the above equations and presented in Figure 3.28. The corresponding Gough plot has been depicted in Figure 3.29. The relationship between $\gamma$ and the spin $\varphi$ follows from Eqn (3.55).

The curves established show good qualitative agreement with measured characteristics. Some details in their features may be different with respect to experimental evidence. In the next section where the simulation model is

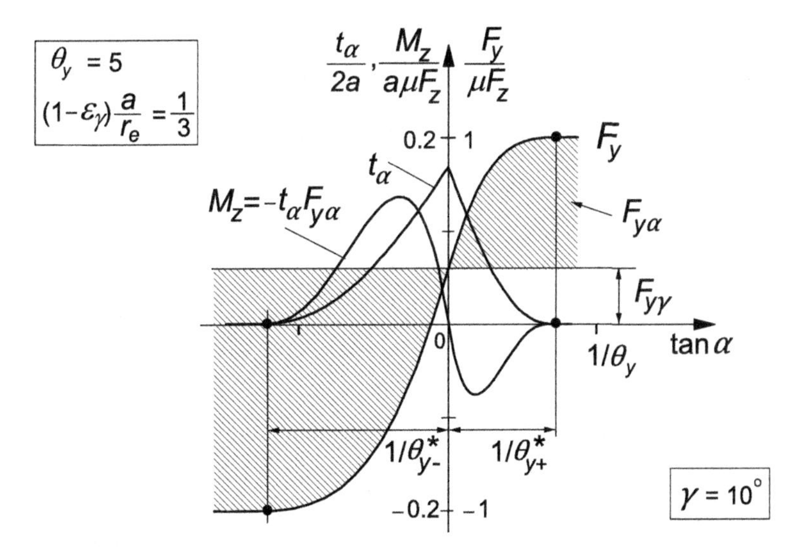


**FIGURE 3.27** Basic configuration of the characteristics vs side slip at camber angle $\gamma = 10°$.

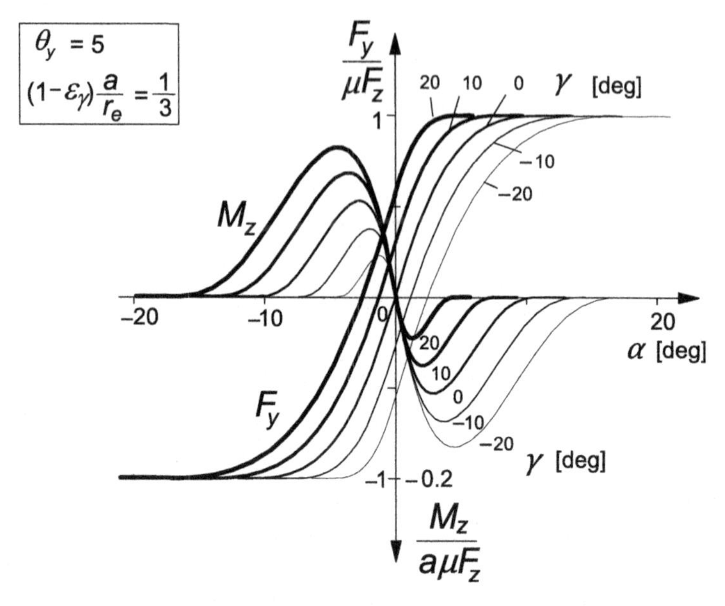


FIGURE 3.28 The calculated side force and moment characteristics at various camber angles.

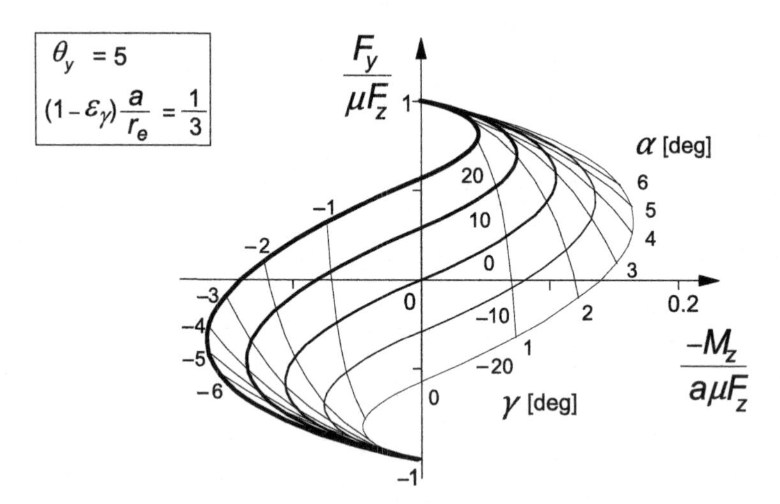


FIGURE 3.29 The corresponding Gough plot.

introduced, the effect of various other parameters like the width of the contact patch and the possibly camber-dependent average friction coefficient on the peak side force will be discussed.

The next item to be addressed is the response to large spin in the presence of side slip. Figures 3.30a,b refer to this situation. Large spin

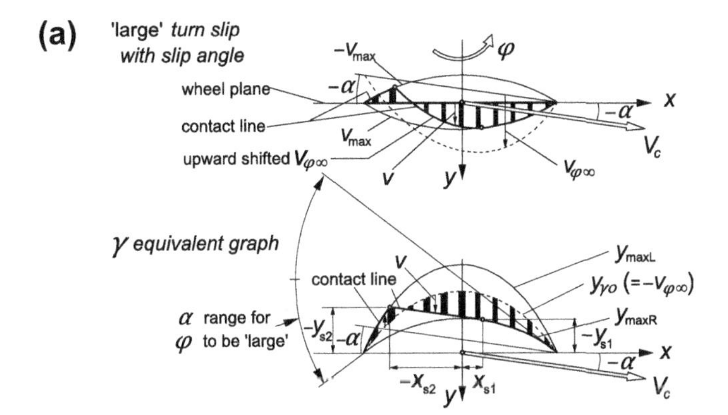


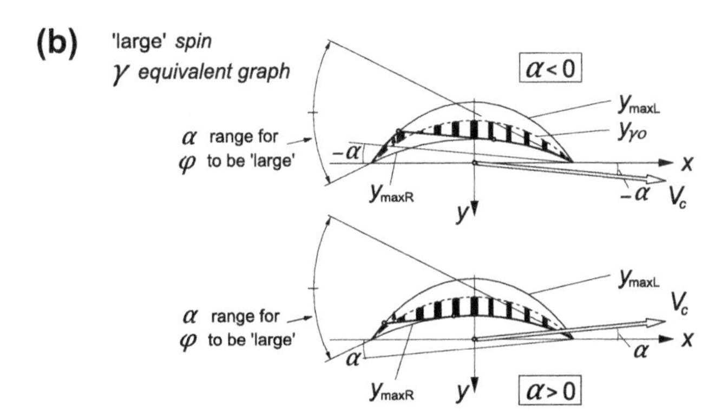


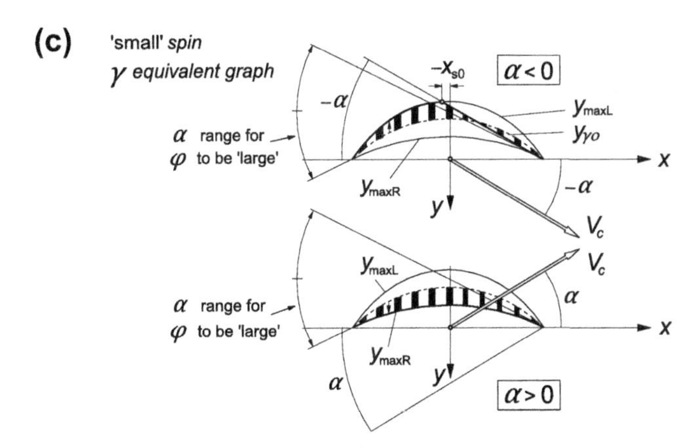


**FIGURE 3.30** The model running at large spin (turning and equivalent camber) at a relatively small (a, b) or large positive or negative slip angle (c).

with two sliding ranges occurs when the following two conditions are fulfilled:

$$|\varphi| \geq \varphi_l = \frac{1}{a\theta_y} \quad \text{and} \quad \tan|\alpha| \leq a|\varphi| - \frac{1}{\theta_y} \; (>0) \tag{3.77}$$

If the second condition is not satisfied, we have a relatively large slip angle and the Eqns (3.74, 3.75) hold again. This situation is illustrated in Figure 3.30c.

For the development of the equations for the deflections, we refer to Figure 3.30a with the camber equivalent graph. First, the distances $y$ will be established and then the deflections $v_{\varphi\infty}$ will be added to obtain the actual deflections $v$. Adhesion occurs in between the two sliding ranges. The straight line runs parallel to the speed vector and touches the boundary $y_{\text{maxR}}$. The tangent point forms the first transition point from sliding to adhesion. More to the rear, the straight line intersects the other boundary $y_{\text{maxL}}$. With the following two quantities introduced,

$$A_1 = \frac{1}{2}\left(a|\varphi| - \frac{1}{\theta_y}\right), \quad A_2 = \frac{1}{2}\left(a|\varphi| + \frac{1}{\theta_y}\right) \tag{3.78}$$

we derive for the $x$-coordinates of the transition points:

$$x_{s1} = \frac{a \tan\alpha \,\text{sgn}\,\varphi}{2A_1} \tag{3.79}$$

and

$$x_{s2} = -\frac{q + \text{sgn}\,\varphi \sqrt{q^2 + 4pr}}{2p} \tag{3.80}$$

with

$$p = \frac{A_2}{a}\text{sgn}\,\varphi\,, \quad q = \tan\alpha, \quad r = aA_2\,\text{sgn}\,\varphi + y_{s1} + x_{s1}\tan\alpha \tag{3.81}$$

The distances $y$ in the first sliding region ($x_{s1} < x < a$) read

$$y_1 = -A_1 \frac{a^2 - x^2}{a}\,\text{sgn}\,\varphi \tag{3.82}$$

in the adhesion region ($x_{s2} < x < x_{s1}$):

$$y_2 = y_{s1} + (x_{s1} - x)\tan\alpha \tag{3.83}$$

with

$$y_{s1} = y_1(x_{s1}) \tag{3.84}$$

and in second sliding range ($-a < x < x_{s2}$):

$$y_3 = -A_2 \frac{a^2 - x^2}{a}\,\text{sgn}\,\varphi \tag{3.85}$$

Integration over the contact length after addition of $v_{\varphi\infty}$ and multiplication with the stiffness per unit length $c_{py}$ gives the side force and after first multiplying with $x$ the aligning torque. We obtain the formulas

$$F_y = c_p \frac{\operatorname{sgn}\varphi}{a}\left\{A_1\left(a^2 x_{s1} - \frac{1}{3}x_{s1}^3\right) - A_2\left(a^2 x_{s2} - \frac{1}{3}x_{s2}^3\right)\right\} + c_p\left\{(y_{s1} + x_{s1}\tan\alpha)(x_{s1} - x_{s2}) - \frac{1}{2}\tan\alpha\left(x_{s1}^2 - x_{s2}^2\right)\right\} \tag{3.86}$$

$$M_z = -\frac{1}{2}c_p\frac{\operatorname{sgn}\varphi}{a}\left\{A_1\left(\frac{1}{2}a^4 - a^2x_{s1}^2 + \frac{1}{2}x_{s1}^4\right) - A_2\left(\frac{1}{2}a^4 - a^2x_{s2}^2 + \frac{1}{2}x_{s2}^4\right)\right\} + c_p\left\{\frac{1}{2}(y_{s1} + x_{s1}\tan\alpha)\left(x_{s1}^2 - x_{s2}^2\right) - \frac{1}{3}\tan\alpha\left(x_{s1}^3 - x_{s2}^3\right)\right\} \tag{3.87}$$

The resulting characteristics have been presented in Figures 3.31 and 3.32. The graphs form an extension of the diagram of Figure 3.28 where the level of camber corresponds to 'small' spin. It can be observed that in accordance with Figure 3.25 the force at zero side slip first increases with increasing spin and then decays. As was the case with smaller spin for the case where spin and side slip have the same sign, the slip angle where the peak side force is reached becomes larger. When the signs of both slip components have opposite signs, the level of side slip where the force saturates may become very large. As can be seen from Figure 3.30b, the deflection pattern becomes more anti-symmetric when with positive spin the slip angle is negative. This explains the

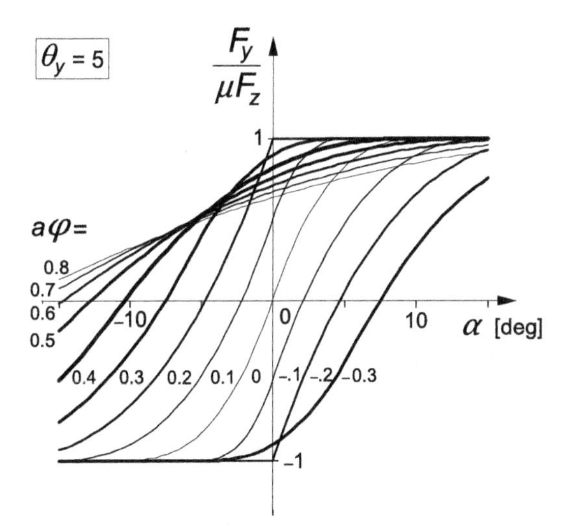


**FIGURE 3.31** Side force characteristics of the single row brush model up to large levels of spin (compare with Figure 3.28 where spin is small and $a\varphi = 0.33 \sin\gamma$).

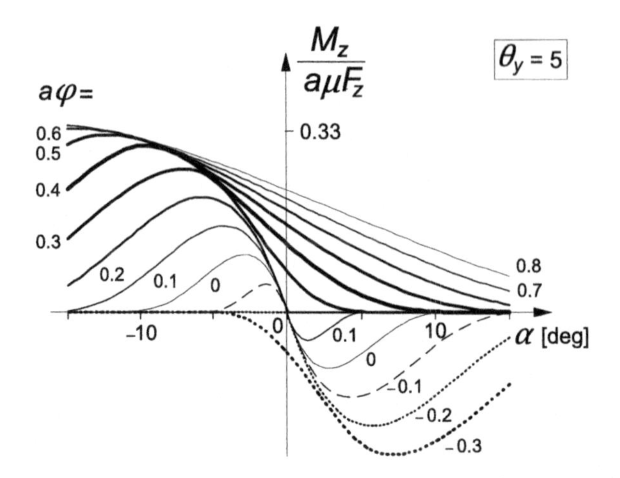


**FIGURE 3.32** Aligning torque characteristics of the single row brush model up to large levels of spin (compare with Figure 3.28 where spin is small and $a\varphi = 0.33 \sin\gamma$).

fact that at higher levels of spin, the torque attains its maximum at larger slip angles with a sign opposite to that of the spin. The observation concerning the peak side force, of course, also holds for the slip angle where the torque reduces to zero.

### *Spin, Longitudinal and Side Slip, the Width Effect*

The width of the contact patch has a considerable effect on the torque and indirectly on the side force because of the consumption of some of the friction by the longitudinal forces involved. Furthermore, for the actual tire with carcass compliance, the spin torque will generate an additional distortion of the carcass which results in a further change of the effective slip angle (beside the distortion already brought about by the aligning torque that results from lateral forces). Among other things, these matters can be taken into account in the tread simulation model to be dealt with in Section 3.3.

In the part that follows now, we will show the complexity involved when longitudinal slip is considered besides spin and side slip. To include the effect of the width of the contact patch, we consider a model with a left and a right row of tread elements positioned at a distance $y_L = -b_{row}$ and $y_R = b_{row}$ from the wheel center plane. In fact, we may assume that we deal with two wheels attached to each other on the same shaft at a distance $2b_{row}$ from each other. The wheels are subjected to the same side slip and turn slip velocities, $V_{sy}$ and $\dot{\psi}$, and show the same camber angle $\gamma$. However, the longitudinal slip velocities are different for the case of camber because of a difference in effective rolling radii. We have, for the longitudinal slip velocity of the left or right wheel positioned at a distance $y_{L,R}$ from the center plane,

$$V_{sxL,R} = V_{sx} - y_{L,R}\{\dot{\psi} - (1 - \varepsilon_\gamma)\Omega \sin\gamma\} \qquad (3.88)$$

This expression is obtained by considering Eqn (2.55) in which conicity is disregarded, steady-state is assumed to occur and the camber reduction factor $\varepsilon_\gamma$ is introduced. The factors $\theta$ are defined as (like in (3.54, 3.55))

$$\theta_{\gamma x}(y) = -\varepsilon_\gamma, \quad \theta_y(x,y) = -\varepsilon_\gamma \frac{x}{r_e} \sin\gamma \tag{3.89}$$

From Eqns (2.55, 2.56) using (3.88), the sliding velocity components are obtained:

$$V_{gxL,R} = V_{sxL,R} - \frac{\partial u_{L,R}}{\partial x} V_r \tag{3.90}$$

$$V_{gy} = V_{sy} - \frac{\partial v}{\partial x} V_r + x\{\dot{\psi} - (1-\varepsilon_\gamma)\Omega \sin\gamma\} \tag{3.91}$$

After introducing the theoretical slip quantities for the two attached wheels

$$\sigma_{xL,R} = -\frac{V_{sxL,R}}{V_r}, \quad \sigma_y = -\frac{V_{sy}}{V_r}, \quad \sigma_\psi = -\frac{\dot{\psi}}{V_r} \tag{3.92}$$

we find, for the gradients of the deflections in the adhesion zone (where $V_g = 0$) if small spin is considered (sliding only at the rear),

$$\frac{\partial u_{L,R}}{\partial x} = -\sigma_{xL,R} \tag{3.93}$$

$$\frac{\partial v}{\partial x} = -\sigma_y - x\left\{\sigma_\psi + (1-\varepsilon_\gamma)\frac{1}{r_e}\sin\gamma\right\} \tag{3.94}$$

which yields after integration for the deflections in the adhesion zone ($x < x_t$):

$$u_{L,R} = (a-x)\sigma_{xL,R} \tag{3.95}$$

$$v = (a-x)\sigma_y + \frac{1}{2}(a^2 - x^2)\left\{\sigma_\psi + (1-\varepsilon_\gamma)\frac{1}{r_e}\sin\gamma\right\} \tag{3.96}$$

The transition point from adhesion to sliding, at $x = x_t$, can be assessed with the aid of the condition

$$c_p e_{L,R} = \mu q_z \tag{3.97}$$

with the magnitude of the deflection

$$e_{L,R} = |\boldsymbol{e}_{L,R}| = \sqrt{u_{L,R}^2 + v^2} \tag{3.98}$$

Solving for $x_t$ and performing the integration over the adhesion range may be carried out numerically. In the sliding range, the direction of the deflections varies with $x$. As an approximation, one may assume that these deflections $\boldsymbol{e}$ (for an isotropic model) are all directed opposite to the slip speed $\boldsymbol{V}_{sL,R}$. Results of such integrations yielding the values of $F_x$, $F_y$ and $M_z$ will not be shown here. We refer to Sakai (1990) for analytical solutions of the single row brush model at combined slip with camber.

## 3.3. THE TREAD SIMULATION MODEL

In this section, a methodology is developed that enables us to investigate effects of elements in the tire model which were impossible to include in the analytical brush model dealt with in the preceding Section 3.2. Examples of such complicating features are: arbitrary pressure distribution; velocity and pressure dependent friction coefficient; isotropic stiffness properties; combined lateral, longitudinal and camber or turn slip; lateral, bending and yaw compliance of the carcass and belt; and finite tread width at turn slip or camber.

The method is based on the time simulation of the deformation history of one or more tread elements while moving through the contact zone. The method is very powerful and can be used either under steady-state or time-varying conditions. In the latter nonsteady situation, we may divide the contact length into a number of zones of equal length in each of which a tread element is followed. In the case of turning or camber, the contact patch should be divided into several parallel rows of elements. While moving through the zones, the forces acting on the elements are calculated and integrated. After having moved completely through a zone, the integration produces the zone forces. These forces act on the belt and the corresponding distortion is calculated. With the updated belt deflection, the next passage through the zones is performed and the calculation is repeated.

Here, we will restrict the discussion to steady-state slip conditions and take a single zone with length equal to the contact length. In Section 2.5, an introductory discussion has been given and reference has been made to a number of sources in the literature. The complete listing of the simulation program *TreadSim* written in Matlab code can be found *online* (cf. App. 2). For details, we may refer to this program.

Figure 3.33 depicts the model with deflected belt and the tread element that has moved from the leading edge to a certain position in the contact zone. In Figure 3.34, the tread element deflection vector $\boldsymbol{e}$ has been shown. The tread element is assumed to be isotropic, thus with equal stiffnesses in the $x$- and $y$-direction. Then, when the element is sliding, the sliding speed vector $\boldsymbol{V}_g$, that has a sense opposite to the friction force vector $\boldsymbol{q}$, is directed opposite to the deflection vector $\boldsymbol{e}$. The figure depicts the deflected element at the ends of two successive time steps $i-1$ and $i$. The first objective is now to find an expression for the displacement $\boldsymbol{g}$ of the tip of the element while sliding over the ground.

The contact length is divided into $n$ intervals. Over each time step $\Delta t$, the base point $B$ moves over an interval length toward the rear. With given length

$$\Delta x = \frac{2a}{n} \tag{3.99}$$

and the linear speed of rolling $V_r = r_e\Omega$, the time step $\Delta t$ is obtained

$$\Delta t = \frac{\Delta x}{V_r} \tag{3.100}$$

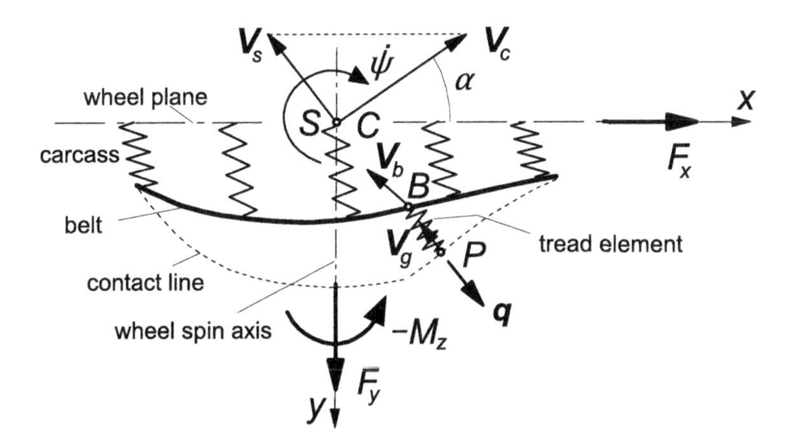


**FIGURE 3.33** Enhanced model with deflected carcass and tread element that is followed from front to rear.

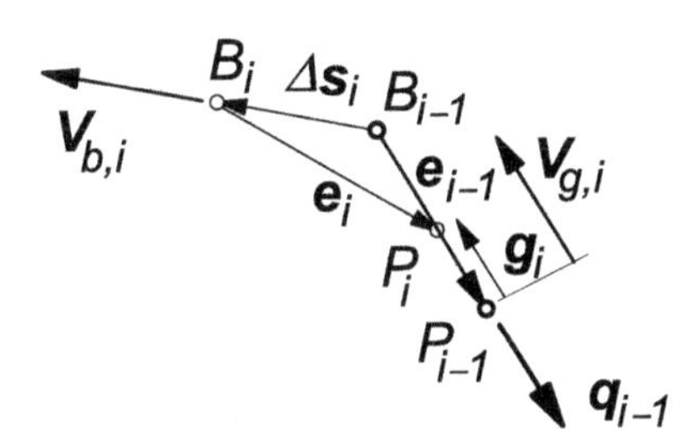


**FIGURE 3.34** The isotropic tread element with deflection $\boldsymbol{e}$ in two successive positions $i-1$ and $i$. Its base point $B$ moves with speed $V_b$ and its tip $P$ slides with speed $V_g$.

with the velocity vector $\boldsymbol{V}_b$ of point $B$, the displacement vector $\Delta\boldsymbol{s}$ of this point over the time step becomes

$$\Delta\boldsymbol{s} = \boldsymbol{V}_b \Delta t \tag{3.101}$$

The base point $B$ moves along the belt peripheral line or a line parallel to this line. With the known lateral coordinate $y_b$ of this line of base points with respect to the wheel center plane, the local slope $\partial y_b/\partial x$ can be assessed. Then, with the slip velocity $\boldsymbol{V}_s$ of the slip point $S$, the yaw rate of the line of intersection $\dot{\psi}$, and the rolling speed $V_r$, the components of $\boldsymbol{V}_b$ can be found.

The velocity of point $B$ may be considered as the sliding velocity of this point with respect to the ground and we may employ equations (2.55, 2.56) for its assessment. In these equations, at steady state, the time derivatives of $u$ and $v$ vanish, the slope $\partial u/\partial x$ is replaced by zero, and for $\partial v/\partial x$ we take the gradient of the belt deflection caused by the external force and moment. We have, with average $x$ position $x_b = x + 0.5\Delta x$,

$$V_{bx} = V_{sx} - y_{bo}(\dot{\psi} - \Omega \sin\gamma) - \theta_{\text{con},x}(y)V_r + \theta_{\gamma x}(y) y_{bo} \Omega \sin\gamma \tag{3.102}$$

and

$$V_{by} = V_{sy} + x_b\dot{\psi} - \frac{\partial y_b}{\partial x_b}V_r \tag{3.103}$$

with the slip and roll velocities

$$V_{sx} = -V_{cx}\kappa, \;\; V_r = r_e\Omega = V_{cx} - V_{sx}, \;\; V_{sy} = -V_{cx}\tan\alpha \tag{3.104}$$

The lateral displacement $y_b$ of the belt at the contact center is attributed to camber, conicity, and the lateral external force (through the lateral compliance of the carcass). The gradient $\partial y_b/\partial x_b$ may be approximately assessed by assuming a parabolic base line $y_b(x_b)$ exhibiting an average slope $c_s$ influenced by the aligning torque (through the yaw compliance) and ply-steer, and a curvature $c_c$ influenced by the side force (through the bending stiffness) and camber and conicity (cf. (3.56)). We have, for the lateral coordinate,

$$y_b = -\frac{a^2}{2r_e}(1-\varepsilon_{\gamma y})\,(\gamma_{\text{con}} + \sin\gamma)\, + \frac{F_y}{c_{\text{lat}}} + c_s x_b + \frac{1}{2}c_c x_b^2 \pm b_{\text{row}} \tag{3.105}$$

and, for its approximation used in (3.102),

$$y_{bo} = \pm b_{\text{row}} \tag{3.106}$$

and the slope:

$$\frac{\partial y_b}{\partial x_b} = c_s + c_c x_b = \frac{M_z'}{c_{\text{yaw}}} - \frac{x_b F_y}{c_{\text{bend}}} + \frac{x_b}{r_e}\sin\gamma + \theta_y(x_b, y_b) \tag{3.107}$$

Conicity and ply-steer will be interpreted here to be caused by 'built-in' camber and slip angles. These equivalent camber and slip angles $\gamma_{\text{con}}$ and $\alpha_{\text{ply}}$ are introduced in the expressions of the quantities $\theta$. We define

$$\theta_{\gamma x}(y_b) = -\varepsilon_{\gamma y}, \;\; \theta_{\text{con},x}(y_b) = -\frac{y_b}{r_e}(1-\varepsilon_{\gamma x})\gamma_{\text{con}} \tag{3.108}$$

and

$$\theta_y(x_b, y_b) = -\varepsilon_{\gamma y}\frac{x_b}{r_e}\sin\gamma + (1-\varepsilon_{\gamma y})\,\frac{x_b}{r_e}\gamma_{\text{con}} + \alpha_{\text{ply}} \tag{3.109}$$

The coefficients $\varepsilon_{\gamma x}$ and $\varepsilon_{\gamma y}$ may be taken equal to each other. The first term of the displacement (3.105) is just a guess. It constitutes the lateral displacement of the base line at the contact center when the tire is pressed on a frictionless surface in the presence of conicity and camber. The displacements at the contact leading and trailing edges are assumed to be zero under these conditions. The approximation $y_{bo}$ (3.106) is used in (3.102) to avoid apparent changes in the effective rolling radius at camber. The actual lateral coordinate $y_b$ (3.105) plus a term $y_{r\gamma}$ is used to calculate the aligning torque. With this additional term, the lateral shift of $F_x$ due to sideways rolling when the tire is being cambered is accounted for. We have $y_{b,\text{eff}} = y_b + y_{r\gamma}$ with $y_{r\gamma} = \varepsilon_{yr\gamma}\, b\sin\gamma$

with an upper limit of its magnitude equal to $b$. The moment $M_z'$ causes the torsion of the contact patch and is assumed to act around a point closer to its center as depicted in Figure 3.19. A reduction parameter $\varepsilon_y'$ is used for this purpose. More refinements may be introduced to better approximate the shape of the base line, especially near the leading and trailing edges (bending back).

With the displacement vector $\Delta\boldsymbol{s}$ (3.101) established, we can derive the change in deflection $\boldsymbol{e}$ over one time step. By keeping the directions of motion of the points $B$ and $P$ in Figure 3.34 constant during the time step, an approximate expression for the new deflection vector is obtained. After the base point $B$ has moved according to the vector $\Delta\boldsymbol{s}$, we have

$$\boldsymbol{e}_i = \boldsymbol{e}_{i-1} + \boldsymbol{g}_i - \Delta\boldsymbol{s}_i \quad \text{with} \quad \boldsymbol{g}_i = -\frac{g_i}{e_{i-1}}\boldsymbol{e}_{i-1} \tag{3.110}$$

Here, $e_{i-1}$ denotes the absolute value of the deflection and $g_i$, the distance $P$, has slided in the direction of $-\boldsymbol{e}_{i-1}$. In the case of adhesion, the sliding distance $g_i = 0$. When the tip slides, the deflection becomes

$$e_i = \frac{\mu_i q_{z,i}}{c_p} \tag{3.111}$$

From (3.110), an approximate expression for $g_i$ can be established, considering that $g$ is small with respect to the deflection $e$. We obtain

$$g_i = \frac{1}{2}e_{i-1}\frac{(e_{x,i-1} - \Delta s_{x,i})^2 + (e_{y,i-1} - \Delta s_{y,i})^2 - e_i^2}{e_{x,i-1}(e_{x,i-1} - \Delta s_{x,i}) + e_{y,i-1}(e_{y,i-1} - \Delta s_{y,i})} \tag{3.112}$$

in which expression (3.111) is to be substituted. If $g_i$ is positive, sliding remains. If not, adhesion commences. In the case of sliding, the deflection components can be found from (3.110) using (3.112). Then, the force vector per unit length is

$$\boldsymbol{q}_i = \mu_i q_{z,i}\frac{\boldsymbol{e}_i}{e_i} \tag{3.113}$$

If adhesion occurs, $g_i = 0$ and with (3.110) the deflection is determined again. The force per unit length now reads

$$\boldsymbol{q}_i = c_p\boldsymbol{e}_i \tag{3.114}$$

As soon as the condition for adhesion

$$q_i = |\boldsymbol{q}_i| \leq \mu_i q_{z,i} \tag{3.115}$$

is violated, sliding begins. The sliding distance $g_i$ is calculated again and its sign checked until adhesion may show up again.

In (3.113), the friction coefficient appears. This quantity may be expressed as a function of the sliding velocity of the tip of the element over the ground. However, this velocity is not available at this stage of the calculation. Through iterations, we may be able to assess the sliding speed at the position considered.

Instead, we will adopt an approximation and use the velocity of the base point (3.102, 3.103) to determine the current value of the friction coefficient. The following functional relationship may be used for the friction coefficient versus the magnitude of the approximated sliding speed $V_b$:

$$\mu = \frac{\mu_0}{1 + a_\mu V_b} \tag{3.116}$$

During the passage of the element through the contact zone, the forces $\Delta \boldsymbol{F}_i$ are calculated by multiplying $\boldsymbol{q}_i$ with the part of the contact length $\Delta x$ covered over the time step $\Delta t$.

Subsequently, the total force components and the aligning moment are found by adding together all the contributions $\Delta F_{xi}$, $\Delta F_{yi}$, and $x_{bi}\Delta F_{yi} - y_{bi,\text{eff}}\Delta F_{xi}$, respectively. A correction to the moment arm may be introduced to account for the side ways rolling of the tire cross section while being cambered and deflected (causing lateral shift of point of action of the resulting normal load $F_z$ and similarly of the longitudinal force $F_x$). Also, the counter-effect of the longitudinal deflection $u_c$ may contribute to this correction factor (cf. Eqn (3.51)).

For details and possible application of the model, we refer to the complete listing of the Matlab program *TreadSim* presented online.

In the sequel, a number of example results of using the tread simulation model have been presented. The following cases have been investigated:

1. Sliding velocity-dependent friction coefficient (rigid carcass but parameter $c$ of Eqn (3.52) is included) (Figure 3.35a).
2. Flexible carcass, without and with camber (Figure 3.35b).
3. Finite tread width (two rows of tread elements, flexible carcass), with and without camber (Figure 3.35c).
4. Combined lateral and turn slip (two rows, flexible carcass) (Figure 3.35d).
5. Pneumatic trail at pure side slip (flexible and rigid carcass) (Figure 3.36).

The computations have been conducted with the set of parameter values listed in Table 3.1.

In the case of a 'rigid' carcass, parameter $c$ ($= 0.01$ m/kN), Eqn (3.52), is used while $c_{\text{lat}}$, $c_{\text{bend}}$, $c_{\text{yaw}} \to \infty$. As indicated, quantities $c_p$ and $\theta$ follow from the model parameters.

The graphs of the lower half of Figure 3.35a relative to the upper half show the most prominent effect of a with sliding velocity decreasing friction coefficient. At $a_\mu = 0.03$, the side force $F_y$ exhibits a clear peak at a slip angle of about seven degrees when the wheel is rolling freely ($\kappa = 0$). Also, the fore-and-aft force $F_x$ tends to decrease after having reached its peak value. The inward endings of the curves at constant values of slip angle shown in the bottom diagram are typical especially for a tire running on wet road surfaces. The peak values themselves will decrease when the speed of travel is increased (not shown) while the initial slopes (slip stiffnesses) remain unchanged. The

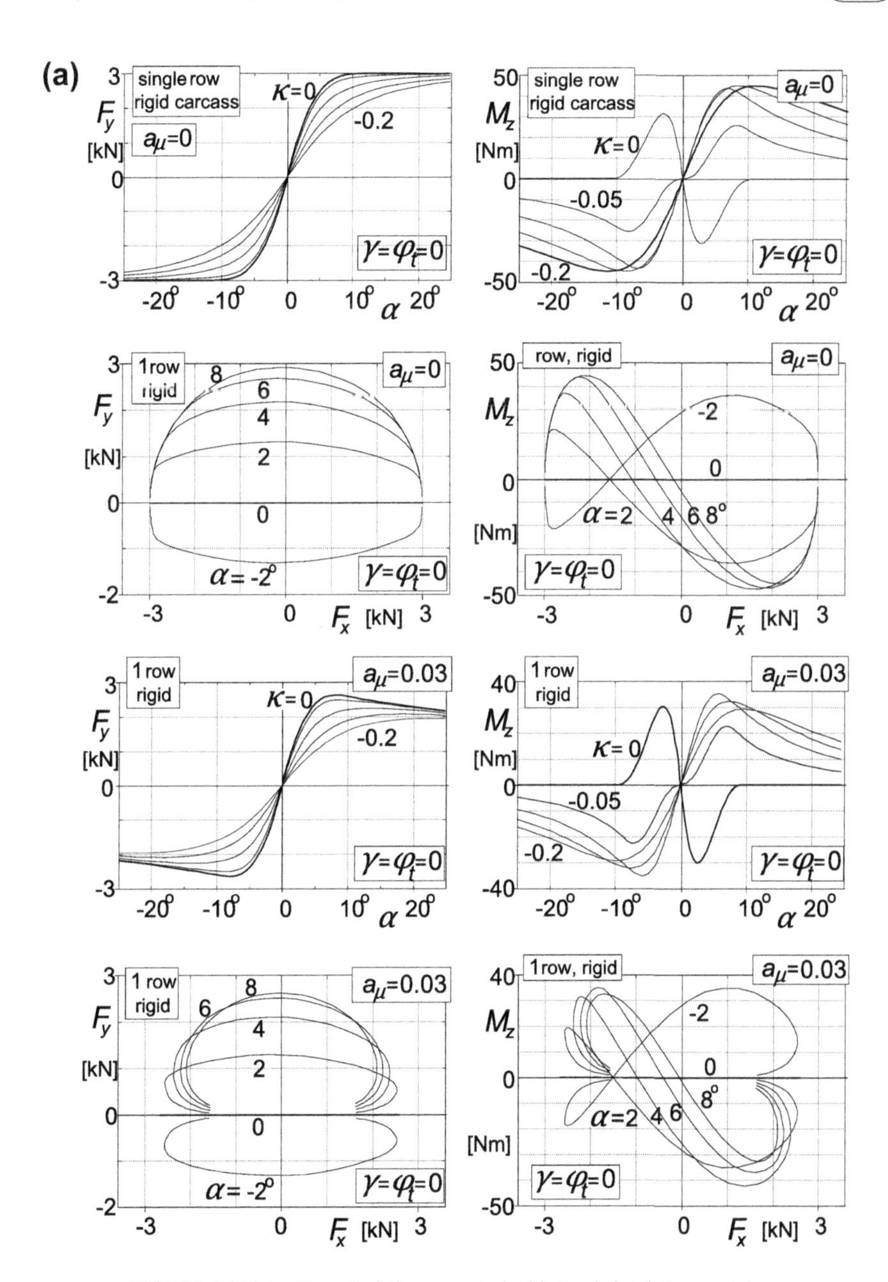


**FIGURE 3.35(a)** Characteristics computed with tread simulation model.

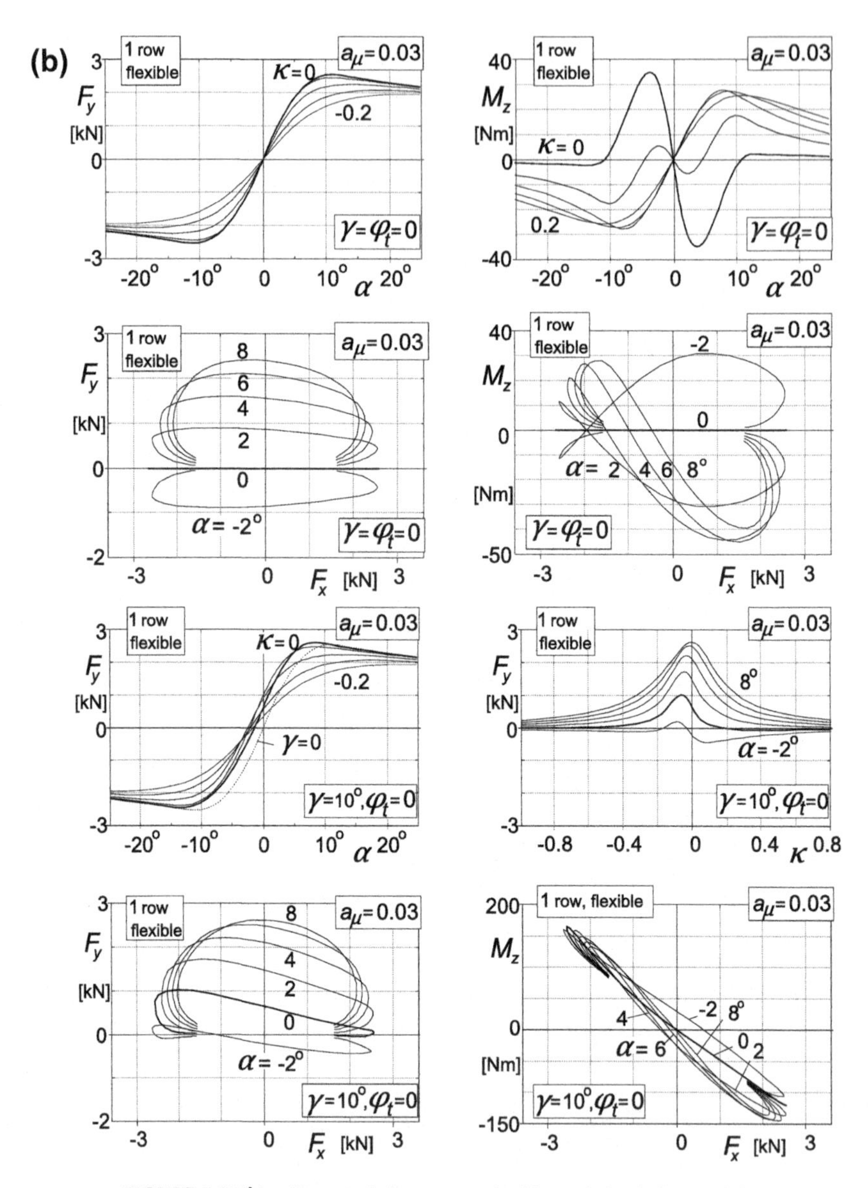


**FIGURE 3.35(b)** Characteristics computed with tread simulation model.

change in sign of the aligning torque in the braking range of its diagram occurs because of the retained parameter $c$ that produces the effect of the flexible carcass according to Eqns (3.51, 3.52). The expected change in sign of the aligning torque curve at pure side slip at higher levels of side slip does not occur

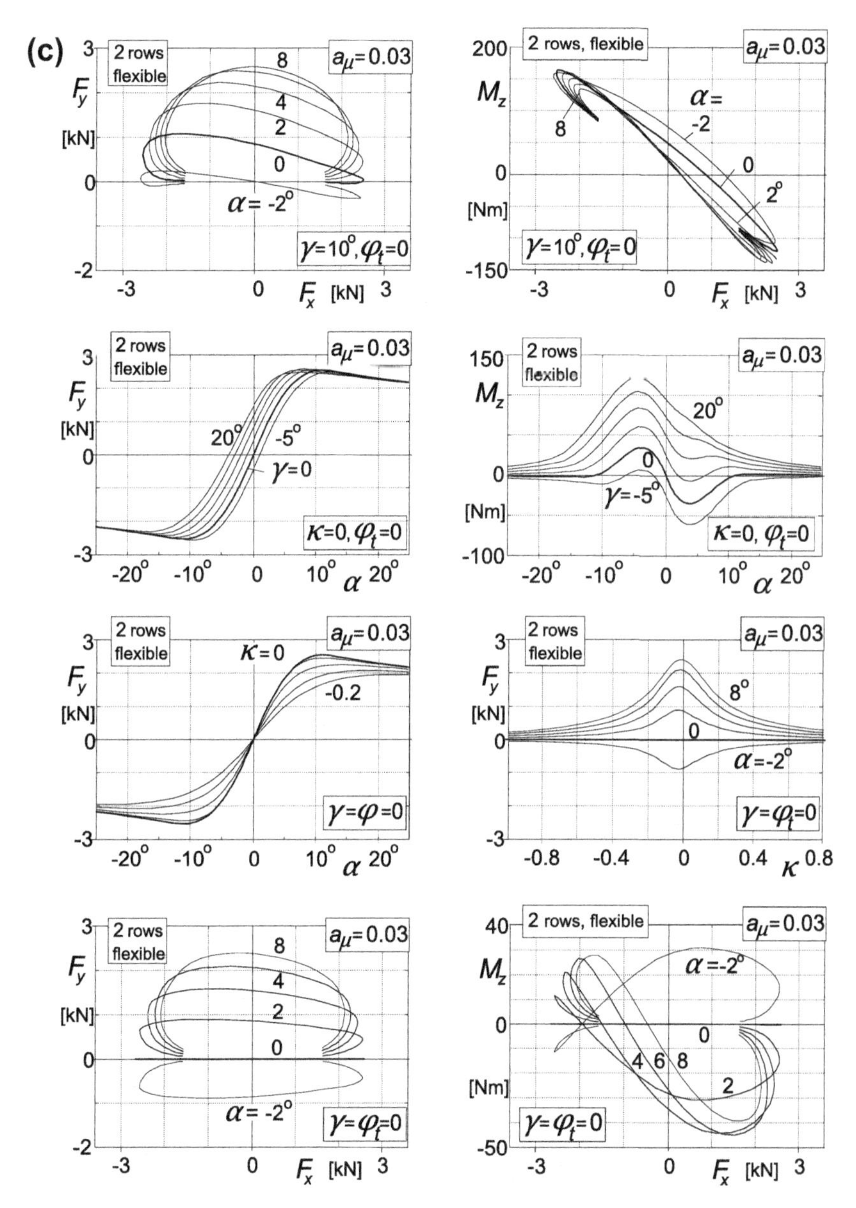


FIGURE 3.35(c) Characteristics computed with tread simulation model.

due to the limitation of the estimation of the sliding speed for which the speed of a point of the belt is taken (here rigid). In Figure 3.35b with the carcass considered flexible, this sign change does show up as illustrated in the upper right diagram.

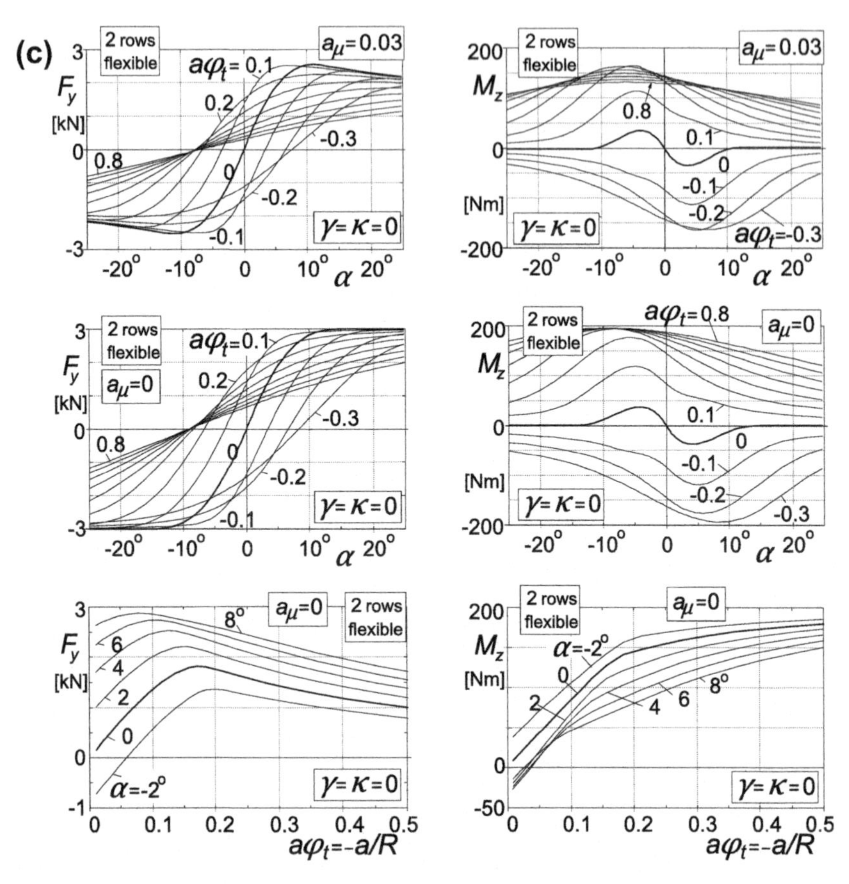


FIGURE 3.35(d) Characteristics computed with tread simulation model.

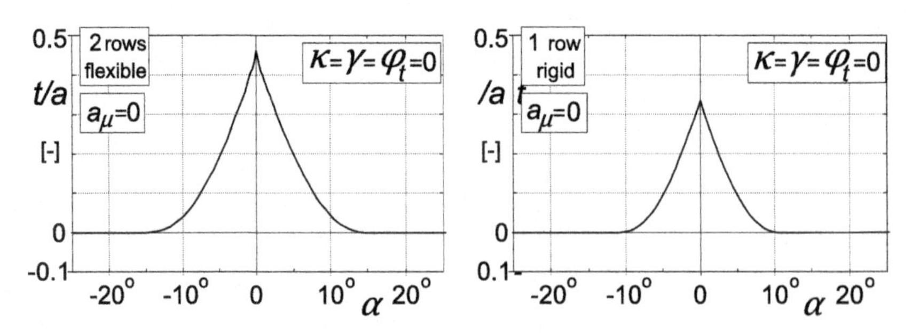


FIGURE 3.36 Pneumatic trail variation as computed with tread simulation model.

A more important effect of the flexible carcass presented in Figure 3.35b is the decreased cornering stiffness while the fore-and-aft slip stiffness remains the same. The lower value of $F_y$ at $\alpha = 2^\circ$ in the $F_y$ vs $F_x$ diagram clearly demonstrates this effect (the curves become less curved). The aligning

**TABLE 3.1** Parameter Values used in the Tread Simulation Model (Figures 3.35–3.37)

| | | | | | | | |
|---|---|---|---|---|---|---|---|
| $a$ | 0.1 m | $F_z$ | 3000 N | $c_{lat}$ | 100 kN/m | $C_{F\kappa}$ | $15F_z$ |
| $b$ | 0.08 m | $V_c$ | 30 m/s | $c_{bend}$ | 4 kNm | $y_o$ | $0 : v_o$ Eqn (3.51) |
| $b_{row}$ | 0.05 m | $\mu_o$ | 1.0 | $c_{yaw}$ | 6 kNm/rad | $\alpha_{ply}$ | 0 |
| $r_e$ | 0.3 m | $a_\mu$ | 0, 0.03 s/m | $c$ | $1/c_{lat}$ | $\gamma_{con}$ | 0 |
| $\varepsilon_{y\gamma}$ | 4.0 | other $\varepsilon$'s = 0 | | $c_p = C_{F\kappa}/(2a^2\, n_{row})$ | | $\theta = C_{F\kappa}/(3\mu_o\, F_z)$ | |

torque, however, is not so much affected. This is due to the larger pneumatic trail which is a result of the curved deflection line of the belt. Figure 3.36 shows the pneumatic trail diagram for the model with and without carcass compliance (the number of rows has no influence in case of zero spin). As expected, the simple brush model with rigid carcass has a trail of $0.33a$ when the slip angle approaches zero. The flexible carcass model considered features a pneumatic trail of about $0.46a$. See Figure 3.38 for the deflection pattern.

The influence on the curves of the introduction of the relatively large camber angle of 10° is indicated in the lower half of Figure 3.35b. The effect is most clearly demonstrated in the lower-left diagram. The camber thrust is accompanied by a lateral deflection that causes a shift of the line of action of the longitudinal force. The resulting torque tends to rotate the lower part of the belt about the vertical axis which now can be accomplished through the yaw compliance of the carcass. At braking, the rotation is such that an apparent slip angle arises that increases the camber side force. At driving, the opposite occurs. As a result, the constant slip angle curves plotted in the diagram show an inclination. The corresponding influence diagram of $\kappa$ on $F_y$ shows distorted curves when compared with those of Figure 3.11. The inversed $S$ shape of the curves at small slip angle is a feature that is commonly encountered in measured characteristics. The aligning torque diagram (lower-right picture) is considerably changed as a result of the action of the torque mentioned above that originates from $F_x$.

In Figure 3.35c, the effect of a finite width of the contact patch is demonstrated. Two rows of tread elements have been considered. At the camber angle of 10°, a spin torque is generated that appears to rotate the lower part of the belt in such a way that an apparent slip angle arises that increases the camber thrust. The upper-left-hand diagram shows the increase in side force. The right-hand diagram indicates the considerable rise in the aligning torque as a result of the spin torque in the range of small longitudinal force $F_x$.

The two diagrams in the second row demonstrate the effects of tread width, carcass flexibility, and friction decay with velocity on the side force and moment vs slip angle characteristics for a series of camber angles. The plots may be compared with those of Figure 3.28. The lower half of Figure 3.35c refers to the case without camber and may be compared with the plots of the upper half of Figure 3.35b that does not include the effects of tread width.

Figure 3.35d presents the force and moment characteristics for a series of levels of turn slip, including large values of spin corresponding to a radius of curvature equal to $a/0.8$. The figures of the second row refer to the case of constant friction ($a_\mu = 0$) and may be compared with Figures 3.31, 3.32. An interesting effect of tread width is the further decrease of the side force at higher levels of turn slip due to longitudinal slip that occurs on both sides of the contact patch which consumes a lot of the available frictional forces. Of course, the aligning torque is now considerably larger. The third row of diagrams represents the spin force and moment characteristics at various levels of side slip. In Figure 3.37, the pure spin characteristics have been drawn for the model with a rigid carcass provided with one row of tread elements (same as Figure 3.25) and with two rows computed with the tread simulation model. Comparison clearly shows the considerable reduction of the peak side force and the much larger level of the aligning torque caused by the finite tread width. The influence of carcass flexibility appears to be very small both for the force and for the moment.

These results indicate that the single row theory developed in Section 3.2 also for high spin in combination with side slip has only limited practical

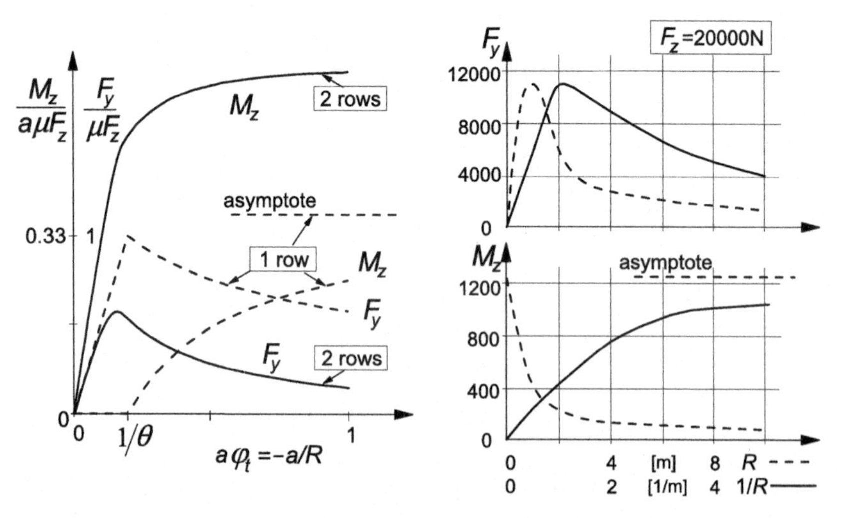


FIGURE 3.37 Pure turn slip characteristics according to the model (left, rigid carcass with one or two rows of elements) and results from experiments with a bias-ply truck tire (right, 9.00–20 eHD, $p_i = 5.5$ bar, $V = 1$–$3$ km/h on dry road, from Freudenstein 1961). Model parameters: $\theta = 5$, $a_\mu = 0$, and $b_{row} = 0.5\,a$.

significance. The influence of small spin, i.e., camber, on the side force vs slip angle characteristic as calculated with the aid of the simple single row, rigid carcass brush model may be considered as reasonable.

Freudenstein (1961) has conducted side-slip and turning experiments with a bias-ply truck tire on a dry road surface. The side-slip measurement results of this tire were already depicted in Figure 3.6. In the right-hand diagram of Figure 3.37, the results from the turning experiments have been presented. As abscissa, both the path radius and the path curvature have been used. Obviously, the calculated two-row model characteristics show good qualitative agreement with the experimental curves. For the values $a = 0.1$ m, $\mu F_z = 20000$ N, and $\theta = 8$, a very reasonable also quantitative correspondence for both the force and the moment characteristics of Figure 3.37 can be obtained. Freudenstein suggests the following formula for the peak moment generated at pure turning at wheel speed $V = 0$:

$$|M_{z,\max}| = M_{z\varphi\infty} \approx \frac{3}{8}\mu F_z\left(a + \frac{2}{3}b\right) \tag{3.117}$$

As Freudenstein did not give the camber characteristics of the truck tires on which the turning behavior was measured, we are not able to compare the responses to camber and turning. According to Hadekel (1952), for aircraft tires the lateral force due to turning is about four times higher than the camber force at equal values of spin. From experiments performed by Higuchi (1997), a factor of about two can be deduced for a radial ply car tire (cf. Chap. 7 discussion above Figure 7.11). This supports the theory of the reduced curvature of the peripheral line of the cambered tire pressed on a frictionless surface due to the

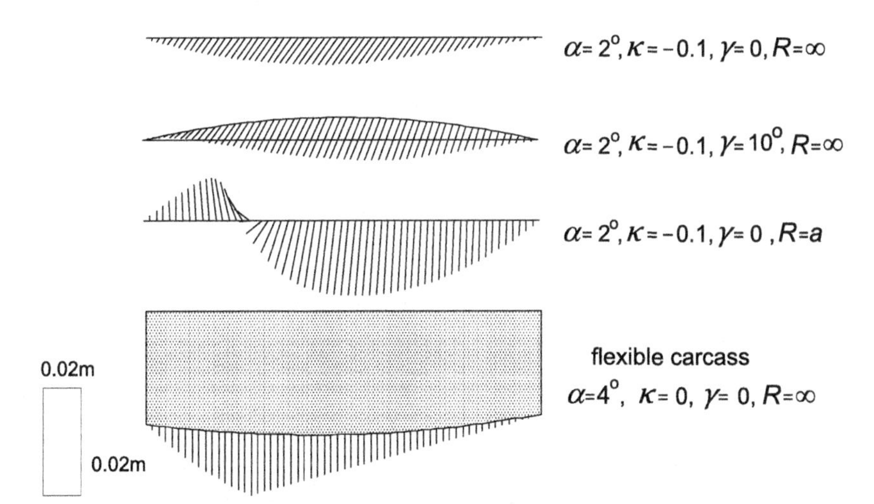


**FIGURE 3.38** Examples of deflection patterns at various combinations of slip for the single row model with rigid and flexible carcass.

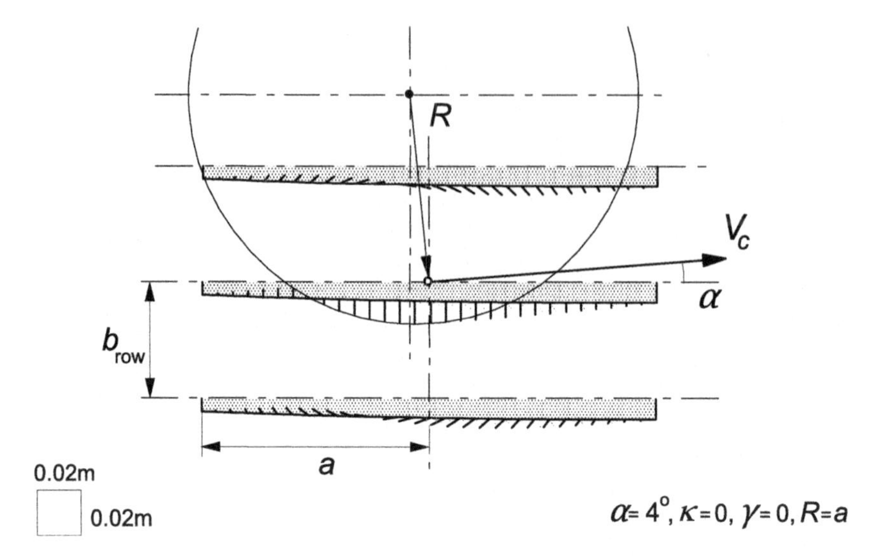


**FIGURE 3.39** Deflections of the tread elements and of the belt for the three row model at side slip and turning.

high lateral bending stiffness of the tread band, as illustrated in Figure 2.10, and the associated coefficient $\varepsilon_\gamma$ in Eqn (3.55). The program *TreadSim* also provides information on the distribution of contact forces and deflections of belt and tread elements. Some examples of the deformation pattern have been depicted in Figures 3.38 and 3.39. In Figure 3.38, the deflections of the single row brush model have been depicted and in addition of the model with carcass lateral, yaw and bending compliance. As indicated, the scale of the drawing has been chosen larger in lateral direction. In the middle two diagrams, the influence of camber and of turning on the deflections at side and brake slip has been shown. Turning was considered with an exceptionally small turn radius $R$ equal to $a$ which is half the contact length. One may note the central section where adhesion occurs. The latter situation also occurs with the middle row of tread elements of the three row brush model depicted in Figure 3.39. In this drawing, the scales are the same. Clearly, the tire generates longitudinal deflections of the outer rows of elements which contribute to the torque.

## 3.4. APPLICATION: VEHICLE STABILITY AT BRAKING UP TO WHEEL LOCK

When the vehicle is being braked forcefully and a possible downward slope does not compensate for the reduction in speed of travel, the situation can no longer be considered as steady state. The influence of the rate of change of the longitudinal speed and of the effect of combined slip on the lateral stability may no longer be neglected. Instead of using the two first-order differential equations

(1.42) for the lateral and yaw motions, we must now consider the complete set, including the equation for the longitudinal motion which is also of the first order.

Because of the complexity involved, the influence of the height of the center of gravity on the vehicle motion will be disregarded. With a finite height, fore-and-aft but also lateral load transfer would occur, the latter causing unequal braking forces on the left- and right-locked wheels that give rise to a stabilizing torque counteracting the effect of the fore and aft load transfer.

The steer angle is kept equal to zero. This two-wheel, single-track, rigid vehicle with zero c.g. height has been depicted in Figure 3.40 where the wheel on axle 2 is considered to be locked. For the three states, we obtain the equations

$$\begin{aligned} m(\dot{u} - vr) &= F_{x1} + F_{x2} \\ m(\dot{v} + ur) &= F_{y1} + F_{y2} \\ mk^2\dot{r} &= aF_{y1} - bF_{y2} \end{aligned} \tag{3.118}$$

with

$$\begin{aligned} \tan\alpha_1 &= -\frac{v + ar}{u} \\ \tan\alpha_2 &= -\frac{v - br}{u} \end{aligned} \tag{3.119}$$

and

$$\begin{aligned} F_{y1} &= F_{y1}(\alpha_1, \kappa_1, F_{z1}) \\ F_{y2} &= F_{y2}(\alpha_2, \kappa_2, F_{z2}) \end{aligned} \tag{3.120}$$

For a proper simulation of the motion, the slip ratio's $\kappa_i$ should result from the wheel speeds of revolution $\Omega_i$ which would require additional degrees of freedom. For this occasion, we will employ an alternative approach that involves the introduction of functions for the side force in which its direct dependence on the braking effort (or brake pressure) is included. This is possible when 'dry' friction is assumed to occur between tire and road and the curves of the right-most diagram of Figure 1.2 do not show inward endings and thus double valued functions are avoided. The characteristics of Figure 3.16

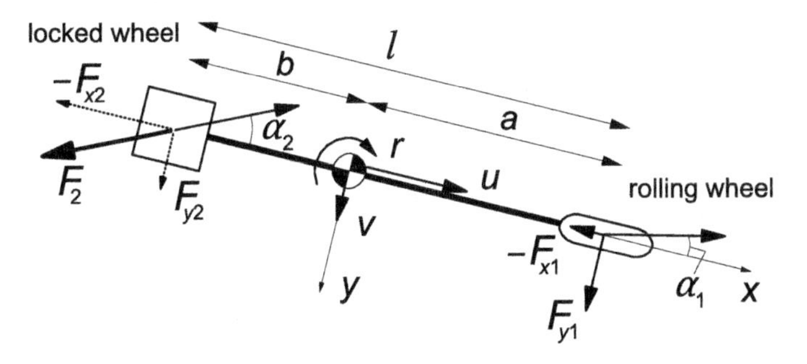


**FIGURE 3.40** 'Bicycle' model with one wheel locked.

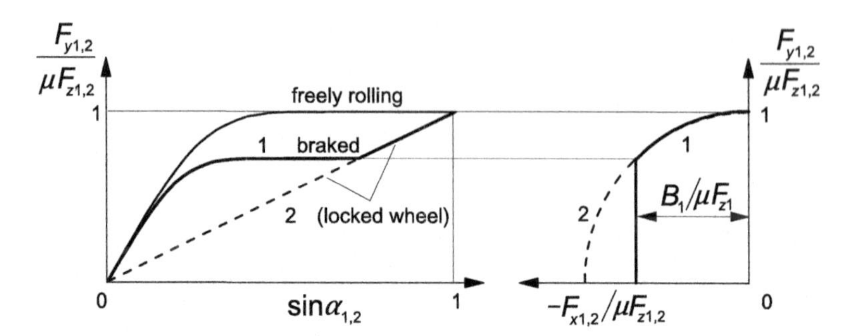


**FIGURE 3.41** Tire side force characteristics for freely rolling and braked wheels.

will be used for our analysis. In Figure 3.41, the characteristics have been reproduced: one curve for the freely rolling wheel, another one for the wheel that is being braked with a braking effort corresponding to a brake force $B_1 = -F_{x1}$ at straight ahead rolling ($\alpha$ small) and finally for a wheel that is locked completely. The second curve shows that at higher slip angles, a lower horizontal plateau is followed until the straight inclined line is reached where also this wheel gets locked (note that sin $\alpha$ has been used as abscissa which makes the relationship linear). A full discussion of this behavior has been given in Subsection 3.2.3 in connection with the treatment of the tire brush model.

As shown in Figure 3.40, we assume that the wheel on axle 2 is locked and that on axle 1 may be braked. The characteristics of Figure 3.41 apply. First, the situation near the undisturbed straight ahead motion will be studied. Linearization with both slip angles assumed small yields for the horizontal wheel forces:

$$\begin{aligned} F_{x1} &= -B_1 \\ F_{y1} &= F_{y1}(\alpha_1, B_1) \;\rightarrow\; C_1(B_1)\alpha_1 \\ F_{x2} &= -\mu F_{z2}\cos\alpha_2 \;\rightarrow\; -\mu F_{z2} \\ F_{y2} &= \mu F_{z2}\sin\alpha_2 \;\rightarrow\; \mu F_{z2}\alpha_2 = C_2\alpha_2 \end{aligned} \tag{3.121}$$

For the locked wheel, we have now an effective side-slip stiffness:

$$C_2 = \mu F_{z2} = \mu\frac{a}{l}mg \tag{3.122}$$

As a result, Eqns (3.118–3.120) reduce to

$$\begin{aligned} m\dot{u} &= -B_1 - \mu F_{z2} \\ m\dot{v} + \frac{1}{|u|}(C_1 + C_2)v &+ \left\{mu + \frac{1}{|u|}(aC_1 - bC_2)\right\}r = 0 \\ mk^2\dot{r} + \frac{1}{|u|}(a^2C_1 + b^2C_2)\,r &+ \frac{1}{|u|}(aC_1 - bC_2)v = 0 \end{aligned} \tag{3.123}$$

Furthermore, we have introduced the absolute value of the forward speed $u$ in the denominators to allow for the consideration of negative values of $u$, while at the same time the sign of $\alpha$ remains unchanged (cf. the corresponding equations (1.46) for the nondriven or braked vehicle). Negative values of $u$ correspond to the case of locked front wheels.

Elimination of the lateral velocity $v$ from Eqn (3.123) yields

$$\begin{aligned} m^2k^2u^2\ddot{r} + m\{-k^2B_1 + (a^2+k^2)C_1 + b^2C_2\}|u|\dot{r} \\ + \{l^2C_1C_2 - mu|u|(aC_1 - bC_2)\}r = 0 \end{aligned} \tag{3.124}$$

When compared with Eqn (1.50), considering (1.48), it is noted that in the second coefficient of (3.124), the term $k^2C_2$ has disappeared and that $-k^2B_1$ has been added. This is due to the differentiation of $u$ in the elimination process of $v$. We had, originally in the second coefficient of (3.124), the term

$$mk^2\dot{u}\,\mathrm{sgn}(u) = -k^2\left(\mu\frac{a}{l}mg + B_1\right) = -k^2(C_2 + B_1) \tag{3.125}$$

which explains the changes observed. In other respects, Eqn (3.124) is similar to the homogeneous version of Eqn (1.50). However, an important difference appears in the coefficients which are now dependent on the time because of the presence of the linearly with time decreasing speed $u$. We have the additional equation for $u$ (by integration of the first of Eqn (3.123)):

$$u = u_o - \frac{1}{m}\left(\mu\frac{a}{l}mg + B_1\right)\mathrm{sgn}(u)\cdot t \tag{3.126}$$

The exact solution of Eqn (3.124) can be found because of the fortunate fact that the equation can be reduced to the differential equation of Bessel, the solution of which is known in tabular form. Before giving an example of such a complete solution, we will analyze the motion just after the application of a slight disturbance which allows us to approximate the equation to one with constant parameters in which $u$ is replaced by its initial value $u_o$. For this substitutive equation, the solution can be found easily. We obtain

$$r = D_1e^{\lambda_1 t} + D_2e^{\lambda_2 t} \tag{3.127}$$

with $D_{1,2}$ denoting the constants of integration (governed by the initial values of the state variables $v$ and $r$) and $\lambda_{1,2}$ representing the eigenvalues, that is, the roots of the characteristic equation of the substitutive differential equation:

$$A\lambda^2 + B\lambda + (C - u_o|u_o|D) = 0 \tag{3.128}$$

Comparison with the coefficients of Eqn (3.124) reveals the expressions for the quantities $A$, $B$, $C$, and $D$. Since, apparently, the first three are always positive and in the case considered $C_2$ is much smaller than $C_1$ which makes $D$ also a positive quantity, we expect that the substitutive system can only become unstable because of a possibly negative third coefficient. Then, one of the roots (lets say $\lambda_1$) becomes positive (while remaining real). Obviously, this can only occur when $u$ is positive which means: when the wheels of the rear axle are locked. Locked front wheels which occur when in our system description $u < 0$ will not destabilize the system but makes the vehicle unsteerable. This is because of the fact that changing the steer angle of a locked wheel cannot effect the orientation of its frictional force vector. The car with locked rear wheels corresponds in behavior with the case of excessive oversteer while locked front wheels would give rise to extreme understeer. The critical speed for the case of locked rear wheels is derived by making the third coefficient of Eqn (3.128), or of Eqn (3.124), equal to zero. We find, by using (3.122),

$$u_{\text{crit}} = \sqrt{\frac{\mu g l}{1 - \dfrac{\mu F_{z1}}{C_1}}} \tag{3.129}$$

From this expression, it can be seen that when a rigid front tire is considered so that $C_1 \to \infty$, we obtain the simple form: $u_{\text{crit}} = \sqrt{(\mu g l)}$. The influence of an elastic front tire is not very great as the front cornering stiffness may range from 6 to 30 times the tire normal load. Considering a dry road with $\mu = 1$, we find a critical speed of about 20 km/h.

In view of this very low speed above which the vehicle with rear wheels locked becomes unstable and the fact that normally the speed will already be a lot larger than the critical speed when the brakes are being applied, it is more useful to consider the degree of instability as a function of speed. We will adopt the root $\lambda_1$ to represent the degree of instability. Figure 3.42 illustrates the

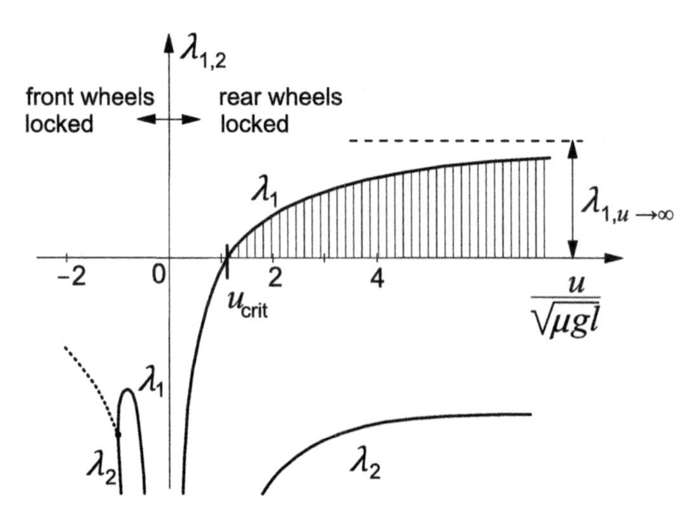


**FIGURE 3.42** Real (part of the) eigenvalues of the substitutive linear system with constant coefficients.

manner in which the roots vary with speed of travel. The diagram includes the case of front wheels locked ($u < 0$), where obviously the roots remain negative and become complex at higher speeds of travel (real part: dotted curve). As a practical indication for the degree of instability, we may employ the height of the horizontal asymptote to which $\lambda_1$ tends when $u \rightarrow \infty$. For that we find

$$\lambda_1\big|_{u\rightarrow\infty} = \sqrt{\frac{g}{l}\frac{ab}{k^2}\left(\frac{C_1}{F_{z1}} - \mu\right)} \tag{3.130}$$

which shows that instability decreases when the front tire cornering stiffness is lower and becomes more in balance with the here very low rear 'cornering stiffness' (3.122). It also becomes clear that when the front wheels are locked as well ($C_1 = \mu F_{z1}$), the degree of instability reduces to zero representing indifferent stability.

The complete solution of Eqn (3.124) can be found by applying the Lommel transformation of the Bessel differential equation and using the modified functions of Bessel in its solution. These functions are available in tabulated form and can be found in the book by Abramowitz and Stegun (1965, cf. Eqn 9.1.53 for the transformed equation and p. 377 for the solution).

In the paper of Koiter and Pacejka (1969), the exact solution has been given together with examples of the numerically computed solution for the complete nonlinear system. For the special case with parameters $a = b = k$, $B_1 = 0$, $C_1/\mu F_z = 6$, the solution of the linear Eqn (3.124) is presented in Figure 3.43. The exact solution is compared with the approximate one (3.127) of the substitutive equation. The solution in the form of a stable and an unstable

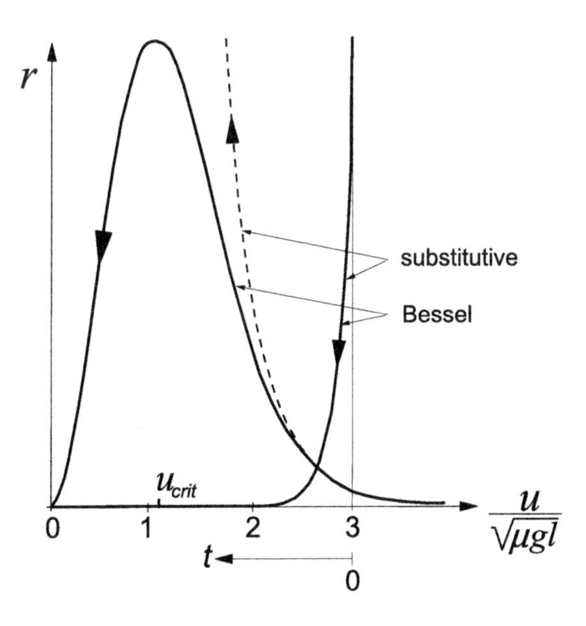


**FIGURE 3.43** Solutions of the exact and the approximated linear systems representing the variation of the yaw rate $r$ with forward velocity $u$ or with time $t$.

branch has been drawn in the $(u, r)$ plane constituting the projection of the complete three-dimensional trajectory in the $(u, v, r)$ space. Along the $u$ axis, we may introduce the time axis considering the relationship (3.126). In the case considered the nondimensional eigenvalues $\lambda_{1,2}\sqrt{(l/\mu g)}$ take the values 1.26 and $-3.43$ respectively. For the sake of comparison, the initial values have been taken the same for both pairs of solutions. No visible difference can be detected between the two decaying stable branches. The 'unstable' branch of the exact solution, however, shows for greater values of time an increasing difference with the corresponding approximate exponential solution. At the start of the motion where the rear axle tends to break away the agreement is very good. It can be shown that when expanded in a series of powers of the time $t$, the solutions are identical up to and including the third term (with $t^2$). This supports our choice of defining the degree of instability.

Along the vertical axis, the yaw rate is plotted. After having attained a maximum, the yaw angle develops at a lesser rate until the motion reaches a complete stop. The area underneath the curves is proportional to the final angle of swing. This finite angle will linearly depend on the initial values of the variables $r$ and $v$. Consequently, the final deviation with respect to the original rectilinear path can be kept within any chosen limit and, strictly speaking, the actual system is always stable.

Finally, the solutions for the nonlinear system governed by the Eqns (3.118, 3.119) and tire characteristics according to Figure 3.41 have been established by numerical integration of the equations of motion. Now, the character of the motion may change with the level of the initial disturbance. In Figures 3.44 and 3.45, the resulting motions for two cases have been depicted: the first without braking the front wheels (Figure 3.44) and the second with also the front wheels being braked but at a lower effort to make sure that they roll at least initially. The initial speed of travel and disturbances has been kept the same for all cases but the coefficient of friction $\mu$ has been varied. At lower friction, the time

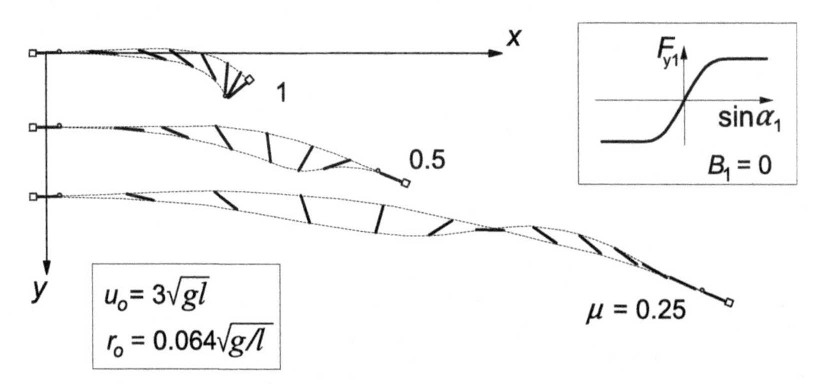


**FIGURE 3.44** Development of the vehicle motion after the rear wheels get locked while the front wheels remain rolling freely.

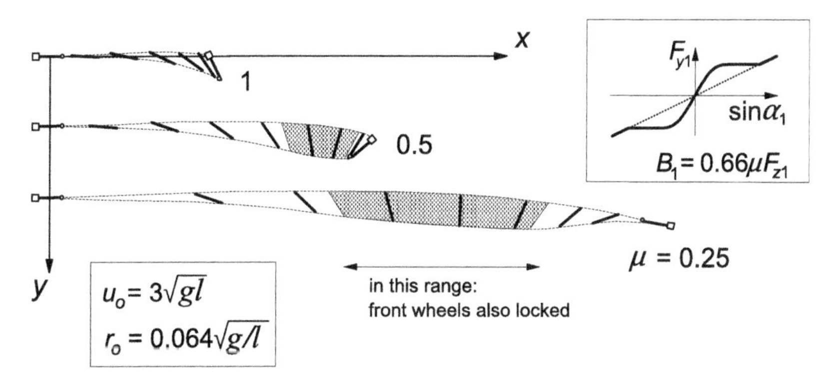


**FIGURE 3.45** Development of the vehicle motion after the rear wheels get locked while also the front wheels are being braked.

available to develop the angle of swing is larger and we see that an angle of more than 180 degrees may be reached. Then the locked wheels are moving at the front ($u < 0$) and we have seen that situation is stable while the motion can become oscillatory (cf. Figure 3.42). For the case of Figure 3.44 with $\mu = 0.25$, we indeed observe that a sign change occurs once for the yaw rate $r$. When the brakes are applied also at the front axle, the deceleration is larger and we have less time to come to a stop. Consequently, the final angles of Figure 3.45 are smaller than those reached in Figure 3.44. The shaded zones indicate the ranges where also the front wheels get locked. This obviously occurs when the slip angle of the initially rolling wheels 1 becomes sufficiently large and exceeds the value indicated in Figure 3.41 where the inclined straight line is reached. It has been found (not surprisingly) that the influence of raising the initial speed is qualitatively the same as the effect of reducing the coefficient of friction.

Chapter 4

# Semi-Empirical Tire Models

**Chapter Outline**



**Tire and Vehicle Dynamics. DOI: 10.1016/B978-0-08-097016-5.00003-6**

## 4.1. INTRODUCTION

In the preceding chapter the theory of the tire force and moment generating properties have been dealt with based on physical tire models. The present chapter treats models that have been specifically designed to represent the tire as a vehicle component in a vehicle simulation environment. The modeling approach is termed 'semi-empirical' because the models are based on measured data but may contain structures that find their origin in physical models like those treated in the preceding chapter. The mathematical descriptions are restricted to steady-state situations. The non-steady-state behavior will be discussed in subsequent chapters.

In the past, several types of mathematical functions have been used to describe the cornering force characteristic. Exponential, arctangent, parabolic (up to its maximum), and hyperbolic tangent functions (difference of two) have been tried with more and less success. Often, only very crude approximations could be achieved. To improve the accuracy, tables of measured data points have been used together with interpolation schemes. Also, higher-order polynomials were popular but proved not to be always suitable in terms of accuracy and the very large deviations that occur outside the ranges of slip covered by the original measurement data used in the fitting process. Mathematical representations of longitudinal force and aligning torque came later, and only relatively recently the combined slip condition was included in the empirical description. The longitudinal slip ratio was introduced as an input variable instead of the braking or driving force which was common practice in the early days of vehicle dynamics analysis. This latter method, however, is still in use for certain applications.

In the following sections of this chapter, first, a relatively simple approach will be discussed that is based on the similarity concept and after that, in the remainder of the chapter, a detailed description will be given of the *Magic Formula* tire model. The two model approaches belong to the second and first categories of Figure 2.11, respectively.

## 4.2. THE SIMILARITY METHOD

The method to be discussed in this section is based on the observation that the pure slip curves remain approximately similar in shape when the tire runs at conditions that are different from the reference condition. The reference condition is defined here as the state where the tire runs at its rated (nominal) load ($F_{zo}$), at camber equal to zero ($\gamma = 0$), at either free rolling ($\kappa = 0$) or at side slip equal to zero ($\alpha = 0$) and on a given road surface ($\mu_o$). A similar shape means that the characteristic that belongs to the reference condition is regained by vertical and horizontal multiplications and shifting of the curve. The similarity method is based on the normalization theory of Fiala (1954) and has been introduced by Pacejka (1958), also cf. Radt and Pacejka (1963) and Pacejka (1971, 1981). A demonstration that, in practice, similarity indeed approximately occurs is given

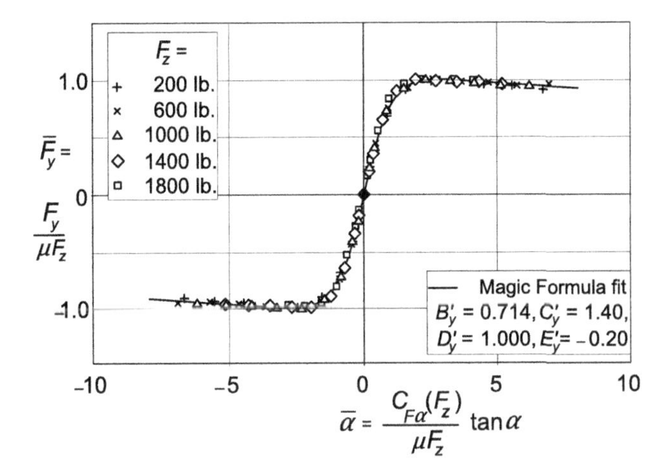


**FIGURE 4.1** Result of Radt's nondimensionalization of tire characteristics showing that the tire side force characteristics measured at different loads reduce to virtually the same curve when the force and the slip angle are normalized as indicated.

by Radt and Milliken (1983), also see Milliken and Milliken (1995). Figures 4.1 and 4.2 present the results when the force and moment as well as the slip angle have been normalized, resulting in the nondimensional quantities shown along the axes. The raw data have been processed to make the characteristics pass through the origin of the graph. The curve results from a Magic Formula fit. The parameters $B'$, $C'$, $D'$, and $E'$ for the nondimensional side force (subscript $y$) and for the nondimensional moment (subscript $z$) have been used in the

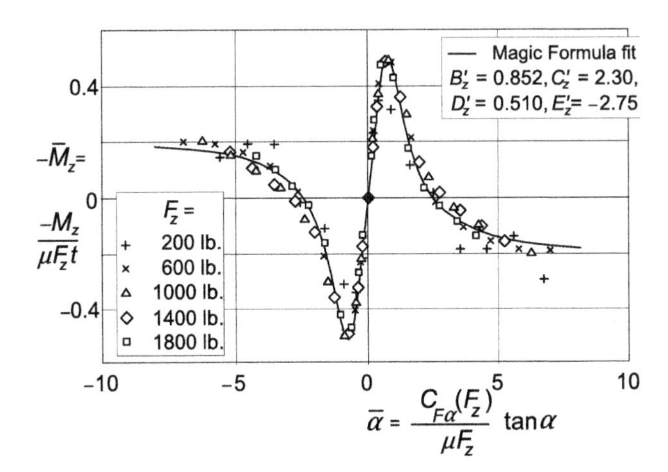


**FIGURE 4.2** Result of Radt's nondimensionalization of the tire aligning torque characteristic after normalizing the moment and the slip angle as indicated; $t$ represents the pneumatic trail at vanishing slip angle.

nondimensional version of the Magic Formula given in Eqn (1.6) and similar for the moment. Further on, the formula will be introduced again, Eqns (4.6, 4.10).

The resulting model is, through its simplicity, relatively fast. It is capable to represent pure slip conditions rather well including the influence of a camber angle. The description of the situation at combined lateral and longitudinal slip is qualitatively satisfactory. Quantitatively, however, deviations may occur at higher levels of the combined two slip values.

### 4.2.1. Pure Slip Conditions

The functions representing the reference curves which are found at pure slip conditions are designated with the subscript o. We have, for instance, the reference function $F_y = F_{yo}(\alpha)$ that represents the side force vs slip angle relationship at nominal load $F_{zo}$, with longitudinal slip and camber equal to zero and the friction level represented by $\mu_o$. We may now try to change the condition to a situation at a different wheel load $F_z$. Two basic changes will occur with the characteristic: (1) a change in level of the curve where saturation of the side force takes place (peak level) and (2) a change in slope at vanishing side slip ($\alpha = 0$). The first modification can be created by multiplying the characteristic both in vertical and horizontal direction with the ratio $F_z/F_{zo}$. The horizontal multiplication is needed to not disturb the original slope. We obtain, for the new function,

$$F_y = \frac{F_z}{F_{zo}} F_{yo}(\alpha_{eq}) \tag{4.1}$$

with the equivalent slip angle:

$$\alpha_{eq} = \frac{F_{zo}}{F_z} \alpha \tag{4.2}$$

Because, obviously, the derivative of $F_{yo}$ with respect to its argument $\alpha_{eq}$ at $\alpha_{eq} = 0$ is equal to the original cornering stiffness $C_{F\alpha o}$, we find, for the derivative of $F_y$ with respect to $\alpha$ the same value for the slope at $\alpha = 0$,

$$\frac{\partial F_y}{\partial \alpha} = \frac{F_z}{F_{zo}} \frac{\mathrm{d}F_{yo}}{\mathrm{d}\alpha_{eq}} \frac{\partial \alpha_{eq}}{\partial \alpha} = \frac{\mathrm{d}F_{yo}}{\mathrm{d}\alpha_{eq}} = C_{F\alpha o} \tag{4.3}$$

which proves that the slope at the origin of the characteristic is not affected by the successive multiplications. The second step in the manipulation of the original curve is the adaptation of the slope. This is accomplished by a horizontal multiplication of the newly obtained characteristic. This is done by multiplying the argument with the ratio of the new and the original cornering stiffness. Consequently, the new argument reads

$$\alpha_{eq} = \frac{C_{F\alpha}(F_z)}{C_{F\alpha o}} \frac{F_{zo}}{F_z} \alpha \tag{4.4}$$

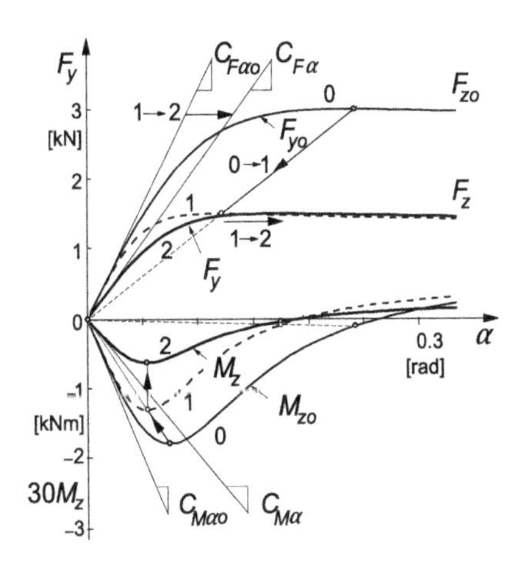


**FIGURE 4.3** Using the similarity method to adapt $F_y$ and $M_z$ curves to new load level.

Together with Eqn (4.1) we have the new formulation for the side force vs slip angle relationship at the new load $F_z$. In Figure 4.3 the two steps taken to obtain the new curve have been illustrated. Here, the nominal load $F_{zo} = 3000$ N and the new load $F_z = 1500$ N.

In the same figure the characteristic of the aligning torque has been adapted to the new condition. For this, we use the knowledge gained when working with the brush model (cf. Figure 3.4). More specifically, we will obey the rule that according to the theory, obviously, the point where the $M_z$ curve reaches the $\alpha$ axis lies below the peak of the $F_y$ curve. In Figure 3.4 this occurs at tan $\alpha = 1/\theta_y$. This requirement means that the same equivalent slip angle (4.4) must be used as for the argument of the $F_y$ function (4.1). In addition, we will use information on the new value of the aligning stiffness $C_{M\alpha}$. The reference curve for $M_{z0}$ that is used is more realistic than the theoretical curve of Figure 3.4. Typically, the moment changes its sign in the larger slip angle range where $F_y$ reaches its peak.

With the same equivalent slip angle and the new aligning stiffness, we obtain the expression for the new value of the aligning torque according to the similarity concept:

$$M_z = \frac{F_z}{F_{zo}} \frac{C_{M\alpha}(F_z)}{C_{M\alpha o}} \frac{C_{F\alpha o}}{C_{F\alpha}(F_z)} M_{zo}(\alpha_{eq}) \tag{4.5}$$

The first and third factors being the inverse of the multiplication factor used in (4.4) are needed to maintain the original slope. The second factor multiplies the intermediate $M_z$ curve (1 in Figure 4.3) in vertical direction to adapt the slope to its new value (2) while not disturbing the $\alpha$ scale. It may be noted that the combined second and third factor equals the ratio of the new and the original values of the pneumatic trail $t/t_o$.

For the calculations connected with Figure 4.3, the reference characteristics for $F_{yo}$, $M_{zo}$, and $C_{F\alpha}$ have been described by means of the *Magic Formula* type

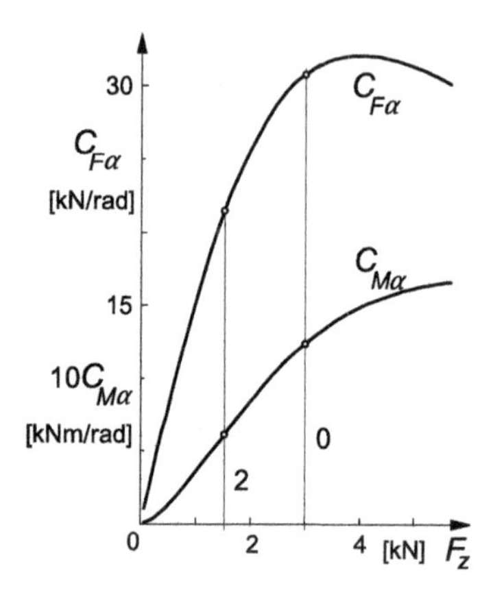


**FIGURE 4.4** Cornering and aligning stiffness vs wheel load.

functions (cf. Eqn (1.6)). For this occasion, the aligning stiffness is modeled as the product of a certain fraction of the contact length $2a$ and the cornering stiffness. The resulting characteristics for these four quantities as shown in Figures 4.3 and 4.4 are realistic. The following formulas have been used:

Side force at nominal load $F_{zo}$

$$F_{yo} = D_{yo} \sin\left[C_y \arctan\left\{B_{yo}\alpha - E_y\left(B_{yo}\alpha - \arctan(B_{yo}\alpha)\right)\right\}\right] \tag{4.6}$$

with stiffness factor

$$B_{yo} = \frac{C_{F\alpha o}}{C_y D_{yo}} \tag{4.7}$$

peak factor for the side force which in general is different from the one for the longitudinal force; consequently we introduce $\mu_{yo}$ besides $\mu_{xo}$:

$$D_{yo} = \mu_{yo} F_{zo} \tag{4.8}$$

and cornering stiffness as function of wheel load $F_z$:

$$C_{F\alpha} = c_1 c_2 F_{zo} \sin\left\{2 \arctan\left(\frac{F_z}{c_2 F_{zo}}\right)\right\} \tag{4.9}$$

Aligning torque at nominal wheel load:

$$M_{zo} = D_{zo} \sin[C_z \arctan\{B_{zo}\alpha - E_z(B_{zo}\alpha - \arctan(B_{zo}\alpha))\}] \tag{4.10}$$

with stiffness factor

$$B_{zo} = -\frac{C_{M\alpha o}}{C_z D_{zo}} \tag{4.11}$$

peak factor ($a_o$ representing half the contact length at nominal load)

$$D_{zo} = c_3 a_o D_{yo} \tag{4.12}$$

and aligning stiffness as function of wheel load $F_z$

$$C_{M\alpha} = tC_{F\alpha} = c_4 a\, C_{F\alpha} \tag{4.13}$$

where apparently the pneumatic trail $t = c_4 a$. Half the contact length $a$ is assumed to be proportional with the square root of the wheel load:

$$a = a_o \sqrt{F_z / F_{zo}} \tag{4.14}$$

The resulting approximation of the pneumatic trail turns out to be quite adequate and might be considered to be used in the set of Magic Formula tire model Eqn (4.E42) to be dealt with later on.

The values of parameters used for the calculations to be conducted have been listed in Table 4.1.

The next two items we will deal with are a change in friction coefficient from $\mu_{yo}$ to $\mu_y$ and the introduction of a camber angle $\gamma$. The first change can be handled by multiplying the curves in radial direction with factor $\mu_y/\mu_{yo}$. Together with the change in load, we find

$$F_y = \frac{\mu_y F_z}{\mu_{yo} F_{zo}} F_{yo}(\alpha_{eq}) \tag{4.15}$$

with

$$\alpha_{eq} = \frac{C_{F\alpha}(F_z)}{C_{F\alpha o}} \frac{\mu_{yo} F_{zo}}{\mu_y F_z} \alpha \tag{4.16}$$

The friction coefficient may be assumed to depend on the slip velocity $V_s$. To model the situation on wet roads, a decaying function $\mu(V_s)$ may be employed, cf. e.g., Eqn (4.E23).

When considering the computed characteristics of Figure 3.35c (2$^{nd}$ row) for small camber angles, a horizontal shift of the $F_y$ curve may give a reasonable result. Then, the peak side force remains unchanged which is a reasonable assumption but not always supported by experimental evidence (where often, but not always, a slight increase in maximum side force is manifested) or by computations with a physical model with constant friction and finite contact

**TABLE 4.1** Parameter Values used in Section 4.2

| | | | | | | | | | | | |
|---|---|---|---|---|---|---|---|---|---|---|---|
| $F_{zo}$ | 3000 N | $C_y$ | 1.3 | $C_x$ | 1.5 | $c_1$ | 8 | $c_5$ | 1 | $c_9$ | 0.3 |
| $a_o$ | 0.08 m | $E_y$ | −1 | $E_x$ | −1 | $c_2$ | 1.33 | $c_6$ | 0.3 | $c_{10}$ | 0 |
| $b$ | 0.07 m | $C_z$ | 2.3 | $\mu_{yo}$ | 1 | $c_3$ | 0.25 | $c_7$ | 100 | $c_{11}$ | 4 |
| $r_e$ | 0.30 m | $E_z$ | −2 | $\mu_{xo}$ | 1.26 | $c_4$ | 0.5 | $c_8$ | 15 | | |

width which show a slight decrease. For small angles, the camber thrust is approximated by the product of the camber stiffness and the camber angle: $F_{y\gamma} = C_{F\gamma}\gamma$. Consequently, the $\alpha$ shift should amount to

$$S_{Hy} = \frac{C_{F\gamma}(F_z)}{C_{F\alpha}(F_z)}\gamma \tag{4.17}$$

so that

$$\alpha^* = \alpha + S_{Hy} \tag{4.18}$$

which gives rise to $\alpha_{eq}$ (4.16) with $\alpha$ replaced by $\alpha^* = \alpha + S_{Hy}$.

For the representation of the aligning moment at the new conditions, the situation is more complex. We observe from Figure 3.35c that for not too large slip angles, the original curve $M_{zo}$ tends to shift both sideways and upward. If the same equivalent slip angle and thus the same horizontal shift (4.17) as for the force is employed, the moment would become zero where also the force curve crosses the $\alpha$-axis. A moment equal to $-C_{M\alpha}\, S_{Hy}$ would arise at $\alpha = 0$. However, the moment should become equal to $C_{M\gamma}\gamma$, with a positive camber moment stiffness $C_{M\gamma}$. This implies that an additional vertical shift required is equal to

$$S_{Vz} = C_{M\gamma}(F_z)\gamma + C_{M\alpha}(F_z)S_{Hy} \tag{4.19}$$

This additional moment corresponds to the so-called residual torque $M_{zr}$, which is the moment that remains when the side force becomes equal to zero. For larger values of the slip angle, the additional moment should tend to zero as can be observed in Figure 3.35c. This may be easily accomplished by dividing (4.19) by a term like $1 + c_7\,\alpha^2$, cf. Eqn (4.26).

The dependencies of the camber stiffnesses on the vertical load may be assumed linear. We have

$$C_{F\gamma} = c_5 F_z, \quad C_{M\gamma} = c_6 \frac{b^2}{r_e} C_{F\kappa} \tag{4.20a}$$

where $b$ denotes half the (assumedly constant) width of the contact patch, $r_e$ the effective rolling radius, and $C_{F\kappa}$ the longitudinal slip stiffness which is considered to be linearly increasing with load:

$$C_{F\kappa} = c_8 F_z \tag{4.20b}$$

In Figure 4.5, the different steps leading to the condition with a (small) camber angle have been demonstrated. First, the original curve (0) is moved to the left over a distance equal to the horizontal shift (4.17) yielding the intermediate curve (1). Then the residual torque is added: vertical shift (4.19), possibly made a function of slip angle by multiplying the shift with a decaying weighting function as suggested above. The load $F_z$ and the friction coefficient $\mu_y$ have been kept equal to their reference values: 3000 N and 1, respectively.

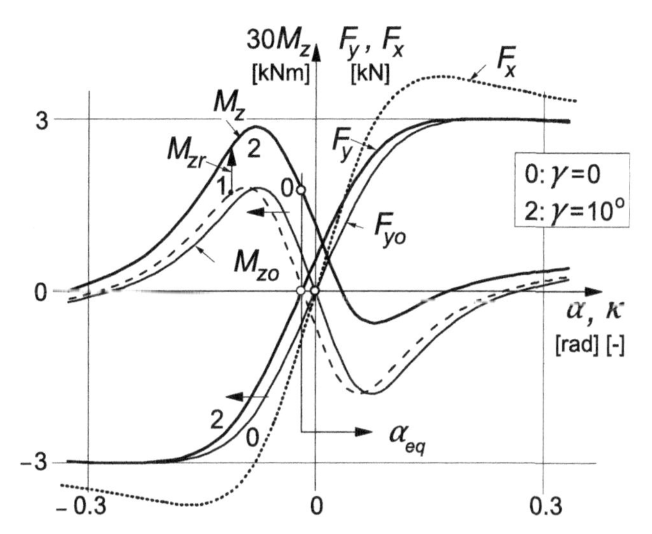


**FIGURE 4.5** Introduction of a camber angle in the side force and the aligning torque vs side slip characteristics according to the similarity method. The longitudinal force vs longitudinal slip characteristic (not affected by camber). ($F_z = F_{zo} = 3000$ N, $\mu_y = \mu_{yo} = 1$, $\mu_x = \mu_{xo} = 1.26$).

The longitudinal force at pure longitudinal slip may be modeled in the same way as we did for the side force. Equations similar to (4.6–4.8) hold for the reference function. An influence of camber on the longitudinal force, however, is assumed not to occur. In the same Figure 4.5, the characteristic for $F_x$ vs $\kappa$ has been drawn. The values of the relevant additional parameters for the longitudinal force and the moment have been listed in Table 4.1.

We finally have the following similarity formulas for pure slip conditions (either at longitudinal or at lateral slip) for the longitudinal force:

$$F_x = \frac{\mu_x F_z}{\mu_{xo} F_{zo}} F_{xo}(\kappa_{eq}) \tag{4.21}$$

with the equivalent longitudinal slip

$$\kappa_{eq} = \frac{C_{F\kappa}(F_z)}{C_{F\kappa o}} \frac{\mu_{xo} F_{zo}}{\mu_x F_z} \kappa \tag{4.22}$$

and, for the side force,

$$F_y = \frac{\mu_y F_z}{\mu_{yo} F_{zo}} F_{yo}(\alpha_{eq}) \tag{4.23}$$

with the equivalent slip angle, containing the horizontal shift (4.17):

$$\alpha_{eq} = \frac{C_{F\alpha}(F_z)}{C_{F\alpha o}} \frac{\mu_{yo} F_{zo}}{\mu_y F_z} \left( \alpha + \frac{C_{F\gamma}(F_z)}{C_{F\alpha}(F_z)} \gamma \right) \tag{4.24}$$

and, for the aligning torque,

$$M_z = \frac{\mu_y F_z}{\mu_{yo} F_{zo}} \frac{C_{M\alpha}(F_z)}{C_{M\alpha o}} \frac{C_{F\alpha o}}{C_{F\alpha}(F_z)} M_{zo}(\alpha_{eq}) + M_{zr} \tag{4.25}$$

with the residual torque corresponding to the vertical shift (4.19) provided with the reduction factor:

$$M_{zr} = \frac{C_{M\gamma}(F_z) + t(F_z) C_{F\gamma}(F_z)}{1 + c_7 \alpha^2} \gamma \tag{4.26}$$

### 4.2.2. Combined Slip Conditions

We will now address the problem of describing the situation at combined slip. The analysis of the brush model has given considerable insight into the mechanisms that play a role and we will use the theoretical slip quantities $\sigma_{x,y}$ (Eqn 3.34 or 3.32) and the magnitude $\sigma$ (3.40) and we will adopt the concept similar to that of assessing the components of the resulting horizontal force according to (3.49) and consider the pneumatic trail $t$ as in (3.50) and include explicit contributions of $F_x$ to the moment as indicated in (3.51). We have, for the theoretical slip quantities with the $\alpha$ shift due to camber included,

$$\sigma_x = \frac{\kappa}{1 + \kappa} \tag{4.27a}$$

$$\sigma_y^* = \frac{\tan \alpha^*}{1 + \kappa} \tag{4.27b}$$

and

$$\sigma^* = \sqrt{\sigma_x^2 + \sigma_y^{*2}} \tag{4.28}$$

with

$$\alpha^* = \alpha + \frac{C_{F\gamma}(F_z)}{C_{F\alpha}(F_z)} \gamma \tag{4.29}$$

In the ensuing theory, we will make use of the theoretical slip quantities $\sigma_x$ and $\sigma_y^*$ (4.27a,b). By using these quantities, the slope in the constant $\alpha$ curves in the $F_y$ vs $F_x$ diagram at $F_x = 0$ does show up, like for the brush model in Figure 3.13. It should be noted, however, that experience shows that the use of practical slip quantities $s_x$ and $s_y^*$ (that is: $\kappa$ and $\tan\alpha^*$ which may be obtained again by omitting the denominators in (4.27a,b)) may already give very good results as will be indicated later on in Figure 4.8. Note that when approaching wheel lock $\sigma_x \to \infty$. This calls for an artificial limitation by adding a small positive quantity to the denominator of (4.27a).

Because we are dealing with an in general nonisotropic tire, we have pure longitudinal and lateral slip characteristics that are not identical (cf. Figure 4.5).

Still, we will take the lateral and longitudinal components but then of the respective pure slip characteristics $F_{yo}$ and $F_{xo}$. If we want to use the theoretical slip quantities $\sigma$, we must have available the pure slip characteristics with $\sigma_x$ and $\sigma_y$ as abscissa. This can be accomplished by fitting the original data after having computed the values of $\sigma_x$ for each value of $\kappa$ by using (4.27a) and similarly for transforming $\alpha$ into tan $\alpha$ and denoting the resulting functions: $F_{xo}(\sigma_x)$, $F_{yo}(\sigma_y)$, and $M_{zo}(\sigma_y)$. Or, simpler, by replacing in the already available pure slip force and moment functions, expressed in terms of the practical slip quantities, the argument $\alpha$ (in radians) by $\tan\alpha$ (probably acceptable approximation) and $\kappa$ by the expression given below (derived from Eqn (4.27a)),

$$\kappa = \frac{\sigma_x}{1 - \sigma_x} \tag{4.30}$$

which would be successful if the $F_x$ characteristic had been fitted for the whole range of positive (driving) and negative (braking) values of $\kappa$. Otherwise if only the braking side is available ($\kappa$ and $\sigma_x < 0$) and the driving side is modeled as the mirror image of the braking side, one might better take

$$\kappa = \frac{\sigma_x}{1 + |\sigma_x|} \tag{4.30a}$$

If the thus obtained function of $F_x$ is plotted vs $\sigma_x$, the resulting curve becomes symmetric. However, if then the abscissa is converted into $\kappa$ by using (4.30), the resulting $F_x(\kappa)$ curve turns out to become asymmetric with the braking side identical to the characteristic we started out with. This asymmetry was already found to occur with the brush model discussed in Chapter 3.

In the combined slip model that we employ, the longitudinal and lateral force components are obtained according to an adapted Eqn (3.49) where for the side force $F(\sigma)$ is replaced by $F_{yo}(\sigma)$ and, for the fore-and-aft force, $F(\sigma)$ is replaced by $F_{xo}(\sigma)$. Figure 4.6 illustrates the procedure. The figure shows how the force vector arises from the individual pure slip characteristics. At small slip, due to slip stiffness differences, the vector does not run opposite with respect to the slip speed vector $\boldsymbol{V}_s$. At wheel lock, however, the force vector does run opposite with respect to the slip speed vector because it is assumed here that the levels of the individual characteristics (asymptotes) are the same for the slip approaching infinity (in the *Magic Formula,* we should have $\mu_{yo}\sin(\frac{1}{2}\pi C_y) = \mu_{xo}\sin(\frac{1}{2}\pi C_x)$).

Next, we should realize that through (4.18) camber has been accounted for and that, as a consequence, $\sigma_y$ is to be replaced by $\sigma_y^*$ defined by Eqn (4.27b). The resulting force and moment functions are denoted by $F_{xo}(\sigma_x)$, $F_{yo}\,(\sigma_y^*)$, and $M_{zo}\,(\sigma_y^*)$. The components now become, with $\sigma^*$ denoting the magnitude of the theoretical slip vector (4.28),

$$F_x = \frac{\sigma_x}{\sigma^*}F_{xo}(\sigma^*), \qquad F_y = \frac{\sigma_y^*}{\sigma^*}F_{yo}(\sigma^*) \tag{4.31}$$

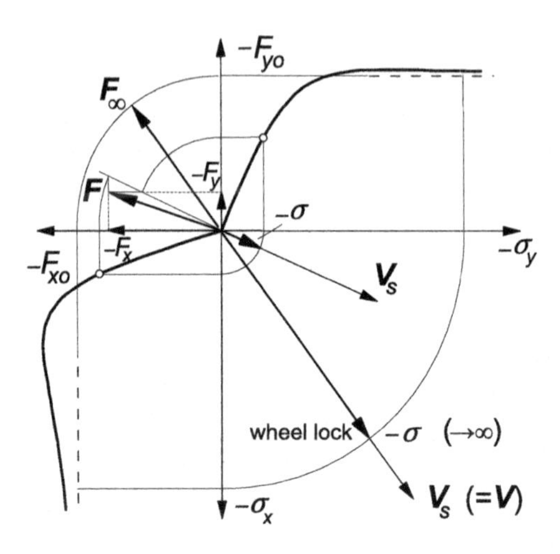


**FIGURE 4.6** Construction of the resulting combined slip force from the pure slip characteristics according to Eqn (4.31) (with camber not considered).

At large slip, the model apparently fails to properly account for the contribution of the camber angle. At wheel lock, one would expect the side force to become zero at vanishing slip angle. This will not exactly be the case in the model due to the equivalent theoretical side slip defined according to Eqn (4.27b) with $\alpha^*$ given by (4.18).

For the aligning torque, we may define

$$M_z = M_z' + M_{zr} - c_9 a_o \frac{F_x F_y}{F_{zo}} - c_{10} a_o F_x - c_{11} b\, \gamma F_x \tag{4.32}$$

with the first term directly attributable to the side force. This term can be written as

$$M_z' = -t(\sigma^*) \cdot F_y \tag{4.33}$$

By considering (4.31), it turns out that (4.33) becomes

$$M_z' = \frac{\sigma_y^*}{\sigma^*} M_{zo}'(\sigma^*) \tag{4.34}$$

The last three terms of Eqn (4.32) result from the moment exerted by the longitudinal force because of the moment arm that arises through the side force-induced deflection, a possibly initial offset of the line of action and the sideways rolling of the tire cross section due to camber. For the sake of simplicity, these effects are assumed not to be influenced by the wheel load.

With the similarity expressions for the pure slip forces $F_{x,yo}$ and moment $M_{zo}'$, we finally obtain the following formulas for combined slip conditions. For the longitudinal force

$$F_x = \frac{\sigma_x}{\sigma^*} \frac{\mu_x F_z}{\mu_{xo} F_{zo}} F_{xo}(\sigma_{eq}^x) \tag{4.35}$$

with the equivalent slip

$$\sigma^x_{eq} = \frac{C_{F\kappa}(F_z)}{C_{F\kappa o}} \frac{\mu_{xo}F_{zo}}{\mu_x F_z} \sigma^* \tag{4.36}$$

and, for the side force,

$$F_y = \frac{\sigma^*_y}{\sigma^*} \frac{\mu_y F_z}{\mu_{yo}F_{zo}} F_{yo}(\sigma^y_{eq}) \tag{4.37}$$

with the equivalent slip

$$\sigma^y_{eq} = \frac{C_{F\alpha}(F_z)}{C_{F\alpha o}} \frac{\mu_{yo}F_{zo}}{\mu_y F_z} \sigma^* \tag{4.38}$$

and, for the aligning torque,

$$\begin{aligned} M_z = & \frac{\sigma^*_y}{\sigma^*} \frac{\mu_y F_z}{\mu_{yo}F_{zo}} \frac{C_{M\alpha}(F_z)}{C_{M\alpha o}} \frac{C_{F\alpha o}}{C_{F\alpha}(F_z)} M'_{zo}(\sigma^y_{eq}) + M_{zr} \\ & - c_9 a_o \frac{F_x F_y}{F_{zo}} - c_{10} a_o F_x - c_{11} b\, \gamma F_x \end{aligned} \tag{4.39}$$

with the residual torque

$$M_{zr} = \frac{C_{M\gamma}(F_z) + t(F_z) C_{F\gamma}(F_z)}{1 + c_7 (\sigma^y_{eq})^2} \gamma \tag{4.40}$$

On behalf of example calculations, the three additional nondimensional parameters $c_9$, $c_{10}$, and $c_{11}$ have been given values listed in Table 4.1.

For the case of a load different from the reference value and the presence of a camber angle, the combined slip characteristics for the tire side force and for the aligning torque have been assessed by using the above equations. The resulting diagrams with slip angle $\alpha$ used as parameter have been presented in Figures 4.7a and b. Due to the use of the theoretical slip quantities (4.27a, 4.27b) in the derivation, we find that the curves of $F_y$ vs $F_x$ show the slight slope near $F_x = 0$ also observed to occur with the physical brush model and sometimes more pronounced in experimentally assessed characteristics (Figure 3.18). The additional slope that arises when a camber angle is introduced (cf. Figure 3.35c, top) for which the torsional compliance is responsible does not appear. A special formulation would be needed to represent this effect. In the next section, we will take care of this in connection with the *Magic Formula* tire model.

It may be further remarked that the similarity model given by Eqns (4.35–4.40) together with (4.27a,b), (4.28) and the reference pure slip functions does represent the actual tire steady-state behavior well in the following extreme cases: (1) in pure slip situations, (2) in the linearized combined slip situation (small $\alpha$ and $\kappa$), and (3) in the case of a locked wheel (at $\gamma = 0$). As illustrated in

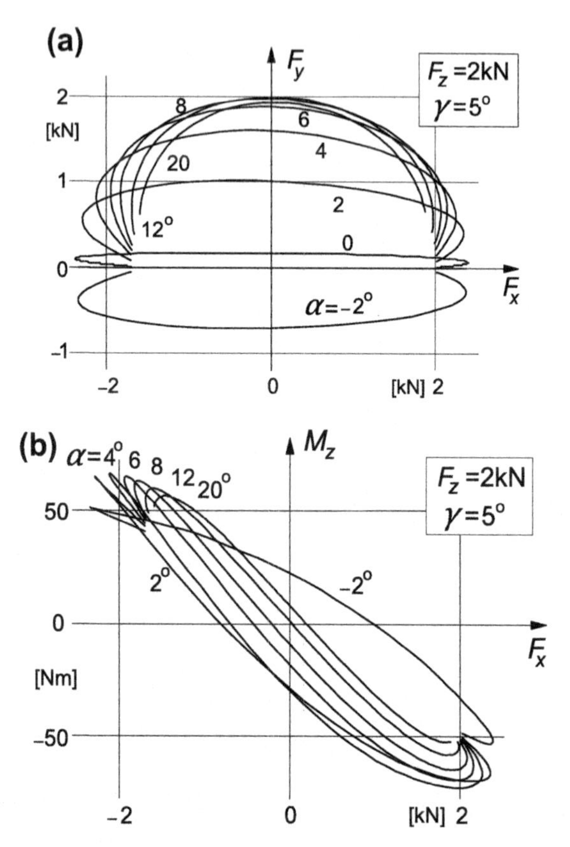


**FIGURE 4.7** (a) Cornering force vs longitudinal force at a new load and camber angle as computed with the similarity method. (b) Aligning torque vs longitudinal force at a new load and camber angle as computed with the similarity method.

Figure 4.6, the model correctly shows that when the wheel is locked, the resulting force acts in a direction opposite to the slip speed vector $\boldsymbol{V}_s$ if, in the original reference pure slip curves, the level of $F_{xo}$ and of $F_{yo}$ tends to the same level when both slip components approach infinity (governed by the parameters $\mu$ and $C$ in the *Magic Formula*). Other combinations of slip may give rise to deviations with respect to measured characteristics. It is noted that when the terms for the aligning torque with parameters $c_9$, $c_{10}$, and $c_{11}$ are disregarded, the combined slip performance can be represented without the necessity to rely on combined slip measurement data.

A comparison with measured data is given in Figure 4.8. Here the similarity method is used where instead of the theoretical slip quantities the practical slip variables $\kappa$ and tan $\alpha$ (that is: $s_x$ and $s_y$) have been employed also in the derivation phase (in Eqns (4.35–4.38)). A relatively good agreement is achieved for this passenger car tire in the combined slip situation.

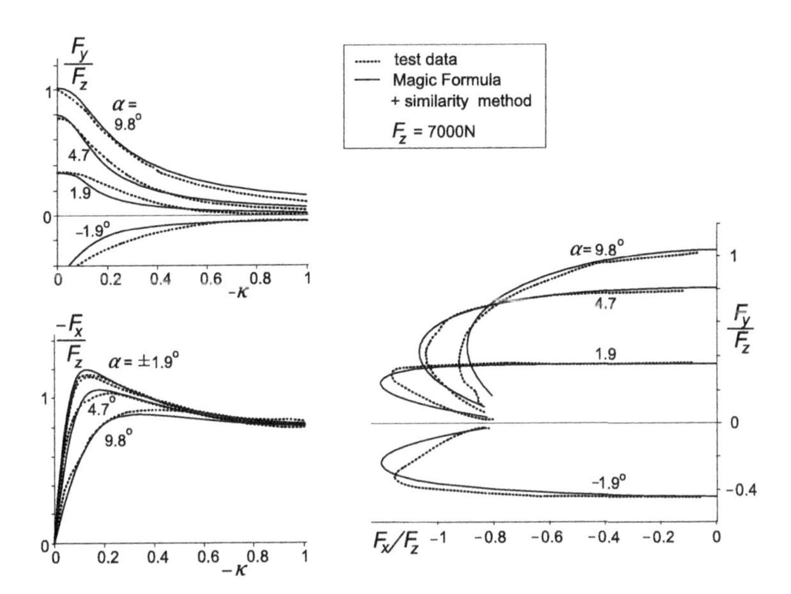


**FIGURE 4.8** Tire combined slip characteristics as obtained from measurements with the Delft tire test trailer compared with the curves calculated with the aid of the similarity method.

## 4.2.3. Combined Slip Conditions with $F_x$ as Input Variable

In simpler vehicle dynamics simulation studies with the wheel spin degree of freedom not included, one may prefer to use the longitudinal force $F_x$ as input quantity instead of the longitudinal slip $\kappa$. This approach has been used, almost exclusively, in early vehicle dynamics research. However, for (quasi) steady-state cornering analysis, notably circuit simulation in the racing world, this option is still popular. An important limitation of using this method is the condition that the $F_x$ vs $\kappa$ characteristic (although not used) is supposed to show a positive slope over the entire range of longitudinal slip while at wheel lock $F_x = -\mu F_z$, or, alternatively, we use only that part of the characteristic that lies in between the two peaks. This entails that the $F_y$ vs $F_x$ curves are single valued in the $F_x$ range employed. Furthermore, it is assumed that the friction coefficient is the same for longitudinal and lateral directions and is denoted by $\mu$.

The obvious main effect of the introduction of $F_x$ on the side force is the lowering of the maximum side force that can be generated. For this to realize, the right-hand expressions of Eqns (4.23) and (4.25) are to be multiplied with the factor

$$\varphi_x = \frac{\sqrt{\mu^2 F_z^2 - F_x^2}}{\mu F_z} \tag{4.41}$$

while expression (4.24) is to be divided by the same factor. The cornering and aligning stiffnesses remain unaffected through this operation. These stiffnesses, however, do depend on the longitudinal force, as becomes obvious from the curves of Figure 3.13 that belong to small side-slip angles. The following functions may serve as a crude approximation of the actual relationships:

$$C_{F\alpha}(\mu, F_z, F_x) = \varphi_{x\alpha}\left\{C_{F\alpha}(F_z) - \frac{1}{2}\mu F_z\right\} + \frac{1}{2}(\mu F_z - F_x) \tag{4.42}$$

with

$$\varphi_{x\alpha} = \left\{1 - \left(\frac{F_x}{\mu F_z}\right)^n\right\}^{1/n} \tag{4.42a}$$

Depending on the user's desire, one may choose for $n$ a value in the range of 2–8 (more or less curved characteristic $C_{F\alpha}(F_x)$). For the aligning stiffness, we may use the formula

$$C_{M\alpha}(\mu, F_z, F_x) = \varphi_x^2\, C_{M\alpha}(F_z) \tag{4.43}$$

and, similar for the camber stiffness,

$$C_{F\gamma}(\mu, F_z, F_x) = \varphi_x^2\, C_{F\gamma}(F_z) \tag{4.44}$$

The camber moment stiffness may be disregarded. The load dependencies $C_{F,M\alpha}(F_z)$ and $C_{F\gamma}(F_z)$ of the freely rolling tire already appeared in Eqns (4.24, 4.25). It may be ascertained that the present model makes sure that the cornering stiffness vanishes when $F_x \rightarrow \mu F_z$ (wheel drive spin, $\kappa \rightarrow \infty$) while at $F_x = -\mu F_z$ (wheel lock, $\kappa = -1$) the side-slip 'stiffness' equals $\mu F_z$, which is correct. The functional relationships (4.21, 4.22) for $F_x$ do not play a role anymore and the friction coefficient $\mu_y$ has been replaced by $\mu$. The imposed longitudinal force should be kept within the boundaries $-\mu F_z \cos\alpha \leq F_x \leq \mu F_z$.

The resulting equations for the side force and the aligning torque with the longitudinal force serving as one of the input quantities now read

For the side force,

$$F_y = \varphi_x \frac{\mu F_z}{\mu_o F_{zo}} F_{yo}(\alpha_{eq}) \tag{4.45}$$

with the equivalent slip angle

$$\alpha_{eq} = \frac{1}{\varphi_x} \frac{C_{F\alpha}(\mu, F_z, F_x)}{C_{F\alpha o}} \frac{\mu_o F_{zo}}{\mu F_z}\left(\alpha + \frac{C_{F\gamma}(\mu, F_z, F_x)}{C_{F\alpha}(\mu, F_z, F_x)}\gamma\right) \tag{4.46}$$

and, for the aligning torque,

$$M_z = \varphi_x \frac{\mu F_z}{\mu_o F_{zo}} \frac{C_{M\alpha}(\mu, F_z, F_x)}{C_{M\alpha o}} \frac{C_{F\alpha o}}{C_{F\alpha}(\mu, F_z, F_x)} M_{zo}(\alpha_{eq})$$
$$+ M_{zr} - c_9 a_o \frac{F_x F_y}{F_{z0}} - c_{10} a_o F_x - c_{11} b\, \gamma F_x \tag{4.47}$$

where the last terms have been taken from (4.39) and the residual torque reads

$$M_{zr} = \varphi_x \frac{C_{M\gamma}(F_z) + t(F_z)C_{F\gamma}(F_z)}{1 + c_7\alpha^2}\gamma \tag{4.48}$$

Exercise 4.1 given at the end of Section 4.3 addresses the problem of assessing the side force characteristics using the similarity technique with the longitudinal force considered as one of the input quantities.

In Section 4.3, the *Magic Formula* tire model will be treated in detail. This complex model is considerably more accurate and will again employ the longitudinal slip $\kappa$ as input variable.

## 4.3. THE *MAGIC FORMULA* TIRE MODEL

A widely used semi-empirical tire model to calculate steady-state tire force and moment characteristics for use in vehicle dynamics studies is based on the so-called *Magic Formula*. The development of the model was started in the mid-eighties. In a cooperative effort, TU-Delft and Volvo developed several versions (Bakker et al. 1987, 1989, Pacejka et al. 1993). In these models, the combined slip situation was modeled from a physical viewpoint. In 1993, Michelin (cf. Bayle et al. 1993) introduced a purely empirical method using *Magic Formula*-based functions to describe the tire horizontal force generation at combined slip. This approach has been adopted in subsequent versions of the *Magic Formula Tire Model*. In the newer version of the model, the original description of the aligning torque has been altered to accommodate a relatively simple physically based combined slip extension. The pneumatic trail is introduced as a basis to calculate this moment about the vertical axis, cf. Pacejka (1996). A complete listing of the model is given in Section 4.3.2. In Section 4.3.3, the model is further extended by introducing formulas for the description of the situation at large camber angle $\gamma$ and turn slip $\varphi_t$.

### 4.3.1. Model Description

The general form of the formula, that holds for given values of vertical load and camber angle, reads

$$y = D\sin[C\arctan\{Bx - E(Bx - \arctan Bx)\}] \tag{4.49}$$

with

$$Y(X) = y(x) + S_V \tag{4.50}$$

$$x = X + S_H \tag{4.51}$$

where $Y$ is the output variable $F_x$, $F_y$ or possibly $M_z$ and $X$ the input variable $\tan\alpha$ or $\kappa$, and $B$ is the stiffness factor, $C$ the shape factor, $D$ the peak value, $E$ the curvature factor, $S_H$ the horizontal shift, and $S_V$ the vertical shift.

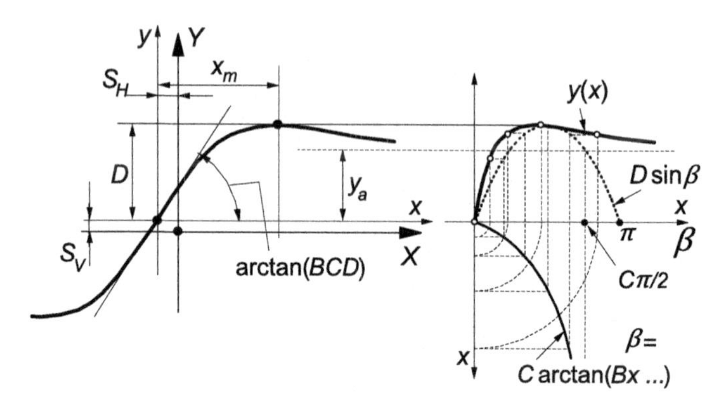


**FIGURE 4.9** Curve produced by the original sine version of the Magic Formula, Eqn (4.49). The meaning of curve parameters has been indicated.

The *Magic Formula y(x)* typically produces a curve that passes through the origin $x = y = 0$, reaches a maximum, and subsequently tends to a horizontal asymptote. For given values of the coefficients *B, C, D,* and *E,* the curve shows an anti-symmetric shape with respect to the origin. To allow the curve to have an offset with respect to the origin, two shifts $S_H$ and $S_V$ have been introduced. A new set of coordinates *Y(X)* arises as shown in Figure 4.9. The formula is capable of producing characteristics that closely match measured curves for the side force $F_y$ (and if desired also for the aligning torque $M_z$) and for the fore-and-aft force $F_x$ as functions of their respective slip quantities: the slip angle $\alpha$ and the longitudinal slip $\kappa$ with the effect of load $F_z$ and camber angle $\gamma$ included in the parameters.

Figure 4.9 illustrates the meaning of some of the factors by means of a typical side force characteristic. Obviously, coefficient $D$ represents the peak value (with respect to the central $x$-axis and for $C \geq 1$) and the product $BCD$ corresponds to the slope at the origin ($x = y = 0$). The shape factor $C$ controls the limits of the range of the sine function appearing in formula (4.49) and thereby determines the shape of the resulting curve. The factor $B$ is left to determine the slope at the origin and is called the stiffness factor. The factor $E$ is introduced to control the curvature at the peak and, at the same time, the horizontal position of the peak.

From the heights of the peak and of the horizontal asymptote, the shape factor $C$ may be computed:

$$C = 1 \pm \left(1 - \frac{2}{\pi} \arcsin \frac{y_a}{D}\right) \tag{4.52}$$

From $B$ and $C$ and the location $x_m$ of the peak, the value of $E$ may be assessed:

$$E = \frac{Bx_m - \tan\{\pi/(2C)\}}{Bx_m - \arctan(Bx_m)} \quad \text{if } C > 1 \tag{4.53}$$

The offsets $S_H$ and $S_V$ appear to occur when ply-steer and conicity effects and possibly the rolling resistance cause the $F_y$ and $F_x$ curves not to pass through the origin. Wheel camber may give rise to a considerable offset of the $F_y$ vs $\alpha$ curves. Such a shift may be accompanied by a significant deviation from the pure anti-symmetric shape of the original curve (cf. Figure 3.35c,d). To accommodate such an asymmetry, the curvature factor $E$ is made dependent on the sign of the abscissa ($x$):

$$E = E_o + \Delta E \cdot \mathrm{sgn}(x) \tag{4.54}$$

Also, the difference in shape that is expected to occur in the $F_x$ vs $\kappa$ characteristic between the driving and braking ranges can be taken care of. In Figure 4.10, the influence of the two shape factors $C$ and $E$ on the appearance of the curves has been demonstrated. The diagrams have been normalized by

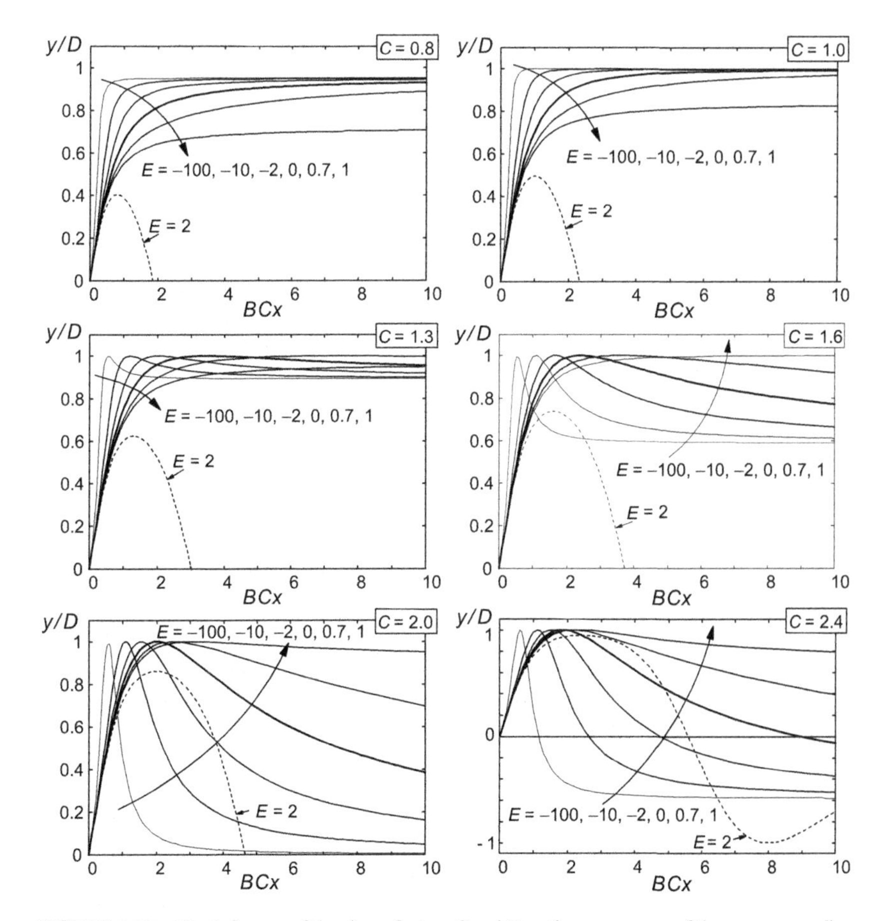


**FIGURE 4.10** The influence of the shape factors $C$ and $E$ on the appearance of the curve according to Eqn (4.49). Note that a value of curvature factor $E > 1$ does not produce realistic curves.

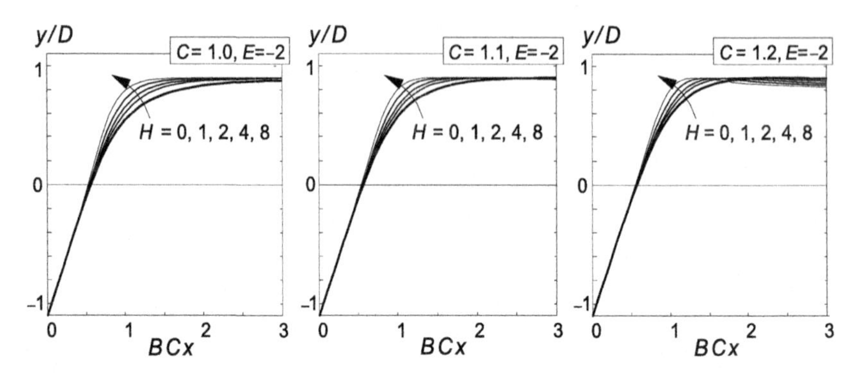


**FIGURE 4.11** Sharpness of curves near the peak may be increased by introducing additional term with sharpness factor $H$ (according to Eqn (4.55)).

dividing $y$ by $D$ and multiplying $x$ with $CB$, making the curve peak level and initial slope independent of the parameters.

In rather extreme cases, the sharpness that can be reached by means of the function given by Eqn (4.49) may not be sufficient. It turns out to be possible to considerably increase the sharpness of the curves by introducing an extra term in the argument of the arctan function. The modified function reads

$$y = D\sin\left[C\arctan\left\{Bx - E(Bx - \arctan Bx) + H\arctan^7 Bx\right\}\right] \qquad (4.55)$$

Figure 4.11 demonstrates the effect of the new coefficient $H$. Too large values may give rise to an upward curvature of the curve near the origin as would also occur at large negative values of $E$ (cf. the lower diagrams of Figure 4.10). In the ensuing text, we will not use this additional coefficient $H$.

It may be furthermore of interest to note that the possibly awkward function arctan($x$) may be replaced by the possibly faster and almost identical pseudo-arctan function psatan($x$) $= x(1 + a|x|)/\{1 + 2(b|x| + ax^2)/\pi\}$ with $a = 1.1$ and $b = 1.6$.

The various factors are functions of normal load and wheel camber angle. Several parameters appear in these functions. A suitable regression technique is used to determine their values from measured data according to a quadratic algorithm for the best fit (cf. Oosten and Bakker 1993). One of the important functional relationships is the variation of the cornering stiffness (almost exactly given by the product of coefficients $B_y$, $C_y$, and $D_y$ of the side force function: $BCD_y = K_{y\alpha} = \partial F_y/\partial\alpha$ at $\tan\alpha = -S_H$) with $F_z$ and $\gamma$:

$$BCD_y = p_1 \sin[2\arctan(F_z/p_2)]/(1 + p_3\gamma^2) \qquad (4.56)$$

For zero camber, the cornering stiffness attains its maximum $p_1$ at $F_z = p_2$. In Figure 4.12, the basic relationship has been depicted. Apparently, for a cambered wheel, the cornering stiffness decreases with increasing $|\gamma|$. Note the difference in curvature left and right of the characteristics at larger values of

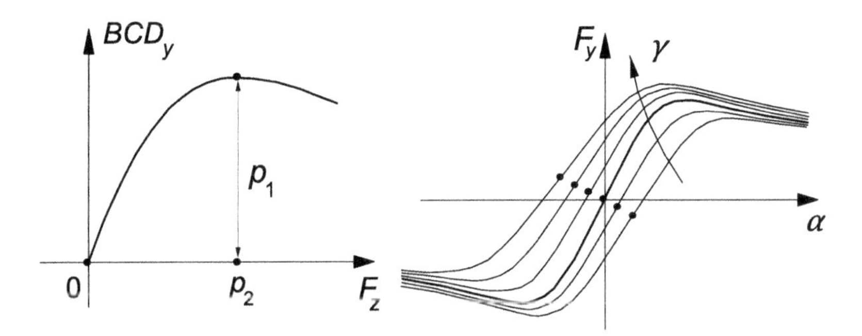


**FIGURE 4.12** Cornering stiffness vs vertical load and the influence of wheel camber, Eqn (4.56).

the camber angle. To accomplish this, a split-$E$ according to Eqn (4.54) has been employed. We refer to Section 4.3.2 for a complete listing of the formulas. Here, nondimensional parameters have been introduced. For example, the parameters in (4.56) will become: $p_1 = F_{zo}\, p_{Ky1}$, $p_2 = F_{zo}\, p_{Ky2}$, and $p_3 = p_{Ky3}$, with $F_{zo}$ denoting the nominal wheel load.

The aligning torque $M_z$ can now be obtained by multiplying the side force $F_y$ with the pneumatic trail $t$ and adding the usually small (except with camber) residual torque $M_{zr}$ (cf. Figure 4.13). We have

$$M_z = -t \cdot F'_y + M_{zr} \tag{4.57}$$

For the differently defined side force $F'_y$, we refer to the remark below Eqn (4.74). The pneumatic trail decays with increasing side slip and is described as follows:

$$t(\alpha_t) = D_t \cos[C_t \arctan\{B_t\alpha_t - E_t(B_t\alpha_t - \arctan(B_t\alpha_t))\}] \tag{4.58}$$

where

$$\alpha_t = \tan\alpha + S_{Ht} \tag{4.59}$$

The residual torque shows a similar decay:

$$M_{zr}(\alpha_r) = D_r \cos[\arctan(B_r\alpha_r)] \tag{4.60}$$

with

$$\alpha_r = \tan\alpha + S_{Hf} \tag{4.61}$$

It is observed that both parts of the moment are modeled using the *Magic Formula*, but instead of the sine function, the cosine function is employed. In that way, a hill-shaped curve is produced. The peaks are shifted sideways.

The residual torque is assumed to attain its maximum $D_r$ at the slip angle where the side force becomes equal to zero. This is accomplished through the horizontal shift $S_{Hf}$. The peak of the pneumatic trail occurs at $\tan\alpha = -S_{Ht}$. This

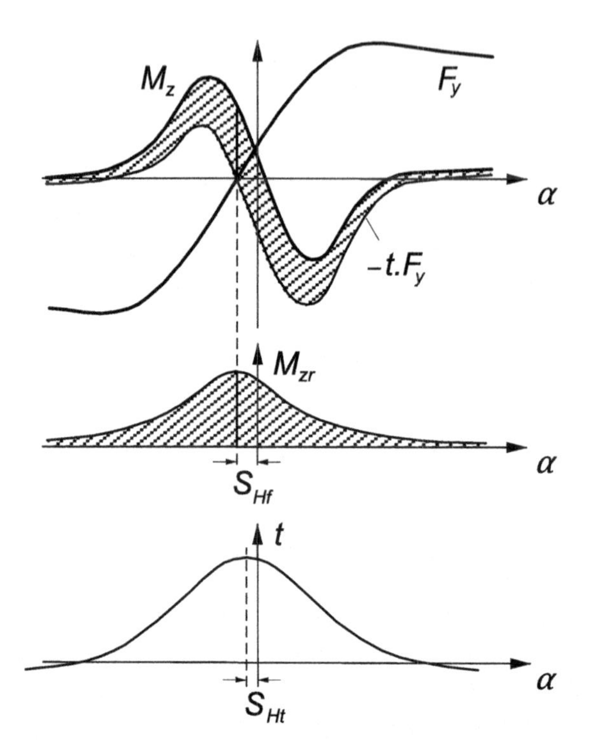


**FIGURE 4.13** The aligning torque characteristic composed of a part directly attributed to the side force and a part due the so-called residual torque (due to tire conicity and camber).

formulation has proven to give very good agreement with measured curves. The advantage with respect to the earlier versions, where formula (4.49) is used for the aligning torque as well, is that we have now directly assessed the function for the pneumatic trail which is needed to handle the combined slip situation.

In Figure 4.14, the basic properties of the cosine based curve have been indicated (subscripts of factors have been deleted again). Again, $D$ is the peak value, $C$ is a shape factor determining the level $y_a$ of the horizontal asymptote, and now $B$ influences the curvature at the peak (illustrated with the inserted parabola).

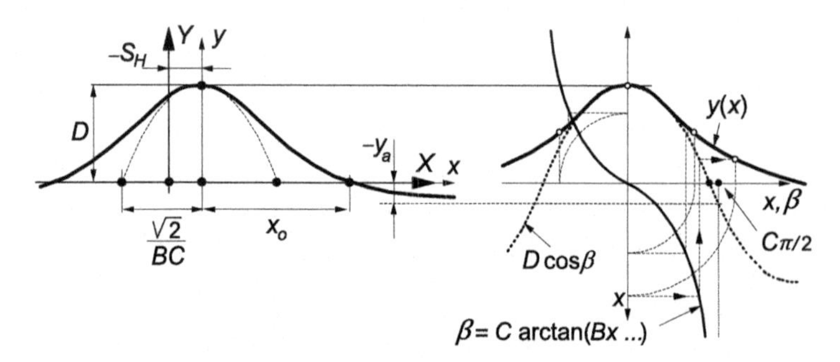


**FIGURE 4.14** Curve produced by the cosine version of the *Magic Formula*, Eqn (4.58). The meaning of curve parameters has been indicated.

Factor $E$ modifies the shape at larger values of slip and governs the location $x_o$ of the point where the curve intersects the $x$-axis. The following formulas hold:

$$C = \frac{2}{\pi} \arccos \frac{y_a}{D} \tag{4.62}$$

$$E = \frac{Bx_o - \tan\{\pi/(2C)\}}{Bx_o - \arctan(Bx_o)} \quad (\text{if } C > 1) \tag{4.63}$$

In the original version of the formula (4.57), cf. 1st or 2nd edition of this book, the force $Fy$ was used instead of $F_y'$. The force $F_y'$ constitutes the side force due to only side slip. That is, with input camber (and turn slip) disregarded. In addition, the different form of the force function (4.49) specially introduced for the possibly at large camber operating motorcycle, has been abandoned. The new formulations are listed in Sec. 4.3.2 which can now be employed successfully for both car and motorcycle tyres. Also see discussion below Eq. (4.74).

In the case of the possible presence of large camber angles (motorcycles), it may be better to use in (4.57) the side force $F_y$ that would arise at $\gamma = 0$. Also, the side force function (4.49) and the cornering stiffness function (4.56) may be modified to better approximate large camber response for motorcycle tires, cf. De Vries (1998a) and Sec.11.6 for a full listing of equations. We refer to Section 4.3.3 for the discussion of the model extension for larger camber and turn slip (path curvature) also applicable in the case of combined slip with braking or driving forces.

In the paper of Pacejka and Bakker (1993), the tire's response to combined slip was modeled by using physically based formulas. A newer more efficient way is purely empiric. This method was developed by Michelin and published by Bayle, Forissier, and Lafon (1993). It describes the effect of combined slip on the lateral force and on the longitudinal force characteristics. Weighting functions $G$ have been introduced which, when multiplied with the original pure slip functions (4.49), produce the interaction effects of $\kappa$ on $F_y$ and of $\alpha$ on $F_x$. The weighting functions have a hill shape. They take the value one in the special case of pure slip (either $\kappa$ or $\alpha$ equal to zero). When, for example, at a given slip angle a from zero increasing brake slip is introduced, the relevant weighting function for $F_y$ may first show a slight increase in magnitude (becoming larger than one) but will soon reach its peak after which a continuous decrease follows. The cosine version of the *Magic Formula* is used to represent the hill shaped function:

$$G = D \cos[C \arctan(Bx)] \tag{4.64}$$

Here, $G$ is the resulting weighting factor and $x$ is either $\kappa$ or $\tan \alpha$ (possibly shifted). The coefficient $D$ represents the peak value (slightly deviating from one if a horizontal shift of the hill occurs), $C$ determines the height of the hill's base, and $B$ influences the sharpness of the hill. Coefficient $B$ constitutes the main factor responsible for the shape of the weighting functions.

As an extension to the original function published by Bayle et al., the part with shape factor $E$ will be added later on. This extension appears to improve the approximation, in particular at large levels of slip, especially in view of the strict condition that the weighting function $G$ must remain positive for all slip conditions.

For the side force, we get the following formulas:

$$F_y = G_{y\kappa} \cdot F_{yo} + S_{Vy\kappa} \tag{4.65}$$

with the weighting function now expressed such that it equals unity at $\kappa = 0$:

$$G_{\kappa y} = \frac{\cos[C_{y\kappa} \arctan(B_{y\kappa}\kappa_S)]}{\cos[C_{y\kappa} \arctan(B_{y\kappa}S_{Hy\kappa})]} \quad (>0) \tag{4.66}$$

where

$$\kappa_S = \kappa + S_{Hy\kappa} \tag{4.67}$$

and further the coefficients

$$B_{y\kappa} = r_{By1}\cos\left[\arctan\{r_{By2}(\tan\alpha - r_{By3})\}\right] \tag{4.68a}$$

$$C_{y\kappa} = r_{Cy1} \tag{4.68b}$$

$$S_{Hy\kappa} = r_{Hy1} + r_{Hy2}\, df_z \tag{4.68c}$$

$$S_{Vy\kappa} = D_{Vy\kappa} \sin[r_{Vy5} \arctan(r_{Vy6}\kappa)] \tag{4.69a}$$

$$D_{Vy\kappa} = \mu_y F_z\, (r_{Vy1} + r_{Vy2}\, df_z + r_{Vy3}\gamma)\cos[\arctan(r_{Vy4} \tan\alpha)] \tag{4.69b}$$

with $df_z$ the notation for the nondimensional increment of the vertical load with respect to the (adapted) nominal load, cf. next Subsection 4.3.2, Eqn (4.E2a). Figure 4.15 depicts the two weighting functions displayed as functions of both $\alpha$ and $\kappa$ and the resulting force characteristics (parameters according to Table 4.2 on p. 190). Below, an explanation is given.

In Eqn (4.65) $F_{yo}$ denotes the side force at pure side slip obtained from Eqn (4.49). The denominator of the weighting function (4.66) makes $G_{y\kappa} = 1$ at $\kappa = 0$. The horizontal shift $S_{Hy\kappa}$ of the weighting function accomplishes the slight increase that the side force experiences at moderate braking before the peak of $G_{y\kappa}$ is reached and the decay of $F_y$ commences.

This horizontal shift may be made dependent on the vertical load. $C_{y\kappa}$ controls the level of the horizontal asymptote. If $C_{y\kappa} = 1$, the weighting function (4.66) will approach zero when $\kappa \to \pm\infty$. This would be the correct value for $C_{y\kappa}$ if $\kappa$ is expected to be used in the entire range from plus to minus infinity. If this is not intended, then $C_{y\kappa}$ may be chosen different from one if $G_{y\kappa}$ is optimized with the restriction to remain positive. The factor $B_{y\kappa}$ influences the sharpness of the hill shaped weighting function. As indicated, the hill becomes more flat (wider) at larger slip angles. Then $B_{y\kappa}$ decreases according to (4.68a). When in an extreme situation $\alpha$ approaches 90°, that is when $V_{cx} \to 0$, $B_{y\kappa}$ will go to zero and, consequently, $G_{y\kappa}$ will remain equal to one unless $\kappa$ goes to infinity which may easily be the case when at $V_{cx} \to 0$ the wheel speed of revolution $\Omega$ and thus the longitudinal slip velocity $V_{sx}$ remain unequal to zero. The quantity $S_{Vy\kappa}$ is the

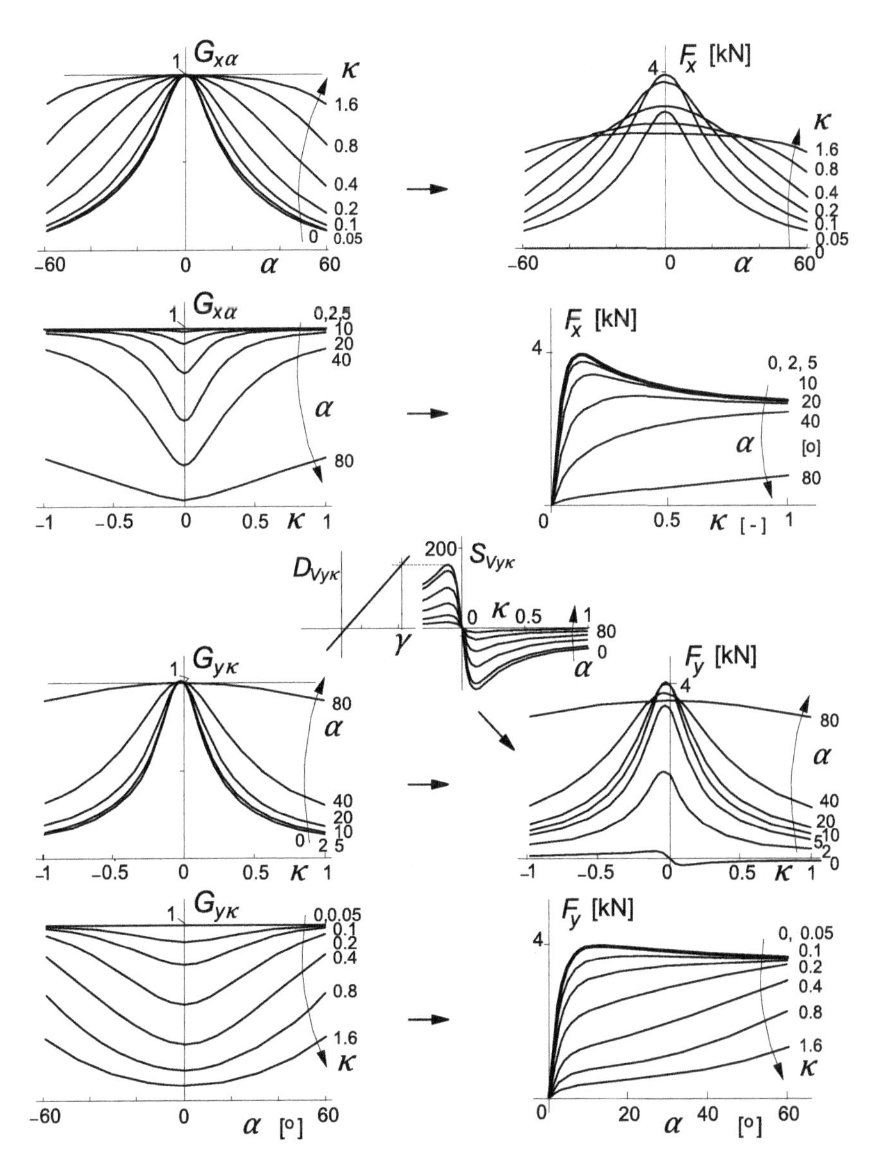

**FIGURE 4.15** Nature of weighting functions and the resulting combined slip longitudinal and lateral forces, the latter also affected by the $\kappa$ induced ply-steer 'vertical shift' $S_{Vy\kappa}$.

vertical 'shift', which sometimes is referred to as the $\kappa$-induced ply-steer. At camber, due to the added asymmetry, the longitudinal force clearly produces a torque that creates a torsion angle comparable with a possibly already present ply-steer angle. This shift function varies with slip $\kappa$ indicated in (4.69a).

As illustrated in Figure 4.15, its peak value $D_{Vy\kappa}$ depends on the camber angle $\gamma$ and decays with increasing magnitude of the slip angle $\alpha$. Figure 4.16

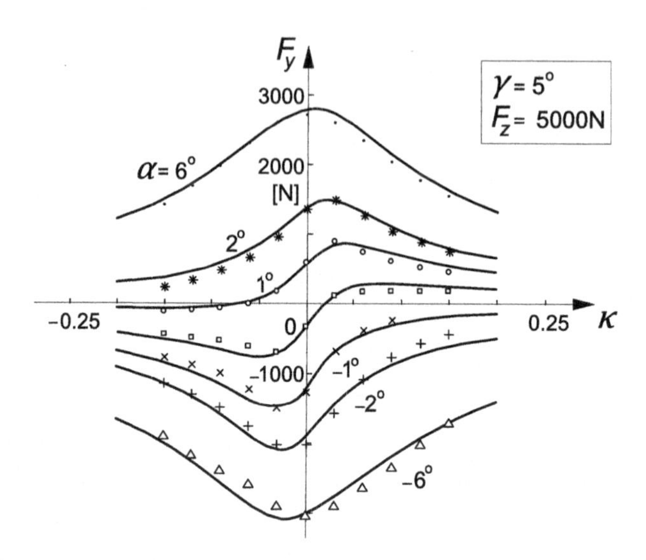


**FIGURE 4.16** The combined side slip force characteristics in the presence of a camber angle, Bayle et al. (1993).

presents measured data together with the fitted curves as published by Bayle et al.

The combined slip relations for $F_x$ are similar to what we have seen for the side force. However, a vertical shift function was not needed. In Figure 4.17, a three-dimensional graph is shown indicating the variation of $F_x$ and $F_y$ with both $\alpha$ and $\kappa$. The initial '*S*' shape of the $F_y$ vs $\kappa$ curves (at small $\alpha$) due to the vertical shift function is clearly visible.

For possible improvement of the general tendency of the model at larger levels of combined slip beyond the range of available test data, one might

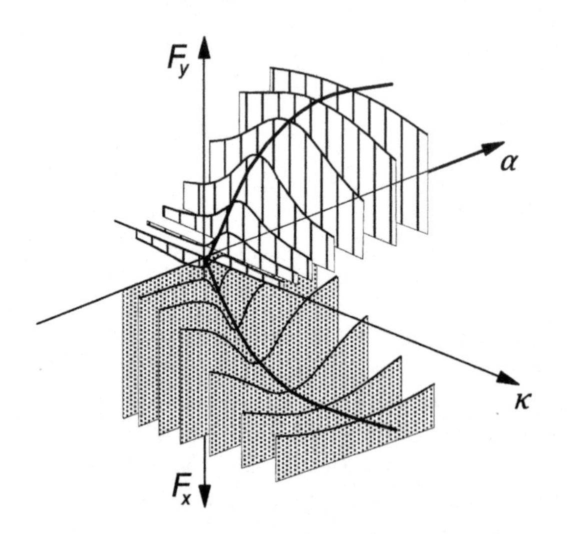


**FIGURE 4.17** Three-dimensional graph of combined slip force characteristics.

include additional 'fabricated' data which are derived from similarity method results at larger values of the slip angle. Another possibility is the usage of the conditions at wheel lock where one might assume that the force and slip vector are colinear. We then have, for the ratio of the components,

$$\frac{F_x}{F_y} = \frac{V_{sx}}{V_{sy}} = \frac{\kappa}{\tan\alpha} = \frac{-1}{\tan\alpha} \tag{4.70}$$

or

$$F_x = F_{xo(\kappa=-1)} \cdot \cos\alpha = F_{xo(\kappa=-1)} \cdot \frac{V_{sx}}{V_s} \tag{4.71}$$

$$F_y = -F_{xo(\kappa=-1)} \cdot \sin\alpha = F_{xo(\kappa=-1)} \cdot \frac{V_{sy}}{V_s} \tag{4.72}$$

Regarding the aligning torque, physical insight is used to model the situation at combined slip. We write

$$M_z = -t(\alpha_{t,eq}) \cdot F'_y + M_{zr}(\alpha_{r,eq}) + s(F_y, \gamma) \cdot F_x \tag{4.73}$$

The arguments $\alpha_t$ and $\alpha_r$ (including a shift) appearing in the functions (4.59, 4.61) for the pneumatic trail and residual torque at pure side slip are replaced by equivalent slip angles, as indicated by Eqn (4.74), incorporating the effect of $\kappa$ on the composite slip:

$$\alpha_{t,eq} = \sqrt{\alpha_t^2 + \left(\frac{K_{x\kappa}}{K_{y\alpha}}\right)^2 \kappa^2} \cdot \operatorname{sgn}(\alpha_t) \tag{4.74}$$

and similar for $\alpha_{r,eq}$. To approximate the same effect on the degree of sliding in the contact patch as would occur with side slip, the longitudinal slip $\kappa$ is multiplied with the ratio of the longitudinal and lateral slip stiffnesses.

Besides, an extra term is introduced in (4.73) to account for the fact that a moment arm $s$ arises for $F_x$ as a result of camber $\gamma$ and lateral tire deflection related to $F_y$. This extra term may give rise to a sign change of the aligning torque in the range of braking as discussed before (cf. Figure 3.20).

With respect to the previous version of the expression for the aligning torque, presented in e.g., the 2nd edition of this book, an important change has been introduced that originates from the special version developed for the motorcycle tire (MC-MF tire) that can handle large camber angles, cf. 2nd edition and De Vries and Pacejka (1998a). In the first term of Eqn (4.73), the full side force $F_y$ has been replaced by the side force $F'_y$ that is generated by only the side-slip angle $\alpha$ (possibly influenced by the fore-and-aft slip $\kappa$). Consequently, $F'_y$ results from $F_y$ calculated with $\gamma = \varphi_t = 0$. With this replacement, the full set of equations, presented in the next Subsection 4.3.2, is now capable of accurately modeling force and moment response for both car and motorcycle tires. Note that this replacement is more logical because $\gamma$ and $\varphi_t$ cause almost symmetrical lateral deflections. The accompanying moment $M_{zr}$ arises through anti-symmetric longitudinal tread deflections. The replacement might, however, be partially disturbed through an $M_{zr}$-induced slip angle (contact patch torsion through carcass torsional compliance). This makes the lateral deflection a little less symmetric.

At the end of this chapter, force and moment characteristics for car, truck, and motorcycle tires as computed with the *Magic Formula* equations have been presented. They are compared with curves that result from steady-state full-scale tire tests.

### 4.3.2. Full Set of Equations

The complete set of steady-state formulas contains nondimensional parameters $p$, $q$, $r$, and $s$. In addition, user scaling factors $\lambda$ have been introduced. With that tool, the effect of changing friction coefficient, cornering stiffness, camber stiffness, etc., can be quickly investigated in a qualitative way without having the need to implement a completely new tire data set. Scaling is done in such a way that realistic relationships are maintained. For instance, when changing the cornering stiffness and the friction coefficient in lateral direction (through $\lambda_{Ky\alpha}$ and $\lambda_{\mu y}$), the abscissa of the pneumatic trail characteristic is changed in a way equal to that of the side force characteristic and in accordance with the similarity method of Section 4.2.

The *Magic Formula* model equations listed below are in accordance with the description given in the MF-Tire/MF-Swift 6.1.2 Equation Manual, cf. TNO Automotive (2010). The equations contain the nondimensional model parameters $p$, $q$, $r$, and $s$ and, in addition, a set of scaling factors $\lambda$. Other parameters and variable quantities used in the equations are:

| | |
|---|---|
| $g$ | acceleration due to gravity, |
| $V_c$ | magnitude of the velocity of the wheel contact center $C$, |
| $V_{cx,y}$ | components of the velocity of the wheel contact center $C$, |
| $V_{sx,y}$ | components of slip velocity $V_s$ (of point $S$) with $V_{sy} \approx V_{cy}$, Eqn (2.13), |
| $V_r$ | $(=R_e\,\Omega = V_{cx} - V_{sx})$ forward speed of rolling, |
| $V_o$ | reference velocity $(=\sqrt{(gR_o)}$ or other specified value), |
| $R_o$ | unloaded tire radius $(=r_o)$, |
| $R_e$ | effective rolling radius $(=r_e)$, |
| $\Omega$ | wheel speed of revolution, |
| $\rho_z$ | tire radial deflection ($>0$ if compression), and |
| $F_z$ | normal load $(=F_N \geq 0)$. Note: in Chaps. 9 and 10 $F_z = -F_N$ $(\leq 0)$: |
| $F_{zo}$ | nominal (rated) wheel load, |
| $F'_{zo}$ | adapted nominal load, |
| $p_i$ | tire inflation pressure, and |
| $p_{io}$ | nominal inflation pressure. |

The effect of having a tire with a different nominal load may be roughly approximated by using the scaling factor $\lambda_{Fzo}$:

$$F'_{zo} = \lambda_{Fzo} F_{zo} \tag{4.E1}$$

Further, we introduce the normalized change in vertical load

$$df_z = \frac{F_z - F'_{zo}}{F'_{zo}} \tag{4.E2a}$$

and similarly the normalized change in inflation pressure:

$$dp_i = \frac{p_i - p_{io}}{p_{io}} \tag{4.E2b}$$

The approximate influence of a change in inflation pressure as assessed through numerous tire tests has been recently investigated and introduced in the Magic Formula model equations, cf. Besselink et al. (2009).

Instead of taking the slip angle $\alpha$ itself (in radians, from Eqn (2.12)) as input quantity, one may, in the case of very large slip angles and possibly backwards running of the wheel, better use the tangent of the slip angle defined as the lateral slip:

$$\alpha^* = \tan\alpha \cdot \operatorname{sgn} V_{cx} = -\frac{V_{cy}}{|V_{cx}|} \tag{4.E3}$$

For the spin due to the camber angle, we introduce

$$\gamma^* = \sin\gamma \tag{4.E4}$$

The longitudinal slip ratio is defined as follows:

$$\kappa = -\frac{V_{sx}}{|V_{cx}|} \tag{4.E5}$$

If the forward speed $V_{cx}$ becomes or is equal to zero, one might add a small quantity $\varepsilon$ in the denominator of (4.E3, 4.E5) to avoid singularity, or, when transient slip situations occur, one should use the transient slip quantities (or deformation gradients) $\tan\alpha'$ and $\kappa'$ as defined and used in Chapters 7 and 8.

To avoid the occurrence of similar singularities in the ensuing equations due to e.g., zero velocity or zero vertical load, a small additional quantity $\varepsilon$ (with same sign as its neighboring main quantity) will be introduced in relevant denominators like in the next equation.

For the factor $\cos\alpha$ appearing in the equations for the aligning torque to properly handle the case of large slip angles and possibly backwards running ($V_{cx} < 0$), we have defined

$$\cos'\alpha = \frac{V_{cx}}{V'_c} \tag{4.E6}$$

with

$$V_c' = V_c + \varepsilon_V \tag{4.E6a}$$

where we may choose $\varepsilon_V = 0.1$.

For the normally encountered situations where turn slip $\varphi_t$ may be neglected (path radius $R \rightarrow \infty$) and camber remains small, the factors $\zeta_i$ appearing in the equations may be set equal to unity:

$$\zeta_i = 1 \quad (i = 0, 1, \ldots, 8)$$

In the following Subsection 4.3.3 where the influence of spin (turn slip and camber) is discussed, the proper expressions will be given for the factors $\zeta_i$, and additional equations will be introduced.

### User Scaling Factors

The following set of scaling factors $\lambda$ is available. The default value of these factors is set equal to one (except $\lambda_{\mu V}$ which equals zero if not used). We have:

Pure slip
- $\lambda_{Fzo}$ nominal (rated) load
- $\lambda_{\mu x,y}$ peak friction coefficient
- $\lambda_{\mu V}$ with slip speed $V_s$ decaying friction
- $\lambda_{Kx\kappa}$ brake slip stiffness
- $\lambda_{Ky\alpha}$ cornering stiffness
- $\lambda_{Cx,y}$ shape factor
- $\lambda_{Ex,y}$ curvature factor
- $\lambda_{Hx,y}$ horizontal shift
- $\lambda_{Vx,y}$ vertical shift
- $\lambda_{Ky\gamma}$ camber force stiffness
- $\lambda_{Kz\gamma}$ camber torque stiffness
- $\lambda_t$ pneumatic trail (effecting aligning torque stiffness)
- $\lambda_{Mr}$ residual torque

Combined slip
- $\lambda_{x\alpha}$ $\alpha$ influence on $F_x(\kappa)$
- $\lambda_{y\kappa}$ $\kappa$ influence on $F_y(\alpha)$
- $\lambda_{Vy\kappa}$ $\kappa$ induced 'ply-steer' $F_y$
- $\lambda_s$ $M_z$ moment arm of $F_x$

Other
- $\lambda_{Cz}$ radial tire stiffness
- $\lambda_{Mx}$ overturning couple stiffness
- $\lambda_{VMx}$ overturning couple vertical shift
- $\lambda_{My}$ rolling resistance moment

To change from a relatively high friction surface to a low friction surface, the factors $\lambda_{\mu x}$ and $\lambda_{\mu y}$ may be given a value lower than unity. In addition, to reflect a slippery surface (wet) with friction decaying with increasing (slip) speed, one may choose for $\lambda_{\mu V}$ a value larger than zero, e.g., Eqns (4.E13, 4.E23). The publications of Dijks (1974) and of Reimpell et al. (1986) may be useful in this respect. Note that the slip stiffnesses are not affected through these changes. For the composite friction scaling factor, in $x$- and $y$-direction respectively, we have

$$\lambda^*_{\mu x,y} = \lambda_{\mu x,y}/(1 + \lambda_{\mu V} V_s/V_o) \tag{4.E7}$$

A special degressive friction factor $\lambda'_{\mu x,y}$ is introduced to recognize the fact that vertical shifts of the force curves do vanish when $\mu \rightarrow 0$ but at a much slower rate:

$$\lambda'_{\mu x,y} = A_\mu \lambda^*_{\mu x,y} / \left\{1 + (A_\mu - 1)\lambda^*_{\mu x,y}\right\} \quad (\text{suggestion: } A_\mu = 10) \tag{4.E8}$$

For the three forces and three moments acting from road to tire and defined according to the diagram of Figure 2.3, the equations, first those for the condition of pure slip (including camber) and subsequently those for the condition of combined slip, read (version 2004).

### *Longitudinal Force (Pure Longitudinal Slip, $\alpha = 0$)*

$$F_{xo} = D_x \sin[C_x \arctan\{B_x\kappa_x - E_x(B_x\kappa_x - \arctan(B_x\kappa_x))\}] + S_{Vx} \tag{4.E9}$$

$$\kappa_x = \kappa + S_{Hx} \tag{4.E10}$$

$$C_x = p_{Cx1} \cdot \lambda_{Cx} \qquad (>0) \tag{4.E11}$$

$$D_x = \mu_x \cdot F_z \cdot \zeta_1 \qquad (>0) \tag{4.E12}$$

$$\mu_x = (p_{Dx1} + p_{Dx2} df_z)(1 + p_{px3} dp_i + p_{px4} dp_i^2)(1 - p_{Dx3}\gamma^2) \cdot \lambda^*_{\mu x} \tag{4.E13}$$

$$E_x = (p_{Ex1} + p_{Ex2} df_z + p_{Ex3} df_z^2)\{1 - p_{Ex4}\operatorname{sgn}(\kappa_x)\} \cdot \lambda_{Ex} \quad (\leq 1) \tag{4.E14}$$

$$K_{x\kappa} = F_z(p_{Kx1} + p_{Kx2} df_z)\exp(p_{Kx3} df_z)(1 + p_{px1} dp_i + p_{px2} dp_i^2)$$

$$\lambda_{Kx\kappa} \quad (= B_x C_x D_x = \partial F_{xo}/\partial \kappa_x \text{ at } \kappa_x = 0) \; (= C_{F\kappa}) \tag{4.E15}$$

$$B_x = K_{x\kappa}/(C_x D_x + \varepsilon_x) \tag{4.E16}$$

$$S_{Hx} = (p_{Hx1} + p_{Hx2} df_z) \cdot \lambda_{Hx} \tag{4.E17}$$

$$S_{Vx} = F_z \cdot (p_{Vx1} + p_{Vx2} df_z) \cdot \lambda_{Vx} \lambda'_{\mu x} \zeta_1 \tag{4.E18}$$

### *Lateral Force (Pure Side Slip, $\kappa = 0$)*

$$F_{yo} = D_y \sin\left[C_y \arctan\left\{B_y\alpha_y - E_y(B_y\alpha_y - \arctan(B_y\alpha_y))\right\}\right] + S_{Vy} \tag{4.E19}$$

$$\alpha_y = \alpha^* + S_{Hy} \tag{4.E20}$$

$$C_y = p_{Cy1} \cdot \lambda_{Cy} \quad (>0) \tag{4.E21}$$

$$D_y = \mu_y \cdot F_z \cdot \zeta_2 \tag{4.E22}$$

$$\mu_y = (p_{Dy1} + p_{Dy2}\, df_z)(1 + p_{py3} dp_i + p_{py4} dp_i^2)(1 - p_{Dy3}\gamma^{*2})\lambda^*_{\mu y} \tag{4.E23}$$

$$E_y = (p_{Ey1} + p_{Ey2}df_z)\{1 + p_{Ey5}\gamma^{*2} - (p_{Ey3} + p_{Ey4}\gamma^*)\text{sgn}(\alpha_y)\}\ \lambda_{Ey} \quad (\leq 1) \tag{4.E24}$$

$$K_{y\alpha} = p_{Ky1}F'_{zo}(1 + p_{py1}dp_i)(1 - p_{Ky3}|\gamma^*|) \cdot \sin\left[p_{Ky4}\arctan\left\{\frac{F_z/F'_{zo}}{(p_{Ky2} + p_{Ky5}\gamma^{*2})(1 + p_{py2}dp_i)}\right\}\right] \cdot \zeta_3\lambda_{Ky\alpha} \tag{4.E25}$$

$$(= B_yC_yD_y = \partial F_{yo}/\partial\alpha_y \text{ at } \alpha_y = 0)$$

$$(\text{if } \gamma = 0\text{: } = K_{y\alpha o} = C_{F\alpha})\ (p_{Ky4} \approx 2)$$

$$B_y = K_{y\alpha}/(C_yD_y + \varepsilon_y) \tag{4.E26}$$

$$S_{Hy} = (p_{Hy1} + p_{Hy2}df_z)\ \lambda_{Hy} + \frac{K_{y\gamma o}\gamma^* - S_{Vy\gamma}}{K_{y\alpha} + \varepsilon_K}\zeta_o + \zeta_4 - 1 \tag{4.E27}$$

$$S_{Vy\gamma} = F_z\cdot(p_{Vy3} + p_{Vy4}df_z)\gamma^*\cdot\lambda_{Ky\gamma}\lambda'_{\mu y}\zeta_2 \tag{4.E28}$$

$$S_{Vy} = F_z\cdot(p_{Vy1} + p_{Vy2}df_z)\cdot\lambda_{Vy}\lambda'_{\mu y}\zeta_2 + S_{Vy\gamma} \tag{4.E29}$$

$$K_{y\gamma 0} = F_z\cdot(p_{Ky6} + p_{Ky7}\ df_z)(1 + p_{py5}dp_i)\cdot\lambda_{Ky\gamma}$$

$$(\approx \partial F_{yo}/\partial\gamma \text{ at } \alpha = \gamma = 0)(= C_{F\gamma}) \tag{4.E30}$$

### *Aligning Torque (Pure Side Slip, $\kappa = 0$)*

$$M_{zo} = M'_{zo} + M_{zro} \tag{4.E31}$$

$$M'_{zo} = -t_o\cdot F_{yo,\gamma=\varphi_t=0} \tag{4.E32}$$

$$t_0 = t(\alpha_t) = D_t\cos[C_t\arctan\{B_t\alpha_t - E_t(B_t\alpha_t - \arctan(B_t\alpha_t))\}]\cdot\cos'\alpha \tag{4.E33}$$

$$\alpha_t = \alpha^* + S_{Ht} \tag{4.E34}$$

$$S_{Ht} = q_{Hz1} + q_{Hz2}df_z + (q_{Hz3} + q_{Hz4}df_z)\gamma^* \tag{4.E35}$$

$$M_{zro} = M_{zr}(\alpha_r) = D_r\cos[C_r\arctan(B_r\alpha_r)]\cdot\cos'\alpha \tag{4.E36}$$

$$\alpha_r = \alpha^* + S_{Hf} \quad (= \alpha_f) \tag{4.E37}$$

$$S_{Hf} = S_{Hy} + S_{Vy}/K'_{y\alpha} \tag{4.E38}$$

$$K'_{y\alpha} = K_{y\alpha} + \varepsilon_K \tag{4.E39}$$

$$B_t = (q_{Bz1} + q_{Bz2}df_z + q_{Bz3}df_z^2)(1 + q_{Bz5}|\gamma^*| + q_{Bz6}\gamma^{*2})\ \lambda_{Ky\alpha}/\lambda^*_{\mu y} \quad (> 0) \tag{4.E40}$$

$$C_t = q_{Cz1} \quad (> 0) \tag{4.E41}$$

$$D_{to} = F_z\cdot(R_o/F'_{zo})\cdot(q_{Dz1} + q_{Dz2}df_z)(1 - p_{pz1}dp_i)\cdot\ \lambda_t\cdot\text{sgn}\ V_{cx} \tag{4.E42}$$

$$D_t = D_{to} \cdot (1 + q_{Dz3}|\gamma^*| + q_{Dz4}\gamma^{*2}) \cdot \zeta_5 \tag{4.E43}$$

$$E_t = (q_{Ez1} + q_{Ez2}df_z + q_{Ez3}df_z^2) \cdot \left\{1 + (q_{Ez4} + q_{Ez5}\gamma^*)\ \frac{2}{\pi}\arctan(B_t C_t \alpha_t)\right\}\ (\leq 1) \tag{4.E44}$$

$$B_r = (q_{Bz9} \cdot \lambda_{Ky\alpha}/\lambda^*_{\mu y} + q_{Bz10}B_y C_y) \cdot \zeta_6 \quad (\text{preferred: } q_{Bz9} = 0) \tag{4.E45}$$

$$C_r = \zeta_7 \tag{4.E46}$$

$$D_r = F_z R_o \left[(q_{Dz6} + q_{Dz7}df_z)\lambda_{Mr}\zeta_2 + \{(q_{Dz8} + q_{Dz9}df_z)(1 + p_{pz2}dp_i) + (q_{Dz10} + q_{Dz11}df_z)|\gamma^*|\}\gamma^*\lambda_{Kz\gamma}\,\zeta_0\right]\lambda^*_{\mu y}\ \mathrm{sgn}V_{cx}\cos'\alpha + \zeta_8 - 1 \tag{4.E47}$$

$$K_{z\alpha o} = D_{to}K_{y\alpha(\gamma=0)}\ (\approx -\partial M_{zo}/\partial\alpha_y \text{ at } \alpha_y = \gamma = 0)\ (= C_{M\alpha}) \tag{4.E48}$$

$$K_{z\gamma o} = F_z R_o(q_{Dz8} + q_{Dz9}df_z)(1 + p_{pz2}dp_i)\lambda_{Kz\gamma}\lambda^*_{\mu y} - D_{to}K_{y\gamma o}\ (\approx \partial M_{zo}/\partial\gamma \text{ at } \alpha = \gamma = 0)(= C_{M\gamma}) \tag{4.E49}$$

### *Longitudinal Force (Combined Slip)*

$$F_x = G_{x\alpha} \cdot F_{xo} \tag{4.E50}$$

$$G_{x\alpha} = \cos[C_{x\alpha}\arctan\{B_{x\alpha}\alpha_S - E_{x\alpha}(B_{x\alpha}\alpha_S - \arctan(B_{x\alpha}\alpha_S))\}]/G_{x\alpha o}\ (> 0) \tag{4.E51}$$

$$G_{x\alpha o} = \cos[C_{x\alpha}\arctan\{B_{x\alpha}S_{Hx\alpha} - E_{x\alpha}(B_{x\alpha}S_{Hx\alpha} - \arctan(B_{x\alpha}S_{Hx\alpha}))\}] \tag{4.E52}$$

$$\alpha_S = \alpha^* + S_{Hx\alpha} \tag{4.E53}$$

$$B_{x\alpha} = (r_{Bx1} + r_{Bx3}\gamma^{*2})\cos[\arctan(r_{Bx2}\kappa)] \cdot \lambda_{x\alpha} \quad (> 0) \tag{4.E54}$$

$$C_{x\alpha} = r_{Cx1} \tag{4.E55}$$

$$E_{x\alpha} = r_{Ex1} + r_{Ex2}df_z \quad (\leq 1) \tag{4.E56}$$

$$S_{Hx\alpha} = r_{Hx1} \tag{4.E57}$$

### *Lateral Force (Combined Slip)*

$$F_y = G_{y\kappa} \cdot F_{yo} + S_{Vy\kappa} \tag{4.E58}$$

$$G_{y\kappa} = \cos\left[C_{y\kappa}\arctan\{B_{y\kappa}\kappa_S - E_{y\kappa}(B_{y\kappa}\kappa_S - \arctan(B_{y\kappa}\kappa_S))\}\right]/G_{y\kappa o}\ (> 0) \tag{4.E59}$$

$$G_{y\kappa o} = \cos\left[C_{y\kappa}\arctan\{B_{y\kappa}S_{Hy\kappa} - E_{y\kappa}(B_{y\kappa}S_{Hy\kappa} - \arctan(B_{y\kappa}S_{Hy\kappa}))\}\right] \tag{4.E60}$$

$$\kappa_S = \kappa + S_{Hy\kappa} \tag{4.E61}$$

$$B_{y\kappa} = (r_{By1} + r_{By4}\gamma^{*2})\cos\left[\arctan\left\{r_{By2}(\alpha^* - r_{By3})\right\}\right]\cdot\lambda_{y\kappa} \quad (>0) \tag{4.E62}$$

$$C_{y\kappa} = r_{Cy1} \tag{4.E63}$$

$$E_{y\kappa} = r_{Ey1} + r_{Ey2}df_z \quad (\leq 1) \tag{4.E64}$$

$$S_{Hy\kappa} = r_{Hy1} + r_{Hy2}df_z \tag{4.E65}$$

$$S_{Vy\kappa} = D_{Vy\kappa}\sin\left[r_{Vy5}\arctan(r_{Vy6}\kappa)\right]\cdot\lambda_{Vy\kappa} \tag{4.E66}$$

$$D_{Vy\kappa} = \mu_y F_z\cdot(r_{Vy1} + r_{Vy2}df_z + r_{Vy3}\gamma^*)\cdot\cos\left[\arctan(r_{Vy4}\alpha^*)\right]\cdot\zeta_2 \tag{4.E67}$$

*Normal Load (see also Eqns (7.48) and (9.217))*

$$F_z = \left\{1 + q_{V2}|\Omega|\frac{R_o}{V_o} - \left(q_{Fcx}\frac{F_x}{F_{zo}}\right)^2 - \left(q_{Fcy}\frac{F_y}{F_{zo}}\right)^2\right\} \cdot\left\{\left(q_{Fz1} + q_{Fz3}\gamma^2\right)\frac{\rho_z}{R_o} + q_{Fz2}\frac{\rho_z^2}{R_o^2}\right\}\cdot(1 + p_{pFz1}dp_i)\cdot F_{zo} \tag{4.E68}$$

*Overturning Couple (also see Section 4.3.5)*

$$M_x = R_o F_z\cdot\left[q_{sx1}\lambda_{VMx} - q_{sx2}\gamma(1 + p_{pMx1}dp_i) + q_{sx3}\frac{F_y}{F_{zo}} + q_{sx4}\cos\left\{q_{sx5}\arctan\left(q_{sx6}\frac{F_z}{F_{zo}}\right)^2\right\}\sin\left\{q_{sx7}\gamma + q_{sx8}\arctan\left(q_{sx9}\frac{F_y}{F_{zo}}\right)\right\} + q_{sx10}\arctan\left(q_{sx11}\frac{F_z}{F_{zo}}\right)\cdot\gamma\right]\cdot\lambda_{Mx} \tag{4.E69}$$

*Rolling Resistance Moment (see Eqns (9.236, 9.230, 9.231))*

$$M_y = F_z R_o\left\{q_{sy1} + q_{sy2}\frac{F_x}{F_{zo}} + q_{sy3}\left|\frac{V_x}{V_o}\right| + q_{sy4}\left(\frac{V_x}{V_o}\right)^4 + \left(q_{sy5} + q_{sy6}\frac{F_z}{F_{zo}}\right)\gamma^2\right\}\left\{\left(\frac{F_z}{F_{zo}}\right)^{q_{sy7}}\cdot\left(\frac{p_i}{p_{io}}\right)^{q_{sy8}}\right\}\cdot\lambda_{My} \tag{4.E70}$$

*Aligning Torque (combined slip)*

$$M_z = M_z' + M_{zr} + s\cdot F_x \tag{4.E71}$$

$$M_z' = -t\cdot F_y' \tag{4.E72}$$

$$t = t(\alpha_{t,eq}) = D_t \cos\left[C_t \arctan\left\{B_t \alpha_{t,eq} - E_t(B_t \alpha_{t,eq} - \arctan(B_t \alpha_{t,eq}))\right\}\right] \cos'\alpha \tag{4.E73}$$

$$F'_y = G_{y\kappa(\gamma=\varphi_t=0)} \cdot F_{yo(\gamma=\varphi_t=0)} \tag{4.E74}$$

$$M_{zr} = M_{zr}(\alpha_{r,eq}) = D_r \cos[C_r \arctan(B_r \alpha_{r,eq})] \tag{4.E75}$$

$$s = R_o \cdot \left\{s_{sz1} + s_{sz2}(F_y/F'_{zo}) + (s_{sz3} + s_{sz4} df_z)\gamma^*\right\} \cdot \lambda_s \tag{4.E76}$$

$$\alpha_{t,eq} = \sqrt{\alpha_t^2 + \left(\frac{K_{x\kappa}}{K'_{y\alpha}}\right)^2 \kappa^2} \cdot \mathrm{sgn}(\alpha_t) \tag{4.E77}$$

$$\alpha_{r,eq} = \sqrt{\alpha_r^2 + \left(\frac{K_{x\kappa}}{K'_{y\alpha}}\right)^2 \kappa^2} \cdot \mathrm{sgn}(\alpha_r) \tag{4.E78}$$

Note that actually $-S_{Hx}C_{F\kappa}$ should equal $M_y/r_l$ making $\kappa = 0$ at free rolling (drive torque $M_D = 0$). If $S_{Hx}$ is set equal to zero, $\kappa$ vanishes at $M_D = M_y$, making $F_x = 0$.

A set of parameter values has been listed in App. 3 for an example tire in connection with the *SWIFT* tire model to be dealt with in Chapter 9. Examples of computed characteristics compared with experimentally assessed curves for both pure slip and combined slip conditions are discussed in Section 4.3.6, cf. Figures 4.28–4.34. In Section 4.3.3, the effect of having turn slip is modeled and some calculated characteristics have been presented for a set of hypothetical model parameters (Figure 4.19). Section 4.3.4 examines the possibility to define the effects of conicity and ply-steer as responses to equivalent camber and slip angles. Section 4.3.5 gives the basis for the expression Eqn (4E.69) of the overturning couple of truck, car, and racing tires due to side slip and provides a discussion on the part due to camber.

## 4.3.3. Extension of the Model for Turn Slip

The model described so far contains the input slip quantities side slip, longitudinal slip, and wheel camber. In the present section, the turn slip or (in steady state) path curvature is added which completes the description of the steady-state force and moment generation properties of the tire.

Turn slip is one of the two components which together form the spin of the wheel, cf. Figure 3.22. The turn slip is defined as (last equality, with $R$ the turn radius, is valid only if the side-slip angle $\alpha$ remains zero or constant)

$$\varphi_t = -\frac{\dot{\psi}}{V'_c} = -\frac{1}{R} \tag{4.75}$$

and the total tire spin, cf. Eqn (3.55):

$$\varphi = -\frac{1}{V_c'}\{\dot{\psi} - (1-\varepsilon_\gamma)\Omega\sin\gamma\} \tag{4.76}$$

with the singularity protected velocity $V_c'$. Again, $\varepsilon_\gamma$ denotes the camber reduction factor for the camber to become comparable with turn slip. For radial-ply car tires, this reduction factor $\varepsilon_\gamma$ may become as high as ca. 0.7 (for some truck tires even slightly over 1.0), while for a motorcycle tire the factor is expected to be close to zero as with a homogeneous solid ball, cf. discussion below Eqn (3.117).

In the previous section, the effect of camber was introduced. For the side force, this resulted in a horizontal and a vertical shift of the $F_y$ vs $\alpha$ curves, while for the aligning torque, besides the small effect of the shifted $F_y$ curve, the residual torque $M_{zr}$ was added in which a contribution of camber occurs. Furthermore, changes in shape appeared to occur which were represented by the introduction of $\gamma$ in factors such as $\mu_y$, $E_y$, $K_{y\alpha}$, $B_t$, $D_t$, etc. These changes may be attributed to changes in tire cross section and contact pressure distribution resulting from the wheel inclination angle.

These shifts and shape changes will be retained in the model extension but will be expanded to cover the complete range of spin in combination with side slip and also longitudinal slip. The spin may change from zero to $\pm\infty$ when path curvature goes to infinity and, consequently, path radius to zero (when the velocity $V_c \rightarrow 0$). Weighting functions will again be introduced to gradually reduce peak side and longitudinal forces with increasing spin. Also, cornering stiffness and pneumatic trail will be subjected to such a reduction.

The theoretical findings which have been gained from the physical model of Section 3.3 presented in Figure 3.35d will be used as basis for the model development. Since these findings hold for only one value of the vertical load and zero longitudinal slip, the influence of load variations and longitudinal slip has been structured tentatively.

In Figure 4.18 for the two extreme cases of $\alpha = 0$ and $\alpha = 90°$, the courses of $F_y$ and $M_z$ vs $a\varphi_t = -a/R$ have been depicted. The pure spin side force characteristic shown in the left-hand diagram (at $\alpha = 0$), also indicated in the lower left diagram of Figure 3.35d, can be fabricated by sideways shifting of the side force curve belonging to zero spin (turn slip $\varphi_t = 0$ and $\gamma = 0$, upper or middle left diagram) while at the same time reducing its peak value $D_y$ and its slope at the curve center $K_{y\alpha}$. For this, we define the reduction functions to be substituted in the previous Eqns (4.E22, 4.E25, 4.E28, 4.E29). The peak side force reduction factor (note: $R_o$ here denotes the unloaded tire radius!)

$$\zeta_2 = \cos\left[\arctan\left\{B_{y\varphi}(R_o|\varphi| + p_{Dy\varphi4}\sqrt{R_o|\varphi|}\,)\right\}\right] \tag{4.77}$$

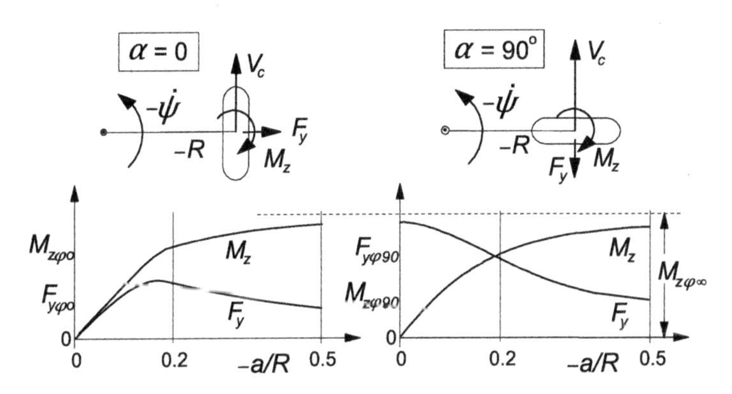


**FIGURE 4.18** Two basic spin force and moment diagrams.

with sharpness factor

$$B_{y\varphi} = p_{Dy\varphi1}(1 + p_{Dy\varphi2}df_z)\cos[\arctan(p_{Dy\varphi3}\tan\alpha)] \tag{4.78}$$

and the slope reduction factor

$$\zeta_3 = \cos[\arctan(p_{Ky\varphi1}R_o^2\varphi^2)] \tag{4.79}$$

Herewith, the condition is satisfied that at $\varphi_t \to \infty$, where the wheel is steered about the vertical axis at a speed of travel equal to zero (that is: velocity of contact center $V_c \to 0$ and path radius $R \to 0$, leading to $\zeta_2 = \zeta_3 = 0$), the side force reduces to zero although a slip angle may theoretically still exist. Considering the upper- or middle-left diagram of Figure 3.35d, it seems that the sideways shift saturates at larger values of spin. To model this phenomenon, which actually says that beyond a certain negative slip angle no spin is large enough to make the side force vanish, the sine version of the *Magic Formula* is used. We have

$$\begin{aligned} S_{Hy\varphi} =& D_{Hy\varphi}\sin\left[C_{Hy\varphi}\arctan\{B_{Hy\varphi}R_o\varphi \right. \\ &\left. - E_{Hy\varphi}(B_{Hy\varphi}R_o\varphi - \arctan(B_{Hy\varphi}R_o\varphi))\}\right]\cdot \operatorname{sgn} V_x \end{aligned} \tag{4.80}$$

The shape factor $C_{Hy\varphi}$ is expected to be equal or smaller than unity. To expression (4.80) is added the shift due to ply-steer and conicity. Finally, we subtract the horizontal displacement of the point of intersection of the side force vs slip angle curve which arises due to the vertical shift of the curve (which for now is thought to be solely attributed to the camber component of spin). This leads to a total horizontal shift:

$$S_{Hy} = (p_{Hy1} + p_{Hy2}df_z)\cdot\lambda_{Hy} + S_{Hy\varphi} - S_{Vy\gamma}/K'_{y\alpha} \tag{4.81}$$

in which the vertical shift due to $\gamma$ reads

$$S_{Vy\gamma} = F_z\cdot(p_{Vy3} + p_{Vy4}df_z)\gamma^*\cdot\zeta_2\cdot\lambda_{Ky\gamma}\cdot\lambda'_{\mu y} \tag{4.82}$$

cf. Eqn (4.E28). The quantity $K'_{y\alpha}$ is the singularity-protected cornering stiffness defined by Eqn (4.E39). Through this manipulation, the camber/spin stiffness is solely attributed to the horizontal shift of the point of intersection $S_{Hy\varphi}$, cf. Eqns (4.88, 4.89). Apparently, the factors $\zeta$ in Eqn (4.E27) now read

$$\zeta_0 = 0 \tag{4.83}$$

$$\zeta_4 = 1 + S_{Hy\varphi} - S_{Vy\gamma}/K'_{y\alpha} \tag{4.84}$$

The various factors appearing in (4.80) are defined as follows:

$$C_{Hy\varphi} = p_{Hy\varphi 1} \quad (> 0) \tag{4.85}$$

$$D_{Hy\varphi} = (p_{Hy\varphi 2} + p_{Hy\varphi 3} df_z) \cdot \mathrm{sgn}(V_{cx}) \tag{4.86}$$

$$E_{Hy\varphi} = p_{Hy\varphi 4} \quad (\leq 1) \tag{4.87}$$

$$B_{Hy\varphi} = K_{yR\varphi o}/(C_{Hy\varphi} D_{Hy\varphi} K'_{y\alpha o}) \tag{4.88}$$

where the spin force stiffness $K_{yR\varphi o}$ is related to the camber stiffness $K_{y\gamma o}$ ($= C_{F\gamma}$) that is given by (4.E30):

$$K_{yR\varphi o} = \frac{K_{y\gamma o}}{1 - \varepsilon_\gamma} \quad (= \partial F_{yo}/\partial(R_o \varphi) \text{ at } \alpha = \gamma = \varphi = 0) \quad (= C_{F\varphi}/R_o) \tag{4.89}$$

for which we may define

$$\varepsilon_\gamma = p_{\varepsilon\gamma\varphi 1}(1 + p_{\varepsilon\gamma\varphi 2} df_z) \tag{4.90}$$

Obviously, this parameter governs the difference of the response to camber with respect to that of turn slip.

For modeling the aligning torque, we will use as before the product of the side force and the pneumatic trail in the first term of Eqns (4.E31, 4.E71), while for the second term the residual torque will be expanded to represent large spin torque.

The middle-right diagram of Figure 3.35d shows that at increasing turn slip the residual or spin torque increases while the moment due to side slip (clearly visible at $\varphi_t = 0$) diminishes. This decay is modeled by means of the weighting function $\zeta_5$ multiplied with the pneumatic trail in Eqn (4.E43):

$$\zeta_5 = \cos[\arctan(q_{Dt\varphi 1} R_o \varphi)] \tag{4.91}$$

The second term of Eqns (4.E31, 4.E71), the residual torque, which in the present context may be better designated as the spin moment, is given by Eqns (4.E36, 4.E75). Its peak value $D_r$ has an initial value due to conicity that is expected to be taken over gradually by an increasing turn slip. In (4.E47), this is accomplished by the weighting function $\zeta_2$ (4.77). The remaining terms will be replaced by the peak spin torque $D_{r\varphi}$. This means that the $\zeta$'s appearing in (4.E47) become, according to (4.83),

$$\zeta_8 = 1 + D_{r\varphi} \tag{4.92}$$

As observed in the middle-right diagram of Figure 3.35d, the peak torque (that will be assumed to occur at a slip angle $\alpha = -S_{Hf}$ where $F_y = 0$) grows with increasing spin and finally saturates. *The Magic Formula* describes this as follows:

$$\begin{aligned} D_{r\varphi} =& D_{Dr\varphi} \sin\left[C_{Dr\varphi} \arctan\left\{B_{Dr\varphi} R_o \varphi \right.\right. \\ & \left.\left. - E_{Dr\varphi}(B_{Dr\varphi} R_o \varphi - \arctan(B_{Dr\varphi} R_o \varphi))\right\}\right] \end{aligned} \tag{4.93}$$

Its maximum value (if $C_{Dr\varphi} \geq 1$) is $D_{Dr\varphi}$. The asymptotic level of the peak spin torque is reached at $\varphi \rightarrow \infty$ or $R \rightarrow 0$, and is denoted as $M_{z\varphi\infty}$. Consequently, with the shape factor $C_{Dr\varphi}$, the maximum value is expressed by

$$D_{Dr\varphi} = M_{z\varphi\infty} / \sin(0.5\pi C_{Dr\varphi}) \tag{4.94}$$

where the moment $M_{z\varphi\infty}$ that occurs at vanishing wheel speed and at constant turning about the vertical axis is formulated as a function of the normal load:

$$M_{z\varphi\infty} = q_{Cr\varphi 1} \mu_y R_o F_z \sqrt{F_z / F'_{zo}} \cdot \lambda_{M\varphi} \quad (> 0) \tag{4.95}$$

This expression may be compared with or replaced by expression (3.117) formulated by Freudenstein (1961). The shape factors in (4.93) are assumed to be given by constant parameters:

$$C_{Dr\varphi} = q_{Dr\varphi 1} \quad (> 0) \tag{4.96}$$

$$E_{Dr\varphi} = q_{Dr\varphi 2} \quad (\leq 1) \tag{4.97}$$

while

$$B_{Dr\varphi} = K_{z\gamma ro} / \{C_{Dr\varphi} D_{Dr\varphi} (1 - \varepsilon_\gamma) + \varepsilon_r\} \tag{4.98}$$

with $K_{z\gamma ro}$ assumed to depend on the normal load as follows:

$$K_{z\gamma ro} = F_z R_o \cdot \{q_{Dz8} + q_{Dz9} df_z + (q_{Dz10} + q_{Dz11} df_z)|\gamma|\} \cdot \lambda_{Kz\gamma} \tag{4.99}$$

As has been indicated with Eqn (4.E49), we have, now for the camber moment stiffness,

$$K_{z\gamma o} = K_{z\gamma ro} - D_{to} K_{y\gamma o} \quad (= C_{M\gamma}) \tag{4.100}$$

and consequently for the moment stiffness against spin:

$$K_{zR\varphi o} = K_{z\gamma o} / (1 - \varepsilon_\gamma) \quad (= C_{M\varphi} / R_o) \tag{4.101}$$

Now that the formulas for the peak of the spin torque $D_{r\varphi}$ have been developed, we must consider the remaining course of $M_{zr}$ with the slip angle $\alpha$. For this, Eqn (4.E36) is employed. According to the middle-right diagram of Figure 3.35d, the curves become flatter as the spin increases. Logically, for $\varphi \to \infty$ the moment should become independent of the slip angle. The sharpness is controlled by the factor $B_r$ which we let gradually decrease to zero with increasing spin. For this, we introduce in Eqn (4.E45) the weighting function:

$$\zeta_6 = \cos[\arctan(q_{Br\varphi 1} R_o \varphi)] \tag{4.102}$$

The factor $C_r$ controls the asymptotic level which corresponds to the torque at $\alpha = 90°$. In that situation, the turn slip $\varphi_t = -\dot{\psi}/|V_{cy}| = 1/R$ which corresponds to the definition given by Eqn (4.76). The moment $M_{z\varphi 90}$ that is generated when the wheel moves at $\alpha = 90°$ increases with increasing curvature $1/R$ up to its maximum value that is attained at $R = 0$ and equals $M_{z\varphi\infty}$. We use the formula

$$M_{z\varphi 90} = M_{z\varphi\infty} \cdot \frac{2}{\pi} \cdot \arctan\left(q_{Cr\varphi 2} R_o |\varphi_t|\right) \cdot G_{y\kappa}(\kappa) \tag{4.103}$$

with parameter $q_{Cr\varphi 2}$. This quantity may be difficult to assess experimentally; the value 0.1 is expected to be a reasonable estimate. (In any case, the argument of (4.104) should remain $<1$.) This moment at 90° is multiplied with the weighting function $G_{y\kappa}$ (4.E59) to account for the attenuation through the action of longitudinal slip.

We obtain, for the factor $\zeta_7 = C_r$ in Eqn (4.E46) using (4.E36) with $|\alpha_r| \to \infty$,

$$\zeta_7 = \frac{2}{\pi} \cdot \arccos\left[M_{z\varphi 90} / \left(|D_{r\varphi}| + \varepsilon_r\right)\right] \tag{4.104}$$

Finally, we must take care of the weighting function $\zeta_1$ which is introduced in the expressions (4.E12, 4.E18) for $F_x$ to let the longitudinal force diminish with increasing spin. We define

$$\zeta_1 = \cos[\arctan(B_{x\varphi} R_o \varphi)] \tag{4.105}$$

with

$$B_{x\varphi} = p_{Dx\varphi 1}(1 + p_{Dx\varphi 2} df_z)\cos[\arctan(p_{Dx\varphi 3}\kappa)] \tag{4.106}$$

With the factors $\zeta_0$, $\zeta_1$,...$\zeta_8$ determined and substituted in the Eqns (4.E9–4.E78), the description of the steady-state force and moment generation has been completed. To show that the formulas produce qualitatively correct results, a collection of computed curves has been presented in the diagrams of Figure 4.19. In Table 4.2 the values of the model parameters have been listed. The diagrams show that the formulas are perfectly capable to at least qualitatively approximate the curves that have been computed with the brush simulation model of Chapter 3 (cf. Figure 3.35). It has not been attempted to find a best fit for the parameters.

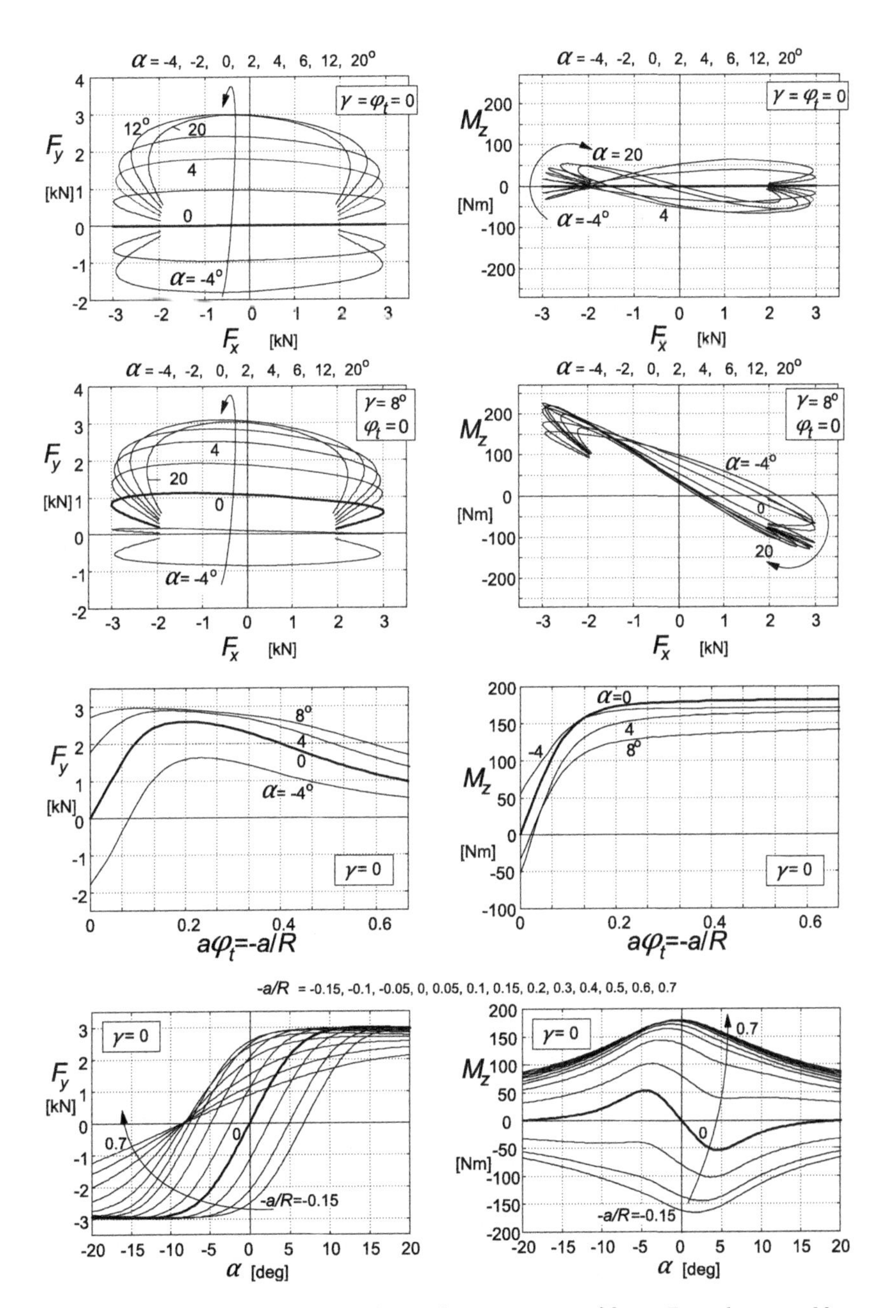


**FIGURE 4.19** *Magic Formula* results for steady-state response of forces $F_{x,y}$ and moment $M_z$ to slip angle $\alpha$, slip ratio $\kappa$, camber angle $\gamma$, and nondimensional path curvature $\varphi_t = -a/R$.

**TABLE 4.2** *Magic Formula* Hypothetical Model Parameter Values for Diagrams of Figure 4.19. Parameters that Govern the Influence of Longitudinal Slip and Changes in Vertical Load have been given the Value Zero

| $F_z = 3000$ N $F_{zo} = 3000$ $R_o = 0.3$ m $R_e = R_o$ (a = 0.1 m) | | | | | | |
|---|---|---|---|---|---|---|
| $p_{Cx1} = 1.65$ | $p_{Dx1} = 1$ | $p_{Dx2} = 0$ | $p_{Ex1} = -0.5$ | $p_{Ex2} = 0$ | $p_{Ex3} = 0$ | $p_{Ex4} = 0$ |
| $p_{Kx1} = 12$ | $p_{Kx2} = 10$ | $p_{Kx3} = -0.6$ | $p_{Hx1} = 0$ | $p_{Hx2} = 0$ | $p_{Vx1} = 0$ | $p_{Vx2} = 0$ |
| $p_{Cy1} = 1.3$ | $p_{Dy1} = 1$ | $p_{Dy2} = 0$ | $p_{Dy3} = 0$ | | | |
| $p_{Ey1} = -1$ | $p_{Ey2} = 0$ | $p_{Ey3} = 0$ | $p_{Ey4} = 0$ | | | |
| $p_{Ky1} = 10$ | $p_{Ky2} = 1.5$ | $p_{Ky3} = 0$ | $p_{Ky4} = 2$ | $p_{Ky5} = 0$ | $p_{Ky6} = 2.5$ | $p_{Ky7} = 0$ |
| $p_{Hy1} = 0$ | $p_{Hy2} = 0$ | $p_{Vy1} = 0$ | $p_{Vy2} = 0$ | $p_{Vy3} = 0.15$ | $p_{Vy4} = 0$ | |
| $q_{Bz1} = 6$ | $q_{Bz2} = -4$ | $q_{Bz3} = 0.6$ | $q_{Bz4} = 0$ | $q_{Bz5} = 0$ | $q_{Bz9} = 0$ | $q_{Bz10} = 0.7$ |
| $q_{Cz1} = 1.05$ | $q_{Dz1} = 0.12$ | $q_{Dz2} = -0.03$ | $q_{Dz3} = 0$ | $q_{Dz4} = -1$ | $q_{Dz6} = 0$ | $q_{Dz7} = 0$ |
| $q_{Dz8} = 0.6$ | $q_{Dz9} = 0.2$ | $q_{Dz10} = 0$ | $q_{Dz11} = 0$ | $q_{Ez1} = -10$ | $q_{Ez2} = 0$ | $q_{Ez3} = 0$ |
| $q_{Ez4} = 0$ | $q_{Ez5} = 0$ | $q_{Hz1} = 0$ | $q_{Hz2} = 0$ | $q_{Hz3} = 0$ | $q_{Hz4} = 0$ | |
| $r_{Bx1} = 5$ | $r_{Bx2} = 8$ | $r_{Bx3} = 0$ | $r_{Cx1} = 1$ | $r_{Hx1} = 0$ | | |
| $r_{By1} = 7$ | $r_{By2} = 2.5$ | $r_{By3} = 0$ | $r_{By4} = 0$ | $r_{Cy1} = 1$ | $r_{Hy1} = 0.02$ | |
| $r_{Vy1} = 0$ | $r_{Vy2} = 0$ | $r_{Vy3} = -0.2$ | $r_{Vy4} = 14$ | $r_{Vy5} = 1.9$ | $r_{Vy6} = 10$ | |
| $s_{sz1} = 0$ | $s_{sz2} = -0.1$ | $s_{sz3} = -1.0$ | $s_{sz4} = 0$ | | | |
| $p_{Dx\varphi 1} = 0.4$ | $p_{Dx\varphi 2,3} = 0$ | $p_{Ky\varphi 1} = 1$ | $p_{Dy\varphi 1} = 0.4$ | $p_{Dy\varphi 2} = 0$ | $p_{Dy\varphi 3,4} = 0$ | |
| $p_{Hy\varphi 1} = 1$ | $p_{Hy\varphi 2} = 0.15$ | $p_{Hy\varphi 3} = 0$ | $p_{Hy\varphi 4} = -4$ | $p_{\varepsilon y\varphi 1} = 0$ | $p_{\varepsilon y\varphi 2} = 0$ | |
| $q_{Dt\varphi 1} = 10$ | $q_{Cr\varphi 1} = 0.2$ | $q_{Cr\varphi 2} = 0.1$ | $q_{Br\varphi 1} = 0.1$ | $q_{Dr\varphi 1} = 1$ | $q_{Dr\varphi 2} = -1.5$ | |

### 4.3.4. Ply-Steer and Conicity

In the formulas for the side force $F_y$ and the aligning torque $M_z$, the vertical shift $S_{Vy}$, the horizontal shift $S_{Hy}$, and the residual torque peak value $D_r$ contain terms which produce the initial values of side force and aligning torque that occur at straight ahead running (at $\alpha = 0$). These initial values are known to be the result of conicity and ply-steer, which are connected with nonsymmetry of the tire construction. These two possible sources result in markedly different behavior of the tire when, at geometrically zero side slip, the tire is rolled forward or backward. If we would have a tire that exhibits ply-steer but no conicity, the generated side force will point to the opposite direction when the wheel is changed from forward to backward rolling. This would also be the case when on a test rig the wheel moves at a small steer angle $\psi$ and the road surface motion is changed from backward to forward. For that reason, ply-steer is sometimes referred to as pseudo-side-slip. If, on the other hand, the tire would show pure conicity, the side force will remain pointing in the same direction when the wheel is rolled in the opposite direction. This behavior is similar to that of a cambered wheel, which explains the term: pseudo-camber. In Figure 4.20 for the different cases, the diagrams for the side force variation resulting from yaw angle variations have been depicted. The curved or skewed foot prints of the tire that due to nonsymmetric construction of carcass and belt would arise on a zero friction surface explains the resulting characteristics. Definitions of conicity and ply-steer follow from the forces found at zero steer angle $\psi$.

The deformations of the tire rolling on a friction surface resemble those that would occur with a tire (free of conicity and ply-steer) that rolls at a small camber and side slip angle, respectively. It is therefore tempting to assume that in these comparable cases the moment $M_z$ and the force $F_y$ respond at the same rate which would mean that the associated pneumatic trails are equal for

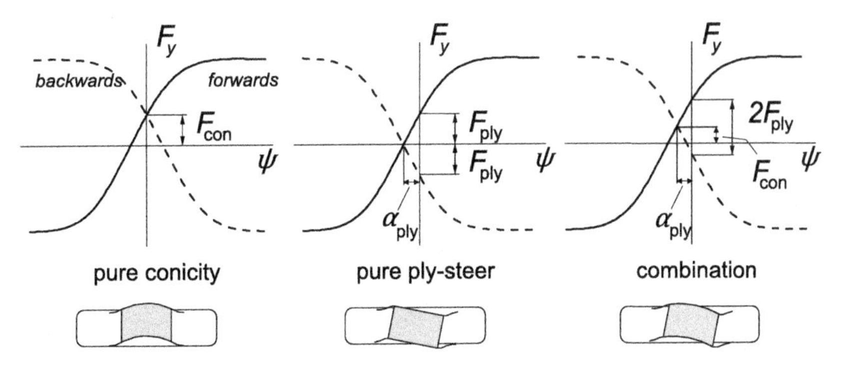


**FIGURE 4.20** Side force vs steer angle characteristics at forward and backward rolling showing the cases of pure conicity, pure ply-steer, and a combination of both situations. Below, the associated foot prints on a frictionless surface have been depicted.

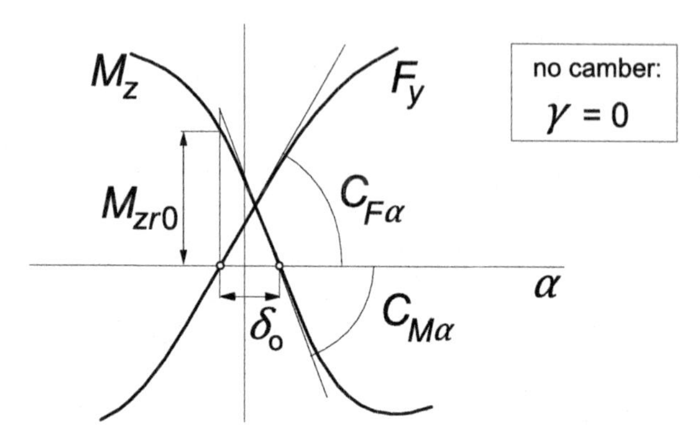


**FIGURE 4.21** The situation near zero slip angle, force, and moment due to conicity and ply-steer.

ply-steer and side slip and for conicity and camber. The assumption is partly supported by an extensive experimental investigation of Lee (2000), which assessed a strong linear correlation between conicity and the difference between slip angles where $F_y$ and $M_z$ become equal to zero. In Figure 4.21, this difference is designated as $\delta_o$. Lee found that at $\delta_o = 0$ conicity is almost zero. This means that for vanishing conicity the remaining ply-steer produces a moment that is approximately equal to minus the pneumatic trail for side slip times the side force. With the introduction of the (small) equivalent slip angle $\alpha_{\text{ply}}$, we have

$$M_{\text{ply}} \approx -C_{M\alpha}\alpha_{\text{ply}} \quad \text{with} \quad C_{M\alpha} = tC_{F\alpha} \tag{4.107}$$

and

$$F_{\text{ply}} \approx C_{F\alpha}\alpha_{\text{ply}} \tag{4.108}$$

Also according to Lee, the residual torque at zero side force, $M_{zr0}$ in Figure 4.21, is strongly correlated with $\delta_o$ and thus with conicity. We find, from the diagram,

$$M_{zr0} \approx C_{M\alpha}\delta_o \tag{4.109}$$

If we may further assume that for the conicity force and moment a similar correspondence with camber response exists, we would have after introducing an equivalent camber angle $\gamma_{\text{con}}$:

$$M_{\text{con}} \approx C_{M\gamma}\gamma_{\text{con}} \quad \text{with} \quad C_{M\gamma} = t_\gamma C_{F\gamma} \tag{4.110}$$

where $t_\gamma$ (>0) represents the distance of the point of application of the resulting camber thrust in front of the contact center (that is: negative trail). The conicity force becomes

$$F_{\text{con}} \approx C_{F\gamma}\gamma_{\text{con}} \tag{4.111}$$

With these assumptions, it is possible to estimate the contributions of ply-steer and conicity in the initial values of side force and aligning torque (at $\alpha = \gamma = 0$) from a set of tire parameters that belongs to one direction of rolling.

This enables the vehicle modeler to switch easily from the set of parameter values of the tire model that runs on the left-hand wheels of a car to the set for the right-hand wheels. To accomplish this, the equivalent camber parts of the left-hand tire model must be changed in sign for both the aligning moment and the side force to make the model suitable for the right-hand tire. Also, it is then easy to omit e.g., all conicity contributions which were originally present in the tire from which the parameters have been assessed through the fitting process.

To develop the theory, we will first consider the case of a tire without ply-steer and conicity and study the situation when a camber angle is applied. This condition is reflected by the diagram of Figure 4.22. Considering the relations entered in this figure, we may write, for the distance $\delta_\gamma$,

$$\delta_\gamma = \left(\frac{C_{M\gamma}}{C_{M\alpha}} + \frac{C_{F\gamma}}{C_{F\alpha}}\right)\gamma \tag{4.112}$$

and consequently we can find the camber angle $\gamma$ from the distance $\delta_\gamma$. In a similar fashion, the conicity will be assessed in terms of the equivalent camber angle and after that, the part attributed to ply-steer can be determined and expressed in terms of the equivalent slip angle. Figure 4.23 illustrates the conversion to equivalent angles. We find, similar to the inverse of (4.112),

$$\gamma_{\text{con}} = \frac{\delta_o}{\dfrac{C_{M\gamma}}{C_{M\alpha}} + \dfrac{C_{F\gamma}}{C_{F\alpha}}} \tag{4.113}$$

The equivalent slip angle can now be obtained from

$$\alpha_{\text{ply}} = \Delta\alpha_o - \frac{C_{F\gamma}}{C_{F\alpha}}\gamma_{\text{con}} \tag{4.114}$$

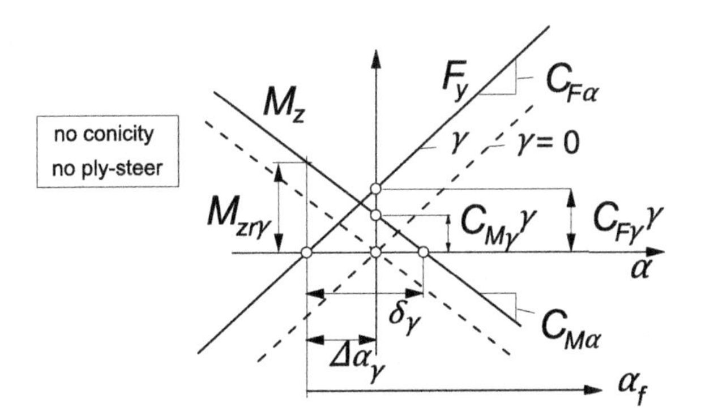


**FIGURE 4.22** Characteristics of a tire without ply-steer or conicity near the origin at $\alpha = 0$, with and without camber.

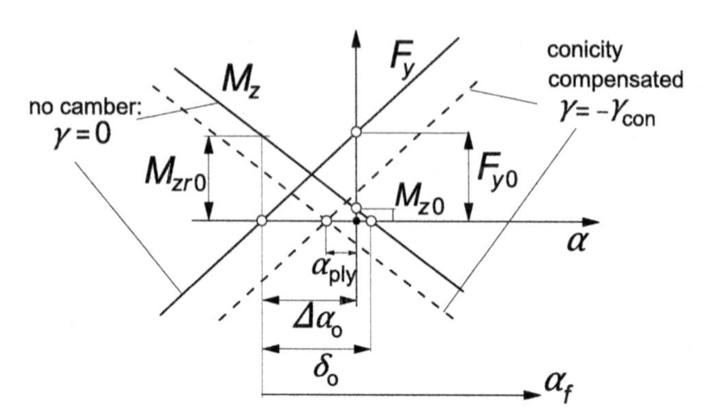


**FIGURE 4.23** Characteristics of a tire with ply-steer and conicity showing conversion to equivalent camber and slip angle.

The four slip and camber stiffnesses are available from Eqns (4.E25, 4.E30, 4.E48, 4.E49). The sign of the forward velocity has been properly introduced.

The next thing we have to do is expressing $\delta_o$ and $\Delta\alpha_o$ in terms of the shifts and the initial residual torque as defined in Section 4.3.2. We obtain, with (4.E38),

$$\Delta\alpha_o = S_{Hf(\gamma=\varphi=0)} \tag{4.115}$$

and, from (4.109) and Figure 4.23,

$$\delta_o = \frac{F_{y0}}{C_{F\alpha}} + \frac{M_{z0}}{C_{M\alpha}} = \frac{M_{zr0}}{C_{M\alpha}} \tag{4.116}$$

with $C_{M\alpha}$ according to Eqn (4.E48) and the initial residual torque from (4.E47):

$$M_{zr0} = D_{ro} = F_z R_o (q_{Dz6} + q_{Dz7} df_z) \cdot \lambda_{Mr} \tag{4.117}$$

Finally, the initial side force and torque are to be removed from the equations by putting the parameters $p_{Hy1}$, $p_{Hy2}$, $p_{Vy1}$, $p_{Vy2}$, $q_{Dz6}$, and $q_{Dz7}$ or the scaling factors $\lambda_{Hy}$, $\lambda_{Vy}$, and $\lambda_{Mr}$ equal to zero and by replacing in Eqns (4.E20, 4.E37) the original side slip input variable $\alpha^* = \tan\alpha \cdot \mathrm{sgn} V_{cx}$ by its effective value:

$$\alpha^*_{\mathrm{eff}} = (\tan\alpha + \alpha_{\mathrm{ply}})\mathrm{sgn} V_{cx} \tag{4.118}$$

and, in Eqns (4.E27, 4.E28, 4.E47) the original camber, $\gamma^* = \sin\gamma$ by the effective total camber:

$$\gamma^*_{\mathrm{eff}} = \sin(\gamma + \gamma_{\mathrm{con}}) \tag{4.119}$$

where it should be realized that both $\alpha_{\mathrm{ply}}$ and $\gamma_{\mathrm{con}}$ are small quantities. The resulting diagram of Figure 4.24 shows characteristics that pass through the origin when the effective camber angle is equal to zero. For the tire running on the other side of the vehicle, the same model can be used but with $\gamma_{\mathrm{con}}$ in (4.119) changed in sign.

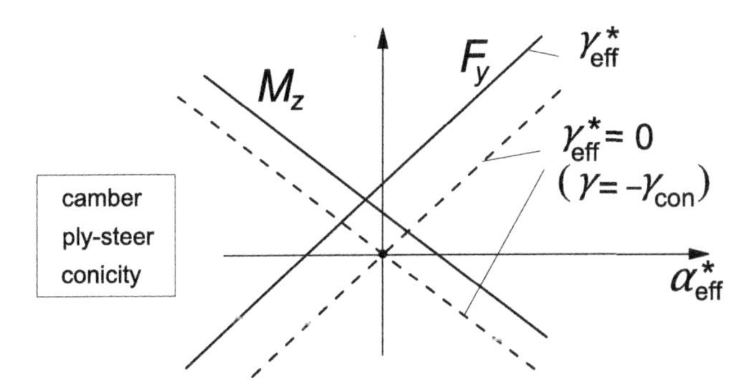


**FIGURE 4.24** The final diagram with effective side slip and camber variables.

The theory above is based on considerations near the origin of the side force and aligning torque vs slip angle diagrams. The introduction of effective slip and camber angles may, however, give rise to slight changes in peak levels of the side force and probably also of the aligning torque. For the former, the situation may be repaired by moving the force characteristic in a direction parallel to the tangent at $F_y = 0$ resulting in an additional vertical shift:

$$\Delta S_{Vy} = F_z \cdot \{(p_{Vy1} + p_{Vy2} df_z) - (p_{Vy3} + p_{Vy4} df_z)\gamma_{con}\} \tag{4.120}$$

and an associated additional horizontal shift:

$$\Delta S_{Hy} = -\Delta S_{Vy}/K'_{y\alpha} \tag{4.121}$$

## *Tire Pull*

If identical tires would be fitted on the front axle of an automobile but with the conicity forces pointing in the same direction, and the vehicle moves along a straight line (that is: side forces are equal to zero), a steering torque must be applied that opposes the residual torques $M_{zr0}$ generated by the front tires (Figure 4.21). This is actually only approximately true because we may neglect the trails connected with ply-steer and conicity with respect to the vehicle wheel base. (In reality we have a small couple acting on the car exerted by the equal but opposite side forces front and rear which counteract the small, mainly conicity, torques front, and rear.) The phenomenon that a steer torque must be applied when moving straight ahead is called tire or vehicle pull. If the steering wheel would be released, the vehicle will deviate from its straight path.

If on the right-hand wheel a tire is mounted that is identical with the left-hand tire, one would actually expect that the conicity forces are directed in opposite directions and neutralize each other (as will occur also with the moments). This is because of the observation that one might compare the condition on the right-hand side with an identical tire rolling backward with respect to condition of the left-hand tire. In contrast, the ply-steer forces of the left and right tires act in the same direction but are compensated by side forces that arise through a small slip angle of the whole vehicle.

If the ply-steer angles, front and rear, are not the same, a small steer angle of the front wheels equal to the difference of the ply-steer slip angles front and rear is required for the vehicle to run straight ahead. The whole vehicle will run at a slip angle equal to that of the rear wheels.

### 4.3.5. The Overturning Couple

The overturning couple is especially important to investigate the vehicle roll-over occurrence and the curving behavior of a motorcycle. In moderate conditions, Eqn (4.E69) may often be sufficient to model the overturning moment. In the present section, we will study cases that require more elaborate modeling. The overturning couple is generated through the action of the side force and due to the camber angle. Eqn (4.E69) gives an expression valid for the combination of both components, further developed by TNO Automotive.

#### *Due to Side Slip*

Experience with measuring truck tires, racing tires, and passenger car tires shows that a positive overturning couple arises especially at low vertical loads as a response to positive side slip (situation depicted in Figure 4.25) while camber remains zero. This is probably in contrast to the behavior of narrow passenger car or motorcycle tires where, due to the lateral deflection connected with the side force $F_y$, the point of application of the resultant vertical force

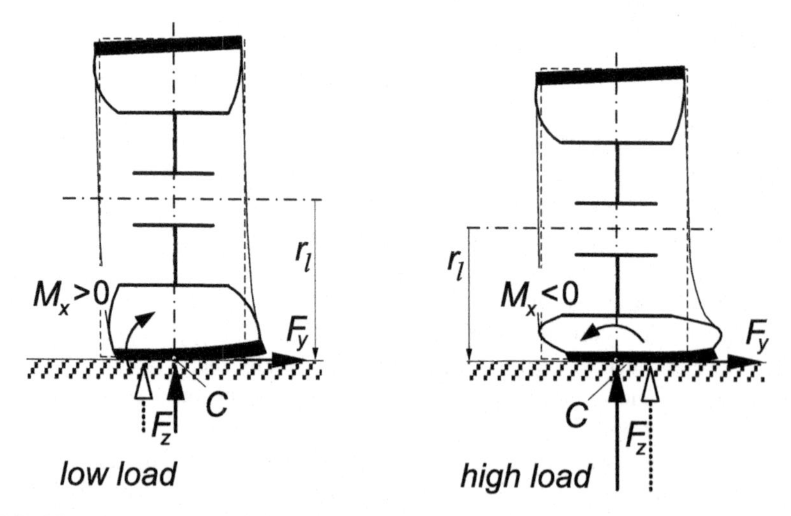


**FIGURE 4.25** Tire at low load and at high load subjected to side slip giving rise to an overturning couple that may be of opposite sign.

$F_z$ moves in the direction of the side force and as a result generates a negative couple $M_x$. The observation that a tire may show a positive moment is expected to be connected with the belt that is stiff in torsion in the contact zone. This high stiffness opposes the flattening of the belt to remain in contact with the road when the belt is tilted through the action of the side force. This will especially be true when the contact patch is short, that is, at low load. As a result, the resultant vertical force vector will move more toward the edge away from the side force which causes the contact pressure to become relatively high on that side of the contact patch, thereby enabling the overturning couple to become positive. At higher loads, the middle portion of the belt contact range will show less resistance to remain horizontal and the overturning couple may turn to the more common negative values. It is of importance to note that the positive overturning couple at low loads and the ensuing sign change is observed to occur when measurements are conducted on, e.g., a flat plank or flat track machine where the axle height is recorded accurately and in the processing stage proper account is given to the standard definition of the overturning couple, that is the moment about the line of intersection of wheel center plane and ground plane. If the change in loaded radius is disregarded as is often done in past practice when measuring with over the road test vehicles, the positive part and the connected sign change is not observed. Consequently, it is of crucial importance that the tire test engineer and the vehicle dynamicist take care of employing, at least, the same definition for the overturning couple.

The phenomenon may in some way be connected with the negative camber stiffness that is observed to occur with some type of tires also especially at low loads turning to positive at higher loads. The negative camber force generally changes to positive values when the camber angle becomes sufficiently large. At moderate loads, this also appears to happen with the overturning couple changing from positive to negative values at larger slip angles, that is: at higher side forces. The overturning couple response to wheel camber is always negative as might be expected. Obviously, the associated camber moment stiffness to be defined as $C_{Mx\gamma} = -\partial M_x/\partial\gamma$ at zero camber is positive.

To model such a sign changing situation, the overturning couple response to side slip of the wheel is taken as an example. We will use a formula that is established by adding two functions:

$$M_x = -F_z(y_1 + y_2) \tag{4.122}$$

with the two contributions to the moment arm:

$$y_1 = R_o q_{x1} \frac{F_y}{F_{zo}} \tag{4.123}$$

$$y_2 = -R_o q_{x2} \cos\left\{ q_{x3} \arctan(q_{x4} F_z/F_{zo})^2 \right\} \cdot \sin\{ q_{x5} \arctan(q_{x6} F_y/F_{zo} \} \tag{4.124}$$

**TABLE 4.3** Parameter Values for Overturning Couple Computation, cf. Eqns (4.120–4.122)

| $F_{zo} = 4.4$ kN | $R_o = 0.3$ m | | | | |
|---|---|---|---|---|---|
| $\mu_y = 1$ | $C_y = 1.4$ | $E_y = -1$ | $p_{Ky1} = 20$ | $p_{Ky2} = 1.5$ | $p_{Ky4} = 2$ |
| $q_{x1} = 0.042$ | $q_{x2} = 0.56$ | $q_{x3} = 0.955$ | $q_{x4} = 2.35$ | $q_{x5} = 1.25$ | $q_{x6} = 0.46$ |

The first term $y_1$ increases with the side force. The second term $y_2$ grows, at least initially, with $-F_y$ and diminishes with vertical load $F_z$. For the parameter values listed in Table 4.3, also containing assumed parameters for the side force characteristic, Eqns (4.E19–4.E26), the overturning couple has been calculated. Figure 4.26 shows the resulting curves. The same tendencies have been found to occur both with truck tires and passenger car tires. For the latter, we may refer to the diagram given in Figure 4.30. The same figure, which has been reproduced from the dissertation of Van der Jagt (2000), contains the diagram showing the decrease of the loaded radius with slip angle. This information is needed to assess the moment about the longitudinal axis through the wheel center. In Chapter 9, a formula is given which assumes that the decrease of the loaded radius is proportional to the squares of the lateral and the longitudinal deflections of the carcass, cf. Eqn (9.222). Experiments indicate that also the tire vertical stiffness is affected by horizontal tire forces, cf. Reimpell (1986), Eqn (9.217).

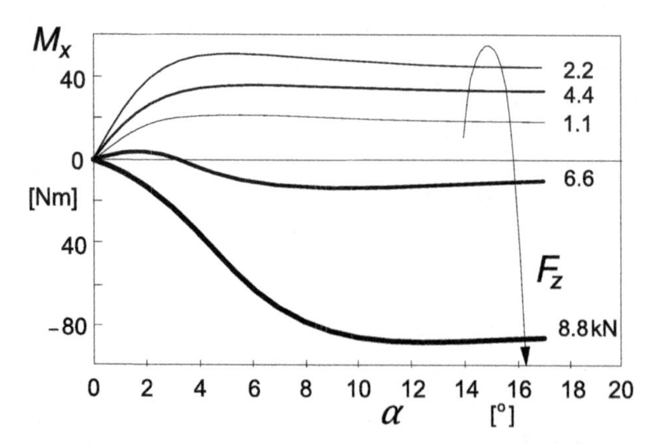


**FIGURE 4.26** Overturning couple calculated with formulas (4.122–4.124), cf. Figure 4.30 for similar measured characteristics.

### Due to Camber

The second case to be dealt with concerns the motorcycle tire. A special problem can be identified that concerns the possibly very large wheel inclination angle. Similar situations may be encountered with a car tire when examining the roll-over event or when the wheel runs over a locally steep transverse slope. In these cases, the geometry of the cross section of the tire with its finite width becomes of importance and is largely responsible for the overturning couple that arises as a result of the wheel inclination angle with respect to the road surface.

In Figure 4.27, the configuration at very large camber of a motorcycle wheel has been depicted. For simplicity, the tire cross section has been assumed to be purely circular. Alternatively, one may take an ellipse (cf. Figure 7.14). Of course, if instead of relying on a possibly less precise measurement, the overturning couple may be derived from a calculation where the location of the point of application of the normal force and its distance to the line of intersection (i.e., to point $C$) which is the moment arm, is of crucial importance, the actual measured cross section contour of the undeformed tire, free radius versus distance to wheel center plan $r_{yo}(y_{co})$, should be utilized. Note that at the point of contact with the not yet deformed tire, we have: $dr_{yo}/dy_{co} = -\tan\gamma$. As a consequence of the definition of the point of intersection or contact center $C$ which is located on the line of intersection of the wheel center plane and the road plane, the loaded radius $r_l$

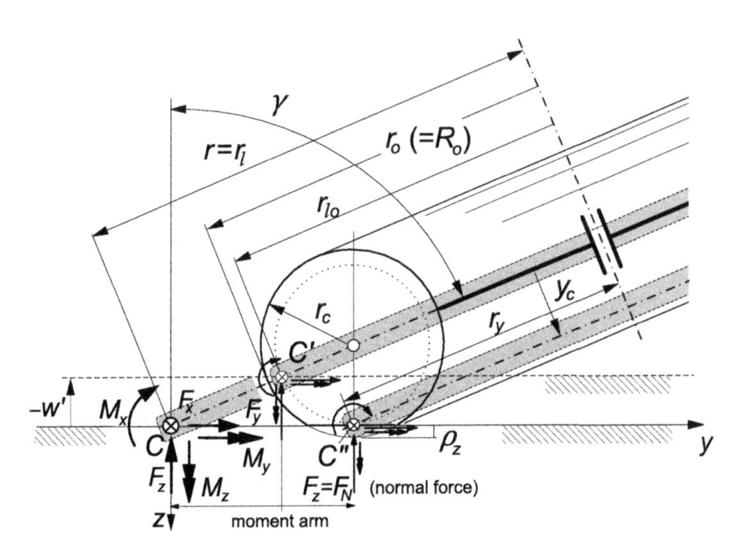


**FIGURE 4.27** The motorcycle wheel at very large camber angle $\gamma$. Two alternative approaches to define the contact center $C$ and the associated 'loaded radius' $r_l$ and as a consequence the overturning couple $M_x$. Instead of the circle, the actual section contour may be taken.

which by definition is equal to the distance between wheel center and contact center may become excessively large when the camber angle becomes large.

Although still perfectly correct, this standard definition may seem less attractive and one might want to choose for an alternative approach. Two possible alternatives have been indicated in Figure 4.27. The first one employs a value for the loaded radius $r_l'$ that remains equal to the value that holds for the upright wheel: $r_{lo}$ which is equal to the free radius $r_o$ minus the normal deflection $\rho_z$. It should then be realized that a concurrent (slight) rise $-w'$ of the effective road plane (cf. Chap.10) may not be negligible. This method to acquire the overturning couple $M_x'$ corresponds to the practical procedure often employed when measuring on the road where the actual height of the wheel center above the road surface is not measured. The second alternative is possibly the physically most realistic one. It defines a contact center located $C''$ above the lowest point of the undeformed tire. A drawback is that an effective wheel plane has to be defined in the vehicle dynamic model that is shifted with respect to the wheel center plane over a distance $y_c$.

It turns out that the overturning couples that would act about the three possible virtual contact centers, defined above, become quite different from each other although the moment about the longitudinal axis through the wheel center remains, of course, the same. We find successively if the resultant normal force is assumed to act along a line that passes through the lowest point of the (supposedly circular) tire cross section and a through $F_y$ with lateral compliance induced additional shift is disregarded:

**1.** *Standard definition, using C* (for general contour and for circular section):

$$r_l = r_{yo} + y_{co}\tan\gamma - \rho_z/\cos\gamma = r_o - r_c + (r_c - \rho_z)/\cos\gamma \tag{4.125}$$

$$M_x = -F_z(y_{co}/\cos\gamma - \rho_z\tan\gamma) = -F_z(r_c - \rho_z)\tan\gamma \tag{4.126}$$

**2.** *Practical definition, using C′*

$$r_l' = r_{lo} = r_o - \rho_z \tag{4.127}$$

$$M_x' = -F_z(r_c - \rho_z)\sin\gamma + F_y w' \tag{4.128}$$

$$w' = -(r_c - \rho_z)(1 - \cos\gamma) \tag{4.129}$$

**3.** *Physically more realistic definition, using C″*

$$r_l'' = r_y = r_o - r_c + (r_c - \rho_z)\cos\gamma \tag{4.130}$$

$$M_x'' = 0 \tag{4.131}$$

$$y_c = (r_c - \rho_z)\sin\gamma \tag{4.132}$$

The forces act in the points $C$, $C'$, and $C''$, respectively.

The overturning couple $M_x$ according to the standard definition (4.126) is directly responsible for the lean angle of the motorcycle in a steady turn to become larger than what would be expected according to the ratio of centrifugal force and vehicle weight. In the practical definition, it would be the combination of $M_x'$ and the effective road surface rise $w'$ that causes the increase in roll angle, while with the third definition it would be due to the lateral shift $y_c$ of the virtual wheel plane.

Irrespective of the definition employed, it must be clear how the measured data have been processed and converted from the wheel axle system of axes (where in general the forces are measured) to the road and line of intersection-based axes system. An additional conversion may be necessary to suit the requirements of the vehicle dynamicist. It may be realized that with respect to the standard definition of the loaded radius $r_l$, Eqn (4.125), the effective rolling radius $r_e$ follows a quite different course. Also at large camber angles, $r_e$ will remain close to the radius $r_l'' = r_y$ defined by (4.130) and shown in Figure 4.27.

To account for the effect of the lateral tire deflection and through that the change in vertical pressure distribution, the part of $M_x$ attributed to wheel camber and defined by one of the above or similar expressions should be extended with the part of the overturning couple described by Eqns (4.122–4.124) or, if appropriate, by a simpler version of that.

### Note on the Aligning Torque at Large Camber

It may be of interest to be aware of another unexpected phenomenon that arises at large camber angles. One might make a mistake when deriving the aligning torque from the wheel axle-oriented measured forces and moments. If these latter quantities are denoted with an additional subscript $a$, the aligning torque is obtained from the moment equilibrium about the vertical axis as follows (cf. Figure 4.27):

$$M_z = M_{za} \cos \gamma + M_{ya} \sin \gamma - F_{xa} r \sin \gamma \tag{4.133}$$

or alternatively from the moment equilibrium about the wheel radius:

$$M_z = M_y \tan \gamma + M_{za} / \cos \gamma \tag{4.134}$$

If the fact is not recognized that at camber a longitudinal force $F_x$ is generated even at free rolling where $M_{ya}$ is zero or very small, and consequently, $F_{xa}$ is disregarded in Eqn (4.133), the calculated aligning torque is incorrect. For a freely rolling tire with $M_{ya} = 0$, the longitudinal force appears to read

$$F_x = -\frac{1}{r} (M_z \sin \gamma + M_y \cos \gamma) \tag{4.135}$$

### 4.3.6. Comparison with Experimental Data for a Car, a Truck, and a Motorcycle Tire

In Figures 4.28, 4.29 and 4.31, 4.32, the computed and measured characteristics of a 195/65 R15 car tire (on the road with Delft Tire Test Trailer,

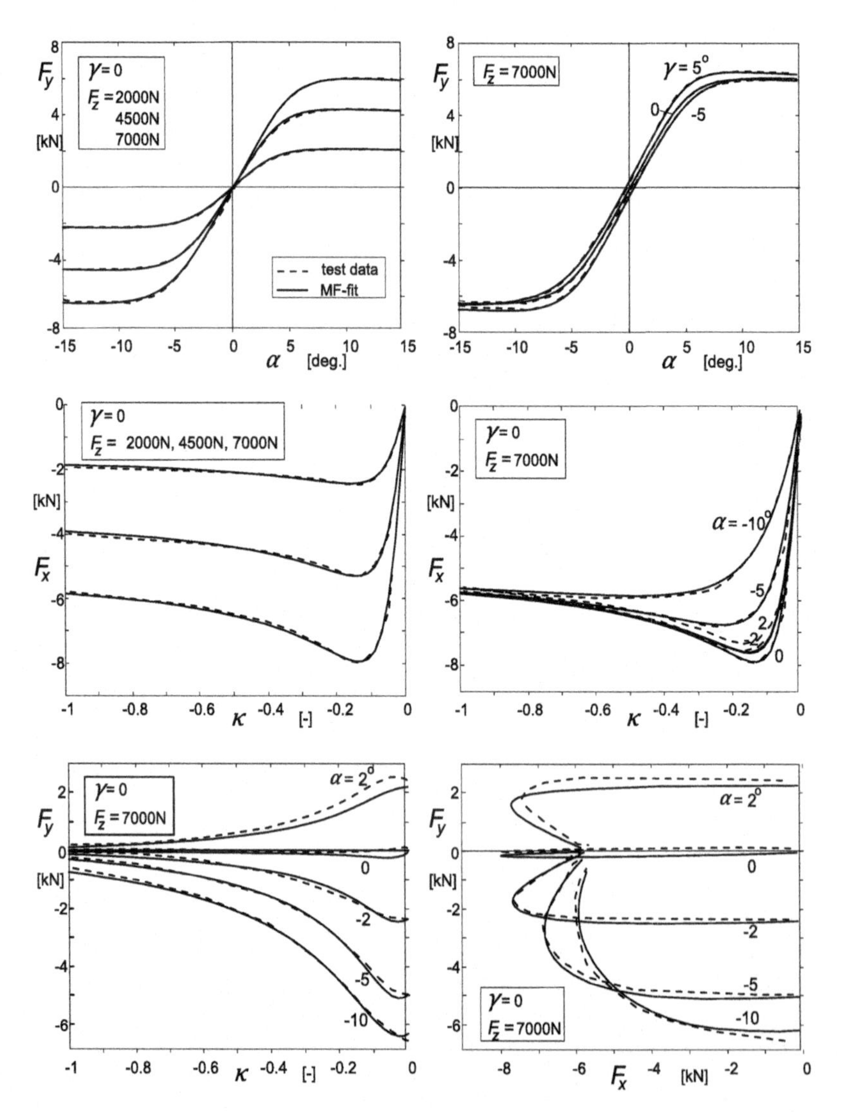


**FIGURE 4.28** Force characteristics of a 195/65 R15 car tire. *Magic Formula* computed results compared with data from measurements (dotted curves) conducted with the Delft Tire Test Trailer (2000).

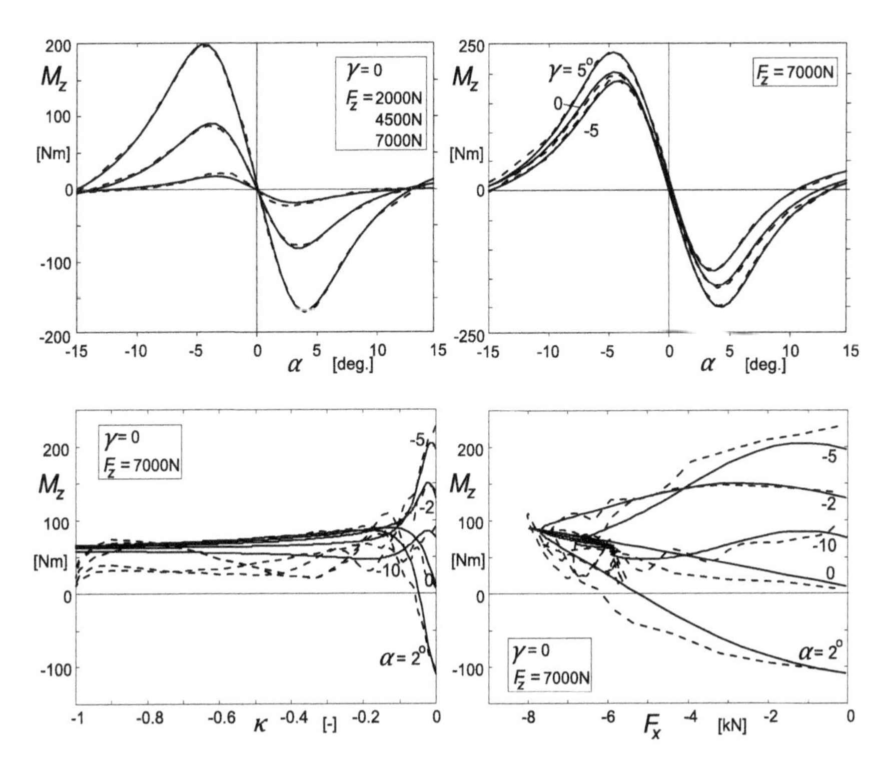


**FIGURE 4.29** Aligning torque characteristics of a 195/65 R15 car tire. *Magic Formula* model-computed results compared with data from measurements (dotted curves) conducted with the Delft Tire Test Trailer (2000).

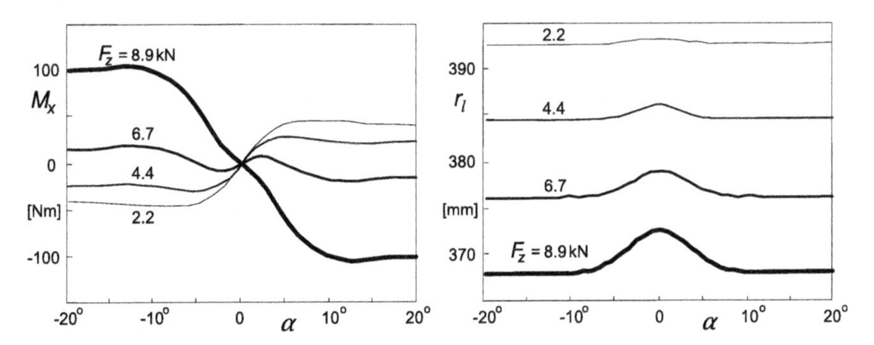


**FIGURE 4.30** Overturning couple and loaded radius of a passenger car tire measured on Ford's MTS Flat-Trac III tire testing machine (V.d. Jagt 2000). Compare with Figure 4.26.

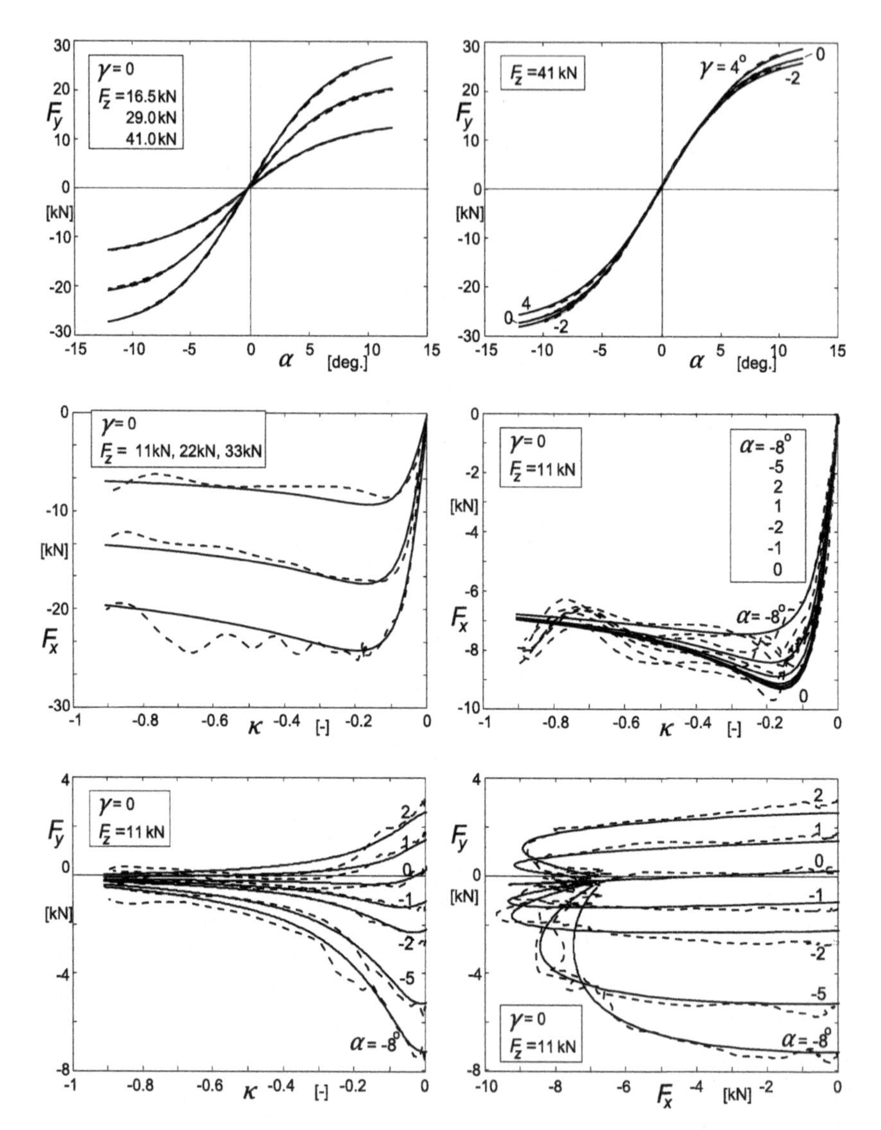


**FIGURE 4.31** Force characteristics of a 315/80 R22.5 truck tire. *Magic Formula*-computed results compared with data from measurements conducted with the Calspan flat track tire test facility (2000) (dotted curves).

averaged over two slip angle and brake pressure sweep cycles) and of a 315/80 R22.5 truck tire (on Calspan flat track test stand) have been presented. Test and model results of a 180/55ZR17 motorcycle tire have been presented in Figures 11.33, 11.34. This tire has been tested with the TNO Tire Test-Semi-Trailer, cf. Figure 12.1 (replacement of the Delft Tire Test Trailer but

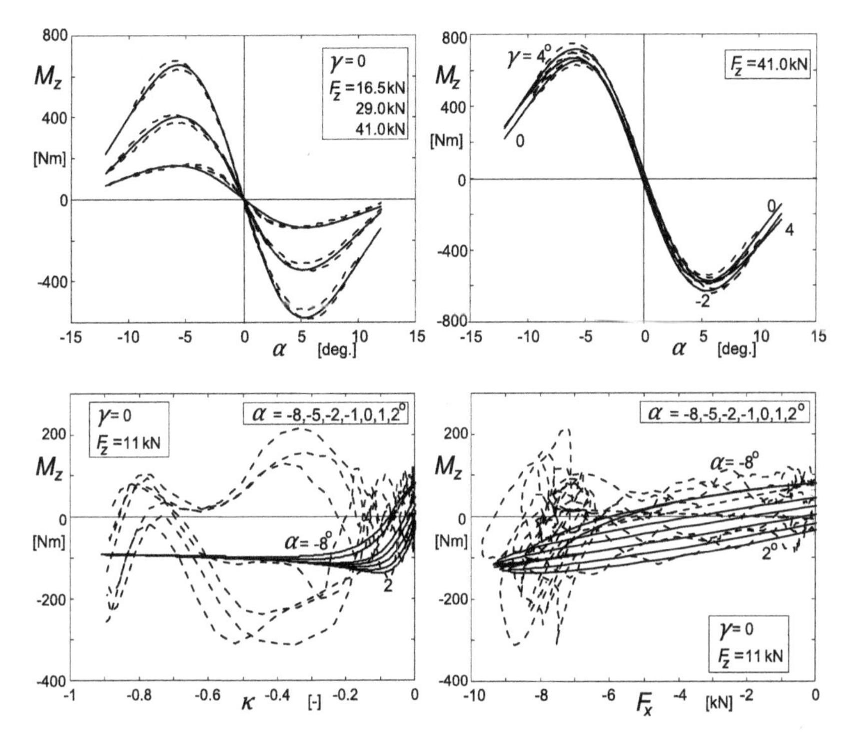


**FIGURE 4.32** Aligning torque characteristics of a 315/80 R22.5 truck tire. *Magic Formula*-computed results compared with data from measurements conducted with the Calspan flat track tire test facility (2000) (dotted curves).

equipped with the same measuring stations) using the new specially developed test device that can handle large camber angles (up to 70 degrees), cf. Figure 12.2. The fitted results have been obtained using the newly adapted *Magic Formula* model that is now suitable for fitting data for large camber angles as well, cf. Section 4.3.2.

In general, it is observed that good agreement between computed and measured curves can be achieved with the model. An exception is the attempt to (physically) describe the measured course of the aligning torque at braking (except for the motorcycle tire). The diagrams concerned, in Figures 4.29 and 4.32, show a rather unpredictable variation of the moment while the brake pressure is increased until wheel lock is reached. This is supposedly due to the large direct contribution of the braking force times the moment arm. This distance between the line of action of $F_x$ and the wheel center plane is suspected to vary due to local road camber variations and possibly tire nonuniformities while the tire rolls. Apparently, the motorcycle tire with narrow contact patch is much less sensitive to this effect.

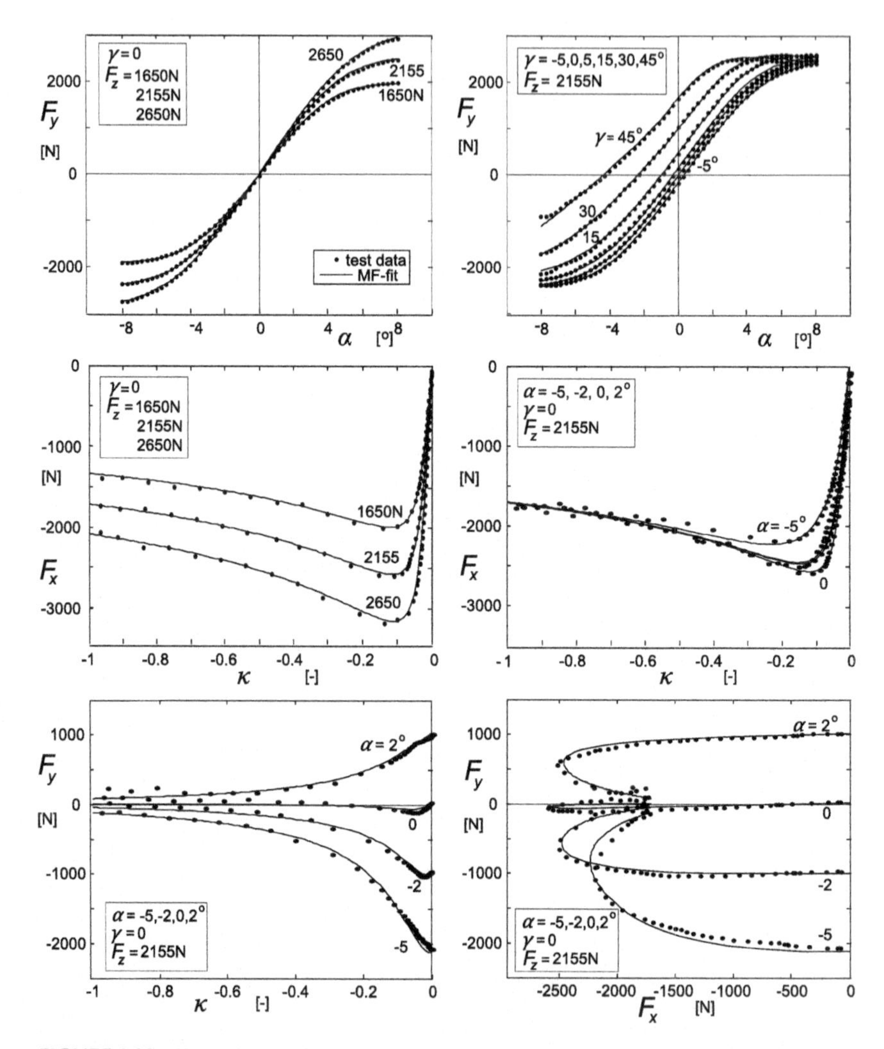


**FIGURE 4.33** Force characteristics of a 180/55ZR17 motorcycle tire. *Magic Formula*-computed results compared with data from measurements conducted with the TNO test-semi-trailer, equipped with large camber test device (cf. Figure 12.2).

Figure 4.30 presents the overturning couple and the loaded radius for a passenger car tire as reported by Van der Jagt (2000) and measured with the Ford MTS Flat-Trac III tire testing facility (cf. Figure 12.3). One may compare the measured characteristics with the calculated curves shown in Figure 4.26 obtained from Eqns (4.122–4.124) with hypothetical (not optimized) parameter values (Table 4.3).

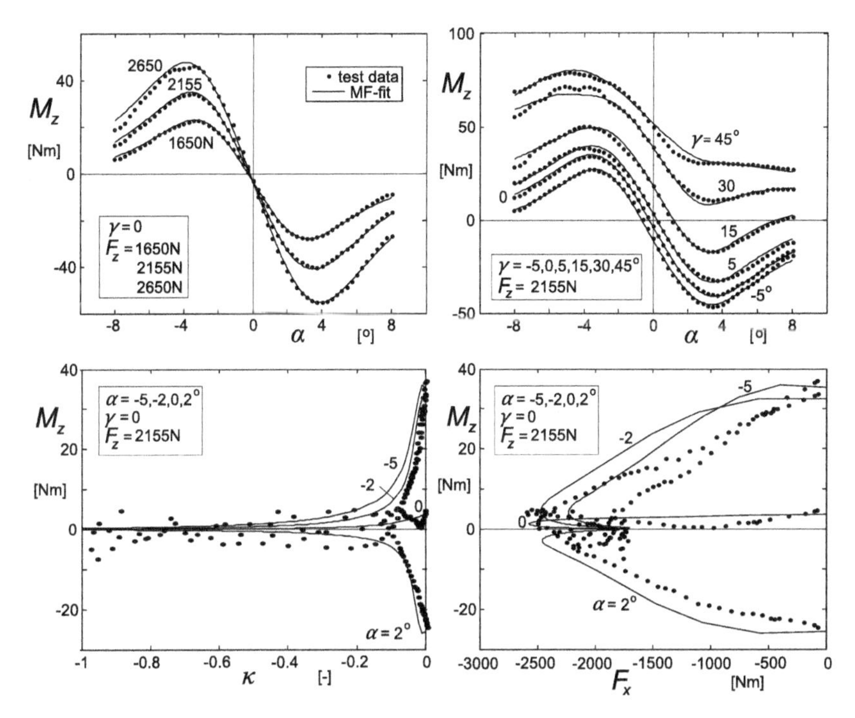


**FIGURE 4.34** Aligning torque characteristics of a 180/55ZR17 motorcycle tire. *Magic Formula*-computed results compared with data from measurements conducted with the TNO test-semi-trailer, equipped with large camber test device (cf. Figure 12.2).

---

## Exercise 4.1 Assessment of Off-Nominal Tire Side Force Characteristics and Combined Slip Characteristics with $F_x$ as Input Quantity

Consider the diagrams of Figure 4.35. For the original (nominal) side force characteristic $F_{yo}(\alpha)$ and for the cornering stiffness vs wheel load characteristic $C_{F\alpha}(F_z)$, employ the formulas as given by Eqns (4.6–4.9) using the data indicated in the figure. The variation of camber stiffness with wheel load $C_{F\gamma}(F_z)$ is assumed to be linear. To cover the combined slip situation, use Eqns (4.41–4.44) (note: $C_{F\alpha}$ is expressed in N/degree, convert this first into N/rad to get in line with Eqn (4.42)). In Eqn (4.42a), choose $n = 4$. In the nominal condition, we have parameter values:

$$F_{zo} = 3000\,\text{N}, \mu_o = 0.8, \; C_{Fao} = 400\,\text{N/deg}, \; C_{F\gamma o} = 50\,\text{N/deg}$$

Use Eqns (4.45, 4.46) and do the following:

1. Derive and plot the function $F_y(\alpha)$ for the following parameter values describing the tire at a different condition:

$$F_z = 4000\,\text{N}, \; \mu = 0.9, \gamma = \; 4^\circ, \; F_x = 2000\,\text{N}$$

2. Plot the graph for $F_y\,(F_x)$ for the constant slip angles $\alpha = 2$, 4, 6, and 8° at the condition $\mu = \mu_o$, $F_z = F_{zo}$, and $\gamma = 0$.

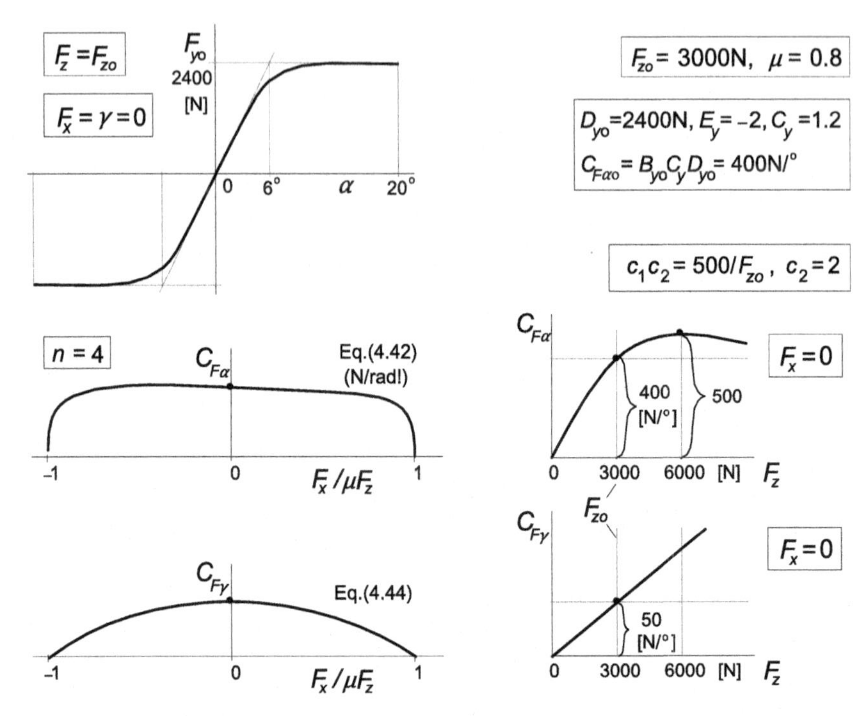


**FIGURE 4.35** On the assessment of off-nominal and combined slip tire side force characteristics with longitudinal force as input quantity using the similarity method (Exercise 4.1).

## Exercise 4.2. Assessment of Force and Moment Characteristics at Pure and Combined Slip using the *Magic Formula* and the *Similarity Method* with $\kappa$ as Input

Given are the pure slip force and moment characteristics at nominal vertical load $F_{zo}$ = 4000 N and the cornering stiffness vs vertical load (cf. Figures 4.36, 4.37). The problems are formulated as follows:

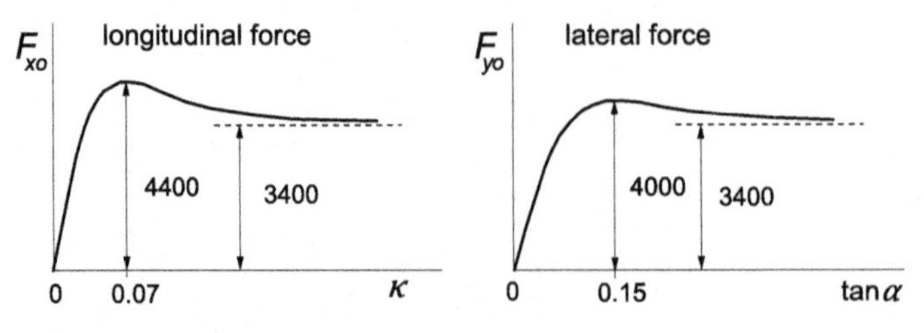


**FIGURE 4.36** Original tire force characteristics at pure slip (Exercise 4.2).

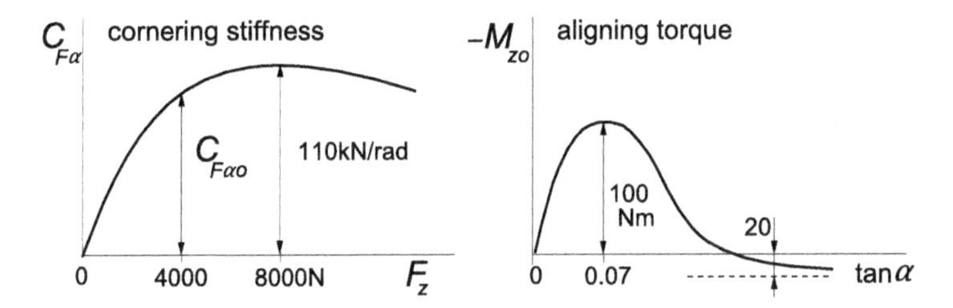


**FIGURE 4.37** Cornering stiffness and original moment characteristics (Exercise 4.2).

1. Determine for the original curves, that is: at the nominal load, the values of the coefficients *B, C, D,* and *E* from the data indicated in the figures below. The vertical and horizontal shifts are disregarded. Use formula (4.49) for the longitudinal force, the side force, and also for the moment. Consider Figure 4.9 and use formulas (4.52, 4.53) and (4.56). Draw the resulting curves.

$\mu_{xo} = 1.1$
$F_{x,\text{peak}} = 4400$ N
$F_{xa} = 3400$ N
$\kappa_m = 0.07$
$C_{F\kappa o} = C_{F\kappa}(4000)$
$p_1 = 110$ kN/rad
$p_2 = 8000$ N
further:
$C_{F\kappa}(F_z) = 40F_z$ [N]
effective compliance parameter for $M_z$, Eqn (4.39):
$c_9 = 2 \times 10^{-5}\ F_{zo}/a_o$

$\mu_{y0} = 1.0$
$F_{y,\text{peak}} = 4000$ N
$F_{ya} = 3400$ N
$\tan\alpha_m = 0.15$
$C_{F\alpha 0} = C_{F\alpha}(4000)$
$-M_{z,\text{peak}} = 100$ Nm
$-M_{za} = -20$ Nm
$\tan\alpha_m = 0.07$
$C_{M\alpha o} = C_{M\alpha}(4000)$
$t(F_z) = 0.5 \times 10^{-5}\ F_z$ [m]
$C_{M\alpha}(F_z) = t.C_{F\alpha}(F_z)$ [Nm/rad]
$c_{10} = 0$

2. For $F_z = 2000$, 4000, and 6000 N with $\mu_x = \mu_{xo}$ and $\mu_y = \mu_{yo}$ while $\gamma = 0$, compute and plot the curves for pure slip $F_x(\kappa)$, $F_y(\alpha)$ and $M_z(\alpha)$. Employ the similarity Eqns (4.21–4.26).
3. For $F_z = 6000$ N with $\mu_x = \mu_{xo}$, $\mu_y = \mu_{yo}$, and $\gamma = 0$ compute and plot the combined slip curves $F_y(F_x)$ and $M_z(F_x)$ at two slip angles: $\alpha = 2°$ and $8°$. The values of the longitudinal slip may range from $\kappa = -1$ to $+1$. Use Eqns (4.27–4.29) and (4.35–4.40).

Chapter 5

# Non-Steady-State Out-of-Plane String-Based Tire Models

Chapter Outline



**Tire and Vehicle Dynamics. DOI: 10.1016/B978-0-08-097016-5.00003-6**

## 5.1. INTRODUCTION

The transient and oscillatory dynamic behavior of the tire will be discussed in this and two ensuing chapters. The present chapter is devoted to the model development of the tire as an integral component. The stretched string model is chosen as the basis for the physical description of the out-of-plane (anti-symmetric) behavior. This model exhibits a finite contact length that allows the study of short path wavelength phenomena. The model is relatively simple in structure and integrates the carcass compliance and contact patch slip properties. For the moment response to yaw variations, the finite width of the contact patch needs to be introduced which is accomplished by connecting to the string the brush model featuring only fore-and-aft tread element compliance. In the more advanced string model, the added tread elements are allowed to also deflect sideways. The behavior of this more complex model is expected to be more realistic. This becomes especially apparent in the treatment of the side force response to a constant slip angle when the wheel runs over an undulated road surface. The inertia of the tire is of importance when running at higher speeds (gyroscopic couple) and when the frequency of lateral and yaw excitation can no longer be considered small. Several approximations of the kinematic and dynamic model will be discussed. In Chapter 6, the theory will be applied in the analysis of the wheel shimmy phenomenon. In Chapter 7, the model will be simplified to the single-point contact model that restricts the application to longer wavelength situations but enables the extension of the application to longitudinal and combined transient slip situations. Chapter 9 treats the more complex model that includes an approximate representation of the effect of the finite contact length, the compliance of the carcass, and the inertia of the belt. This more versatile model is able to consider both out-of-plane and in-plane tire dynamic behavior that can be extended to the nonlinear slip range also at relatively short wavelengths. Moreover, rolling-over road unevennesses will be included in Chapter 10.

## 5.2. REVIEW OF EARLIER RESEARCH

In the theories describing the horizontal non-steady-state behavior of tires, one can identify two trends of theoretical development. One group of authors assumes a bending stiffness of the carcass and the other bases its theory on the string concept.

In principle, the string theory is simpler than the beam theory, since with the string model the deflection of the foremost point alone determines the path of the tread for given wheel movements, whereas with the beam model the slope at the foremost point also has to be taken into account as an additional variable. The latter leads to an increase in the order of the system by one.

Von Schlippe (1941) presented his well-known theory of the kinematics of a rolling tire and introduced the concept of the stretched string model. For the first time, a finite contact length was considered. In the same paper, Dietr applied this theory to the shimmy problem. Later on, two papers of Von Schlippe and Dietrich (1942,1943) were published in which the effect of the width of the contact area is also considered. Two rigidly connected coaxial wheels, both approximated by a one-dimensional string model, are considered. The strings and their elastic supports are also supposed to be elastic in the circumferential direction.

Segel (1966) derived the frequency response characteristics for the one-dimensional string model. These appear to be similar to response curves which arise in Saito's approximate theory for the beam model (1962). For the same string model, Sharp and Jones (1980) developed a digital simulation technique which is capable of generating the exact response of the model. Earlier, Pacejka (1966) employed an analog computer and tape recorder (as a memory device) for the simulation according to the excellent von Schlippe approximation in which the contact line is approximated by a straight line connecting the two end points.

Smiley (1958) gave a summary theory resembling the one-dimensional theory of von Schlippe (1941). He has correlated various known theories with several systematic approximations to his summary theory.

Pacejka (1966) derived the non-steady-state response of the string model of finite width provided with tread elements. The important gyroscopic effect due to tire inertia has been introduced and the nonlinear behavior of the tire due to partial sliding has been discussed. Applications of the tire theory to the shimmy motion of automobiles were presented. In Pacejka (1972), the effect of mass of the tire has been investigated with the aid of an exact analysis of the behavior of a rolling stretched string tire model provided with mass. This complicated and cumbersome theory not suitable for dynamic vehicle studies was then followed by an approximate more convenient theory (Pacejka, 1973a), taking into account the inertial forces only up to the first harmonic of its distribution along the tire circumference. An outline of the theory together with experimental results will be given in the present chapter.

Rogers derived empirical differential equations (1971) which later were given a theoretical basis (1972). As a result of Rogers' research, shimmy response of tires in the low frequency range (mass effect not included!) can be described satisfactorily up to rather high reduced frequencies (i.e. short wavelengths) by using simple second-order differential equations.

Sperling (1977) conducted an extensive comparative study of different kinematic models of rolling elastic wheels. Recently, Besselink (2000) investigated and compared a number of interesting earlier models (Von Schlippe, Moreland, Smiley, Rogers, Kluiters, Keldysh, Pacejka) in terms of frequency response functions and step responses to side slip, path curvature (turn slip), and yaw angle, and judged their performances also in connection with the shimmy phenomenon.

Sekula et al. (1976) derived transfer functions from random slip angle input test data in the range of 0.05–4.0 Hz. From this information, cornering force responses were deduced for both radial and bias-ply tires to slip angle step inputs. Ho and Hall (1973) conducted an impressive experimental investigation using relatively small aircraft tires tested on a 120-inch research road wheel up to an oscillating yaw frequency of 3 Hz. A critical correlation study with theoretical results revealed that reasonable or good fit of the experimental frequency response plots can be achieved by using the theoretical functions (5.32, 5.92, 5.93) presented hereafter. It should be pointed out that in testing small-scale tires, certain similarity rules, cf. Pacejka (1974), should be obeyed.

Full-scale experimental tire frequency response tests have been carried out by several researchers, e.g., by Meier-Dörnberg, up to a yaw frequency of 20 Hz. Some of the latter investigations have been reported on by Strackerjan (1976). This researcher developed a dynamic tire model based on a somewhat different modeling philosophy compared to the model described by Pacejka (1973a) and discussed hereafter. Both types of models show good agreement with measured behavior. The straightforward approach employed by Strackerjan is similar to the method followed in Chapter 9.

In 1977, Fritz reported on an extensive experimental investigation concerning the radial force and the lateral force and aligning torque response to vertical axle oscilla at different constant yaw (side slip) angles of the wheel. Also, the mean values of the side force and of the moment have been determined. Earlier, Metcalf (1963) conducted small-scale tire experiments and Pacejka (1981) produced theoretical results in terms of mean cornering stiffness using a taught string tire model provided with elastic tread elements. Laerman (1986) conducted extensive tests on both car and truck tires and compared results with a theoretical model in terms of average side force and aligning torque and frequency response characteristics. The model includes tire mass (based on Strackerjan 1976) and allows sliding of tread elements with respect to the ground. Similar experiments on car tires have been carried out by Takahashi and Pacejka (1987). They developed a relatively simple mathematical model suitable for vehicle dynamics studies, including cornering on uneven roads (cf. Chapter 8).

In 2000 Maurice published experimental data up to over 60 Hz and a dynamic and short wavelength nonlinear model development that will be dealt with in Chapter 9. The model is connected with the in-plane dynamic model of Zegelaar (1998, cf. Chapter 9) and forms a versatile combined slip model that works with the *Magic Formula* steady-state functions presented in Chapter 4.

Another aspect of tire behavior is the non-steady-state response to wheel camber variations. Segel and Wilson (1976) found for a specific motorcycle tire that after a step change in camber angle about 20 percent of the ultimately attained side force responds instantaneously to this input. The remainder responds in a way similar to the response of the side force to a step change in slip angle, although with a larger relaxation length. Higuchi (1997) conducted comprehensive research on the nonlinear response to camber based on the string model and experimentally assessed responses to stepwise changes of the camber angle. In Chapter 7, the response to camber changes will be discussed.

The next section treats the essentials of the theory of lateral and yaw tire non-steady-state and dynamic behavior on the basis of the stretched string model.

## 5.3. THE STRETCHED STRING MODEL

The tire model, depicted in Figure 5.1, consists of an (assumedly endless) string which is kept under a certain pretension by a uniform radial force distribution (comparable with inflation pressure in real tires). In the axial direction, the string is elastically supported with respect to the wheel-center-plane but is prevented from moving in the circumferential direction. The string contacts the horizontal smooth road over a finite length. It is assumed that the remaining free portion of the string maintains its circular shape (in side view).

The model may be extended with a number of parallel strings keeping a constant mutual distance. As a result of this extension, a finite contact width

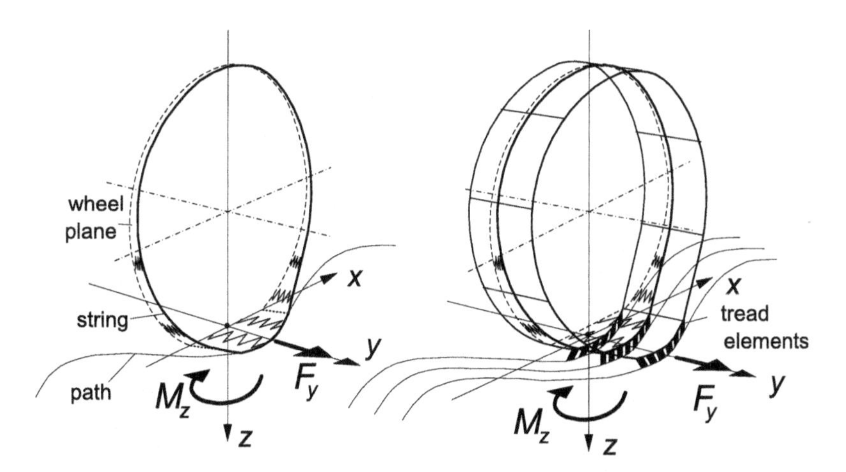


**FIGURE 5.1** Tire model with single stretched string and model extended with more parallel strings provided with tread elements which are flexible in the longitudinal direction.

arises. The strings are thought to be provided with a large number of elastic tread elements which, for reasons of simplification, are assumed to be flexible in the circumferential direction only. From the above assumptions, it follows that longitudinal deformations which arise at both sides of the wheel-center-plane when the wheel axle is subjected to a yaw rate are supposed to be taken up by the tread elements only.

For the theory to be linear, we must restrict ourselves to small lateral deformations and assume complete adhesion in the contact area. The wheel-center-plane is subjected to motions in the lateral direction (lateral displacement $\bar{y}$) and about the axis normal to the road (yaw angle $\psi$). These motions constitute the input to the system. The excitation frequency is denoted by $\omega$ ($=2\pi n$). With $V$ representing the assumed constant speed of travel, we obtain for the spatial or path frequency $\omega_s = \omega/V$ and the wavelength of the path of contact points $\lambda = V/n = 2\pi/\omega_s$. The distance traveled becomes $s = Vt$.

Alternative input quantities may be considered which are not related to the position of the wheel with respect to the road but to its rate of change characterized by the slip angle $\alpha = \psi - \mathrm{d}\bar{y}/\mathrm{d}s$ and the path curvature or turn slip $\varphi = -\mathrm{d}\psi/\mathrm{d}s$. The force $F_y$ and the moment $M_z$ which act from the ground on the tire in the $y$-direction and about the $z$-axis respectively form the response to the imposed wheel plane motion.

### 5.3.1. Model Development

To obtain an expression for the lateral deflection $v$ of the string, we consider the lateral equilibrium of an element of the tread band shown in Figure 5.2. The element is of full tread width and contains elements of the parallel strings with the rubber in between. In the lateral direction ($y$), the equilibrium of forces

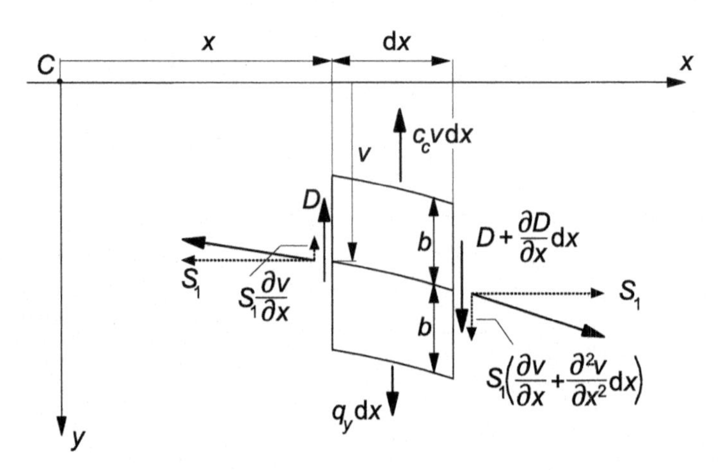


**FIGURE 5.2** Lateral equilibrium of deflected tread band element.

acting on the element of length d$x$ and width $2b$ results in the following equation:

$$q_y \mathrm{d}x - c_c v \,\mathrm{d}x - D + D + \frac{\partial D}{\partial x}\mathrm{d}x - S_1\frac{\partial v}{\partial x} + S_1\left(\frac{\partial v}{\partial x} + \frac{\partial^2 v}{\partial x^2}\mathrm{d}x\right) = 0 \quad (5.1)$$

where $c_c$ denotes the lateral carcass stiffness per unit length, $S_1$ the circumferential (in the $x$-direction) component of the total tension force acting on the set of strings, and $D$ the shear force in the cross section of the tread band acting on the rubber matrix. The shear force is assumed to be a linear function of the shear angle according to formula

$$D = S_2\frac{\partial v}{\partial x} \quad (5.2)$$

With the introduction of the effective total tension $S = S_1 + S_2$, we deduce from Eqn (5.1)

$$S\frac{\partial^2 v}{\partial x^2} - c_c v = -q_y \quad (5.3)$$

In the part of the tire not making contact with the road, the contact pressure vanishes so that $q_y = 0$ and

$$S\frac{\partial^2 v}{\partial x^2} - c_c v = 0 \quad \text{if } |x| > a \quad (5.4)$$

The lateral behavior of the model with several parallel strings and of the model with a single string will be identical if parameter $S$ is the same. In Figure 5.3, a top view of the single string model is depicted. The length $\sigma$,

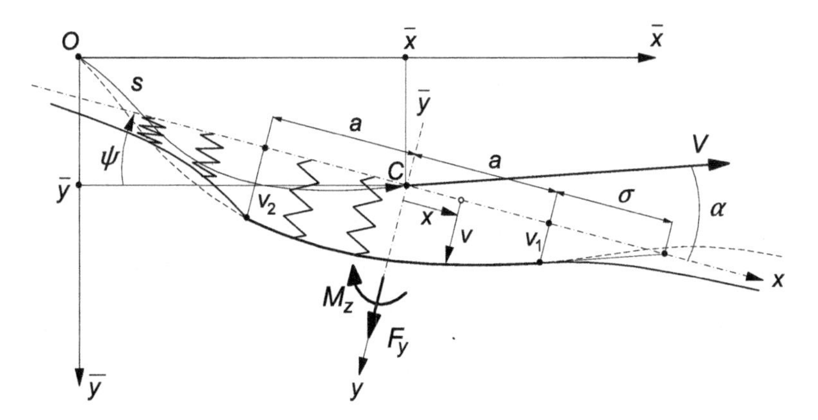


**FIGURE 5.3** Top view of the single string model and its position with respect to the fixed frame.

designated as the relaxation length, has been indicated in the figure. The relaxation length equals:

$$\sigma = \sqrt{\frac{S}{c_c}} \tag{5.5}$$

With this quantity introduced, Eqn (5.4) for the free portion of the string becomes

$$\sigma^2 \frac{\partial^2 v}{\partial x^2} - v = 0 \quad \text{if } |x| > a \tag{5.6}$$

If we consider the circumference of the tire to be much longer than the contact length, we may assume that the deflection $v_2$ at the trailing edge has a negligible effect on the deflection $v_1$ at the leading edge. The deflections of the free string near the contact region may then be considered to be the result of the deflections of the string at the leading edge and at the trailing edge respectively and not of a combination of both. The deflections in these respective regions now read approximately:

$$\begin{aligned} v &= C_1 e^{-x/\sigma} \quad \text{if } x > a \\ v &= C_2 e^{x/\sigma} \quad \text{if } x < -a \end{aligned} \tag{5.7}$$

which constitute the solution of Eqn (5.6) considering the simplifying boundary condition that for large $|x|$ the deflection tends to zero. At the edges $x = \pm a$ where $v = v_1$ and $v = v_2$ respectively, we have, for the slope,

$$\begin{aligned} \frac{\partial v}{\partial x} &= -\frac{v_1}{\sigma} \quad \text{for } x \downarrow a \\ \frac{\partial v}{\partial x} &= \frac{v_2}{\sigma} \quad \text{for } x \uparrow -a \end{aligned} \tag{5.8}$$

Because we do not consider the possibility of sliding in the contact zone, a kink may show up in the shape of the deflected string at the transition points from the free range to the contact zone. It seems a logical assumption that through the rolling process, the string forms a continuously varying slope around the leading edge while at the rear, because of the absence of bending stiffness, a discontinuity in slope may occur. An elegant proof of this statement follows by considering the observation that in vanishing regions of sliding at the transition points, cf. Figure 5.4, the directions of sliding speed of a point of the string with respect to the road and the friction force exerted by the road on the string that is needed to maintain a possible kink are compatible with each other at the trailing edge but incompatible at the leading edge. Therefore, it must be concluded that a kink may only arise at the trailing edge of the contact line. Consequently, the equation for the slope at the leading edge (first of (5.8)) can be rewritten as

$$\frac{\partial v}{\partial x} = -\frac{v_1}{\sigma} \quad \text{if } x \uparrow a,\ x = a,\ x \downarrow a \tag{5.9}$$

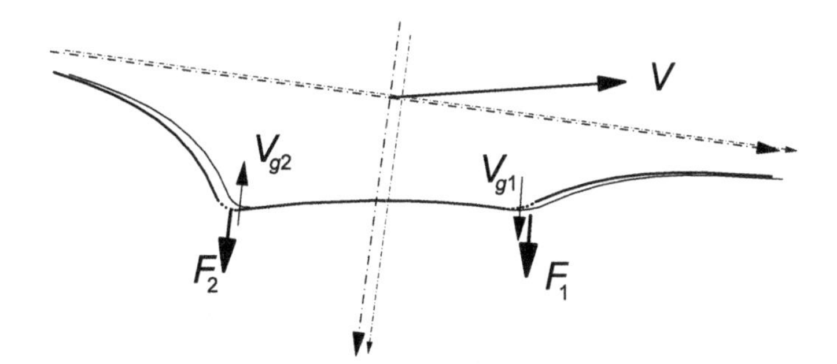


**FIGURE 5.4** Two successive positions of the string model. Vanishing regions of sliding at the leading and trailing edges and the compatibility of sliding speed $V_g$ and frictional forces $F$ required to maintain a possible kink which apparently can only exist at the trailing edge (2).

This equation constitutes an important relationship for the development of the ultimate expression for the deflection of the string as we will see a little later.

In Chapter 2, we have derived the general differential equations for the longitudinal and lateral sliding velocity of a rolling body which is subjected to lateral slip and spin. These equations (2.58, 2.59) become, if the sliding speed equals zero as is our assumption here,

$$\frac{\partial u}{\partial x} - \frac{\partial u}{\partial s} = y\varphi \tag{5.10}$$

$$\frac{\partial v}{\partial x} - \frac{\partial v}{\partial s} = -a - x\varphi \tag{5.11}$$

In these equations, $s$ denotes the distance traveled by the wheel center (or better: the contact center) and $x$ and $y$ the coordinates the considered particle would have with respect to the moving axes system ($C$, $x$, $y$) in the horizontally undeformed state.

These partial differential equations will be solved by using Laplace transformation. The transforms will be written in capitals. The transformation will not be conducted with respect to time but with respect to the distance traveled $s = Vt$ where the speed $V$ is assumed to be a constant. The Laplace transform of a variable quantity, generally indicated by $q$, is defined through

$$L[q(s)] = Q(p) = \int_0^\infty e^{-ps} q(s) \mathrm{d}s \tag{5.12}$$

where $p$ is the Laplace variable.

With initial conditions $u(x, 0) = v(x, 0) = 0$ at $s = 0$, we obtain, from (5.10, 5.11) the transformed equations,

$$\frac{\mathrm{d}U}{\mathrm{d}x} - pU = y\Phi \tag{5.13}$$

$$\frac{\mathrm{d}V}{\mathrm{d}x} - pV = -A - x\Phi \tag{5.14}$$

The solutions of these ordinary first-order differential equations read

$$U = C_u e^{px} - \frac{1}{p} y\Phi \tag{5.15}$$

$$V = C_v e^{px} + \frac{1}{p} A + \frac{1}{p}\left(\frac{1}{p} + x\right)\Phi \tag{5.16}$$

In Eqns (5.15, 5.16), the coefficients $C_u$ and $C_v$ are constants of integration. They are functions of $p$ and depend on the tire construction, which is the structure of the model. For our string model with tread elements which can be deformed in the longitudinal direction only, we have the boundary conditions at the leading edge $x = a$:

$$u = 0 \quad \text{or} \quad U = 0 \tag{5.17}$$

and

$$\frac{\partial v}{\partial x} = -\frac{v_1}{\sigma} \quad \text{or} \quad \frac{\mathrm{d}V}{\mathrm{d}x} = -\frac{V_1}{\sigma} \tag{5.18}$$

with the latter Eqn (5.18) corresponding to Eqn (5.9).

The constant $C_u$ now obviously becomes

$$C_u = \frac{1}{p} y\Phi e^{-pa} \tag{5.19}$$

For the determination of $C_v$, we have to differentiate Eqn (5.16) with respect to $x$:

$$\frac{\mathrm{d}V}{\mathrm{d}x} = C_v p e^{px} + \frac{1}{p}\Phi \tag{5.20}$$

With (5.18), we obtain

$$C_v = -\frac{1}{p}\left(\frac{1}{\sigma} V_1 + \frac{1}{p}\Phi\right) e^{-pa} \tag{5.21}$$

which with (5.16) yields for the deflection at the leading edge

$$V_1 = \frac{\sigma}{1 + \sigma p}(A + a\Phi) \tag{5.22}$$

and for the deflection in the contact zone

$$V = \frac{1}{p}\left\{ -\frac{A + \left(\sigma + a + \frac{1}{p}\right)\Phi}{1 + \sigma p} e^{p(x-a)} + A + \left(x + \frac{1}{p}\right)\Phi \right\} \tag{5.23}$$

The terms containing $e^{p(x-a)}$ point to a retardation behavior, which corresponds to delay terms in the original expressions. Note that a memory effect exists due to the fact that the nonsliding contact points retain information about their location with respect to the inertial system of axes $(\bar{x}, O, \bar{y})$ as long as they are in the contact zone.

Eqn (5.22) transformed back yields the first-order differential equation for the deflection of the string at the leading edge:

$$\frac{\mathrm{d}v_1}{\mathrm{d}s} + \frac{1}{\sigma} v_1 = \alpha + a\varphi \quad \left( = \psi - \frac{\mathrm{d}\bar{y}}{\mathrm{d}s} - a\frac{\mathrm{d}\psi}{\mathrm{d}s} \right) \tag{5.24}$$

This equation which is of fundamental importance for the transient behavior of the tire model (note the presence of the relaxation length $\sigma$ in the left-hand member) might have been found more easily by considering a simple trailing wheel system with trail equal to $\sigma$ and swivel axis located in the wheel plane a distance $\sigma + a$ in front of the wheel axis, cf. Figure 5.3. The equation may also be found immediately from Eqn (5.14) by taking $x = a$ and considering condition (5.9). Eqn (5.16) may also be transformed back which produces the delay terms mentioned before. However, we prefer to maintain the expression in the transformed state since we would like to obtain the result, that is: the force and moment response, in the form of transfer functions.

### *The Force and Moment Transfer Functions*

For the calculation of the lateral force $F_y$ and the moment $M'_z$ due to lateral deflections acting on the string two methods may be employed. According to the first method used by von Schlippe and Dietrich (1941) and by Segel (1966), the internal (lateral) forces acting on the string are integrated along the length of the string extending from minus to plus infinity taking into account the circular shape of the string from side view. The latter is important for the moment acting about the vertical axis. A correct result for the moment is obtained if not only the lateral forces acting on the string are taken into account but also the radial forces (the air pressure) which arise due to the string tension and which act along lines out of the center plane due to the lateral deflection of the string. Surprisingly, a simpler configuration where the string lies in horizontal plane (without considering the circular shape) appears to produce the same result. This is proven by considering the second method.

The second method which has been used by Temple (cf. Hadekel 1952) is much simpler and leads to the same correct result. The equilibrium of only that portion of the string is considered which makes contact with the road surface. On this piece of string, the internal lateral forces, the string tension force, and the external forces, constituting the force $F_y$ and the moment $M'_z$, are acting (cf. Figure 5.5).

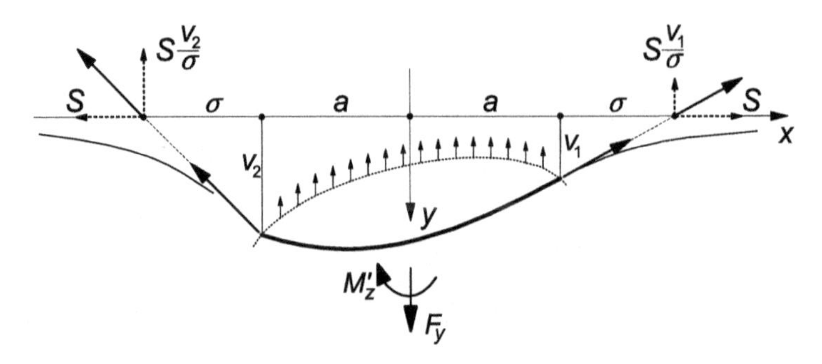


FIGURE 5.5 Equilibrium of forces in the contact region.

According to Temple's method, we obtain, for the lateral force,

$$F_y = c_c \int_{-a}^{a} v\mathrm{d}x + S(v_1 + v_2)/\sigma \tag{5.25}$$

and, for the moment due to lateral deformations,

$$M'_z = c_c \int_{-a}^{a} vx\mathrm{d}x + S(a + \sigma)(v_1 - v_2)/\sigma \tag{5.26}$$

with $v_1$ and $v_2$ denoting the deflections at $x = a$ and $-a$ respectively and $S = \sigma^2 c_c$ according to Eqn (5.5). For a first application of the theory, we refer to Exercise 5.1.

The moment $M_z^*$ due to longitudinal deformations which in our model are performed by only the tread elements is derived from the deflection $u$ distributed over the contact area. With $c'_{px}$ denoting the longitudinal stiffness of the elements per unit area, we obtain

$$M_z^* = -c'_{px} \int_{-a}^{a} \int_{-b}^{b} u\, y\, \mathrm{d}x\, \mathrm{d}y \tag{5.27}$$

By adding up both contributions, the total moment is formed:

$$M_z = M'_z + M_z^* \tag{5.28}$$

The Laplace transforms of $F_y$, $M'_z$, and $M_z^*$ are now readily obtained using Eqns (5.23) and (5.15) with (5.19) and the transformed versions of Eqns (5.25), (5.26), and (5.27). In general, the transformed responses may be written as

$$\begin{aligned} L\{F_y\} &= H_{F,\alpha} A + H_{F,\varphi} \Phi \\ &= H_{F,y} Y + H_{F,\psi} \Psi \end{aligned} \tag{5.29}$$

etc. The coefficients of the transformed input variables constitute the transfer functions. The formulas, for convenience written in vectorial form, for the responses to $\alpha$, $\varphi$, and $\psi$ read (since $\alpha = -\mathrm{d}\bar{y}/\mathrm{d}s$ expressions for the response to $\bar{y}$ have been omitted)

$$\boldsymbol{H}_{F,(\alpha,\varphi,\psi)}(p) = c_c\frac{1}{p}\left[\ 2(\sigma+a)\left(1,\frac{1}{p},0\right)\right. \tag{5.30}$$

$$\left. -\frac{1}{p}\left(1+\frac{\sigma p-1}{\sigma p+1}e^{-2pa}\right)\left(1,\sigma+a+\frac{1}{p},-(\sigma+a)p\right)\right]$$

$$\boldsymbol{H}_{M',(\alpha,\varphi,\psi)}(p) = c_c\frac{1}{p}\left[\ 2a\left\{\sigma(\sigma+a)+\frac{1}{3}a^2\right\}(0,1,-p)\right.$$

$$\left. -\frac{a(1+e^{-2pa})+p\{\sigma(\sigma+a)-1/p^2\}(1-e^{-2pa})}{(\sigma p+1)p}\left(1,\sigma+a+\frac{1}{p},-(\sigma+a)p\right)\right] \tag{5.31}$$

and furthermore

$$\boldsymbol{H}_{M^*,(\alpha,\varphi,\psi)}(p) = \frac{\kappa^*}{ap}\left\{1-\frac{1}{2ap}(1-e^{-2pa})\right\}(0,1,-p) \tag{5.32}$$

in which the parameter

$$\kappa^* = \frac{4}{3}a^2b^3c'_{px} \tag{5.33}$$

has been introduced.

The transfer functions of the responses to $\bar{y}$ and $\psi$ are obtained by considering the relations between the transformed quantities

$$\Phi = -p\Psi \tag{5.34}$$

$$A = \Psi - pY \tag{5.35}$$

and inserting these into (5.29). We find in general, for the transfer function conversion,

$$H_y = -pH_\alpha \tag{5.36}$$

$$H_\psi = H_\alpha - pH_\varphi \tag{5.37}$$

By transforming back the expressions such as (5.23, 5.29), the deflection, the force, and the moment can be found as a function of distance traveled $s$ for a given variation of $\alpha$ and $\varphi$ or of $\bar{y}$ and $\psi$.

An interesting observation may be made when considering the situation depicted in Figure 5.6. Here a yaw oscillation of the wheel plane is considered around an imaginary vertical steering axis located at a distance $\sigma + a$ in front of the wheel center. When yaw takes place about this particular point, the contact line remains straight and is positioned on the line along which the steering axis

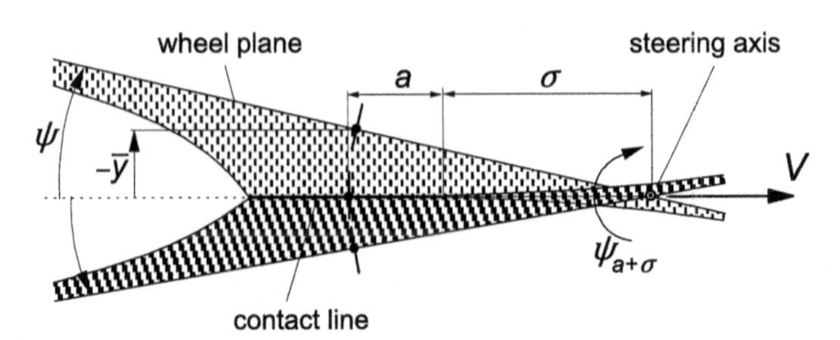


**FIGURE 5.6** Steady-state response that occurs when swivelling about the $\sigma + a$ point.

moves. Consequently, the response of the model to such a yaw motion is equal to the steady-state response. That is, for the force $F_y$ and the moment $M_z'$, the transfer functions become equal to the cornering stiffness $C_{F\alpha}$ and minus the aligning stiffness $C_{M\alpha}$ respectively. As we realize that the angular motion about the $\sigma + a$ point is composed of the yaw angle and the lateral displacement $\bar{y} = -(\sigma + a)\psi$ and furthermore that the response to $d\bar{y}/ds$ is equal to the response to $-\alpha$, we find, e.g., for the transfer function of $F_y$ to $\psi_{\sigma+a}$,

$$H_{F,\psi_{\sigma+a}} = C_{F\alpha} = H_{F,\psi} + (\sigma + a)pH_{F,\alpha} \tag{5.38}$$

With the aid of (5.37), the fundamental relationship between the responses to $\alpha$ and $\varphi$ can be assessed. We have, in general,

$$H_{\varphi} = \frac{1}{p}(H_{\alpha} - H_{\psi ss}) + (\sigma + a)H_{\alpha} \tag{5.39}$$

where for the responses to $F_y$ and $M_z'$ we have the steady-state response functions denoted by $H_{\psi ss}$:

$$H_{F,\psi_{\sigma+a}} = H_{F,\psi ss} = C_{F\alpha} \tag{5.40}$$

and

$$H_{M',\psi_{\sigma+a}} = H_{M',\psi ss} = -C_{M\alpha} \tag{5.41}$$

The important conclusion is that we may suffice with establishing a single pair of transfer functions, e.g. $H_{\alpha}$ for $F_y$ and $M_z'$, and derive from that the other functions by using the relations (5.36, 5.37, 5.39) together with (5.40, 5.41). Since in practice the frequency response functions are often assessed experimentally by performing yaw oscillation tests, we give below the conversion formulas to be derived from the transfer functions $H_{\psi}$. Later on, we will address the problem of first subtracting $M_z^*$ from the measured total moment $M_z$ to retrieve $M_z'$ for which the conversion is valid:

$$H_{\alpha} = -\frac{1}{(\sigma + a)p}(H_{\psi} - H_{\psi ss}) \tag{5.42}$$

$$H_{\varphi} = \frac{1}{p}(H_{\alpha} - H_{\psi}) \tag{5.43}$$

$$H_{y} = -pH_{\alpha} \tag{5.44}$$

Strictly speaking, the above conversion formulas only hold exactly for our model. The actual tire may behave differently especially regarding the effect of the moment $M_z^*$ that in reality may slightly rotate the contact patch about the vertical axis and thus affects the slip angle seen by the contact patch. As a consequence, the observation depicted in Figure 5.6 may not be entirely true for the real tire.

In the following, first the step response functions will be assessed and after that the frequency response functions.

### 5.3.2. Step and Steady-State Response of the String Model

An important characteristic aspect of transient tire behavior is the response of the lateral force to a stepwise variation of the slip angle $\alpha$. The initial condition at $s = 0$ reads: $v(x) = 0$; for $s > 0$, the slip angle becomes $\alpha = \alpha_o$. From Eqn (5.23), we obtain, for small slip angles by inverse transformation for the lateral deflection of the string in the contact region,

$$\frac{v}{\alpha_o} = a - x + \sigma\left\{1 - e^{-(s-a+x)/\sigma}\right\} \quad \text{for } x > a - s \tag{5.45}$$

while for the old points which are still on the straight contact line, the simple expression holds:

$$\frac{v}{\alpha_o} = s \quad \text{for } x \leq a - s \tag{5.46}$$

Expression (5.45) is composed of a part $(a - x)\alpha_o$ which is the lateral displacement of the wheel plane during the distance rolled $a - x$ and an exponential part. The point which at the distance rolled $s$ considered is located at coordinate $x$ was the point at the leading edge when the wheel was rolled a distance $a - x$ ago that is at $s - a + x$. At that instant, we had a deflection at the leading edge $v_{1(s-a+x)}$. The new $v = v_{(s)}$ equals the old $v_1 = v_{1(s-a+x)}$ plus the subsequent lateral displacement of the wheel $(a - x)\alpha_o$. Obviously, the exponential part of (5.45) is the $v_1$ at the distance rolled $s - a + x$. This can easily be verified by solving Eqn (5.24) for $v_1$.

Finally, with (5.25) the expression for the force can be calculated for the two intervals, with and without the old contact points:

$$F_y = \Gamma_{F,\alpha}\alpha_o = c_c\left\{2(\sigma + a)s - \frac{1}{2}s^2\right\}\alpha_o \quad \text{if } s \leq 2a \tag{5.47}$$

$$F_y = \Gamma_{F,\alpha}\alpha_o = 2c_c\left\{(\sigma + a)^2 - \sigma^2 e^{-(s-2a)/\sigma}\right\}\alpha_o \quad \text{if } s > 2a \tag{5.48}$$

For the aligning torque, we obtain, by using (5.26),

$$M_z' = \Gamma_{M',\alpha}\alpha_o = c_c\left\{\frac{1}{6}s^3 - \frac{1}{2}(\sigma + a)s^2\right\}\alpha_o \quad \text{if } s \le 2a \tag{5.49}$$

$$M_z' = \Gamma_{M',\alpha}\alpha_o = -2c_c\left\{\frac{1}{3}a^3 + \sigma a(\sigma + a) - \sigma^2 a e^{-(s-2a)/\sigma}\right\}\alpha_o \quad \text{if } s > 2a \tag{5.50}$$

where the quantities $\Gamma(s)$ designate the unit step response functions. These functions correspond to the integral of the inverse Laplace transforms of the transfer functions (5.30, 5.31) given above.

The graph of Figure 5.7 shows the resulting variation of $F_y$ and $M_z'$ vs traveled distance $s$. As has been indicated, the curves are composed of a parabola (of the second and third degree respectively) and an exponential function. The step responses have been presented as a ratio to their respective steady-state values.

The steady-state values of $F_y$ and $M_z'$ (or $M_z$) are directly obtained from (5.25) and (5.26) by considering the shape of the deflected string at steady-state side slip (Figure 5.8), i.e. a straight contact line at an angle $\alpha$ with the wheel plane and a deflection at the leading edge $v_1 = \sigma\alpha$ through which the condition to avoid a kink in the string at that point is obeyed, or from Eqns (5.48, 5.50) by letting $s$ approach infinity. We have

$$F_y = C_{F\alpha}\alpha \tag{5.51}$$

$$M_z = -C_{M\alpha}\alpha \tag{5.52}$$

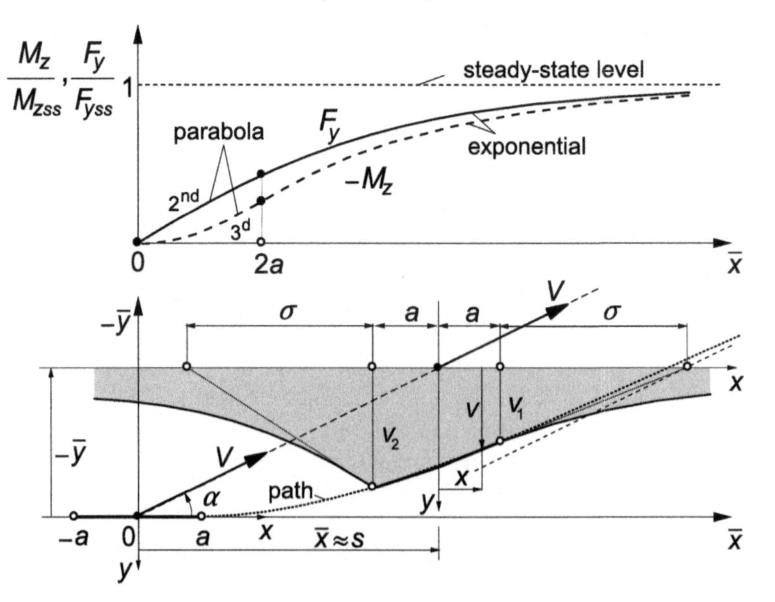


**FIGURE 5.7** The response of the lateral force $F_y$ and the aligning torque $M_z$ to a step input of the slip angle $\alpha$, calculated for a relaxation length $\sigma = 3a$.

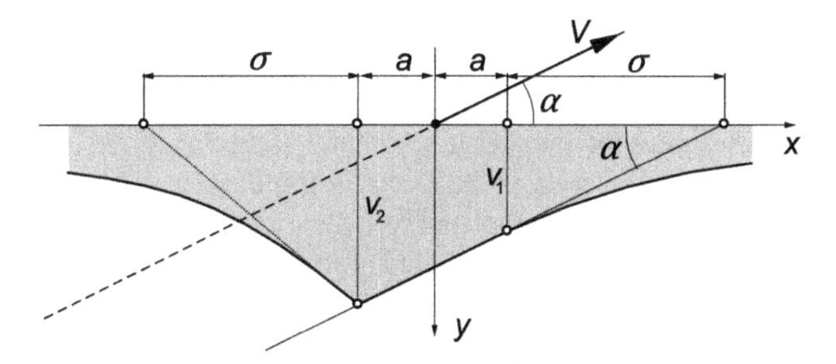


**FIGURE 5.8** Steady-state deflection of a side slipping string tire model (complete adhesion).

with the cornering and aligning stiffnesses

$$C_{F\alpha} = 2c_c(\sigma + a)^2 \tag{5.53}$$

$$C_{M\alpha} = 2c_c a\left\{\sigma(\sigma + a) + \frac{1}{3}a^2\right\} \tag{5.54}$$

and the pneumatic trail

$$t = \frac{C_{M\alpha}}{C_{F\alpha}} = \frac{a\left\{\sigma(\sigma + a) + \frac{1}{3}a^2\right\}}{(\sigma + a)^2} \tag{5.55}$$

The variation of these quantities and of the pneumatic trail $t = C_{M\alpha}/C_{F\alpha}$ and also of the relaxation length with wheel load $F_z$, the latter being assumed to vary proportionally with $a^2$, turns out to be quite unrealistic when compared with experimental evidence. A variation much closer to reality would be obtained if tread elements were attached to the string. For more details, also concerning the nonlinear characteristics and the nonsteady analysis of that enhanced but much more complicated model, we refer to Section 5.4.3.

The unit step response functions $\Gamma(s)$ to the other input variable $\varphi$ and the associated variables $\bar{y}$ and $\psi$ are of interest as well. They can be derived by integration of the unit impulse response functions $\Pi(s)$, which are found by inverse transformation of the transfer functions (5.30, 5.31, 5.36, 5.37). Alternatively, the other step response functions may be directly established by considering the associated string deflections similar to (5.45, 5.46) or from the unit step responses to the slip angle, corresponding to the coefficients of $\alpha_0$ shown in (5.47–5.50), by making use of the following relationships analogous to Eqns (5.39, 5.36, 5.37):

$$\Gamma_\varphi = \int(\Gamma_\alpha - H_{\psi ss})\mathrm{d}s + (\sigma + a)\Gamma_\alpha \tag{5.56}$$

$$\Gamma_y = -\frac{\mathrm{d}}{\mathrm{d}s}\Gamma_\alpha \tag{5.57}$$

$$\Gamma_{\psi} = \Gamma_{\alpha} - \frac{\mathrm{d}}{\mathrm{d}s}\Gamma_{\varphi} \tag{5.58}$$

Figure 5.9 illustrates the manner in which the deflection of the string model reacts to a step change of each of the four wheel motion variables (slip angle, lateral displacement of the wheel plane, turn slip, and yaw angle). Figure 5.10 presents the associated responses of the side force and the aligning torque. The responses have been divided by either the ultimate steady-state value of the transient response or the initial value, if relevant. For the moment response to turn slip and lateral displacement, both the initial and the final values vanish and a different reference value had to be chosen to make them nondimensional. The various steady-state coefficients and the lateral and torsional stiffnesses read in terms of the model parameters:

Lateral stiffness of the standing tire:

$$C_{Fy} = 2c_c(\sigma + a) \tag{5.59}$$

Cornering or lateral slip stiffness

$$C_{F\alpha} = 2c_c(\sigma + a)^2 \tag{5.60}$$

Aligning stiffness

$$C_{M\alpha} = 2c_c a\left\{\sigma(\sigma + a) + \frac{1}{3}a^2\right\} \tag{5.61}$$

Torsional stiffness of (thin) standing tire

$$C_{M\psi} = 2c_c a\left\{\sigma(\sigma + a) + \frac{1}{3}a^2\right\} \tag{5.62}$$

Turn slip stiffness for the force

$$C_{F\varphi} = 2c_c a\left\{\sigma(\sigma + a) + \frac{1}{3}a^2\right\} \tag{5.63}$$

Note that the steady-state response of $M_z'$ to $\varphi$ equals zero. The responses to side slip have already been presented in Figure 5.7. It now appears that the response of the side force $F_y$ to turn slip $\varphi$ is identical to the response of the aligning torque due to lateral deflections $M_z'$ to $\alpha$. This reciprocity property is also reflected by the equality of the slip stiffnesses given by (5.61, 5.63). It furthermore appears that the responses of $F_y$ to $\psi$ and $\varphi$ and of $M_z'$ to $\alpha$ tend to approach the steady-state condition at the same rate. This will be supported by the later finding that the corresponding frequency responses at low frequencies (large wavelengths) are similar. The frequency responses at short wavelengths are mainly governed by the step response behavior shortly after the step change has commenced. As appears from the graphs, at distances rolled smaller than

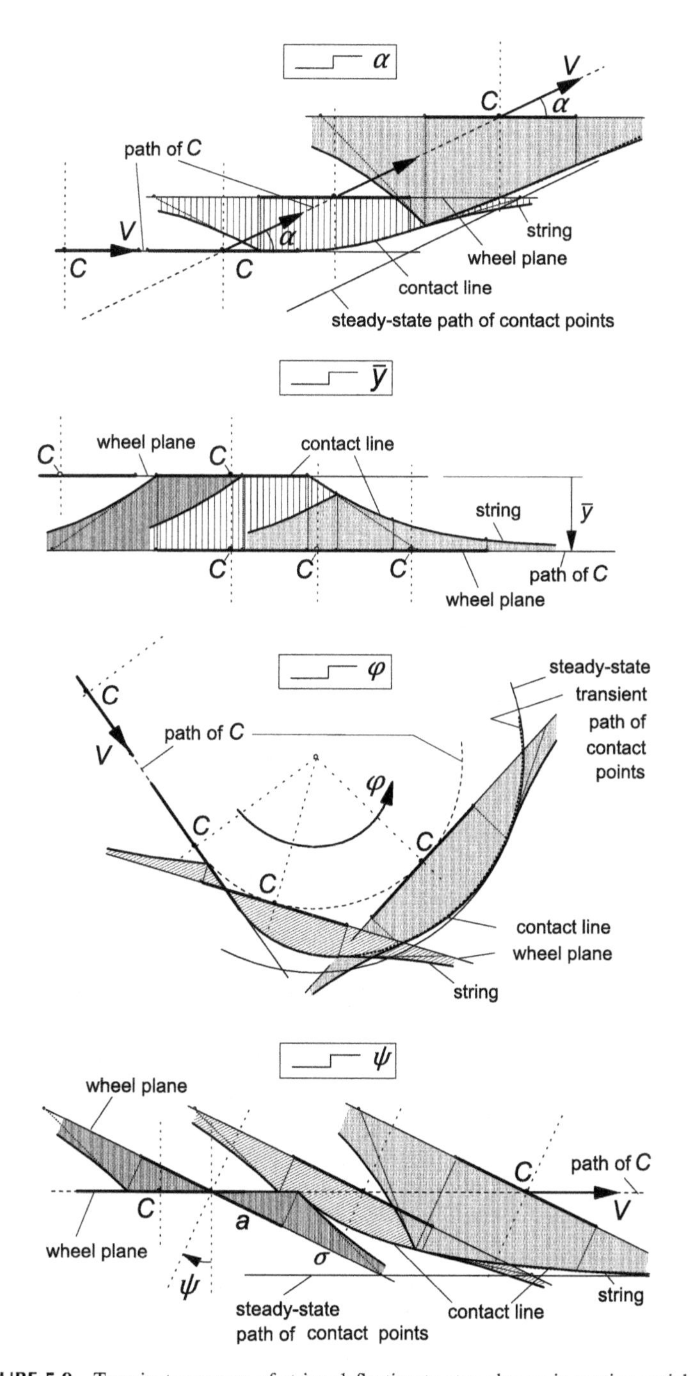


FIGURE 5.9 Transient response of string deflection to step change in motion variables.

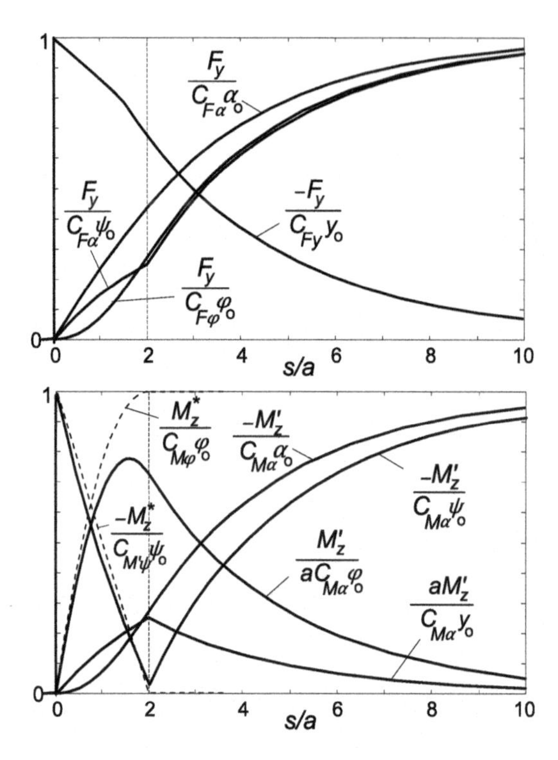


**FIGURE 5.10** Unit step response of side force $F_y$, aligning torque $M_z'$ and tread width moment $M_z^*$ to slip angle $\alpha$, lateral displacement $\bar{y}$, turn slip $\varphi$, and yaw angle $\psi$. Computed for string model with relaxation length $\sigma = 3a$.

the contact length large differences in transient behavior occur. As expected, the initial responses of $F_y$ to $\bar{y}$ and of $M_z'$ to $\psi$ are immediate and associated with the respective stiffnesses (5.59, 5.62).

The response of the moment $M_z^*$ due to tread width modeled with the brush model that deflects only in the longitudinal direction may be derived by considering the Laplace transform of the longitudinal deflection $u$ according to Eqn (5.15) with (5.19). Through inverse transformation or simply by inspection of the development of this deflection while the element moves through the contact range, the following expressions are obtained:

$$\frac{u}{\varphi_o} = -(a-x)y \quad \text{for } x > a-s \tag{5.64}$$

$$\frac{u}{\varphi_o} = -ys \quad \text{for } x \leq a-s \tag{5.65}$$

By using Eqn (5.27), the following expressions for the step response of $M_z^*$ to $\varphi$ result:

$$M_z^* = \Gamma_{M^*,\varphi}\varphi_o = \frac{4}{3}c'_{px}\left(as - \frac{1}{4}s^2\right)b^3\varphi_o \quad \text{if } s \leq 2a \tag{5.66a}$$

$$M_z^* = \Gamma_{M^*,\varphi}\varphi_o = \frac{4}{3}\, c'_{px}a^2b^3\varphi_o \quad \text{if } s > 2a \tag{5.66b}$$

The graphical representation of these formulas is given in Figure 5.10. The slip stiffness and stiffness coefficients employed read

Turn slip stiffness for the moment, cf. (5.66b):

$$C_{M\varphi} = \kappa^* = \frac{4}{3}\, c'_{px}a^2b^3 \tag{5.67a}$$

Torsional stiffness of standing tire due to tread width, cf. (5.32) with $p \to \infty$:

$$C_{M^*\psi} = \frac{1}{a}C_{M\varphi} = \frac{4}{3}\, c'_{px}ab^3 \tag{5.67b}$$

Note that the steady-state response of $M'_z$ to $\varphi$ equals zero.

As a result of a step change in turn slip, longitudinal slip at both sides of the contact patch occurs. The transient response extends only over a distance rolled equal to the contact length, at the end of which the steady-state response has been reached. As indicated in the graph, the approach curve has a parabolic shape. The response to a step change in yaw angle is immediate (like that of $M'_z$) after which a decline occurs which for $M_z^*$ is linear (derivative of response to $\varphi$).

In Figures 5.11 and 5.12, experimentally obtained responses to small step changes of the input have been presented. The diagrams show very well the exponential nature of the force response to side slip. Especially the aircraft tire exhibits the 'delayed' response of the aligning torque to side slip as predicted by the theory (Figure 5.10). Figure 5.12 shows a similar delay in the responses of $F_y$ to turn slip also found in the theoretical results. The peculiar response of the moment to yaw and turn slip is clearly formed by the sum of the responses of $M'_z$ and of $M_z^*$, although their ratio differs from the assumption adopted in Figure 5.10.

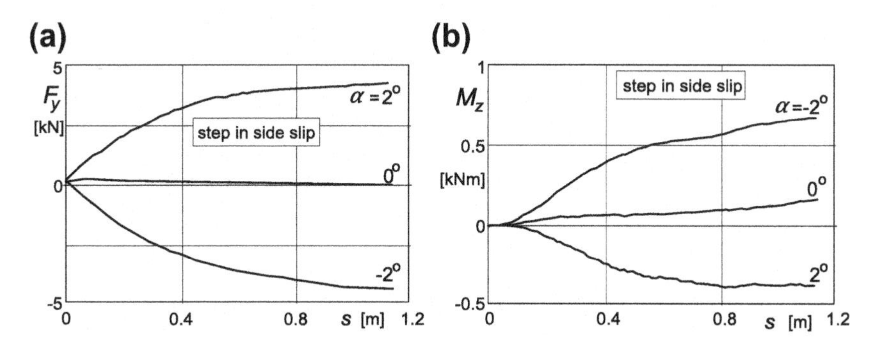


**FIGURE 5.11** Step response of side force $F_y$ and aligning torque $M_z$ to slip angle $\alpha$ as measured on an aircraft tire with vertical load $F_z = 35$ kN. (From Besselink 2000; test data is provided by Michelin Aircraft Tire Corporation).

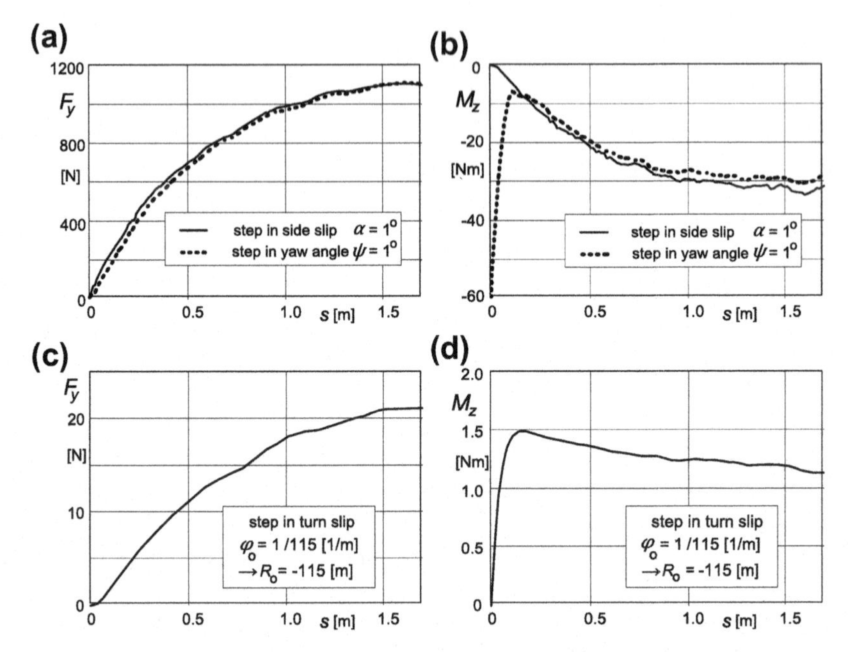


**FIGURE 5.12** Step response of side force $F_y$ and aligning torque $M_z$ to slip angle $\alpha$, yaw angle $\psi$, and turn slip $\varphi$ ($\alpha = 0$) as measured on a passenger car tire at load $F_z = 4$ kN. Tests were conducted on the flat plank machine of TU-Delft, cf. Figure 12.6 (Higuchi 1997, p. 44).

The response to a small step change in turn slip $\varphi_o = -1/R_o$ has been obtained by integration of the response to a small turn slip impulse = – step yaw angle $\psi_o$ (while $\alpha$ remains zero) and division by $R_o\psi_o$:

$$\Gamma_{F,\varphi} = \int \Pi_{F,\varphi}\,\mathrm{d}s = -\int \Gamma_{F,\psi}\,\mathrm{d}s \tag{5.68a}$$

$$F_{\text{step}\varphi} = \frac{1}{R_o\psi_o}\int F_{\text{impulse}\varphi}\,\mathrm{d}s = -\frac{1}{R_o\psi_o}\int F_{\text{step}\psi}\,\mathrm{d}s \tag{5.68b}$$

For $\psi_o = -0.5° = -1/115$ rad and by choosing $R_o = 115$ m, we divide by unity. For more details, we refer to Pacejka (2004).

Graphs of step response functions may serve to compare the performance of different models and approximations with each other. This will be done in Section 5.4. First we will discuss the frequency response functions.

### 5.3.3. Frequency Response Functions of the String Model

The frequency response functions for the force and the moment constitute the response to sinusoidal motions of the wheel and can be easily obtained by replacing in the transfer functions (5.30, 5.31, 5.32) the Laplace variable $p$ by

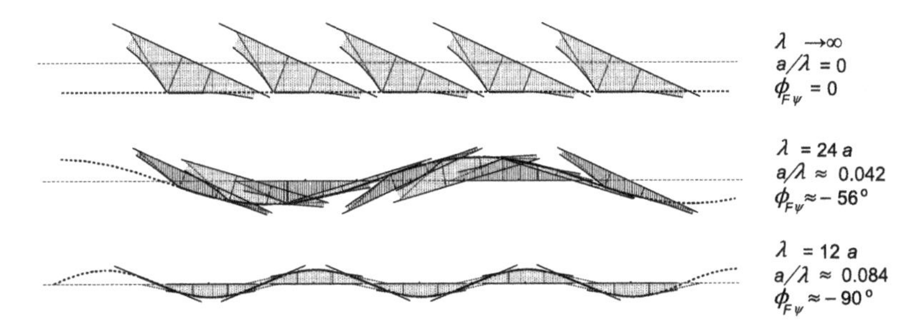


**FIGURE 5.13** The string model at steady-state side slip and subjected to yaw oscillations with two different wavelengths $\lambda$.

$i\omega_s$. The path frequency $\omega_s$ (rad/m) is equal to $2\pi/\lambda$, where $\lambda$ denotes the wavelength of the sinusoidal motion of the wheel. Figure 5.13 illustrates the manner in which the string deflection varies with traveled distance when the model is subjected to a yaw oscillation with different wavelengths.

The frequency response functions such as $H_{F,\psi}(i_{\omega s})$ are the complex ratios of output, e.g. $F_y$, and input, e.g. $\psi$. In Figures 5.14 and 5.15, the various frequency response functions have been plotted as a function of the nondimensional path frequency $a/\lambda = {}^1\!/_2\, \omega_s a/\pi$. The functions are represented by their absolute value $|H|$ and the phase angle $\phi$ of the output with respect to the input (if negative then output lags behind input), e.g.

$$H_{F,\psi} = \left|H_{F,\psi}\right| e^{i\phi_{F\psi}} = \frac{|F_y|}{|\psi|} e^{i\phi_{F\psi}} \tag{5.69}$$

If the variables are considered as real quantities, one gets $\alpha = |\alpha| \cos(\omega_s s)$ and, for its response, $F_y = |F_y| \cos(\omega_s s + \phi)$. In the figure, the absolute values have been made nondimensional by showing the ratio to their values at $\omega_s = 0$, the steady-state condition. Three different ways of presentation have been used, each with its own advantage.

The force response to slip angle very much resembles a first-order system behavior, as can be seen in the upper graph with a log–log scale. The cutoff frequency that is found by considering the steady-state response and the asymptotic behavior at large path frequencies appears to be equal to

$$\omega_{s,F\alpha,co} = \frac{1}{\sigma + a} \tag{5.70}$$

However, the phase lag at frequencies tending to zero is not equal to $\omega_s(\sigma + a)$, as one would expect for a first-order system, but somewhat smaller. Analysis reveals that the phase lag tends to

$$-\phi_{F,\alpha} \rightarrow \sigma_{F\alpha}\omega_s \quad \text{for } \omega_s \rightarrow 0 \tag{5.71}$$

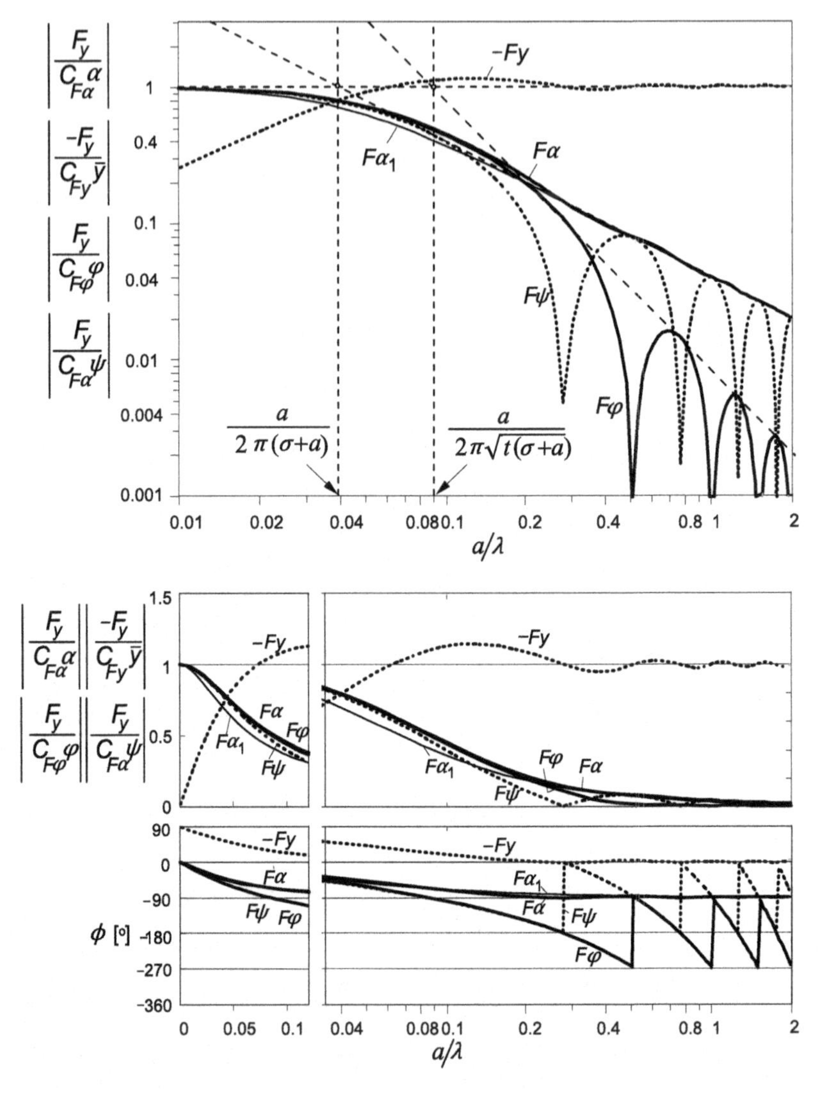


**FIGURE 5.14** String model frequency response functions for the side force with respect to various motion input variables. The nondimensional path frequency is equal to half the contact length divided by the wavelength of the motion: $a/\lambda = \omega_s/2\pi$. For the response to $\alpha$ a first-order approximation with the same cutoff frequency, $\omega_s = 1/(\sigma + a)$, has been added ($F\alpha_1$). Three ways of presentation have been used: log–log, lin–log, and lin–lin. The model parameter $\sigma = 3a$, which leads to $t = 0.77a$.

with the relaxation length for the side force with respect to the slip angle

$$\sigma_{F\alpha} = \sigma + a - t \tag{5.72}$$

which with (5.55) becomes equal to $3.23a$ if $\sigma = 3a$. The phase lag does approach 90° for frequencies going to infinity. The first-order approximation

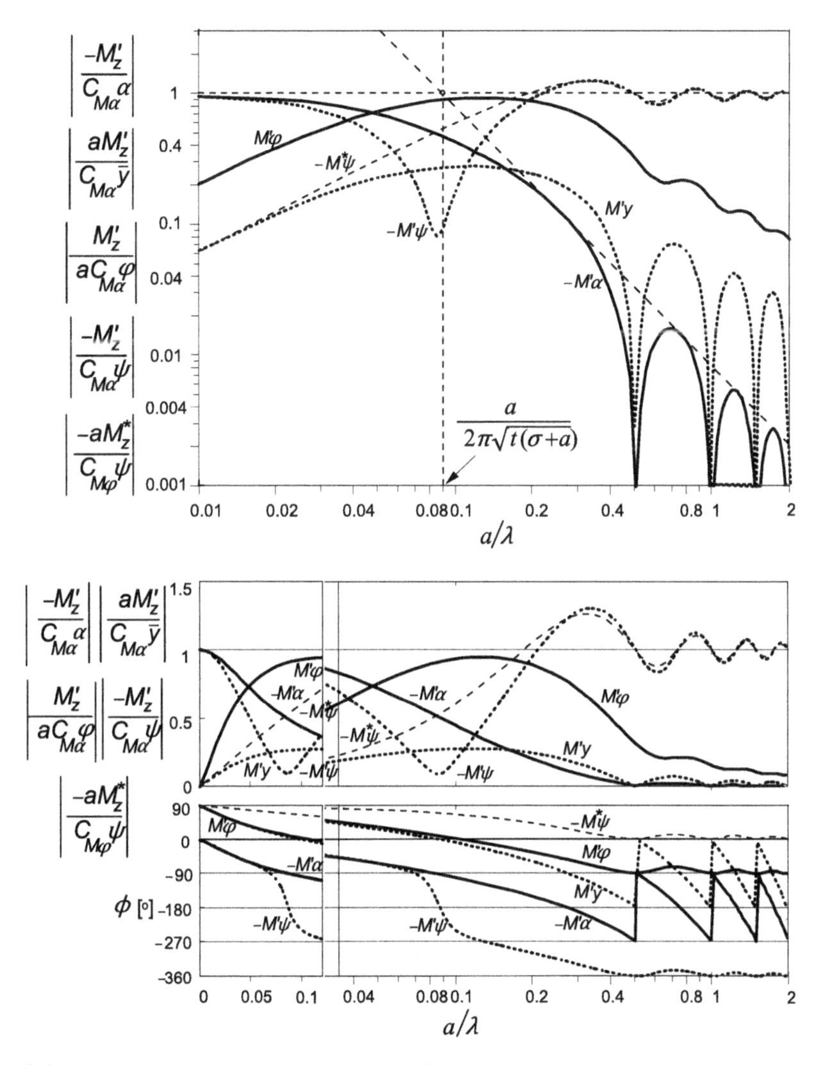


**FIGURE 5.15** String model frequency response functions for the aligning torque with respect to various motion input variables. Same conditions as in Figure 5.14. The response functions of the moment due to tread width $M_z^*$ have been added.

with the same cutoff frequency has been added in the graph for comparison. The corresponding frequency response function reads

$$H_{F\alpha_1} = \frac{C_{F\alpha}}{1 + i\omega_s(\sigma + a)} \tag{5.73}$$

The frequency response of the force to yaw shows a wavy curve for the amplitude at higher frequencies (at wavelengths smaller than ca. two times the contact length). The decline of the peaks occurs according to the same asymptote as found for the slip angle response. Consequently, the same cutoff frequency applies:

$$\omega_{s,F\psi,c0} = \frac{1}{\sigma + a} \tag{5.74}$$

When analyzing the behavior at small frequencies, it appears that here the phase lag does tend to

$$-\phi_{F,\psi} \rightarrow \sigma_{F\psi}\omega_s \quad \text{for } \omega_s \rightarrow 0 \tag{5.75}$$

with the relaxation length for the side force with respect to the yaw angle:

$$\sigma_{F\psi} = \sigma + a \tag{5.76}$$

Further analysis reveals that, when developing the frequency response function $H_{F,\psi}$ in a series up to the second degree in $i\omega_s$,

$$\begin{aligned} H_{F,\psi}(i\omega_s) &= C_{F\alpha}\{1 - \sigma_{F\psi} i\omega_s(1 - b_{F2} i\omega_s)\} + \cdots \\ &= C_{F\alpha}(1 - \sigma_{F\psi} b_{F2}\omega_s^2 - \sigma_{F\psi} i\omega_s) + \cdots \end{aligned} \tag{5.77}$$

and subsequently employing the fundamental relationship (5.42) between $\alpha$ and $\varphi$ responses, the frequency response function $H_{F,\alpha}$ up to the first degree in $i\omega_s$ becomes

$$\begin{aligned} H_{F,\alpha}(i\omega_s) &= C_{F\alpha}(1 - \sigma_{F\alpha} i\omega_s) + \cdots \\ &= C_{F\alpha}(1 - b_{F2} i\omega_s) + \cdots \end{aligned} \tag{5.78}$$

which shows that

$$\sigma_{F\alpha} = b_{F2} \tag{5.79}$$

which is an important result in view of assessing $\sigma_{F\alpha}$ from yaw oscillation measurement data and checking the correspondence with (5.72).

The aligning torque ($-M'_z$, Figure 5.15) shows a response to the slip angle which is closer to a second-order system with a phase lag tending to a variation around 180° and a 2:1 asymptotic slope of the amplitude with a cutoff frequency equal to

$$\omega_{s,M\alpha,co} = \frac{1}{\sqrt{t(\sigma + a)}} \tag{5.80}$$

where $t$ denotes the pneumatic trail, cf. (5.55). Again, the response of $F_y$ to $\varphi$ turns out to be the same as the response of $-M'_z$ to $\alpha$. As the graph of Figure 5.15 shows, the amplitude of $M'_z$ as a response to yaw oscillations $\psi$ exhibits a clear dip at (with parameter $\sigma = 3a$) a wavelength $\lambda = \sim 12a$. This condition corresponds to the situation depicted in Figure 5.13 (third case) and is referred to as the meandering phenomenon or as kinematic shimmy which occurs in practice when the wheel is allowed to swivel freely about the vertical axis through the wheel center and the system is slowly moved forward. The nearly symmetric string deformation explains why the amplitude of the aligning torque almost vanishes at this wavelength. At higher frequencies, the amplitude remains finite and approaches at $\omega_s \rightarrow \infty$ the same value as it had at $\omega_s = 0$ that is: $-M_z' = C_{M\alpha}\psi = C_{M'\psi}\psi$. The phase angle approaches $-360°$. It is interesting that analysis at frequencies approaching zero shows that the phase lag both for the response of $-M_z'$ to $\alpha$ and to $\psi$ (and thus also for $F_y$ to $\varphi$) approaches the value $\omega_s(\sigma + a)$ that also appeared to be true for the response of $F_y$ to $\psi$. So we have

$$\sigma_{M'\psi} = \sigma_{M\alpha} = \sigma_{F\varphi} = \sigma_{F\psi} = \sigma + a \tag{5.81}$$

Expressions equivalent to (5.70, 5.72, 5.74, 5.79–5.81) appear to hold for the enhanced model with tread elements attached to the string, cf. Section 5.4.3.

The torque due to tread width $-M_z^*$ shows a response to yaw angle $\psi$ that increases in amplitude with path frequency and starts out with a phase lead of 90° with respect to $\psi$. At low frequencies, the moment acts like the torque of a viscous rotary damper with damping rate inversely proportional with the speed of travel *V*. We find, with Eqn (5.67),

$$M_z^* = -\frac{1}{V}\kappa^*\dot{\psi} \quad \text{for } \omega_s \rightarrow 0 \tag{5.82}$$

At high frequencies $\omega_s \rightarrow \infty$, that is at vanishing wavelength $\lambda \rightarrow 0$ where the tire is standing still, the tire acts like a torsional spring and the moment $-M_z^*$ approaches $C_{M^*\psi}\psi = C_{M\varphi}/a$, cf. (5.68). The cutoff frequency appears to become

$$\omega_{s,M^*\varphi,co} = \frac{1}{a} \tag{5.83}$$

The total moment about the vertical axis is obtained by adding the components due to lateral and longitudinal deformations:

$$M_z = M_z' + M_z^* \tag{5.84}$$

For the standing tire, one finds from a yaw test the total torsional stiffness $C_{M\psi}$ which relates to the aligning stiffness and the stiffness due to tread width as follows:

$$C_{M\psi} = \frac{1}{a}C_{M\varphi} + C_{M\alpha} \tag{5.85}$$

Obviously, this relationship offers a possibility to assess the turn slip stiffness for the moment $C_{M\varphi}$.

Due to the action of $M_z^*$, the phase lag of $-M_z$ will be reduced. This appears to be important for the stabilization of wheel shimmy oscillations (cf. Chapter 6). The total moment and its components can best be presented in a Nyquist plot

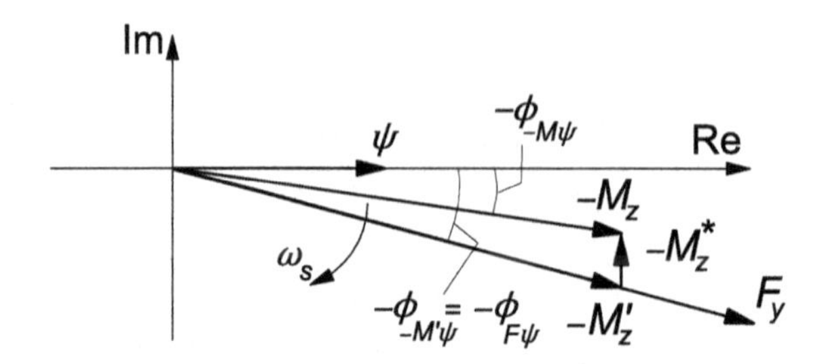


**FIGURE 5.16** Complex representation of side force and moment response to yaw oscillations $\psi$.

where the moment components and the resulting total moment appear as vectors in a polar diagram. Figure 5.16 depicts the vector diagram at a low value of the path frequency $\omega_s$.

By considering the various phase angles at low frequencies, we may be able to extract the moment response due to tread width from the total (measured) response and find the response of the moment for a 'thin' tire. Since at low frequencies the moment vector $-M_z^*$ tends to point upward, we find, while considering (5.75, 5.81) and (5.82),

$$|M_z^*| = |M_z|(\sigma_{F\psi}\omega_s + \phi_{-M\psi}) = \kappa^*\omega_s|\psi| \tag{5.86}$$

With known $\sigma_{F\psi}$ and $\phi_{-M\psi}$ to be determined from the measurement at low frequency, the moment turn slip stiffness $C_{M\varphi} = \kappa^*$ may be assessed in this way.

In the diagrams of Figures 5.17 and 5.18, the nondimensional frequency response functions $H_{F,\psi}(i\omega_s)/C_{F\alpha}$ and $-H_{M,\psi}(i\omega_s)/C_{M\alpha}$ with its components $-H_{M',\psi}(i\omega_s)/C_{M\alpha}$ and $-H_{M^*,\psi}(i\omega_s)/C_{M\alpha}$ have been presented as a function of the nondimensional path frequency $a/\lambda$. The parameter values are $\sigma = 3a$ and $C_{M\varphi} = aC_{M\alpha}$.

The diagram of Figure 5.17 clearly shows the increase in phase lag and decrease of the amplitude of $F_y$ with decreasing wavelength $\lambda$. The wavy behavior and the associated jumps from 180° to 0 of the phase angle displayed in Figure 5.14 become clear when viewing the loops that appear to occur when the wavelength becomes smaller than about two times the contact length.

The aligning torque vector, Figure 5.18, for the 'thin' tire turns over 360° before from a wavelength of about the contact length the loops begin to show up. At about $\lambda = 12a$, the curve gets closest to the origin. This corresponds to the frequency where the dip occurs in Figure 5.15 and is illustrated as the last case of Figure 5.13.

For values of $\kappa^*$ sufficiently large, the total moment curve does not circle around the origin anymore. The curve stretches more to the right and ends where the wheel does not roll anymore and the tire acts as a torsional spring with stiffness expressed by (5.85). In reality, the tire will exhibit some damping due to hysteresis. That will result in an end point located somewhat above the horizontal axis.

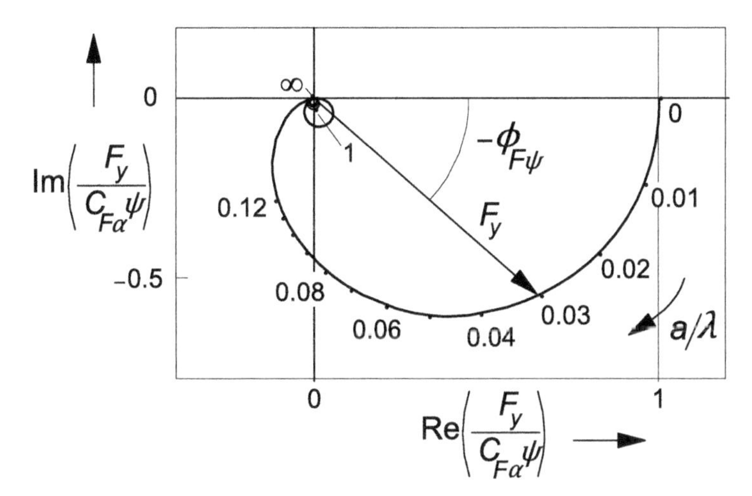


**FIGURE 5.17** Nyquist plot of the nondimensional frequency response function of the side force $F_y$ with respect to the wheel yaw angle $\psi$. Parameter: $\sigma = 3a$.

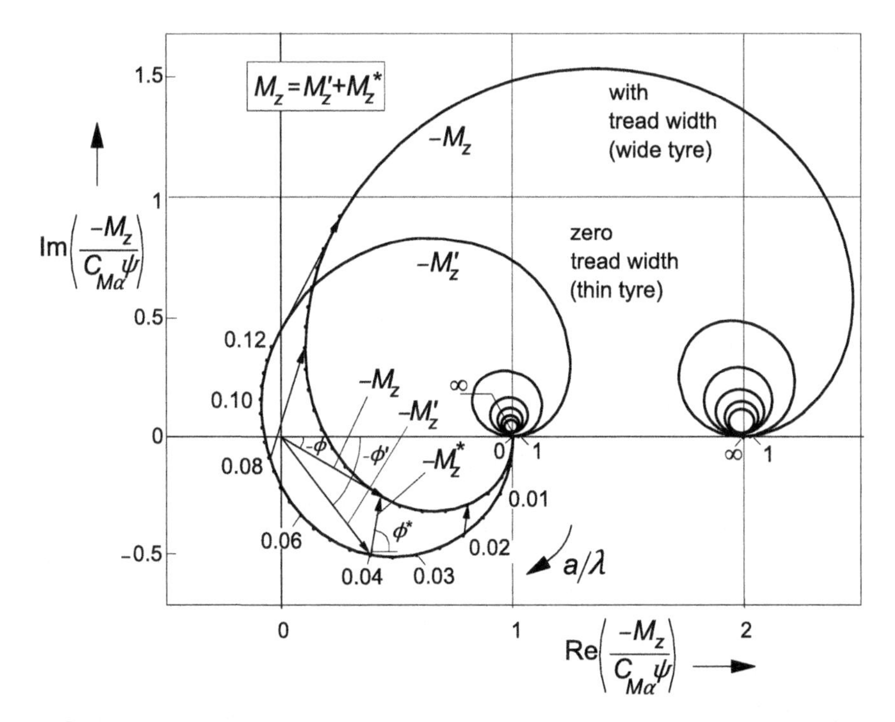


**FIGURE 5.18** Nyquist plot of the nondimensional frequency response function of the aligning moment $M_z$ with respect to the wheel yaw angle $\psi$. The contributing components $M_z'$ due to lateral deformations and $M_z^*$ due to tread width and associated longitudinal deformations. Parameters: $\sigma = 3a$ and $C_{M\varphi} = aC_{M\alpha}$.

The calculated behavior of the linear tire model has unmistakable points of agreement with results found experimentally at low values of the yaw frequency. At higher frequencies and higher speeds of rolling, the influence of the tire inertia and especially the gyroscopic couple due to tire lateral deformation rates is no longer negligible. In Section 5.5 and Chapter 9, these matters will be addressed.

---

**Exercise 5.1. String Model at Steady Turn Slip**

Consider the single stretched string tire model running along a circular path with radius $R$ anti-clockwise ($\varphi = 1/R$) and without side slip ($\alpha = 0$) as depicted in Figure 5.19.

Derive the expression for the lateral force $F_y$ acting upon the model under these steady-state circumstances. First find the expression for the lateral deflection $v(x)$ using Eqn (2.61) which leads to a quadratic approximation of the contact line. Then use Eqn (5.25) for the calculation of the side force.

Now consider in addition some side slip and determine the value of $\alpha$ required to neutralize the side force generated by the path curvature $1/R$. Make a sketch of the resulting string deformation and wheel-plane orientation with respect to the circular path for the following values of radius and relaxation length: $R = 6a$ and $\sigma = 2a$.

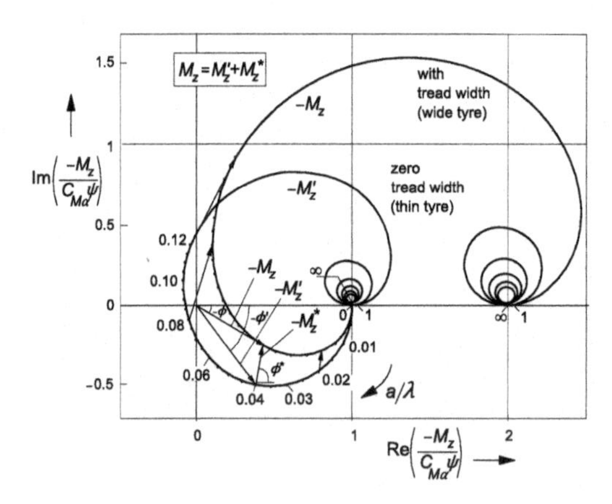


**FIGURE 5.19** The string tire model in steady turning state (Ex. 5.1).

---

## 5.4. APPROXIMATIONS AND OTHER MODELS

In the present section, approximations to the exact theory will be treated to make the theory more accessible to applications. Subsection 5.4.2 discusses a number of other models known in the literature. After that in Subsection 5.4.3, a more complex model showing tread elements flexible also in the lateral direction will be treated to provide a reference model that is closer in performance to the real tire.

### 5.4.1. Approximate Models

In the literature, several simpler models have been proposed. Not all of these are based on the string model but many are. Figure 5.20 depicts a number of approximated contact lines as proposed by several authors. The most well-known and accurate approximation is that of Von Schlippe (1941) who approximated the contact line by forming a straight connection between the leading and trailing edges of the exact contact line. Kluiters (1969) gave a further approximation by introducing a Padé filter to approximately determine the location of the trailing edge. Smiley (1957) proposes an alternative approximation by considering a straight contact line that touches the exact contact line in its center in a more or less approximate way. Pacejka (1966) considered a linear or quadratic approximation of the contact line touching the exact one at the leading edge; the first and simplest approximation is referred to as the straight tangent approximation. A further simplification, completely disregarding the influence of the length of the contact line, results in the first-order approximation referred to as the point contact approximation.

In the sequel, we will discuss Von Schlippe's and Smiley's second-order approximation as well as the straight tangent and point contact approximations. The performance of these models will be shown in comparison with the exact 'bare' string model and with the enhanced model with laterally compliant tread elements. Figures 5.21–5.27 give the results in terms of step response, frequency response Bode plots, and Nyquist diagrams.

For some of the other approximate models (Rogers 1972, Kluiters 1969, Keldysh 1945 and Moreland 1954), only the governing equations will be provided with some comments on their behavior. For more information, we refer to the original publications or to the extensive comparative study of Besselink (2000).

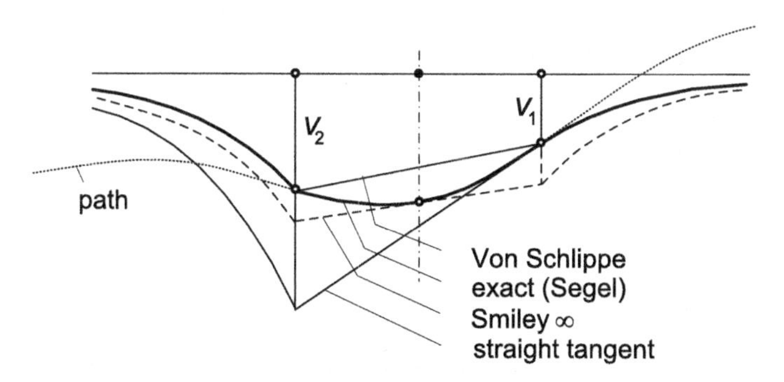


**FIGURE 5.20** Several approximative shapes of the contact line of the string model.

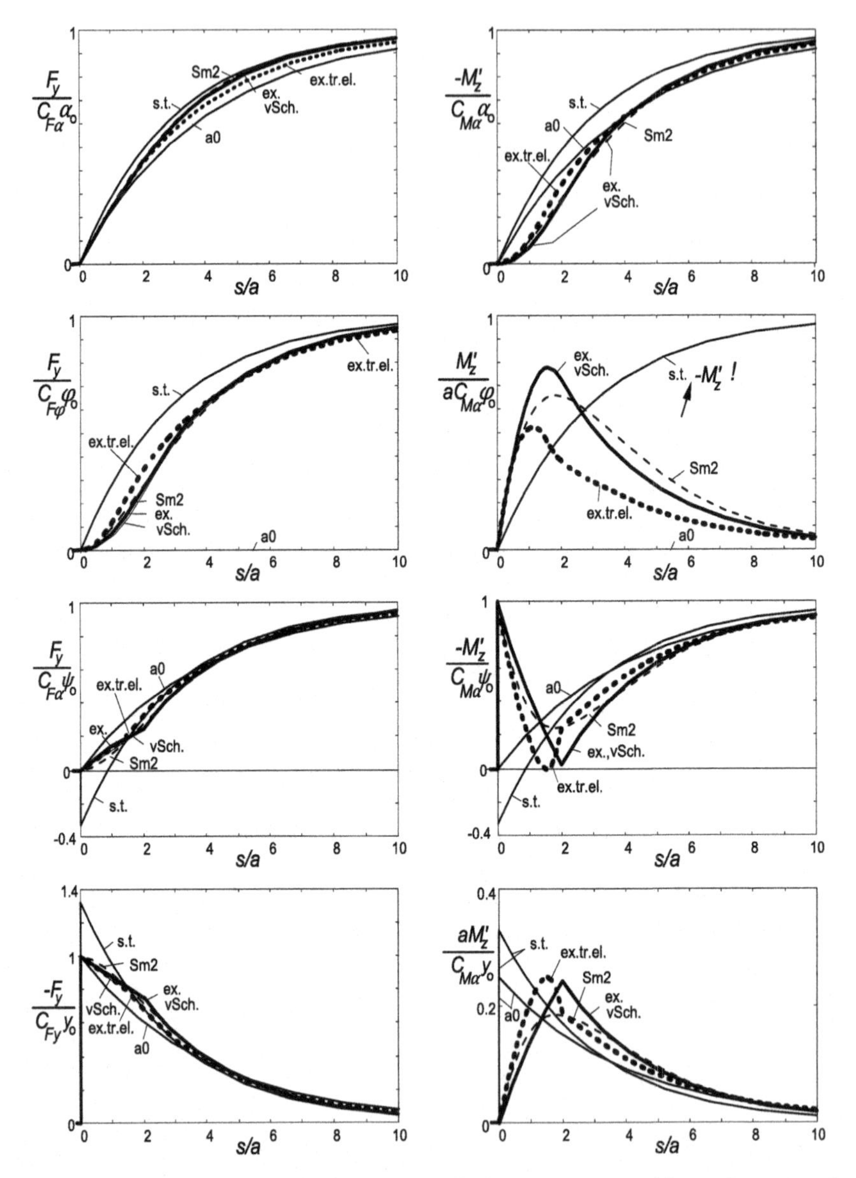


**FIGURE 5.21** Step responses of side force and aligning torque to several inputs for exact and approximate models: ex.: exact 'bare' string model; vSch.: Von Schlippe; Sm2: Smiley second order; s.t.: straight tangent; a0: point contact; ex.tr.el.: exact with tread elements.

The simplest models – straight tangent and point contact – can easily be extended into the nonlinear regime. In Chapter 6, this will be demonstrated for the straight tangent model in connection with the nonlinear analysis of the shimmy phenomenon. In Chapter 7, the nonlinear single-point contact model

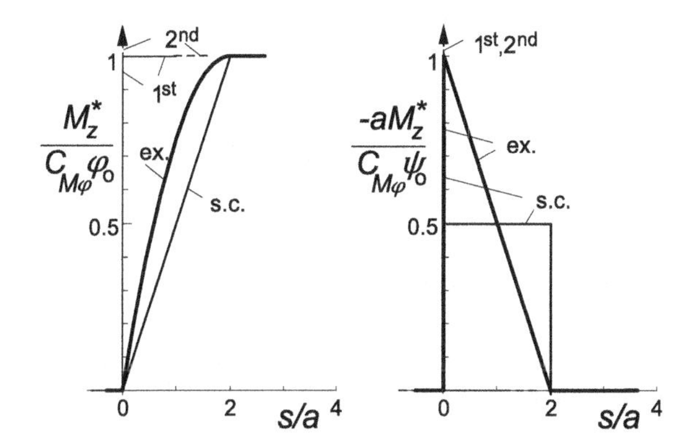


**FIGURE 5.22** Step response of the moment due to tread width to turn slip (path curvature) and yaw angle for exact and approximations: ex.: exact (brush model); s.c.: straight connection (linear interpolation); 2nd: second-order approximation; 1st: first-order approximation.

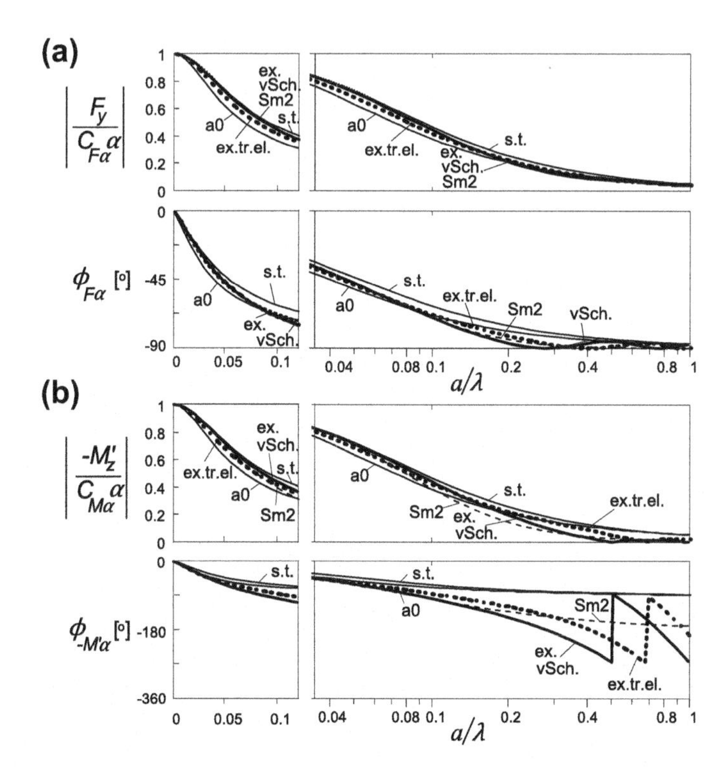


**FIGURE 5.23** a,b,c,d,e,f. Frequency response functions of side force and aligning torque to several inputs for exact and approximate models: ex.: exact 'bare' string model; vSch.: Von Schlippe; Sm2: Smiley second order; s.t.: straight tangent; a0: point contact; ex.tr.el.: exact with tread elements.

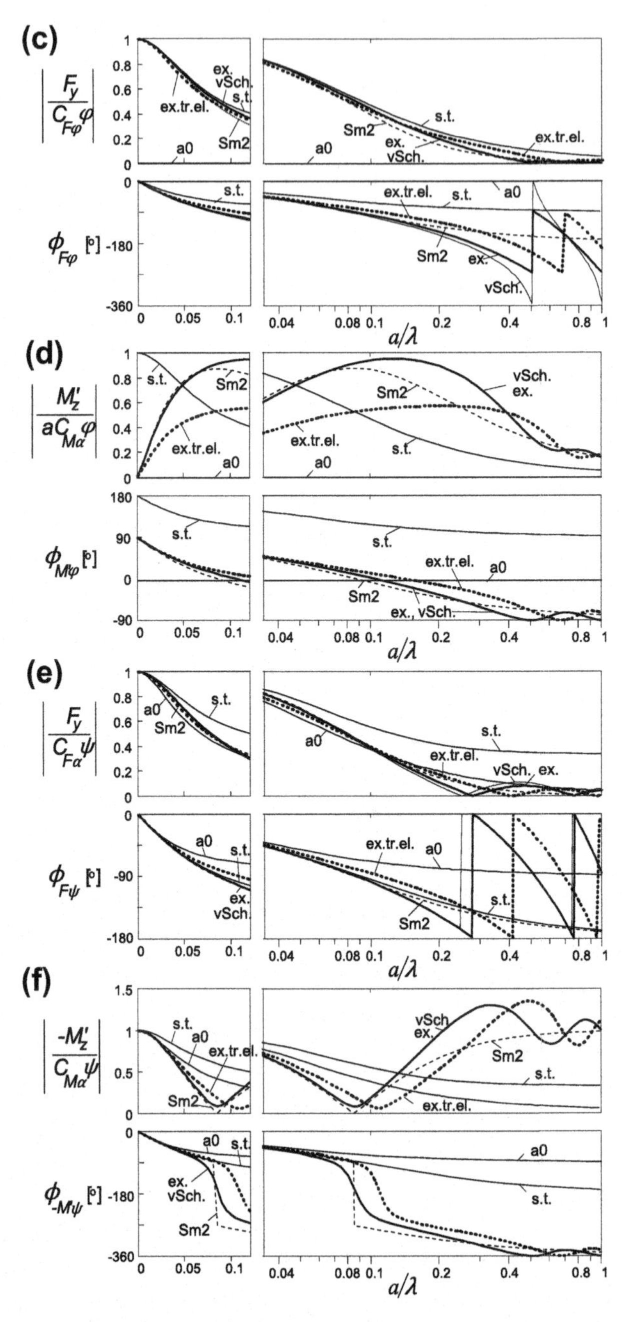


FIGURE 5.23 (*Continued*).

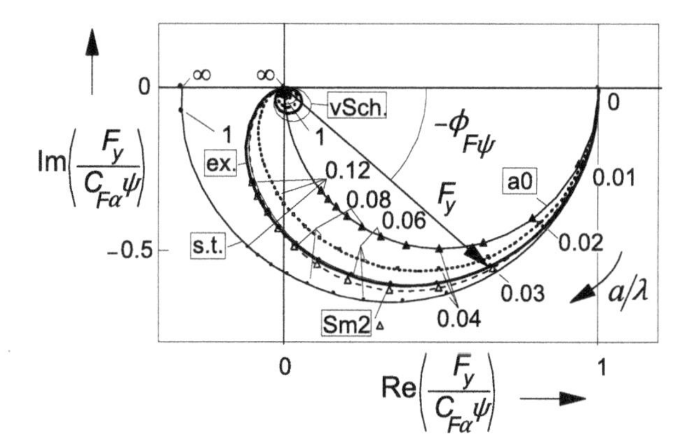


**FIGURE 5.24** Nyquist plot of the frequency response function of the side force $F_y$ with respect to the wheel yaw angle $\psi$. Parameter: $\sigma = 3a$. ex.: exact 'bare' string model; vSch.: Von Schlippe; Sm2: Smiley second order; s.t.: straight tangent; a0: point contact; ex.tr.el.: exact with tread elements.

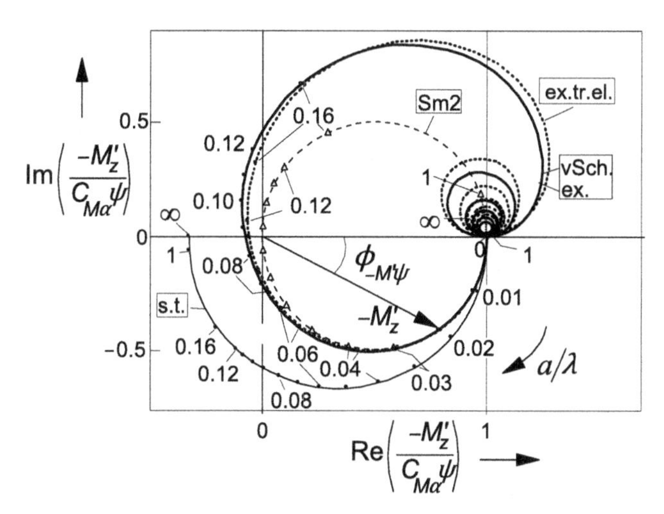


**FIGURE 5.25** Nyquist plot of the frequency response function of the aligning torque $-M_z^*$ with respect to the wheel yaw angle $\psi$. Curve a0 is hidden by Sm2. Same conditions as in Figure 5.24.

will be fully exploited. These models only show an acceptable accuracy for wheel oscillations at wavelengths which are relatively large with respect to the contact length. Chapter 9 is especially devoted to the development of a model that can operate at smaller wavelengths and nonlinear (combined slip) conditions which requires the inclusion of the effect of the length of the contact zone.

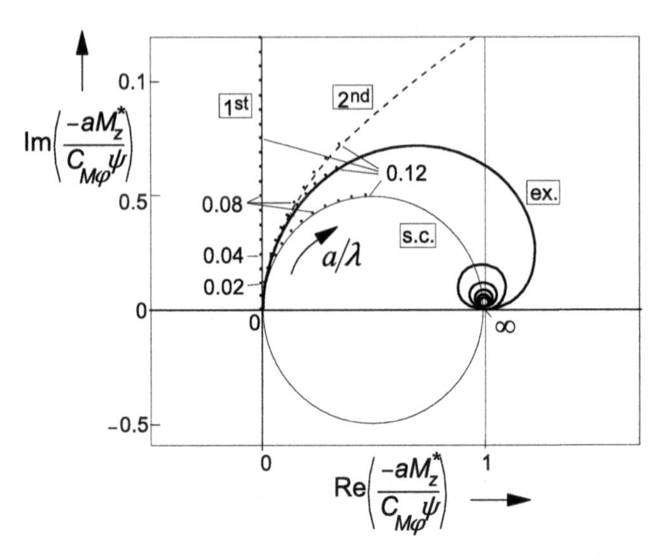


**FIGURE 5.26** Nyquist plot of the frequency response function of the torque due to tread width $-M_z^*$ with respect to the wheel yaw angle $\psi$. ex.: exact (brush); s.c.: straight connection (linear interpolation); 2nd: second-order app.; 1st: first-order approx.

### *Von Schlippe's Straight Connection Model*

This model which shows results that can hardly be distinguished from the exact ones only requires the string deflections at the leading and trailing edges $v_1$ and $v_2$. We find, for the Laplace transforms of these deflections as derived from the expressions (5.22, 5.23) in vectorial form,

$$\boldsymbol{H}_{v1,(\alpha,\varphi,\psi)}(p) = \frac{\sigma}{1+\sigma p}(1, a, 1-ap) \tag{5.87}$$

$$\boldsymbol{H}_{v2,(\alpha,\varphi,\psi)}(p) = -\frac{1}{p}\frac{e^{-2ap}}{1+\sigma p}\left(1, \sigma+a+\frac{1}{p},\ -(\sigma+a)p\right) + \frac{1}{p}\left(1, -a+\frac{1}{p}, ap\right) \tag{5.88}$$

where the exponential function refers to the retardation effect over a distance equal to the contact length. The transfer functions to the alternative set of input variables ($y$, $\psi$) may be obtained by using the conversion formulas (5.36, 5.37).

The responses to a step change in slip angle become, cf. (5.45, 5.46),

$$v_1(s) = \Gamma_{v1,\alpha}\alpha_o = \sigma(1-e^{-s/\sigma})\alpha_o \tag{5.89}$$

$$v_2(s) = \Gamma_{v2,\alpha}\alpha_o = s\alpha_o \quad \text{if } s \le 2a \tag{5.90}$$

$$v_2(s) = \Gamma_{v2,\alpha}\alpha_o = \left[2a + \sigma\left\{1 - e^{-(s-2a)/\sigma}\right\}\right]\alpha_o \quad \text{if } s > 2a \tag{5.91}$$

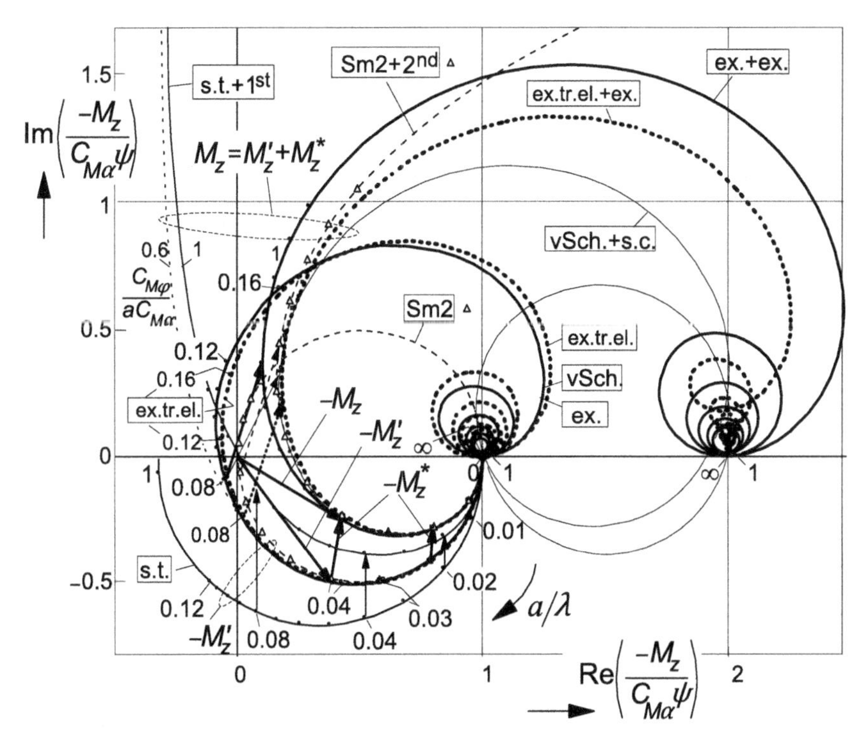


**FIGURE 5.27** Nyquist plot of the nondimensional frequency response function of the aligning moment $M_z$ with respect to the wheel yaw angle $\psi$. Contributing components $M_z'$ due to lateral deformations and $M_z^*$ due to tread width and associated longitudinal deformations. Parameters: $\sigma = 3a$ and $C_{M\varphi} = aC_{M\alpha}$. Exact and approximate models: ex.: exact 'bare' string model; vSch.: Von Schlippe; Sm2: Smiley second order; s.t.: straight tangent; a0: point contact (hidden by Sm2); ex.tr.el.: exact with tread elements; ex.: exact (brush ); s.c.: straight connection (linear interpolation); 2nd : second-order approximation; 1st : first-order approx. (also $C_{M\varphi} = 0.6aC_{M\alpha}$).

Responses to step changes of other input variables may be determined by using the conversion formulas (5.56–5.58).

The side force and the aligning torque are obtained as follows:

$$F_y = c_c(\sigma + a)(v_1 + v_2) = \frac{1}{2} C_{F\alpha} \frac{v_1 + v_2}{\sigma + a} \tag{5.92}$$

$$M_z' = c_c\left\{\sigma(\sigma + a) + \frac{1}{3}a^2\right\}(v_1 - v_2) = \frac{1}{2} C_{M\alpha} \frac{v_1 - v_2}{a} \tag{5.93}$$

The first equation shows that the force is obtained by multiplying the average lateral deflection with the lateral stiffness, cf. (5.59) with (5.60), and the moment by multiplying the slope of the connection line with the torsional stiffness of the 'thin' tire (5.62) or (5.61). It may be noted that due to the adopted straight contact line, the turn slip stiffness for the force becomes slightly less than the exact one (5.63):

$$C_{F\varphi} = 2c_c a\sigma(\sigma + a) \tag{5.94}$$

The various diagrams show that only in some particular cases, a (small) difference between 'exact' and 'Von Schlippe' can be observed. The model can be easily used in vehicle simulation studies by remembering the $(\overline{x}, \overline{y})$ coordinates of the leading edge with respect to the global axis system (cf. Figure 5.3) and use these coordinates again after the wheel is rolled a distance $2a$ further when the trailing edge has assumed this location.

### Smiley's and Roger's Approximations

Assuming in Figure 5.3 that the wheel moves along the $\overline{x}$ axis with only small deviations in the lateral direction and in yaw, the lateral coordinate $\overline{y}_1$ of the leading edge follows from Eqn (5.24). After realizing that

$$\overline{y}_1 = \overline{y} + a\psi + v_1 \tag{5.95}$$

the differential equation for $\overline{y}_1$ becomes

$$\sigma\frac{d\overline{y}_1}{ds} + \overline{y}_1 = \overline{y} + (\sigma + a)\psi \tag{5.96}$$

The problem Smiley has addressed is the assessment of the location of the center of the contact line. The lateral coordinate $\overline{y}_o$ of this point may be approximated by using a Taylor series. Starting out from the position of the center contact point, the slope and the curvature etc. of the path of contact points, the relationship between the lateral coordinate of the foremost contact point and that of the center point can be written as follows:

$$\overline{y}_1 = \overline{y}_o + a\frac{d\overline{y}_0}{ds} + \frac{a^2}{2!}\frac{d^2\overline{y}_0}{ds^2} + \cdots \tag{5.97}$$

Its derivative becomes

$$\frac{d\overline{y}_1}{ds} = \frac{d\overline{y}_0}{ds} + a\frac{d^2\overline{y}_0}{ds^2} + \frac{a^2}{2!}\frac{d^3\overline{y}_0}{ds^3} + \cdots \tag{5.98}$$

After substitution of these series in Eqn (5.96), the following generic formula is obtained:

$$\sum_{i=1}^{n}\frac{(i\sigma + a)a^{i-1}}{i!}\frac{d^i\overline{y}_0}{ds^i} + \overline{y}_o = \overline{y} + (\sigma + a)\psi \tag{5.99}$$

The side force and the aligning torque can be found by multiplying the lateral deflection $v_0 = \bar{y}_0 - \bar{y}$ and the torsion angle $\beta = d\bar{y}_0/ds - \psi$ with the respective stiffnesses $C_{Fy}$ and $C_{M'\psi}$, cf. Eqns (5.59, 5.62). We obtain

$$F_y = C_{Fy}(\bar{y}_0 - \bar{y}) \tag{5.100}$$

$$M'_z = -C_{M'\psi}\left(\frac{d\bar{y}_0}{ds} - \psi\right) \tag{5.101}$$

Using the conversion formulas (5.36, 5.37), the transfer functions with respect to $\alpha$ and $\varphi$ can be obtained. For the order $n$ of (5.99) equal to 2, the model becomes of the second order and one finds the following transfer functions in vector form:

$$\boldsymbol{H}_{F,(\alpha,\varphi,\psi)}(p) = C_{Fy}\frac{\left(a\left(\sigma+\frac{1}{2}a\right)p+\sigma+a,\ a\left(\sigma+\frac{1}{2}a\right),\sigma+a\right)}{a\left(\sigma+\frac{1}{2}a\right)p^2+(\sigma+a)p+1} \tag{5.102}$$

and

$$\boldsymbol{H}_{M',(\alpha,\varphi,\psi)}(p) = -C_{M'\psi}\frac{\left(1,\ -a\left(\sigma+\frac{1}{2}a\right)p,\ a\left(\sigma+\frac{1}{2}a\right)p^2+1\right)}{a\left(\sigma+\frac{1}{2}a\right)p^2+(\sigma+a)p+1} \tag{5.103}$$

It may be noted that the side force turn slip stiffness for this approximate model is somewhat larger than according to the exact expression. We have, for Smiley's model,

$$C_{F\varphi} = 2c_c a\left(\sigma+\frac{1}{2}a\right)(\sigma+a) \tag{5.104}$$

The step responses of Figure 5.21 show reasonable to very good correspondence for the different inputs. Also, the frequency response functions, shown in Figure 5.23, indicate that the accuracy of this relatively simple Smiley2 approximation is quite good. The dip in the $M'_z$ response to $\psi$ is well represented and is located at the 'meandering' path frequency (zero of (5.103)) $a\omega_s = 2\pi a/\lambda = \sqrt{\left\{a/\left(\sigma+\frac{1}{2}a\right)\right\}}$. The approximation may be considered to be acceptable for wavelengths larger than about 4 times the contact length ($a/\lambda = 0.125$). In his publication, Smiley recommends to use the order $n = 3$.

The initial empirically assessed formulas of Rogers (1972) are almost the same as the functions (5.102, 5.103); the terms $\frac{1}{2}a$ do not appear in his expressions but in the numerator of the moment response to turn slip (second element of 5.103) the empirically assessed term $-\varepsilon$ is added. Also, a connection with Kluiters' approximation appears to exist.

### Kluiters' Approximation

Kluiters (1969) (cf. Besselink 2000) adopted a Padé filter to approximate the position of the trailing edge. The transfer function of $y_2$ with respect to $y_1$ reads, if the order of the filter is taken equal to 2,

$$H_{y_2,y_1}(p) = \frac{\frac{1}{3}a^2p^2 - ap + 1}{\frac{1}{3}a^2p^2 + ap + 1} \tag{5.105}$$

with

$$\bar{y}_1 = \bar{y} + a\psi + v_1 \tag{5.106}$$

$$v_2 = \bar{y}_2 - \bar{y} + a\psi \tag{5.107}$$

We obtain, with (5.87),

$$\boldsymbol{H}_{v2,(\alpha,\varphi,\psi)}(p) = -\frac{1}{p}\frac{\frac{1}{3}a^2p^2 - ap + 1}{(\sigma p + 1)\left(\frac{1}{3}a^2p^2 + ap + 1\right)}\left(1,\ \sigma + a + \frac{1}{p},\ -(\sigma + a)p\right) + \frac{1}{p}\left(1,\ -a + \frac{1}{p},\ ap\right) \tag{5.108}$$

The deflection at the foremost point is, of course, governed by transfer function (5.87). The correspondence of this third-order model with Von Schlippe's model is very good for wavelengths larger than about 1.5 times the contact length. The accuracy is better than even Smiley's third-order approach. It turns out that when a first-order Padé filter is used, the formulas of Rogers without $\varepsilon$ arise.

### Straight Tangent Approximation

For this very simple approximation, the contact line is solely governed by the deflection $v_1$ at the leading edge. The approximated shape of the deflected string corresponds to the steady-state deflection depicted in Figure 5.8 with deflection angle $\alpha' = v_1/\sigma$. The side force and aligning torque are found by multiplying the deflection angle with the cornering stiffness and the aligning stiffness, respectively.

Using (5.87), we obtain, for the transfer functions,

$$\boldsymbol{H}_{F,(\alpha,\varphi,\psi)}(p) = C_{F\alpha}\frac{1}{1 + \sigma p}(1,\ a,\ 1 - ap) \tag{5.109}$$

$$\boldsymbol{H}_{M',(\alpha,\varphi,\psi)}(p) = -C_{M\alpha}\frac{1}{1 + \sigma p}(1,\ a,\ 1 - ap) \tag{5.110}$$

The step responses equal the respective slip stiffnesses multiplied with the responses of $v_1/\sigma$ according to (5.89). As expected, the accuracy becomes much less and from the frequency response functions we may conclude that acceptable agreement is attained when the path frequency is limited to about $a/\lambda = 0.04$ or a wavelength larger than about 12 times the contact length. The response of the aligning torque to path curvature appears to be far off when compared with the exact responses. Since this particular response is the least important in realistic situations, the straight tangent model may still be acceptable. In the analysis of the shimmy phenomenon, this will be demonstrated to be true for speeds of travel which are not too low (where the wavelength becomes too short).

Figures 5.24 and 5.25 clearly show that at least up to $a/\lambda = 0.04$, the phase lag closely follows the exact variation with frequency. The amplitude, however, is too large, especially for the moment. A combination with a fictive $M_z^*$ (first-order approximation, see further on) with properly chosen parameter $\kappa^* = C_{M\varphi}$ may help to reduce the amplitude of the moment to a more acceptable level also for nondimensional path frequencies higher than 0.04. Figure 5.27 shows a considerable improvement if we would choose $C_{M\varphi} = 0.6C_{M\alpha}$.

Differential Eqn (5.125), to be shown later on, governs the straight tangent approximation. Because we have a deflection shape equal to that occurring at steady-state side-slip motion, the extension of the model to nonlinear large slip conditions is easy to establish. We may employ the steady-state characteristics, e.g. *Magic Formulas*, and obtain

$$F_y = F_y(\alpha'), \; M_z = M_z(\alpha') \tag{5.111}$$

with the deflection angle

$$\alpha' = \frac{v_1}{\sigma} \tag{5.112}$$

The amplitude of the self-excited shimmy oscillation appears to become limited due to the nonlinear, degressive, characteristics of the side force and the aligning torque versus slip angle.

## Single-point Contact Model

This simplest approximation disregards the influence of the length of the contact line. The lateral deflection at the contact center at steady-state side slip should be taken equal to that of the exact model. This requires a model relaxation length $\sigma_0$ equal to the sum of the string deflection relaxation length $\sigma$ and half the contact length $a$. The corresponding transfer function for the deflection $v_0$ becomes

$$\boldsymbol{H}_{v0,(\alpha,\varphi,\psi)}(p) = \frac{\sigma_0}{1+\sigma_0 p}\,(1,0,1) \tag{5.113}$$

with

$$\sigma_0 = \sigma + a \tag{5.114}$$

Apparently, the response to turn slip vanishes. A deflection angle may still be defined: $\alpha' = v_0/\sigma_0$. As a result, response functions for the force and the moment are obtained:

$$\boldsymbol{H}_{F,(\alpha,\varphi,\psi)}(p) = C_{F\alpha}\frac{1}{1+\sigma_0 p}\,(1,0,1) \tag{5.115}$$

$$\boldsymbol{H}_{M',(\alpha,\varphi,\psi)}(p) = -C_{M\alpha}\frac{1}{1+\sigma_0 p}\,(1,0,1) \tag{5.116}$$

This function corresponds to the frequency response function (5.73) that holds for a first-order system with the same cutoff frequency as the one for the exact model. With relaxation length $\sigma_0 = \sigma + a$, the model produces a correct phase lag for the response to yaw oscillations at large wavelengths as was already assessed before, cf. (5.81). Except for the moment response to yaw, the amplitudes appear to become somewhat too small in the probably acceptable wavelength range $\lambda > \sim 24a$. In the Nyquist plot of Figure 5.25, the curve for the moment response coincides with that of Smiley's approximation (not the frequency marks!). The response of the side force to turn slip may be artificially introduced by putting an '$a$' instead of the '0' in the input vector of (5.115). This would, however, require an additional first-order differential equation for the side force.

Differential Eqn (5.130) governs the single-point contact model.

## *Approximations of Tread Width Moment Response*

The transfer function (5.32) for $M_z^*$ may be simplified by following the approach of Von Schlippe but now for the longitudinal deflections, that is: by assuming a linear interpolation of the longitudinal deflection $u$ between the exact deflections at the leading and trailing edges. As the deflection at the first point is equal to zero, the linear interpolation would lead to an average deflection equal to half the deflection at the rear most point $u_2$. The corresponding transfer function becomes

$$\boldsymbol{H}_{M^*,(\alpha,\varphi,\psi)}(p) = \frac{\kappa^*}{2ap}\left(1-e^{-2pa}\right)(0,1,-p) \tag{5.117}$$

Figures (5.22, 5.26) show the performance of this linear interpolation model together with the exact response and other approximated model responses. The approach may be used in connection with the Von Schlippe lateral deflection approximation where the location of the contact point at the leading edge is remembered over a distance rolled equal to the contact length when this information is used to calculate the deflection at the rear edge.

Expansion of the exact response function (5.32) in a series of powers of $p$ yields when limiting to the second power (for the response to $\psi$):

$$\boldsymbol{H}_{M^*,(\alpha,\varphi,\psi)}(p) = \kappa^*\left(1 - \frac{2}{3}ap\right)(0,\ 1, -p) \tag{5.118}$$

where $\kappa^* = C_{M\varphi}$. The first-order (only first term in 5.118) and the second-order (both terms) approximations show responses as presented in the figures. The first-order response has been indicated in combination with the straight tangent response for $M'_z$ in Figure 5.27. The vectors for $-M^*_z$ point upward (vertically). The case of $C_{M\varphi} - 0.6aC_{M\alpha}$ generates a response which improves on the straight tangent bare string response. Especially compared with the enhanced model provided with tread elements, the agreement is satisfactory. To model the foot print, Chapter 9 uses a first-order differential equation with the relaxation length equal to $a$. This corresponds to the cutoff frequency of the exact model (5.83).

## *Differential Equations*

From the various approximate transfer functions, the governing differential equations can be easily established by replacing the Laplace variable $p$ with the differentiation operator d/d$s$. This is not the case with the exact and the Von Schlippe transfer functions since these are of infinite order. We may expand these functions in series of powers of $p$ and truncate at a certain power, after which $p$ is replaced by d/d$s$. Truncation after the first degree of the series expansion of the exact functions (5.30–5.32) yields the differential equations written in terms of input variables $\alpha$ and $\varphi$ ($t$ denoting the pneumatic trail):

$$\sigma\frac{dF_y}{ds} + F_y = C_{F\alpha}\alpha + C_{F\varphi}\varphi - C_{F\alpha}(a-t)\frac{d\alpha}{ds} - C_{F\varphi}a\frac{d\varphi}{ds} + \cdots \tag{5.119}$$

$$\sigma\frac{dM'_z}{ds} + M'_z = -C_{M\alpha}\alpha + aC_{M\alpha}\frac{d\alpha}{ds} + a\left\{\left(\sigma + \frac{1}{3}a\right)C_{M\alpha} + \frac{1}{15}a(C_{M\alpha} - a\sigma C_{Fy})\right\}\frac{d\varphi}{ds} + \cdots \tag{5.120}$$

with coefficients according to (5.55, 5.59–5.63). When using the conversion formulas (5.34, 5.35), equations in terms of input variables $\bar{y}$ and $\psi$ are obtained:

$$\sigma\frac{dF_y}{ds} + F_y = C_{F\alpha}\left(\psi - \frac{d\bar{y}}{ds} - a\frac{d\psi}{ds} + (a-t)\frac{d^2\bar{y}}{ds^2}\right) + \cdots \tag{5.121}$$

$$\sigma\frac{dM'_z}{ds} + M'_z = -C_{M\alpha}\left(\psi - \frac{d\bar{y}}{ds} - a\frac{d\psi}{ds} + a\frac{d^2\bar{y}}{ds^2}\right) + \cdots \tag{5.122}$$

From Eqn (5.118), we, finally, obtain the differential equations for $M_z^*$:

$$M_z^* = C_{M\varphi}\left(\varphi - \frac{2}{3}a\frac{\mathrm{d}\varphi}{\mathrm{d}s}\right) + \cdots \tag{5.123}$$

or, in terms of the yaw angle,

$$M_z^* = -C_{M\varphi}\left(\frac{\mathrm{d}\psi}{\mathrm{d}s} - \frac{2}{3}a\frac{\mathrm{d}^2\psi}{\mathrm{d}s^2}\right) + \cdots \tag{5.124}$$

The straight tangent approximation remains when the last term of the right-hand member of both Eqns (5.121, 5.122) is omitted. The same result is obtained when the single first-order differential Eqn (5.24) is used in combination with the deflection angle $\alpha'$ (5.112) and the cornering and aligning stiffnesses. We have the following set of equations for the straight tangent approximation:

$$\frac{\mathrm{d}v_1}{\mathrm{d}s} + \frac{1}{\sigma}v_1 = \psi - \frac{\mathrm{d}\bar{y}}{\mathrm{d}s} - a\frac{\mathrm{d}\psi}{\mathrm{d}s} \quad (= \alpha + a\varphi) \tag{5.125}$$

$$\alpha' = \frac{v_1}{\sigma} \tag{5.126}$$

$$F_y = C_{F\alpha}\alpha', \quad M_z = -C_{M\alpha}\alpha' \tag{5.127}$$

or, in case of larger slip angles,

$$F_y = F_y(\alpha'), \quad M_z = M_z(\alpha') \tag{5.128}$$

The set may be used in combination with the first-order version of (5.124):

$$M_z^* = -C_{M\varphi}\frac{\mathrm{d}\psi}{\mathrm{d}s} \quad (= C_{M\varphi}\varphi) \tag{5.129}$$

or in the nonlinear case $\varphi$ may be used as input in the steady-state model description extended with turn slip (cf. Section 4.3.3).

The single-point contact model with transfer functions (5.115, 5.116) is governed by the single first-order differential equation:

$$\frac{\mathrm{d}v_0}{\mathrm{d}s} + \frac{1}{\sigma_0}v_0 = \psi - \frac{\mathrm{d}\bar{y}}{\mathrm{d}s} \quad (= \alpha) \tag{5.130}$$

with the relaxation length

$$\sigma_0 = \sigma + a = \frac{C_{F\alpha}}{C_{Fy}} \tag{5.131}$$

The 'transient' slip angle

$$\alpha' = \frac{v_0}{\sigma_0} \tag{5.132}$$

completes the description together with Eqns (5.127 or 5.128).

By considering only terms up to the first degree in $p$ of the (transformed) Eqns (5.119–122), the following individual first-order differential equations can be defined which hold exactly for $\lambda \to \infty$ ($p \to 0$), if use is made of the notation for the individual relaxation lengths (5.72) and (5.81):

$$\sigma_{F\alpha}\frac{dF_y}{ds} + F_y = C_{F\alpha}\alpha \tag{5.133}$$

$$\sigma_{M\alpha}\frac{dM'_z}{ds} + M'_z = -C_{M\alpha}\alpha \tag{5.134}$$

$$\sigma_{F\varphi}\frac{dF_y}{ds} + F_y = C_{F\varphi}\varphi \tag{5.135}$$

$$\sigma\frac{dM'_z}{ds} + M'_z = a\left\{\left(\sigma + \frac{1}{3}a\right)C_{M\alpha} + \frac{1}{15}a(C_{M\alpha} - a\sigma C_{Fy})\right\}\frac{d\varphi}{ds} \tag{5.136}$$

and, in terms of the alternative set of inputs,

$$\sigma_{F\alpha}\frac{dF_y}{ds} + F_y = -C_{F\alpha}\frac{d\bar{y}}{ds} \tag{5.137}$$

$$\sigma_{M\alpha}\frac{dM'_z}{ds} + M'_z = C_{M\alpha}\frac{d\bar{y}}{ds} \tag{5.138}$$

$$\sigma_{F\psi}\frac{dF_y}{ds} + F_y = C_{F\alpha}\psi \tag{5.139}$$

$$\sigma_{M'\psi}\frac{dM'_z}{ds} + M'_z = -C_{M\alpha}\psi \tag{5.140}$$

where the respective individual relaxation lengths read (according to the exact bare string model)

$$\sigma_{F\alpha} = \sigma + a - t \tag{5.72}$$

$$\sigma_{M'\psi} = \sigma_{M\alpha} = \sigma_{F\varphi} = \sigma_{F\psi} = \sigma + a \tag{5.81}$$

If also the relaxation length for $F_y$ with respect to $\alpha$ is taken equal to $\sigma_0 = \sigma + a$, with respect to turn slip improved, the single contact point approximation arises.

Table 5.1 presents the values of the individual relaxation lengths as defined by Eqns (5.71, 5.75, etc.) and assessed numerically for the various models discussed above, with $\sigma = 3a$. For the corresponding enhanced model with tread elements, we refer to Section 5.4.3. For all models, $\sigma_{F\psi} = C_{F\alpha}/C_{Fy}$ has been kept the same.

In practical cases, shimmy occurs at path frequencies below the meandering frequency, even below about half this value. This may correspond to a practical upper limit of $a/\lambda = 0.04$ or for the wavelength $\lambda$ a lower limit of ca. 12.5 times the contact length ($25a$). It turns out that in this practical range, the agreement between the various models is good or reasonable.

**TABLE 5.1** Individual Relaxation Lengths for the Various Models, Defined for $\omega_s \to 0$

| | Exact | vSchl. | Smiley2 | str.tang. | a = 0 | ex.tr.el. |
|---|---|---|---|---|---|---|
| $\sigma_{F\alpha}/a$ | 3.23 | 3.25 | 3.12 | 3.00 | 4.00 | 3.51 |
| $\sigma_{M\alpha}/a$ | 4.00 | 4.00 | 4.00 | 3.00 | 4.00 | 4.00 |
| $\sigma_{F\varphi}/a$ | 4.00 | 4.11 | 4.00 | 3.00 | – | 4.00 |
| $\sigma_{F\dot{\psi}}/a$ | 4.00 | 4.00 | 4.00 | 4.00 | 4.00 | 4.00 |
| $\sigma_{M'\dot{\psi}}/a$ | 4.00 | 4.00 | 4.00 | 4.00 | 4.00 | 4.00 |

## 5.4.2. Other Models

Two more models will be briefly discussed. These are the models of Keldysh, cf. Goncharenko et al. (1981) and Besselink (2000), and of Moreland (1954). The models are not based on the stretched string concept. They feature two degrees of freedom – the lateral deflection and the torsion deflection – and disregard the finite length of the contact zone. The single-point or straight tangent approximations of the string model have a single degree of freedom where the lateral and the angular deflections at the leading edge are linked through an algebraic relationship. The additional degree of freedom of Keldysh's and Moreland's single-point models with a nonholonomic constraint (first-order differential equation) raises the order of the description from one to two.

### *Keldysh's Model*

This model is known to be used in the Russian aircraft industry and was published in 1945. Goncharenko employed the model for his research in shimmy of landing gears. In addition to lateral and torsional stiffnesses, damping has been included between contact patch and rim. The side force and aligning torque become expressed in terms of lateral deflection $v$ and torsion angle $\beta$:

$$F_y = k_y \dot{v} + c_y v \tag{5.141}$$

$$M_z = k_\psi \dot{\beta} + c_\psi \beta \tag{5.142}$$

With the assumption of full adhesion, the following relation must hold regarding the direction of the path of the contact point of the rolling tire:

$$\psi + \beta = \frac{\mathrm{d}(v + \bar{y})}{\mathrm{d}s} \tag{5.143}$$

Regarding the path curvature, a linear equation is introduced:

$$\frac{\mathrm{d}(\psi+\beta)}{\mathrm{d}s} = -q_\alpha v - q_\beta \beta + q_\gamma \gamma \tag{5.144}$$

Interesting is that Keldysh did take into account the wheel camber angle $\gamma$ as an additional input variable. The transfer functions with respect to $\alpha$, $\varphi$, $\psi$, and $\gamma$ become

$$\boldsymbol{H}_{F,(\alpha,\varphi,\psi,\gamma)}(p) = (c_y + k_y V p)\,\frac{(p+q_\beta, 1, q_\beta, q_\gamma)}{p^2 + q_\beta p + q_\alpha} \tag{5.145}$$

and

$$\boldsymbol{H}_{M,(\alpha,\varphi,\psi,\gamma)}(p) = -(c_\psi + k_\psi V p)\,\frac{(q_\alpha,\ -p,\ p^2+q_\alpha, -q_\gamma p)}{p^2 + q_\beta p + q_\alpha} \tag{5.146}$$

Surprisingly, it turns out that these transfer functions (except the one with respect to $\gamma$) correspond to the functions of Rogers (with $\varepsilon = 0$) (Eqns 5.102, 5.103 with $\frac{1}{2}a$ omitted) if, as Besselink indicated, the following equivalence conditions hold:

$$q_\alpha = \frac{1}{a\sigma},\ q_\beta = \frac{a+\sigma}{a\sigma},\ c_y = C_{Fy} = \frac{C_{F\alpha}}{a+\sigma},\ c_\psi = C_{M'\psi} = C_{M\alpha} \tag{5.147}$$

Furthermore, comparison with the Von Schlippe approximation at steady-state shows that

$$q_\beta = \frac{C_{F\alpha}}{C_{F\varphi}},\ q_\gamma = \frac{C_{F\gamma}}{C_{F\varphi}} \tag{5.148}$$

where, as before, $C_{F\gamma}$ designates the camber force stiffness.

## *Moreland's Model*

This model, first published in 1954 in the paper "The story of shimmy," used to be a popular tool in the U.S. for the analysis of aircraft shimmy. The structure is similar to that of Keldysh's model. The differences are that the damping coefficient $k_\psi$ is omitted and, more important, turn slip and camber are not considered and instead of Eqn (5.144) the following relation is used:

$$\frac{\mathrm{d}\beta}{\mathrm{d}t} = -\frac{1}{\tau}\left(\frac{F_y}{C_{F\alpha}} + \beta\right) \tag{5.149}$$

The transfer functions with respect to the slip angle that were established read

$$H_{F,\alpha}(p) = C_{F\alpha}\frac{k_y \tau V^2 p^2 + (k_y + c_y \tau) V p + c_y}{C_{F\alpha}\tau V p^2 + (k_y V + C_{F\alpha}) p + c_y} \tag{5.150}$$

$$H_{M,\alpha}(p) = -C_{M\alpha}\frac{k_y V p + c_y}{C_{F\alpha}\tau V p^2 + (k_y V + C_{F\alpha}) p + c_y} \tag{5.151}$$

The fact that we are dealing here with a *time* constant $\tau$ gave rise to criticism and caused difficulties in attempts to fit experimental data. To make the model pure path dependent, $\tau$ should be made inversely proportional with speed $V$; this in addition to omitting the damping coefficient $k_y$. It may be observed that when considering the relations (5.147), the substitution for $\tau$

$$\tau = \frac{a\sigma}{(a+\sigma)V} \tag{5.152}$$

leads to a transfer function with respect to $\alpha$ that is identical to the corresponding functions established by Keldysh and Rogers.

### 5.4.3. Enhanced String Model with Tread Elements

Although the string model with its simplicity and essential features served us well in gathering insight and establishing proper mathematical descriptions and useful approximations, the string model shows a behavior that in several views differs considerably from the properties of the real tire. The relaxation length $\sigma$ for the tire deflection does not change with vertical load, and thus the relaxation length for the side force to yaw $\sigma_{F\psi} = \sigma + a$ does not vary sufficiently and does not approach zero when $F_z \rightarrow 0$ as would occur with the real tire. The same discrepancy holds for the cornering stiffness. The pneumatic trail of the string model is often too large. When considering larger values of slip, that is, when sliding is allowed to occur in the contact zone, the calculated steady-state characteristics for the bare string model do not appear to be adequate as they exhibit a kink at the slip angle where the force has reached its maximum and the moment becomes equal to zero where full sliding commences. Also, when at side slip the lateral force distribution along the contact line is analyzed, it appears that a discontinuity occurs at the point of transition from adhesion to sliding. This is caused by the fact that in the range of adhesion the contact line (that is: the string) is straight and the external frictional forces are equal to the internal elastic forces and remain relatively small. From the transition point onward, the string slides over the ground and the side forces are governed by the vertical force distribution and the friction coefficient. Here the contact line gets curved to let the string tension force $S$ contribute to the transmission of the ground forces to the tire. Finally, it can be easily assessed that the relaxation length of the bare string model does not change with increasing slip angle as is observed to occur in practice. These reasons lead to the desire to develop a model that shows a better behavior and when too complex may serve as a reference for understanding and for the development of simpler models. In the sequel, we will briefly discuss the development of the string model provided with elastic tread elements; for more information, cf. Pacejka (1966, 1981). The results concerning this model's transient and oscillatory out-of-plane behavior have already been included in the various graphs of the preceding section.

### Steady-State Characteristics

Figure 5.28 depicts the structure of the model. The total deflection $v$ is made up of the lateral deflection of the string (the carcass) $v_c$ and the lateral deflection of the tread element $v_p$. Associated stiffnesses of carcass and tread elements are $c_c$ and $c_p$ both per unit length and the latter integrated over the tread width $2b$. The string effective tension force is $S$, cf. Section 5.3.1. The following model constants are introduced:

$$\sigma = \sqrt{\frac{S}{c_c}}, \quad \sigma_c = \sqrt{\frac{S}{c_c + c_p}}, \quad \varepsilon = \frac{\sigma_c}{\sigma} = \sqrt{\frac{c_c}{c_c + c_p}} \tag{5.153}$$

A parabolic vertical pressure distribution is assumed according to Eqn (3.4). At the trailing edge, a sliding range starts to build up when side slip increases. Also at the leading edge, sliding may occur when the tread element stiffness is relatively large. The bare string model does show a relatively short sliding range that grows and ultimately meets the rear sliding range at increasing side slip. We will assume that the tread element stiffness is sufficiently small to disregard front sliding.

The total deflection $v$ at steady-state side slip in the range $x_t < x < a$, where the tips of the elements adhere to the ground, follows the differential Eqn (5.11) with $\varphi = 0$:

$$\frac{\mathrm{d}v}{\mathrm{d}s} = -\alpha \quad \text{if } x_t < x < a \tag{5.154}$$

The deflection of the string outside the contact range is governed by the differential equation:

$$\sigma^2 \frac{\mathrm{d}^2 v_c}{\mathrm{d}\,x^2} - v_c = 0 \quad \text{if } |x| > a \tag{5.155}$$

and in the contact zone:

$$\sigma_c^2 \frac{\mathrm{d}^2 v_c}{\mathrm{d}\,x^2} - v_c = -(1 - \varepsilon^2)v \quad \text{if } |x| \le a \tag{5.156}$$

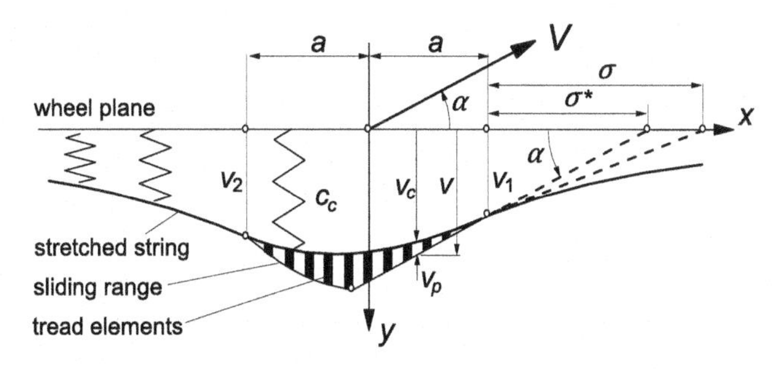


**FIGURE 5.28** Enhanced string model provided with tread elements.

To solve for the coordinate of the transition point $x_t$ and the five constants of integration in the solutions of Eqn (5.154) and of Eqn (5.156) for the two zones, adhesion and sliding, we need six boundary conditions. These concern zero tread element deflection at the leading edge and continuity of deflection and slope of the string and continuity of the element deflection at the transition point. The conditions read

$$x = a: \quad v_c = v\,(= v_1), \quad \frac{\mathrm{d}v_c}{\mathrm{d}x} = -\frac{v_c}{\sigma} \tag{5.157}$$

$$x = x_t: \quad \lim_{x \uparrow x_t}\left(\frac{\mathrm{d}v_c}{\mathrm{d}x}, v_c, v\right) = \lim_{x \downarrow x_t}\left(\frac{\mathrm{d}v_c}{\mathrm{d}x}, v_c, v\right) \tag{5.158}$$

$$x = -a: \quad \frac{\mathrm{d}v_c}{\mathrm{d}x} = \frac{v_c}{\sigma} \tag{5.159}$$

Once expressions for the deflection $v_p = v - v_c$ and the transition point $x_t$ have been established, the force and moment can be assessed by integration along the contact length of $q_y$ over the sliding range and of $c_p v_p$ over the adhesion range. The resulting formulas can be found in the original publication. We will restrict ourselves here with showing the characteristics for the values $\sigma = 3.75a$ and $c_p/c_c = 55$. Figure 5.29 depicts the deflected model at a number of slip angles. In Figure 5.30, the force and moment characteristics have been presented with along the abscissa the quantity $\theta_c \alpha$. The composite model parameter $\theta_c$ is defined as

$$\theta_c = \frac{2}{3}\frac{c_c a^2}{\mu F_z} \tag{5.160}$$

For transient tire analysis, the relaxation length for the leading edge, $\sigma$ for the bare string model and $\sigma^*$ for the model with tread elements here defined as

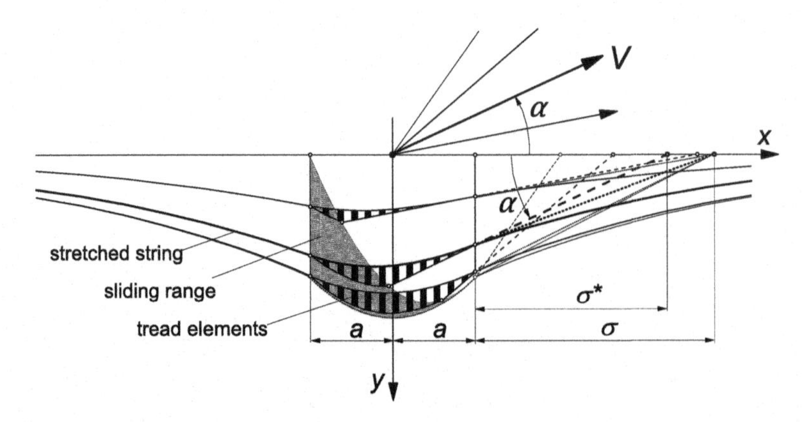


**FIGURE 5.29** The enhanced string tire model at increasing slip angles showing growing sliding range and decreasing 'intersection' length $\sigma^*$.

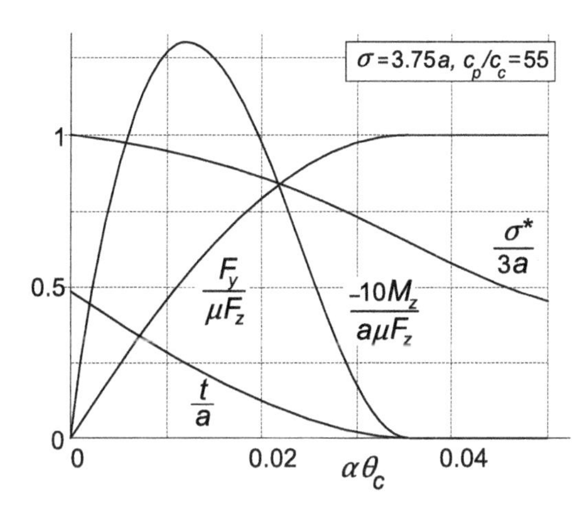


**FIGURE 5.30** Steady-state side-slip characteristics of the string model with tread elements.

the distance between leading edge and the point of intersection of wheel plane and elongation of the straight contact line (Figures 5.8, 5.28, and 5.29), is of importance. We have

$$\sigma^* = \frac{v_1}{\alpha} \tag{5.161}$$

The variation of this 'intersection length' $\sigma^*$ with slip angle has been plotted as well. The parameters have been chosen such that at $\alpha = 0$, we have $\sigma^* = 3a$.

The diagram also contains the curve for the pneumatic trail $t = -M_z/F_y$. At $\alpha = 0$, the value of the pneumatic trail becomes $t = 0.49a$ which is much more in accordance with experimental findings than what we had with the bare string model. The value $0.49a$ also supports the magnitude that had been found for the relaxation length for the force to side slip $\sigma_{F\alpha} = \sigma^* + a - t = 3.51a$ as indicated in Table 5.1, cf. Eqn (5.72).

Expressions for the slip stiffnesses, the pneumatic trail and the relaxation length, and their relationship with vertical load will be presented in the subsequent subsection, where the linear model is discussed.

### *Non-Steady-State Behavior at Vanishing Sliding Range (Linear Analysis)*

We will restrict ourselves to the derivation of the response to the slip angle $\alpha$ and use conversion formula (5.39), with $\sigma$ replaced by $\sigma^*$ (5.171) to obtain the response functions to turn slip $\varphi$ and (5.36, 5.37) for the responses to $y$ and $\psi$. For the total deflection $v$, or transformed $V$, Eqns (5.11, 5.14) and thus (5.16) hold in general. With $A$ representing the transformed $\alpha$, we get

$$V = C_v e^{px} + \frac{1}{p} A = (C_A e^{px} + 1) \frac{1}{p} A \tag{5.162}$$

With the Laplace transformed version of Eqn (5.156), the differential equation for $V_c$ is obtained:

$$\sigma_c^2 \frac{\mathrm{d}^2 V_c}{\mathrm{d}\,x^2} - V_c = -(1-\varepsilon^2)(C_A e^{px} + 1)\,\frac{1}{p}\,A \tag{5.163}$$

with solution:

$$V_c = \left\{ C_+\, e^{\frac{a-x}{\sigma_c}} - C_-\, e^{-\frac{a-x}{\sigma_c}} + (1-\varepsilon^2)\left(\frac{C_A\, e^{px}}{1-\sigma_c^2 p^2} + 1\right)\right\}\frac{1}{p}\,A \tag{5.164}$$

The three integration constants can be determined by using the boundary conditions at the leading and trailing edges (5.157, 5.159):

$$C_\pm = \frac{1}{2}\varepsilon(1\pm\varepsilon)\left(\frac{1+\sigma p}{1\pm\sigma_c p}\,C_A e^{pa} + 1\right) \tag{5.165}$$

$$C_A = -\frac{2 + B_+ + B_-}{B} \tag{5.166}$$

with

$$B = (1+\sigma p)\left(\frac{B_+}{1+\sigma_c p} + \frac{B_-}{1-\sigma_c p}\right)e^{pa} + 2\,\frac{1-\sigma p}{1-\sigma_c^2 p^2}\,e^{-pa} \tag{5.167}$$

$$B_\pm = \frac{1\pm\varepsilon}{1\mp\varepsilon}\,e^{\pm 2\frac{a}{\sigma_c}} \tag{5.168}$$

Integration of $V_p = V - V_c$ times the stiffness $c_p$ over the contact length yields the transformed expressions for the force $F_y$ and, after multiplication with $x$, the moment $M_z'$. The following formulas are obtained for their transfer functions to the slip angle $\alpha$:

$$\begin{aligned} H_{F,\alpha}(p) = {} & c_p\varepsilon\frac{1}{p}\left\{2\varepsilon a + \frac{1-\sigma^2 p^2}{1-\sigma_c^2 p^2}\,\frac{e^{pa}-e^{-pa}}{p}\,\varepsilon C_A \right.\\ & \left. + \sigma(1-e^{\frac{2a}{\sigma_c}})C_+ + \sigma(1-e^{-\frac{2a}{\sigma_c}})C_-\right\} \end{aligned} \tag{5.169}$$

$$\begin{aligned} H_{M',\alpha}(p) = {} & c_p\varepsilon\frac{1}{p}\left[\frac{1-\sigma^2p^2}{1-\sigma_c^2p^2}\left(a\frac{e^{pa}+e^{-pa}}{p} - \frac{e^{pa}-e^{-pa}}{p^2}\right)\varepsilon C_A \right.\\ & + \sigma\left\{a\left(1+e^{\frac{2a}{\sigma_c}}\right) + \sigma_c\left(1-e^{\frac{2a}{\sigma_c}}\right)\right\}C_+ + \sigma\left\{a\left(1+e^{-\frac{2a}{\sigma_c}}\right)\right.\\ & \left.\left. - \sigma_c\left(1-e^{-\frac{2a}{\sigma_c}}\right)\right\}C_-\right] \end{aligned} \tag{5.170}$$

By letting $p \rightarrow 0$ in the expressions for $\sigma^* = V_1/A$ (cf. (5.162), $x = a$) and the transfer functions (5.169, 5.170), the relaxation length and the slip stiffnesses may be assessed. We find

$$\sigma^* = \frac{\sigma\{(1+\varepsilon)e^{2a/\sigma_c} + (1-\varepsilon)e^{-2a/\sigma_c} - 2\} - 4a}{\frac{1+\varepsilon}{1-\varepsilon}e^{2a/\sigma_c} + \frac{1-\varepsilon}{1+\varepsilon}e^{-2a/\sigma_c} + 2} \tag{5.171}$$

$$\begin{aligned} C_{F\alpha} = {} & 2c_p\varepsilon^2\Big[a(\sigma^* + a) - \frac{1}{4}\sigma\sigma^*\{(1+\varepsilon)e^{2a/\sigma_c} + (1-\varepsilon)e^{-2a/\sigma_c} - 2\} \\ & + \frac{1}{4}\sigma^2(1-\varepsilon^2)(e^{2a/\sigma_c} + e^{-2a/\sigma_c} - 2)\Big] \end{aligned} \tag{5.172}$$

$$\begin{aligned} C_{M\alpha} = {} & 2c_p\varepsilon^2\Big[\frac{1}{3}a^3 - \frac{1}{4}\sigma\{\sigma^*(1+\varepsilon) - \sigma(1-\varepsilon^2)\}\{a(1+e^{2a/\sigma_c}) \\ & + \sigma_c(1-e^{2a/\sigma_c})\} - \frac{1}{4}\sigma\{\sigma^*(1-\varepsilon) - \sigma(1-\varepsilon^2)\}\{a(1+e^{-2a/\sigma_c}) \\ & - \sigma_c(1-e^{-2a/\sigma_c})\}\Big] \end{aligned} \tag{5.173}$$

$$C_{F\varphi} = C_{M\alpha} \tag{5.174}$$

and the pneumatic trail:

$$t = \frac{C_{M\alpha}}{C_{F\alpha}} \tag{5.175}$$

It is of interest to examine how these quantities vary with vertical load $F_z$. For this purpose, we assume that the contact length varies with the square root of the radial tire compression and that the load is a linear function of the compression. Then we find that, if at the nominal or static load $F_{z0}$ half the contact length is denoted by $a_o$, the $F_z$ vs $\alpha$ relationship is described by

$$\frac{a}{a_o} = \sqrt{\frac{F_z}{F_{zo}}} \tag{5.176}$$

This, introduced in the above expressions, yields the characteristics shown in the diagram of Figure 5.31. The various nondimensional quantities have been presented both for the bare string model and for the enhanced model. As has been expected, we now see that the relaxation length for the force and moment responses to yaw, $\sigma^* + a$, vanishes when the load tends to zero. The pneumatic trail's ratio to half the contact length is a lot more realistic when the tread elements are added. The same holds for the cornering and aligning stiffnesses shown in Figure 5.32. With the enhanced model, the curves both start at the origin and exhibit a realistic shape. The feature, usually exhibited by passenger

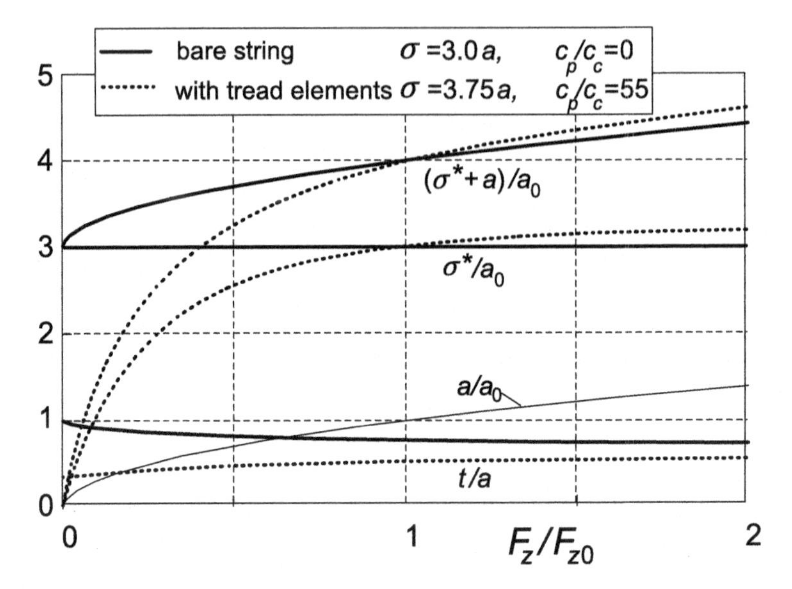


**FIGURE 5.31** The variation of the relaxation length and half the contact length as a ratio to the static value of the latter and the ratio of the pneumatic trail to half the contact length with normalized vertical load both for the bare string model and for the model with tread elements.

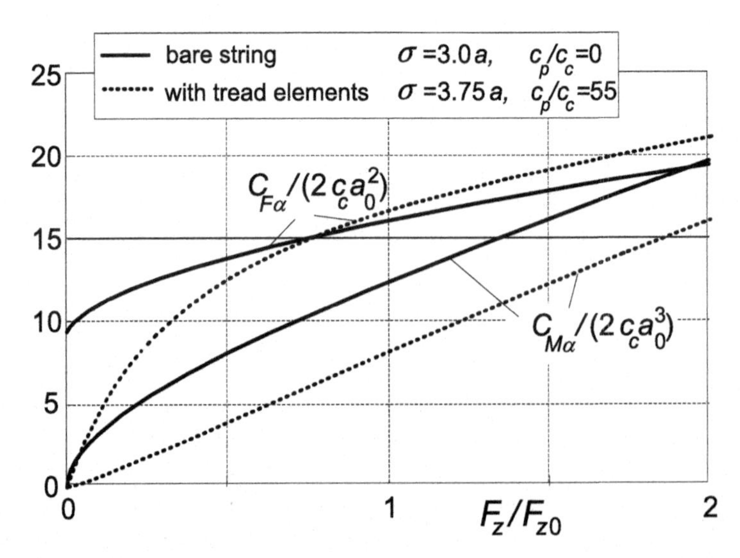


**FIGURE 5.32** The variation of nondimensionalized cornering and aligning stiffnesses with normalized vertical load both for the bare string model and the model with tread elements.

car tires, that the cornering stiffness bends downwards after it has reached the peak (cf. Figure 1.3) may be attributed to the reduced lateral stiffness of the tire cross section at higher vertical compression, cf. Pacejka (1981, p. 729 or 1971, p. 698) for an analytical formula.

The frequency response functions for the enhanced model are presented in Figures 5.33 and 5.34. They may be compared with the plots of Figures 5.14

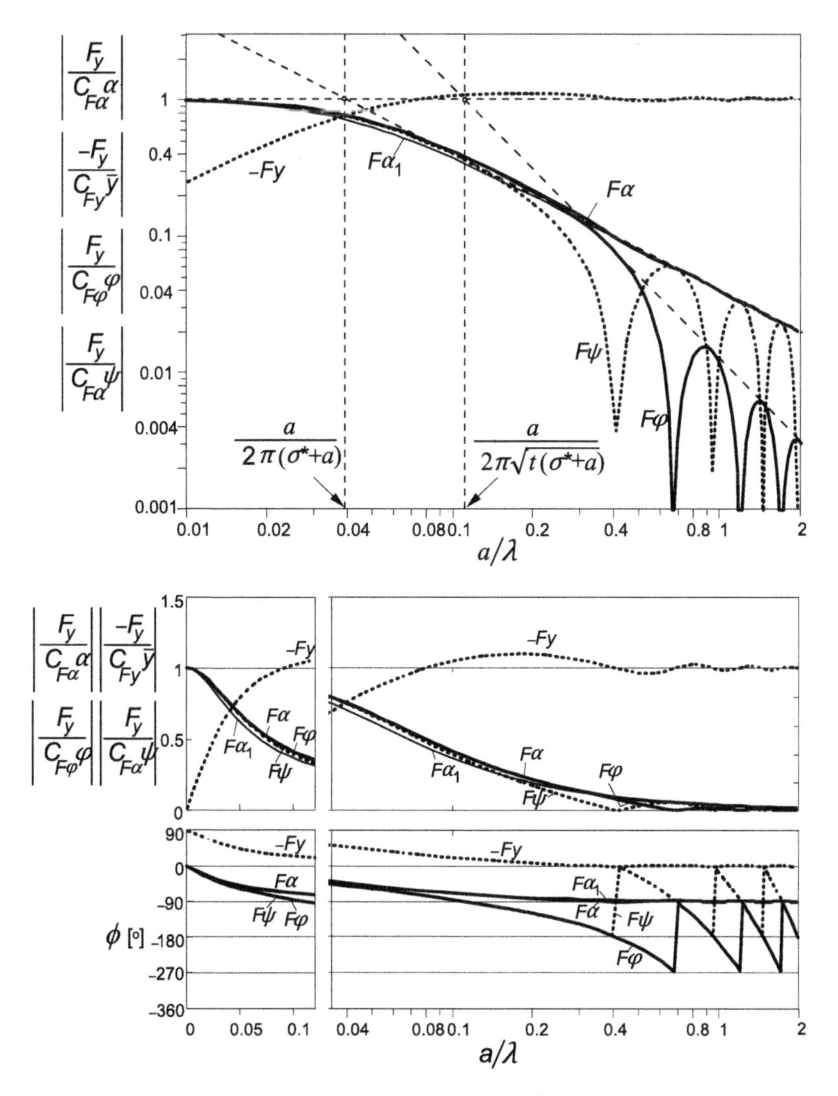


**FIGURE 5.33** Frequency response functions of the string model with tread elements for the side force with respect to various motion input variables. For the response to $\alpha$, a first-order approximation with the same cutoff frequency, $\omega_s = 1/(\sigma^* + a)$, has been added ($F\alpha_1$). Three ways of presentation have been used: log–log, lin–log, and lin–lin. The model parameters are: $\sigma = 3.75a$, $c_p/c_c = 55$ which leads to $\sigma^* = 3a$, $t = 0.49a$.

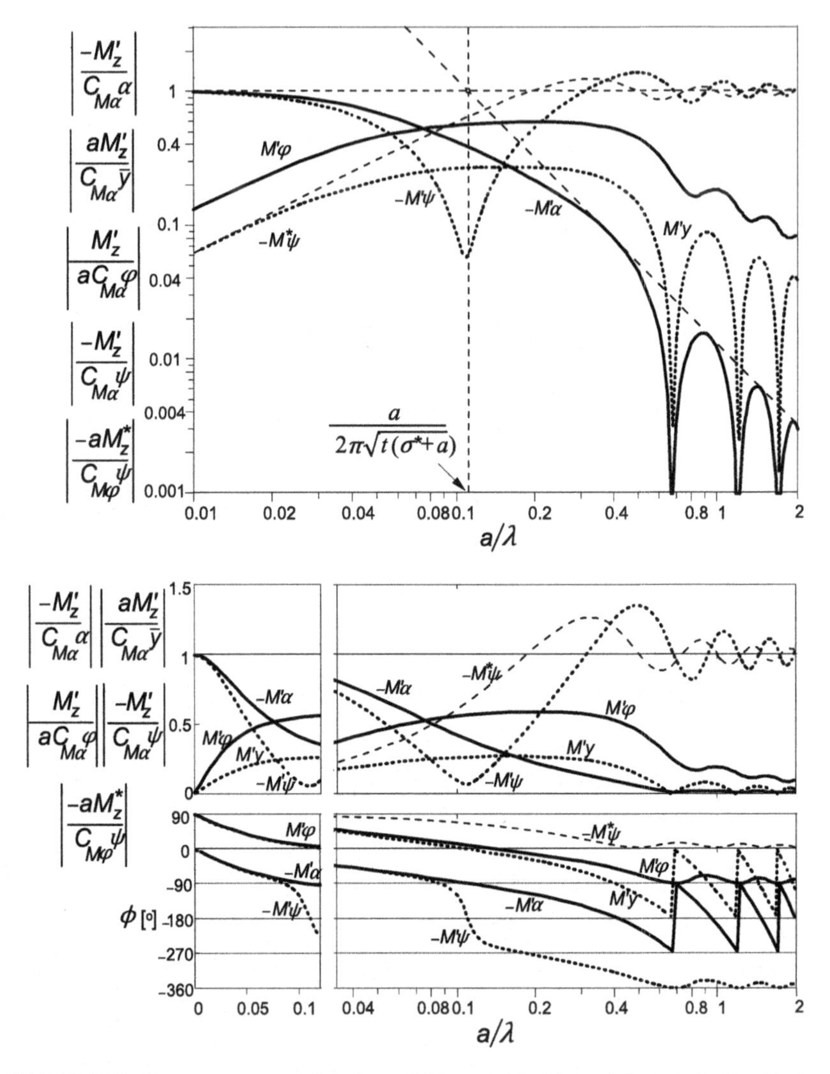

**FIGURE 5.34** Frequency response functions of string model with tread elements for the aligning torque with respect to various motion input variables. Same conditions as in Figure 5.33. The response functions of the moment due to tread width $M_z^*$ have been added.

and 5.15 which hold for the bare string model. The model parameters are again: $\sigma = 3.75a$, $c_p/c_c = 55$. An important result of the analysis is that expressions (5.70, 5.72, 5.74, 5.79–5.81) for relaxation lengths and cutoff frequencies are the same for the bare string model and the enhanced model if, in these expressions, $\sigma$ is replaced by $\sigma^*$ defined by (5.171).

For comparison, the plots for the step responses and for the frequency response functions have been presented in Figures 5.21–5.27 together with results of other models. Also, Table 5.1 (p. 256) contains data computed with

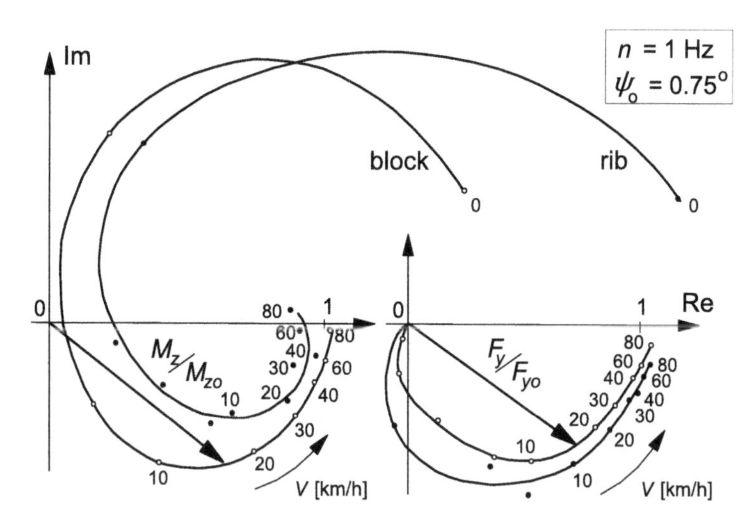


**FIGURE 5.35** Frequency response plots of $F_y$ and $M_z$ to $\psi$ for two light truck tires with same bias-ply carcass but different tread patterns. Measurements were performed on a 2.5-m drum at a vertical load of 10 kN, a yaw amplitude of 0.75° at a frequency $n = 1$ Hz.

the new model. The general course of the curves is similar. Important differences with respect to the bare string model are the considerable shift of the cutoff frequency of the $M_z'$ response to $\alpha$ and similarly of $F_y$ to $\varphi$ and, together with these, the meandering frequency where the dip of the $M_z'$ response to $\psi$ occurs. The reduction of the pneumatic trail $t$ is responsible for the increase of the cutoff frequency.

Experimental results assessed on a 2.5-m diameter steel drum have been presented in Figure 5.35. The responses are depicted in the Nyquist diagram (cf. Figures 5.17 and 5.18) as a ratio to their values assessed at steady-state side slip. Two light truck tires have been tested at a vertical load of $F_z = 10$ kN. Both tires possess the same cross-ply construction but exhibit a different tread design. The original tire shows a block tread design (central rib with transverse blocks, oval foot print), which is relatively weak in the longitudinal direction at the sides of the contact patch. The modified tire is provided with circumferential ribs (three ribs, almost rectangular foot print) making the tread relatively stiff in the longitudinal direction. As expected, the ribbed tire generates more resistance against turning which is reflected by a larger moment due to tread width $M_z^*$ and consequently a considerably smaller phase lag of the resulting self-aligning torque $-M_z$. The side force does not appear to be much affected at least in the larger wavelength range.

In Table 5.2, parameters estimated from the measurements have been listed. The intersection length or deflection relaxation length $\sigma^*$ was determined from the formula:

$$\sigma^* = \frac{C_{F\alpha}}{C_{Fy}} - a \tag{5.177}$$

**TABLE 5.2** Parameters for Two 9.00–16 Tires with Different Tread Patterns

| | Blocks | Ribs |
|---|---|---|
| $r_o$ | 0.5 m | 0.5 m |
| $a$ | 0.115 m | 0.08 m |
| $b$ | 0.092 m | 0.10 m |
| $t$ | 0.53$a$ | 0.82$a$ |
| $\sigma^*$ | 2.4$a$ | 2.8$a$ |
| $F_z$ | 10 kN | 10 kN |
| $C_{F\alpha}$ | 64 kN/rad | 53 kN/rad |
| $C_{M\alpha}$ | 3.9 kNm/rad | 3.5 kNm/rad |
| $C_{Fy}$ | 165 kN/m | 175 kN/m |
| $\kappa^*$ | 0.25 $C_{M\alpha}a$ | 1.0 $C_{M\alpha}a$ |

When considering the relation $a/\lambda = an/V$ with $V_{\text{km/h}} = 3.6\ V_{\text{m/s}}$, the value for the nondimensional path frequency at $V = 10$ km/h and an excitation frequency of 1 Hz becomes $a/\lambda = 0.0414$ which roughly corresponds with that indicated on the thin tire model curve of Figure 5.18.

## 5.5. TIRE INERTIA EFFECTS

It has been found experimentally that higher frequencies and greater speeds of travel bring about an increasingly important deviation of the response with respect to the kinematic representation with dynamic influences disregarded.

Figure 5.36 presents an example of response curves obtained experimentally. The response curves for the moment $M_z$ and the side force $F_y$ to the yaw angle $\psi$ have been shown as a function of the speed of travel $V$ for fixed values of the excitation frequency $n$ of the imposed yaw oscillation. The force and moment outputs are given as a ratio to their steady-state values (at $\omega_s = \omega = 0$) indicated by the subscript 0.

We observe that for higher frequencies the curves of the moment response appear to shift upward. The force response curves, on the other hand, show an increase in amplitude while the phase lag remains approximately unchanged for constant wavelength $\lambda = V/n$.

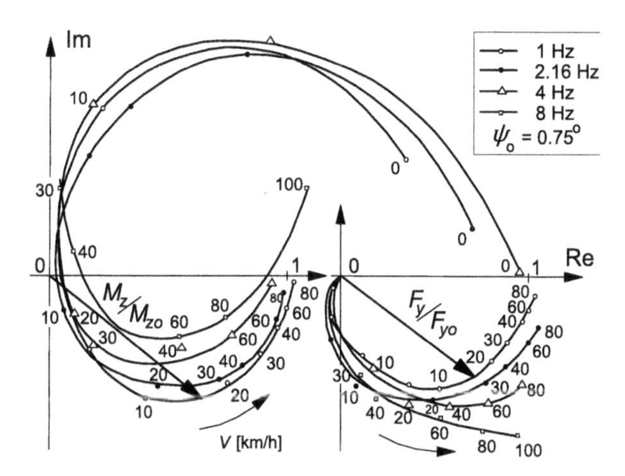


**FIGURE 5.36** Frequency response plots of $F_y$ and $M_z$ to $\psi$ for light truck tire with block tread pattern. Measurements were performed on a 2.5-m drum at a vertical load of 10 kN, a yaw amplitude of 0.75° and at four different time frequencies.

## 5.5.1. First Approximation of Dynamic Influence (Gyroscopic Couple)

In the low frequency range, $n < 10$ Hz, the aligning torque appears to be affected most by the inertia of the tire. A simple addition to the kinematic model may approximately describe the dynamic changes in the response of the moment. In this theory, the gyroscopic couple due to the lateral deformation rate of the lower portion of the tire has been introduced as the only dynamic influence. It is assumed here that the dynamic forces have a negligible effect on the distortion of the tire.

The gyroscopic couple about the vertical axis is proportional to the wheel rotational speed and the time rate of change of the tilt (camber) distortion of the tire, cf. Figure 5.37. Since this distortion is connected with the lateral deflection

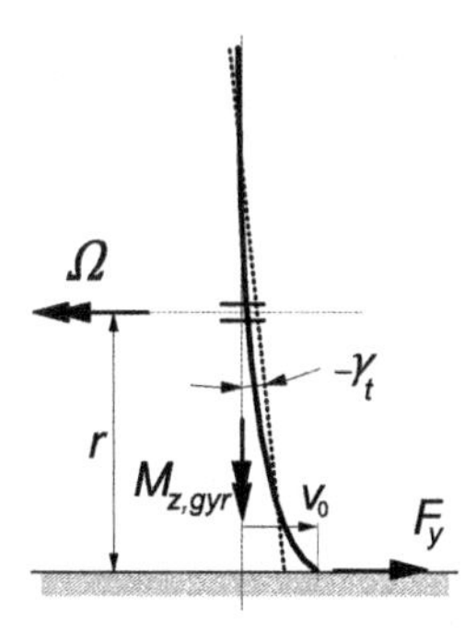


**FIGURE 5.37** The average tilt angle of the peripheral line of the deflected tire.

of the lower part of the tire and the side force $F_y$ is proportional to this deflection, we have, for the gyroscopic couple,

$$M_{z,\mathrm{gyr}} = C_{\mathrm{gyr}} V \frac{\mathrm{d}F_y}{\mathrm{d}t} \tag{5.178}$$

in which $t$ denotes the time. The tire constant $C_{\mathrm{gyr}}$ is proportional to the tire mass $m_t$ and inversely proportional to the lateral stiffness $C_{Fy}$ of the standing tire. We have, with the nondimensional quantity $c_{\mathrm{gyr}}$ introduced,

$$C_{\mathrm{gyr}} = c_{\mathrm{gyr}} \frac{m_t}{C_{Fy}} \tag{5.179}$$

The total moment which acts upon the wheel is now composed of the moment due to lateral tire deformation $M_z'$ and that due to longitudinal anti-symmetric tread deformations $M_z^*$ and finally the gyroscopic couple $M_{z,\mathrm{gyr}}$:

$$M_z = M_z' + M_z^* + M_{z,\mathrm{gyr}} \tag{5.180}$$

With this simple addition to the enhanced stretched string tire model (provided with tread elements distributed over the assumedly rectangular contact area), calculations have been performed. Figure 5.38 shows the Nyquist plot of the responses along with the curves $\omega_s\, a = 2\pi\, na/V = 2\pi\, a/\lambda$. It should be noted that the vector of the gyroscopic couple is directed perpendicularly to the vector of the lateral force. For the range of frequencies considered, the simple approximate dynamic extension theory gives a good representation of the effect on the moment response curves. However, the theory does not account for the dynamic changes in the force response.

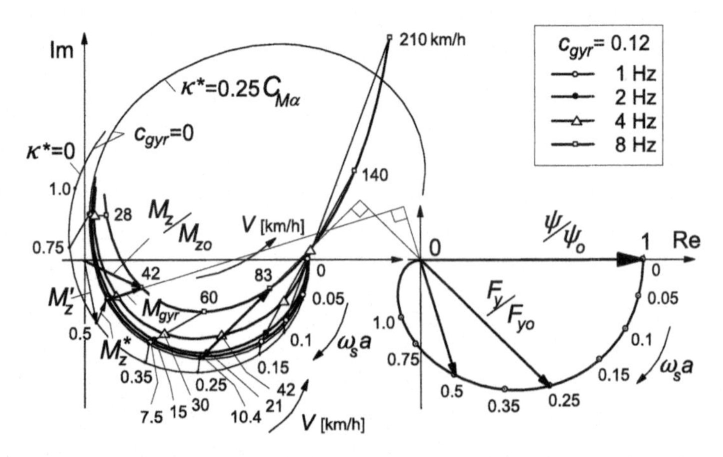


**FIGURE 5.38** The influence of the frequency of the yaw oscillation on the dynamic response according to the simple approximate model extension using the gyroscopic couple due to the lateral tire deflection time rate of change.

### 5.5.2. Second Approximation of Dynamic Influence (First Harmonic)

More complex theories as developed by Pacejka (1973a) and by Strackerjan (1976) and more recently by Maurice (2000), cf. Chapter 9, yield better results also with respect to the force response and extends to higher frequencies covering the first natural frequencies of the out-of-plane tire dynamic behavior. In these theories, the tire circular belt is considered to move not only about a longitudinal axis but also about a vertical axis and in the lateral direction. The belt rotation about the vertical axis gives rise to a gyroscopic couple about the longitudinal axis which on its turn results in a change in slip angle which influences $F_y$ (and also $M_z$).

The interesting feature of the approach followed in the first reference mentioned above is that the kinematic theory and response equations as treated in Sections 5.3 and 5.4 with tire inertia equal to zero can be employed. The kinematic part of the model will be considered to be subjected to a different, effective, set of input motion variables. Instead of taking the motion of the actual wheel plane defined by $\bar{y}$ and $\psi$ (or $\alpha$ and $\varphi$), an effective input is introduced that is defined as the lateral and the yaw motion of the line of intersection of the 'dynamic' belt plane and the road surface $(\bar{y}_e, \psi_e)$. Figure 5.39 depicts the situation. The dynamic belt plane is defined as the virtual center plane of the belt of the tire that is displaced with respect to the rim through the action of lateral inertial forces only. These inertial forces distributed along the circumference of the belt are approximated in such a way that only the forces resulting from the zeroth and first modes of vibration are taken

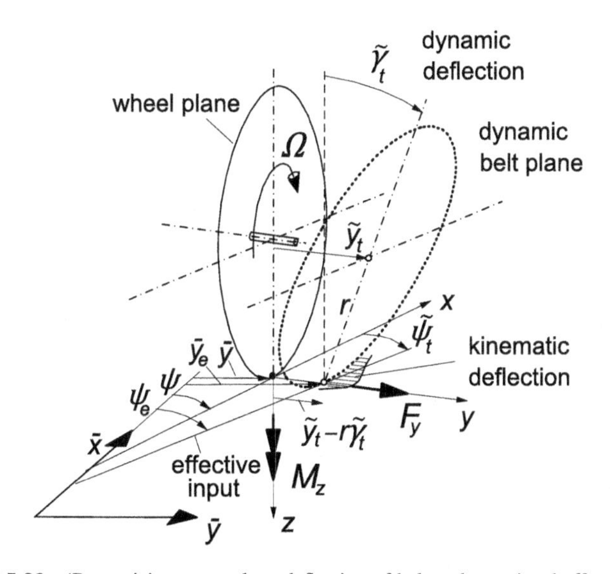


**FIGURE 5.39** 'Dynamic' average plane deflection of belt and associated effective input.

into account, that is: as if the belt would remain circular. Harmonics of second and higher order are neglected in the dynamic part of the model. We have, for the approximate dynamic lateral, camber, and yaw deflections, respectively,

$$\begin{aligned} \tilde{y}_t &= -m_t(\ddot{\bar{y}} + \ddot{y}_t)/C_y \\ \tilde{\gamma}_t &= -\{I_t\ddot{\gamma}_t + I_{tp}\Omega(\dot{\psi} + \dot{\psi}_t)\}/C_\gamma \\ \tilde{\psi}_t &= -\{I_t(\ddot{\psi} + \ddot{\psi}_t) - I_{tp}\Omega\dot{\gamma}_t\}/C_\psi \end{aligned} \tag{5.181}$$

with the total tire deflections composed of dynamic and static parts (the latter resulting from external ground forces):

$$\begin{aligned} y_t &= \tilde{y}_t + F_y/C_y \\ \gamma_t &= \tilde{\gamma}_t - F_y r/C_\gamma \\ \psi_t &= \tilde{\psi}_t + M_z'/C_\psi \end{aligned} \tag{5.182}$$

and stiffnesses of the whole belt with respect to the rim $C_y$, $C_\gamma$, and $C_\psi$. Furthermore, the equations contain effective tire inertia parameters $m_t$, $I_t$, and polar moment of inertia $I_{tp}$. The effective inputs to the kinematic model are defined by

$$\begin{aligned} \bar{y}_e &= \bar{y} + \tilde{y}_t - r\tilde{\gamma}_t \\ \psi_e &= \psi + \tilde{\psi}_t \end{aligned} \tag{5.183}$$

With the aid of these equations plus the kinematic tire equations (e.g. Von Schlippe's approximate Eqns (5.87, 5.88, 5.92, 5.93) with (5.36, 5.37) and (5.32) for $M_z^*$ with inputs $(\bar{y}, \psi)$ replaced by effective inputs $(\bar{y}_e, \psi_e)$, the frequency response functions of the ground force $F_y$ and moment $M_z = M_z' + M_z^*$ with respect to the actual inputs $(\bar{y}, \psi)$ may be readily established.

For applications in vibratory problems associated with wheel suspension and steering systems, it is of greater interest to know the response of the equivalent force and moment that would act on the assumedly rigid wheel with tire inertia included at the contact center. The equivalent force and moment are obtained as follows:

$$\begin{aligned} F_{y,eq} &= F_y - m_t\ddot{y}_t \\ M_{z,eq} &= M_z' + M_z^* - I_z\ddot{\psi}_t + I_{tp}\Omega\dot{\gamma}_t \end{aligned} \tag{5.184}$$

This result should correspond to the force and moment measured in the wheel hub, $F_{y,\mathrm{hub}}$ and $M_{z,\mathrm{hub}}$, after these have been corrected for the inertial force and moment acting on the assumedly rigid wheel plus tire (inertia between load cells and ground surface). With inertia parameters of wheel plus tire, we find, for this correction,

$$\begin{aligned} F_{y,eq} &= F_{y,\text{hub}} + m_{wt}\ddot{y} \\ M_{z,eq} &= M_{z,\text{hub}} + I_{wt}\ddot{\psi} \end{aligned} \tag{5.185}$$

When lifted from the ground, the equivalent ground to tire forces vanish if the frequency of excitation is not too high (low with respect to the first natural frequency of the tire).

For the set of parameter values measured on a radial-ply steel-belted tire listed in Table 5.3, the amplitude and phase of the equivalent moment as a response to yaw angle have been computed and are presented in Figure 5.40 as a function of the excitation frequency $n = \omega/2\pi$. Experimentally obtained curves up to a maximum of 8 Hz (on 2.5 m drum) have been added. The phase lag that occurs in the lower frequency range is largely responsible for the creation of self-excited wheel shimmy. The influence of tire inertia is evident: the phase lag changes into phase lead at higher values of speed (similar as in Figure 5.38). In contrast to the dynamic response depicted in Figure 5.41, the kinematic model response (without the inclusion of tire inertia) would produce the same response at equal wavelengths, $\lambda = V/n$, irrespective of the value of the speed of travel $V$.

The resonance frequencies, which at speeds close to zero approach a value somewhat larger than the natural frequency of the free non-rotating tire, $n_{00}$, appear to separate at increasing speed. Vibration experiments performed on a free tire (not contacting the road) for the purpose of obtaining the values of certain tire parameters gave a clear indication that a reduction of the lowest resonance frequency arises when the wheel is rotated around the wheel spin axis at higher speeds of revolution.

**TABLE 5.3 Parameters for Car Tire Model with Dynamic Belt Extension (2nd Approximation)**

| Bias-Ply Tire | | | | Radial-Ply Steel-Belted | | | |
|---|---|---|---|---|---|---|---|
| $a$ | 0.066 m | $C_\psi = C_\gamma$ | 29 Nm | $a$ | 0.063 m | $C_\psi = C_\gamma$ | 18.1 Nm |
| $\sigma$ | 0.230 m | $C_y$ | 270 kN/m | $\sigma$ | 0.377 m | $C_y$ | 360 kN/m |
| $r$ | 0.324 m | $m_t$ | 4.5 kg | $r$ | 0.322 m | $m_t$ | 5.7 kg |
| $C_{F\alpha}$ | 35.5 kN | $I_t$ | 0.235 kg m$^2$ | $C_{F\alpha}$ | 49.0 kN | $I_t$ | 0.295 kg m$^2$ |
| $C_{M\alpha}$ | 1.22 kNm | $I_{tp}$ | 0.47 kg m$^2$ | $C_{M\alpha}$ | 1.4 kNm | $I_{tp}$ | 0.59 kg m$^2$ |
| $\kappa^*$ | 80 Nm$^2$ | ($m_{\text{tire}}$ = 8.9 kg) | | $\kappa^*$ | 104 Nm$^2$ | ($m_{\text{tire}}$ = 8.5 kg) | |

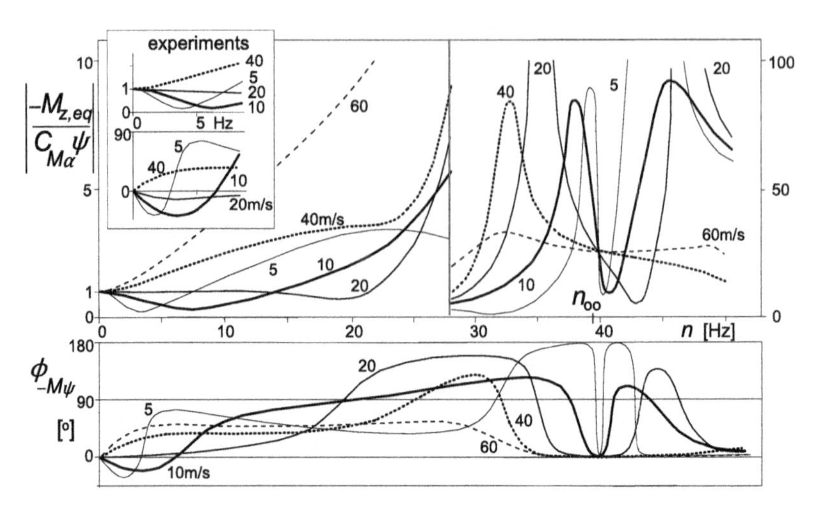


**FIGURE 5.40** Amplitude and phase response of equivalent aligning torque (as if acting from ground on rigid wheel tire) to yaw angle as a function of time frequency of excitation [Hz] computed using Von Schlippe approximation and exact $M_z^*$ extended with dynamic theory according to second approximation with dynamic belt. Model parameters are listed in Table 5.3 (radial steel belted tire). Curves hold for given speed of travel $V$ [m/s]. Experimental results are given up to a frequency of 8 Hz. The yaw natural frequency of the free non-rotating tire with fixed wheel spindle is $n_{00}$.

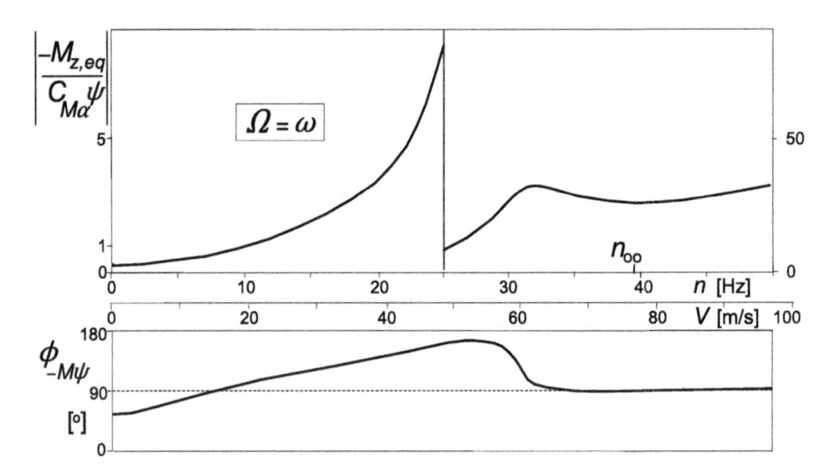


**FIGURE 5.41** Amplitude and phase response as a function of frequency with the frequency of excitation equal to that of wheel revolution.

Nyquist plots for the $M_{z,eq}$ and $F_{y,eq}$ responses to $\psi$ have been presented in Figures 5.42 and 5.43. Both for the bias-ply and for the radial steel-belted tire of Figure 5.40, theoretical and experimental results are shown. Parameters are according to Table 5.3. For each of the curves, the excitation frequency remains constant while the speed varies along the curve.

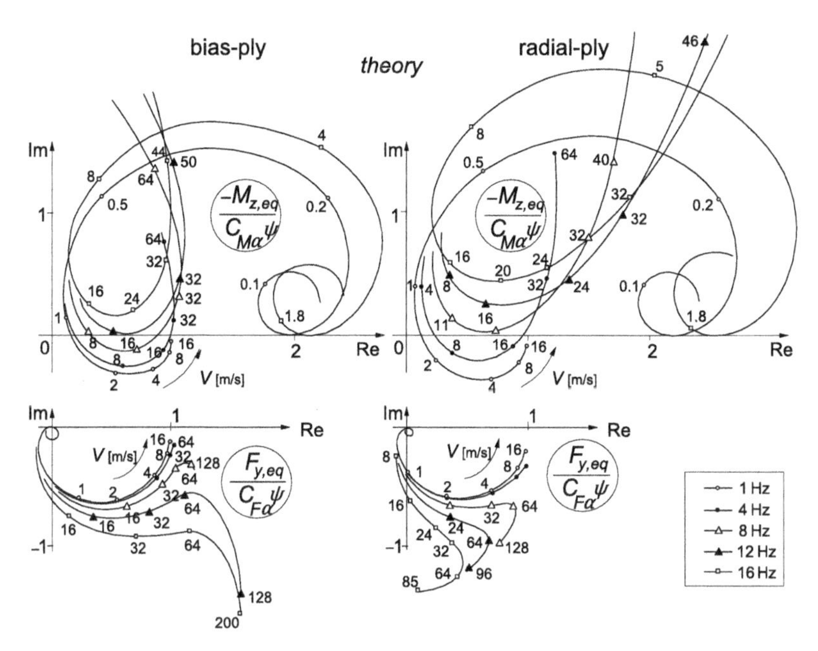


FIGURE 5.42 Theoretical frequency response Nyquist plots of side force and aligning torque to yaw angle for different time frequencies. Same conditions as in Figure 5.40.

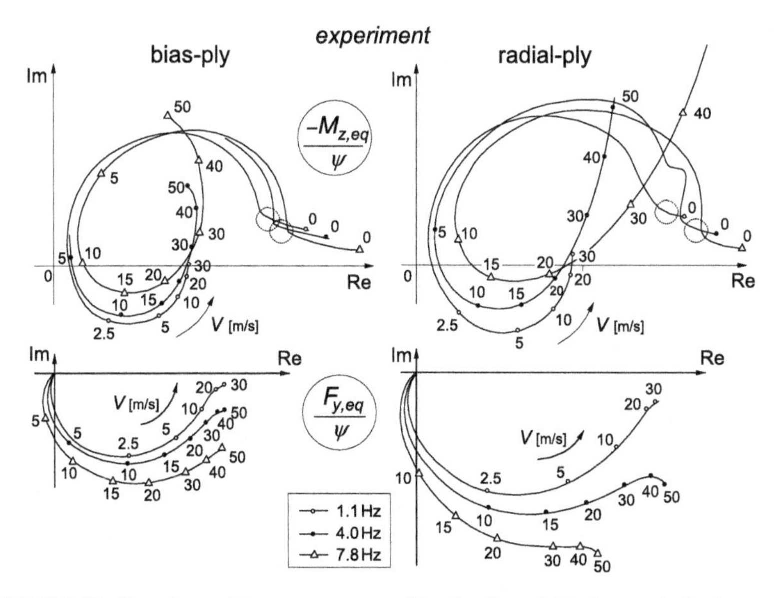


FIGURE 5.43 Experimental frequency response Nyquist plots of side force and aligning torque to yaw oscillations for a radial and a bias-ply car tire measured on a 2.5-m drum.

An increase in frequency tends to raise the moment curves while the force response curves appear to expand with the phase angle at a given wavelength remaining not much changed. The typical differences between the two types of tires, notably in the shape of the upward ends of the moment curves where the speed assumes high values, are well represented in the theoretical diagrams.

We may conclude that the simple first approximation appears to give a considerable improvement for the moment response in the lower frequency range, while for the force response and the higher frequency excitations the inclusion of the dynamics of the belt up to the first modes of vibration is required.

Since we have here a dynamic model extension that allows the use of existing kinematic (mass less) models, we have been able to investigate the effect of tire inertia on these models. However, for the sake of simplicity these models have been restricted to a linear description allowing straightforward frequency response analysis. Extension into the nonlinear regime in the time domain is cumbersome for models including a finite contact length (retardation) effect (although an approximate nonlinear Von Schlippe model may be established by attaching nonlinear springs to the wheel plane at the leading and trailing edges the tips of which follow the path as if we had full adhesion) and also because in the dynamic extension, Eqns (5.181–5.183), the function $F_y = F_y(\alpha')$ would be needed in an inverse form.

To further develop mathematical models that can be used under conditions that include short wavelengths, higher frequencies, and large (combined) slip, a different route appears to be more promising. The adopted principle is: separate modeling of the flexible belt with carcass and the contact patch with its transient slip properties. The theory of the non-steady-state rolling and slipping properties of the brush model helps in this modeling effort. In Chapter 9, this advanced model will be discussed.

For less demanding cases (large wavelength, low frequency), the straight tangent or single-point contact models may be employed. Such models may be used in combination with the approximate dynamic extension that accounts for the gyroscopic couple due to the lateral deflection velocity. Furthermore, these simple models can be easily adapted to cover large slip nonlinear behavior, cf. Eqns (5.125, 5.126, 5.128) or (5.130–5.132) and Eqn (5.178) with $F_y$ replaced by $C_{Fy}v_0$. Chapter 7 is devoted to the treatment of the very useful and relatively simple, possibly nonlinear, single contact point transient slip models. With that single-point contact model, the problem of cornering (and braking) on undulated road surfaces can be made more accessible for vehicle simulation studies. First, we will address this difficult subject in the subsequent section using exact string models with and without tread elements which shows the deeper cause of the associated loss in cornering power.

## 5.6. SIDE FORCE RESPONSE TO TIME-VARYING LOAD

In studies by Metcalf (1963), Pacejka (1971), and more recently by Laermann (1986) and by Takahashi (1987) and Pacejka (1992), it was found that the average value of the side force generated by a constant slip angle is reduced as a result of a periodical change of the vertical load. This reduction may be considerable and appears to depend on the amplitude of the vertical load variation as a ratio to the static load, on the wavelength of the motion, the speed of travel, and on the magnitude of the slip angle. In Chapter 7, a more practical approach based on the single contact point model (that means: for accuracy limited to relatively large wavelengths) is adopted to model the non-steady-state behavior at varying vertical load, including high levels of slip. To lay the basis for that theory, we will employ the advanced string model with tread elements that was developed in Section 5.4.3 for the case of a constant vertical load. The presence of tread elements turns out to be crucial to properly model the loss of side force when cornering on an undulated road surface. As in Section 5.4.3, only small slip angles will be considered allowing the theory to remain linear. First, we will develop the model based on the advanced string model with tread elements. Then, to simplify the model description, an adapted bare string model will be introduced and used in the further calculations.

### 5.6.1. String Model with Tread Elements Subjected to Load Variations

Figure 5.44 shows the tire model in two successive positions while the wheel moves at a constant (small) slip angle over an undulated road surface which causes the contact length to vary with distance traveled $s$. When the contact

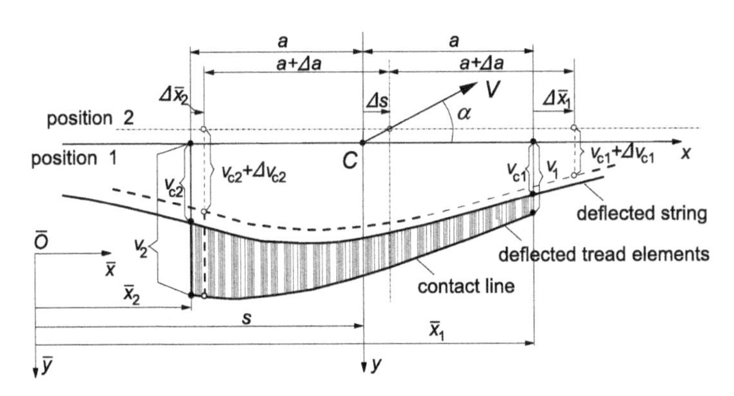


**FIGURE 5.44** Two successive positions of tire model with varying contact length at constant slip angle $\alpha$ (sliding not considered).

center $C$ is moved a distance $\Delta s$, the following changes occur simultaneously: lateral shift of the wheel plane, forward rolling, and change in contact length. These changes influence the position of the leading and trailing edges of the string. The $\bar{x}$ coordinates of these points are:

$$\begin{aligned} \bar{x}_1 &= s + a \\ \bar{x}_2 &= s - a \end{aligned} \tag{5.186}$$

the changes of which become

$$\begin{aligned} \Delta\bar{x}_1 &= \Delta s + \Delta a \\ \Delta\bar{x}_2 &= \Delta s - \Delta a \end{aligned} \tag{5.187}$$

The changes in front and rear lateral string deflections $v_{c1,2}$ are composed of various contributions. For the increment of the deflection at the leading edge $\Delta v_{c1}$ we obtain

**1.** due to lateral wheel displacement $\alpha\Delta s$:

$$\Delta v_{c1} = A(a)\cdot\alpha\Delta s \tag{5.188a}$$

**2.** due to loss of contact at the trailing edge ($\Delta\bar{x}_2 > 0$):

$$\Delta v_{c1} = -B(a)\frac{v_2 - v_{c2}}{\sigma_c}\,\Delta\bar{x}_2 \tag{5.188b}$$

**3.** due to loss of contact at the leading edge ($\Delta\bar{x}_1 < 0$):

$$\Delta v_{c1} = -B^*(a)\frac{v_1 - v_{c1}}{\sigma_c}\,\Delta\bar{x}_1 \tag{5.188c}$$

**4.** due to longitudinal displacement of leading edge ($\Delta\bar{x}_1$):

$$\Delta v_{c1} = -\frac{v_{c1}}{\sigma}\,\Delta\bar{x}_1 \tag{5.188d}$$

Similar expressions are obtained for the contributions to the change of the lateral deflection at the rear $v_{c2}$. The contact length-dependent coefficients appearing in the equations are

$$A(a) = \left\{-2\varepsilon + (1+\varepsilon)e^{2a/\sigma_c} - (1-\varepsilon)e^{-2a/\sigma_c}\right\}\Big/P(a) \tag{5.189}$$

$$B(a) = 2/P(a) \tag{5.190}$$

$$B^*(a) = \left\{(1+\varepsilon)e^{2a/\sigma_c} + (1-\varepsilon)e^{-2a/\sigma_c}\right\}\Big/P(a) \tag{5.191}$$

$$P(a) = \frac{1+\varepsilon}{1-\varepsilon}e^{2a/\sigma_c} - \frac{1-\varepsilon}{1+\varepsilon}e^{-2a/\sigma_c} \tag{5.192}$$

They are derived from solutions of the differential equations for the shape of the string deflection (Eqns (5.155, 5.156)). For the parameters employed, we refer to Eqn (5.153). From the increments (5.187, 5.188), the following differential equations are established for the unknown string deflections $v_{c1}$ and $v_{c2}$:

$$\frac{dv_{c1}}{ds} = -\left(1+\frac{da}{ds}\right)\frac{v_{c1}}{\sigma} - B(a)\left(1-\frac{da}{ds}\right)\frac{v_2 - v_{c2}}{\sigma_c} + B^*(a)\left(1+\frac{da}{ds}\right)\frac{v_1 - v_{c1}}{\sigma_c} + A(a)\alpha \tag{5.193}$$

$$\frac{dv_{c2}}{ds} = \left(1-\frac{da}{ds}\right)\frac{v_{c2}}{\sigma} + B(a)\left(1+\frac{da}{ds}\right)\frac{v_1 - v_{c1}}{\sigma_c} - B^*(a)\left(1-\frac{da}{ds}\right)\frac{v_2 - v_{c2}}{\sigma_c} + A(a)\alpha \tag{5.194}$$

The two remaining unknowns $v_1$ and $v_2$ representing the total lateral deflections and consequently the distances between wheel plane and contact line at the leading and trailing edges are found by considering the following. In the rolling process of a side slipping tire with a continuously changing contact length, in general, three intervals can be distinguished during the interval of the loading cycle where contact with the road exists. In Figure 5.45, the load

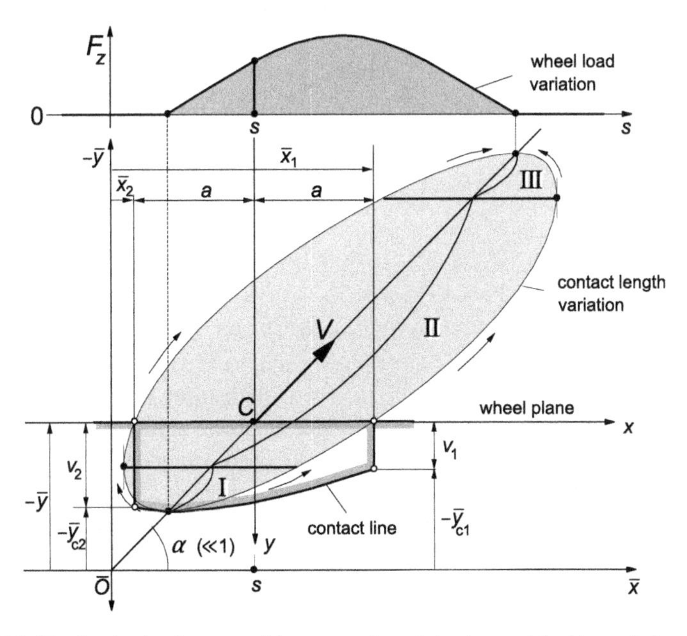


**FIGURE 5.45** On the development of the contact line of the tire model with tread elements that periodically loses contact with the road.

variation together with the corresponding variation of the contact length $2a$ of a side slipping tire that periodically jumps from the road has been shown in the road plane $(\bar{x}, \bar{y})$ for the interval that contact exists. Immediately after the first point touches the road, the contact line grows in two directions (interval I). This growth continues until the second interval II is reached, where growth of contact takes place only at the front, and at the rear loss of contact occurs. Finally, in the third interval III, loss of contact at both ends takes place until the tire leaves the road. When the tire does not leave the road, an additional interval II occurs before interval I is reached again and the cycle starts anew. In less severe cases, intervals I and III may not occur. We then have the relatively simple situation of continuous growth of contact at the front and loss of contact at the rear.

The unknowns $v_1$ and $v_2$ and the coordinate of the contact points in the $(\bar{x}, \bar{y})$ plane are obtained as follows (cf. Figure 5.45):

Interval I: $\dot{\bar{x}}_1 > 0, \ \dot{\bar{x}}_2 < 0 \quad \left(\frac{da}{ds} > 1\right)$

$$\begin{aligned} v_1 &= v_{c1}, \quad v_2 = v_{c2} \\ -\bar{y}_{c1} &= s\alpha - v_{c1}, \qquad -\bar{y}_{c2} = s\alpha - v_{c2} \end{aligned} \tag{5.195a}$$

Interval II: $\dot{\bar{x}}_1 > 0, \ \dot{\bar{x}}_2 > 0 \quad \left(-1 < \frac{da}{ds} < 1\right)$

$$\begin{aligned} v_1 &= v_{c1}, v_2 = s\alpha + \bar{y}_{c2} \\ -\bar{y}_{c1} &= s\alpha - v_{c1} \end{aligned} \tag{5.195b}$$

Interval III: $\dot{\bar{x}}_1 < 0, \ \dot{\bar{x}}_2 > 0 \quad \left(\frac{da}{ds} < -1\right)$

$$v_1 = s\alpha + \bar{y}_{c1}, \quad v_2 = s\alpha + \bar{y}_{c2} \tag{5.195c}$$

By means of numerical integration, Eqns (5.193, 5.194) have been solved with the aid of the Eqns (5.195) and the expressions (5.189–5.192) for the case formulated below in which the vertical tire deflection $\rho$ has been given a periodic variation with amplitude $\hat{\rho}$ greater than the static deflection $\rho_o$, so that periodic loss of contact between tire and road occurs.

The static situation will be indicated with subscript $_o$. The tire radius is denoted with $r$. The vertical deflection and the contact length are governed by the equations

$$\rho = \rho_o - \hat{\rho} \sin \omega_s s \tag{5.196}$$

$$a^2 = 2r\rho - \rho^2 \tag{5.197}$$

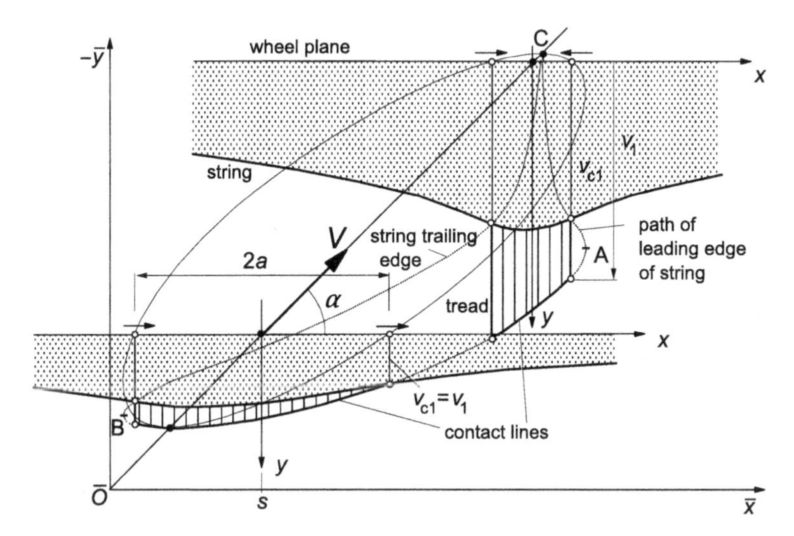


**FIGURE 5.46** The deflected tire model in two positions during the interval of contact with the road.

The numerical values of the parameters of the system under consideration are

$$a_o = 0.25r, (\rho_o = 0.03175r), \hat{\rho} = 0.1r, \omega_s = \pi/(4a_o), \ (\lambda = 8a_o)$$
$$\sigma = 3a_o, \ \varepsilon = 0.25, (\sigma_c = 0.75a_o, \ c_p = 15c_c) \qquad (5.198)$$

In Figure 5.46, the calculated variation of contact length $2a$ (oval curve), the path of the contact points (lower curve AB), and the course of the points of the string at the leading and trailing edge (AC and BC) have been indicated. Two positions of the tire have been depicted, one in interval II and the other in interval III.

It is necessary that the calculations in the numerical integration process be carried out with great accuracy. A small error gives rise to a rapid build up of deviations from the correct course. In the case considered above, the integration time is limited and accurate results can be obtained. In cases, however, where continuous contact between tire and road exists, a long integration time is needed before a steady-state situation is attained, which is due to the fact that the exact initial conditions of a periodic loading cycle are not known. For this sort of situation, the exact method described above is difficult to apply due to strong drift of $v_{c2}$ in particular.

For further investigation of the effect of a time-varying load, we will turn to an approximate method based on the string model without tread elements.

## 5.6.2. Adapted Bare String Model

The use of this model is much simpler since the deflections at both edges are independent of each other. Drift does not occur in the calculation process.

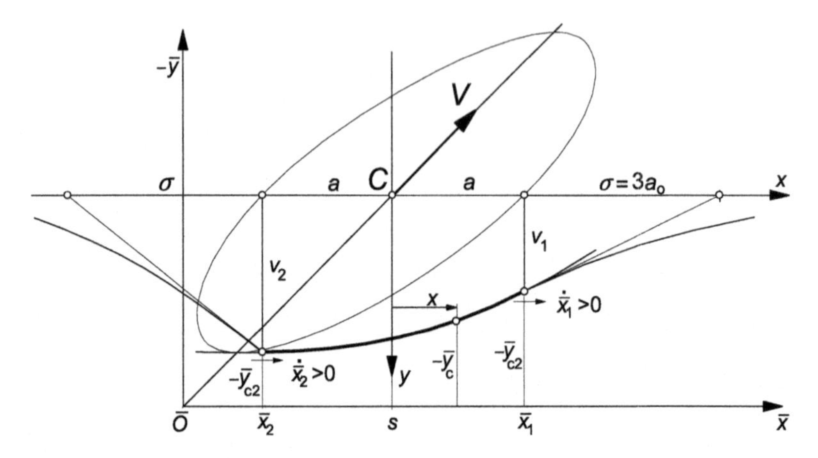


FIGURE 5.47 On the development of the contact line of the bare string model.

In Figure 5.47, the deflected model has been shown in interval II. For the three intervals, the following sets of equations apply:
*growth of contact at leading edge*:

$$\dot{\bar{x}}_1 > 0 \quad \left(\frac{da}{ds} > -1\right) \tag{5.199a}$$

$$\frac{dv_1}{ds} = \alpha - \left(1 + \frac{da}{ds}\right)\frac{v_1}{\sigma} \tag{5.199b}$$

$$\bar{x}_1 = s + a \tag{5.199c}$$

$$\bar{y}_{c1} = -s\alpha + v_1 \tag{5.199d}$$

*growth of contact at trailing edge*:

$$\dot{\bar{x}}_2 < 0 \quad \left(\frac{da}{ds} > 1\right) \tag{5.200a}$$

$$\frac{dv_2}{ds} = \alpha - \left(1 - \frac{da}{ds}\right)\frac{v_2}{\sigma} \tag{5.200b}$$

$$\bar{x}_2 = s - a \tag{5.200c}$$

$$\bar{y}_{c2} = -s\alpha + v_2 \tag{5.200d}$$

*loss of contact at leading edge*:

$$\dot{\bar{x}}_1 < 0 \quad \left(\frac{da}{ds} < -1\right) \tag{5.201a}$$

$$\bar{y}_{c1} = \bar{y}_c(\bar{x} = s + a) \tag{5.201b}$$

$$v_1 = s\alpha + \bar{y}_{c1} \tag{5.201c}$$

*loss of contact at trailing edge*:

$$\dot{x}_2 > 0 \quad \left(\frac{\mathrm{d}a}{\mathrm{d}s} < 1\right) \tag{5.202a}$$

$$\bar{y}_{c2} = \bar{y}_c(\bar{x} = s - a) \tag{5.202b}$$

$$v_2 = s\alpha + \bar{y}_{c2} \tag{5.202c}$$

Solutions of the above equations show considerable differences from the results obtained using the more advanced model with tread elements. The most important difference is the fact that with the simple model the lateral force does not gradually drop to zero as the tire leaves the ground.

In order to get better agreement, we will introduce a nonconstant relaxation length $\sigma = \sigma^*(a)$ that is a function of the contact length $2a$ and thus of the vertical load. We take $\sigma^*$ equal to the relaxation length of the more advanced model according to Eqn (5.171).

Figure 5.48 presents the variation of relaxation length and contact length as a function of the vertical load as calculated for the current parameter values (5.198).

In Figure 5.49, a comparison is made between the results of the three models: without tread elements, with tread elements (exact), and according to the approximation with varying $\sigma = \sigma^*(a)$. The calculations have been carried

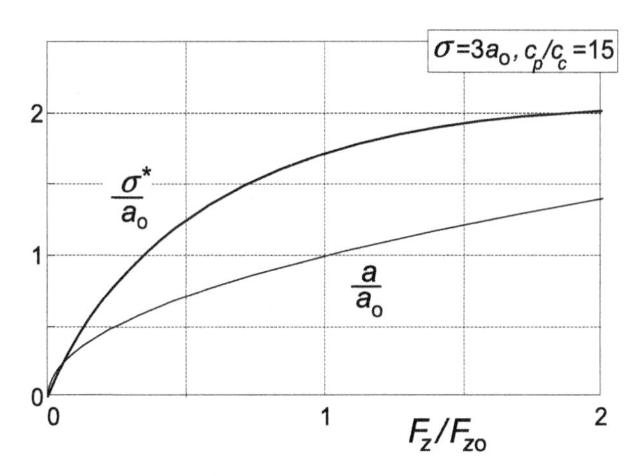


**FIGURE 5.48** The calculated variation of half the contact length and the relaxation length as a ratio to the static half contact length versus the vertical load ratio for the model with tread elements using Eqn (5.171).

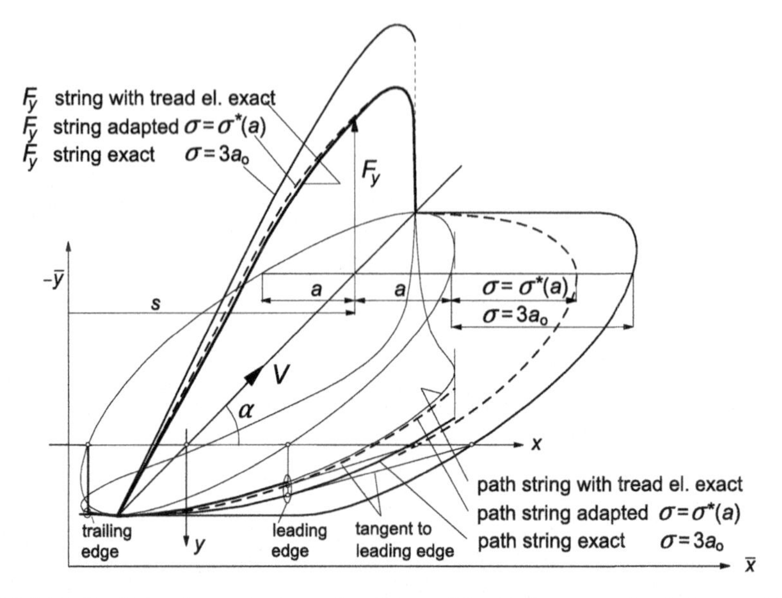


**FIGURE 5.49** Paths of contact points and variations of the side force according to three models.

out for the parameter values (5.198). The approximate path of contact points shows good agreement with the path for the exact model with tread elements. When the tread elements are omitted, we have the bare string model for which the path becomes wider and the lateral deflections greater.

## 5.6.3. The Force and Moment Response

The lateral force $F_y$ and the moment $M_z$ which act on the tire can be assessed with good approximation through the following simplified formulas. In their derivations, we have replaced the contact line by the straight line connecting the beginning and end points of the contact line (Von Schlippe's approach). For the model with tread rubber elements, we obtain

$$F_y = \frac{2c_c c_p}{c_c + c_p}\{a + \sigma A(a)\}v_0 \tag{5.203}$$

$$M_z = -\frac{2c_c c_p a}{c_c + c_p}\left\{\frac{1}{3}a^2 + \sigma(\sigma + a)C(a)\right\}\alpha_0 \tag{5.204}$$

and, for the model without tread elements,

$$F_y = 2c_c(a + \sigma)v_0 \tag{5.205}$$

$$M_z = -2c_c a\left\{\frac{1}{3}a^2 + \sigma(\sigma + a)\right\}\alpha_0 \tag{5.206}$$

In these expressions, the following quantities have been introduced:

$$v_0 = \frac{1}{2}(v_1 + v_2), \quad \alpha_0 = \frac{1}{2}(v_1 - v_2)/a \tag{5.207}$$

and

$$C(a) = A(a) - \frac{\sigma}{a}\varepsilon B^*(a) + \left(2 + \frac{\sigma}{a}\right)\varepsilon B(a) \tag{5.208}$$

in which $A$, $B$, and $B^*$ are given by expressions (5.189–5.192).

For the sake of completeness, the formulas for the moment have been added. However, we will restrict ourselves to the discussion of the variation of the force. For the three models considered, the variation of the cornering force has been shown in the same Figure 5.49. As expected, the approximate and exact theories drop to zero at the point of lift-off, whereas the force acting on the simple string model remains finite, at least under the here assumed condition of no sliding. The correspondence between exact and approximate solutions of the path and of the side force is satisfactory, and we will henceforth use the approximate method according to the adapted bare string model for the investigation of the model with tread elements.

For a series of deflection amplitudes $\hat{\rho}$ and wavelengths $\lambda$, the variation of the cornering force, or rather of the cornering stiffness, has been calculated. Since during loss of contact the vertical load remains zero, the period of the total loading cycle must become longer in order to keep the average vertical load unchanged. This change in period has been taken into account in the calculation of the average side force or cornering stiffness $C_{F\alpha}$.

Figures 5.50 and 5.51 show the final result of this investigation, viz. the cornering stiffness averaged over one complete loading cycle as a function of the vertical deflection amplitude $\hat{\rho}$ and the path wavelength $\lambda$ for both models with and without tread elements, that is: with the adapted bare string model

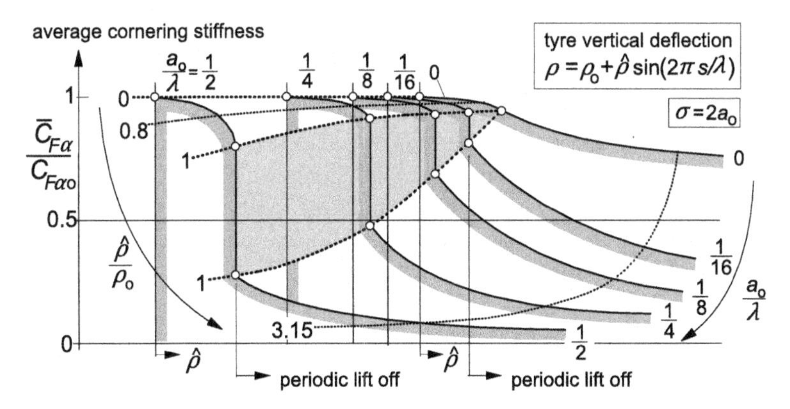


**FIGURE 5.50** Computed variation of average cornering stiffness with amplitude of the vertical tire deflection, $\rho$, and path frequency, $1/\lambda$, for the bare string model.

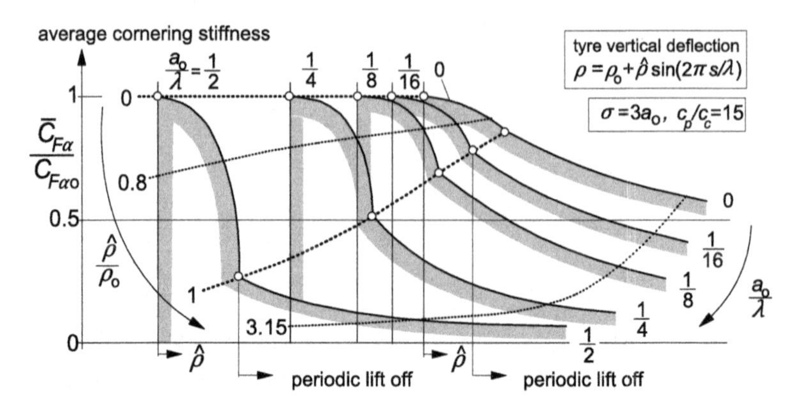


**FIGURE 5.51** Computed variation of average cornering stiffness with amplitude of the vertical tire deflection ($\rho$) and path frequency ($1/\lambda$) using the adapted bare string model that approximates the string model with tread elements.

and with the original bare string model. With the parameter values $\sigma = 3a_o$ and $c_p = 15c_c$, the relaxation length of the more advanced model with tread elements at the static load becomes $\sigma_o^* = 1.7a_o$. For the model without tread rubber, the relaxation length has been taken equal to $\sigma = 2a_o$.

The figures clearly illustrate the unfavorable effect of increasing the amplitude $\hat{\rho}$ and the path frequency (decreasing the wavelength $\lambda$). The curve at zero path frequency is purely due to the nonlinear variation of the cornering stiffness $C_{F\alpha}$ versus vertical load $F_z$ shown in Figure 5.32 (cf. Figure 1.3 and discussion on the effect of load transfer in Chapter 1, Figure 1.7) and reflects the 'static' drop in average cornering stiffness. A pronounced difference between the response of both types of models is the sudden drop in average cornering stiffness that shows up in the curves of the bare string model when the frequency is larger than zero and the deflection amplitude reaches the value of the static deflection. This is in contrast to the gradual variation in average cornering stiffness which is observed to occur with the more advanced model, and no doubt also with the real tire. Furthermore, the more advanced model with tread elements already shows a noticeable decrease in average cornering stiffness before the tire starts to periodically leave the ground. The 'dynamic' drop is represented by the dotted curves along which the path frequency changes and the deflection amplitude remains the same. The very low level, to which the average stiffness is reduced, once loss of contact occurs, is of the same order of magnitude for both tire models.

Chapter 6

# Theory of the Wheel Shimmy Phenomenon

Chapter Outline



## 6.1. INTRODUCTION

As an application of the theory developed in the previous chapter, we will study the self-excited oscillatory motion of a wheel about (an almost) vertical steering axis. This type of unstable motion is usually designated as the wheel shimmy oscillation. Shimmy is a violent and possibly dangerous vibration that may occur with front wheels of an automobile and with aircraft landing gears. Wobble of the front wheel of a motorcycle is an oscillatory unstable mode similar to shimmy. This steering vibration will be discussed in Chapter 11.

To investigate the shimmy phenomenon, the boundary of stability is established for system models of different degrees of complexity and we will examine the effect of employing different tire models. The linear stability analysis is extended to nonlinear systems which enables us to determine the


**Tire and Vehicle Dynamics. DOI: 10.1016/B978-0-08-097016-5.00003-6**

limit of the shimmy amplitude (limit cycle) and the magnitude of initial disturbance or unbalance to initiate the self-excited oscillation. Much of the contents of the present chapter is based on the work of Von Schlippe and Dietrich (1941), Pacejka (1966,1981) and Besselink (2000).

Fromm, cf. Becker, Fromm and Maruhn (1931), is one of the first investigators who developed a theory for the shimmy motion of automobiles. Besides his advanced theoretical work, as a result of which the gyroscopic coupling between the angular motions about the longitudinal axis and about the steering axis was found to be the main factor causing shimmy, he has carried out and described, together with his co-workers Becker and Maruhn, some tests on a system with a rigid front axle. Also, Den Hartog (1940) and Rocard (1949) have treated this 'gyroscopic' shimmy for systems with live axles. The phenomenon was experimentally examined by Olley (1947).

Another kind of shimmy is closely related to tire and suspension lateral compliance and has been observed to occur with aircraft landing gears and automobiles equipped with independent front wheel suspensions. Many of the authors mentioned in connection with the development of non-steady-state out-of-plane tire models in Chapter 5 have used their model in connection with the analysis of the shimmy problem.

## 6.2. THE SIMPLE TRAILING WHEEL SYSTEM WITH YAW DEGREE OF FREEDOM

First, the simplest system that can generate shimmy is discussed. Consider the trailing wheel system depicted in Figure 6.1. The vertical swivel axis moves along the $\bar{x}$ axis with speed $V$. The motion variable is the yaw angle $\psi$ of the wheel plane about the swivel axis. This axis intersects the road plane at a distance $e$ (mechanical trail or caster length) in front of the contact center. The system is provided with a rotation damper with viscous damping coefficient $k$ and possibly with a torsional spring with stiffness $c_\psi$. The moment of inertia about the swivel axis is denoted with $I$. This quantity varies with caster length $e$. However, we may consider a constant value of $I$ if caster is accomplished by

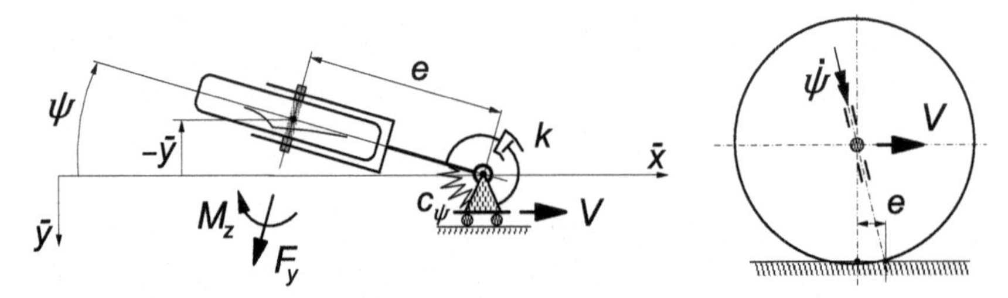


**FIGURE 6.1** Simple trailing wheel system that is capable of showing shimmy. Two ways of realizing the mechanical caster length $e$ of the contact center with respect to the steering axis (kingpin) have been indicated.

inclining the swivel axis about the wheel spin axis. If the caster angle remains relatively small, the camber angle that arises through steering may be neglected and the equations assuming a vertical axis will hold with good approximation.

The tire will be considered massless and the contact line is approximated by a straight line tangent to the actual contact line at the leading edge (straight tangent approximation) for which the Eqns (5.125–5.127) hold or by a single point, Eqns (5.130–5.132), with and without the effect of tread width. The results will be compared with those obtained when using the almost exact Von Schlippe tire model. We have the equations:

$$I\ddot{\psi} + k\dot{\psi} + c_{\psi}\psi = -F_y e + M_z \tag{6.1}$$

$$F_y = C_{F\alpha}\alpha' \tag{6.2}$$

$$M_z = M'_z + M^*_z \tag{6.3}$$

$$M'_z = -C_{M\alpha}\alpha' \tag{6.4}$$

$$M^*_z = -\kappa^* \frac{\mathrm{d}\psi}{\mathrm{d}s} \tag{6.5}$$

$$\alpha' = \frac{v_1}{\sigma} \tag{6.6}$$

$$\frac{\mathrm{d}v_1}{\mathrm{d}s} + \frac{v_1}{\sigma} = \psi - a\frac{\mathrm{d}\psi}{\mathrm{d}s} - \frac{\mathrm{d}\bar{y}}{\mathrm{d}s} \tag{6.7}$$

$$s = \bar{x} = Vt \tag{6.8}$$

$$\bar{y} = -e\psi \tag{6.9}$$

where $a = 0$ when the single contact point tire model is considered. To reduce the number of governing system parameters we will introduce the following nondimensional quantities (in bold letters) with the reference length $a_o$ representing the actual or the nominal half contact length:

$$\boldsymbol{a} = \frac{a}{a_o},\ \boldsymbol{s} = \frac{s}{a_o},\ \boldsymbol{e} = \frac{e}{a_o},\ \boldsymbol{t} = \frac{C_{M\alpha}}{a_o C_{F\alpha}},\ \boldsymbol{\sigma} = \frac{\sigma}{a_o},\ \boldsymbol{v}_1 = \frac{v_1}{a_o},\ \boldsymbol{\omega}_s = \omega_s a_o$$

$$\boldsymbol{V} = V\sqrt{\frac{I}{C_{F\alpha}a_o^3}},\ \boldsymbol{\kappa}^* = \frac{\kappa^*}{C_{F\alpha}a_o^2} = \frac{\kappa^* t}{C_{M\alpha}a_o^2},\ \boldsymbol{k} = \frac{k}{\sqrt{IC_{F\alpha}a_o}},\ \boldsymbol{c}_{\psi} = \frac{c_{\psi}}{C_{F\alpha}a_o} \tag{6.10}$$

which includes the nondimensional pneumatic trail $\boldsymbol{t} = t/a_o$. After elimination of the time and all the variables except $\psi$ and $v_1$ in Eqns (6.1–6.9) and using the conversions (6.10) we obtain the nondimensional differential equations:

$$\begin{aligned} &\boldsymbol{V}^2\frac{\mathrm{d}^2\psi}{\mathrm{d}\boldsymbol{s}^2} + (\boldsymbol{kV} + \boldsymbol{\kappa}^*)\frac{\mathrm{d}\psi}{\mathrm{d}\boldsymbol{s}} + \boldsymbol{c}_{\psi}\psi + (\boldsymbol{e} + \boldsymbol{t})\frac{\boldsymbol{v}_1}{\boldsymbol{\sigma}} = 0 \\ &\frac{\mathrm{d}\boldsymbol{v}_1}{\mathrm{d}\boldsymbol{s}} + \frac{\boldsymbol{v}_1}{\boldsymbol{\sigma}} = (\boldsymbol{e} - \boldsymbol{a})\frac{\mathrm{d}\psi}{\mathrm{d}\boldsymbol{s}} + \psi \end{aligned} \tag{6.11}$$

The system is linear and of the third order. It is assumed that the tire deflection relaxation length $\sigma = 3a$ or, with $a = a_o$, nondimensionally: $\boldsymbol{\sigma} = 3$. In the case of the single contact point tire model $\boldsymbol{a}$ must be taken equal to zero and $\boldsymbol{\sigma}$ must be given the value of $\boldsymbol{\sigma} + a = 4a$. In the nondimensional notation $\boldsymbol{\sigma}$ takes the value of $\boldsymbol{\sigma} + \boldsymbol{a} = \boldsymbol{\sigma} + 1 = 4$. For the pneumatic trail we assume $t = 0.5a$ or $\boldsymbol{t} = 0.5$.

The characteristic equation of system (6.11) becomes

$$\begin{vmatrix} \boldsymbol{V}^2p^2 + (\boldsymbol{kV} + \boldsymbol{\kappa}^*)p + \boldsymbol{c}_\psi & \boldsymbol{e} + \boldsymbol{t} \\ (\boldsymbol{e} - \boldsymbol{a})p + 1 & -(\boldsymbol{\sigma}p + 1) \end{vmatrix} = 0 \tag{6.12}$$

or

$$\begin{aligned} &\boldsymbol{\sigma V}^2p^3 + \{\boldsymbol{V}^2 + \boldsymbol{\sigma}(\boldsymbol{kV} + \boldsymbol{\kappa}^*)\}p^2 \\ &\quad + \{\boldsymbol{kV} + \boldsymbol{\kappa}^* + \boldsymbol{\sigma c}_\psi + (\boldsymbol{e} + \boldsymbol{t})(\boldsymbol{e} - \boldsymbol{a})\}p + \boldsymbol{e} + \boldsymbol{t} + \boldsymbol{c}_\psi = 0 \end{aligned} \tag{6.13}$$

In general, we may write

$$a_0p^3 + a_1p^2 + a_2p + a_3 = 0 \tag{6.14}$$

According to the criterion of Hurwitz the conditions for stability of the motion of the third-order system read as follows:

- all coefficients $a_i$ of the characteristic equation must be positive:

$$a_0 > 0, \;\; a_1 > 0, \;\; a_2 > 0, \;\; a_3 > 0 \tag{6.15}$$

- the Hurwitz determinants $H_{n-1}$, $H_{n-3}$, etc. must be positive, which yields for the third order ($n = 3$) system:

$$H_2 = \begin{vmatrix} a_1 & a_0 \\ a_3 & a_2 \end{vmatrix} > 0 \tag{6.16}$$

The first two coefficients of Eqn (6.13) are always positive. For the remaining coefficients, the conditions for stability become

$$a_2 = \boldsymbol{kV} + \boldsymbol{\kappa}^* + \boldsymbol{\sigma c}_\psi + (\boldsymbol{e} + \boldsymbol{t})(\boldsymbol{e} - \boldsymbol{a}) > 0 \tag{6.17}$$

$$a_3 = \boldsymbol{e} + \boldsymbol{t} + \boldsymbol{c}_\psi > 0 \tag{6.18}$$

while according to (6.16):

$$H_2 = a_2\{\boldsymbol{V}^2 + \boldsymbol{\sigma}(\boldsymbol{kV} + \boldsymbol{\kappa}^*)\} - \boldsymbol{\sigma V}^2(\boldsymbol{e} + \boldsymbol{t} + \boldsymbol{c}_\psi) > 0 \tag{6.19}$$

It can be easily seen that, if the last two conditions are satisfied, the first one is satisfied automatically. If the condition $a_n = a_3 > 0$ is the first to be violated then the motion turns into a monotonous unstable motion (divergent instability, that is, without oscillations). Consequently, if $\boldsymbol{e} < -(\boldsymbol{t} + \boldsymbol{c}_\psi)$ meaning that the steering axis lies a distance $-e$ behind the contact center that is larger than $t + c_\psi/C_{F\alpha}$, the wheel swings around over 180 degrees to the new stable situation. If $H_{n-1} = H_2$ is the first to become negative, the motion becomes oscillatorily unstable. The boundaries of the two unstable areas are: $a_3 = 0$ and $H_2 = 0$.

In the case of vanishing damping ($k = \kappa^* = 0$) condition (6.19) reduces to

$$(\boldsymbol{e} + \boldsymbol{t})(\boldsymbol{e} - \boldsymbol{a} - \boldsymbol{\sigma}) > 0 \tag{6.20}$$

Apparently, in the ($\boldsymbol{e}$, $V$) parameter plane the boundaries of (6.19) reduce to two parallel lines at $\boldsymbol{e} = \boldsymbol{\sigma} + \boldsymbol{a}$ and $\boldsymbol{e} = -\boldsymbol{t}$. When the caster lies in between these two values, the yaw angle performs an oscillation with exponentially increasing amplitude at any speed of travel. Apparently, when the damping is zero, the speed and the torsional stiffness do not influence the extent of the unstable area. They will, however, change the degree of instability and the natural frequency (eigenvalues).

If we have damping, the limit situation may be considered where the speed $V$ tends to zero. The condition for stability then becomes

$$\kappa^* + \boldsymbol{\sigma c_\psi} + (\boldsymbol{e} + \boldsymbol{t})(\boldsymbol{e} - \boldsymbol{a}) > 0 \tag{6.21}$$

which shows that the values for $e$ become complex if $\sigma c_\psi + \kappa^* > {}^1\!/_4 (\boldsymbol{t} - \boldsymbol{a})^2 + \boldsymbol{ta}$. This implies that in that case the unstable area becomes detached from the $\boldsymbol{e}$ axis.

In Figure 6.2 the unstable areas have been shown for different values of the viscous steer damping $\boldsymbol{k}$ and with steer stiffness $\boldsymbol{c}_\psi = 0$. Damping reduces the size of the unstable area and pushes the extreme right-hand edge to lower values of speed. The curves resulting from the application of the straight tangent and the single contact point tire model approximations have been displayed together

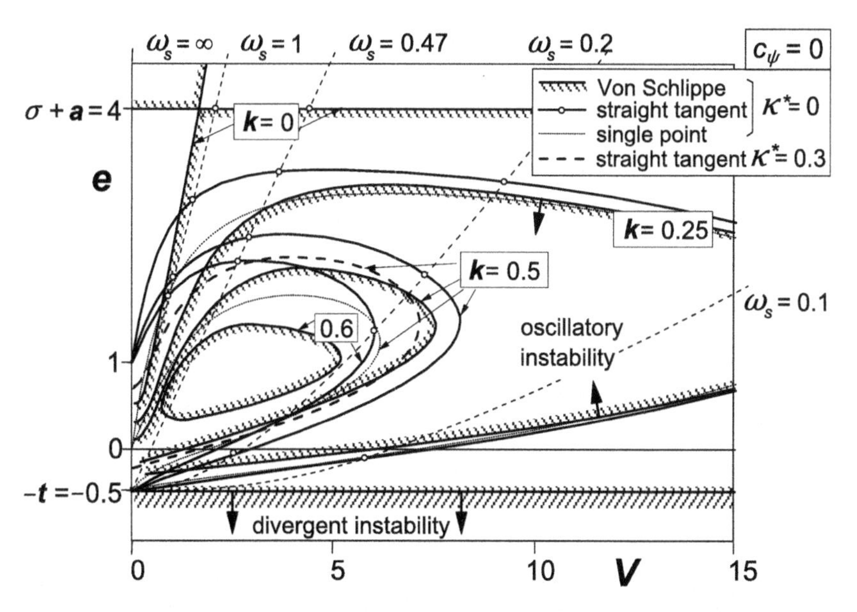


**FIGURE 6.2** Areas of instability for the trailing wheel system of Figure 6.1 without steer stiffness at different levels of viscous steer damping. Consequences of using different tire models can be observed.

with the shaded curves representing the boundaries according to the almost exact Von Schlippe approximation. For a more detailed study of the situation at small values of $\boldsymbol{V}$ (< ca. 2), where at very low damping alternative stable and unstable ranges appear to arise when the exact theory is applied, we refer to Stepan (1997).

Appreciable deviations appear to occur at low values of speed $\boldsymbol{V}$ where the path frequency $\boldsymbol{\omega}_s$ is relatively high and thus the wavelength $\lambda$ relatively short. This is in accordance with findings in Figure 5.23–5.27 where the frequency response functions according to various approximations have been compared. The straight tangent approximation shows a tendency to predict a shimmy instability that is somewhat too strong. On the other hand, the single point contact model turns out to be too stable: for $\boldsymbol{k} = 0.6$ the unstable area vanishes all together. When employing the straight tangent approximation we at least appear to be on the safe side. For the case $\boldsymbol{k} = 0.5$, a curve has been added in Figure 6.2 to show the effect of artificially introducing a proper value of $\kappa^*$ to improve the performance of the straight tangent model. The value $\kappa^* = C_{M\varphi} = 0.6aC_{M\alpha}$ that corresponds to $\boldsymbol{\kappa}^* = 0.3$ indeed produces a better result. This is in accordance with the curve added in the Nyquist diagram of Figure 5.27 for $C_{M\varphi} = 0.6aC_{M\alpha}$ that shows an improved aligning torque frequency response.

The nondimensional path frequencies shown in the figure occur on the boundaries computed for the system with straight tangent approximation. Here the motion shows an undamped oscillation. Then, the solution of (6.13) contains a pair of purely imaginary roots. By replacing $p$ by $i\omega_s$ in the characteristic equation and then separating the imaginary and the real terms we get two equations which can be reduced to one by eliminating the total damping coefficient $\boldsymbol{kV} + \boldsymbol{\kappa}^*$.

The resulting relationship between $\boldsymbol{V}$ and $\boldsymbol{e}$ has been shown for a number of values of $\boldsymbol{\omega}_s$. At path frequency $\boldsymbol{\omega}_s = 1/\sqrt{\{\boldsymbol{\sigma}(1 + \boldsymbol{t})\}} = 0.47$ a special case arises where the frequency curve reduces to a straight line originating from (0,$-\boldsymbol{t}$). By using one of the two equations, an expression for the nondimensional path frequency may be obtained:

$$\boldsymbol{\omega}_s^2 = \frac{a_3}{a_1} = \frac{\boldsymbol{e} + \boldsymbol{t} + \boldsymbol{c}_\psi}{\boldsymbol{V}^2 + \boldsymbol{\sigma}(\boldsymbol{kV} + \boldsymbol{\kappa}^*)} \tag{6.22}$$

Figure 6.3 presents the diagram indicating the influence of damping due to tread width $\kappa^*$. This type of damping is especially effective at low speed. This is understandable because we have seen that the equivalent viscous damping coefficient decreases inversely proportionally with the forward velocity. The boundaries now have a chance to become closed at the left-hand side, cf. Eqn (6.21).

As Figure 6.4 depicts, the steer torsional stiffness reduces the unstable shimmy area in size especially when sufficient damping is provided. At vanishing damping, the area of oscillatory instability appears to remain limited by the horizontal lines obtained from (6.20). The area of divergent instability is reduced through the restoring action of the torsional spring about the steering axis. Its upper boundary shifts downwards according to the negative trail $-\boldsymbol{e} = \boldsymbol{t} + \boldsymbol{c}_\psi$.

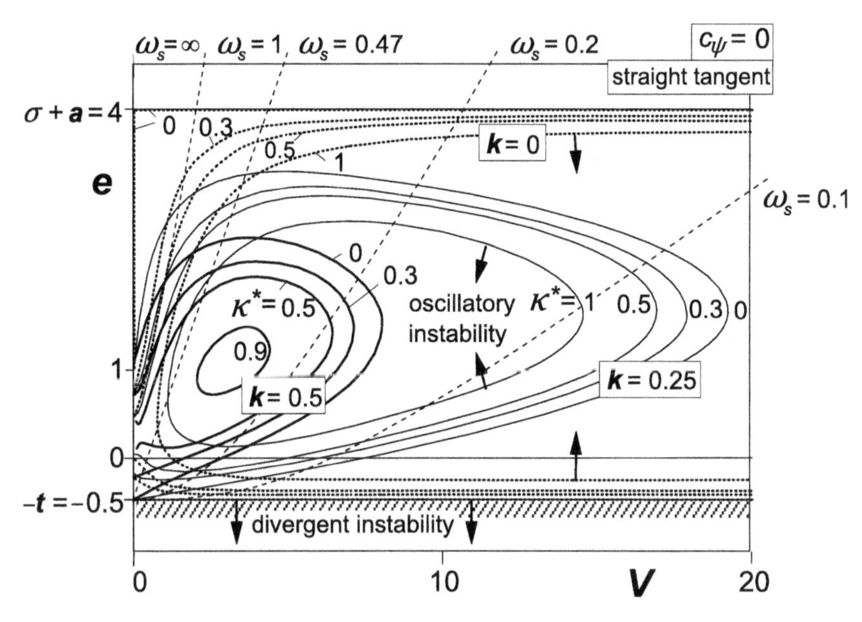


**FIGURE 6.3** Areas of instability for the trailing wheel system of Figure 6.1 without steer stiffness. Influence of tread width damping at different levels of viscous steer damping.

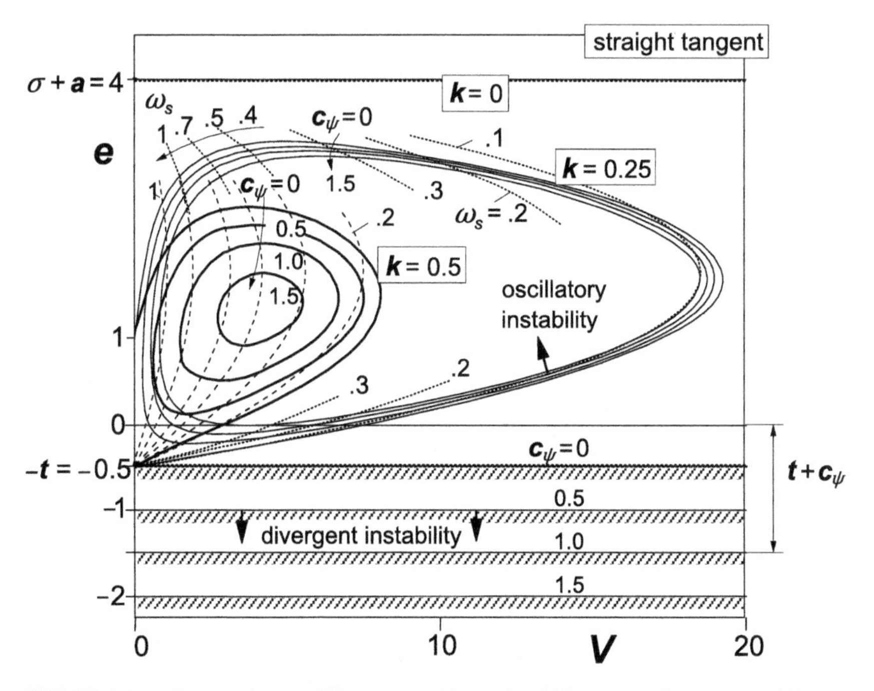


**FIGURE 6.4** Influence of steer stiffness $c_\psi$ on shimmy instability and on divergent instability at different levels of steer damping without tread width damping.

The influence of tire inertia has not been indicated in the diagrams. As may be expected from the diagram of Figure 5.38, which shows that the phase lag of the moment response is reduced through the action of the gyroscopic couple $M_{z,gyr}$ especially at higher speeds, the area becomes limited at the right-hand side even in the absence of damping. Exercise 6.1 given below addresses this problem.

If the moment of inertia about the steering axis cannot be considered as a constant but rather as a function of the trail $e$, e.g., replace $\boldsymbol{I}$ by $I_z + me^2$, then, considering the nondimensionalization defined by (6.10), the curves of the various diagrams must be reinterpreted: at a constant $\boldsymbol{k}$ the actual damping coefficient $k$ is not a constant along the curve but increases with increasing $\boldsymbol{e}$ while at a given $\boldsymbol{V}$ the actual speed $V$ diminishes at increasing $\boldsymbol{e}$.

*Example*

Consider the system with parameters:

$$a = a_3 = 0.14\,\mathrm{m},\ \ \boldsymbol{t} = 0.5,\ \ \boldsymbol{\sigma} = 3\ \ \boldsymbol{e} = 0,$$
$$I = 5.4\,\mathrm{kgm}^2,\ \ C_{F\alpha} = 70000\,\mathrm{N/rad},\ \ \kappa^* = 0.25\,\boldsymbol{t} = 0.125.$$

Speed of travel:

$$\boldsymbol{V} = V\sqrt{\frac{I}{C_{F\alpha}a_o^3}} \rightarrow V \approx 6\boldsymbol{V}\ [\mathrm{m/s}] \approx 21\boldsymbol{V}\ [\mathrm{km/h}]$$

Shimmy wavelength:

$$\boldsymbol{\omega}_s = a\omega_s \rightarrow \omega_s = 7.14\boldsymbol{\omega}_s\quad [\mathrm{rad/m}]$$
$$\lambda = 2\pi/\omega_s = 0.88/\boldsymbol{\omega}_s\quad [\mathrm{m}]$$

so that at

$$\boldsymbol{\omega}_s = 0.3 \rightarrow \lambda = 2.9\mathrm{m} = 21a$$

and at

$$\boldsymbol{\omega}_s = 0.1 \rightarrow \lambda = 8.8\mathrm{m} = 63a$$

Shimmy frequency:

$$\omega_s = \omega_s V\quad [\mathrm{rad/s}]$$
$$n = V/\lambda = \omega/(2\pi)\quad [\mathrm{Hz}]$$

at $\boldsymbol{V} = 5.6$ we have $V = 33$ m/s and according to the stability graph of Figure 6.3 ($\boldsymbol{c}_\psi = 0$) we have at this speed and caster $e = 0$ a nondimensional path frequency $\boldsymbol{\omega}_s \approx 0.12$ so that $\omega_s \approx 0.85$ rad/m, $\lambda \approx 7.3$ m, $\omega \approx 28$ rad/s, and $n \approx 4.5$ Hz. This would be the case at the boundary of stability where $\boldsymbol{k} \approx 0.25$ with $\kappa^* = 0.3$. At about the same point in the diagram of Figure 6.4 we would have with a nondimensional steer stiffness $\boldsymbol{c}_\psi = 1$ and damping coefficient $\boldsymbol{k} = 0.25$ a nondimensional path frequency $\boldsymbol{\omega}_s \approx 0.2$ which yields $\omega_s \approx 1.4$ rad/m, a wavelength $\lambda \approx 4.4$ m and a frequency $n \approx 7.5$ Hz.

**Exercise 6.1. Influence of the Tire Inertia on the Stability Boundary**

Consider the wheel system of Figure 6.1. Derive the stability conditions for this system with the inclusion of the effect of tire inertia approximated by the introduction of $M_{z,gyr}$ (5.178). In nondimensional form, the gyroscopic couple becomes

$$\boldsymbol{M}_{z,\mathrm{gyr}} = \frac{M_{z,\mathrm{gyr}}}{C_{F\alpha}a} = -\boldsymbol{C}_{\mathrm{gyr}}\boldsymbol{V}^2\frac{\mathrm{d}\boldsymbol{F}_y}{\mathrm{d}\boldsymbol{s}} \tag{6.23}$$

with

$$\boldsymbol{C}_{\mathrm{gyr}} = C_{\mathrm{gyr}}\frac{aC_{F\alpha}}{I} = c_{\mathrm{gyr}}\frac{m_t}{C_{Fy}}\frac{aC_{F\alpha}}{I} = c_{\mathrm{gyr}}\frac{m_t}{I}a(\sigma + a) \tag{6.24}$$

cf. Eqns (5.59, 5.60) and (5.179).

Determine how the stability boundaries for the system without damping ($\kappa^* = k = c_\psi = 0$) are changed due to the inclusion of the gyroscopic couple. Consider the following parameter values

$$\boldsymbol{t} = 0.5, \quad \boldsymbol{\sigma} = 3, \quad \boldsymbol{C}_{\mathrm{gyr}} = 0.04$$

Draw in the ($\boldsymbol{e}$, $\boldsymbol{V}$) diagram the boundaries for $C_{\mathrm{gyr}} = 0$ and sketch the new stability boundary by calculating at least the boundary values for $\boldsymbol{V}$ at $\boldsymbol{e} = -\boldsymbol{t}$, $\boldsymbol{e} = 0$ and $\boldsymbol{e} = 1$ and, moreover, the boundary on the $\boldsymbol{e}$-axis. Indicate where the system is stable.

## 6.3. SYSTEMS WITH YAW AND LATERAL DEGREES OF FREEDOM

The additional degree of freedom that allows the kingpin to move sideways leads to a system which is of the fifth order. Figure 6.5 depicts the system with lateral compliance of the wheel suspension possibly associated with a torsional (camber) compliance with (virtual) axis of rotation assumed to be located at a height $h$ above the ground (right-hand diagram). The inertia of the system ($m$, $I_z$) may be represented by two point masses $m_A$ and $m_B$ and a residual moment of inertia $I^*$ about the vertical axis. The two mass points are connected by a massless rod with length $e$, and are located on the wheel axis and the steer axis, respectively. Lateral damping will not be considered.

The following set of equations hold for the dynamics of the system of Figure 6.5 assuming a vertical steering axis and adopting the straight tangent tire model, Eqns (5.125–5.127, 6.7):

$$-m_A(e\ddot{\psi} - \ddot{y}) + m_B\ddot{y} + c_y y = F_y \tag{6.25}$$

$$m_A e(e\ddot{\psi} - \ddot{y}) + I^*\ddot{\psi} + k\dot{\psi} + c_\psi\psi = -F_y e + M_z \tag{6.26}$$

$$F_y = C_{F\alpha}\alpha' \tag{6.2}$$

$$M_z = M_z' + M_z^* \tag{6.3}$$

$$M_z' = -C_{M\alpha}\alpha' \tag{6.4}$$

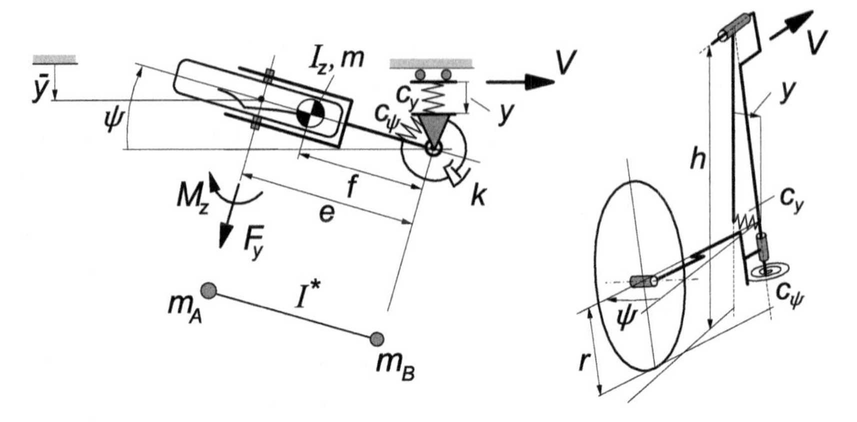


FIGURE 6.5 Lateral compliance introduced in the system.

$$M_z^* = -\kappa^* \frac{d\psi}{ds} \tag{6.5}$$

$$\alpha' = \frac{v_1}{\sigma} \tag{6.6}$$

$$\frac{dv_1}{ds} + \frac{v_1}{\sigma} = \alpha - a\frac{d\psi}{ds} \tag{6.27}$$

$$s = \bar{x} = Vt \tag{6.8}$$

where the slip angle $\alpha$ is obtained from

$$\alpha = \psi - \frac{1}{V}(\dot{y} - e\dot{\psi}) \tag{6.28}$$

Before further studying the fifth-order system, let us try to reduce the system to a lower order while retaining the lateral compliance.

### 6.3.1. Yaw and Lateral Degrees of Freedom with Rigid Wheel/Tire (Third Order)

It can be shown that the lateral compliance of the wheel suspension has an effect that is similar to the effect of tire lateral compliance. The basic effect of lateral compliance of the suspension may be isolated by considering the wheel and tire as a rigid disk. As a result, slip angle $\alpha = 0$ and a nonholonomic constraint equation arises that reduces the order of the system to three. Through this first-order differential equation, the condition that the wheel is unable to perform lateral slip is obeyed. We obtain the constraint equation from (6.28) with wheel slip angle $\alpha$ forced to be equal to zero:

$$\alpha = \psi - \frac{1}{V}(\dot{y} - e\dot{\psi}) = 0 \tag{6.29}$$

In addition, we assume that for the rigid disk the aligning torque and the contact length vanish:

$$M_z = 0, \quad a = 0 \tag{6.30}$$

After elimination of $F_y$ from the Eqns (6.25, 6.26) and expressing $\dot{y}$ in terms of $\psi$ and its derivative we find a third-order differential equation. Its characteristic equation with variable $s$ reads as

$$(I^* + m_B e^2)s^3 + (k + m_B eV)s^2 + (c_\psi + c_y e^2)s + c_y eV = 0 \tag{6.31}$$

The trivial condition against (divergent) instability that results from the last term of (6.31) which must remain positive becomes

$$e > 0 \tag{6.32}$$

The condition for stability $H_2 > 0$, cf. (6.16), reads as

$$(k + m_B eV)(c_\psi + c_y e^2) > (I^* + m_B e^2)c_y eV \tag{6.33}$$

For the undamped case ($k = 0$) this reduces to the simple condition for stability, if $e > 0$:

$$m_B c_\psi > I^* c_y \tag{6.34}$$

Apparently, a negative residual moment of inertia ($I^* < 0$) will ensure stability. When $f$ defines the location of the center of gravity behind the steer axis and $i^*$ denotes the radius of inertia of the combination ($m_A$, $m_B$, $I^*$), a negative residual moment of inertia would occur if $i^{*2} < f(e - f)$. For such a system, an increase of the lateral stiffness strengthens the stability.

If, on the other hand, $I^* > 0$ a sufficiently large steer stiffness is needed to stabilize the system. It is surprising that a larger lateral stiffness may then cause the system to become unstable again.

The conclusion that the freely swiveling wheel system (that is, $c_\psi = k = 0$) is stable if the residual moment of inertia $I^* < 0$ entails that when caster is realized by inclining the steering axis backwards (like in Figure 6.1 right-hand diagram) we have the situation that $m_A = 0$ so that $I^* = I$ which is always positive. Consequently, for such a freely swiveling system equipped with a rigid thin tire the motion is always oscillatorily unstable.

## 6.3.2. The Fifth-Order System

The gyroscopic coupling terms that arise due to the angular camber velocity of the wheel system about the longitudinal axis located at a height $h$ above the ground will be added to the Eqns (6.25, 6.26). For this, the coefficient $\beta_{\text{gyr}}$ is introduced:

$$\beta_{\text{gyr}} = \frac{I_p}{mrh} \tag{6.35}$$

where $I_p$ denotes the polar moment of inertia of the wheel. Furthermore, the ratio of distances with respect to the camber torsion center is introduced:

$$\zeta_h = \frac{h}{h-r} \tag{6.36}$$

Moreover, to fully account for the effects of the angular displacement about the longitudinal axis, which also arises when steering around an inclined kingpin (to provide caster), one may include the camber force in Eqn (6.2), and the small (negative) stiffness effects resulting from the lateral shift of the normal force $F_z$. Because of the fact that these effects are relatively small and appear to partly compensate each other, we will neglect the extra terms.

The differential equations that result when returning to the set $I_z$ and $m$ that replaces the equivalent set $m_A$, $m_B$, and $I^*$, and eliminating all variables except $y$, $\psi$ and $\alpha'$ read as

$$m(\ddot{y} - f\ddot{\psi}) + k_y\dot{y} + c_y y - m\beta_{\text{gyr}}V\dot{\psi} = \zeta_h C_{F\alpha}\alpha' \tag{6.37}$$

$$I_z\ddot{\psi} + (k + \kappa^*/V)\dot{\psi} + c_\psi\psi + m\beta_{\text{gyr}}V\dot{y} + fk_y\dot{y} + fc_y y = -C_{F\alpha}(e + t - f)\alpha' \tag{6.38}$$

$$\sigma\dot{\alpha}' + V\alpha' = -\zeta_h\dot{y} + (e-a)\dot{\psi} + V\psi \tag{6.39}$$

In the equations, the lateral suspension damping, $k_y$, has been included. In the further analysis, this quantity will be left out of consideration. To study the pure gyroscopic coupling effect, the ratio $\zeta_h$ will be taken equal to unity which would represent the case that the center of gravity and the lateral spring are located at ground level. For the special case that the center of gravity lies on the (inclined) steer axis, distance $f=0$, the system description is considerably simplified. Our analysis will mainly be limited to this configuration. Besselink (2000) has carried out an elaborate study on the system of Figure 6.5 with the mass center located behind the (vertical) steering axis ($f>0$), even behind the wheel center ($f>e$). For the complete results of this for aircraft shimmy important analysis we refer to the original publication. One typical result, however, where $f=e$ will be discussed here.

The complete set of nondimensional quantities, extended and slightly changed with respect to the set defined by Eqn (6.10), reads as

$$\boldsymbol{a} = \frac{a}{a_o},\ \boldsymbol{e} = \frac{e}{a_o},\ \boldsymbol{f} = \frac{f}{a_o},\ \boldsymbol{t} = \frac{C_{M\alpha}}{a_o C_{F\alpha}},\ \boldsymbol{\sigma} = \frac{\sigma}{a_o},\ \boldsymbol{i}_z = \frac{i_z}{a_o},\ \boldsymbol{y} = \frac{y}{a_o},$$

$$\boldsymbol{\omega}_s = \omega_s a_o,\ \boldsymbol{\kappa}^* = \frac{\kappa^*}{C_{F\alpha}a_o^2},\ \boldsymbol{k} = \frac{k}{\sqrt{IC_{F\alpha}a_o}},\ \boldsymbol{c}_\psi = \frac{c_\psi}{C_{F\alpha}a_o},\ \boldsymbol{c}_y = \frac{c_y a_o}{C_{F\alpha}}, \tag{6.40}$$

$$\boldsymbol{m} = \frac{ma_o^2}{I_z} = \frac{1}{i_z^2},\ \boldsymbol{V} = V\sqrt{\frac{I_z}{C_{F\alpha}a_o^3}},\ \boldsymbol{\omega} = \omega\sqrt{\frac{I_z}{C_{F\alpha}a_o}},\ \boldsymbol{s} = s\sqrt{\frac{I_z}{C_{F\alpha}a_o}}$$

Here, $s$ denotes the nondimensional Laplace variable. The nondimensional differential equations follow easily from the original Eqns (6.37–6.39). For the stability analysis, we need the characteristic equation which reads in nondimensional form:

$$a_0 s^5 + a_1 s^4 + a_2 s^3 + a_3 s^2 + a_4 s + a_5 = 0 \tag{6.41}$$

The coefficients of (6.41) with $f = k_y = 0$ are

$$\begin{aligned}
a_0 &= m\sigma \\
a_1 &= mV + m\sigma(k + \kappa^*/V) \\
a_2 &= m\sigma c_\psi + \sigma c_y + \zeta_h + m(e+t)(e-a) + \sigma(m\beta_{\rm gyr}V)^2 + m(kV + \kappa^*) \\
a_3 &= mc_\psi V + c_y V + m(e+t)V + (\zeta_h + \sigma c_y)(k + \kappa^*/V) - (a + \zeta_h t \\
&\quad - m\beta_{\rm gyr}V^2)m\beta_{\rm gyr}V \\
a_4 &= (\sigma c_y + \zeta_h)c_\psi + (e+t)(e-a)c_y + m\beta_{\rm gyr}V^2 + c_y(kV + \kappa^*) \\
a_5 &= c_y(c_\psi + e + t)V
\end{aligned} \tag{6.42}$$

The influence of a number of nondimensional parameters will be investigated by changing their values. The other parameters will be kept fixed. The values of the fixed set of parameters are

$$a = 1,\ \sigma = 3,\ t = 0.5,\ m = 0.5,\ \zeta_h = 1 \tag{6.43}$$

For the fifth-order system the Hurwitz conditions for stability read as:

- all coefficients $a_i$ of the characteristic equation must be positive:

$$a_0 > 0,\ a_1 > 0,\ a_2 > 0,\ a_3 > 0,\ a_4 > 0, a_5 > 0 \tag{6.44}$$

- the Hurwitz determinants $H_{n-3}$ and $H_{n-1}$ must be positive, which yields:

$$H_2 = \begin{vmatrix} a_1 & a_0 \\ a_3 & a_2 \end{vmatrix} > 0, \quad H_4 = \begin{vmatrix} a_1 & a_0 & 0 & 0 \\ a_3 & a_2 & a_1 & a_0 \\ a_5 & a_4 & a_3 & a_2 \\ 0 & 0 & a_5 & a_4 \end{vmatrix} > 0 \tag{6.45}$$

It turns out that for the case without damping and gyroscopic coupling terms ($k = \kappa^* = \beta_{\rm gyr} = 0$), a relatively simple set of analytical expressions can be derived for the conditions of stability. It can be proven that when $a_5 > 0$, $H_2 > 0$, $H_4 > 0$ the other conditions are satisfied as well. Consequently, the governing conditions for stability of the system without damping and gyroscopic coupling read as:

$$\frac{a_5}{c_y V} = (c_\psi + e + t) > 0 \tag{6.46}$$

$$\frac{H_2}{m^2 V} = (e+t)(e-a-\sigma) + \frac{1}{m} > 0 \tag{6.47}$$

$$\frac{H_4}{m^3V^2} = (e+t)\left(c_\psi - \frac{c_y}{m}\right)\left\{\left(c_\psi - \frac{c_y}{m}\right)(e-a-\sigma) + \frac{H_2}{m^2V}\right\} > 0 \quad (6.48)$$

We may ascertain that for the special case of a rigid wheel where $C_{F\alpha} \rightarrow \infty$ and $a, \sigma, t \rightarrow 0$ the condition (6.46) corresponds with (6.32) and the condition

$$c_\psi - \frac{c_y}{m} > 0 \quad (6.49)$$

that results from (6.48) corresponds with condition (6.34) if we realize that in the configuration studied here we have: $m_A = 0$ and $I^* = I_z$.

For the case with an elastic tire with stability conditions (6.46–6.48) we will present stability diagrams and show the effects of changing the stiffnesses. In addition, but then necessarily starting out from Eqns (6.42, 6.44, 6.45), the influence of the steer and tread damping and that of the gyroscopic coupling terms will be assessed.

First, let us consider the system with lateral suspension compliance but without steer torsional stiffness, $c_\psi = 0$, that is, a freely swiveling wheel possibly damped (governed by the combined coefficient $k + \kappa^*/V$) and subjected to the action of the gyroscopic couple (that arises when the lateral deflection is connected with camber distortion as governed by a nonzero coefficient $\beta_{gyr}$). From the above conditions it can be shown that for the simple system without damping and camber compliance stability may be achieved only at the academic case of large caster $e > a + \sigma$ when the lateral stiffness of the wheel suspension is sufficiently large:

$$c_y > m(e+t) + \frac{1}{e-a-\sigma} \qquad \text{while} \quad e > a + \sigma \quad (6.50)$$

The minimum stiffness where stability may occur is

$$c_{y,\text{crit}} = m(a + \sigma + t + 2i_z) \quad (6.51)$$

with $i_z$ representing the radius of inertia:

$$i_z^2 = \frac{1}{m} = \frac{I_z}{ma^2} \quad (6.52)$$

The value of the caster length $e_c$ where stability commences (at $c_{y,\text{crit}}$) is

$$e_c = a + \sigma + i_z \quad (6.53)$$

Consequently, it may be concluded that with respect to the third-order system of Figure 6.2 the introduction of lateral suspension compliance shifts the upper boundary of the area of instability to values of caster length $e$ larger than $a + \sigma$. At the same time, a new area of instability emerges from above. When the lateral stiffness becomes lower than the critical value (6.51) the stable area vanishes altogether at $e = e_c$.

The case with steer and tread damping shows ranges of stability that vary with speed of travel $V$. In Figure 6.6 areas of instability have been depicted for

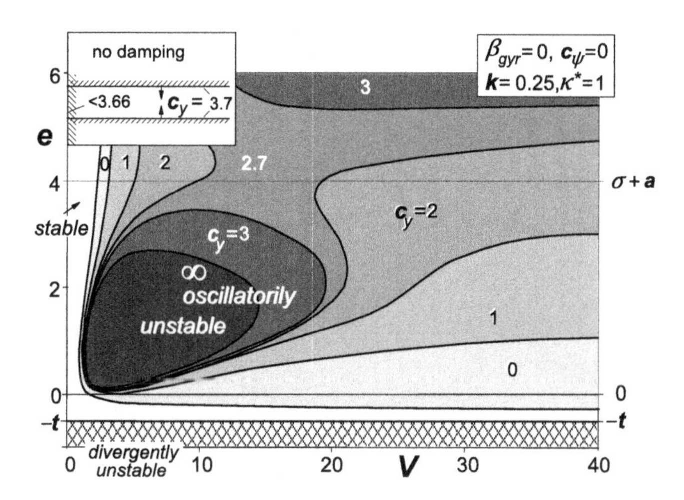


**FIGURE 6.6** Influence of lateral suspension compliance with nondimensional stiffness $\boldsymbol{c}_y$ on stability. Gyroscopic couple and steer stiffness not regarded. No-dimensional parameters: $f=0$, $\boldsymbol{m}=0.5$, $\boldsymbol{t}=0.5$, $\boldsymbol{\sigma}=3$.

different values of the lateral stiffness. The upper left inset illustrates the case without damping. At vanishing lateral stiffness a narrow range of stability appears to remain in the negative trail range with a magnitude smaller than the pneumatic trail $\boldsymbol{t}$.

When a finite height $h$ of the longitudinal torsion axis is considered, parameter $\beta_{\text{gyr}}$ defined by (6.35) obtains a value larger than zero. For $\beta_{\text{gyr}}=0.2$ the diagram of Figure 6.7 arises. An important phenomenon appears to occur that is essential for this two degree of freedom system (not counting the 'half' degree of freedom stemming from the flexible tire). A new area of instability appears to show up at higher values of speed. The area increases in size when the lateral (camber) stiffness decreases. At the same time, the original area of oscillatory instability shrinks and ultimately vanishes. Around zero caster stable motions appear to become possible for all values of speed when a (near) optimum lateral stiffness is chosen. The high-speed shimmy mode exhibits a considerably higher frequency than the one occurring in the lower speed range of instability. This is illustrated by the nondimensional frequencies indicated for the case $\boldsymbol{c}_y=2$ at $\boldsymbol{e}=0.5$. The corresponding nondimensional wavelength $\boldsymbol{\lambda}=\lambda/a$ is obtained by using the formula: $\boldsymbol{\lambda}=2\pi\boldsymbol{V}/\boldsymbol{\omega}$. In Table 6.1 the nondimensional frequencies $\boldsymbol{\omega}$, wavelengths $\boldsymbol{\lambda}$, and path frequencies $1/\boldsymbol{\lambda}=a/\lambda$ have been presented as computed for the four different values of nondimensional velocities where the transition from stable to unstable or vice versa occurs. Also, the mode of vibration is quite different for this 'gyroscopic' shimmy exhibiting a considerably larger lateral motion amplitude of the wheel contact center with respect to the yaw angle amplitude than in the original 'lateral (tire) compliance' shimmy mode. To illustrate this, the table shows the

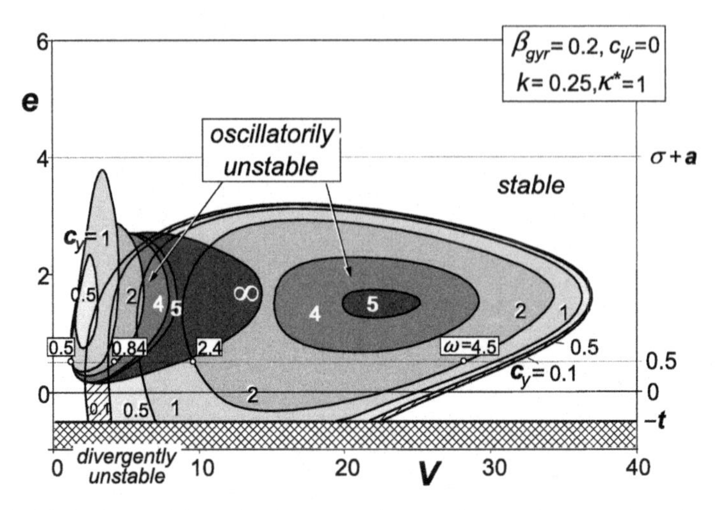


**FIGURE 6.7** Influence of lateral (camber) suspension compliance with gyroscopic coupling. Steer stiffness is equal to zero. At increasing compliance $1/c_y$, a new area of oscillatory instability shows up while the original area shrinks. Frequencies have been indicated for the case $c_y = 2$ at $e = 0.5$.

**TABLE 6.1** Nondimensional Frequencies, Wavelengths and Ratio of Amplitudes of Lateral Displacement at Contact Center and Yaw Angle $\psi$ Times Half-Contact Length *a* for Speeds at Stability Boundaries as Indicated in Figure 6.7 at Trail e = 0.5

| $V = 1.24$ | $\omega = 0.50$ | $\lambda = 15.5$ | $1/\lambda = a/\lambda = 0.065$ | $\lvert \bar{y}/a\psi \rvert = 0.45$ |
|---|---|---|---|---|
| 4.45 | 0.84 | 33.3 | 0.030 | 0.14 |
| 9.60 | 2.40 | 25.2 | 0.040 | 2.51 |
| 28.50 | 4.50 | 39.8 | 0.025 | 1.64 |

ratio of amplitudes of the lateral displacement of the contact center (cf. Figure 6.5) and the yaw angle multiplied with half the contact length: $a\psi$. For the two lower values of speed, the lateral displacement appears to lag behind the yaw angle with about 135° while for the two higher speeds the phase lag amounts to about 90°.

Figure 6.8 depicts the special case of vanishing damping for a number of values of the lateral stiffness. At caster near or equal to zero limited ranges of speed with stable motions appear to occur. These speed ranges correspond with the stable ranges in between the unstable areas of Figure 6.7 (if applicable). For the indicated points A and B and for a number of neighboring points the

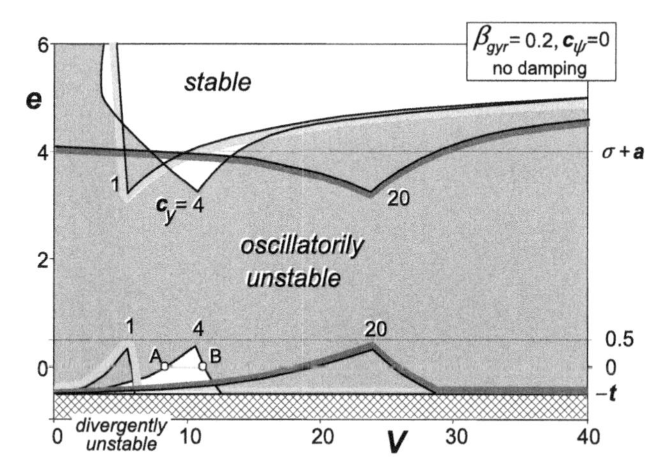


**FIGURE 6.8** Influence of lateral (camber) suspension compliance with gyroscopic coupling. Steer stiffness is equal to zero. Steer and tread damping equal to zero. Eigenvalues and eigenvectors of points A and B are analyzed in Figures 6.22, 6.23.

eigenvalues and eigen-vectors have been assessed and shown in Figures 6.22, 6.23 in Section 6.4.1 where the energy flow is studied.

The system will now be provided with torsional steer stiffness $c_\psi$. For the undamped case, the conditions (6.46–6.48) hold to ensure stability. It can be observed that except for the condition (6.46) $\boldsymbol{c}_\psi$ appears in combination with $\boldsymbol{c}_y$ in the factor $\boldsymbol{c}_\psi - \boldsymbol{c}_y/\boldsymbol{m}$. We may introduce an effective yaw stiffness and an effective lateral stiffness:

$$\boldsymbol{c}^*_\psi = \boldsymbol{c}_\psi - \frac{\boldsymbol{c}_y}{\boldsymbol{m}} \quad \text{and} \quad \boldsymbol{c}^*_y = \boldsymbol{c}_y - \boldsymbol{m}\boldsymbol{c}_\psi \tag{6.54}$$

and establish a single diagram for the ranges of stability for the two effective stiffnesses as presented in Figure 6.10. The way in which the boundaries are formed may be clarified by the diagram of Figure 6.9. The parabola represents the variation of the second Hurwitz determinant $H_2$ divided by $\boldsymbol{m}^2\boldsymbol{V}$, cf. Eqn (6.47), as a function of the caster length $\boldsymbol{e}$. This same term also appears in the expression for the fourth Hurwitz determinant $H_4$ according to (6.48). For both cases $\boldsymbol{c}^*_\psi > 0$ and $\boldsymbol{c}^*_y > 0$ the straight line originating from the point $(\boldsymbol{\sigma} + \boldsymbol{a}, 0)$ at a slope $\boldsymbol{c}^*_\psi$ and $\boldsymbol{c}^*_y/\boldsymbol{m}$, respectively, have been depicted. The points of intersection with the parabola correspond to values of $\boldsymbol{e}$ where $H_4 = 0$. Depending on the signs of factor $\boldsymbol{e} + \boldsymbol{t}$ and of $H_2$ and $a_5$ the ranges of stability can be found. Five typical cases A–E have been indicated. They correspond to similar cases indicated in Figure 6.10. Case D at small negative $\boldsymbol{e}$ lies inside the small stable triangle of Figure 6.10. Here $\boldsymbol{c}^*_\psi > 0$, $\boldsymbol{e} + \boldsymbol{t} > 0$ and $H_2/\boldsymbol{m}^2\boldsymbol{V} > \boldsymbol{c}^*_\psi\,(\sigma + \boldsymbol{a} - \boldsymbol{e})$. Case C shows a range of $\boldsymbol{e}$ where $H_4 > 0$ but at the same time $H_2 < 0$ so that instability prevails. For larger negative values of the caster stability arises

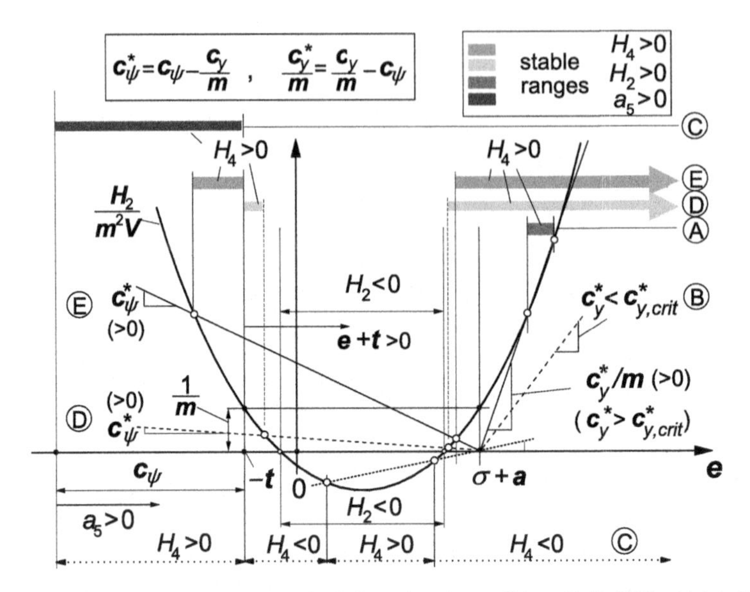

**FIGURE 6.9** Establishing the ranges of stability using the conditions (6.46–6.48) which hold for the fifth-order system without damping and gyroscopic coupling. The cases A–E correspond to those indicated in the stability diagram of Figure 6.10.

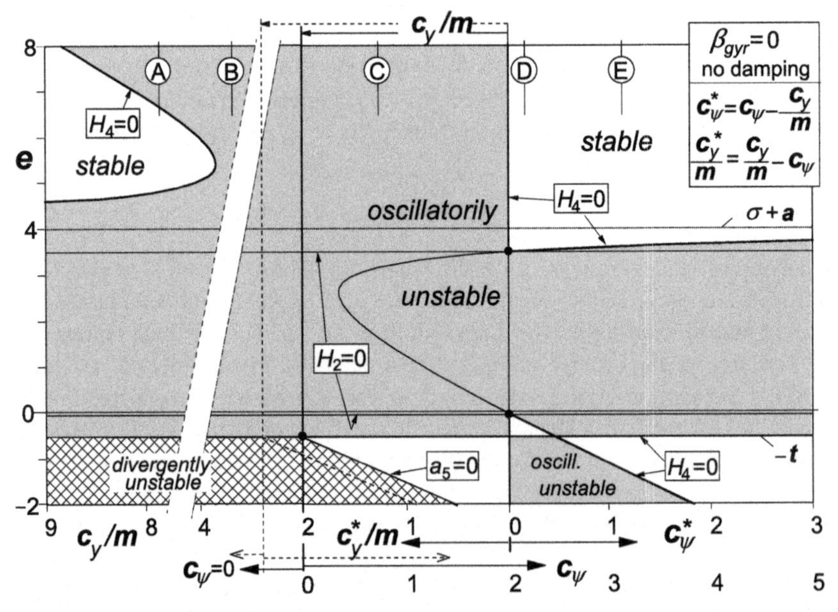

**FIGURE 6.10** Basic stability diagram according to conditions (6.46–6.48) for the fifth-order system without damping and gyroscopic coupling. The cases A–E correspond to those indicated in the stability diagram of Figure 6.9. The diagram holds in general, the case with $c_y/m = 2$ is an example, note: $m = 0.5$.

because $\boldsymbol{e}+\boldsymbol{t}$ changes sign while $a_5$ remains positive. Also for larger values of $\boldsymbol{c}_{\psi}^{*}$ a stable area appears to show up for $\boldsymbol{e} < -\boldsymbol{t}$. For cases like D and E, stability occurs for large values of the mechanical trail $\boldsymbol{e}$ (not very realistic for our present assumption that $f=0$ which means: c.g. on (inclined) steering axis). In a similar large range of $\boldsymbol{e}$ stability occurs for large effective lateral stiffness, e.g., case A. This case has already been referred to previously when discussing the situation without steer stiffness, cf. Figure 6.6 upper left inset, Eqn (6.50). Case B shows instability for all values of $\boldsymbol{e}$. As has been indicated in Figure 6.10, we can establish the abscissa $\boldsymbol{c}_{\psi}$ by shifting the vertical axis to the left over a distance $\boldsymbol{c}_y/\boldsymbol{m}$. At the left of the shifted axis the steer stiffness must be considered equal to zero so that $\boldsymbol{c}_y^{*} = \boldsymbol{c}_y$. The stability boundary $a_5=0$ shifts to the left together with the $\boldsymbol{c}_{\psi}=0$ axis.

In Figure 6.11 the diagram of Figure 6.10 is repeated but only for the version with $\boldsymbol{c}_{\psi}$ as abscissa and for one fixed value of $\boldsymbol{c}_y$. Figures 6.12 and 6.13 clearly show the influence of damping. The stable areas increase in size, while the unstable area tends to split itself into two parts, the higher stiffness part of which ($\boldsymbol{c}_{\psi} > \sim\boldsymbol{c}_y/\boldsymbol{m}$) appears to vanish at sufficient damping. Also in the low yaw stiffness range the instability is suppressed especially in the lower range of speed. For a caster length close to zero it appears possible to achieve stability through adapting the steer stiffness (making it a bit larger than $\boldsymbol{c}_y/\boldsymbol{m}$, cf. Case D of Figure 6.10) and/or by supplying sufficient damping.

The more complex situation where gyroscopic coupling is included has been considered in Figures 6.14–6.16. For vanishing steer stiffness, $\boldsymbol{c}_{\psi} \rightarrow 0$, the situation of Figures 6.8, 6.7 is reached again. In Figure 6.14 where the damping is zero, the curve at low speed resembles the graph of Figure 6.11. As was observed to occur in the zero yaw stiffness case, we see that in the lower yaw stiffness range ($\boldsymbol{c}_{\psi} < \sim\boldsymbol{c}_y/\boldsymbol{m}$) an optimum value of the speed appears to exist

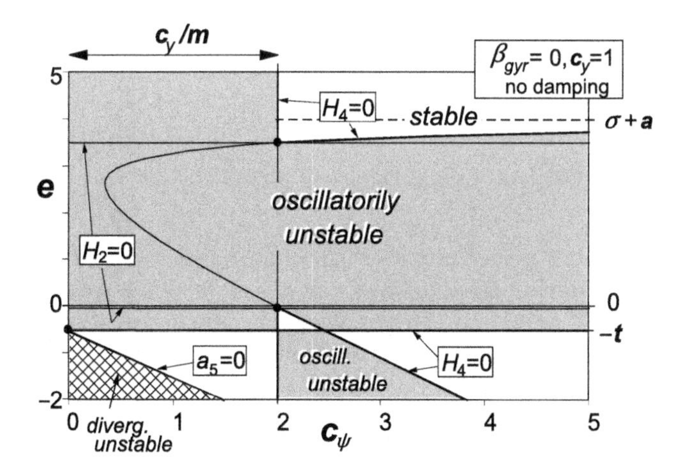


**FIGURE 6.11** Stability diagram according to Figure 6.10 with steer stiffness along abscissa, at a fixed value of the lateral stiffness, without damping or gyroscopic coupling.

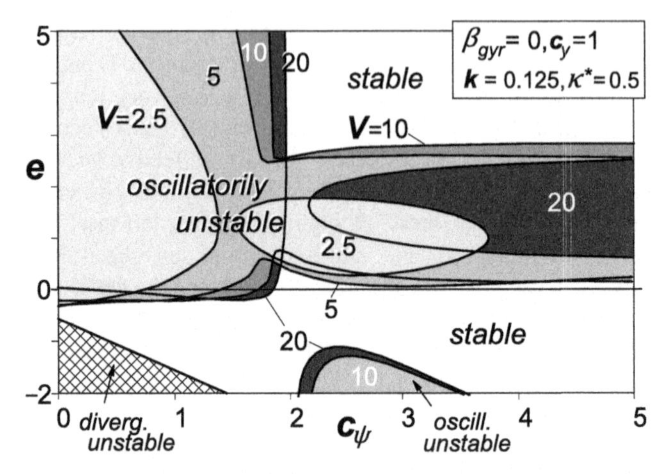


**FIGURE 6.12** Same as Figure 6.11 but with some damping. Areas of instability at different levels of speed.

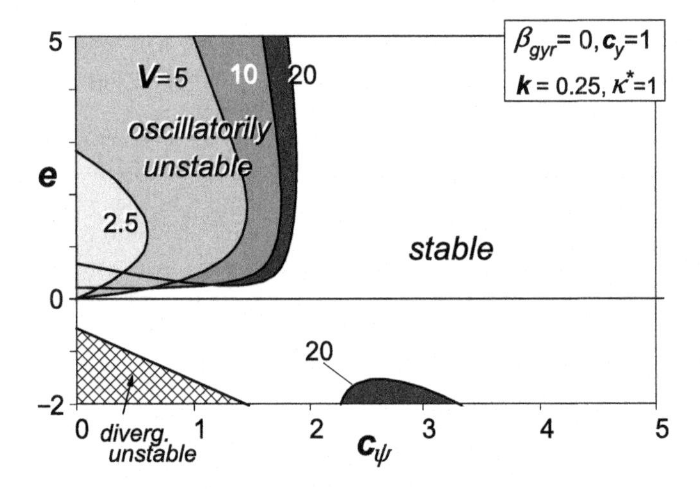


**FIGURE 6.13** Same as Figure 6.12 but with more damping.

where the unstable $e$ range becomes smallest or vanishes altogether (at sufficient damping). In the higher stiffness range, increasing the speed promotes shimmy except in the negative $e$ range where the opposite may occur. In this same range of yaw stiffness an increase of the stiffness appears to lower the tendency to shimmy.

Besselink (2000) has analyzed the more general and for aircraft landing gears more realistic configuration where the center of gravity of the swiveling part is located behind the steering axis. Notably, the case where $f = e$ and the peculiar case with $f > e$ which may be beneficial to suppress shimmy. For the

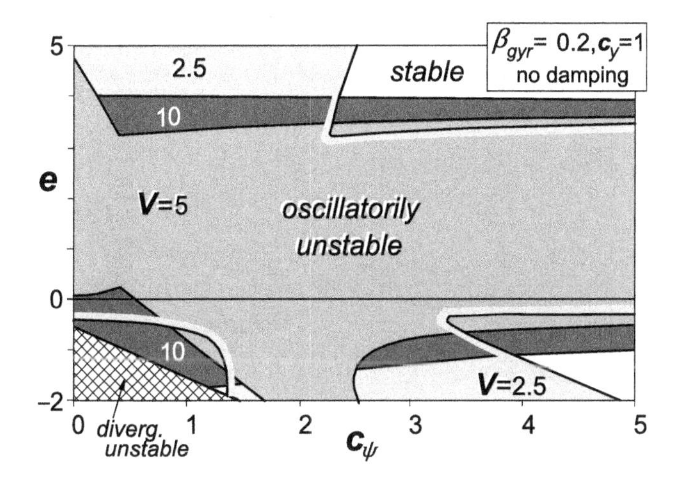


**FIGURE 6.14** Diagram for the case of Figure 6.11 but with gyroscopic coupling. Areas of instability at different levels of speed.

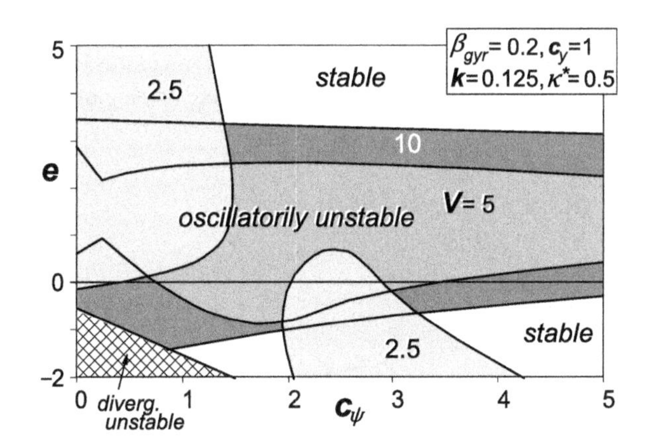


**FIGURE 6.15** Same as Figure 6.14 but with some damping.

configuration that $f = e$ and damping and gyroscopic coupling are disregarded, the following conditions for stability hold:

$$a_5 > 0: \quad c_\psi > c_{\psi 1} = -(e + t) \tag{6.55}$$

$$H_2 > 0: \quad i_z^2 = \frac{1}{m} > (\sigma + a)t \tag{6.56}$$

$$H_4 > 0: \quad \min(c_{\psi 2}, c_{\psi 3}) < c_\psi < \max(c_{\psi 2}, c_{\psi 3}) \qquad 6.57)$$

with the two parabolic functions of $e$:

$$c_{\psi 2} = \frac{1 - ec_y}{m(\sigma + a)} + \frac{c_y}{m} + c_y e(\sigma + a) - c_y e^2 - t \tag{6.58}$$

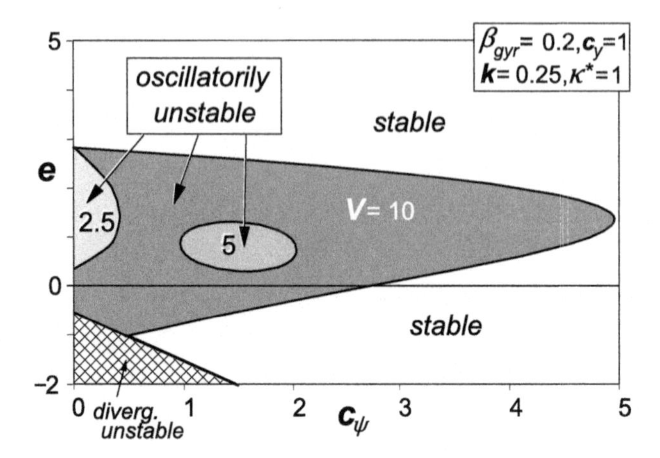


**FIGURE 6.16** Same as Figure 6.15 but with more damping.

$$c_{\psi 3} = \frac{c_y(e+t)(1-emt)}{mt} \tag{6.59}$$

Figure 6.17 depicts the boundaries of stability according to the above functions. The ordinates of two characteristic points have already been given in the figure. For the three other points, we have the $e$ values:

$$e_1 = -\frac{i_z^2}{\sigma + a} \tag{6.60}$$

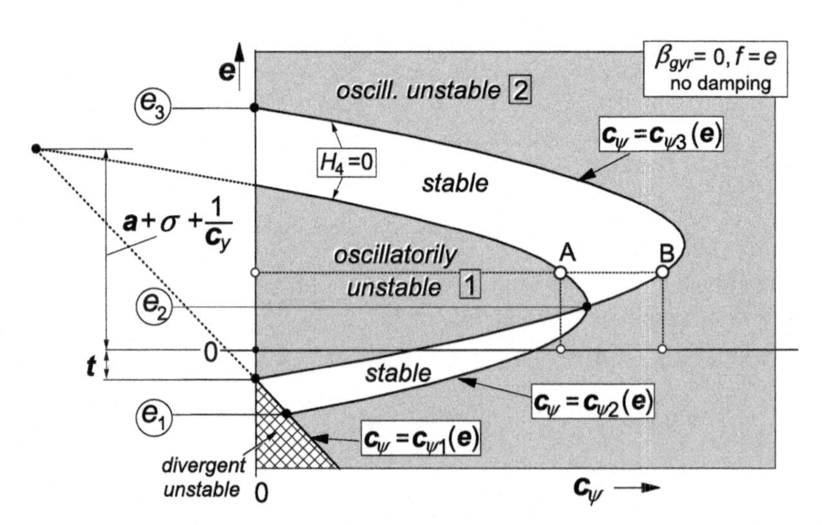


**FIGURE 6.17** Basic stability diagram for the case $f = e$ (c.g. distance $e$ behind steer axis). Steer and tread damping are equal to zero. Gyroscopic coupling is disregarded. The functions $c_{\psi i}(e)$ are defined by Eqns (6.55, 6.58, 6.59). Eigenvalues and eigen-vectors of points A and B are analyzed in Figures 6.18, 6.19. (From: Besselink 2000).

$$e_2 = \frac{t}{c_y(\sigma + a + t)} \tag{6.61}$$

$$e_3 = \frac{i_z^2}{t} \tag{6.62}$$

The appearance of the stability diagram is quite different with respect to that of Figure 6.11 where $f = 0$. Still, an area of instability exists in between the levels of the trail: $-t < e < \sigma + a$. Like in the $f = 0$ case (clearly visible in Figure 6.12), we have two different ranges of instability: area 1 and area 2 where different modes of unstable oscillations occur. In the range of the caster length $-t < e < \sigma + a$ these areas correspond to the lower and the higher yaw stiffness regions.

For the points A and B which lie on the boundaries of the lower and the higher stiffness instability areas, respectively, the modes of vibration have been analyzed. The conditions for stability (6.55–6.57) of the undamped system without gyros are independent of the speed. When changing the speed of travel the damping of the mode that is just on a boundary of instability remains equal to zero. The other mode shows a positive damping which changes with speed as does the frequency of this stable mode. In Figures 6.18 and 6.19 (from Besselink 2000), the damping, frequency, amplitude ratio, and phase lead of the lateral displacement of the

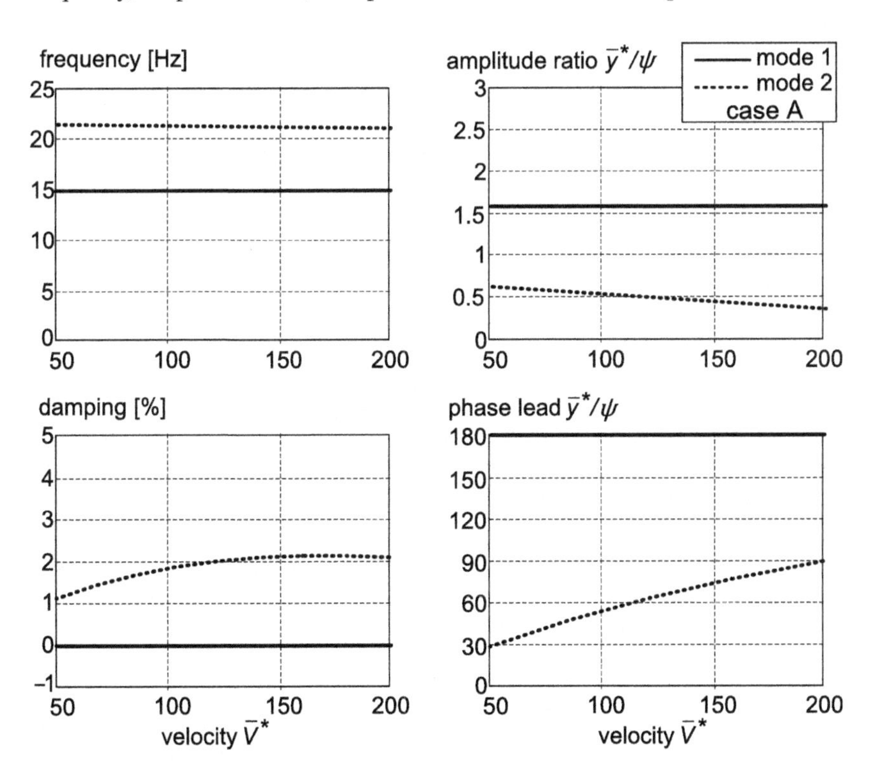


**FIGURE 6.18** Natural frequencies, damping and mode shapes for case A of Figure 6.17.

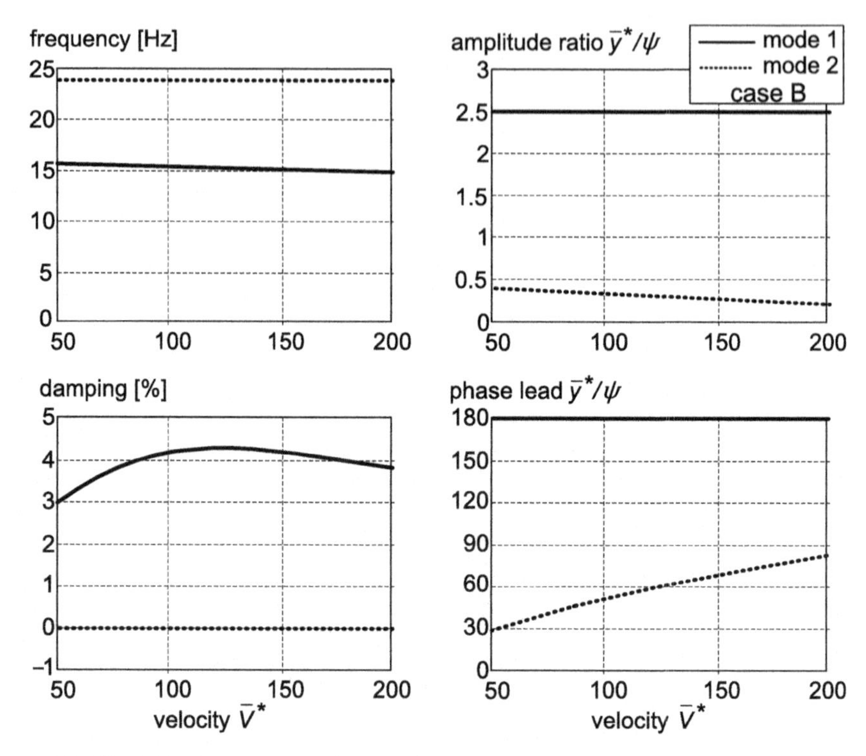


**FIGURE 6.19** Natural frequencies, damping and mode shapes for case B of Figure 6.17.

contact center $\bar{y}$ with respect to the yaw angle $\psi$ have been presented for the two modes of vibration – in Figure 6.18 for case A and in Figure 6.19 for case B of Figure 6.17. Using a reference tire radius $r_{ref}$ the quantities $\bar{y}^*$ and $\bar{V}^*$ have been defined as follows: $\bar{y}^* = y/r_{ref}$ [−], $\bar{V}^* = V/r_{ref}$ [1/sec]. From the graphs can be observed that the lower yaw stiffness case A shows a mode on the verge of instability (mode 1) that has a lower frequency and a larger amplitude ratio $|\bar{y}^*/\psi|$ than mode 2 in case B. Evidently, case B exhibits a more pronounced yaw oscillation, whereas the motion performed by mode 1 of case A comes closer to a lateral translational oscillation. The next section discusses the mode shapes of these periodic motions in relation with the energy flow into the unstable system.

Besselink has analyzed the effects of changing a number of parameters of a realistic two-wheel, single axle main landing gear model configuration (of civil aircraft, also see Van der Valk 1993). Degrees of freedom represent axle yaw angle, strut lateral deflection, axle roll angle, and tire lateral deflection (straight tangent). This makes the total order of the system equal to seven. Important conclusions regarding the improvement of an existing system aiming at avoidance of shimmy have been drawn. Two mutually different configurations have been suggested as possible solutions to the problem, with the c.g. position $f$ assumed to remain equal to the trail $e$.

- *large positive trail* ($e > 0$):
  'An increase of the mechanical trail and a reduction of the lateral stiffness are required to improve stability at high forward velocity. Increasing the yaw stiffness will generally improve the shimmy stability of a gear with a positive mechanical trail provided that some structural damping is present. A limited reduction of the roll stiffness improves the stability at high forward velocities. The yaw moment of inertia should be kept as small as possible in case of a large positive mechanical trail'.
- *small negative trail* ($e < 0$):
  'For a landing gear with a negative trail an upper boundary exists for the yaw stiffness. The actual value of this stiffness may be quite low: for the baseline configuration, the yaw stiffness would have to be reduced almost by a factor 3 to obtain a stable configuration with a negative trail. A relatively small increase in track width relaxes this requirement considerably. If the absolute value of the negative trail equals the pneumatic trail, there exists no lower limit for the yaw stiffness to maintain stability. The effect of increasing the lateral stiffness is quite similar to increasing the track width and improves the shimmy stability in case of negative trail'.

In comparison with the baseline configuration, the proposed modifications result in a major gain in stability. This illustrates that it may be possible to design a stable conventional landing gear, provided that the combination of parameter values is selected correctly.

For detailed information also on the results of simulations with a multi-body complex model in comparison with full-scale experiments on a landing aircraft exhibiting shimmy we refer to the original work of Besselink (2000).

## 6.4. SHIMMY AND ENERGY FLOW

To sustain the unstable shimmy oscillation, energy must be transmitted 'from the road to the wheel'. We realize that, ultimately, the power can only originate from the vehicle's propulsion system. The relation between unstable modes and the self-excitation energy generated through the road to tire side force and aligning moment is discussed in detail in Subsection 6.4.1. The transition of driving energy (or the vehicle forward motion kinetic energy) to self-excitation energy is analyzed in Subsection 6.4.2. Obviously, besides, for the supply of self-excitation energy, the driving energy is needed to compensate for the energy that is dissipated through partial sliding in the contact patch.

### 6.4.1. Unstable Modes and the Energy Circle

To start out with the problem matter let us consider the simple trailing wheel system of Figure 6.1 with the yaw angle $\psi$ representing the only degree of freedom of the wheel plane. If the mechanical trail is equal to zero it is only the aligning torque that exerts work when the wheel plane is rotated about the steering axis. If the yaw angle varies harmonically

$$\psi(s) = \psi_o \sin(\omega_s s) \tag{6.63}$$

The moment response becomes for small path frequency $\omega_s$ approximately:

$$M_z'(s) = -C_{M\alpha}\psi_o \sin(\omega_s s + \phi) \tag{6.64}$$

where the phase angle follows, for instance, from transfer functions (5.110) or (5.113) or Table 5.1 (below Figure 5.27):

$$\phi = \phi_{-M'\psi} = -\sigma_{M'\psi}\omega_s = -(\sigma + a)\omega_s \tag{6.65}$$

The negative value of which corresponds to the fact that the moment lags behind the yaw angle. The energy transmitted during one cycle can be found as follows ($\theta = \omega_s s$):

$$\begin{aligned} W &= \int_{s=0}^{\lambda} M_z' \, d\psi = C_{M\alpha}\psi_o^2 \int_0^{2\pi} \cos(\theta + \phi) \sin\theta \, d\theta \\ &= -\pi C_{M\alpha}\psi_o^2 \sin\phi \quad (> 0) \end{aligned} \tag{6.66}$$

Apparently, because of the negative phase angle the work done by the aligning torque is positive which means that energy is fed into the system. As a result, the undamped wheel system will show unstable oscillations which is in agreement with our earlier findings. It is of interest to note that in contrast to the above result that holds for a tire, a rotary viscous damper (with a positive phase angle of 90°) shows a negative energy flow which means that energy is being dissipated.

If we would consider a finite caster length $e$, more self-excitation will arise due to the contribution of the side force that also responds to the yaw angle with phase lag. At the same time, however, a part of $F_y$ will now respond to the slip angle and, consequently, damps the oscillation. The slip angle arises as a result of the lateral slip speed which is equal to the yaw rate times the caster length. Finally, at $e = \sigma + a$, the moment and the side force are exactly in phase with the yaw angle (cf. situation depicted in Figure 5.6) and the stability boundary is reached. At $e = -t$ the moment about the steer axis vanishes if the simplified straight tangent or single contact point model is used. This also implies that the boundary of oscillatory instability is attained.

We will now turn to the much more complicated system with lateral compliance of the suspension included. The two components of the periodic oscillatory motion of the wheel that occurs on the boundary of (oscillatory) instability may be described as follows:

$$\bar{y}(t) = a_m \eta \sin(\theta + \xi) \tag{6.67}$$

$$\psi(t) = a_m \sin\theta \tag{6.68}$$

The quantity $\eta$ represents the amplitude ratio, $\xi$ indicates the phase lead of the lateral motion with respect to the yaw motion and $\theta$ equals $\omega t$ or $\omega_s s$. The energy $W$ that flows from the road to the wheel during one cycle is equal to the work executed by the side force and the aligning torque that act on the moving wheel plane. Hence, we have

$$W = \int_0^{2\pi} \left( F_y \frac{\mathrm{d}\bar{y}}{\mathrm{d}\theta} + M_z' \frac{\mathrm{d}\psi}{\mathrm{d}\theta} \right) \mathrm{d}\theta \tag{6.69}$$

when $W > 0$ energy flows into the system and if the system is considered to be undamped ($k = \kappa^* = 0$) the conclusion must be that the motion is unstable. Ultimately, the energy must originate from the power delivered by the vehicle propulsion system. How this transfer of energy is realized will be treated in the next section.

To calculate the work done, the force and moment are to be expressed in terms of the wheel motion variables (6.67, 6.68). For this, a tire model is needed and we will follow the theory developed by Besselink (2000) and take the straight tangent approximation of the stretched string model, which is governed by the Eqns (6.2, 6.4, 6.6, 6.7). For the periodic response of the transient slip angle $\alpha'$ to the input motion (6.67, 6.68) the following expression is obtained:

$$\begin{aligned} \alpha'(\theta) &= \frac{a_m}{1 + \sigma^2 \omega_s^2} \{ -(\sigma + a)\omega_s \cos\theta - \eta\omega_s \cos(\theta + \xi) \\ &\quad + (1 - a\sigma\omega_s^2) \sin\theta - \eta\sigma\omega_s^2 \sin(\theta + \xi) \} \end{aligned} \tag{6.70}$$

with the expressions (6.2, 6.4) for the force and the moment the integral (6.69) becomes

$$\begin{aligned} W &= \frac{\pi a_m^2 C_{F\alpha} \omega_s}{1 + \sigma^2 \omega_s^2} \Bigg\{ -\eta^2 + (\sigma + a)t - (\sigma + a - t)\eta \cos\xi \\ &\quad - \left( \frac{1}{\omega_s} - (a + t)\sigma\omega_s \right) \eta \sin\xi \Bigg\} \end{aligned} \tag{6.71}$$

On the boundary of stability $W = 0$ and for a given path frequency $\omega_s$ or wavelength $\lambda = 2\pi/\omega_s$ a relationship between $\eta$ and $\xi$ results from (6.71). To establish the functional relationship it is helpful to switch to Cartesian coordinates:

$$x_p = \eta \cos\xi, \quad y_p = \eta \sin\xi \tag{6.72}$$

The expression for the energy now reads as

$$\begin{aligned} W &= \frac{\pi a_m^2 C_{F\alpha} \omega_s}{1 + \sigma^2 \omega_s^2} \Bigg\{ x_p^2 + y_p^2 + (\sigma + a - t)x_p \\ &\quad + \left( \frac{1}{\omega_s} - (a + t)\sigma\omega_s \right) y_p - (\sigma + a)t \Bigg\} \end{aligned} \tag{6.73}$$

Apparently, for a given value of $W$ this represents the description of a circle. For $W = 0$ and after the introduction of the wavelength $\lambda$ as parameter the function takes the form

$$x_p^2 + y_p^2 + (\sigma + a - t)x_p + \left(\frac{\lambda}{2\pi} - \frac{2\pi}{\lambda}(a + t)\sigma\right)y_p - (\sigma + a)t = 0 \quad (6.74)$$

When the mode shape of the motion of the wheel plane defined by $\eta$ and $\xi$ or by $x_p$ and $y_p$ is such that (6.74) is satisfied, the motion finds itself on the boundary of stability. The center of Besselink's energy circle is located at

$$(x_{pc}, y_{pc}) = -\frac{1}{2}\left((\sigma + a - t), \left\{\frac{\lambda}{2\pi} - \frac{2\pi}{\lambda}(a + t)\sigma\right\}\right) \quad (6.75)$$

From (6.74) it can be ascertained that independent of $\lambda$ the circles will always pass through the two points on the $x_p$ axis: $x_p = -(\sigma + a)$ and $x_p = t$. We have now sufficient information to construct the circles with $\lambda$ as parameter. In Figure 6.20, these circles have been depicted together with circles that arise when different tire models are used. As expected, the circles for the more exact tire models deviate more from those resulting from the straight tangent tire model when the wavelength becomes smaller. These circles that arise also for

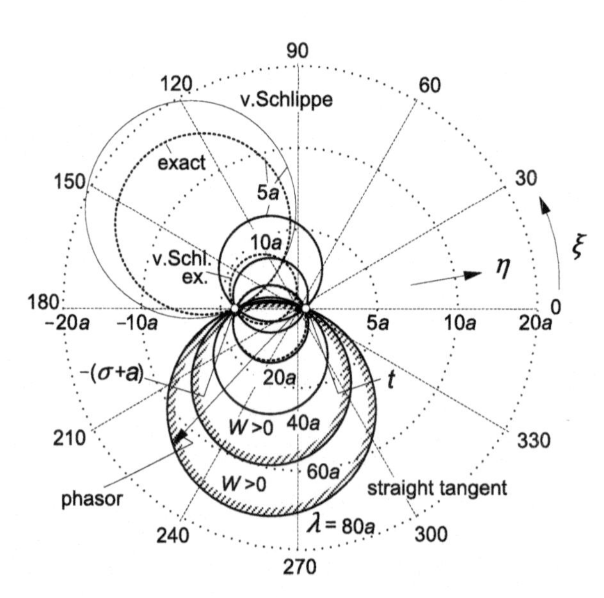


**FIGURE 6.20** Mode shape plot (amplitude ratio $\eta$, phase $\xi$) with Besselink's zero energy circles for different wavelengths and tire model approximations, indicating mode shapes for which shimmy begins to show up (instability occurs if at a particular value of the motion wavelength $\lambda$ the phasor end point gets inside the corresponding circle). At smaller wavelengths the circle belonging to the straight tangent approximation begins to appreciably deviate from the exact or Von Schlippe representation.

other massless models of the tire can be easily proven when considering the fact that at a given wavelength we have a certain amplitude and phase relationship of the force $F_y$ and the moment $M'_z$ with respect to the two input motion variables $\bar{y}$ and $\psi$. That the circles pass through the point $(-\sigma - a, 0)$ follows from the earlier finding that the response of all 'thin' tire models when the wheel is swiveled around a vertical axis that is positioned a distance $\sigma + a$ in front of the wheel axle (cf. Figure 5.6) are equal and correspond to the steady-state response.

When parameters of the system are changed and the mode shape is changed so that the end point of the phasor with length $\eta$ and phase angle $\xi$ moves from outside the circle to the inside, the system becomes oscillatory unstable. It becomes clear that depending on the wavelength a whole range of possible mode shapes of the wheel plane motion is susceptible to forming unstable shimmy oscillations.

As an example, the situation that arises with the cases A and B of Figures 6.17–6.19 has been depicted in Figure 6.21. In case A mode 1 is on the verge of becoming unstable. The mode shape corresponds to the motion indicated in Figure 5.6 and remains unchanged when the speed is varied. The circles, however, change in size and position when the speed is changed which correspond to a change in wavelength. Mode 2 appears to remain outside the circles indicating that this particular mode is stable. In case B mode 1 is stable and mode 2 sits on the boundary of stability. This is demonstrated in the right-hand diagram where the end point of the phasor of mode 2 remains located on the (changing) circles when the speed is changed.

As a second example we apply this theory to the configuration with $f=0$ with the gyroscopic coupling included. In Figure 6.8 the cases A and B have been indicated at a trail $e=0$. Because of the gyroscopic action the condition

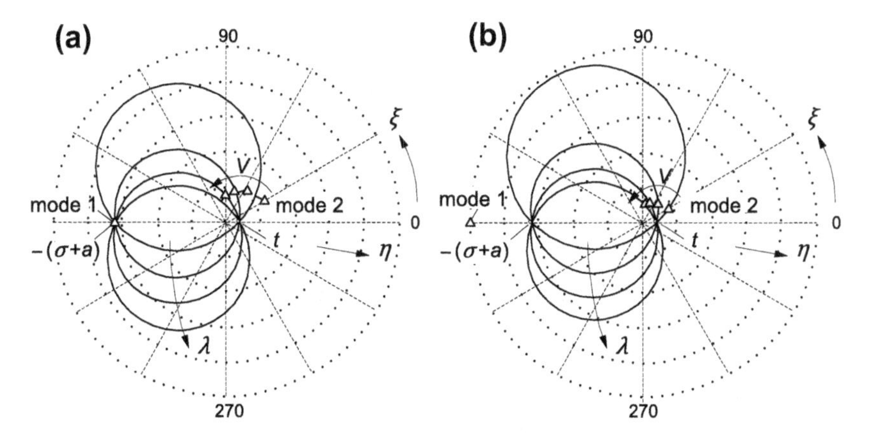


**FIGURE 6.21** Zero energy circles for the cases A and B of Figure 6.17, indicating the mode shape of mode 1 at the boundary of stability in case A and the same for mode 2 in case B.

for stability is speed dependent, although damping is assumed equal to zero. This makes it possible to actually see that the zero energy circle is penetrated or exited when the speed is increased.

In Figures 6.22 and 6.23 the phasors (here considered as complex quantities) have been indicated for a number of values of the nondimensional speed **V**. The nondimensional wavelength and eigenvalue of the mode considered change with speed as have been indicated in the respective diagrams. Cases A and B (with respectively mode 1 and 2 on the boundary of stability) have been marked. When the speed is increased from 6 to 14, two boundaries of stability (A and B) are passed in between which according to Figure 6.8 a stable range of speed exists. In Figure 6.22 we see that in accordance with this the phasor of mode 1 first lies inside the circle (belonging to $V = 6$ with frequency $\omega = 0.66$ and corresponding wavelength $\lambda = 57a$) then crosses the circle at $V = 9.05$ (A) and finally ends up at a speed $V = 14$ outside the circle meaning that this mode is now stable. At the same time, following mode 2 in Figure 6.23, we observe that first the mode point lies outside the circle at $V = 6$ (now for the different frequency $\omega = 2.99$ and wavelength $\lambda = 12.6a$), crosses the circle at $V = 11.2$ (B) and moves to the inside of the circle into the unstable domain. From the eigenvalues $s$ the change in damping and frequency can be observed. The mode shapes are quite different for the two modes. Mode 1 shows a relatively small lateral displacement $\bar{y}$ with respect to the yaw angle multiplied with half the contact length $a\psi$. At first $\bar{y}$ slightly lags behind $\psi$ which turns into a lead at higher speeds. Mode 2 shows a higher frequency and smaller wavelength while $\bar{y}$ is relatively large and lags behind $\psi$.

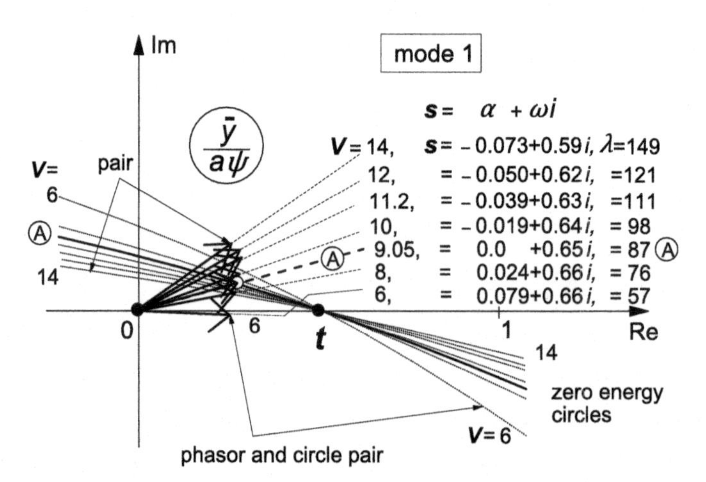


**FIGURE 6.22** System of Figure 6.8 with gyroscopic coupling making stability speed dependent. Circle segments and phasors of mode shapes of mode 1 for a series of speed values covering the range in which points A and B of Figure 6.8 are located. At increasing speed the phasor crosses the circle at $V = 9.05$ into the stable domain.

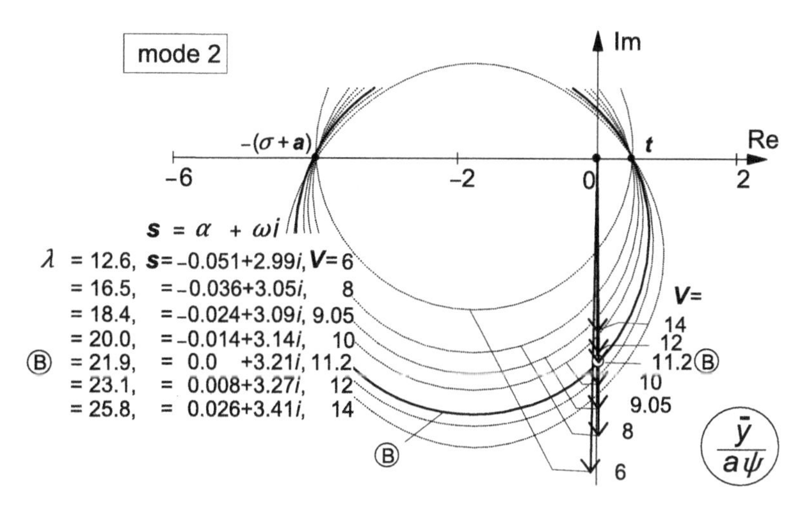


FIGURE 6.23 Circles and phasors of mode shapes of mode 2 for same series of speed values as in Figure 6.22. At increasing speed, the phasor crosses the circle at $V = 11.2$ toward its inside which means: entering the unstable domain.

---

**Exercise 6.2. Zero Energy Circle Applied to the Simple Trailing Wheel System**

Consider the third-order system of Figure 6.1 described by the Eqns (6.1–6.9) but without damping: $\boldsymbol{k} = \kappa^* = 0$. Establish the reduced expression for the work $W$ from the general formula (6.69). Show that the stability condition $H_2 > 0$ corresponds to $W < 0$ which means that the mode shape point lies outside the zero energy circle.

---

## 6.4.2. Transformation of Forward Motion Energy into Shimmy Energy

The only source that is available to sustain the self-excited shimmy oscillations is the vehicle propulsion unit. The tire side force has a longitudinal component that when integrated over one shimmy cycle must be responsible for an average drag force that is balanced by (a part of) the vehicle driving force.

In the previous section, we have seen that the work done by the side force and the aligning torque is fed into the system to generate the shimmy oscillation. The link that apparently exists between self-excitation energy and driving energy will be examined in the present section. For this analysis, dynamic effects may be left out of consideration. The wheel is considered to roll freely (without braking or driving torques).

During a short span of traveled distance d$s$ of the wheel center, the wheel plane moves sideways over a distance d$\bar{y}$ and rotates about the vertical axis over the increment of the yaw angle d$\psi$. At the same time, the tire lateral deflection is changed and, when sliding in the contact patch is considered, some elements will slide over the ground. Consequently, we expect that the supplied driving energy d$E$ is equal to the sum of the changes of (self-)excitation energy d$W$, tire potential energy d$U$ and tire dissipation energy d$D$.

In Figure 6.24 the tire is shown in a deflected situation. We have the pulling force $F_d$ acting in forward ($\bar{x}$) direction from vehicle to the wheel axle and furthermore the lateral force $F_e$ and yaw moment $M_e$ from rim to tire. These forces and moment are in equilibrium with the side force $F_y$ and aligning torque $M_z$ which act from road to tire. A linear situation is considered with small angles and a vanishing length of the sliding zone. For the sake of simplicity, we consider an approximation of the contact line according to the straight tangent concept.

Energy is lost due to dissipation in the sliding zone at the trailing edge of the contact zone. To calculate the dissipated energy, we need to know the frictional force and the sliding velocity distance. For the bare string model, we have the concentrated force $F_g$ that acts at the rear edge and that is needed to maintain the kink in the deformation of the string. This force is equal to the difference of the slopes of the string deflection just behind and in front of the rear edge times the string tension force $S = c_c\sigma^2$. For the straight tangent string deformation with transient slip angle or deformation gradient $\alpha'$, the slope difference is equal to $\varphi_g = 2(\sigma + a)\ \alpha'/s$. The force becomes: $F_g = 2c_c\sigma(\sigma + a)\ \alpha'$. The average sliding speed $V_g$ over the vanishing sliding range is equal to half the difference in slope of the string in front and just passed the kink times the forward speed $V$. This multiplied with the time

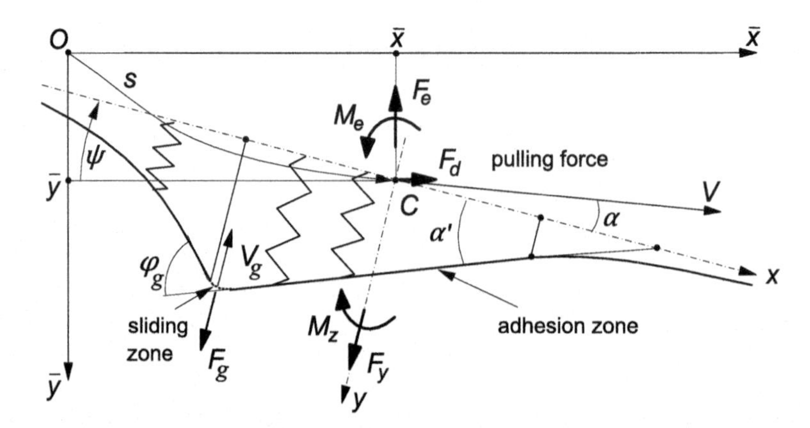


**FIGURE 6.24** On the balance of driving, self-excitation, and dissipation energy.

needed to travel the distance d$s$ gives the sliding distance $s_g$. We find for the associated dissipation energy:

$$dD = F_g s_g = \frac{1}{2} F_g \varphi_g ds = 2c_c \sigma(\sigma + a)\alpha' \frac{\sigma + a}{\sigma} \alpha' ds = C_{F\alpha} \alpha'^2 ds = F_y \alpha' ds \tag{6.76}$$

The same result appears to hold for the brush model. When considering the theory of Chapter 3, we find for small slip angle $\alpha'$ a length of the sliding zone at the trailing edge: $2a\theta\ \alpha'$ and a slope of the contact line in this short sliding zone with respect to the wheel plane $\beta_g = 1/\theta$. The average sliding speed becomes: $V_g = V(\beta_g + \alpha')$ which makes the sliding distance equal to $s_g = (\beta_g + \alpha')ds$. The average side force acting in the sliding range appears to be: $F_g = C_F\alpha\theta\ \alpha'^2$. This results in the dissipated energy (considering that $\beta_g$ is of finite magnitude and thus much larger than $\alpha'$):

$$dD = F_g s_g = C_{F\alpha} \theta \alpha'^2 \beta_g ds = C_{F\alpha} \alpha'^2 ds = F_y \alpha' ds \tag{6.77}$$

which obviously is the same expression as found for the string model, Eqn (6.76).

At steady-state side slipping conditions with constant yaw angle $\psi = \alpha' = \alpha$ we obviously have a drag force $F_d = F_y \psi$. The dissipation energy after a distance traveled $s$ equals $D = F_y^2 / C_{F\alpha} s = F_y \alpha' s = F_y \psi s = F_d s$ which is a result that was to be expected.

The increment of the work $W$ done by the side force and aligning torque acting on the sideways moving and yawing wheel plane (considered in the previous subsection) is

$$dW = F_e d\bar{y} + M_e d\psi = \left( F_y \frac{d\bar{y}}{ds} + M_z \frac{d\psi}{ds} \right) ds \tag{6.78}$$

The change in tire deflection (note that the tire almost completely adheres to the ground) arises through successively moving the wheel plane sideways over the distances d $\bar{y}$ and $\psi$ d$s$ and rotating over the yaw angle d$\psi$ and finally rolling forward in the direction of the wheel plane over the distance d$s$. These contributions correspond with the successive four terms in the expression for the change in potential energy:

$$dU = F_y \left( -\frac{d\bar{y}}{ds} + \psi \right) ds - M_z \frac{d\psi}{ds} ds - F_y \alpha' ds \tag{6.79}$$

The increase in driving energy is

$$dE = F_y \psi\, ds \tag{6.80}$$

In total we must have the balance of energies:

$$dE = dW + dU + dD \tag{6.81}$$

which after inspection indeed appears to be satisfied considering the expressions of the three energy components (6.76–6.79).

Integration over one cycle of the periodic oscillation will show that $U = 0$ so that for the energy consumption over one cycle remains:

$$E = W + D \tag{6.82}$$

At steady-state side slip with slip angle $\alpha = \psi$ the excitation energy $W = 0$ and the driving energy becomes

$$E = D = \frac{F_y^2}{C_{F\alpha}} s = F_y \psi \, s = F_d \, s \tag{6.83}$$

The result presented through Eqn (6.81) demonstrates that the propulsion energy is partly used to compensate the energy lost by dissipation in the contact patch and partly to provide the energy to sustain the unstable shimmy oscillations.

## 6.5. NONLINEAR SHIMMY OSCILLATIONS

Nonlinear shimmy behavior may be investigated by using analytical and computer simulation methods. The present section first gives a brief description of the analytical method employed by Pacejka (1966) that is based on the theory of the harmonic balance of Krylov and Bogoljubov (1947). The procedure that is given by Magnus (1955) permits a relatively simple treatment of the oscillatory behavior of weakly nonlinear systems. Further on in the section, results obtained through computer simulation will be discussed.

Due to the degressive nature of the tire force and moment characteristics, the system which was found to be oscillatory unstable near the undisturbed rectilinear motion (result of linear analysis) will increase in amplitude until a periodic motion is approached which is designated as the limit cycle. The maximum value of the steer angle reached during this periodic motion is referred to as the limit amplitude. For weakly nonlinear systems this oscillation can be approximated by a harmonic motion. At the limit cycle, a balance is reached between the self-excitation energy and the dissipated energy.

In the analytical procedure, an equivalent linear set of equations is established in which the coefficient of the term that replaces the original nonlinear term is a function of the amplitude and the frequency of the oscillation. For example, the moment exerted by the tire side force and the aligning torque about the steering axis, $f^*(\alpha') = F_y e - M_z$, may be replaced by the equivalent total aligning stiffness multiplied with the transient slip angle: $C^*\alpha'$. If dry friction is considered in the steering system, the frictional couple $m^*(\dot{\psi}) = K \,\mathrm{sgn}(\dot{\psi})$ may be replaced by the equivalent viscous damping couple $k^*\dot{\psi}$. The equivalent coefficients $C^*$ and $k^*$ are functions of the amplitudes of $\alpha'$ and $\dot{\psi}$ (or of $\omega\psi$), respectively. The functions are determined by considering a harmonic variation of $\alpha'$ and $\psi$ and then taking the first harmonic of the

Fourier series of the periodic response of the corresponding original nonlinear terms.

We have with $\alpha' = \alpha_o' \sin \tau$ and $\psi = \psi_o \sin \theta$:

$$C^* = \frac{1}{\pi \alpha_o} \int_0^{2\pi} f^*(\alpha') \sin\tau \, d\tau \tag{6.84}$$

and

$$k^* = \frac{1}{\pi \omega \psi_o} \int_0^{2\pi} m^*(\dot{\psi}) \cos\theta \, d\theta = \frac{4K}{\pi \omega \psi_o} \tag{6.85}$$

As expected, the function of the equivalent total aligning stiffness $C^*(\alpha_0')$ starts for vanishing amplitude at the value $C_{F\alpha} e + C_{M\alpha}$ at a slope equal to zero and then gradually decays to zero with the amplitude $\alpha_o'$ tending to infinity. The equivalent damping coefficient $k^*(\omega\psi_o)$ (6.85) appears to vary inversely proportionally with the steer angle amplitude, which means that it begins at very large damping levels and decreases sharply with increasing swivel amplitude according to a hyperbola. If clearance in the wheel bearing about the steer axis would be considered (in series with the steer damper) the matter becomes more complex and an equivalent steer stiffness $c^*$ must be introduced. If the total angle of play is denoted by $2\delta$ the equivalent coefficients $k^*$ and $c^*$ are found to be expressed by the functions

$$k^* = \frac{4K}{\pi \omega \psi_o} \left(1 - \frac{\delta}{\psi_o}\right) \quad (\psi_o > \delta) \tag{6.86}$$

and

$$c^* = \frac{4K}{\pi \psi_o} \sqrt{\frac{\delta}{\psi_o} \left(1 - \frac{\delta}{\psi_o}\right)} \quad (\psi_o > \delta) \tag{6.87}$$

If $\psi_o < \delta$ the oscillation takes place inside the free space of the clearance and the equivalent damping and stiffness become equal to zero. Obviously, the clearance alleviates the initial strong damping effect of the dry friction.

If not all nonlinearities are considered, we may, e.g., replace $k^*$ by $k$ and $c^*$ by zero or $c_\psi$.

For the third-order system of Section 6.2, the nonlinear version of Eqn (6.11) is replaced by equivalent linear differential equations containing the coefficients $C^*$, $c^*$, and $k^*$. The damping due to tread width is kept linear. The characteristic equation, which is similar to (6.13), becomes

$$\begin{aligned} &\boldsymbol{\sigma V^2 p^3 + \{V^2 + \sigma(k^* V + \kappa^*)\}p^2} \\ &\quad \boldsymbol{+ \{k^* V + \kappa^* + \sigma c^* + C^*(e - a)\}p + C^* + c^*} = 0 \end{aligned} \tag{6.88}$$

with nondimensional quantities according to (6.10): $C^*$, $c^*$, and $k^*$ being treated like $C_M\alpha$, $c\delta$, and $k$, respectively. When the shimmy motion has been fully developed and the limit cycle is attained, the solution of the equivalent linear system represents a harmonic oscillation with path frequency $\omega_s$ and amplitudes $\alpha'_o$ and $\psi_o$. The amplitudes are obtained by using the condition that at this sustained oscillation the Hurwitz determinant $H_{n-1} = 0$. Consequently, we have for our third-order system at the limit cycle: $H_2 = 0$. Hence, the equation that is essential for finding the magnitude of the limit cycle reads as

$$\{V^2 + \sigma(k^* V + \kappa^*)\}\{k^* V + \kappa^* + \sigma c^* + C^*(e - a)\} = \sigma V^2 (C^* + c^*) \tag{6.89}$$

The frequency of the periodic oscillation that occurs when the relation (6.89) is satisfied can easily be found when in (6.88) $p$ is replaced by $i\omega_s$. We obtain

$$\omega_s^2 = \frac{C^* + c^*}{V^2 + \sigma(k^* V + \kappa^*)} \tag{6.90}$$

Moreover, we need the ratio of the amplitudes $\alpha'_o$ and $\psi_o$. From the second equation of (6.11), which is not changed in the linearization process, we get with $\alpha' = v_1/\sigma$

$$\frac{\alpha'_o}{\psi_o} = \sqrt{\frac{1 + (a - e)^2 \omega_s^2}{1 + \sigma^2 \omega_s^2}} \tag{6.91}$$

Equations (6.89–6.91) provide sufficient information to compute the amplitude and the frequency of the limit cycle.

The stability of the limit cycle of the weakly nonlinear system may be assessed by taking the derivative of the Hurwitz determinant with respect to the amplitude. If $dH_2/d\psi_o > 0$, the limit cycle is stable and attracts the trajectories; if negative, the limit cycle is unstable and the oscillations deviate more and more from this periodic solution. In the original publication, more information can be found on the analytic assessment of the solutions.

For the four different cases investigated, Figure 6.25 shows the basic motion properties. Figure 6.26 gives for the three nonlinear cases the variation of the limit amplitudes with damping parameter. The left-hand diagram shows that, with just nonlinear tire behavior, in the range of the damping coefficient below the critical value as obtained in the linear analysis, the motion is unstable and the shimmy oscillation develops after a minute disturbance. The degressive shape of the tire force and moment characteristics causes an increase in self-excitation energy that is less than the increase of the dissipation energy from the viscous damper. When the limit amplitude is reached, the two energies become equal to each other. If through some external disturbance the amplitude has become larger than the limit amplitude, the dissipation energy exceeds the self-excitation energy and the oscillation reduces in amplitude until the limit cycle is reached again but now from the other side.

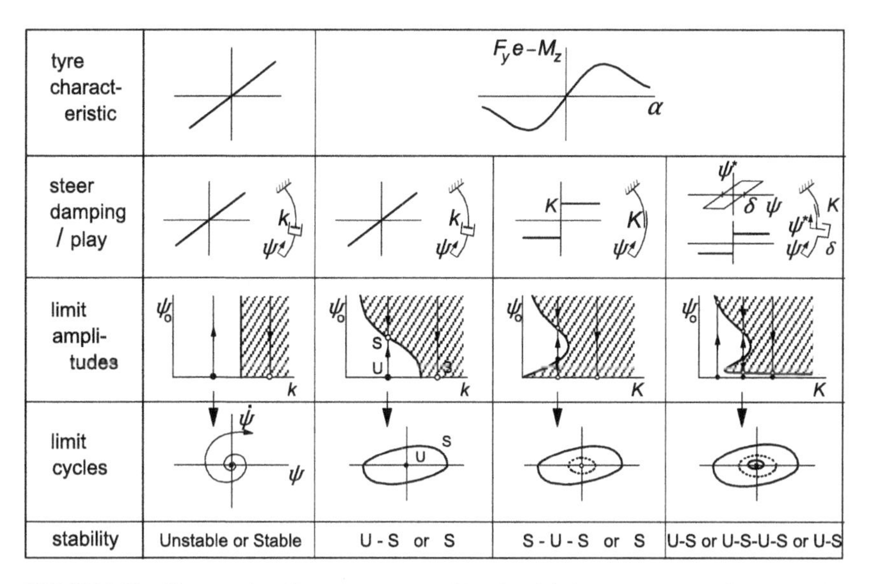


**FIGURE 6.25** Linear and nonlinear systems considered and their autonomous motion properties in terms of stability and possible limit cycles.

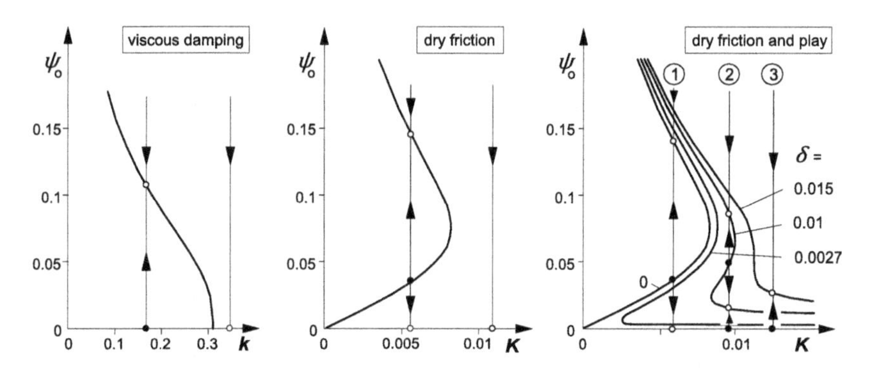


**FIGURE 6.26** Limit amplitudes as a function of damping for increasing number of nonlinear elements: nonlinear tire, dry friction, and play ($V = 6.66$, $\sigma = 3$, $e = 0$, $\kappa^* = 0$).

If dry friction is considered instead of the viscous damping, we observe that the center position is stable. In fact, the system may find its rest position away from the center (at a small steer angle) if the dry frictional torque is sufficiently large. We now need a finite external disturbance (running over an asymmetric obstacle) to overcome the dry friction. If that has happened, the motion develops itself further and the stable limit cycle is reached. If the initial conditions are chosen correctly we may spend a while near the unstable limit cycle before either the rest position or the large stable limit cycle is approached.

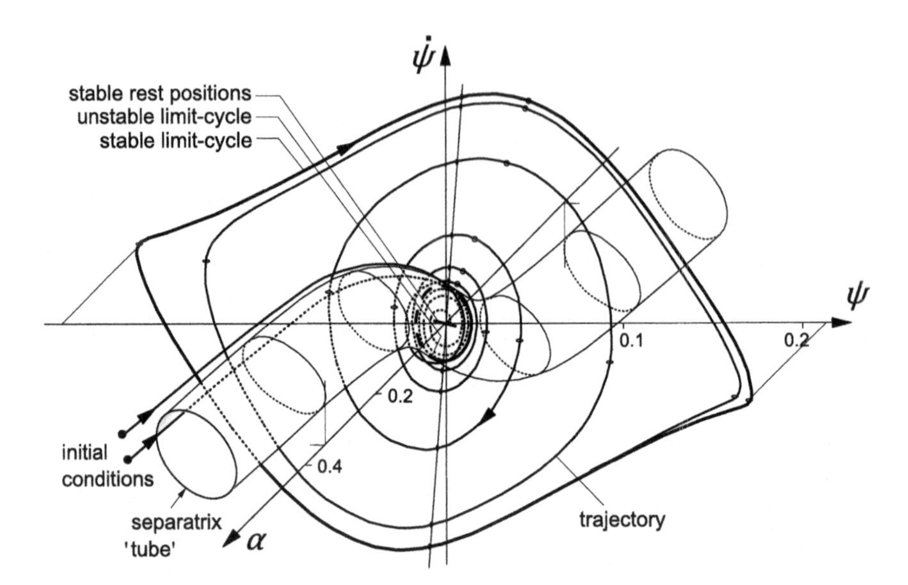


**FIGURE 6.27** Solutions in the three-dimensional state space. System with nonlinear tire and dry friction ($V = 6.66$, $\sigma = 3$, $e = 0$, $\kappa^* = k = \delta = 0$, $K = 0.0035$). Line piece for possible rest positions, unstable and stable limit cycle. Tube-shaped separatrix separating space of trajectories leading to a rest position from the space of solution curves leading to the stable limit cycle.

Figure 6.27 shows for this case the solutions in the three-dimensional state space. The unstable limit cycle appears to lie on a tube-shaped surface which is here the separatrix in the solution space. We see that when the initial conditions are taken outside the tube, the stable limit cycle is reached. When the starting condition is inside the tube, one of the indicated possible rest positions will be ultimately attained.

The introduction of play about the steering axis, with $\delta$ being half the clearance space, appears indeed to be able to relax the action of the dry friction. For small play and enough dry friction (case 2 in Figure 6.26) the small stable limit cycle is reached automatically. An additional disturbance may cause the motion to get over the 'nose' and reach the large stable limit cycle. The plot of Figure 6.28 shows for this case the three limit cycles and trajectories as projected on the $(\dot{\psi}, \psi)$ plane. If the play is larger or the damping less, the large limit cycle is reached without an external disturbance.

It is of interest to see how the limit amplitudes $\psi_o$ vary with the speed of travel $V$. Figure 6.29 depicts two cases: viscous damping and dry friction. The area inside which the amplitude increases due to instability may be designated as the area of self-excitation. The diagrams show the courses of the limit amplitude for the two configurations with linear damping ($K = \delta = 0$) with unstable ranges of speed indicated in Figure 6.3.

The right-hand diagram also gives the (smaller) area of self-excitation when dry friction is added. When the damping is linear, shimmy arises when the

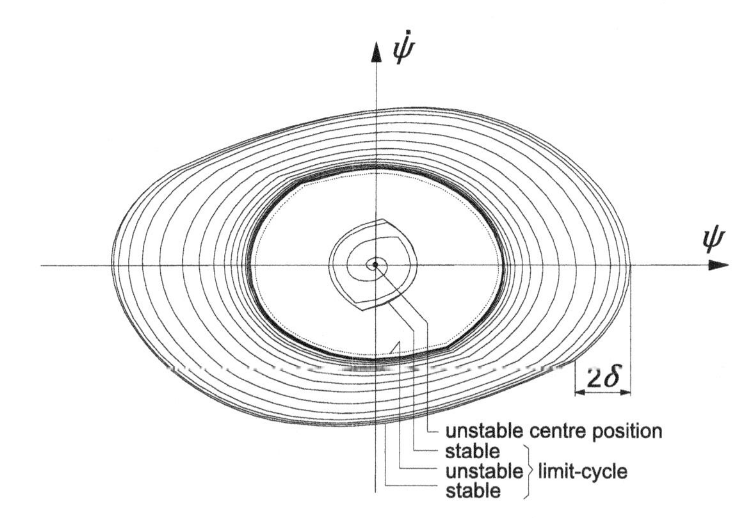


**FIGURE 6.28** Limit cycles and trajectories for the system with nonlinear tire, dry friction, and play in the wheel bearings (Case 2 of Figure 6.26).

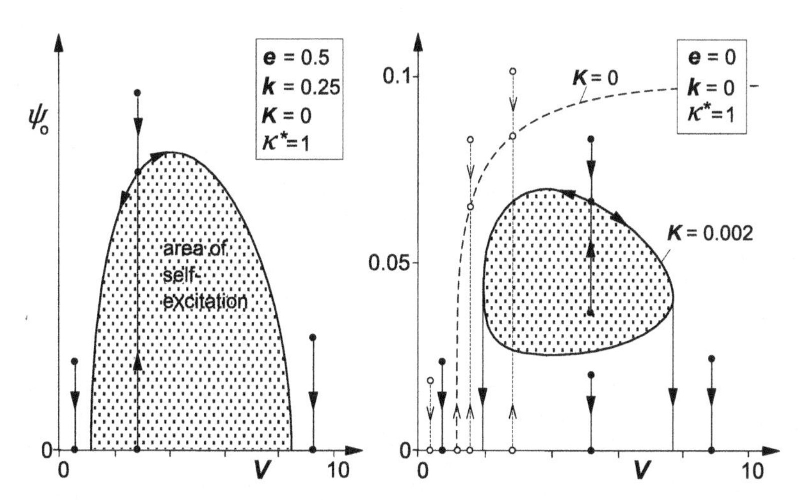


**FIGURE 6.29** Boundary of area of self-excitation representing the course of limit amplitudes of steer angle $\psi$ as a function of speed for system with nonlinear tire and viscous damping (left and right) and dry friction (right) ($\sigma = 3$, cf. Figure 6.3).

critical speed is exceeded. The amplitude grows as the vehicle speeds up. In the left-hand diagram, a maximum is reached beyond which the amplitude decreases and finally, at the higher boundary of stability, the oscillation dies out. With dry friction, the stable limit cycle cannot be reached automatically. A sufficiently strong external disturbance may get the shimmy started.

Another way of initiating the shimmy in the case of dry friction may be the application of wheel unbalance. Beyond a certain speed, the imposed unbalance

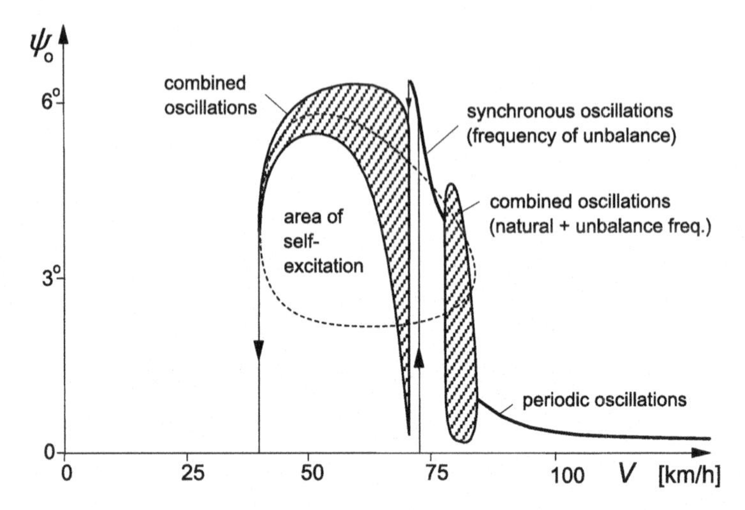


**FIGURE 6.30** The response of the tenth-order system, representing a light truck with nonlinear tire and dry friction, to wheel unbalance moment. The area of self-excitation of the autonomous system has been indicated. At nearly 75 km/h violent shimmy develops. Then a range of speed with the so-called synchronous oscillations occurs. When the system is detuned too much, combined oscillations (beats) show up.

couple may have become large enough to overcome the dry friction. Then, when the forcing frequency is not too much apart from the shimmy natural frequency, the amplitude rises quickly and a state of synchronous motion may arise as depicted in Figure 6.30. In that state, the system oscillates with a single frequency which corresponds to the wheel speed of revolution. When from that point onwards the vehicle speed is increased or decreased, the synchronous oscillation may persist until the difference between the two frequencies becomes too large (in other words, until the difference between free shimmy wavelength and wheel circumference is too great, that is, the system is detuned too much). Then, the unbalance torque is no longer able to drag the free shimmy motion along. The picture of the oscillation is now changed considerably. We have a motion with a beat character that consists of oscillations with two frequencies: one is the unbalance forcing frequency and the other will be close to the free shimmy frequency at the current speed. The shaded areas shown in the diagram indicate the speed ranges where these combination vibrations show up. The upper and lower boundaries of these areas represent the limits in between which the amplitude of the motion varies. When at decreasing speed the point at the vertical tangent to the area of self-excitation is reached, the shimmy oscillation disappears. At increasing speed, the combination oscillations may pass to a forced vibration with a single frequency. This occurs when the degree of self-excitation has become too low. A similar phenomenon of synchronous motions and combined oscillations has been treated by Stoker (1950, p.166). He uses an approximate analytical method to investigate the second-order nonlinear system of Van der Pol that is provided with a forcing member.

The diagram of Figure 6.30 represents the result of a computer simulation study with a relatively complex model of the 10th order. The model is developed to investigate the violent shimmy vibration generated by a light military truck equipped with independent trailing arm front-wheel suspensions. For details, we refer to the original publication, Pacejka (1966). The model features degrees of freedom represented by the following motion variables: lateral displacement and roll of the chassis, lateral and camber deflection of the suspension, steer angle of the front wheel (same left/right), rotation of the steering wheel, and lateral tire deflection. The degree of freedom of the steering wheel has been suppressed in the depicted case by clamping the steering system in the node that appears to occur in the free motion with the front wheels and the steering wheel moving in counter phase. The other lower frequency mode with front wheels and steering wheel moving in phase occurs at lower values of speed and partly overlaps the range of the counter-phase mode. It is expected that the in-phase mode is easily suppressed by loosely holding the steering wheel. This was more or less confirmed by experiments on a small mechanical wheel suspension/steering system model. The full-scale truck only showed shimmy with counter oscillating front wheels and steering wheel.

Finally, Figure 6.31 presents the results of measurements performed on the truck moving over a landing strip. The front wheels were provided with unbalance weights. Shimmy appeared to start at a speed of ca. 75 km/h. Synchronous oscillations were seen to occur as can be concluded by

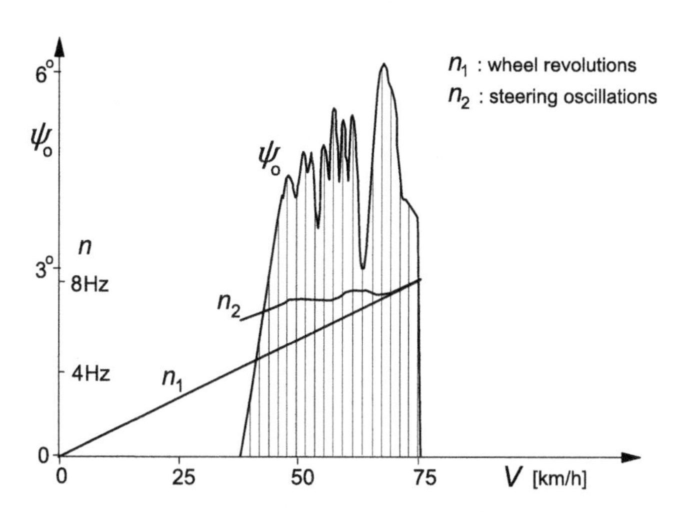


**FIGURE 6.31** Front wheel steer oscillation amplitude as a response to wheel unbalance as measured on a light truck. The speed is gradually reduced after violent shimmy was developed at ca. 75 km/h. After a short range of speed exhibiting synchronous oscillations (single frequency), combined vibrations occur with two distinct frequencies.

considering the frequencies that appear to coincide in the small range of speed just below the speed of initiation of 75 km/h. Further downwards, the frequencies get separated and follow independent courses. This strengthens the impression that here the motion may be able to sustain itself. Afterwards, another test was conducted with only one wheel provided with an unbalance weight. The weight was attached to a cable that made it possible to remove the unbalance during the test run. After having the unbalance detached at the instant that the shimmy was fully developed, the shimmy remained to manifest itself with about the same intensity. This constituted the proof that we dealt with a self-sustained oscillation. The correspondence between the diagrams of Figures 6.30 and 6.31 is striking. Also, the frequency of the autonomous vibration of the model (along the upper boundary of the area of self-excitation shown in Figure 6.30) was close to that according to the test results. The model frequency appears to vary from 6.1 Hz at the low end of the speed (40–45 km/h) to 7 Hz at the initiation velocity of 70–75 km/h.

Chapter 7

# Single-Contact-Point Transient Tire Models

**Chapter Outline**



## 7.1. INTRODUCTION

For relatively low-frequency and large-wavelength transient and oscillatory vehicle motions, tire inertia and the effect of the finite length of the contact patch may be neglected or taken care of in an approximative manner. In Chapter 5 a thorough treatment has been given of the out-of-plane stretched string tire model together with a number of approximate models. One of these models did ignore the contact length. For the aligning torque, the effect of tread width and the gyroscopic couple was introduced.

The present chapter deals with the further development of this type of model which in its simplest form has been and still is very popular in vehicle dynamics studies. Both in-plane and out-of-plane models will be discussed for small-slip, linear and for large-slip, nonlinear conditions. The concept of the relaxation length is central in the model structure. The development of the single-point-contact models follows an essentially different and much simpler line compared with the theoretical approach on which the string model is based. Because of its simplicity, it is possible to enhance the model to cover the full nonlinear combined slip range including rolling from standstill or even change direction from forward to backward rolling. Camber and turn slip may be included.

**Tire and Vehicle Dynamics. DOI: 10.1016/B978-0-08-097016-5.00007-3**

With the linear and nonlinear models, to be developed in this chapter, various vehicle dynamics problems may be approximately analyzed such as the shimmy phenomenon (cf. curve for single-point model in Figure 6.2), transient vehicle motions with oscillatory steer inputs, motion over undulated road surfaces at side slip and camber, steering vibrations induced by wheel imbalance, and tire out-of-roundness. In these studies, the effect of tire lag may be ascertained. In a number of applications to be treated in Chapter 8, we will address these problems (except wheel shimmy, which was the subject of Chapter 6).

## 7.2. MODEL DEVELOPMENT

The model consists of a contact patch (point) that is suspended with respect to the wheel rim by a longitudinal (circumferential) and a lateral spring. These springs represent the compliance of the tire carcass. The contact point may move (slip) with respect to the ground in lateral and longitudinal directions. Through this relative motion the side force, longitudinal force, and aligning torque are generated. To determine these forces and moments, contact patch lateral slip (possibly including ply-steer) and longitudinal slip are defined. In addition, contact line curvature, due to camber possibly including conicity and due to turn slip, is assumed to be detected. These contact patch slip quantities may then be used as input in the steady-state tire slip model, e.g. the *Magic Formula*, to calculate the transient force and moment variation that act upon the contact patch.

### 7.2.1. Linear Model

Figure 7.1 depicts the model in top view. At the instant considered, the wheel slip point $S$ (attached to the wheel rim at a level near the road surface) and the contact point $S'$ are defined to be located in the plane through the wheel axis and normal to the road. These points (which may be thought to lie on two parallel slip circles) move over the road surface with the wheel and contact patch slip velocities,

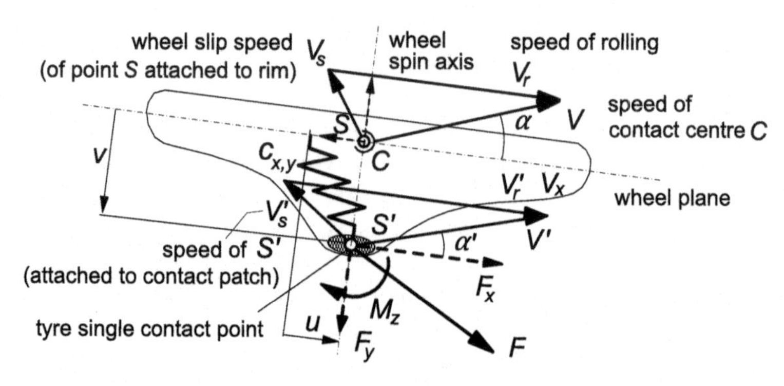


**FIGURE 7.1** Single-contact-point tire model showing lateral and longitudinal carcass deflections, $u$ and $v$ (top view).

respectively. In the figure, the $x$ and $y$ components of the slip velocities have been shown as negative quantities. The difference of the velocities of the two points causes the carcass springs to deflect. Consequently, the time rates of change of the longitudinal and lateral deflections $u$ and $v$ read

$$\frac{\mathrm{d}u}{\mathrm{d}t} = -(V_{sx} - V'_{sx}) \tag{7.1}$$

and

$$\frac{\mathrm{d}v}{\mathrm{d}t} = -(V_{sy} - V'_{sy}) \tag{7.2}$$

If we assume small values of slip, we may write, for the side force acting from road to contact patch with $C_{F\alpha}$ denoting the cornering or side slip stiffness,

$$F_y = C_{F\alpha}\alpha' = -C_{F\alpha}\frac{V'_{sy}}{|V_x|} \tag{7.3}$$

It is assumed here that the difference between the wheel center longitudinal velocity $V_x$ and the longitudinal velocity $V_{cx}$ of the contact center is negligible:

$$V_{cx} \approx V_x \tag{7.4}$$

Consequently, we may employ $V_x$ in the present chapter. With the lateral tire stiffness at road level $C_{Fy}$, we have, for the elastic internal force that balances the side slip force,

$$F_y = C_{Fy}v \tag{7.5}$$

and we can write, for Eqn (7.2) with (7.3, 7.5) after having introduced the relaxation length for side slip $\sigma_\alpha$,

$$\sigma_\alpha = \frac{C_{F\alpha}}{C_{Fy}} \tag{7.6}$$

the differential equation for the lateral deflection due to side slip $v_\alpha$ (later on we will also have a lateral deflection due to camber):

$$\frac{\mathrm{d}v_\alpha}{\mathrm{d}t} + \frac{1}{\sigma_\alpha}|V_x|v_\alpha = |V_x|\alpha = -V_{sy} \tag{7.7}$$

where $\alpha$ is the wheel slip angle: $\alpha \approx -V_{sy}/|V_x|$. The side force is obtained by multiplying $v_\alpha$ with $C_{Fy}$.

In a similar way, we can deal with the longitudinal force response. With the longitudinal tire stiffness $C_{Fx}$ at road level and the longitudinal slip stiffness $C_{F\kappa}$, we obtain, for the relaxation length $\sigma_\kappa$,

$$\sigma_\kappa = \frac{C_{F\kappa}}{C_{Fx}} \tag{7.8}$$

and we can derive with Eqn (7.1) the differential equation for the fore and aft deflection $u$:

$$\frac{du}{dt} + \frac{1}{\sigma_\kappa}|V_x|u = |V_x|\kappa = -V_{sx} \tag{7.9}$$

with $\kappa$ the longitudinal wheel slip ratio: $\kappa \approx -V_{sx}/|V_x|$. The longitudinal force may be obtained by multiplying $u$ with the stiffness $C_{Fx}$.

Next, we consider wheel camber as input. For a suddenly applied camber angle $\gamma$ (about the line of intersection at ground level!), we assume that a contact line curvature and thus the camber thrust $C_{F\gamma}\gamma$ are immediately felt at the contact patch. As a reaction, a contact patch side-slip angle $\alpha'$ is developed that builds up the lateral carcass deflection $v = v_\gamma$. Again Eqn (7.5) applies. The side force that acts on the wheel now becomes

$$F_y = C_{Fy}v_\gamma = C_{F\gamma}\gamma + C_{F\alpha}\alpha' \tag{7.10}$$

With wheel side slip kept equal to zero, $V_{sy} = 0$, and $V'_{sy} = -|V_x|\alpha'$, Eqn (7.2) can be written in the form

$$\frac{dv_\gamma}{dt} + \frac{1}{\sigma_\alpha}|V_x|v_\gamma = \frac{C_{F\gamma}}{C_{F\alpha}}|V_x|\gamma \tag{7.11}$$

This equation shows that according to this simple model the camber force relaxation length $\sigma_\gamma$ is equal to the relaxation length for side slip $\sigma_\alpha$. This theoretical result is substantiated by careful step response experiments, performed by Higuchi (1997) on a flat plank test rig (cf. Section 7.2.3).

A similar equation results for the total spin $\varphi$ including turn slip and camber:

$$\frac{dv_\varphi}{dt} + \frac{1}{\sigma_\alpha}|V_x|v_\varphi = \frac{C_{F\varphi}}{C_{F\alpha}}|V_x|\varphi \tag{7.12}$$

with, according to (4.76),

$$\varphi = -\frac{1}{V_x}\{\dot{\psi} - (1 - \varepsilon_\gamma)\Omega \sin\gamma\} \tag{7.13}$$

that shows that the turn slip velocity $\dot{\psi}$ can be converted into an equivalent camber angle.

The forces and moment are obtained from the deflections $u$ and $v$ by first assessing the transient slip quantities $\alpha'$, $\kappa'$ and $\gamma'$ and from these with the slip stiffnesses the forces and moment.

According to the steady-state model employed here, different from Eqn (4.E71,72), the moment response to camber (and turn slip) is the sum of the residual torque, $M_{zr}$, supposedly mainly due to finite tread width, and $-t_\alpha F_y$, supposedly caused by camber induced side slip, cf. discussion later on (below Eqn (7.40)). A first-order approximation for the response of $M_{zr}$ with the short relaxation length equal to half the contact length $a$ may be employed. This, however, will be saved for Chapter 9 where short wavelength responses are considered. Here we suffice with the assumption that the moment due to tread width responds instantaneously to camber and turn slip.

For the linear, small-slip condition, we find

$$\alpha' \approx \tan\alpha' = \frac{v_\alpha}{\sigma_\alpha}, \quad F_{y\alpha} = C_{F\alpha}\alpha', \quad M_{z\alpha} = -C_{M\alpha}\alpha' = -t_\alpha F_{y\alpha} \tag{7.14}$$

$$\kappa' = \frac{u}{\sigma_\kappa}, \quad F_x = C_{F\kappa}\kappa' \tag{7.15}$$

$$\gamma' = \frac{C_{F\alpha}}{C_{F\gamma}}\frac{v_\gamma}{\sigma_\alpha}, \quad F_{y\gamma} = C_{F\gamma}\gamma', \quad M_{zr\gamma} = (C_{M\gamma} + t_\alpha C_{F\gamma})\gamma \tag{7.16}$$

and similar for $\varphi$. The pneumatic trail due to side slip is denoted here by $t_\alpha$. The total aligning torque becomes (cf. Eqn (4.E71) and Figure 4.21)

$$M_z = -t_\alpha(F_{y\alpha} + F_{y\gamma}) + M_{zr\gamma} \tag{7.17}$$

In an alternative model, used for motorcycle dynamics studies, the terms with $t_\alpha C_{F\gamma}$ or $t_\alpha F_{y\gamma}$ in (7.16, 7.17) are omitted, cf. discussion below Eqn (7.40).

Eqns (7.7, 7.9, 7.11) may be written in terms of the transient slip quantities. For example, we may express $v_\alpha$ in terms of $\alpha'$ by using the first equation of (7.14). Insertion in (7.7) gives

$$\sigma_\alpha\frac{d\alpha'}{dt} + |V_x|\alpha' = |V_x|\alpha = -V_{sy} \tag{7.18}$$

If we recognize the fact that the relaxation length is a function of the vertical load and if the average slip angle is unequal to zero, an additional term shows up in the linearized equation (variation of $\alpha'$ and of $F_z(t)$ are small!) which results from the differentiation of $v_\alpha = \sigma_\alpha\alpha'$ with respect to time. Then, Eqn (7.7) becomes

$$\sigma_\alpha\frac{d\alpha'}{dt} + \left(|V_x| + \frac{d\sigma_\alpha}{dF_z}\frac{dF_z}{dt}\right)\alpha' = |V_x|\alpha = -V_{sy} \tag{7.19}$$

Obviously, when using Eqn (7.18) a response to a variation of the vertical load cannot be expected. If the load varies, Eqn (7.18) is inadequate and the original Eqn (7.7) should be used or the corresponding Eqn (7.19).

With (7.5), Eqn (7.7) may be written directly in terms of $F_y$. If we may consider the carcass lateral stiffness $C_{Fy}$ virtually independent of the wheel load $F_z$, we obtain, by using Eqn (7.6),

$$\sigma_\alpha\frac{dF_y}{dt} + |V_x|F_y = |V_x|F_{yss} \tag{7.20}$$

Since we have the same relaxation length for both the responses to side slip and camber, this equation appears to hold for the combined linear response to the inputs $\alpha$ and $\gamma$ or $\varphi$. In the right-hand member, $F_{yss}$ denotes the steady-state response to these inputs and possibly a changing vertical load at a given slip condition. Multiplication with the pneumatic trail produces the moment $-M'_z$ (if $\gamma = \varphi = 0$). A similar differential equation may be written for the $F_x$ response to $\kappa$.

Eqns (7.7, 7.9) have been written in the form ($V_x$ not in denominator and $V_{sx,y}$ used as right-hand member) that makes them applicable for simulations of stopping and starting from zero speed occurrences. At speed $V_x = 0$, Eqn (7.9) turns into an integrator: $u = -\int V_{sx}\mathrm{d}t$. With Eqn (7.15), the longitudinal force becomes $F_x = C_{F\kappa}\kappa' = C_{F\kappa}u/\sigma_\kappa$ which with (7.8) is equal to $C_{Fx}u$. This is the correct expression for the tire that at standstill acts like a longitudinal or tangential spring. When the wheel starts rolling, the tire gradually changes into a damper with rate: $C_{F\kappa}/|V_x|$. Figure 7.2 depicts a corresponding mechanical model with spring and damper in series. It shows that at low speed the damper becomes very stiff and the spring dominates. At higher forward velocities, the spring becomes relatively stiff and the damper part dominates the behavior of the tire. A similar model may be drawn for the transient lateral behavior. It may be noted that Eqn (7.20) is not suited for moving near or at $V_x = 0$. In Section 8.6, the use of the transient models for the response to lateral and longitudinal wheel slip speed at and near zero speed will be demonstrated.

Apparently, Eqn (7.12) with (7.13) fails to describe the response to variations in wheel yaw angle $\psi$ at vanishing speed. Then, the lateral deflection becomes $v_\varphi = -(C_{F\varphi}/C_{F\alpha})\int \mathrm{d}\psi$. With $\varphi' = (C_{F\alpha}/C_{F\varphi})v_\varphi/\sigma_\alpha$, we have $\varphi' = -\psi/\sigma_\alpha$ indicating an instantaneous response of $F_y$ which, however, should remain zero! We refer to Chapter 9, Section 9.2.1, Eqn (9.56 etc.), that suggests a further developed model that can handle this situation correctly. Consequently, it must be concluded that the present transient model cannot be employed to simulate parking maneuvers unless $F_y$ is suppressed artificially in the lower speed range.

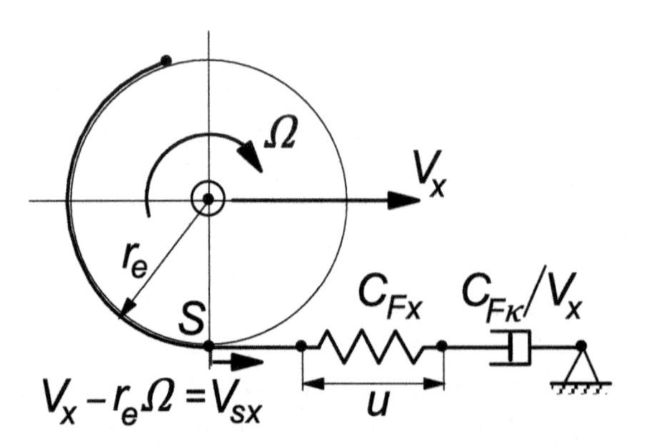


**FIGURE 7.2** Mechanical model of transient tangential tire behavior.

### 7.2.2. Semi-Non-Linear Model

For the extension of the linear theory to cover the non-linear range of the slip characteristics, it may be tempting to employ Eqn (7.20) and use the instantaneous nonlinear force response as input in the differential equation. The input steady-state side force is calculated, e.g. with the *Magic Formula*, using the current wheel slip angle $\alpha$. This method, however, may lead to incorrect results as, due to the phase lag in side force response, the current (varying) wheel load may not correspond to the calculated magnitude attained by the side force. In limit conditions, the tire may then be predicted to be still in adhesion while in reality full sliding occurs. A better approach is to use the original Eqn (7.7) and to calculate the side force afterward by using the resulting transient slip angle $\alpha'$ as input in the *Magic Formula*.

In general, we have the three Eqns (7.7, 7.9, 7.11) and possibly (7.12) and the first equations of (7.14–7.16) producing $\alpha'$, $\kappa'$ and $\gamma'$ or $\varphi'$ which are used as input in the nonlinear force and moment functions ($\gamma$ or $\varphi$ directly in the expressions for $M_r$), e.g. the equations of the *Magic Formula* tire model (Chapter 4):

$$F_x = F_x(\kappa', \alpha', F_z) \tag{7.21}$$

$$F_y = F_y(\alpha', \gamma', \kappa', F_z) \tag{7.22}$$

$$M_z' = -t_\alpha F_y \tag{7.23}$$

$$M_{zr} = M_{zr}(\gamma, \alpha', \kappa', F_z) \tag{7.24}$$

$$M_z = M_z' + M_{zr} + s \cdot F_x \tag{4.E71}$$

where, if required, $\gamma$ may be replaced by $\varphi$ as the spin argument.

This nonlinear model is straightforward and is often used in transient or low-frequency vehicle motion simulation applications. Starting from zero speed or stopping to standstill is possible. However, as has been mentioned before, at $V_x$ equal or close to zero, Eqns (7.7, 7.9) act as integrators of the slip speed components $V_{sx,y}$, which may give rise to possibly very large deflections. The limitation of these deflections may be accomplished by making the derivatives of the deflections $u$ and $v$ equal to zero, if (1) the forward wheel velocity has become very small ($<V_{\text{low}}$) and (2) the deflections take values larger than physically possible. This can be seen to correspond with the combined equivalent side slip value exceeding the level $\alpha_{sl}$ where the peak horizontal force occurs. Approximately, we may adopt the following limiting algorithm with the equivalent slip angle according to Eqn (4.E78):

if: $\quad \left|\alpha'_{r,eq}\right| > \alpha_{sl}$ and $|V_x| < V_{\text{low}}$

then: $\quad$ if: $(V_{sx} + |V_x| u/\sigma_\kappa) u < 0: \quad \dot{u} = 0$ $\quad$ else: Eqn (7.9) applies

$\quad\quad\quad$ if: $(V_{sy} + \left|V_x\right| v/\sigma_\alpha) v < 0: \quad \dot{v} = 0$ $\quad$ else: Eqn (7.7) applies

else: $\quad$ Eqns (7.7, 7.9) apply $\quad\quad$ (7.25)

with roughly

$$\alpha_{sl} = 3D_y/C_{F\alpha}$$

Experience in applying the model has indicated that starting from standstill gives rise to oscillations which are practically undamped. Damping increases when speed is built up. To artificially introduce some damping at very low speed, which with the actual tire is established through material damping, one might employ the following expression for the transient slip $\kappa'$, as suggested by Besselink, instead of Eqn (7.15):

$$\kappa' = \left(\frac{u}{\sigma_\kappa} - \frac{k_{V\text{low}}}{C_{F\kappa}}V_{sx}\right) \tag{7.26}$$

The damping coefficient $k_{V\text{low}}$ should be gradually suppressed to zero when the speed of travel $V_x$ approaches a selected low value $V_{\text{low}}$. Beyond that value the model should operate as usual. In Chapter 8, Section 8.6, an application will be given. A similar equation may be employed for the lateral transient slip.

Another extreme situation is the condition at wheel lock. At steady state, Eqn (7.9) reduces to

$$\frac{1}{\sigma_\kappa}|V_x|u = -V_x \tag{7.27}$$

which indicates that the deflection $u$ according to the semilinear theory becomes as large as the relaxation length $\sigma_\kappa$. Avoiding the deflections from becoming too large, which is of importance at e.g. repetitive braking, calls for an enhanced nonlinear model. Another shortcoming of the model that is to be tackled concerns the experimentally observed property of the tire that its relaxation length depends on the level of slip. At higher levels of side slip, the tire shows a quicker response to additional changes in side slip. This indicates that the relaxation lengths decrease with increasing slip.

### 7.2.3. Fully Nonlinear Model

Figure 5.29, repeated here as Figure 7.3, shows the deflected string model provided with tread elements at various levels of steady-state side slip. Clearly, this model predicts that the 'intersection' length $\sigma^*$ decreases with increasing $\alpha$ that is: when the sliding range grows. In the single-point-contact model, we may introduce a similar reduction of the corresponding length. The intersection length $\sigma^*_\alpha$ is defined here as the ratio of the lateral deflection $v_\alpha$ and the transient lateral slip tan $\alpha'$:

$$\sigma^*_\alpha = \frac{v_\alpha}{\tan\alpha'} \tag{7.28}$$

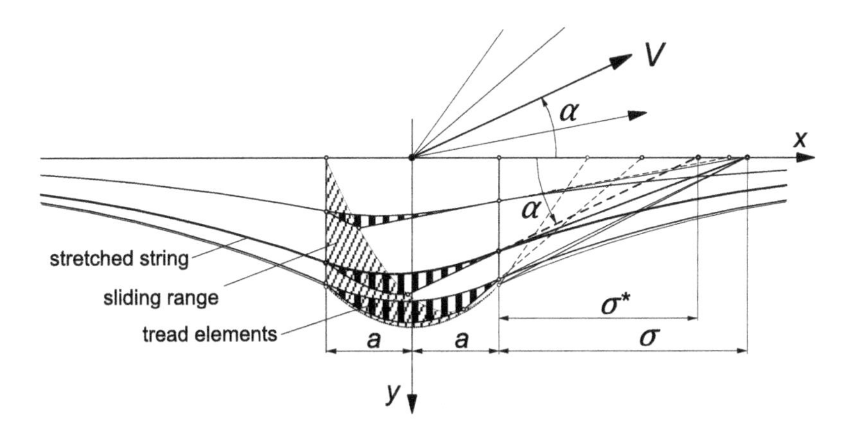


**FIGURE 7.3** The string tire model with tread elements at increasing slip angles showing growing sliding range and decreasing 'intersection' length $\sigma^*$.

In Eqn (7.7), the relaxation length $\sigma_\alpha$ is replaced by $\sigma^*_\alpha$ and we get

$$\frac{dv_\alpha}{dt} + \frac{1}{\sigma^*_\alpha}|V_x|v_\alpha = |V_x|\tan\alpha = -V_{sy} \tag{7.29}$$

with apparently

$$\sigma^*_\alpha = \frac{1}{C_{Fy}}\frac{F_y}{\tan\alpha'} = \frac{\sigma_{\alpha o}}{C_{F\alpha}}\frac{F_y}{\tan\alpha'} \approx \frac{\sigma_{\alpha o}}{C_{F\alpha}}\frac{|F'_y| + C_{F\alpha}\varepsilon_F}{|\tan\alpha'_f| + \varepsilon_F} \tag{7.30}$$

with the initial relaxation length (at $\alpha' = 0$):

$$\sigma_{\alpha o} = \frac{C_{F\alpha}}{C_{Fy}} \tag{7.31}$$

to which $\sigma^*_\alpha$ approaches when $\alpha' \to 0$. To avoid singularity, one may use the last expression of (7.30) with small $\varepsilon_F$ and add to $\alpha'$ the shift $\Delta\alpha$ to arrive at $\alpha'_f$ as indicated in Figure 4.22. Figure 7.4 presents the characteristic of the

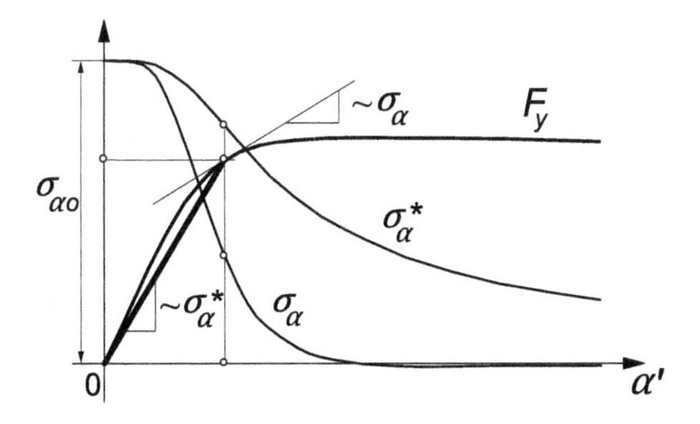


**FIGURE 7.4** Characteristics of tire side force, 'intersection length' $\sigma^*_\alpha$, and relaxation length $\sigma_\alpha$.

intersection length together with the side force characteristic from which it is derived. Also in the equation for the camber deflection response (7.11), the relaxation length $\sigma_\alpha$ is replaced by $\sigma^*_\alpha$.

A more direct way to write Eqn (7.29) is yielded by eliminating $\sigma^*_\alpha$ with the use of Eqn (7.28):

$$\frac{dv_\alpha}{dt} + |V_x|\tan\alpha' = |V_x|\tan\alpha = -V_{sy} \tag{7.32}$$

The transient slip angle $\alpha'$ is obtained from the deflection $v_\alpha$ by using the inverse possibly adapted $F'_y(\alpha')$ characteristic, cf. Higuchi (1997):

$$v_\alpha = \frac{F'_y(\alpha')}{C_{Fy}} \tag{7.33}$$

To avoid double valued solutions for this purpose an adapted characteristic for $F'_y(\alpha')$ in (7.30) or (7.33) may be required showing a positive slope in the slip range of interest. The value for $\alpha'$ or for $\sigma^*_\alpha$, when Eqn (7.29) is used, is obtained through iterations or by using information from the previous time step, cf. Higuchi (1997), Pacejka and Takahashi (1992), and Takahashi and Hoshino (1996). The ultimate value of the force $F_y$ is finally obtained from Eqn (7.22) by using the computed $\tan\alpha'$ $(= v_\alpha/\sigma^*_\alpha)$.

Writing Eqn (7.32) entirely in terms of the transient slip angle by using (7.5) with $v = v_\alpha$ and remembering that $F_y = F_y(\alpha', F_z)$ directly yields

$$\frac{1}{C_{Fy}}\frac{\partial F_y}{\partial\tan\alpha'}\frac{d\tan\alpha'}{dt} + |V_x|\tan\alpha' = -V_{sy} - \frac{1}{C_{Fy}}\frac{\partial F_y}{\partial F_z}\frac{dF_z}{dt} \tag{7.34}$$

The additional input $dF_z/dt$ requires information of the slope $\partial F_y/\partial F_z$ at given values of the slip angle. If the vertical load remains constant, the last term vanishes and we have the often used equation of the restricted fully nonlinear model:

$$\sigma_\alpha\frac{d\tan\alpha'}{dt} + |V_x|\tan\alpha' = -V_{sy} \tag{7.35}$$

with

$$\sigma_\alpha = \frac{1}{C_{Fy}}\frac{\partial F_y}{\partial\tan\alpha'} \tag{7.36}$$

If we consider an average slip angle $\alpha_o$ and a small variation $\tilde{\alpha}$ of $\tan\alpha$ and the corresponding lateral slip velocities, Eqn (7.35) becomes, after having subtracted the average part,

$$\sigma_\alpha\frac{d\tilde{\alpha}'}{dt} + |V_x|\tilde{\alpha}' = -\tilde{V}_{sy} \tag{7.37}$$

which indicates that the structure of (7.35) is retained and that $\sigma_\alpha$ (7.36) represents the actual relaxation length of the linearized system at a given load and slip angle. Its characteristic has been depicted in Figure 7.4 as well. Obviously, the relaxation length is associated with the slope of the side force characteristic. It also shows that the relaxation length becomes negative beyond the peak of the side force characteristic which makes the solution of (7.35) but also of the original Eqn (7.29) or (7.32) unstable if the point of operation lies in that range of side slip. We may, however, limit $\sigma_\alpha$ downward to avoid both instability and excessive computation time: $\sigma_\alpha = \max(\sigma_\alpha, \sigma_{\min})$. The transient response of the variation of the force proceeds in proportion with the variation of $\tilde{\alpha}'$ as $\tilde{F}_y = (\partial F_y/\partial \tan\alpha)\tilde{\alpha}'$.

When Eqn (7.35) is used, an algebraic loop does not occur as is the case when Eqn (7.29) is employed. The relaxation length $\sigma_\alpha$ can be directly determined from the already available $\tan\alpha'$. However, since the last term of (7.34) has been omitted, Eqn (7.35) has become insensitive to $F_z$ variations.

Similar functions and differential equations can be derived for the transient response to longitudinal slip. We obtain, for the distance factors,

$$\sigma_\kappa^* = \frac{1}{C_{Fx}}\frac{F_x}{\kappa'} = \frac{\sigma_{\kappa o}}{C_{F\kappa}}\frac{F_x}{\kappa'} \approx \frac{\sigma_{\kappa o}}{C_{F\kappa}}\frac{|F_x| + C_{F\kappa}\varepsilon_F}{|\kappa'| + \varepsilon_F} \tag{7.38}$$

and

$$\sigma_\kappa = \frac{1}{C_{Fx}}\frac{\partial F_x}{\partial \kappa'} \tag{7.39}$$

It is of interest to note that the extreme case of wheel lock can now be handled correctly. As we have seen below Eqn (7.27), the deflection then becomes equal to minus the relaxation length which according to the fully nonlinear model becomes $u = -\sigma_\alpha^*$. With (7.38) and $\kappa' = -1$, the deflection takes the value that would actually occur at wheel lock: $u = F_x/C_{Fx}$.

The use of Eqns (7.29, 7.32) is attractive for simulation purposes because the solution contains the transient effect of changing vertical load. However, we may have computational difficulties to be reckoned with which may require some preparations. Using Eqn (7.35) proceeds in a straightforward manner if the derivative of the force vs slip characteristic is known beforehand. This requires some preparation and may become quite complex if the general combined slip situation is to be considered. Approximations using equivalent total slip according to Eqn (4.E78) may be realized. A drawback, of course, is the fact that Eqn (7.35) does not respond to changes in $F_z$. In Chapter 9 this equation is used to handle the transient response of the tread deflections in the contact patch.

Figures 7.5–7.12 present results obtained by Higuchi (1997) for a passenger car 205/60R15 tire tested on a flat plank machine (TU-Delft, cf. Figure 12.6). Figure 7.5 shows the response of the side force to step changes in slip angle at

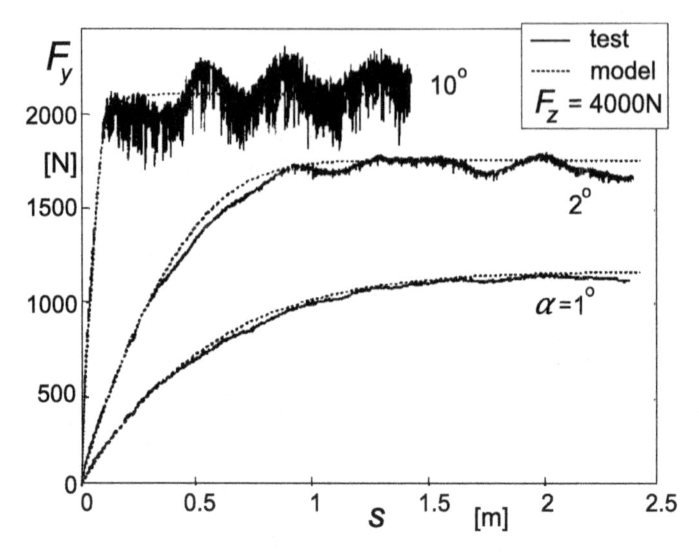


**FIGURE 7.5** Side force response to small and large step change in slip angle as assessed by flat plank experiments and computed with the model defined by Eqns (7.32) or (7.29).

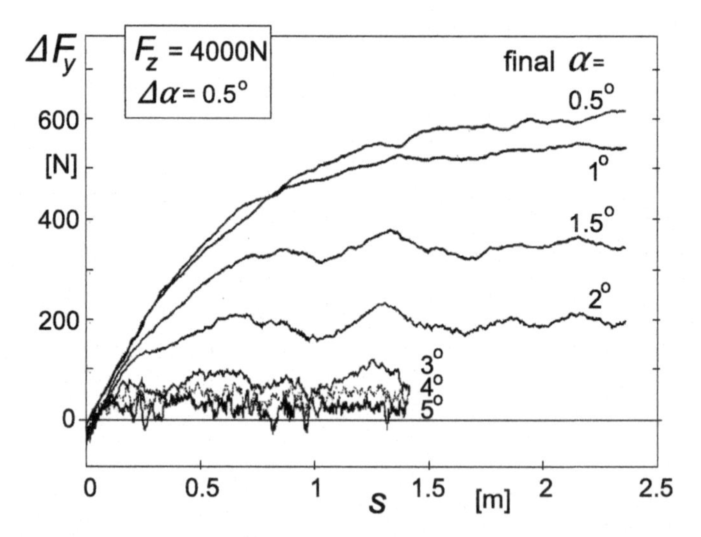


**FIGURE 7.6** Measured side force response to small increment of slip angle $\Delta\alpha$ at different levels of slip angle indicating the decrease of relaxation length at increasing side slip level which is in agreement with expression (7.36).

the nominal load of 4000 N. The diagram clearly indicates the difference in behavior at the different levels of side slip with a very rapid response occurring at $\alpha = 10°$. The fully nonlinear model according to Eqns (7.32) or (7.29) or as in this case, with constant vertical load, Eqn (7.35), in conjunction with

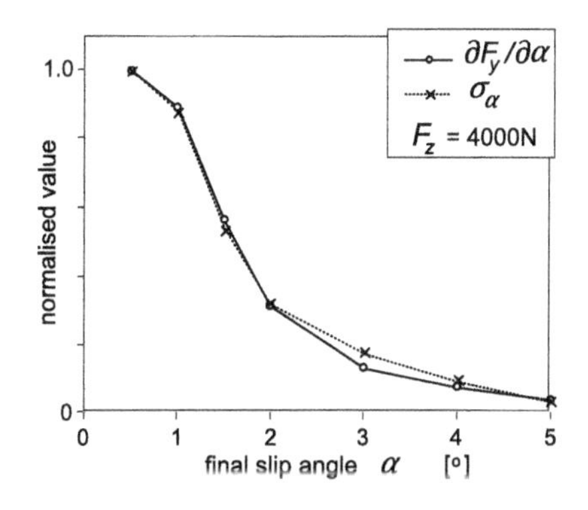


**FIGURE 7.7** Correspondence between $F_y$ vs $\alpha$ slope and relaxation length $\sigma_\alpha$ as assessed by experiments.

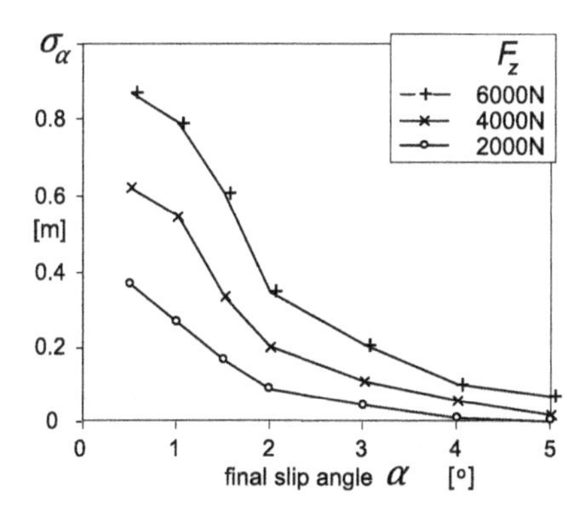


**FIGURE 7.8** Relaxation length measured at various wheel loads, decaying with increasing slip angle.

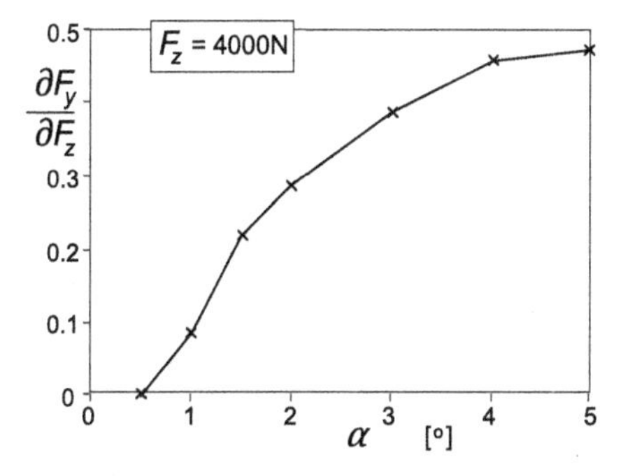


**FIGURE 7.9** The slope of $F_y$ vs $F_z$ as a function of slip angle, appearing in Eqn (7.34), as assessed by tests.

Eqns (7.22, 7.23) gives satisfactory agreement. The measured response curves have been corrected for the side force variation that arises already at zero slip angle. The test is performed by loading the tire after the slip angle (steer angle) has been applied. Subsequently, the plank is moved. The aligning torque behaves in a similar manner although a little initial delay in response occurs as

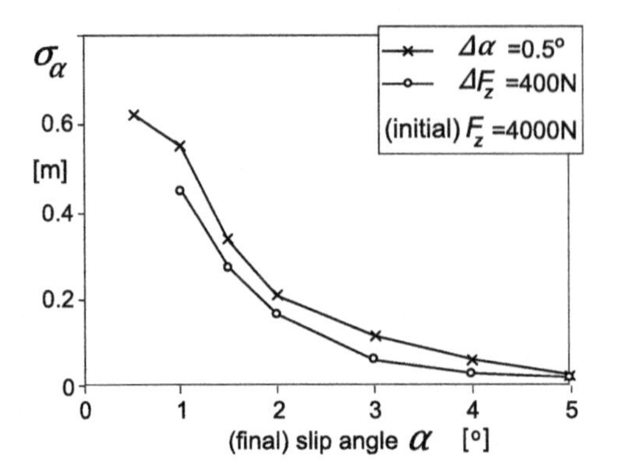


**FIGURE 7.10** Relaxation lengths resulting from small step changes in slip angle or wheel load, as a function of slip angle level.

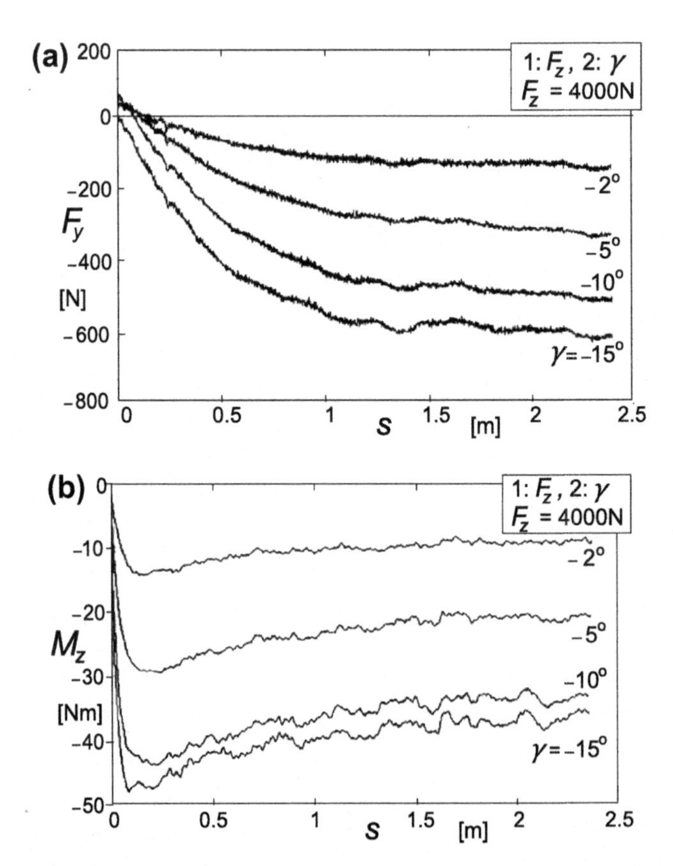


**FIGURE 7.11** (a) Side force response to step change in camber angle. (b) Moment response to step change in camber angle.

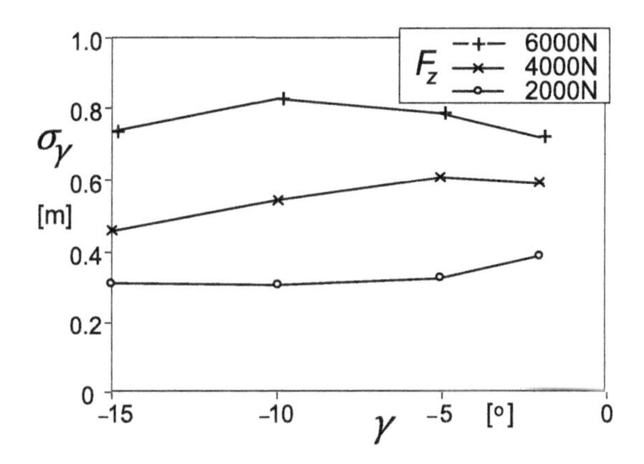


**FIGURE 7.12** Relaxation length for step responses to different camber angles (abscissa) at various wheel loads (camber after loading).

correctly predicted by the string model and shown in Figure 5.12. At larger slip angles (beyond the level where the $M_z$ peak occurs), the aligning torque first shows a peak in the response after which the moment decays to its steady-state value. This behavior is nicely followed by the model. Figure 7.6 shows the side force response to small increments in slip angle. The experiment is conducted by, after having stopped the plank motion, steering the wheel half a degree further and continue the forward motion. It was first ascertained that the response of the side force to a step steer input was hardly influenced by having first lifted the tire from the road surface or not. Of course, the aligning moment response, in the latter case, is quite different because of initial torsion about the vertical axis. The resulting side force responses as depicted in Figure 7.6 have been compared with the exponential response of the side force increment as predicted by the solution of Eqn (7.37) with (7.36). Fitting with a least-square procedure gave the relaxation length $\sigma_\alpha$ as presented in Figure 7.7. The resulting variation of $\sigma_\alpha$ with slip angle level is compared with the slope of the steady-state side force characteristic, both normalized making their values equal to unity at vanishing slip angle.

The excellent agreement supports the theoretical findings. Additional experiments have been conducted by Higuchi investigating the response to small increments in vertical load at constant slip angle. As shown by Eqn (7.34) the partial derivative $\partial F_y/\partial F_z$ plays the role as input parameter. Figure 7.9 shows its variation with slip angle level. Figure 7.10 compares the resulting relaxation length values with those associated with changes in slip angle at constant load. Apparently, the relaxation length for slip angle change is larger than the one belonging to load change. However, due to the finite magnitude of the increments (0.5° and 400 N respectively), the actual differences are expected to be smaller. In the diagram, the curve for $\Delta\alpha$ may be better shifted to the left over 0.25° while the curve for $\Delta F_z$ may be reduced a little in height using information from Figure 7.8 accounting for a decrease of the average load level of 200 N.

Higuchi also investigated the responses to changes in camber angle, and to a limited extent also to the related turn slip. The change in camber was correctly established by rotating the road surface (the plank) about the line of intersection of wheel center plane and plank surface (Figure 12.5, steer angle is kept equal to zero). The step response to turn slip $\varphi_t = -1/R$ is obtained by integrating the response to a pulse change in turn slip, that is: load tire, twist ($\psi$) and then roll in the new wheel plane direction, and multiply the result with $1/(R\psi)$. The responses to camber and turn slip show quantitatively similar responses, cf. Pacejka (2004). The small initial delay of the response of the car tire side force to camber and turn slip (cf. Figures 5.12, 5.10) will effectively increase the relaxation length a little. Figures 7.11a and 7.11b show the responses of side force and moment to step changes in camber angle. The side force behaves in a manner similar to the response to side slip, Figure 7.5. Since we have a camber stiffness much smaller than the cornering stiffness, the camber force characteristic remains almost linear over a larger range of the camber angle (meaning: there is less sliding in the contact patch). Therefore, the difference in step responses remains relatively small. This is reflected by the diagram of Figure 7.12 showing the variation of the resulting relaxation length. Comparison with Figure 7.8 reveals that the relaxation lengths assessed for the responses to side slip (at small side slip) and camber have comparable magnitudes. This supports the theory of Eqn (7.11).

Comparison of the Figures 5.12 (lower left diagram) and 7.11a may reveal the quantitative difference in steady-state side force response to turn slip and camber angle. According to Eqn (3.55), the relationship between turn slip stiffness and camber stiffness will be: $C_{F\gamma} = (1 - \varepsilon_\gamma)C_{F\varphi}/r_e$. With the tire effective rolling radius equal to approximately 0.3 m and the estimated steady-state levels reached in Figures 5.12 and 7.11a for $R = 115$ m and $\gamma = 2°$ respectively, we find, for the reduction factor approximately, $\varepsilon_\gamma = 0.5$.

The response of the aligning torque to camber is similar to the response to turn slip (cf. Figure 5.12). The moment quickly reaches a maximum after which a slower decay to the steady-state level occurs. The string model with tread width effect predicts the same for the response to turn slip as can be concluded by adding the curves for $M_z'$ and $M_z^*$ (response to $\varphi$, Figure 5.10) in an appropriate proportion. The model developed in the present chapter generates a similar response but through a different mechanism. Eqn (4.E71) that for $F_x = 0$ takes the form

$$M_z = M_z' + M_{zr} = -t_\alpha F_y + M_{zr} \tag{7.40}$$

is used after having computed the transient slip and camber angles $\alpha'$ and $\gamma'$.

As can be seen in Figure 4.13 or 4.21, the aligning torque caused by camber at zero slip angle is attributed to the residual torque $M_{zr}$ and to the counteracting moment equal to the aligning stiffness $C_{M\alpha}$ times the camber induced slip angle $\Delta\alpha_\gamma$ which is the same as $-t_\alpha F_y$ at $\alpha = 0$. The residual torque responds quickly with a relatively short relaxation length (about equal to half the contact length) or as has been suggested above for the present model instantaneously. As the side force responds slowly with relaxation length $\sigma_\alpha$ and $-t_\alpha F_y$ is opposite in sign with respect to the residual torque, a similar response as depicted in Figure 7.11b will be developed by the model represented by Eqn (7.40).

The mechanism behind this response is supported by the physical reasoning that the moment due to tread width, that responds relatively fast to changes in camber and turn slip, gives rise to a yaw torsion of the carcass/belt in the contact zone which acts as a slip angle. This side slip generates a side force that acts in the same direction as the camber force generated by the camber induced spin. In addition, an aligning moment is generated that acts in a sense opposite to that of the tread-width camber (spin) moment. Ideally, in Figure 4.21, the moment due to camber spin is equal to the residual torque $M_{zr}$ while $-C_{M\alpha}$ $\Delta\alpha_\gamma = t_\alpha F_y$ is equal to the moment due to the yaw torsion-induced side slip. However, in reality, only a part of the camber force $F_y$ results from camber spin-induced side slip. It is expected that the above analysis is partly true, perhaps even for an appreciable part. The remaining part that is responsible for the 'hump' in the moment response must then be due to the transient asymmetric lateral tire deformation that vanishes when the steady-state condition is reached. Figure 5.9 depicts the nature of this transient deflection. This concept has been followed in Chapter 9.

In the model of Chapter 4 (used in Chapter 9), that was originally introduced for motorcycle dynamics studies dealing with possibly large camber angles, the other extreme is used and the term $t_\alpha F_y$ is replaced by $t_\alpha F_{y\alpha}$, that is: with the side force attributed to the (transient) side slip angle alone (cf. Section 4.3.2 where $F_{y\alpha} = F'_y$).

## 7.2.4. Nonlagging Part

The force response to a change in camber exhibits a peculiar feature. From low-velocity experiments conducted on the flat plank machine (cf. Figure 12.5), it turns out that directly after that the wheel is cambered (about the line of intersection), a side force is developed instantaneously. This, obviously, is caused by the nonsymmetric distortion of the cross section of the lower part of the tire. This initial 'nonlagging' side force that occurs at a distance rolled $s = 0$, as shown in Figure 7.11a, appears, for the tire considered, to act in a direction opposite to the steady-state side force. Equal directions turn out to occur also, e.g., for a motorcycle tire, cf. Segel and Wilson (1976). Also in these reported experiments, the camber angle is applied after the tire has been loaded. Figure 7.13 gives the percentage of the nonlagging part with respect to the steady-state side force for three wheel loads and for three different ways of reaching the loaded and cambered condition before rolling has started. In the first case ($Z$), the free wheel is first cambered and then loaded by moving it toward the horizontal road surface in vertical direction. This sequence appears to result in a somewhat larger nonlagging part. In the second case ($R$), the road

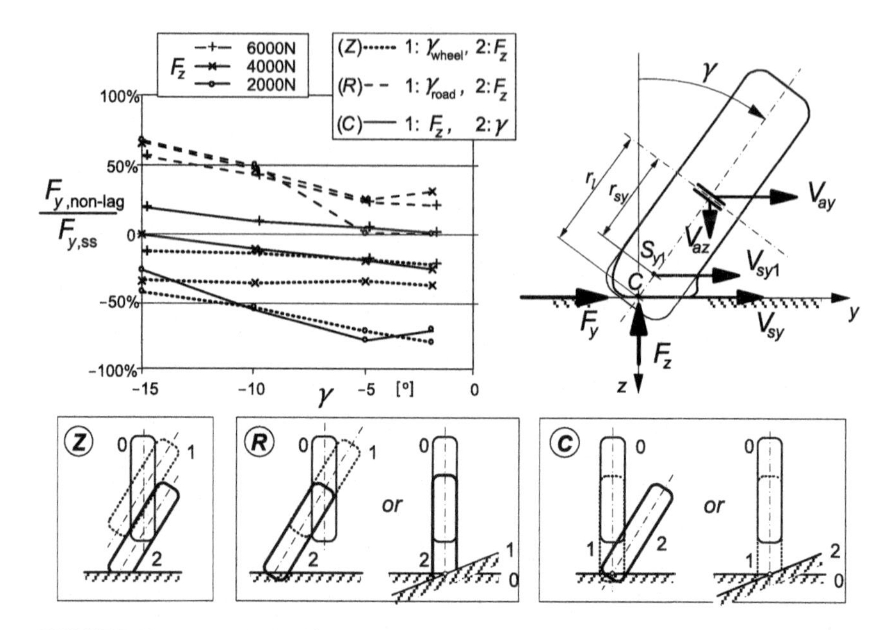


**FIGURE 7.13** Nonlagging part of side force response for the cases: (*Z*) loading the tire vertically after having cambered the wheel, (*R*) loading the tire radially after having cambered the wheel, or loading the tire vertically after having cambered the road, and (*C*) applying wheel or road camber after having loaded the tire. *Upper-right diagram*: New additional side-slip point $S_{y1}$ with side-slip velocity component $V_{sy1}$.

surface is cambered first after which the tire is loaded vertically. It turns out that now a response arises with the same sign as the steady-state response. The diagram indicates that also in case (*C*), the sign may remain unchanged if the wheel load is sufficiently high.

To simulate the development of the nonlagging side force, the following model is suggested. The tire response that arises due to loading and/or tilting of the wheel while $V_x = 0$ is considered to be the result of the integrated lateral horizontal velocity of the lower part of the wheel. Besides the lateral velocity $V_{sy}$ of the contact center *C*, the lateral velocity $V_{sy1}$ of a newly introduced slip point $S_{y1}$ that is thought to be attached to the wheel at a radius $r_{sy}$, is used as an additional component of the effective lateral slip speed $V_{sy,\text{eff}}$. The upper-right diagram of Figure 7.13 depicts the situation. We define

$$V_{sy,\text{eff}} = \varepsilon_c V_{sy} + (1 - \varepsilon_c) V_{sy1} \tag{7.41}$$

with $\varepsilon_c$ being the participation factor. The effective lateral slip speed replaces $V_{sy}$ in Eqn (7.7). In the case of a horizontal flat road surface, the velocities of the two points *C* and $S_{y1}$ become

$$V_{sy} = V_{ay} + \left( V_{az} \sin\gamma - r_l \frac{d\gamma}{dt} \right) \frac{1}{\cos\gamma} \tag{7.42a}$$

$$V_{sy1} = V_{ay} - r_{sy} \frac{d\gamma}{dt} \cos\gamma \tag{7.42b}$$

The location of $S_{y1}$ is defined through the proposed function for the slip radius:

$$r_{sy} = r_l - p_{NL1}\, \rho_z (1 - p_{NL2}\sqrt{F_z/F_{zo}} - p_{NL3}|\gamma| - p_{NL4}\gamma^2) \tag{7.43}$$

The participation factor $\varepsilon_c$ is defined by the proposed function:

$$\varepsilon_c = \frac{p_{NL5} - p_{NL6}|\gamma|}{1 + p_{NL7}F_z/F_{zo} + p_{NL8}(F_z/F_{zo})^2} \tag{7.44}$$

It may be seen from Eqns (7.42a, 7.42b) that at constant camber angle and a purely vertical axle motion (case $Z$), $V_{sy1} = V_{ay} = 0$, and $V_{sy} = V_{az} \tan\gamma$ is the governing component of the effective slip speed (7.41). If the loading is conducted by a radial approach of the road surface (case $R$), we have $V_{sy} = 0$, and the governing part is $V_{sy1} = V_{ay} = -V_{az} \tan\gamma$. The third case $C$ is achieved by first vertical loading of the upright tire, $V_{sy} = V_{sy1} = 0$, and subsequently tilting the wheel about the line of intersection that is: about point $C$. In the latter phase, the lateral slip velocity components become $V_{sy} = 0$ and $V_{sy1} = (r_l - r_{sy})(d\gamma/dt) \cos\gamma$.

For the tire parameter values ($C_{Fz}$ and $C_{Fy}$ possibly considered functions of $\gamma$): $C_{Fz} = 200$ kN/m, $C_{Fy} = 130$ kN/m, $F_{zo} = 4$ kN, $r_o = 0.3$ m, $r_c = 0.15$ m, and for Eqns (4.E19–4.E30) with $\zeta$'s and $\lambda$'s $= 1$: $p_{Cy1} = 1.3$, $p_{Dy1} = 1$, $p_{Ey1} = -1$, $p_{Ky1} = 15$, $p_{Ky2} = 1.5$, $p_{Ky3} = 6$, $p_{Ky4} = 2$, $p_{Ky6} = 1$, $p_{Vy3} = 1$ and remaining $p$'s $= 0$, the following values for $p_{NL1}$. $p_{NL8}$ were assessed through a manual fitting process: $p_{NL1} = 2.5$, $p_{NL2} = 0.8$, $p_{NL3} = 0$, $p_{NL4} = 3$, $p_{NL5} = 1$, $p_{NL6} = 2$, $p_{NL7} = -2.5$, $p_{NL8} = 10$. With these values, the responses computed for the nonlagging part of the side force show reasonable correspondence with the experimental results of Figure 7.13. See Exercise 7.1 for parameters of a motorcycle tire.

The vertical load has been calculated using a tire model with a circular contour of the cross section with radius $r_c$. For the more general case of an elliptic contour, cf. Figure 7.14, the following equations apply. We have, for the coordinates of the lowest point,

$$\begin{aligned} \zeta &= b/\sqrt{1 + (a/b)^2 \tan^2\gamma} \\ \eta &= a(a/b)\tan\gamma/\sqrt{1 + (a/b)^2 \tan^2\gamma} \end{aligned} \tag{7.45}$$

The vertical compression $\rho_z$, which is the distance of the lowest point of the ellipse to the road surface if this distance is non-negative, now reads with $r_o$ the free tire radius and $r_l$ the loaded tire radius:

$$\rho_z = \max((r_o - r_l - b + \zeta)\cos\gamma + \eta \sin\gamma,\ 0) \tag{7.46}$$

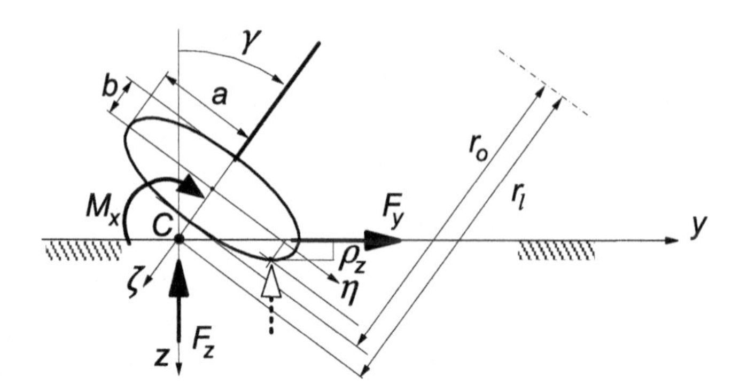


**FIGURE 7.14** Cross-section contour with elliptical shape.

In the simpler case of a circular contour ($r_c = a = b$), the formula reduces to

$$\rho_z = \max((r_o - r_l - r_c)\cos\gamma + r_c\ ,\ 0) \tag{7.47}$$

The normal load is now calculated as

$$F_z = C_{Fz}\,\rho_z \tag{7.48}$$

For a wheel subjected to oscillations, the observed tire properties will be of importance especially when loss of road contact occurs. In Exercise 7.1 given at the end of this chapter and in Section 8.2, this problem is addressed. The more complex case of moving over short obstacles, exhibiting forward and transverse road slope variations, is treated in Section 10.1.6, Eqns (10.33, 10.34).

## 7.2.5. The Gyroscopic Couple

In Chapter 5 the gyroscopic couple that arises as a result of the time rate of change of the average tire tilt deflection angle has been introduced. This angle is considered to be proportional with the lateral tire deflection $v$ or the side force $F_y$. Equation (5.178) may be written in terms of the deflection. With (5.179) and with the $F_z/F_{zo}$ factor added to give smoother results for a jumping tire, we get

$$M_{z,gyr} = c_{gyr} m_t r_e \Omega \frac{\mathrm{d}v}{\mathrm{d}t}\frac{F_z}{F_{zo}} \tag{7.49}$$

where at free rolling the wheel speed of revolution equals the forward velocity divided by the effective rolling radius of the tire. In general, we have

$$\Omega = \frac{V_x - V_{sx}}{r_e} \tag{7.50}$$

For a radial ply steel-belted car tire, the nondimensional coefficient $c_{gyr}$ has been estimated to take the value 0.5. The quantity $m_t$ represents the mass of the tire. Extending the expression (7.40) yields for the total aligning moment:

$$M_z = M'_z + M_{zr} + M_{z,gyr} \tag{7.51}$$

## 7.3. ENHANCED NONLINEAR TRANSIENT TIRE MODEL

A totally different approach to model the transient rolling properties of the tire is based on the separation of contact patch slip properties and carcass compliance not through the use of relaxation lengths but by incorporating the carcass springs in the model explicitly. The contact patch is given some inertia to facilitate the computational process (computational causality). This has the drawback that a relatively high natural frequency is introduced, possibly making the computation slower. We may, however, employ alternative methods to avoid the inclusion of the small mass. The model to be discussed automatically accounts for the property that the lag in the response to wheel slip and load changes diminishes at higher levels of slip that in the previous section was realized by decreasing the relaxation length. This latter approach, however, appeared to possibly suffer from computational difficulties (at load variations). Also, combined slip was less easy to model. In developing the enhanced model, we should, however, try to maintain the nice feature of the relaxation length model to adequately handle the simulation at speeds near or equal to zero.

Figure 7.15 depicts the structure of the enhanced transient model. The contact patch can deflect in circumferential and lateral direction with respect to the lower part of the wheel rim. Only translations are allowed to ensure that the slip angle seen by the contact patch at steady state is equal to that of the wheel plane. To enable straightforward computations, a mass point is thought to be attached to the contact patch. That mass point coincides with point $S^*$ the velocity of which constitutes the slip speed of the contact point. This slip velocity is used, together with the (supposedly at the contact patch detected) path curvature due to turn slip and wheel camber, to compute the forces $F_x$, $F_y$

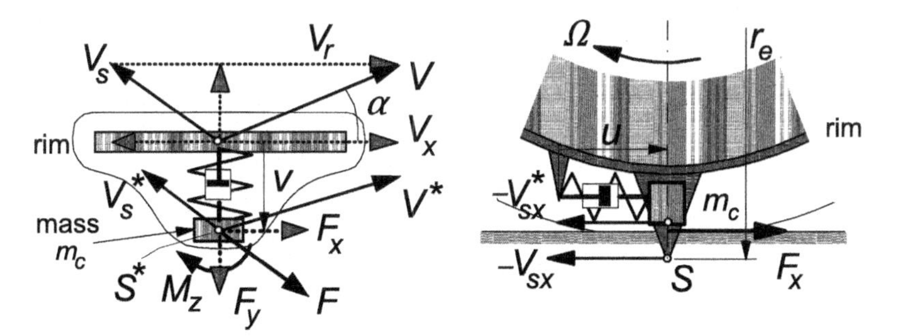


**FIGURE 7.15** Enhanced transient tire model in top and side view showing carcass compliances and contact patch mass.

and the moment $M_z$ which act from road to contact patch. We may add a simple relaxation length model to enable computations near zero speed. The model was first employed by Van der Jagt et al. (1989) and later generalized by Pacejka and Besselink (1997). In Chapter 8, the model will be applied to investigate the effect of road undulations on the efficiency of anti-lock brake control.

With mass $m_c$ and carcass stiffnesses $c_{cx,y}$ and damping ratios $k_{cx,y}$ introduced, the equations of motion for the contact patch mass point with longitudinal and lateral speed components $V^*_{sx,y}$ read (yaw rate terms in accelerations disregarded)

$$m_c \dot{V}^*_{sx} + k_{cx}\dot{u} + c_{cx}u = F_x(\kappa', \alpha', F_z) \tag{7.52}$$

$$m_c \dot{V}^*_{sy} + k_{cy}\dot{v} + c_{cy}v = F_y(\alpha', \kappa', \gamma, F_z) - F_{y,NL} \tag{7.53}$$

The forces acting from ground to contact patch shown on the right-hand sides are computed from the steady-state formulas. The nonlagging camber force part $F_{y,NL}$ is assumed to act directly on the wheel rim. We may approximate the nonlagging force part by a linear relation with $\gamma$ using the camber thrust stiffness $C_{F\gamma}$, the nonlagging fraction $\varepsilon_{NL}$ (cf. Section 7.2.4), and the weighting function $G_{y\kappa}$ (4.E59) to take care of the presence of a fore-and-aft force $F_x$:

$$F_{y,NL} = G_{y\kappa}\varepsilon_{NL}C_{F\gamma}\gamma \tag{7.54}$$

It is noted that $\varepsilon_{NL}$ changes with $F_z$ and $\gamma$, cf. Figure 7.13. For this reason we may better employ the method treated in Section 7.2.4 with $V_{sy,\text{eff}}$ replacing $V_{sy}$ in (7.62).

To enable calculations near or at standstill, we may add additional first-order differential equations with relaxation length $\sigma_c$. From these equations, the transient slip quantities result which act as input in the steady-state slip force formulas. We have

$$\sigma_c \frac{d\alpha'}{dt} + |V_x|\alpha' = -V^*_{sy} \tag{7.55}$$

$$\sigma_c \frac{d\kappa'}{dt} + |V_x|\kappa' = -V^*_{sx} \tag{7.56}$$

If needed, we may in the right-hand member of (7.53) replace argument $\gamma$ with the transient spin slip $\varphi'$ that results from an equation similar to (7.55):

$$\sigma_c \frac{d\varphi'}{dt} + |V_x|\varphi' = |V_x|\varphi \tag{7.57}$$

where in the right-hand member $\varphi$ is to be replaced by expression (7.13).

The contact relaxation length $\sigma_c$ may be given a small but finite value, for instance equal to half the contact length, $a_0$, at nominal load that corresponds to our findings in Chapter 9. Eqns (7.55–7.57) do not respond to load changes. For this, we rely on the effect of the carcass compliance (in conjunction with the load dependent cornering stiffness) which gives adequate results. Effectively, the resulting lateral compliance of the standing tire is

$$\frac{1}{C_{Fy}} = \frac{1}{c_{cy}} + \frac{\sigma_c}{C_{F\alpha}} \tag{7.58}$$

and the effective resulting tire relaxation length (for side slip response):

$$\sigma_\alpha = \frac{C_{F\alpha}}{C_{Fy}} = \frac{C_{F\alpha}}{c_{cy}} + \sigma_c \tag{7.59}$$

This equation may, in fact, be employed to assess the lateral carcass stiffness at ground level $c_{cy}$. From the measured tire relaxation length $\sigma_\alpha$ and cornering stiffness $C_{F\alpha}$, the lateral stiffness of the standing tire $C_{Fy}$ follows. By taking $\sigma_c$ equal to half the contact length $a$, the carcass lateral stiffness $c_{cy}$ can be determined. The relaxation length for load variations remains equal to

$$\sigma_{\alpha,Fz} = \frac{C_{F\alpha}}{c_{cy}} \tag{7.60}$$

which is a little smaller than $\sigma_\alpha$. Although this results from practical modeling considerations, we may in fact come close to the measurement results of Figure 7.10.

If one is not interested to include the ability to simulate near or at zero forward speed, the contact relaxation length $\sigma_c$ may be disregarded and taken equal to zero. In Chapter 9 the contact relaxation length $\sigma_c$ forms an essential element in the model developed for short wavelength behavior.

The deflection rates needed in Eqns (7.52, 7.53) are equal to the difference in slip velocities:

$$\dot{u} = V_{sx}^* - V_{sx} \tag{7.61}$$

$$\dot{v} = V_{sy}^* - V_{sy} \tag{7.62}$$

As has been mentioned before, the wheel slip velocity $V_s$ with components $V_{sx,y}$ is defined as the horizontal velocity of the slip point $S$ that is thought to be attached to the wheel rim a distance $r_e$, the effective rolling radius, below the wheel center in the wheel center plane:

$$V_{sx} = V_x - r_e\Omega \tag{7.63}$$

$$V_{sy} = V_y - r\dot{\gamma} \tag{7.64}$$

where $V_{x,y}$ denote the horizontal (parallel to road plane) components of the wheel center velocity. In Eqn (7.64), the camber angle is assumed to be small and $r_e$ is replaced by the loaded radius $r$.

The forces acting on the wheel rim, finally, become

$$F_{xa} = k_{cx}\dot{u} + c_{cx}u \tag{7.65}$$

$$F_{ya} = k_{cy}\dot{v} + c_{cy}v + F_{y,NL} \tag{7.66}$$

The aligning torque becomes

$$M'_z = -t_\alpha F_y \tag{7.67}$$

$$M_{zr} = M_{zr}(\gamma, \alpha', \kappa', F_z) \tag{7.68}$$

$$M_z = M'_z + M_{zr} + s \cdot F_x + M_{z,gyr} \tag{7.69}$$

The last equation forms an extension of Eqn (4.E71) through the introduction of the gyroscopic couple that follows from Eqn (7.49).

To illustrate the performance of the model, the calculated side force response to successive step changes in camber angle, slip angle, and brake slip has been presented in Figure 7.16. The contact patch relaxation length $\sigma_c$ has not been used in this example. In Chapter 8 the model is applied in the problems of controlled braking on an uneven road and on starting from standstill where in the latter case $\sigma_c$ is employed. The less demanding case of the response of side force and aligning torque to successive steps in pure side slip, shown in Figure 7.17, was calculated with the transient model of Section 7.2.3, Eqn (7.29). Very similar results are found when the enhanced model is used, that is, with the inclusion of carcass compliance and the contact mass $m_c$.

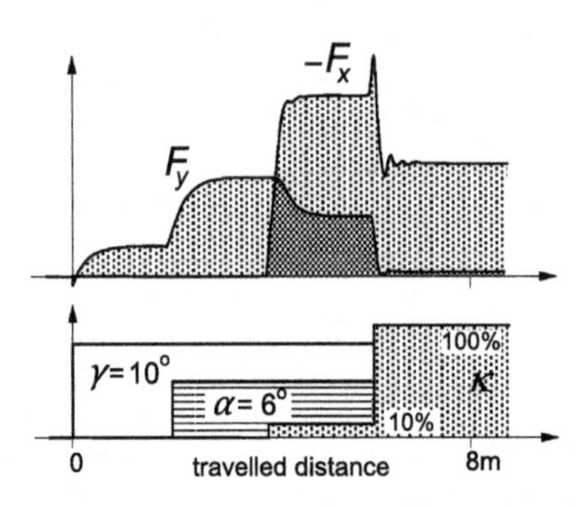


**FIGURE 7.16** Side force and brake force response to step changes in combined slip, computed with the enhanced transient tire model of the present section.

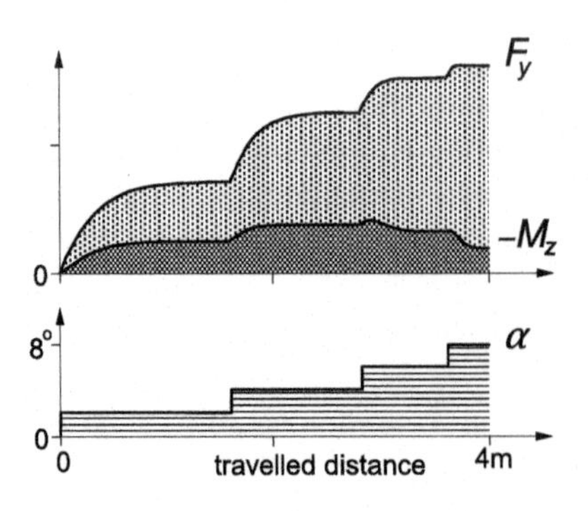


**FIGURE 7.17** Side force and aligning torque response to successive steps in side slip as computed with the transient fully nonlinear model of Section 7.2.4, Eqn (7.29).

### Exercise 7.1. Wheel Subjected to Camber, Lateral, and Vertical Axle Oscillations

Consider the wheel assembly system depicted in Figure 7.18 the axle of which is constrained to move around a longitudinal hinge that is assumed to be positioned at a constant vertical height $h$ above the smooth horizontal road surface. The hinge point moves forward in longitudinal direction with velocity $V_x$. This forward speed is assumed to increase linearly from zero at $t = t_0$ until $t = t_1$, after which the speed remains $V_x = V_{max}$. The wheel axle is subjected to forced camber variation:

$$\gamma = \gamma_o + \hat{\gamma} \sin \omega t \tag{7.70}$$

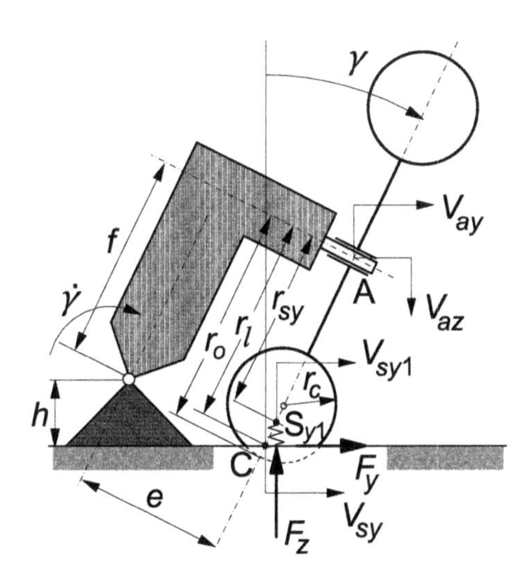


**FIGURE 7.18** Rear view of wheel assembly subjected to combined camber, vertical and lateral oscillations (Exercise 7.1).

Compute the response of the side force $F_y$ by using Eqns (7.7) and (7.11) in the Semi-Non-Linear Model (Section 7.2.2) and the Magic Formula Model of Section 4.3.2. The first for the matter of simplicity because the Fully Non-Linear Model would be more correct:

$$\frac{dv_\alpha}{dt} + \frac{1}{\sigma'} V_x v_\alpha = -V_{sy,eff} \tag{7.71}$$

$$\frac{dv_\gamma}{dt} + \frac{1}{\sigma'} V_x v_\gamma = \frac{C_{F\gamma}}{C'_{F\alpha}} V_x \sin \gamma \tag{7.72}$$

where, protected against singularity,

$$\sigma' = \max\left(\frac{C_{F\alpha}}{C_{Fy}}, 0.01\right) \tag{7.73}$$

and

$$C'_{F\alpha} = \max(C_{F\alpha}, 0.01) \tag{7.74}$$

Furthermore, we have the transient slip quantities, cf. (7.14, 7.16), that replace $\alpha^*$ and $\gamma^*$ in Eqns (4.E19–4.E30):

$$\alpha^* = \tan\alpha' = \frac{v_\alpha}{\sigma'}, \quad \gamma^* = \sin\gamma' = \frac{C_{F\alpha}}{C'_{F\gamma}}\frac{v_\gamma}{\sigma'} \tag{7.75}$$

with

$$C'_{F\gamma} = \max(C_{F\gamma}, 0.01) \tag{7.76}$$

Now take Eqns (7.41–7.44, 7.47, 7.48) for the effective side slip velocity and Eqns (4.E19–4.E30) for the calculation of the side force.
*Parameters*:
For the assembly and motion parameters, we have

$$\begin{aligned} e &= 0.15\text{ m}, \; f = 0.2\text{ m}, \; h = 0.145\text{ m} \\ t_o &= 0.1\text{ s}, \; t_1 = 0.2\text{ s}, \; t_{end} = 0.4\text{ s}, \; V_{max} = 2, 5, 10 \text{ and } 20\text{ m/s}, \\ \gamma_o &= 0.4\text{ rad}, \; \hat{\gamma} = 0.12\text{ rad}, \; \omega = 20\pi\text{ rad/s} \end{aligned}$$

for the tire parameters:

$$r_o = 0.3\text{ m}, \; r_c = 0.1\text{ m}, \; F_{zo} = 2000\text{ N}, \; C_{Fz} = 180\text{ kN/m},$$

$$C_{Fy} = 100 \text{ and } 200\text{ kN/m}$$

and for Eqns (4.E19–4.E30) with $\zeta$'s and $\lambda$'s = 1 and remaining $p$'s = 0:

$$p_{Cy1} = 1.3, \; p_{Dy1} = 1, \; p_{Ey1} = -1, \; p_{Ky1} = 20, \; p_{Ky2} = 1.5,$$

$$p_{Ky3} = 1, \; p_{Ky4} = 2, \; p_{Ky6} = 0.8, \; p_{Vy3} = 0.2$$

and for Eqns (7.43, 7.44):

$$p_{NL1} = -0.13, \; p_{NL2} = -1.5, \; p_{NL3} = -6, \; p_{NL4} = 0, \; p_{NL5} = -0.025,$$

$$p_{NL6} = 0.5, \; p_{NL7} = -0.75, \; p_{NL8} = 0.25.$$

Plot for the eight cases the computed $F_y$, $F_z$, $\rho_z$, $\gamma$, $V_x$ and for comparison the steady-state values of the side force $F_{y,st.st.}$ that arises if in Eqns (4.E19–4.E30) $\alpha^* = -V_{sy}/V_x$ and $\gamma^* = \sin\gamma$ are used directly as input variables.

Chapter 8

# Applications of Transient Tire Models

## Chapter Outline



This chapter is devoted to the application of the single contact point transient tire models as developed in the preceding chapter. The applications demonstrate the effect of the tire relaxation length on vehicle dynamic behavior. In Chapter 6 a similar first-order tire model was used to study the wheel shimmy phenomenon, cf. Eqn (6.7) with $a = 0$.


**Tire and Vehicle Dynamics. DOI: 10.1016/B978-0-08-097016-5.00008-5**

## 8.1. VEHICLE RESPONSE TO STEER ANGLE VARIATIONS

In Chapter 1, Section 1.3.2, the dynamic response of the two-degree of freedom vehicle model depicted in Figures 1.9, 1.11 to steer angle input has been analyzed. As an extension to this model, we will introduce tires with lagged side force response. The system remains linear and we may use Eqn (7.18). The relaxation length will be denoted by $\sigma$. We have the new set of equations:

$$m(\dot{v} + Vr) = C_{F\alpha 1}\alpha_1' + C_{F\alpha 2}\alpha_2' \tag{8.1}$$

$$I\dot{r} = aC_{F\alpha 1}\alpha_1' - bC_{F\alpha 2}\alpha_2' \tag{8.2}$$

$$\frac{\sigma}{V}\dot{\alpha}_1' + \alpha_1' = \alpha_1 \tag{8.3}$$

$$\frac{\sigma}{V}\dot{\alpha}_2' + \alpha_2' = \alpha_2 \tag{8.4}$$

$$\alpha_1 = \delta - \frac{1}{V}(v + ar) \tag{8.5}$$

$$\alpha_2 = -\frac{1}{V}(v - br) \tag{8.6}$$

Figure 8.1 presents the computed frequency response functions, which hold for the extended system, for different values of the relaxation length. In Figure 8.2 the corresponding step response functions are depicted. The diagrams show that the influence of the tire lag on the vehicle motion is relatively small. At higher values of speed, the effect diminishes and may become negligible.

However, for closed-loop vehicle control systems, the additional phase lag caused by the relaxation lengths may significantly affect the performance.

An interesting effect can be seen to occur near the start of the step response functions. With the relaxation length introduced, the responses appear to exhibit a more relaxed nature. The yaw rate response shows some initial lag.

## 8.2. CORNERING ON UNDULATED ROADS

When a car runs along a circular path over an uneven road surface, the wheels move up and down and may even jump from the road while still the centripetal forces are to be generated. Under these conditions, the wheels run at slip angles which may become considerably larger than on a smooth road. Consequently, on average, the cornering stiffness diminishes. This phenomenon was examined in Chapter 5, Section 5.6, where the stretched string model was found to be suitable to explain the decrease in cornering stiffness if the string relaxation length is made dependent on the varying wheel load to imitate the model provided with tread elements, cf. Figure 5.51.

These tire models, however, are considered to be too complicated to be used in vehicle simulation studies. Instead, we will try the very simple transient tire model as discussed in Section 7.2.1 and compare the results with experimental

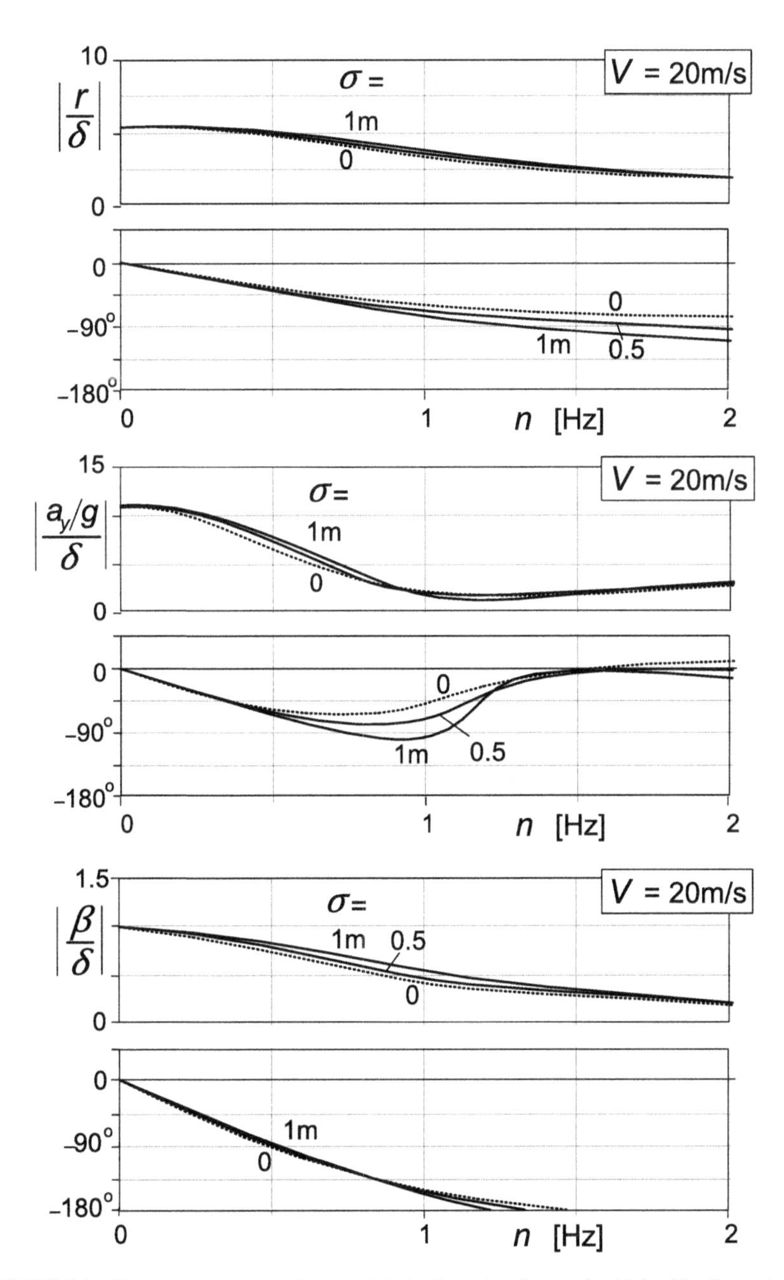


**FIGURE 8.1** Frequency response of yaw rate, lateral acceleration, and vehicle side slip angle to steer angle of vehicle model featuring tires with lagging side force response, cf. Figure 1.15.

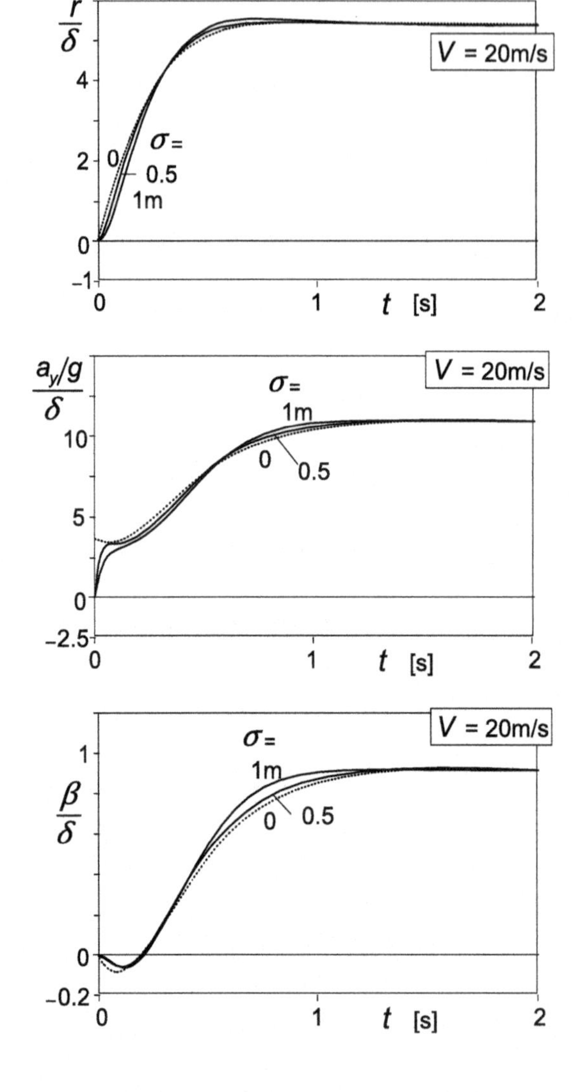


**FIGURE 8.2** Step response of yaw rate, lateral acceleration, and vehicle side slip angle to steer angle of vehicle model featuring tires with lagging side force response.

findings. First, the small slip angle linear model will be considered. We use Eqn (7.20) which may be given the form:

$$\sigma \frac{dF_y}{ds} + F_y = F_{yss} = C_{F\alpha}\alpha \tag{8.7}$$

where $s$ denotes the traveled distance and $ds = V_x dt$. The slip angle is assumed to be constant while the vertical load varies periodically.

$$F_z = F_{zo} + \Delta F_z = F_{zo} + \hat{F}_z \sin 2\pi \frac{s}{\lambda} \tag{8.8}$$

Essential is that both the cornering stiffness and the relaxation length vary with the vertical load:

$$C_{F\alpha} = C_{F\alpha}(F_z) \text{ and } \sigma = \sigma(F_z) \tag{8.9}$$

Figure 8.3 illustrates the situation.

The loss in cornering power can be divided into the 'static' loss and the 'dynamic' loss. The static loss arises due to the curvature of the cornering stiffness vs wheel load characteristic. A similar loss has been experienced to occur as a result of lateral load transfer in the analysis of steady-state cornering of an automobile, cf. Figure 1.7. Figure 8.4 explains the situation at sinusoidally changing wheel load.

As will be shown in the analysis below, the dynamic loss appears to be attributed to the rate of change of the relaxation length with wheel load. This, of course, can only occur because of the fact that the cornering stiffness varies

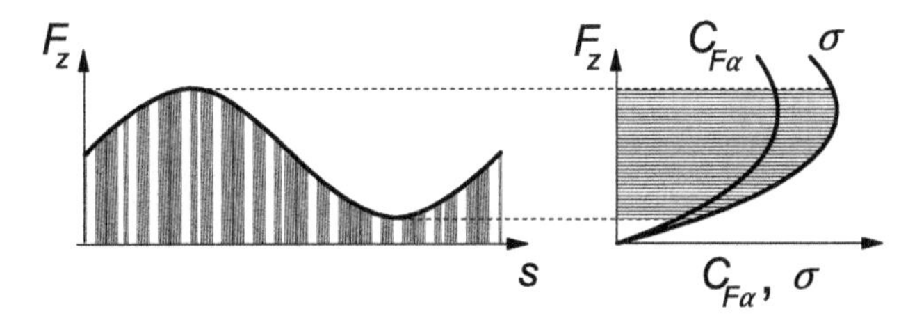


**FIGURE 8.3** Cornering stiffness and relaxation length varying with wheel load.

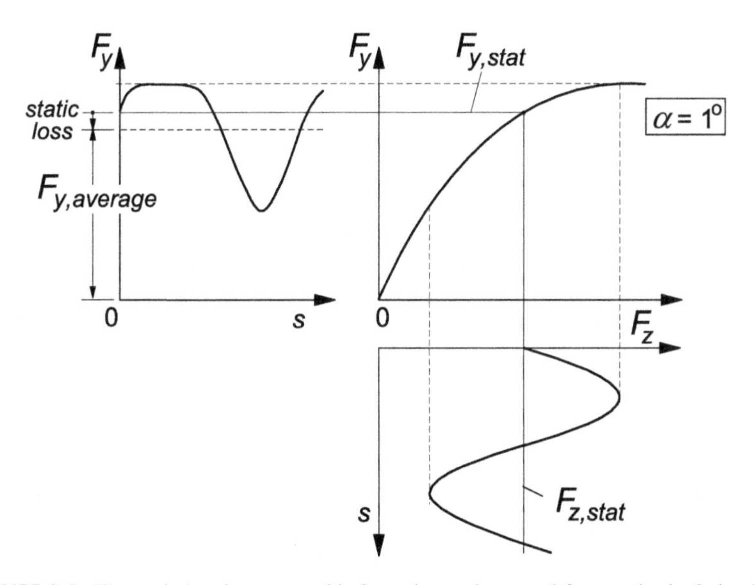


**FIGURE 8.4** The static loss in average side force due to the curved force vs load relationship.

with wheel load since that rate of change forms the input in Eqn (8.7). Apparently, with both $F_y$ and $\sigma$ being dependent on $F_z$ while $\alpha$ in the present analysis only affects $F_y$, the phenomenon is nonlinear in $F_z$ and linear in $\alpha$.

If we assume a quadratic function for the $C_{F\alpha}$ vs $F_z$ relationship with the curvature $bC_{F\alpha o}$

$$C_{F\alpha} = C_{F\alpha o}\left(1 + a\Delta F_z - \frac{1}{2}b\Delta F_z^2\right) \tag{8.10}$$

the static loss can be simply found to become

$$C_{F\alpha,\text{stat,loss}} = \frac{1}{4}b\hat{F}_z^2 C_{F\alpha o} \tag{8.11}$$

The occurrence of the dynamic loss may be explained by assuming a linear variation of both the cornering stiffness and the relaxation length with load:

$$\begin{aligned}\sigma &= cF_z \\ C_{F\alpha} &= eF_z\end{aligned} \tag{8.12}$$

Equation (8.7) becomes herewith:

$$cF_z\frac{\mathrm{d}F_y}{\mathrm{d}s} + F_y = eF_z\alpha \tag{8.13}$$

or simplified:

$$z\frac{\mathrm{d}y}{\mathrm{d}s} + y = \beta z \tag{8.14}$$

with $z = cF_z$, $y = F_y$, $\beta = e\alpha/c$. Consider a truncated Fourier series approximation of the periodic solution:

$$y = y_o + y_{s1}\sin\omega_s s + y_{c1}\cos\omega_s s + y_{s2}\sin^2\omega_s s + \cdots \tag{8.15}$$

and its derivative:

$$\frac{\mathrm{d}y}{\mathrm{d}s} = y_{s1}\omega_s\cos\omega_s s - y_{c1}\omega_s\sin\omega_s s + 2y_{s2}\omega_s\cos 2\omega_s s - \cdots \tag{8.16}$$

while the input

$$z = z_o + \hat{z}\sin\omega_s s \quad \text{with} \quad \omega_s = \frac{2\pi}{\lambda} \tag{8.17}$$

Substitution of (8.15, 8.16, 8.17) in (8.14) and subsequently making the coefficients of corresponding terms in the left and right members equal to each other yields for the average output:

$$y_o = \beta z_o\left(1 - \frac{1}{2}\frac{\omega_s^2\hat{z}^2}{1 + \omega_s^2 z_o^2}\right) \tag{8.18}$$

The average side force $F_{y,\mathrm{ave}}$ expressed in terms of the original quantities now reads as

$$F_{y,\mathrm{ave}} = C_{F\alpha o}\left(1 - \frac{1}{2}\,\frac{\omega_s^2 c^2 \hat{F}_z^2}{1 + \omega_s^2 c^2 F_{zo}^2}\right)\alpha \quad \text{with } c = \frac{\mathrm{d}\sigma}{\mathrm{d}F_z} \tag{8.19}$$

For large $F_z$ amplitudes and short wavelengths $\lambda = 2\pi/\omega_s$ the formula shows that the dynamic loss approaches 50% of the side force generated on smooth roads. As we will see, this finding agrees very well with simulation and test results.

The aligning torque variation may be computed as well. In their study, Takahashi and Pacejka (1987) employed the equation based on the string concept and added the gyroscopic couple. We obtain according to Eqns (5.135) and (5.178), the latter expression corresponding with (7.49):

$$\sigma\frac{\mathrm{d}M_z'}{\mathrm{d}s} + M_z' = -C_{M\alpha}\alpha \tag{8.20}$$

and

$$M_{z,\mathrm{gyr}} = C_{\mathrm{gyr}}V^2\frac{\mathrm{d}F_y}{\mathrm{d}s} \tag{8.21}$$

The total moment responding to the slip angle becomes

$$M_z = M_z' + M_{z,\mathrm{gyr}} \tag{8.22}$$

Experiments have been conducted to assess the relevant parameters of a 195/60R14-87H tire at various values of the vertical load. The diagrams of Figure 8.5 present the characteristics for the relaxation length, the cornering stiffness, and the aligning stiffness. The relaxation length was determined by curve fitting of the data obtained from a swept sine steering oscillation test at a speed of 30 km/h in the frequency range of 0–2 Hz. The gyroscopic

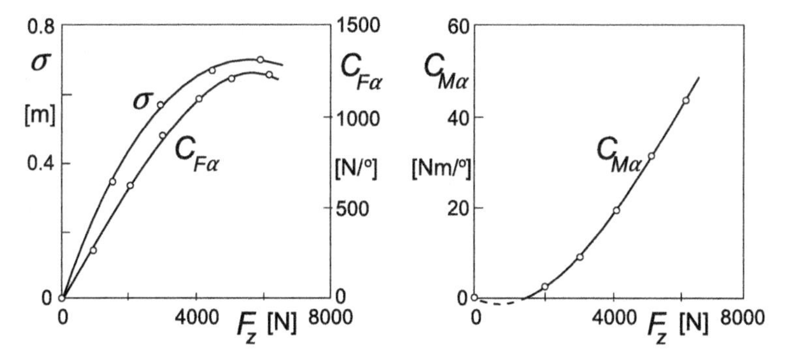


**FIGURE 8.5** Load dependency of tire parameters according to measurements and third-degree polynomial fits.

coefficient $C_{gyr}$ was assessed in a similar test but conducted at a speed of 170 km/h. The resulting value turns out to be $C_{gyr} = 3.62 \times 10^{-5}$ s$^2$ that corresponds to the nondimensional coefficient $c_{gyr} \approx 0.5$, cf. Eqn (5.179).

To assess the validity of the calculated force and moment response to load variations experiments have been carried out on a 2.5-m drum with a special test rig at the Delft University of Technology. The wheel axle can be moved vertically while set at a given slip angle. The side force and moment measured at that slip angle have been corrected for the responses belonging to a slip angle equal to zero.

In Figure 8.6, in the left-hand diagrams, the test results are presented. The wheel axle is vertically oscillated with frequencies up to 8 Hz and with an amplitude that corresponds to the ratio of amplitude and mean value of the wheel load equal to 0.9. The wheel runs at a constant small slip angle of one degree with a speed of 30 km/h. The response at the low frequency of 0.5 Hz may be considered as quasi-static and varies periodically, practically in phase with the axle motion. To enable a proper comparison at the different frequencies, the responses have been plotted against traveled distance divided by the current wavelength. The right-hand diagrams show the corresponding calculated responses. In view of the very simple set of equations that have been used, the agreement between test and model can be considered to be very good. Calculations conducted with the stretched string model of Section 5.6 were not much better. Only the moment responses turned out to be a little closer to the experimental findings. In the responses, two major features can be identified.

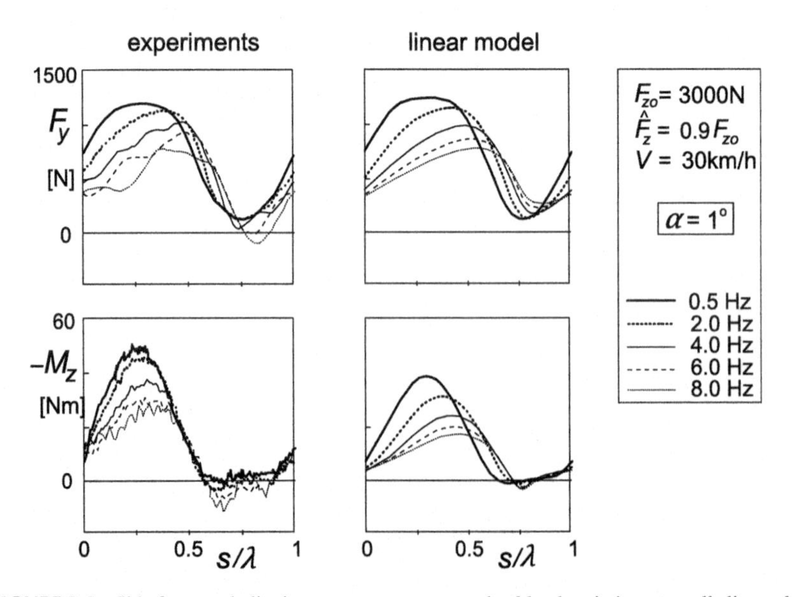


**FIGURE 8.6** Side force and aligning torque response to wheel load variation at small slip angle.

First, we observe that the responses show some lag with respect to the input that grows with increasing frequency $n$ which means: with decreasing wavelength $\lambda$. Second, and more important, we have the clear drop in average value of the side force which becomes larger at increasing frequency. After subtracting the static drop that occurs at the lowest frequency, the dynamic loss remains.

In Figure 8.7 the static and the dynamic parts of the decrease of the average side force, as a ratio to the side force at zero vertical load amplitude, have been plotted versus frequency or wavelength for different input amplitude ratios. The model shows reasonable agreement with the experimental results. It has been found that the correspondence becomes a lot better when the speed is increased to 90 km/h. Then, the wavelength is three times larger and a better performance of the model prevails. The curve of Figure 8.7 that holds for the amplitude ratio 0.95, where the tire almost loses contact, appears to approach the value of ca. 0.5 when the frequency increases to large values. This indeed corresponds to the analytical result established by Eqn (8.19). On the other hand, the more advanced model of Section 5.6 predicts a possibly even larger reduction at the verge of periodic lift-off; cf. Figure 5.51 where the wavelength could be diminished to levels as small as the static contact length $2a_o$.

It may be obvious that an analogous model, Eqn (7.9) with $\kappa' = u/\sigma_\kappa$, can be used for the in-plane non-steady-state response to vertical load variations. Without special measures it is virtually impossible to maintain a certain level of the longitudinal slip $\kappa$ at varying contact conditions. At a fixed braking torque, the average longitudinal braking force remains constant when the wheel keeps rolling. When the wheel runs over an uneven road surface and the wheel load changes continuously, the effect of the tire lag is an increased average level of

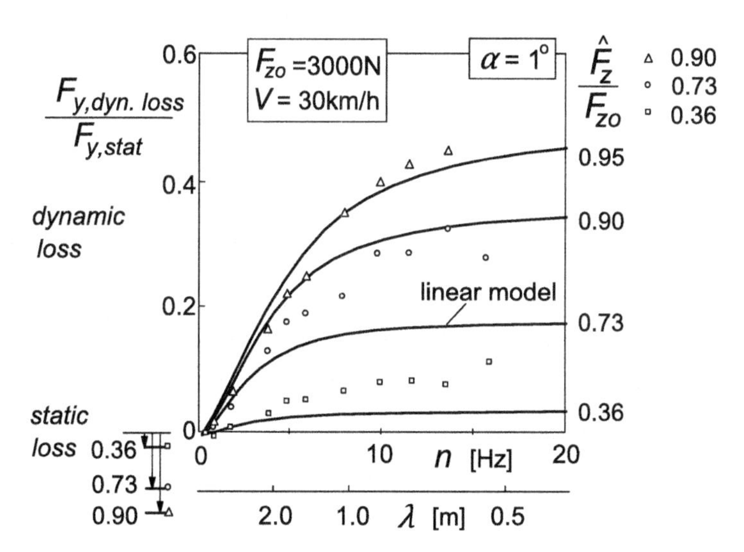


**FIGURE 8.7** Dynamic side force loss at a small slip angle for different load amplitude ratios.

the longitudinal slip ratio. Applications of the single contact point in-plane transient model follow in Sections 8.3 and 8.5 where problems related with tire out of roundness and braking on undulated roads are discussed.

The theory can be extended to larger values of slip if the nonlinear model according to Eqns (7.25, 7.26) is employed. The equations may be used here in the form:

$$\sigma^* \frac{dv}{ds} + v = \sigma^* \alpha \tag{8.23}$$

with

$$\sigma^* = \frac{\sigma_o}{C_{F\alpha}} \frac{F_y}{\alpha'} \tag{8.24}$$

which quantity is a function of both the wheel load $F_z$ and the transient slip angle or deformation gradient $\alpha'$. It reduces to the relaxation length $\sigma_0$ ($=\sigma$ at $\alpha' = 0$) when $\alpha'$ tends to zero and $F_y/\alpha'$to $C_{F\alpha}$. Obviously, offsets of the side force vs slip angle characteristic near the origin have been disregarded here. The transient slip angle follows from

$$\alpha' = \frac{v}{\sigma^*} \tag{8.25}$$

The side force and moment are obtained by using the steady-state characteristics:

$$F_y = F_{y,ss}(\alpha', F_z) \tag{8.26}$$

$$M'_z = M'_{z,ss}(\alpha', F_z) \tag{8.27}$$

With the gyroscopic couple, cf. Eqns (8.21), (7.5) and (5.179)

$$M_{z,gyr} = c_{gyr} m_t V^2 \frac{dv}{ds} \tag{8.28}$$

the total moment becomes

$$M_z = M'_z + M_{z,gyr} \tag{8.29}$$

The steady-state characteristics as measured on the road at a number of wheel loads are presented in Figure 8.8.

As has been noted in Section 7.2.3 the assessment of $\sigma^*$ requires some attention. In the investigation done by Pacejka and Takahashi (1992) and Higuchi (1997), the solution was obtained through iterations. Alternatively, we may use information from the previous time step to assess $\sigma^*$. When the load is rapidly lowered, violent vibrations may occasionally show up in the solution.

With this nonlinear model, simulations have been done and experiments have been carried out for a slip angle equal to six degrees. Plots of the resulting responses are shown in Figure 8.9. The model appears to be successful in predicting the side force variation caused by the wheel load excitation at

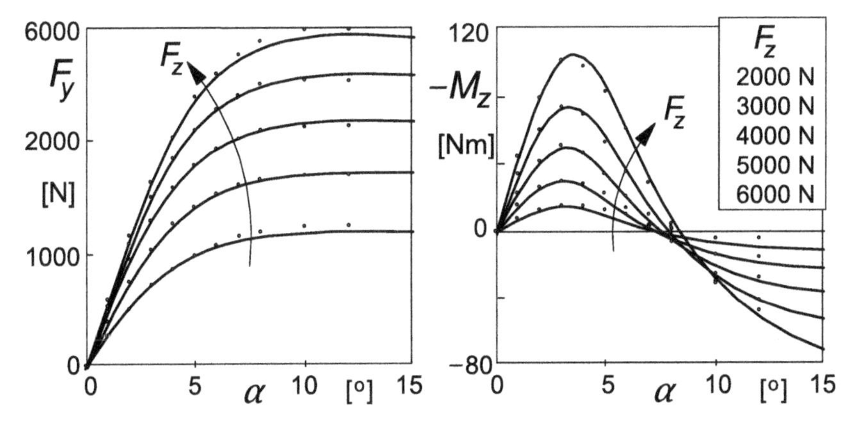


**FIGURE 8.8** Steady-state side force and aligning torque characteristics as measured on the road.

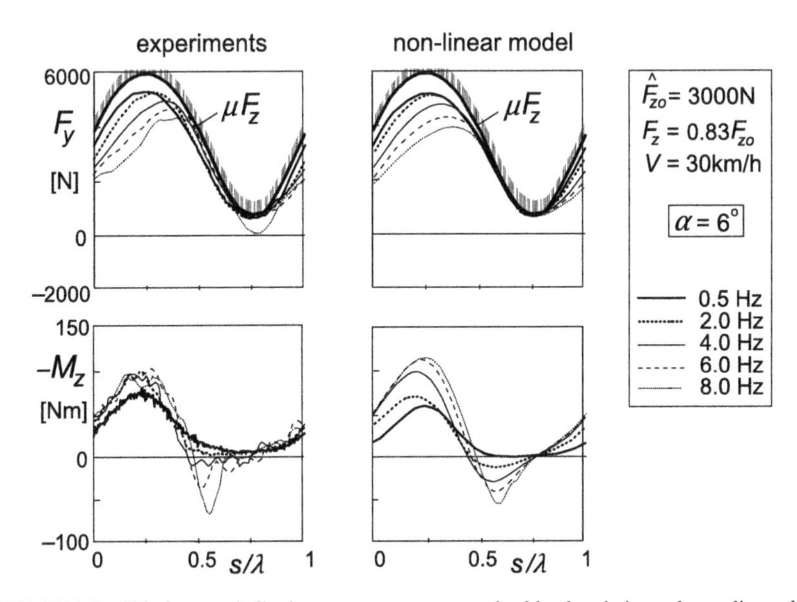


**FIGURE 8.9** Side force and aligning torque response to wheel load variation at large slip angle.

different frequencies. The load amplitude ratio of 0.83 gave rise to full sliding in certain intervals of the periodic motion. This can be seen to occur when the force signal touches the curve representing the maximally achievable side force $\mu F_z$. In this respect, the model certainly performs well which cannot be said of the alternative model governed by Eqn (7.20) with $F_{y,ss}$ calculated with Eqn (8.26) where $\alpha'$ is replaced by $\alpha$. This model gives rise to a calculated force response that exceeds this physical limit when the load approaches its minimal value due to the phase lag of the response. The dynamic loss of the average side

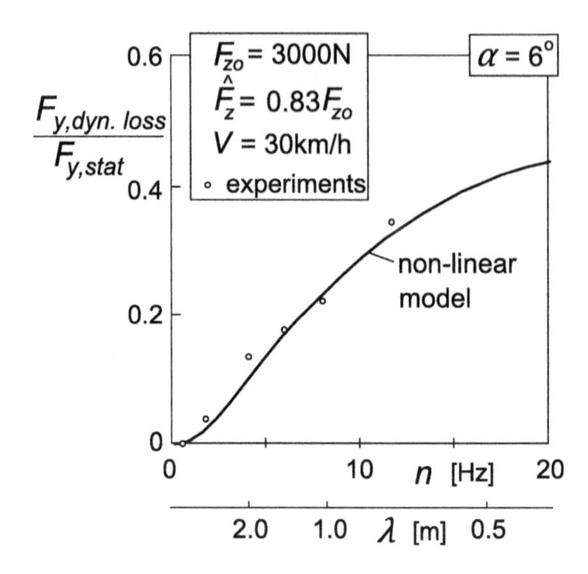


**FIGURE 8.10** The dynamic loss at a large constant slip angle and a large load amplitude ratio.

force follows a tendency similar to what was found with the linear model at small slip angles. The result obtained at a constant slip angle of six degrees is shown in the diagram of Figure 8.10. The correspondence with the experimental outcome becomes even better than in the case of a slip angle equal to one degree, Figure 8.7.

Very similar results have been obtained with the more robust enhanced model of Section 7.3. The difference is that a small dynamic force acts on the little mass of the contact patch used in the model. This causes the ground force to be slightly different from the force felt at the wheel rim. It is, of course, the ground force that cannot exceed the physical limit mentioned above.

## 8.3. LONGITUDINAL FORCE RESPONSE TO TIRE NONUNIFORMITY, AXLE MOTIONS, AND ROAD UNEVENNESS

In this section, the response of the axle forces $F_x$ and $F_z$ to in-plane axle motions ($x$, $z$), road waviness and tire nonuniformities will be discussed. For a given tire–wheel combination, the response depends on rolling speed and frequency of excitation. This dependency, however, appears to be of much greater significance for the fore and aft force variation $F_x$ than for the vertical load variation $F_z$.

The vertical force response has an elastic component and a component due to hysteresis. A similar mechanism is responsible for the generation of (at least an important portion of) the rolling resistance. The hysteresis (damping) part of the vertical load response appears to rapidly decrease to almost negligible levels at from zero increasing forward speed, cf. Pacejka (1981a) and Jianmin et al. (2001). We simplify the relationship between normal load $F_z$ and radial tire deflection $\rho$ using the radial stiffness $C_{Fz}$:

$$F_z = C_{Fz}\rho \tag{8.30}$$

In Chapter 9, tire inertia and other influences will be introduced to improve the relationship at higher frequencies and shorter wavelengths of road unevennesses.

A coupling between vertical deflection variation and longitudinal force appears to exist that is due to the resulting variation of the effective rolling radius of the tire.

For the non-steady-state analysis, we will employ the linear Eqn (7.9) that may take the form with $V = V_x$ (assumed positive):

$$\frac{du}{dt} + \frac{1}{\sigma_\kappa} Vu = V\kappa = -V_{sx} \tag{8.31}$$

The input is the longitudinal slip velocity that is defined as

$$V_{sx} = V - r_e \Omega \tag{8.32}$$

Obviously, the effective rolling radius $r_e$ plays a crucial role here.

## 8.3.1. Effective Rolling Radius Variations at Free Rolling

For a wheel with tire that is uniform and rolls freely at constant speed over an even horizontal road surface, the tractive force required is due to rolling resistance alone. Under these conditions, the effective rolling radius $r_e$ is defined to relate speed of rolling $\Omega$ with forward speed $V$:

$$V = r_e \Omega \tag{8.33}$$

The center of rotation of the wheel body $S$ lies at a distance $r_e$ below the wheel spin axis. As has been discussed in Chapter 1, by definition, this point which can be imagined to be attached to the wheel rim, is stationary at the instant considered if the wheel rolls freely (cf. Figure 8.11 with slip speed $V_{sx} = 0$). In general, the effective rolling radius changes with tire deflection $\rho$. We may write

$$r_e = r_f - f(\rho) \tag{8.34}$$

with $r_f$ denoting the free (undeformed) radius that may vary along the tire circumference due to tire nonuniformity. We have for the loaded radius:

$$r = r_f - \rho \tag{8.35}$$

Figure 8.12 shows variations of the effective rolling radius $r_e$ and the loaded radius $r$ as functions of wheel load $F_z$ for two different tires at different stages of wear and at different speeds as measured on a drum surface of 2.5-m diameter.

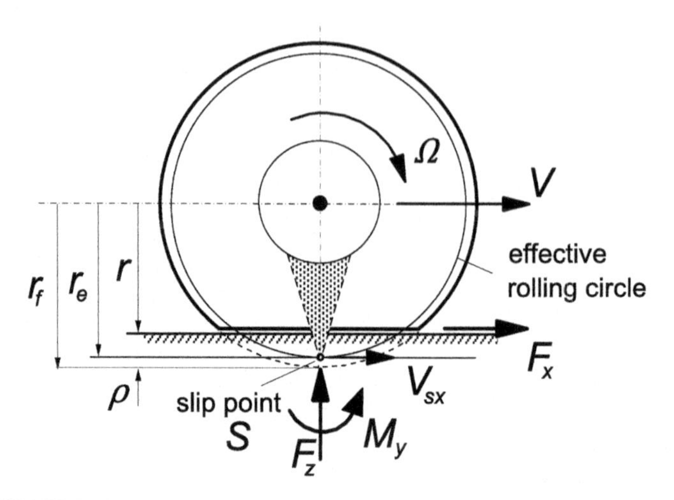


**FIGURE 8.11** The rolling wheel subjected to longitudinal slip (slip speed $V_{sx}$).

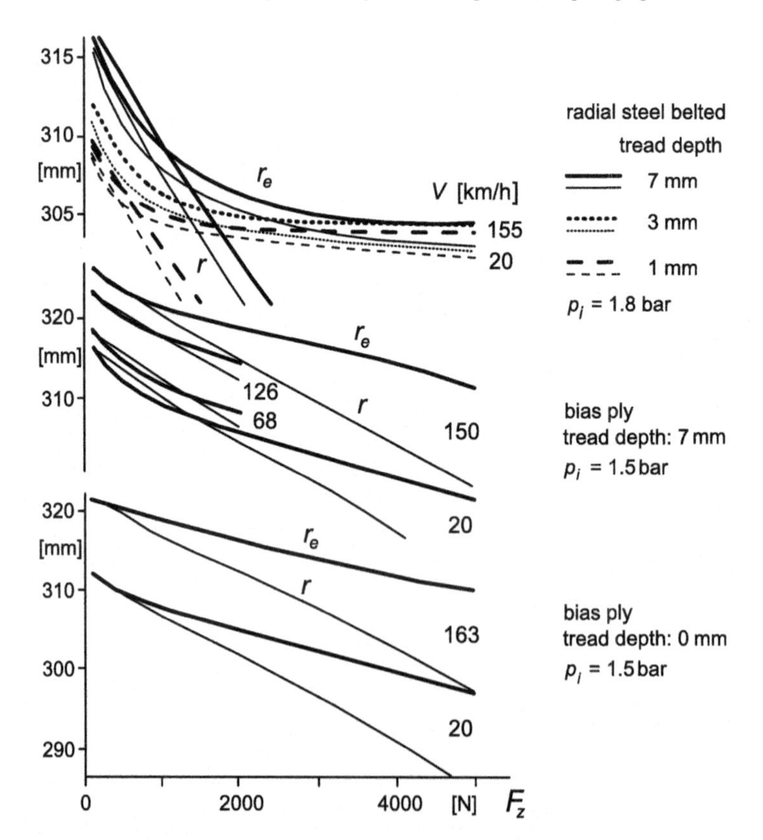


**FIGURE 8.12** Tire effective rolling radius and loaded radius as a function of wheel load, measured on 2.5-m drum at different speeds and for different tread depths and tire design. Note the different scales for the radial and bias-ply tire radii.

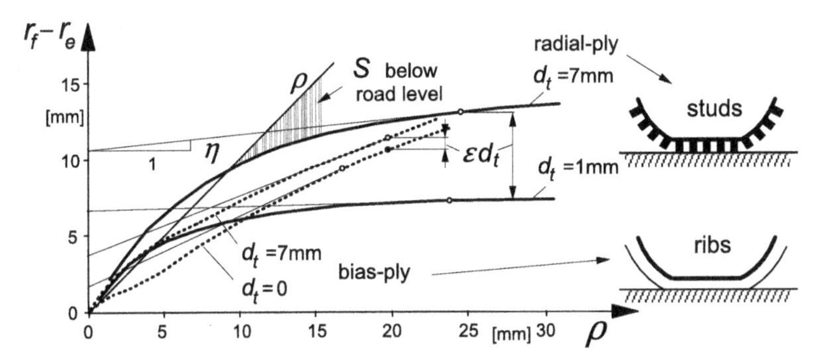


FIGURE 8.13 Reduction of effective rolling radius $r_e$ with respect to tire free radius $r_f$ due to tire radial deflection $\rho$ and tread depth change $d_t$. Situation at nominal load, $F_{zo} = 3000$ N, has been indicated by circles.

The reduction of the effective rolling radius $f(\rho)$ with respect to its theoretical original value $r_f$ at zero wheel load ($\rho = 0$) that has been deduced from diagram 8.12 has been presented in Figure 8.13. The influence of a change in tread depth has been shown. The slope $\eta$ and the coefficient $\varepsilon$ indicate the influence of changes in deflection and tread depth near the nominal load.

Figure 8.14 tries to explain the observation regarding the location of the slip point $S$: possible locations have been indicated. At zero wheel load, $S$ lies at road level or a little lower (due to air and bearing resistance which slows the wheel down at the expense of some slip). A small wheel load causes a number of the originally radially directed tread elements to assume a vertical orientation. The rotation of these tread elements is lost which causes the effective radius to become smaller. Point $S$ may then turn out to lie above road level! At higher loads, we get the more commonly known situation of $S$ located below road level. It has been assumed here that the belt with radius $r_c$ is inextensible. Then, when the lower tread elements have a vertical orientation and translate backwards with respect to the wheel center, the wheel will in one revolution cover a distance equal to the circumference of the belt. This means that the effective rolling radius equals the radius of the belt. When the tire possesses a ribbed tread pattern, it is

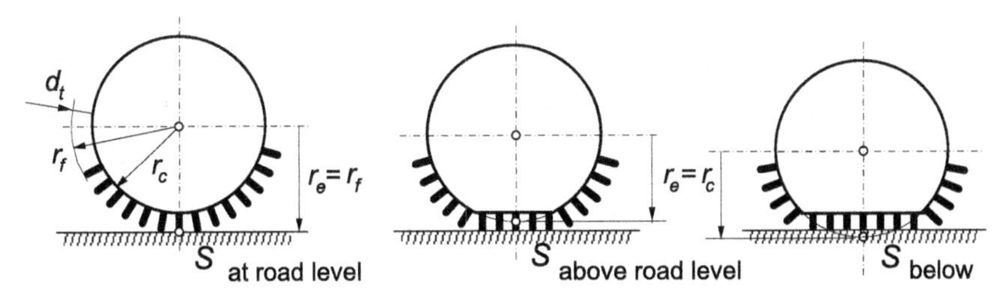


FIGURE 8.14 Location of the center of rotation, the slip point $S$, at three different radial deflections.

expected that due to the greater coherence between rib elements in comparison to the independent studs of the radial-ply tire, some rotation still occurs in the ribs in the contact zone which increases the effective radius and pushes point S earlier below the road surface. This theory seems to be supported by the curves of the new bias-ply tire. The degree of effective incoherence $\varepsilon$ is estimated for the radial-ply tire to be equal to ca. 0.9 and for the bias-ply tire equal to 0.2 at the rated load. Because we have plotted in Figure 8.13 the reduction of the effective radius with respect to the free unloaded radius: $r_f - r_e$, the decrease of $r_e$ due to the wearing off of the tread rubber is not shown directly. Only the indirect effect of tread wear is shown through the tread incoherence parameter $\varepsilon$. If due to the vertical tire deflection the tread band is compressed and has decreased in circumference, the effective radius diminishes. This property, which is more apparent to happen with the bias-ply tire, is partly responsible for the slope of the curves shown at larger vertical deflections. This slope, denoted with $\eta$, is estimated to be equal to about 0.1 for the new radial-ply tire with an almost inextensible belt and 0.4 for the new bias-ply tire at their nominal load.

The analysis will be kept linear and we assume small deviations from the undisturbed condition. In the neighborhood of this average state, we write for the effective radius, the loaded radius, and the radial deflection:

$$r_e = r_{e,o} + \tilde{r}_e, \quad r = r_o + \tilde{r}, \quad \rho = \rho_o + \tilde{\rho} \tag{8.36}$$

Small deviations from the undisturbed condition are indicated by a tilde. Variation of the free radius may occur along the circumference of the tire. This kind of nonuniformity (out of roundness) is composed of two contributions: one is due to variations of the carcass radius $r_c$ and the other due to variations of the tread thickness $d_t$. We have

$$\tilde{r}_f = \tilde{r}_c + \tilde{d}_t \tag{8.37}$$

The value of out of roundness is considered to be present at the moment considered along the radius pointing to the contact center. The following linear relation between the effective radius variation and the imposed disturbance quantities is found to exist:

$$\tilde{r}_e = \tilde{r}_f - \varepsilon\tilde{d}_t - \eta\tilde{\rho} = \tilde{r}_c + (1-\varepsilon)\tilde{d}_t - \eta\tilde{\rho} \tag{8.38}$$

The observation that a reduction in thickness of the tread rubber of a radial-ply tire produces a decrease in effective rolling radius (at $F_{zo}$), amounting to only a small fraction of the reduction in tread depth, is reflected by the factor $(1-\varepsilon)$ in the right-hand expression of (8.38). The coupling coefficient $\eta$ is relatively small for belted tires. In contrast to bias-ply tires, the effective rolling radius of a belted tire does not change very much with deflection once the initial tread thickness effect (at small contact lengths with tread elements still oriented almost radially) has been overcome.

## 8.3.2. Computation of the Horizontal Longitudinal Force Response

The horizontal force response on undulated roads is composed of the following three contributions: the horizontal component of the normal load, the variation of the rolling resistance, and the longitudinal force response to the variation of the wheel longitudinal slip. The horizontal in-plane force designated with $X$ becomes, with linearized variations, cf. Figure 8.15,

$$X = X_o + \tilde{X} = -F_{ro} + \tilde{F}_x + F_{zo}\frac{\mathrm{d}w}{\mathrm{d}s} = -F_{ro} - \tilde{F}_r + \tilde{F}_\kappa + F_{zo}\frac{\mathrm{d}w}{\mathrm{d}s} \quad (8.39)$$

with $F_{ro}$ denoting the average rolling resistance force. For this occasion, it is assumed that the variation in rolling resistance force is directly transmitted to the tread through changes in vertical load. We write

$$\tilde{F}_r = A_r\tilde{F}_z \quad (8.40)$$

where the variation of the vertical load results from variations in deflection and possibly from changes in radial stiffness along the tire circumference:

$$\tilde{F}_z = C_{Fzo}\tilde{\rho} + \rho_o\tilde{C}_{Fz} \quad (8.41)$$

or expressed in terms of the variation in radial static deflection:

$$\tilde{\rho}_s = -\rho_o\frac{\tilde{C}_{Fz}}{C_{Fzo}} \quad (8.42)$$

we may write

$$\tilde{F}_z = C_{Fzo}(\tilde{\rho} - \tilde{\rho}_s) \quad (8.43)$$

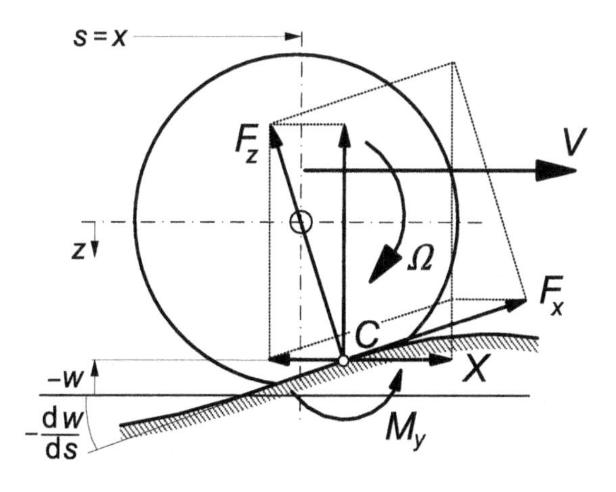


**FIGURE 8.15** In-plane excitation by road unevenness and by vertical and fore and aft axle motions.

The variation of the tangential slip force becomes, with the transient longitudinal slip $\kappa'$,

$$\tilde{F}_\kappa = C_{F\kappa}\kappa' = C_{F\kappa}\frac{u}{\sigma_\kappa} \tag{8.44}$$

where the tire deflection $u$ follows from Eqn (8.31). The longitudinal slip velocity appearing as input in (8.31) is governed by Eqn (8.32). With the variations in forward velocity, speed of revolution, and effective rolling radius, we find in linearized form

$$V_{sx} = \tilde{V} - \Omega_o\tilde{r}_e - r_{eo}\tilde{\Omega} \tag{8.45}$$

The variation of the effective rolling radius follows from (8.38), where the deflection is a result of the combination of wheel vertical displacement, tire out of roundness, and road height changes:

$$\tilde{\rho} = z + \tilde{r}_c + \tilde{d}_t - w \tag{8.46}$$

The wheel angular velocity results from the dynamic equation

$$I_w\dot{\Omega} = -rF_x - M_y \tag{8.47}$$

in which $I_w$ denotes the wheel polar moment of inertia (including a large part of the tire that vibrates along with the wheel rim if the frequency lies well below the in-plane first natural frequency, cf. Chap. 9).

The moment $M_y$ acts about the transverse axis through the contact center $C$, Figure 8.15, and is composed of the part due to rolling resistance and a part due to eccentricity of the tread band which constitutes the first harmonic of the tire out of roundness. This latter part runs 90 degrees ahead in phase with respect to the loaded radius variation that is sensed at the contact center. With subscript 1 referring to the first harmonic we have

$$M_y = M_r + F_z\tilde{r}_{1,(\Omega t+\pi/2)} \tag{8.48}$$

The rolling resistance moment $M_r$ is here supposed to be directly connected with the rolling resistance force $F_r$ which is a simplifying assumption that has a negligible effect on the resulting responses. As a result, the rolling resistance components cancel out in the right-hand member of Eqn (8.47). This equation now reduces to

$$I_w\dot{\Omega} = -r_oF_\kappa - F_{zo}\tilde{r}_{1,(\Omega t+\pi/2)} \tag{8.49}$$

Collection of the relevant Eqns (8.31, 8.38–8.43, 8.45) and elimination of all variables, except the input quantities

$$\tilde{V}(=\dot{\tilde{x}}), z, w, \tilde{r}_c, \tilde{d}_t, \tilde{\rho}_s, \tilde{r}_1(\tilde{r}_1 = \text{1st harmonic of } \tilde{r}_f = \tilde{r}_c + \tilde{d}_t \tag{8.50}$$

and the output $\tilde{X}$ and, for now, also $\tilde{r}_e$ and $\tilde{\rho}$, yields the following Laplace transformed representation of the horizontal force variation with $s$ designating the Laplace variable or $i\omega$:

$$\tilde{X} = C_{F\kappa}\frac{I_w\Omega_o s\cdot\tilde{r}_e - I_w s^2\cdot\tilde{x} - F_{zo}r_o\exp(1/2\pi s/\Omega_o)\cdot\tilde{r}_1}{\sigma_\kappa I_w s^2 + I_w V_o s + C_{F\kappa}r_o^2} - A_r C_{Fz}\cdot(\tilde{\rho} - \tilde{\rho}_s) + F_{z0}s\cdot w/V_o \tag{8.51}$$

where with (8.38) and (8.46) we have the variations of effective radius and tire deflection:

$$\tilde{r}_e = \eta(w - z) + (1 - \eta - \varepsilon)\tilde{d}_t + (1 - \eta)\tilde{r}_c \tag{8.52}$$

and

$$\tilde{\rho} = z + \tilde{r}_c + \tilde{d}_t - w \tag{8.53}$$

From this expression (8.51) of the combined response, the individual transfer functions or the frequency response functions to the various input quantities may be written explicitly. For the latter functions to obtain, we simply have to take the coefficients of one of the input variables (8.50) in (8.51), after (8.52) and (8.53) have been inserted in (8.51), and replace $s$ by $i\omega$ or $i\Omega_o$ depending on the type of input variable.

Figure 8.16 presents both theoretical and experimental data, the latter being obtained from tests performed on a 2.5-m drum test stand provided with

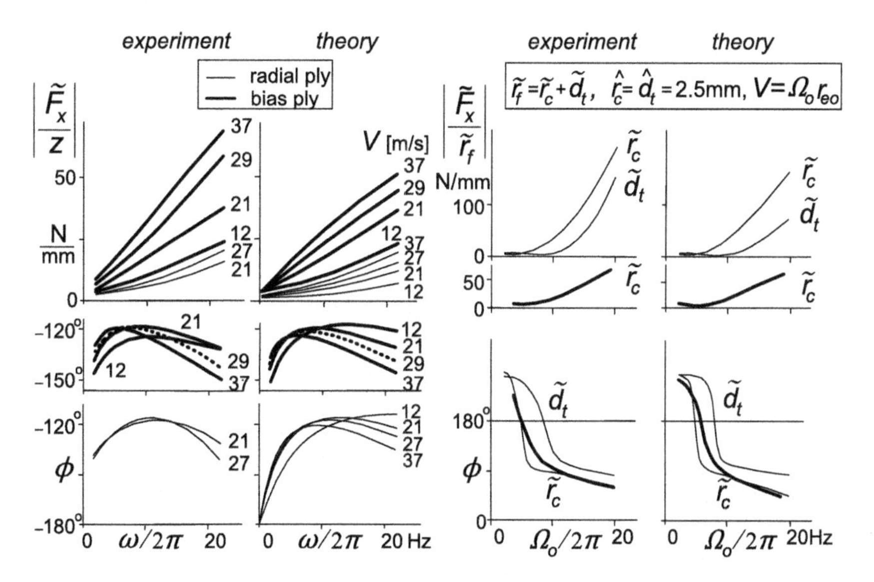


**FIGURE 8.16** Experimental and theoretical frequency response plots of the longitudinal force $F_x$ to vertical axle oscillations $z$ and to out of roundness (1st harmonic of variations in carcass radius $r_c$ or in tread thickness $d_t$). Average vertical load $F_{zo} = 3000$ N.

**TABLE 8.1** Tire In-plane Parameter Values for the Radial and Bias-ply Tires

| |
|---|
| ***Radial steel-belted*** **($p_i$ = 1.8 bar, tread depth (at $\hat{d}_t = 0$) = 6.5 mm)** |
| $I_w = 0.8$ ($z$) or 1.1 ($\tilde{r}_f$) kgm$^2$, $r_{fo} = 0.315$ m, $r_o \approx 0.3$ m, $\eta = 0.1$, $\varepsilon = 0.6$, $\sigma_\kappa = C_{Fx}/C_{F\kappa}$, $C_{Fx} = 565$ kN/m, $C_{F\kappa} = 80$ kN (500 kN at $\hat{d}_t = 2.5$ mm), $F_{zo} = 3000$ N, $F_{ro} = 25$ N, $C_{Fz} = 133$ kN/m, $A_r = 0.0083$ |
| ***Bias-ply*** **($p_i$ = 1.5 bar, tread depth = 6.5 mm)** |
| $I_w = 0.8$ ($z$) or 1.1 ($\tilde{r}_f$) kgm$^2$, $r_{fo} = 0.325$ m, $r_o \approx 0.3$ m, $\eta = 0.4$, $\sigma_\kappa = C_{Fx}/C_{F\kappa}$, $C_{Fx} = 1000$ kN/m, $C_{F\kappa} = 45$ kN, $F_{zo} = 3000$ N, $F_{ro} = 60$ N, $C_{Fz} = 190$ kN/m, $A_r = 0.02$ |

a specially designed rig (Pacejka, Van der Berg and Jillesma 1977). The parameters of the tires tested have been listed in Table 8.1.

The response to $z$ has been found by moving the axle up and down by means of a hydraulic actuator. The variation in carcass radius $r_c$ has been achieved by mounting a reasonably uniform wheel and tire with 2.5-mm eccentricity on the shaft and subsequently balancing the rotating system. In this test, the axle was held fixed. The special arrangement resulted in a larger polar moment of inertia $I_w$ (1.1 kgm$^2$). The response to $d_t$ variations could be obtained by partly buffing off the tire tread. On one side 5 mm less tread rubber remained with respect to the other side of the tire. As a side effect, the average slip stiffness $C_{F\kappa}$ increased. The new value was estimated. All tire parameters have been determined through special tests. Sometimes, slight changes of these values could result in a better match of the response data.

In general, a good correspondence with experimental results could be established. Also, the phase relationship, which is very sensitive to the structure of the model and changes in parameters, turns out to behave satisfactorily. A negative phase angle $\phi$ indicates that the response lags behind the input.

To explain these findings which turn out to be quite different for the two types of tires, we shall now analyze the responses to vertical axle motions $z$ and out of roundness $\tilde{r}_c$ $(= \tilde{r}_{c1}$ first harmonic) in greater detail.

### 8.3.3. Frequency Response to Vertical Axle Motions

After replacing in Eqn (8.51) the Laplace variable $s$ by $i\omega$, with $\omega$ denoting the frequency of the excitation, we find for the response to vertical axle motions the following frequency response function with $X = F_x$:

$$\frac{\tilde{F}_x}{z} = -A_r C_{Fz} - \frac{\eta C_{F\kappa}}{r_o} \frac{2\zeta i\nu}{1 - \nu^2 + 2\zeta i\nu} \tag{8.54}$$

with the nondimensional frequency:

$$\nu = \frac{\omega}{\omega_{\Omega o}} \tag{8.55}$$

and the speed-dependent damping ratio:

$$\zeta = \frac{1}{2}\frac{I_w V}{C_{F\kappa} r_o^2}\,\omega_{\Omega o} = \frac{1}{2}\frac{V}{\sigma_\kappa \omega_{\Omega o}} \tag{8.56}$$

where the natural frequency of the tire–wheel rotation with respect to the foot print has been introduced:

$$\omega_{\Omega o} = \sqrt{\frac{C_{Fx} r_o^2}{I_w}} = \sqrt{\frac{C_{F\kappa} r_o^2}{\sigma_\kappa I_w}} \tag{8.57}$$

An interesting observation is that an increase in slip stiffness or relaxation length decreases the nondimensional damping coefficient $\zeta$. A higher speed $V$, obviously, produces more damping.

For the extreme case, $\omega \to 0$, we find as expected only the contribution to rolling resistance:

$$\left.\frac{\tilde{F}_x}{z}\right|_{\omega \to 0} = -A_r C_{Fz} \tag{8.58}$$

If we neglect the small contribution from the rolling resistance, the first term of (8.54) vanishes. The magnitude of the remaining term reduces to a simple expression if the frequency is considered to be small with respect to the natural frequency:

$$\left|\frac{\tilde{F}_x}{z}\right|_{\omega \ll \omega_{\Omega o}} = 2\zeta \frac{\eta C_{F\kappa}}{r_o}\nu = \frac{\eta I_w}{r_o^3} V\omega = 2\pi \frac{\eta I_w}{r_o^3} Vn \tag{8.59}$$

with $V$ the speed of travel in m/s and $n$ the frequency of the vertical axle motion in Hz. It is noted from Eqn (8.59) that the rise of the response in the lower frequency range with the frequency $n$ and the speed $V$ is in particular influenced by the magnitude of the factor $\eta$ multiplied with the polar moment of inertia $I_w$. A small value of $\eta$, as with radial tires, is favorable which is obviously caused by the fact that then the change of effective rolling radius with deflection is very small so that rotational accelerations and accompanying longitudinal forces are almost absent. The coefficient of $Vn$ in (8.59) for the radial-ply tire becomes 18.6 and for the bias-ply tire 74.5. This would mean that if the wheel with bias-ply tire that rolls at a speed of 10 m/s is oscillated vertically with an amplitude of 1 mm at a frequency of 10 Hz, the resulting longitudinal force gets an amplitude of 7.45 N.

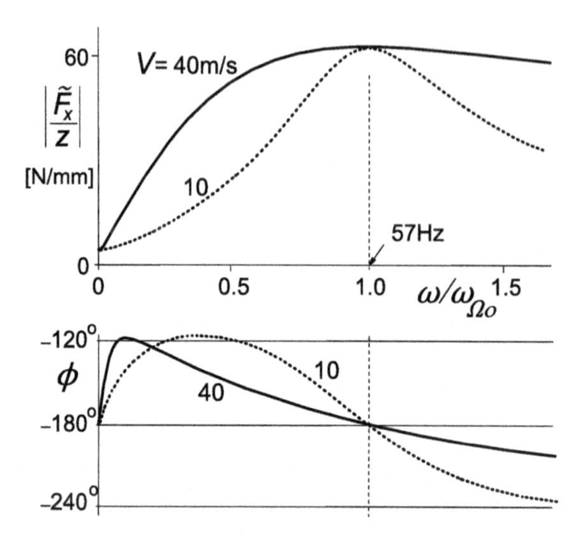


**FIGURE 8.17** Theoretical frequency response function of the longitudinal force $F_x$ to vertical axle motions $z$ for the bias-ply car tire.

At higher frequencies it can be shown that the curves for the magnitude of expression (8.54) reach a common maximum (for all $V$) at $\omega = \omega_{\Omega o}$. The maximum response amplitude amounts to

$$\left|\frac{\tilde{F}_x}{z}\right|_{\max} = \eta \frac{C_{F\kappa}}{r_o} + A_r C_{Fz} \tag{8.60}$$

This maximum is determined again primarily by the effective rolling radius gradient $\eta$ but now multiplied with slip stiffness $C_{F\kappa}$. For the radial tire, the maximum response becomes ca. 28 N/mm at the natural frequency 43 Hz and for the bias-ply tire this becomes ca. 64 N/mm at 57 Hz. Figure 8.17 shows both the amplitude and phase (lag) response calculated for the bias-ply tire. Figure 8.16 has already indicated that for the radial tire a much lower response is expected to occur.

## 8.3.4. Frequency Response to Radial Run-out

From Eqn (8.51) we derive for the frequency response to radial (carcass) run-out considering only its first harmonic $\tilde{r}_f = \tilde{r}_1 = \tilde{r}_c$ with frequency equal to the wheel speed of revolution $\omega = \Omega$ (nondimensional frequency $\nu = \Omega/\omega_{\Omega o}$):

$$\frac{\tilde{F}_x}{\tilde{r}_c} = -A_r C_{Fz} + i\,\frac{(1-\eta)C_{Fx}\nu^2 - F_{zo}/r_o}{1-\nu^2 + i\nu^2 r_o/\sigma_\kappa} \tag{8.61}$$

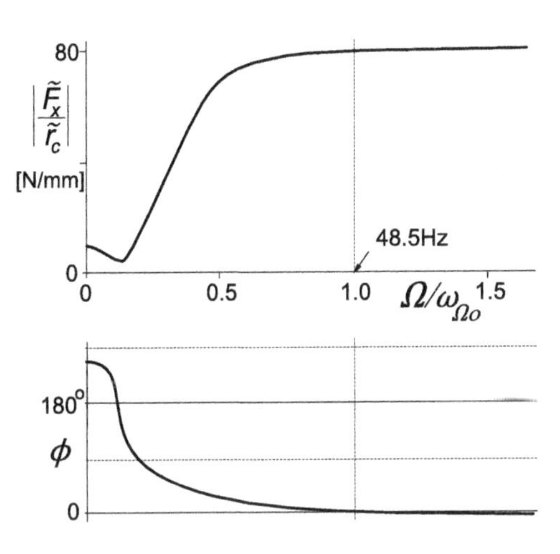


**FIGURE 8.18** Theoretical frequency response curve of longitudinal force to tire out of roundness (1st harmonic) for the bias-ply tire. The speed of travel $V = r_e\Omega$.

This expression differs in two main respects from the one for the response to axle vertical motions (8.54). First, we have now the factor $1 - \eta$ instead of $\eta$ and, second, an additional term with $F_{zo}$ shows up which is due to the eccentricity, cf. Eqn (8.49). Figure 8.18 presents the response plots for the bias-ply tire.

For speeds very low, $\Omega \rightarrow 0$ the expression reduces to

$$\left.\frac{\tilde{F}_x}{\tilde{r}_c}\right|_{\omega \rightarrow 0} = -A_r C_{Fz} - i\frac{F_z}{r_o} \tag{8.62}$$

The amplitude ratio attains a minimum near

$$\nu = \sqrt{\frac{F_{zo}/r_o}{(1-\eta)C_{Fx}}} \tag{8.63}$$

which means for the radial tire, near the frequency 0.14 $n_{\Omega o}$ and, for the bias-ply tire, near 0.13 $n_{\Omega o}$. With parameters corresponding to the experiment conducted ($I_w = 1.1$ kgm$^2$ resulting in natural frequencies of 36.5 and 48.5 Hz) we obtain for these frequencies 5.1 and 6.3 Hz, respectively. The minimum value appears to become very close to $A_r C_{Fz} = 1100$ and 3800 N/m, respectively. As can be seen in the upper right-hand diagram of Figure 8.16, the experimentally assessed curve also features such a dip in the low-frequency range and the accompanying sharp decrease in phase lead. At the natural frequency $n_{\Omega 0}$ ($\nu = 1$), the amplitude ratio comes close to its maximum value. This maximum is approximately equal to

$$\left|\frac{\tilde{F}_x}{\tilde{r}_c}\right|_{\max} \approx -A_r C_{Fz} + (1-\eta)\frac{C_{Fx}}{r_o} - \frac{\sigma_\kappa F_{zo}}{r_o^2} \tag{8.64}$$

The second term by far dominates this expression. The radial tire attains 220 N/mm (at 36.5 Hz) and the bias-ply tire attains 80 N/mm (at 48.5 Hz). Compared with the maximum longitudinal force response to vertical axle motions, Eqn (8.60), we note a much stronger sensitivity of the radial tire to out of roundness (ca. 220 N/mm) than to vertical axle motions (ca. 28 N/mm). This is due to the factor $1-\eta$ occurring in (8.64) which in (8.60) appears as $\eta$. For the bias-ply tire with a much more compressible tread band, we have 80 N/mm versus 64 N/mm.

### Exercise 8.1. Response to Tire Stiffness Variations

Derive from Eqns (8.51) to (8.53) the frequency response function of $F_x$ to the static deflection variation $\tilde{\rho}_s$, that is, to variations in radial tire stiffness, for fixed axle height, $z = 0$. Then release the vertical axle motion and find the frequency response function for $z$ to $\tilde{\rho}_s$. Consider Figure 8.19 that represents the simplified axle/wheel-suspension system. Small motions are considered so that only linear terms can be taken into account. Use the parameter values of the bias-ply tire of Table 8.1 and furthermore the values for the mass, the suspension stiffness, and the damping:

$$m = 50\text{ kg}, \quad c = 40000\text{ N/m}, \quad k = 2000\text{ Ns/m}$$

Now find the total frequency response of $F_x$ to $\tilde{\rho}_s$ for the released axle motion. It may be noted that the excitation by a variation of the radial stiffness is equivalent to the excitation by a vertical force:

$$\tilde{F}_{zs} = \rho_o \tilde{C}_{Fz} = -C_{Fzo}\tilde{\rho}_s$$

representing the variation of the 'static' tire load at constant axle height while the tire rolls.

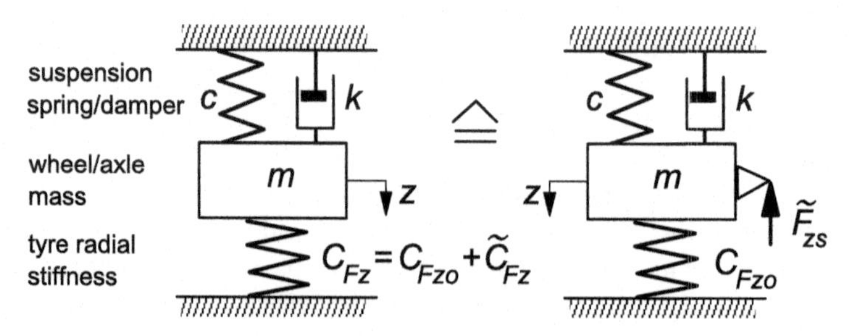


**FIGURE 8.19** Wheel suspension/mass system excited by tire stiffness variations (Exercise 8.1).

### Exercise 8.2. Self-excited Wheel Hop

Investigate the stability of the fourth-order system of Figure 8.20 using the Hurwitz criterion (cf. Eqn (1.118)). Employ Eqns (8.31, 8.38, 8.45–8.47), assume $M_y = 0$, and consider a flat road and a uniform tire. Assess the range of the slope $\beta$ of the wheel axle guidance surface where instability will occur for the undamped wheel suspension system. The effective rolling radius coupling coefficient $\eta$ is the decisive parameter. Plot the stability boundary in the tan $\beta$ versus $\eta$ diagram. Vary $\eta$ between the values 0 and 1. Indicate the areas where stability prevails. Take the parameter values:

$$r_o = r_{eo} = 0.3 \text{ m}, \sigma_\kappa = C_{F\kappa}/C_{Fx} = 0.3 \text{ m}$$

$$C_{Fz} + c_z = C_{Fx} = 2 \times 10^5 \text{ N/m}$$

$$m = 30 \text{ kg}, I_w = 0.5\, mr_o^2 \text{ kgm}^2$$

That a moderate oscillatory instability may indeed show up with such a system has been demonstrated experimentally in professor S.K. Clark's laboratory at the University of Michigan in 1970.

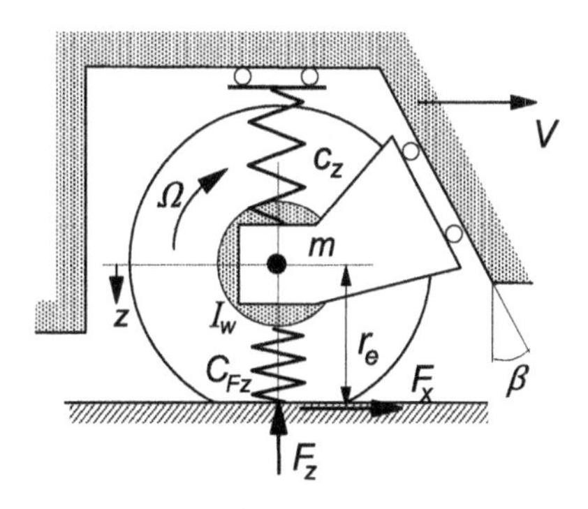


**FIGURE 8.20** On the possibly unstable wheel hop motion (Exercise 8.2).

## 8.4. FORCED STEERING VIBRATIONS

A steering/suspension system of an automobile exhibits a rather complex configuration and possesses many degrees of freedom. A simplification is necessary to conduct a sensible analysis to gain insight into its general dynamic behavior and into the influence of important parameters of the system. Investigation of the steering mode of vibration requires at least the steering degree of freedom of the front wheel, possibly extended with the rotation degree of freedom of the steering wheel. For the sake of simplicity, one degree of freedom may be suppressed by holding the steering spring clamped at the node of the natural mode of vibration. Furthermore, we will consider the influence of two more degrees of freedom: the vertical axle motion and the longitudinal deflection of the suspension with respect to the steadily moving vehicle mass.

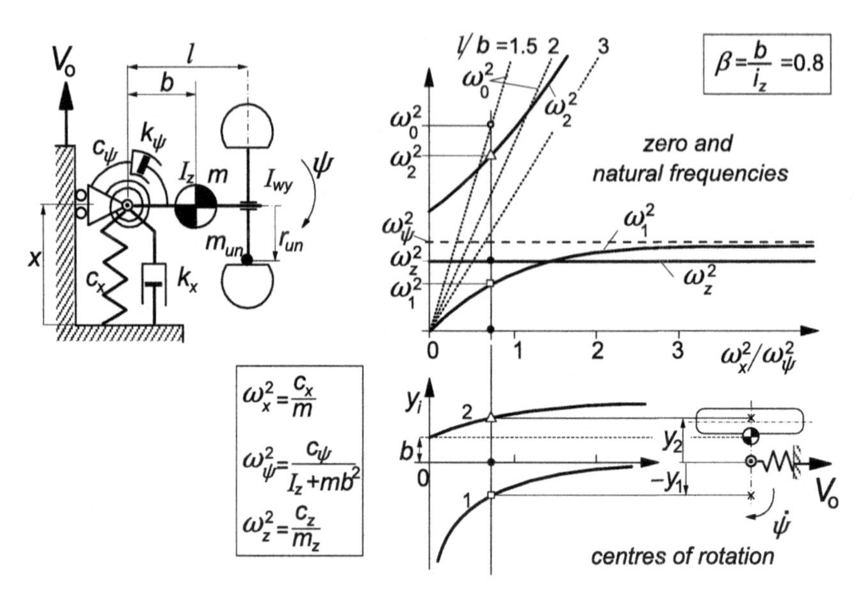


**FIGURE 8.21** Configuration of a simple steering/suspension system and the resulting natural frequencies and vibrational modes.

The picture of Figure 8.21 shows the layout of system. Due to the assumed orthogonality of the system (wheel axis, kingpin, and road plane) the dynamically coupled horizontal motions ($x$, $\psi$) are not coupled with the vertical axle motion when small displacements are considered and tire contact forces are disregarded.

First, we will examine the dynamics of the free system not touching the road. After that, the tire is loaded and tire transient models for the in-plane and out-of-plane behavior are introduced and the system response to wheel unbalance will be assessed and discussed.

## 8.4.1. Dynamics of the Unloaded System Excited by Wheel Unbalance

The simple system depicted in Figure 8.21 possesses two horizontal degrees of freedom: the rotation about the vertical steering axis, $\psi$, and the fore and aft suspension deflection, $x$. The figure provides details about the geometry, stiffnesses, damping, and inertia. The mass $m$ represents the total mass of the horizontally moving parts. The length $i_z$ denotes the radius of inertia: $i_z^2 = I_z/m$.

The wheel rim that revolves with a speed $\Omega$ is provided with an unbalance mass $m_{un}$ (in the wheel center plane at a radius $r_{un}$). The centrifugal force has a component in forward direction:

$$F_{un,x} = -m_{un} r_{un} \Omega^2 \sin \Omega t \tag{8.65}$$

The equations of motion of this fourth-order system read as

$$m\ddot{x} - mb\ddot{\psi} + k_x\dot{x} + c_x x = F_{un,x} \tag{8.66}$$

$$m(i_z^2 + b^2)\ddot{\psi} - mb\ddot{x} + k_\psi\dot{\psi} + c_\psi = -lF_{un,x} \tag{8.67}$$

With damping disregarded, the magnitude of the frequency response function becomes

$$\frac{\hat{\psi}}{m_{un}r_{un}} = \frac{l}{c_\psi}\frac{\Omega^2\left\{1-\left(\frac{\Omega}{\omega_0}\right)^2\right\}}{\left\{1-\left(\frac{\Omega}{\omega_1}\right)^2\right\}\left\{1-\left(\frac{\Omega}{\omega_2}\right)^2\right\}} \tag{8.68}$$

in which we have the zero frequency:

$$\omega_0^2 = \frac{l}{l-b}\omega_x^2 \tag{8.69}$$

and the two natural frequencies:

$$\omega_{1,2}^2 = \frac{1}{2}(1+\beta)\left\{\left(\omega_x^2+\omega_\psi^2\right) \pm \left(\omega_x^2-\omega_\psi^2\right)\sqrt{1+\frac{4\beta}{1+\beta}\frac{\omega_x^2\omega_\psi^2}{\left(\omega_x^2-\omega_\psi^2\right)^2}}\right\} \tag{8.70}$$

where we have introduced the 'uncoupled' natural frequencies:

$$\omega_x^2 = \frac{c_x}{m} \text{ and } \omega_\psi^2 = \frac{c_\psi}{m(i_z^2+b^2)} \tag{8.71}$$

and the coupling factor:

$$\beta = \frac{b^2}{i_z^2} \tag{8.72}$$

For the analysis, we are interested in the influence of the fore and aft compliance of the suspension. In the right-hand diagram of Figure 8.21 the two natural frequencies have been plotted as a function of the longitudinal natural frequency ratio (squared), which is proportional to the longitudinal stiffness $c_x$, together with the constant vertical natural frequency and the zero frequency for three different values of *f/b*. In addition, the location of the two centers of rotation according to the two modes of the undamped vibration has been indicated.

The two steer natural frequencies $\omega_1$ and $\omega_2$ increase with increasing longitudinal suspension stiffness. The lower natural frequency with a center of rotation located at the inside of the kingpin approaches the uncoupled natural frequency $\omega_\psi$.

From (8.69) it is seen that a zero does not occur if the unbalance arm length $l$ lies in the range $0 < l < b$ which does not represent a usual configuration. For the normal situation with $b > 0$ the zero frequency line may cross the second natural frequency curve if $l$ is not too large. If the two frequencies coincide, the second resonance peak of the steer response to unbalance will be suppressed. In that case, the unbalance force line of action passes through the center of rotation of the higher vibration mode.

It may be noted that the situation with contact between wheel and road can be simply modeled if the wheel is assumed to be rigid. The same equations apply with inertia parameters adapted according to the altered system with a point mass attached to the axle in the wheel plane. The point mass has the value $I_w/r^2$ where $I_w$ denotes the wheel polar moment of inertia and $r$ the wheel radius.

### 8.4.2. Dynamics of the Loaded System with Tire Properties Included

In a more realistic model, the in-plane and out-of-plane slip, compliance, and inertia parameters should be taken into account. A possible important aspect is the interaction between vertical tire deflection and longitudinal slip which may cause the appearance of a third resonance peak near the vertical natural frequency of the wheel system. The longitudinal carcass compliance gives rise to an additional natural frequency around 40 Hz of the wheel rotating against the foot print (cf. discussion in Section 8.3.3). Due to damping, originating from tangential slip of the tire, a supercritical condition will arise beyond a certain forward velocity. This causes the additional natural frequency to disappear.

For the extended system with road contact, the following complete set of linear equations applies:

$$m(\ddot{x} + 2\zeta_x\omega_x\dot{x} + \omega_x^2 x) - mb\ddot{\psi} - F_x = F_{un,x} \tag{8.73}$$

$$m_z(\ddot{z} + 2\zeta_z\omega_z\dot{z} + \omega_z^2 z) = F_{un,z} \tag{8.74}$$

$$I_\psi(\ddot{\psi} + 2\zeta_\psi\omega_\psi\dot{\psi} + \omega_\psi^2\psi) - mb\ddot{x} + lF_x - M_z = -lF_{un,x} \tag{8.75}$$

$$I_w\dot{\Omega} + r_oF_x = 0 \tag{8.76}$$

$$\sigma_\kappa\dot{F}_x + V_oF_x = -C_{F\kappa}V_{sx} \tag{8.77}$$

$$\sigma_\alpha\dot{F}_y + V_oF_y = -C_{F\alpha}V_{sy} \tag{8.78}$$

$$V_{sx} = \dot{x} - l\dot{\psi} - r_o(\Omega - \Omega_o) + \Omega_o\eta z \tag{8.79}$$

$$V_{sy} = -V_o\psi \tag{8.80}$$

$$M_z = -t_\alpha F_y - \frac{1}{V_o}\kappa^* \dot{\psi} - C_{gyr} V_o \dot{F}_y \tag{8.81}$$

$$F_{un,x} = -m_{un} r_{un} \Omega_o^2 \sin \Omega_o t \tag{8.82}$$

$$F_{un,y} = m_{un} r_{un} \Omega_o^2 \cos \Omega_o t \tag{8.83}$$

$$V_0 = r_o \Omega_o \tag{8.84}$$

The tire side force differential Eqn (7.20) has been used and similar for the longitudinal force. The longitudinal slip speed follows from (8.45) with (8.52). Note that mechanical caster has not been considered so that the lateral slip speed is simply expressed by (8.80). The aligning torque is based on Eqn (7.51) with for $M_{zr}$ the spin moment $M_z^*$ according to Eqn (5.82) and the gyroscopic couple from Eqns (7.49) or (5.178). The rolling resistance moment has been neglected in (8.76) and the average effective rolling radius has been taken equal to the average axle height or loaded radius $r_o$ (in reality $r_e$ is usually slightly larger than $r_o$).

In Table 8.2 the set of parameter values used in the computations are listed. The moment of inertia about the steering axis is denoted with $I_\psi$ and equals $m(b^2 + i_z^2)$. The amplitude of the steer angle that occurs as a response to a wheel unbalance mass of 0.1 kg has been plotted as a function of the wheel speed of revolution in Figure 8.22. To examine the influence of the longitudinal suspension compliance a series of values of the longitudinal stiffness $c_x$ has been considered. In Figure 8.23, these values are indicated by marks on the stiffness axis.

Clearly, in agreement with the variation of the natural frequencies assessed in this figure, the two resonance peaks move to higher frequencies when the stiffness is raised. A third resonance peak may show up belonging to the vertical natural frequency. This peak remains at the same frequency. It is of interest to observe that when the lowest steer natural frequency $n_1$ coincides with $n_z$ the interaction between vertical and horizontal motions causes the peak to reach relatively high levels. As we have seen before, this interaction is brought about by the slope $\eta$ indicated in Figure 8.13. The curves of the

**TABLE 8.2** Parameter Values of Wheel Suspension System and Tire Considered

| | | | | | | | |
|---|---|---|---|---|---|---|---|
| $m$ | 30 kg | $\omega_z$ | 70 rad/s | $\sigma_\kappa$ | 0.15 m | $F_{zo}$ | 3500 N |
| $m_z$ | 40 kg | $\omega_\psi$ | 85 rad/s | $\sigma_\alpha$ | 0.3 m | $C_{Fz}$ | 160 kN/m |
| $I_\psi$ | 1.2 kgm$^2$ | $\zeta_x$ | 0.01 | $t_\alpha$ | 0.03 m | $C_{F\kappa}$ | 60 kN |
| $I_w$ | 0.8 kgm$^2$ | $\zeta_z$ | 0.06 | $\kappa^*$ | 80 Nm$^2$ | $C_{F\alpha}$ | 40 kN/rad |
| $m_{un}$ | 0.1 kg | $\zeta_\psi$ | 0.08 | $C_{gyr}$ | $2 \times 10^{-5}$ s$^2$ | $\eta$ | 0.4 |

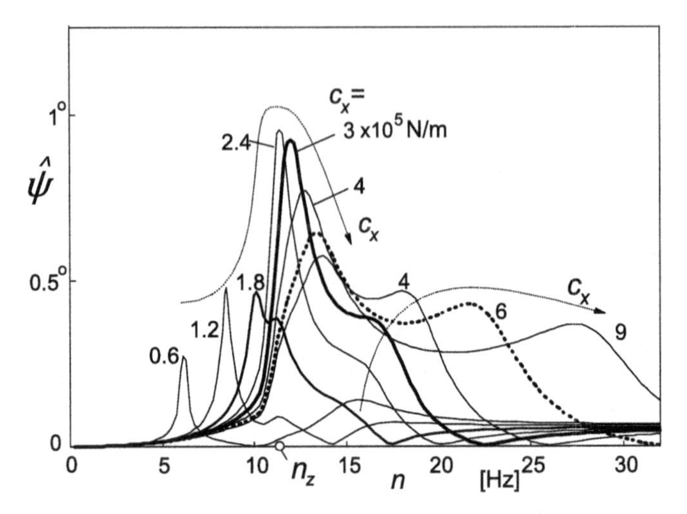


**FIGURE 8.22** Steer vibration amplitude due to wheel unbalance as a function of wheel frequency of revolution $n = \Omega_o/2\pi$ for various values of the longitudinal suspension stiffness.

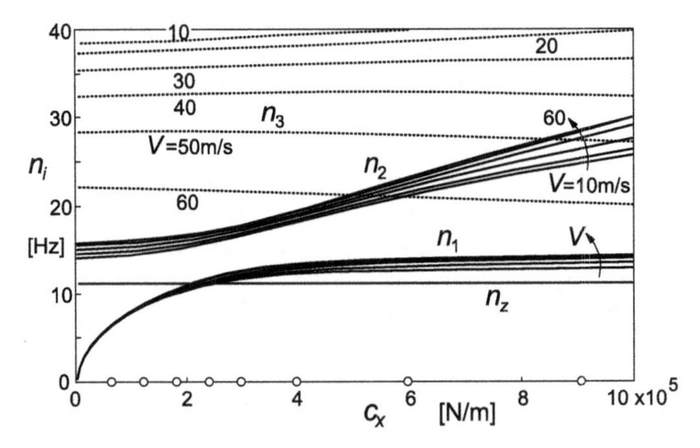


**FIGURE 8.23** Variation of the three natural frequencies with longitudinal suspension stiffness at different speeds together with the constant vertical natural frequency. The circles on the horizontal axis mark the stiffness cases of Figure 8.22.

diagram of Figure 8.22 also exhibit the dip in the various curves that correspond with the zero's ($\omega_0$) of Figure 8.23. The zero frequency closely follows the formula (8.69) of the free system. At the lowest stiffness, the zero frequency $n_0$ almost coincides with the second natural frequency $n_2$ and suppresses the second peak.

It may be of interest to find out how the several tire parameters affect the response characteristics of Figure 8.22. In Figure 8.24 the result of making

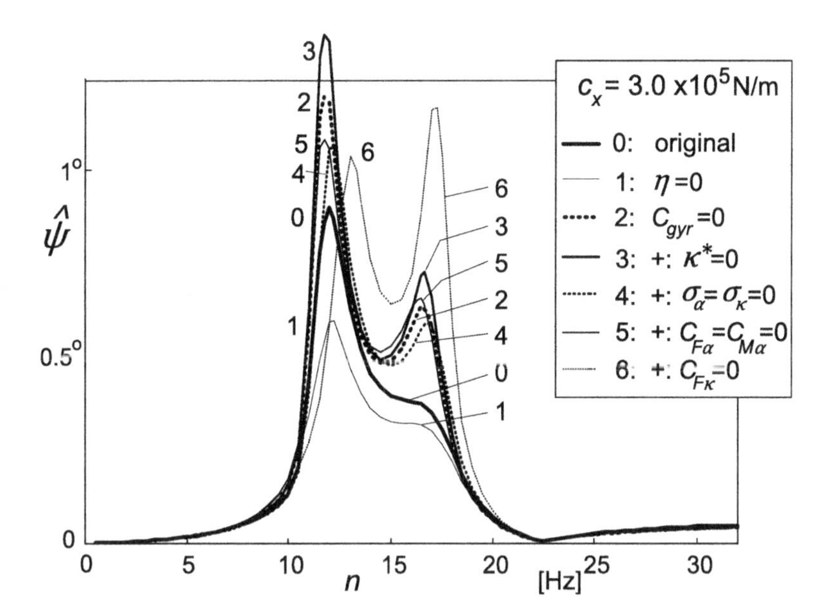


**FIGURE 8.24** Influence of various tire parameters on the steer angle amplitude response curve.

these parameters equal to zero has been depicted for the stiffness case $c_x = 3 \times 10^5$ N/m. Neglecting the factor $\eta$ (case 1), meaning that the effective rolling radius would not change with load, appears to have a considerable effect indicating that, with $\eta$, the vertical motion does amplify the steering oscillation. Omitting the gyroscopic tire moment (2), while reinstating $\eta$, appears, as expected, to effectively decrease the steer damping. This is strengthened by additionally deleting the moment due to tread width (3). Omitting the relaxation lengths (4) lowers the peaks, thus removing the negative damping due to tire compliance. Deleting, in addition, the side force and the aligning torque (5) raises the peaks again, indicating that some energy is lost through the side slip. Disregarding the horizontal tire forces altogether (6) brings us back to the (horizontally) free system treated in the previous subsection. As predicted by the analysis, two sharp resonance peaks arise as well as the dip at the zero frequency.

## 8.5. ABS BRAKING ON UNDULATED ROAD

The aim of this section is to investigate the influence of dynamic effects due to vertical and longitudinal wheel vibrations excited by road irregularities upon the braking performance of the tire and antilock braking system. These disturbing factors affect the angular velocity of the wheel and consequently may introduce disinformation in the signals transmitted to the ABS system upon which its proper functioning is based.

One may take the simple view that the primary function of the antilock device is to control the brake slip of the wheel to confine the wheel slip within a narrow range around the slip value at which the longitudinal tire force peaks. In the same range, it fortunately turns out that the lateral force that the tire develops as a response to the wheel slip angle is usually sufficient to keep the vehicle stable and steerable.

In order to keep the treatment simple, we will restrict the vehicle motion to straight-line braking and consider a quarter vehicle with wheel and axle that is suspended with respect to the vehicle body through a vertical and a longitudinal spring and damper. The analysis is based on the study of Van der Jagt et al. (1989).

### 8.5.1. In-Plane Model of Suspension and Wheel/Tire Assembly

The vehicle is assumed to move along a straight line with speed $V$. The forward acceleration of the vehicle body is considered to be proportional with the longitudinal tire force $F_x$. Vertical and horizontal vehicle body parasitic motions will be neglected with respect to those of the wheel axle. Consequently, the role of the wheel suspension is restricted to axle motion alone. These simplifications enable us to concentrate on the influence of the complex interactions between motions of the axle and the tire upon the braking performance of the tire. Figure 8.25 depicts the system to be studied. We have axle displacements $x$ and $z$ and a vertical road profile described by $w$ and its slope d$w$/d$s$. To suit the limitations of the tire model employed, the wavelength $\lambda$ of the road undulation is chosen relatively large. The wheel angular velocity is $\Omega$ and the braking torque is denoted with $M_B$. The unsprung mass and the polar moment of inertia of the wheel are lumped with a large part of those of the tire and are denoted with $m$ and $I_w$.

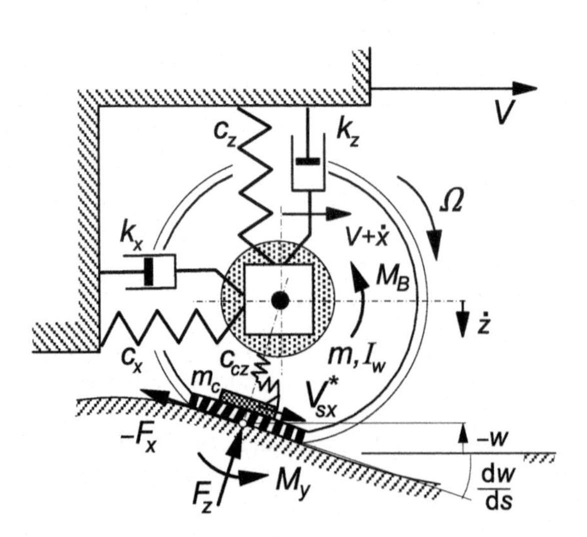


**FIGURE 8.25** System configuration for the study of controlled braking on undulated road surface.

Through the tire radial deflection the normal load is generated. Since we intend to consider possibly large slip forces, the analysis has to account for nonlinear tire characteristic properties. To adequately establish the fore and aft tire contact force, we will employ the enhanced transient tire model discussed in Section 7.3. The contact relaxation length $\sigma_c$ will be disregarded here. In the model, a contact patch mass $m_c$ is introduced that can move with respect to the wheel rim in tangential direction, thereby producing the longitudinal carcass deflection $u$. The contact patch mass may develop a slip speed with respect to the road denoted by $V_{sx}^*$. The transient slip is defined by ($V > 0$)

$$\kappa' = -\frac{V_{sx}^*}{V} \tag{8.85}$$

This longitudinal slip ratio is used as input in the steady-state longitudinal tire characteristic. The internal tire force that acts in the carcass and on the wheel axle will be designated as $F_{xa}$, while the tangential contact force is denoted with $F_x$. According to the theory, this slip force is governed by the steady-state force vs slip relationship $F_{x,ss}(\kappa)$ which may be modeled with the *Magic Formula*. Since we have to consider the influence of a varying normal load, the relationship must contain the dependency on $F_z$. To handle this, the similarity method of Chapter 4 is used. We have the following equations:

$$F_{x,ss} = \frac{F_z}{F_{zo}} F_{xo}(\kappa_{eq}) \tag{8.86}$$

with argument

$$\kappa_{eq} = \frac{F_{zo}}{F_z} \frac{C_{F\kappa}(F_z)}{C_{F\kappa o}} \kappa' \tag{8.87}$$

The master characteristic $F_{xo}(\kappa)$ described by the *Magic Formula* holds at the reference load $F_{zo}$ which is taken equal to the average load. Its argument $\kappa$ is replaced by the equivalent slip value $\kappa_{eq}$. Also, the longitudinal slip stiffness $C_{F\kappa o}$ is defined at the reference load. In Figure 8.26, the steady-state force characteristics are presented for a set of vertical loads.

### *In-Plane Braking Dynamics Equations*

The system has as input the road profile $w$ and the brake torque $M_B$. The level $w$ and the forward slope d$w$/d$s$ are given as sinusoidal functions of the traveled distance $s$. To partly linearize the equations, the road forward slope is assumed to be small. The brake torque ultimately results from a control algorithm but may in the present analysis be considered as a given function of time.

The following equations apply for the wheel rotational dynamics, the horizontal and vertical axle motions, and the tangential motion of the contact patch mass:

$$I_w\dot{\Omega} + rF_{xa} + M_y = -M_B \tag{8.88}$$

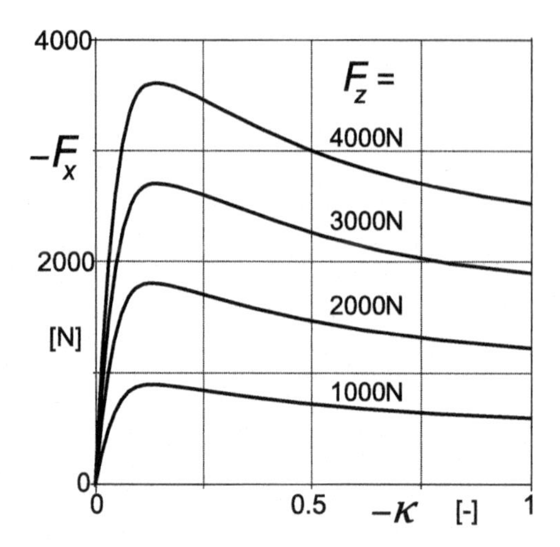


**FIGURE 8.26** Steady-state tire brake force characteristics at different loads as computed with the aid of the *Magic Formula* and the similarity technique.

$$m\ddot{x} + k_x\dot{x} + c_x x - F_{xa} - F_z\frac{\mathrm{d}w}{\mathrm{d}s} = 0 \tag{8.89}$$

$$m\ddot{z} + k_z\dot{z} + c_z z + F_z - F_{xa}\frac{\mathrm{d}w}{\mathrm{d}s} = 0 \tag{8.90}$$

$$m_c\dot{V}^*_{sx} + F_{xa} - F_x = 0 \tag{8.91}$$

with auxiliary equations for the tire loaded radius

$$r = r_o - \rho \tag{8.92}$$

the radial deflection

$$\rho = z - w \tag{8.93}$$

the effective rolling radius

$$r_e = r_e(\rho) \tag{8.94}$$

the longitudinal carcass deflection rate

$$\dot{u} = V^*_{sx} - V_{sx} \tag{8.95}$$

the wheel slip velocity

$$V_{sx} = V + \dot{x} - r_e\Omega + \dot{z}\frac{\mathrm{d}w}{\mathrm{d}s} \tag{8.96}$$

and the transient slip

$$\kappa' = -\frac{V^*_{sx}}{V} \tag{8.97}$$

Moreover, the following constitutive relations are to be accounted for

$$F_z = F_z(\rho, u) \tag{8.98}$$

$$F_{xa} = F_{xa}(u, \dot{u}) \tag{8.99}$$

$$M_y = M_y(F_z, F_x) \tag{8.100}$$

$$F_x = F_{x,ss}(\kappa', F_z) \tag{8.101}$$

The function (8.101) is represented by Eqn (8.86) and illustrated in Figure 8.26. The remaining constitutive relations are simplified to linear expressions.

## *Frequency Response of Wheel Speed to Road Unevenness*

As an essential element of the analysis we will assess the response of the wheel angular velocity variation to road undulations at a given constant brake torque. For this, the set of equations may be linearized around a given point of operation characterized by the average load $F_{zo}$, the constant brake torque $M_{Bo}$, and the average slip ratio $\kappa_o$. We find for the constitutive relations,

$$\tilde{F}_z = c_{cz}(\tilde{\rho} + e_{zx}\tilde{u}) \tag{8.102}$$

$$\tilde{F}_{xa} = c_{cx}\tilde{u} + k_{cx}\dot{\tilde{u}} + e_{xz}\tilde{F}_z \tag{8.103}$$

$$\tilde{M}_y = A_r\tilde{F}_z + B_r\tilde{F}_{xa} \tag{8.104}$$

$$\tilde{F}_x = C_{xz}\tilde{F}_z + C_{x\kappa}\tilde{\kappa}' \tag{8.105}$$

where with the use of Eqns (8.86, 8.87)

$$C_{xz} = \frac{\partial F_{x,ss}}{\partial F_z} = \frac{F_{xo}}{F_{zo}} + \kappa_o\left(\frac{1}{C_{F\kappa o}}\frac{\mathrm{d}C_{F\kappa}}{\mathrm{d}F_z} - \frac{1}{F_{zo}}\right)\frac{\mathrm{d}F_{xo}}{\mathrm{d}\kappa_{eq}} \tag{8.106}$$

and

$$C_{x\kappa} = \frac{\mathrm{d}F_{xo}}{\mathrm{d}\kappa_{eq}} \tag{8.107}$$

For the variation in effective rolling radius, we have

$$\tilde{r}_e = -\eta\tilde{\rho} \tag{8.108}$$

The linearized equations of motion are obtained from the Eqns (8.88–8.91) by subtracting the average, steady-state values of the variable quantities, and neglect products of small variations.

The frequency response of $\Omega$ to $w$ has been calculated for different values of the longitudinal suspension stiffness $c_x$. The average condition is given by the vehicle speed $V_o = 60$ km/h, a vertical load $F_{zo} = 3000$ N, and a brake torque $M_{Bo} = 300$ Nm that corresponds to an average slip ratio $\kappa_o = -0.021$. The road undulation amplitude $\hat{w} = 0.001$ m.

**TABLE 8.3** Parameter Values for Wheel/Tire/Suspension System

| | | | | | | | |
|---|---|---|---|---|---|---|---|
| $I_w$ | 0.95 kgm$^2$ | $c_x$ | 100 kN/m | $c_z$ | 20 kN/m | $F_{z0}$ | 3000 N |
| $m$ | 35 kg | $k_x$ | 2 kNs/m | $k_z$ | 2 kNs/m | $r_0$ | 0.29 m |
| $m_c$ | 1 kg | $c_{cx}$ | 10$^3$ kN/m | $c_{cz}$ | 170 kN/m | $k_{cx}$ | 0.8 kNs/m |

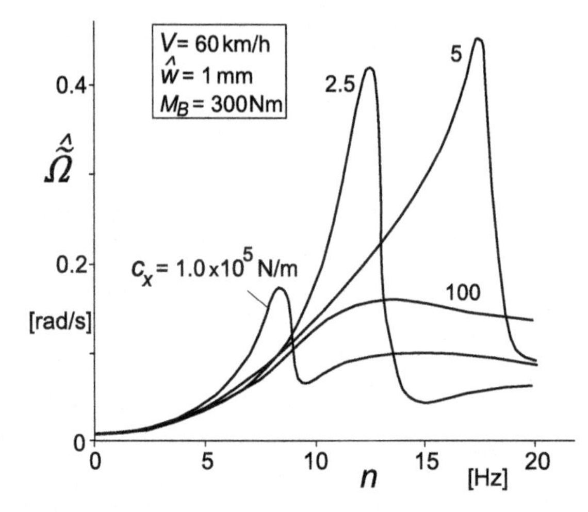


**FIGURE 8.27** Frequency response of amplitude of wheel spin fluctuations to road undulations $w$ at a constant speed $V$ and brake torque $M_B$ and for different values of the longitudinal suspension stiffness $c_x$.

Some of the tire constitutive relations were determined directly from measurements while for some of the parameters values have been estimated. The small interaction parameters, $\eta$, $e_{xz}$, and $e_{zx}$, and also the rolling resistance parameters $A_r$ and $B_r$ have been disregarded. Table 8.3 lists the parameter values used in the computations.

Figure 8.27 displays the amplitude of the wheel angular velocity variation as a function of the imposed frequency that is inversely proportional with the wavelength of the undulation. It is seen that the resonance peak height and the resonance frequency strongly depend on the fore and aft suspension stiffness.

## 8.5.2. Antilock Braking Algorithm and Simulation

The actual control algorithms of ABS devices marketed by manufacturers can be quite complex and are proprietary items which incorporate several practical considerations. However, almost all algorithms make use of raw data pertaining to the angular speed and acceleration of the wheels. A detailed discussion of the various algorithms for predicting wheel motions and modulating the brake torque accordingly has been given by Guntur (1975). The operational characteristic of a control algorithm used here for the purpose

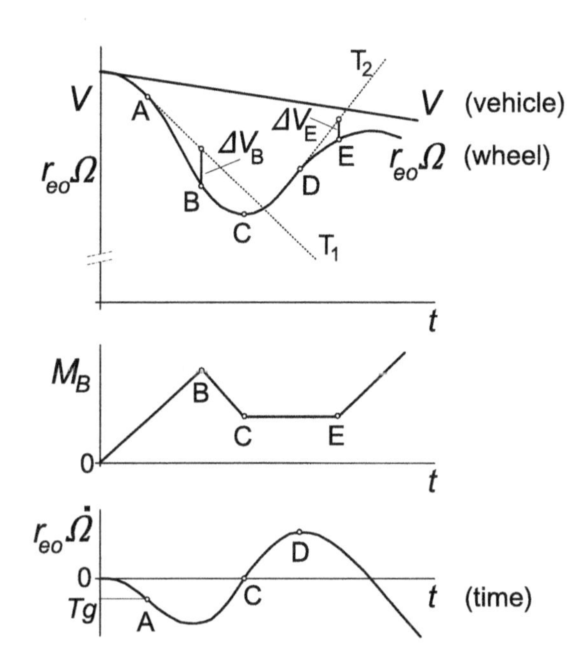


**FIGURE 8.28** Example of an ABS algorithm used to modulate the applied brake torque.

of illustration is shown in Figure 8.28. The salient features of the criteria used to increase or decrease applied brake torque (pressure) during a braking cycle depend critically upon both the momentary and threshold values of $\Omega$ and of $\dot{\Omega}$. It is assumed that the brake torque rate can be controlled at three different levels. Upon application of the brake, the torque rises at a constant rate until point B is reached. After that, the torque is reduced linearly until point C. The brake torque remains constant in the interval between points C and E. Compound criteria are considered at points A and B for predicting when braking should be reduced to prevent the wheel from locking. Thus, we have from the starting of brake application

Increase of torque for $0 < t < t_B$ according to $dM_B/dt = R_1$

The preliminary prediction time $t_A$ is given by

$$r_{eo}\dot{\Omega}(t_A) = Tg$$

where $r_{eo}$ stands for the nominal effective rolling radius and $T$ for a nondimensional constant which sets the threshold for $\dot{\Omega}$.

From time $t_B$ we have

Decrease of torque for $t_B < t < t_C$ according to $dM_B/dt = R_2$

The instant of prediction and action $t_B$ is defined by

$$r_{eo}\{\Omega(t_A) - \Omega(t_B)\} - Tg(t_B - t_A) = \Delta V_B$$

The time $t_C$ marks the beginning of the constant torque phase CE, i.e.

Constant torque for $t_C < t < t_E$ that is $dM_B/dt = 0$

Time $t_C$ is found from

$$\dot{\Omega}(t_C) = 0$$

Similarly, the criterion for increasing the torque at E is:

Increase of torque for $t > t_E$ according to $dM_B/dt = R_1$ is given by the compound reselection conditions generated at points D and E, with $t_D$ and $t_E$ being found from

$$\ddot{\Omega}(t_D) = 0$$

and

$$r_{eo}[\{\Omega(t_D) - \Omega(t_E)\} + \dot{\Omega}(t_D)\cdot(t_E - t_D)] = \Delta V_E$$

For the simulation, the following values have been used:
$T = -1$, $\Delta V_B = 0.1 r_{eo}\Omega\ (t_B)$ and $\Delta V_E = 1$ m/s
The brake torque rates have been set to

$$R_1 = -R_2 = 19000\,\mathrm{Nm/s}$$

In order to simulate the braking maneuver of a quarter vehicle equipped with the antilock braking system, the deceleration of the vehicle is taken to be proportional to $F_x$. The results of the simulations performed are presented in Figure 8.29. The left-hand diagrams refer to the case of braking on an ideally flat road, while the diagrams on the right-hand side depict the results obtained with the same system on a wavy road surface. For the purpose of simulation, the model was extended to include a hydraulic subsystem interposed between the brake pedal and the wheel cylinder. However, this extension is not essential to our discussion. In both cases, the same hydraulic subsystem was used and the same control algorithm as the one discussed above was implemented.

The further parameters used in the simulation were

Initial speed of the vehicle: 60 km/h, the road input with a wavelength of 0.83 m and an amplitude of 0.005 m.

The results of the simulation show that brake slip variations occur both on a flat as well as on undulated road surfaces. However, the very large variations of the transient slip value $\kappa'$ in the latter case lead to a further deterioration of the braking performance. The average vehicle deceleration drops down from 7.1 m/s$^2$ to approximately 6 m/s$^2$.

Although large fluctuations in the vertical tire force are mainly responsible for this reduction, it is equally clear that severe perturbation occurring in both $\Omega$ and $\dot{\Omega}$ may be an additional source of misinformation for the antilock control algorithm. The important reduction in braking effectiveness resulting from vertical tire force variations may be attributed to the term $C_{cz}$ in the linearized constitutive relation (8.106) of the rolling and slipping tire, and in particular to the contribution of the variation of the longitudinal slip stiffness with wheel

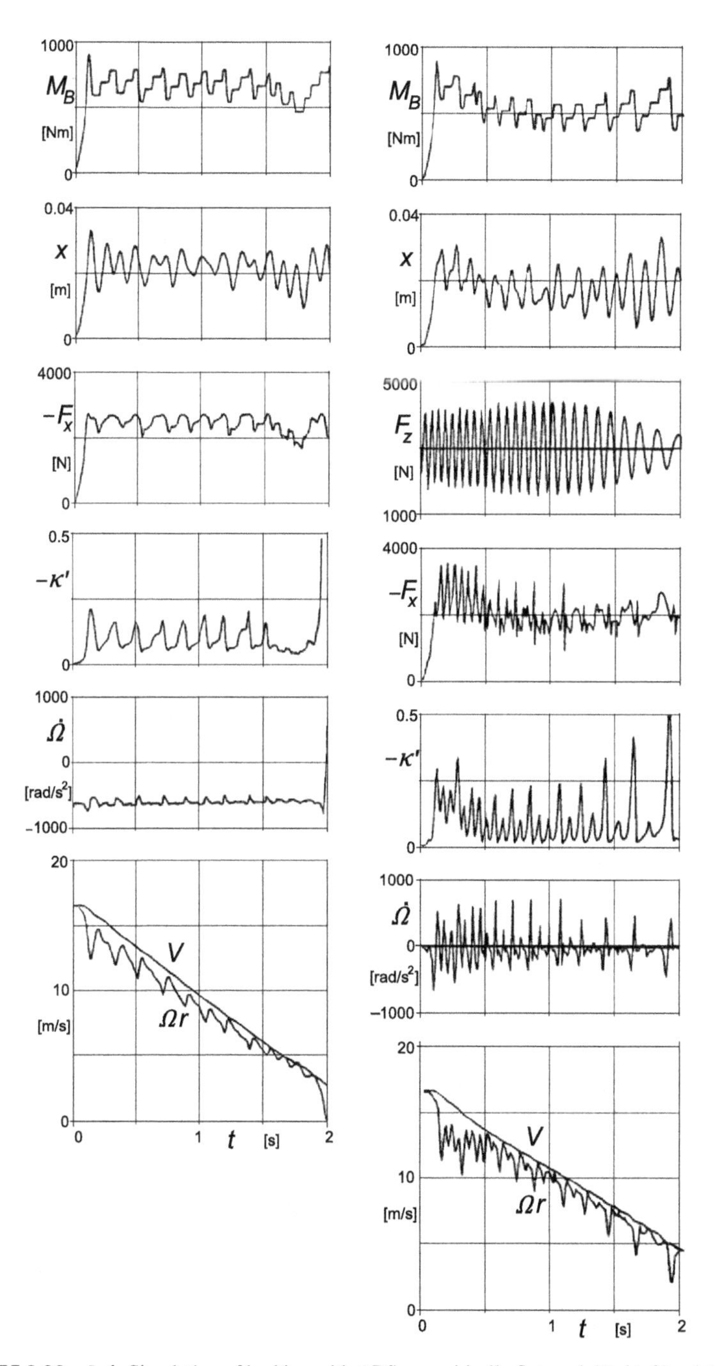


**FIGURE 8.29** *Left*: Simulation of braking with ABS on an ideally flat road. *Right*: Simulation of braking with ABS on a wavy road.

load $dC_{F\kappa}/dF_z$. Both vertical and horizontal vibrations of the axle and the vertical load fluctuations on wavy roads appear to adversely influence the braking performance of the tire as well as that of the antilock system.

In 1986, Tanguy made a preliminary study of such effects using a different control algorithm. He pointed out that wheel vibrations on uneven roads can pose serious problems of misinformation for the control logic of the antilock system. The results of the simulations discussed above and reported by Van der Jagt (1989) confirm Tanguy's findings.

## 8.6. STARTING FROM STANDSTILL

In this section, the ability of the transient models to operate at and near-zero speed conditions will be demonstrated. The four models treated in Sections 7.2.2, 7.2.3, and 7.3 will be employed in the simulation of the longitudinal motion of a quarter vehicle model on an upward slope of 5%, cf. Figure 8.30.

The following three different maneuvers are considered:

- Standing still on slope, stepwise application of drive torque and subsequently rolling at constant speed.
- From standstill on slope, freely rolling backwards, then powerful propulsion followed by free rolling.
- From standstill, rolling backwards, then braking to wheel lock, followed by free rolling again after which a short phase of drive torque is applied. Finally, the quarter vehicle slows down on the slope.

The steady-state longitudinal force function of the transient longitudinal slip $\kappa'$ is described by the formula

$$F_x = D_x \sin\left[C_x \arctan\left\{B_x\kappa' - E_x(B_x\kappa' - \arctan(B_x\kappa'))\right\}\right] \tag{8.109}$$

For the successive transient tire models the equations will be repeated below.

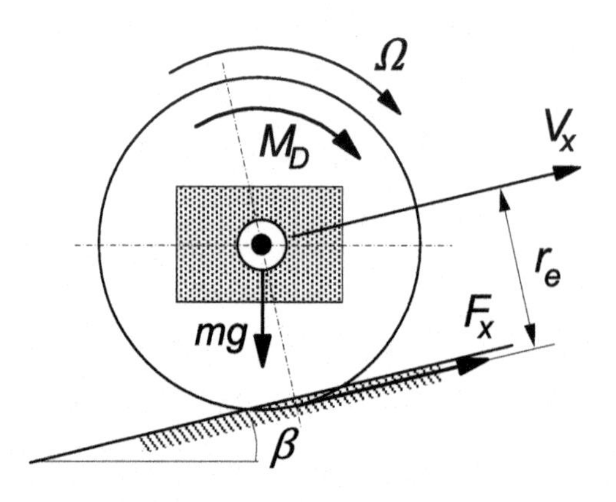


**FIGURE 8.30** On the problem of starting from standstill on a slope.

*Semi-non-linear transient model:*

First-order differential equation for longitudinal tire deflection $u$ according to Eqn (7.9):

$$\frac{\mathrm{d}u}{\mathrm{d}t} + \frac{1}{\sigma_\kappa}|V_x|u = -V_{sx} \tag{8.110}$$

where $\sigma_\kappa$ only depends on a possibly varying vertical load $F_z$. The transient slip reads as

$$\kappa' = \frac{u}{\sigma_\kappa} \tag{8.111}$$

At low values of speed $|V_x| < V_{\mathrm{low}}$ the deflection $u$ may have to be restricted as was formulated in general by Eqn (7.25). For our present problem, we have

if $|\kappa'| > A\kappa_{\mathrm{sl}}$ and $|V_x| < V_{\mathrm{low}}$ and $(V_{sx} + |V_x|u/\sigma_\kappa)u < 0$:

$$\frac{\mathrm{d}u}{\mathrm{d}t} = 0 \tag{8.112}$$

else

$$\frac{\mathrm{d}u}{\mathrm{d}t} = -V_{sx} - \frac{1}{\sigma_\kappa}|V_x|u \tag{8.113}$$

The slip where the peak force occurs is roughly

$$\kappa_{sl} = 3D_x/C_{F\kappa} \tag{8.114}$$

with the slip stiffness

$$C_{F\kappa} = B_x C_x D_x \tag{8.115}$$

For the factor $A$ the value 1 is suggested but a higher value may improve the performance, especially when the force characteristic exhibits a pronounced peak.

*Fully nonlinear transient model:*

The same Eqns (8.110, 8.111) hold but with $\sigma_\kappa$ replaced by the $\kappa'$-dependent and downwards limited quantity:

$$\sigma_\kappa^* = \max\left(\frac{\sigma_{\kappa o}}{C_{F\kappa}} \cdot \frac{|F_x| + C_{F\kappa}\varepsilon_F}{|\kappa'| + \varepsilon_F}, \sigma_{\min}\right) \tag{8.116}$$

where $\sigma_{\kappa o}$ represents the value of the relaxation length at $\kappa' = 0$. We have the equation

$$\frac{\mathrm{d}u}{\mathrm{d}t} + \frac{1}{\sigma_\kappa^*}|V_x|u = -V_{sx} \tag{8.117}$$

The transient slip now reads as

$$\kappa' = \frac{u}{\sigma_\kappa^*} \tag{8.118}$$

A deflection limitation is not necessary for this model. However, because of the algebraic loop that arises, the quantity (8.116) must be obtained from the previous time step.

*Restricted fully nonlinear model*:

Analogous to Eqn (7.35) we have the equation for $\kappa'$

$$\sigma_\kappa \frac{d\kappa'}{dt} + |V_x|\kappa' = -V_{sx} \tag{8.119}$$

with the $\kappa'$-dependent relaxation length

$$\sigma_\kappa = \frac{\sigma_{\kappa o}}{C_{F\kappa}} \frac{\partial F_x}{\partial \kappa'} \tag{8.120}$$

The model is not sensitive to wheel load variations which constitutes the restriction of the model. For the problem at hand, this restriction is not relevant and the model can be used. The great advantage of the model is the fact that an algebraic loop does not occur, and again, a $u$ limitation is not needed. A straightforward simulation can be conducted. For the relation (8.120) the following approximate function is used:

$$\sigma_\kappa = \max\left[\sigma_{\kappa o}\left(1 - \frac{C_{F\kappa}}{3D_x}|\kappa'|\right), \sigma_{\min}\right] \tag{8.121}$$

where $\sigma_{\min}$ represents the minimum value of the relaxation length that is introduced to avoid numerical difficulties.

*Enhanced nonlinear transient model*:

Following Section 7.3 and Figure 7.15, we have for the differential equation for the longitudinal motion of the contact patch mass $m_c$

$$m_c \dot{V}^*_{sx} + k_{cx}\dot{u} + c_{cx}u = F_x(\kappa') \tag{8.122}$$

where $F_x$ denotes the contact force governed by Eqn (8.109).

The transient slip value is obtained from the differential Eqn (7.56)

$$\sigma_c \frac{d\kappa'}{dt} + |V_x|\kappa' = -V^*_{sx} \tag{8.123}$$

an equation similar to (8.119) but here with a constant (small) contact relaxation length. The longitudinal carcass stiffness $c_{cx}$ in the contact zone should be found by satisfying the equation

$$\sigma_{\kappa o} = \frac{C_{F\kappa}}{C_{Fx}} = \frac{C_{F\kappa}}{c_{cx}} + \sigma_c \tag{8.124}$$

The deflection rate needed in Eqn (8.122) is equal to the difference in slip velocities of contact patch and wheel rim:

$$\dot{u} = V^*_{sx} - V_{sx} \tag{8.125}$$

The longitudinal force acting on the wheel rim results from

$$F_{xa} = k_{cx}\dot{u} + c_{cx}u \tag{8.126}$$

*Low-speed additional damping*:

At zero forward speed with each of the first three tire models, a virtually undamped vibration is expected to occur. An artificial damping may be introduced at low speed by replacing $\kappa'$ in (8.109) by

$$\kappa' - \frac{k_{V,\mathrm{low}}}{C_{F\kappa}} V_{sx} \tag{8.127}$$

which corresponds with the suggested Eqn (7.26). The gradual reduction to zero at $V_x = V_{\mathrm{low}}$ is realized by using the formula

$$\begin{aligned} k_{V,\mathrm{low}} &= \frac{1}{2}k_{V,\mathrm{low0}}\left\{1 + \cos\left(\pi\frac{|V_x|}{V_{\mathrm{low}}}\right)\right\} && \text{if } |V_x| \le V_{\mathrm{low}} \\ k_{V,\mathrm{low}} &= 0 && \text{if } |V_x| > V_{\mathrm{low}} \end{aligned} \tag{8.128}$$

Also with the enhanced model, the additional damping will be introduced while $k_{cx}$ is taken equal to zero to better compare the results.

*Vehicle and wheel motion*:

For the above tire models, the wheel slip speed follows from

$$V_{sx} = V_x - r_e\Omega \tag{8.129}$$

The vehicle velocity and the wheel speed of revolution are governed by the differential equations:

$$m\dot{V}_x = F_{xa} - mg\tan\beta \tag{8.130}$$

$$I_w\dot{\Omega} = M_D - r_eF_{xa} \tag{8.131}$$

where $r_e$ is the effective moment arm which turns out to correspond with experimental evidence, cf. Chap. 9, Figure 9.34. For the first three models with contact patch mass not considered, we have

$$F_{xa} = F_x \tag{8.132}$$

When at braking the wheel becomes locked, we have $\Omega = 0$.

*Results*:

The three different longitudinal maneuvers listed above have been simulated with each of the four transient tire models. In Figures 8.31–8.34, some of the results are presented. The parameter values are listed in Table 8.4.

Figures 8.31 and 8.32 depict the process of standing still on a 5% upward slope (appropriate brake or drive torque), followed by a step drive torque input and subsequently rolling at constant speed (back to equilibrium drive torque). Tire/wheel wind-up oscillations occur both at start and at end of step change in propulsion torque $M_D$. Results have been shown of the computations using the semi-non-linear transient tire model (Figure 8.31)

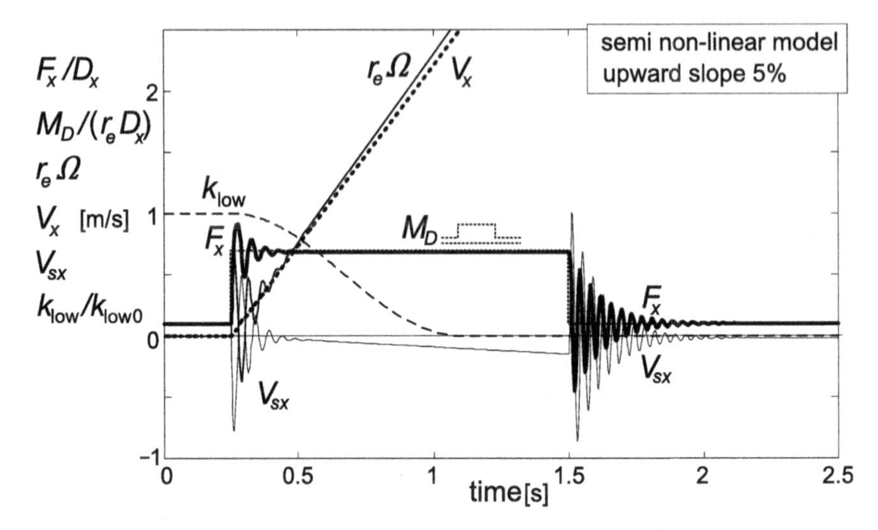


**FIGURE 8.31** Standing still on slope, acceleration, and subsequently rolling at constant speed. Tire/wheel wind-up oscillations occur both at start and at end of step change in propulsion torque $M_D$. Computations have been performed with the semi-non-linear transient tire model with constant relaxation length ($V_{low} = 2.5$ m/s).

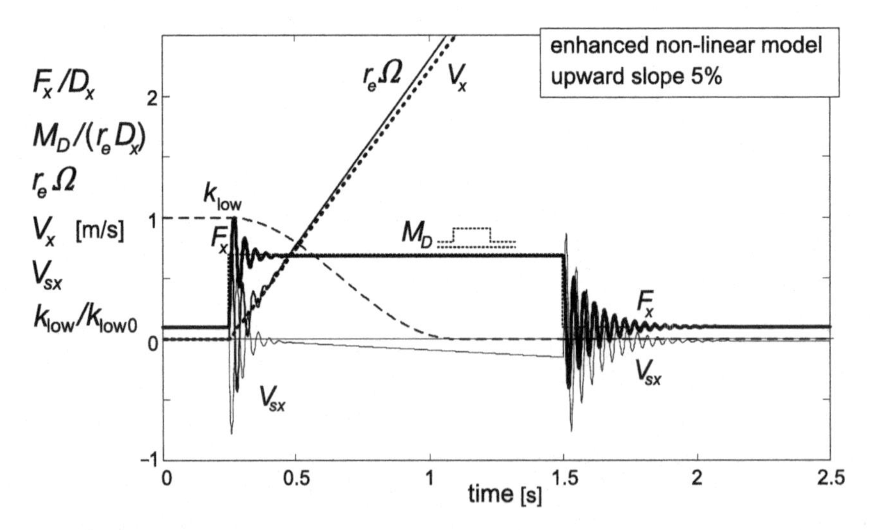


**FIGURE 8.32** Same simulation as in Figure 8.31 but now using the enhanced nonlinear transient tire model with carcass compliance ($V_{low} = 2.5$ m/s). Only at low speed a frequency difference can be observed to occur. Very similar results are obtained with the fully nonlinear transient models with changing relaxation length.

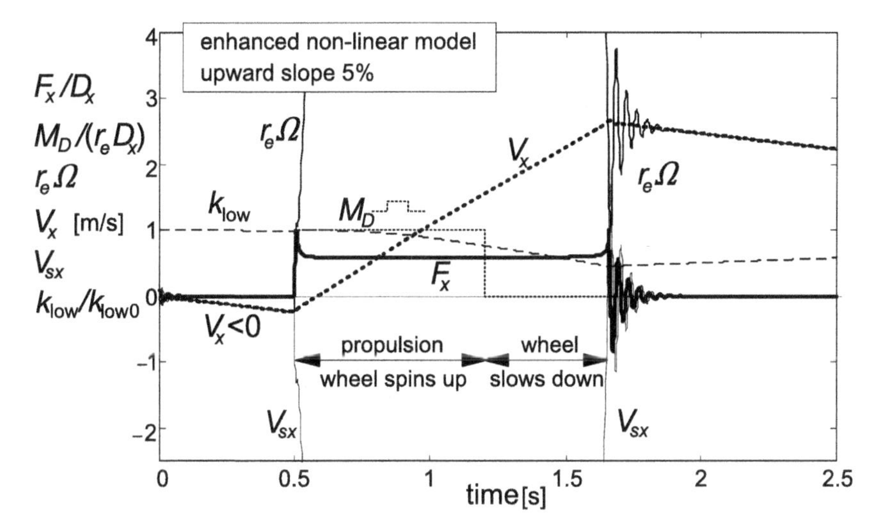


**FIGURE 8.33** From standstill on slope, freely rolling backward, then powerful propulsion followed by free rolling. The tire longitudinal force passes its peak and reaches the lower end of its characteristic which enables the wheel to spin up. Next, when the wheel slows down rapidly, the peak is reached again and a damped wheel/tire wind-up oscillation occurs. Calculations using enhanced transient tire model ($V_{low} = 5$ m/s). Together with the restricted fully nonlinear model, the enhanced model turns out to perform best.

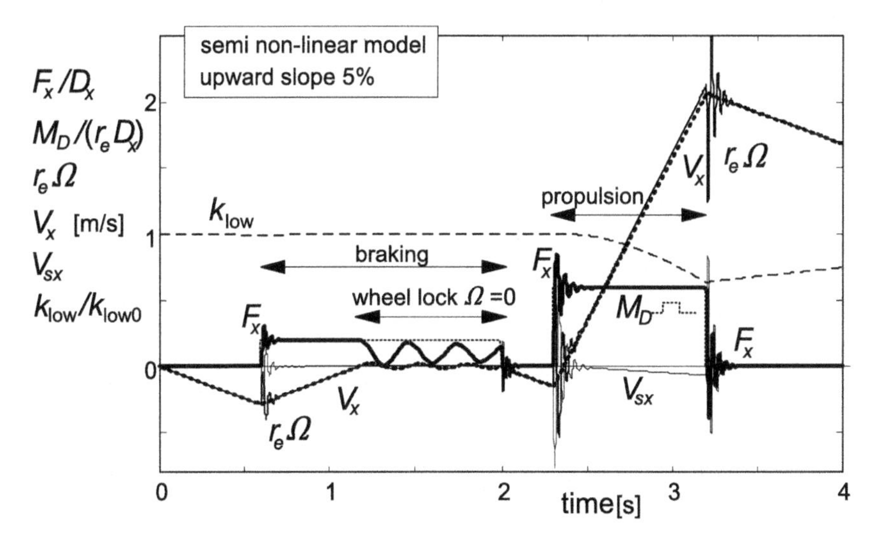


**FIGURE 8.34** From standstill rolling backward, then braking to wheel lock, followed by free rolling again after which a short phase of acceleration occurs. Finally, the quarter vehicle slows down on the slope. During wheel lock, the vehicle vibrates at low frequency with respect to the 'contact patch'. Calculations with the simple semi-non-linear transient tire model. Practically the same results are obtained when using one of the other three transient tire models.

**TABLE 8.4** Parameter Values

| | | | | | | | |
|---|---|---|---|---|---|---|---|
| $m$ | 600 kg | $B_x$ | 12.5 | $\tan\beta$ | 0.05 | $\sigma_{\kappa(0)}$ | 0.2 m |
| $I_w$ | 1 kgm$^2$ | $C_x$ | 1.6 | $k_{V\text{low}0}$ | 770 Ns/m | $\sigma_{min}$ | 0.02 m |
| $g$ | 9.81 m/s$^2$ | $D_x$ | 3000 N | $k_{cx}$ | 0 | $\sigma_c$ | 0.02 m |
| $r_e$ | 0.3 m | $E_x$ | 0 | $m_c$ | 1 kg | $\varepsilon_F$ | 0.01 |

and the enhanced model (Figure 8.32). Virtually the same results as depicted in Figure 8.32 have been achieved by using either the fully nonlinear or the restricted nonlinear transient tire model. Only at low speed, a frequency difference can be observed to occur when using the semi-non-linear model.

Figure 8.33 shows the results of standing on the slope, then freely rolling backward, subsequently applying powerful propulsion, which is followed by free rolling. The tire longitudinal force passes its peak and reaches the lower end of its characteristic, which enables the wheel to spin up. Next, when the wheel slows down rapidly, the peak is passed again and a damped wheel/tire wind-up oscillation occurs. Calculations have been performed using the enhanced transient tire model of Section 7.3 (with contact mass). The restricted fully nonlinear tire model, Eqn (7.35) not responding to load changes, turns out to yield almost equal results. The other two models, especially the semi-non-linear model of Section 7.2.3 (with $\sigma$ constant), appear to perform less good under these demanding conditions (less $F_x$ reduction in spin-up phase which avoids an equally rapid spinning up of the wheel). The low-speed limit value $V_{low}$ was increased to 5 m/s to restrict the $u$ deflection of the semi-non-linear model over a larger speed range. The factor $A$ was increased to 4 which improved the performance of this model. Omission of the limitation of the $u$ deflection would lead to violent back and forth oscillations.

Finally, Figure 8.34 depicts the simulated maneuver: from standstill rolling backwards then braking to wheel lock, which is followed by free rolling again after which a short phase of acceleration occurs; finally, the quarter vehicle slows down on the slope. During wheel lock the vehicle mass appears to vibrate longitudinally at low frequency with respect to the 'contact patch'. Calculations have been conducted with the simple semi-non-linear transient tire model. Practically equal results have been obtained when using one of the other three transient tire models.

It may be concluded that the restricted fully nonlinear model and the enhanced model with contact patch mass show excellent performance. It should be kept in mind that the restricted model does not react on wheel load variations. The semi-non-linear model can be successfully used under less demanding conditions.

The next chapter deals with the development of a more versatile tire model which can handle higher frequencies and shorter wavelengths. The enhanced model and the restricted fully nonlinear transient tire model will be used as the basis of this development.

Chapter 9

# Short Wavelength Intermediate Frequency Tire Model

## Chapter Outline



Tire and Vehicle Dynamics. DOI: 10.1016/B978-0-08-097016-5.00009-7

## 9.1. INTRODUCTION

In Chapter 5 the stretched string model, possibly featuring tread elements and tire inertia (gyroscopic couple), was used to study the dynamic response of the side force and aligning torque to lateral, vertical, and yaw motion variations of the wheel axle. These motions, however, had to be restricted in magnitude to allow the theory to remain linear. Several approximations were introduced to simplify the model description, which made it possible to consider ranges of relatively large slip through a simple nonlinear extension. The limitation of the use of this type of simplified models is that only phenomena with relatively large wavelength (>ca. 1.5 m) and low frequency (<ca. 8 Hz) can be approximately handled.

The present chapter describes a model that is able to cover situations where the wavelength is relatively short (>ca. 10 cm and even shorter for modeling road obstacle enveloping properties) and the frequency is relatively high (<ca. 80 Hz) while the level of slip can be high. Situations in which combined slip occurs can be handled and the *Magic Formula* model can be used as the basis for the nonlinear force and moment description. As a result, a continuous transition from time-varying slip situations to steady-state conditions is realized. The original model development was restricted to the more important responses to variations of longitudinal and side slip. Also, the possibility to traverse distinct road irregularities (cleats) was included in the tire model. The model is based on the work of Zegelaar (1998) and Maurice (2000) conducted at the Delft University of Technology and supported by TNO Automotive and a consortium of industries. The model is referred to as the *SWIFT* model (corresponding to the title of the present chapter). Subsequent developments of the model made it possible to also include variations of camber and turn slip.

The crucial step that was taken to reach further than one can by using the string model is the separation of modeling the carcass and the contact patch. In this way a much more versatile model can be established that correctly describes slip properties at short wavelengths and at high levels of slip. The model achieved can be seen as a further development of the enhanced single contact point model of Section 7.3, Figure 7.15.

Five elements of the model structure can be distinguished: (1) The inertia of the belt that has been taken into account to properly describe the dynamics of the tire. The restriction to frequencies of about 60 Hz allows the belt to be considered as a rigid circular ring. (2) The so-called residual stiffnesses, see explanation below, that have been introduced between contact patch and ring to ensure that the total static tire stiffnesses in vertical, longitudinal, lateral, and yaw directions are correct. The total tire model compliance is made up of the carcass compliance, the residual compliance (in reality a part of the total carcass compliance), and the tread compliance. (3) The brush model that represents the contact patch featuring horizontal tread element compliance and partial sliding. On the basis of this model, the effects of the finite length and width of the footprint are approximately included. This element of the model is the most complex part and accomplishes the reduction of the allowed input wavelength to ca. 10 cm. (4) Effective road inputs to enable the simulation of the tire moving over an uneven road surface with the enveloping behavior of the tire properly represented (Chapter 10). The actual three-dimensional profile of the road is replaced by a set of four effective inputs: the effective height, the effective forward and transverse slopes of the road plane, and the effective forward road curvature that is largely responsible for the variation of the tire effective rolling radius. (5) The *Magic Formula* tire model to describe the nonlinear slip force and moment properties. In Figure 9.1 the model structure has been depicted.

Similar more physically oriented models have been developed. A notable example is the *BRIT* model of Gipser. He employs a brush-ring model featuring tread elements distributed over a finite contact patch with realistic pressure distributions and demonstrates the use of the model with simulations of the tire traversing a sinusoidal road surface also at a slip angle and at braking, cf. Gipser et al. (1997). Other models: *FTire* and *RMOD-K*, have been developed by Gipser (cf. Section 13.2) and Oertel and Fandre (cf. Section 13.1), respectively. In Chapter 13 outlines of these two models and of the *SWIFT* model are presented.

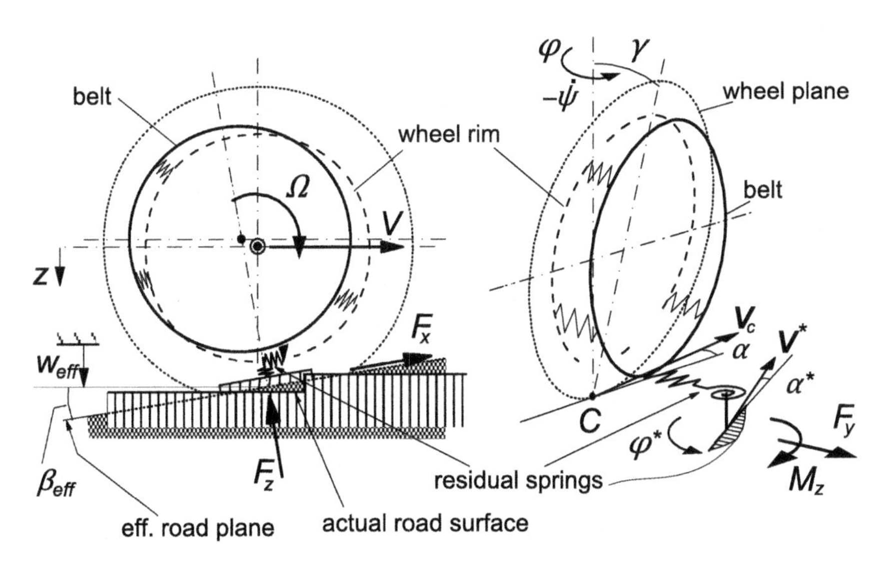


**FIGURE 9.1** General configuration of the *SWIFT* model featuring rigid belt ring, residual stiffnesses, contact patch slip model, and effective road inputs.

In the ensuing theoretical treatment, firstly, in Section 9.2, the slip model of the contact patch covering small wavelengths and large slip will be dealt with. Secondly, the model for the description of the dynamic behavior of the rigid belt ring will be added, Section 9.3. Thirdly, the feature of the model that takes care of running over uneven roads will be addressed, cf. Chapter 10. Full-scale tire test results demonstrate the validity of the model.

## 9.2. THE CONTACT PATCH SLIP MODEL

In this section, we will first represent the contact patch with tread elements by the brush model. Because of its relative complexity, the analytical model that describes the non-steady-state response to slip variations is approximated by a set of first-order differential equations. This contact model is tested by attaching the base line of the brush model to the wheel plane through a compliant carcass. For reasons of practical use, we finally introduce the *Magic Formula* to handle realistic nonlinear slip behavior of the model.

### 9.2.1. Brush Model Non-Steady-State Behavior

The steady-state characteristics of the brush model have been discussed in Chapter 3. As a first step, we will derive the equations that govern the response of the forces and moment to small variations of the wheel slip with respect to a given level of wheel slip indicated with subscript $o$.

#### *Longitudinal Slip*

For the case of pure longitudinal slip $\kappa_{co}$ of the contact patch, the point of transition from adhesion to sliding is located according to Eqn (3.44) with (3.34, 3.40) at a distance $x_t$ from the contact center:

$$x_t = a\left(\frac{2\theta|\kappa_{co}|}{1+\kappa_{co}} - 1\right) \tag{9.1}$$

The composite parameter $\theta$ has been defined in Chapter 3, Eqn (3.46). The length of the adhesion range $2am = a - x_t$ that begins at the leading edge is characterized by the fraction $m$ that in the present chapter replaces the symbol $\lambda$ to avoid confusion with the notation for the wavelength. We have

$$m = 1 - \theta\frac{|\kappa_{co}|}{1+\kappa_{co}} \quad \text{if} \quad \frac{|\kappa_{co}|}{1+\kappa_{co}} < \frac{1}{\theta} \quad \text{else} \quad m = 0 \tag{9.2}$$

An important observation is that at small variations of slip we may assume that only in the adhesion range changes in deflection occur.

For the development of the transient model we start out from the basic rolling contact differential Eqns (2.55, 2.56). In the adhesion range we find for the longitudinal tread element deflection $u_c$ in the case that only longitudinal slip is considered:

$$\frac{\partial \tilde{u}_c}{\partial t} - V_{rco}\frac{\partial \tilde{u}_c}{\partial x} = \tilde{V}_{sxc} = -V_{xco}\tilde{\kappa}_c \tag{9.3}$$

where the tilde designates the variation with respect to the steady-state level. The average linear speed of rolling of the contact patch is equal to that of the wheel rim and reads

$$V_{rco} = V_{ro} = V_x - V_{sxo} = V_x(1+\kappa_{co}) \tag{9.4}$$

The Fourier transform of the deflection becomes with $U$ and $K$ denoting the transformed quantities of $u$ and $\kappa$, respectively, and $\omega_s$ the path frequency:

$$\tilde{U}_c = \frac{1}{i\omega_s}(1 - e^{-i\omega_s(a-x)V_x/V_{rco}})\tilde{K}_c \tag{9.5}$$

Here, the boundary condition that says that the deflection vanishes at the leading edge, is satisfied. By integrating over the range of adhesion and multiplying with the tread element stiffness $c_p$, the frequency response function of the variation of $F_x$ to the variation of $\kappa_c$ is obtained. With the reduced frequency

$$\omega_s' = \frac{m\omega_s}{1+\kappa_{co}} \tag{9.6}$$

and the local derivative of $F_x$ to $\kappa_c$

$$C_{F\kappa c} = \frac{2c_p m^2 a^2}{1+\kappa_{co}} \quad \text{if} \quad \frac{|\kappa_{co}|}{1+\kappa_{co}} < \frac{1}{\theta} \quad \text{else} \quad C_{F\kappa c} = 0 \tag{9.7}$$

the expression for the response function becomes, when still adhesion occurs in the contact patch (first condition of (9.2)),

$$H_{F,\kappa c}(i\omega_s') = \frac{C_{F\kappa c}}{i\omega_s' a}\left\{1 - \frac{1}{2i\omega_s' a}(1 - e^{-2i\omega_s' a})\right\} \tag{9.8}$$

In Figure 9.2 the resulting amplitude and phase characteristics have been shown together with those of an approximate first-order system. Especially the phase curve appears to exhibit a wavy pattern in the higher frequency range. These waves are considerably attenuated when the contact model is

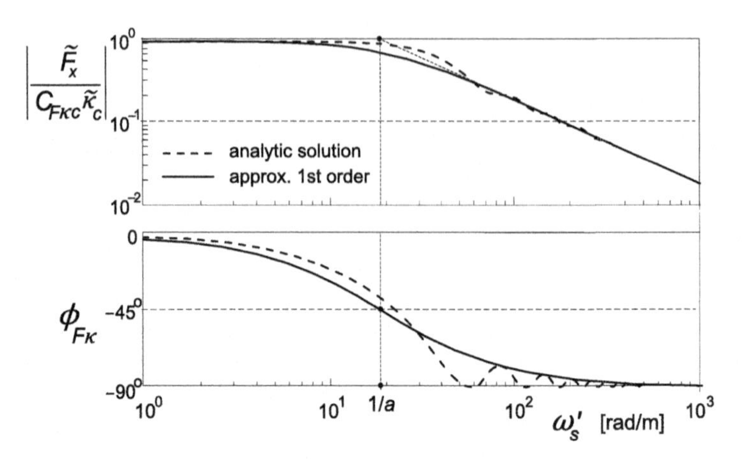


**FIGURE 9.2** Frequency response function of the longitudinal force variation to longitudinal slip variation of the brush model versus reduced path frequency according to the exact analytical solution and that of an approximate substitute first-order system at a given level of longitudinal slip.

incorporated in a more complete tire model including carcass compliance, as will be shown below.

The first-order substitute model has a frequency response function that reads

$$H_{F,\kappa c}(i\omega_s') = \frac{C_{F\kappa c}}{1 + i\omega_s' a} = \frac{C_{F\kappa c}}{1 + i\omega_s \sigma_c} \tag{9.9}$$

The approximation shows the same high frequency asymptote and steady-state level as the exact model. Apparently, the actual cutoff path frequency reads

$$\omega_{s,co} = \frac{1 + \kappa_{co}}{ma} \tag{9.10}$$

that, obviously, reduces to $1/a$ when the average slip $\kappa_{co}$ vanishes and, as a consequence, $m$ (9.2) becomes equal to unity. When the average slip is chosen larger, the length of the adhesion range decreases and the cutoff frequency becomes higher. The relaxation length of the approximate contact model reads, according to Eqn (9.9) with (9.6),

$$\sigma_c = \frac{ma}{1 + \kappa_{co}} \tag{9.11}$$

which reduces to zero when total sliding occurs.

The performance of the first-order system is reasonable but appears to improve when the filtering action of the carcass compliance is taken into account. In that configuration, we have the wheel slip velocity $V_{sx}$ that acts as the input quantity. Adding the time rate of change of the carcass deflection produces the slip velocity of the brush model. The carcass deflection $u$ equals (with fore-and-aft carcass stiffness $c_x$ introduced)

$$u = \frac{F_x}{c_x} \tag{9.12}$$

The feedback loop in the augmented system apparently contains a gain equal to $i\omega_s/c_x$. The resulting complete frequency response function reads

$$H_{F,\kappa}(i\omega_s) = \frac{H_{F,\kappa c}(i\omega_s)}{1 + H_{F,\kappa c}(i\omega_s)i\omega_s/c_x} \tag{9.13}$$

When the approximate first-order system is employed for the contact model, the frequency response function for the complete model becomes, using (9.13) with (9.9),

$$H_{F,\kappa}(i\omega_s) = \frac{C_{F\kappa c}}{1 + i\omega_s\sigma} \tag{9.14}$$

where the total relaxation length has been introduced which apparently reads

$$\sigma = \sigma_c + \frac{C_{F\kappa c}}{c_x} = am + \frac{C_{F\kappa}}{c_x} \tag{9.15}$$

The local longitudinal slip stiffness of the contact patch is here equal to the local longitudinal slip stiffness of the complete model, which obviously is due to the assumed inextensibility of the base line of the brush model. So we have at steady state:

$$\frac{\partial F_x}{\partial \kappa} = C_{F\kappa} = C_{F\kappa c} \tag{9.16}$$

At vanishing slip we will add a subscript 0. When total sliding occurs, both $\sigma_c$ and $C_{F\kappa}$ reduce to zero and the total relaxation length $\sigma$ vanishes.

Introduction of the longitudinal carcass compliance may be accomplished by considering the feedback loop similar to the one ($c_y$) for the more complex configuration for lateral slip shown in Figure 9.6. Figure 9.3 presents the frequency response function for this total model at vanishing slip, $\kappa_0 = 0$.

It is noted that compared with the curves of Figure 9.2 the wavy pattern is considerably reduced. The approximate system performs very well. A wavelength of 10 cm occurs at the path frequency $\omega_s =$ ca. 60 rad/m.

The differential equation that governs the transient slip response of the contact patch and through that the longitudinal force response becomes for the approximate system

$$\sigma_c \frac{d\tilde{\kappa}_c'}{dt} + |V_x|\tilde{\kappa}_c' = |V_x|\tilde{\kappa}_c = -\tilde{V}_{sxc} \tag{9.17}$$

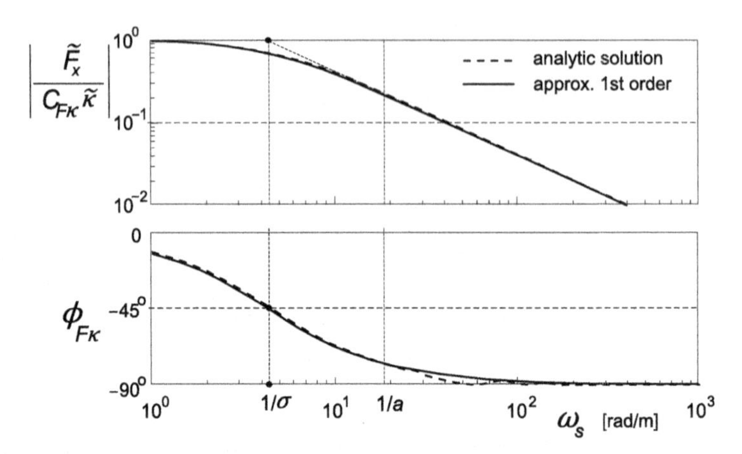


**FIGURE 9.3** Frequency response function of the longitudinal force variation to longitudinal slip variation of the brush model attached to a flexible carcass versus path frequency according to the exact analytical solution and that of the approximate first-order system at zero longitudinal slip.

The variation of the force becomes

$$\tilde{F}_x = C_{F\kappa c}\tilde{\kappa}'_c \tag{9.18}$$

The structure of Eqn (9.17) corresponds with that of Eqn (7.37) of Chapter 7. The response, however, is insensitive to load variations but shows a nice behavior. The transient response to load variations is sufficiently taken care of through the effect of carcass compliance. As a result, the relaxation length for the response to load variations becomes equal to $C_{F\kappa c}/c_x$, which is somewhat smaller than $\sigma$ (9.15) that holds for the response to slip variations.

When the average steady-state relation for the longitudinal slip

$$|V_x|\kappa_{co} = -V_{sxco} \tag{9.19}$$

is added to Eqn (9.17), the equation for the total transient slip is obtained:

$$\sigma_c\frac{\mathrm{d}\kappa'_c}{\mathrm{d}t} + |V_x|\kappa'_c = |V_x|\kappa_c = -V_{sxc} \tag{9.20}$$

that completely corresponds to Eqn (7.54) of the enhanced nonlinear transient tire model. The transient slip $\kappa'_c$ is subsequently used as input into the steady-state longitudinal force characteristic, as will be explained later on, and has already been indicated in Chapter 7, Section 7.3.

### *Lateral Slip*

The lateral slip condition is more complex to handle because we have to deal with both the side force and the aligning torque. In addition, in the test condition, the carcass is allowed to not only deflect in the lateral direction but also about the vertical axis. This is in accordance with the ultimate *SWIFT* configuration. The connected turn slip behavior of the contact patch will be dealt with further on.

As with the longitudinal model development, first, the analytic response functions of the brush model will be assessed, in this case to side slip variations. Eqn (2.56) gives rise to the following equation for the lateral tread deflection variations:

$$\frac{\partial \tilde{v}_c}{\partial t} - V_{rco}\frac{\partial \tilde{v}_c}{\partial x} = \tilde{V}_{syc} = -V_{xco}\tilde{\alpha}_c \tag{9.21}$$

Note that for the sake of simplification in the present chapter the notation $\tan\alpha$ is replaced by $\alpha$, with or without a subscript. As for the longitudinal deflection we find for the Fourier transform $V_c$ of the lateral deflection responding to slip angle variations, $A_c$ denoting the brush model slip angle's transform (or actually of $\tan\alpha_c$):

$$\tilde{V}_c = \frac{1}{i\omega_s}(1 - e^{-i\omega_s(a-x)V_x/V_{rco}})\tilde{A}_c \tag{9.22}$$

Again, the responses to variations of the side slip only occur in the range of adhesion. The transition point from adhesion to sliding now occurs at

$$x_t = a(2\theta|\alpha_{co}| - 1) \tag{9.23}$$

and the corresponding adhesion fraction becomes, cf. Eqn (3.8),

$$m = 1 - \theta|\alpha_{co}| \quad \text{if} \quad |\alpha_{co}| < \frac{1}{\theta} \quad \text{else} \quad m = 0 \tag{9.24}$$

As the slip angle of the contact patch remains small in the range where adhesion still occurs, $\cos\alpha_{co}$ has been replaced by unity. By integration of the transformed deflection over the range of adhesion the frequency response functions of the force and the moment variations are established. They read, at pure lateral slip ($V_{rco} = V_x$, cf. (9.4)),

$$H_{F,\alpha c}(i\omega_s) = \frac{C_{F\alpha c}}{i\omega_s am}\left\{1 - \frac{1}{2i\omega_s am}(1 - e^{-2i\omega_s am})\right\} \tag{9.25}$$

and

$$H_{M,\alpha c}(i\omega_s) = \frac{c_p}{i\omega_s}\left\{-\frac{1}{\omega_s^2}(1 - e^{-2i\omega_s am}) - \frac{a}{i\omega_s}(1 + (2m-1)e^{-2i\omega_s am}) + 2a^2 m(1-m)\right\} \tag{9.26}$$

The local slope of the steady-state side force characteristic of the brush model is given by

$$C_{F\alpha c} = 2c_p m^2 a^2 \quad \text{if} \quad |\alpha_{co}| < \frac{1}{\theta} \quad \text{else} \quad C_{F\alpha c} = 0 \tag{9.27}$$

The normalized response function of the side force is identical to that of the longitudinal force if the factor $1 + \kappa_{co}$ is omitted in (9.6). The approximate first-order description with response function

$$H_{F,\alpha c}(i\omega_s) = \frac{C_{F\alpha c}}{1 + i\omega_s \sigma_c} \tag{9.28}$$

where

$$\sigma_c = am \tag{9.29}$$

shows the same very good agreement with the exact result, as assessed with the longitudinal force response. Similarly we have the differential equation for the variation of the transient side slip:

$$\sigma_c \frac{d\tilde{\alpha}'_c}{dt} + |V_x|\tilde{\alpha}'_c = |V_x|\tilde{\alpha}_c = -\tilde{V}_{syc} \tag{9.30}$$

The variation of the side force becomes

$$\tilde{F}_y = C_{F\alpha c}\tilde{\alpha}'_c \tag{9.31}$$

After adding the steady-state equation

$$|V_x|\alpha_{co} = -V_{syco} \tag{9.32}$$

to Eqn (9.30), the equation for the total transient side slip is obtained:

$$\sigma_c \frac{d\alpha'_c}{dt} + |V_x|\alpha'_c = |V_x|\alpha_c = -V_{syc} \tag{9.33}$$

As before, the resulting $\alpha'_c$ is used as the input of the steady-state side force function.

We may follow the same procedure to assess the aligning torque equations as was done in Section 7.3. The resulting first-order response, however, does not always agree with the analytically found tendency. The phase lag and the slope of the high-frequency amplitude asymptote indicate that a second-order behavior prevails at zero average slip angle while at larger slip angles the response gradually changes into a first-order nature. Apparently, the mechanism is more complex and we should account for the transient response of the pneumatic trail. The variation of the aligning torque may be written as follows:

$$\tilde{M}_z = -t_{co}\tilde{F}_y - \tilde{t}_c F_{yo} \tag{9.34}$$

The analysis conducted by Maurice (2000) shows that the analytically assessed response function of the pneumatic trail variation to slip angle variations can be approximated by the first-order system (9.33) and a so-called phase leading network in series. The frequency response function of the latter reads

$$H_p(i\omega_s) = C_{t\alpha c}\frac{1 + i\omega_s\sigma_1}{1 + i\omega_s\sigma_2} \tag{9.35}$$

The factor in this formula represents the local slope of pneumatic trail characteristic:

$$C_{t\alpha c} = \frac{\partial t_c}{\partial|\alpha_c|} \tag{9.36}$$

which, apparently, is a negative quantity. According to Maurice, adequate values for the parameters $\sigma_1$ and $\sigma_2$ can be obtained through the formulas

$$\sigma_2 = \frac{1}{3}a(1 - \theta|\alpha_{co}|) \tag{9.37}$$

or alternatively

$$\sigma_2 = t_c \tag{9.38}$$

and

$$\frac{\sigma_1}{\sigma_2} = \frac{1}{1 - m^2} \tag{9.39}$$

The block diagram of the current system governed by Eqns (9.28, 9.34, 9.35) is presented in the upper diagram of Figure 9.4. The lower diagram shows an alternative structure of the same system, thereby displaying the extra moment $\Delta M_z$, which is governed by the ratio of parameters (9.39). This ratio tends to infinity when full adhesion occurs. Then, $m = 1$ and $\alpha_{co} = 0$, as becomes clear from Eqn (9.24). The singularity involved has been circumvented by Maurice through the introduction of a function that limits the value of (9.39) around zero lateral slip, $\alpha_{co} = 0$. This, however, will slightly disturb the proper response at zero slip angle and thus degrades the linear analysis around zero slip. An alternative way of avoiding the singularity, which obviously is caused by the fact that we actually may have a moment without the simultaneous presence of a force, is the consideration of the extra transient moment $\Delta M_z$. This moment is obtained by multiplication of the difference of the transient slip quantities for the force and for the pneumatic trail with three factors the combination of which may be designated with $C_{\Delta M}$:

$$C_{\Delta M} = -\frac{\sigma_1}{\sigma_2}C_{t\alpha c}F_{yo} \tag{9.40}$$

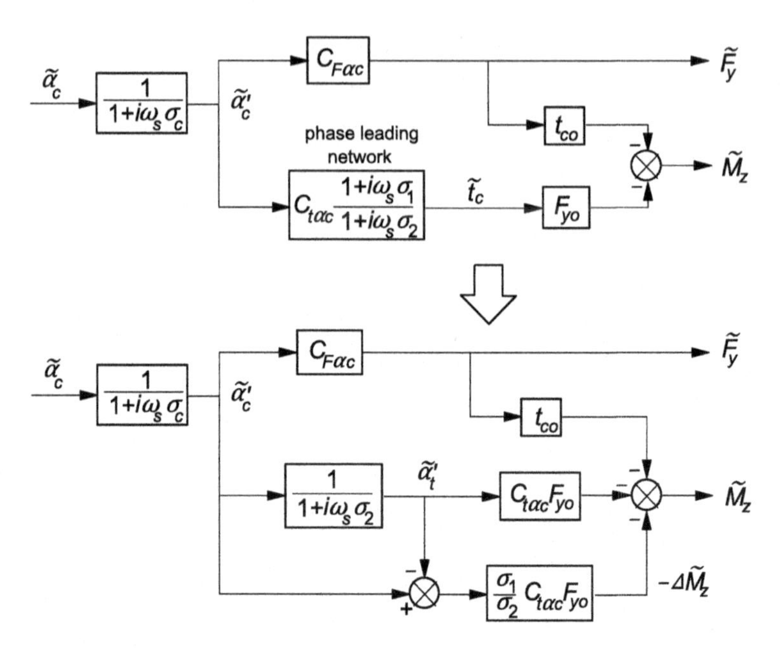


**FIGURE 9.4** Block diagram of the contact patch model to generate short wavelength transient responses of the side force and the aligning torque to small slip angle variations. The original upper diagram can be replaced by the lower diagram, thereby avoiding singularity at zero slip.

It turns out that now the singularity does not show up because both $F_{yo}$ and $1/\sigma_1$ become zero at the same time. This indicates that indeed a moment may arise although at that instant of time the side force is zero. After writing out the factors in (9.40) by using expressions for the side force (3.11), the pneumatic trail (3.13) and further (9.36, 9.39) while in (3.11) $\lambda$ is replaced by $m$ and $\theta_y \sigma_y$ by $\theta \alpha_{co} = z$ we find, for $C_{\Delta M}$,

$$C_{\Delta M} = \frac{1}{3} a C_{F\alpha co} \xi = C_{M\alpha co} \xi \tag{9.41}$$

where $C_{M\alpha co}$ denotes the aligning stiffness of the brush model at zero slip angle and the nondimensional factor $\xi$ is introduced:

$$\xi = \frac{(1-z)^2}{2-z} \left\{ 3 - \frac{(1-z)(1-2z/3)}{1-z+z^2/3} \right\} \quad \text{if} \quad z < 1 \quad \text{else } \xi = 0 \tag{9.42}$$

with

$$z = \theta |\alpha_{co}| \tag{9.43}$$

Evaluation of (9.42) reveals that for $z < 1$ the factor $\xi$ may possibly be approximated by

$$\xi = 1 - z \tag{9.44}$$

This approximation will be introduced later on when we will deal with the application of the *Magic Formula*.

The differential equations that apply for the contact patch model subjected to small side slip variations with respect to a given side slip level now read

$$\sigma_c \frac{d\tilde{\alpha}'_c}{dt} + |V_x|\tilde{\alpha}'_c = |V_x|\tilde{\alpha}_c = -\tilde{V}_{syc} \tag{9.45}$$

$$\sigma_2 \frac{d\tilde{\alpha}'_t}{dt} + |V_x|\tilde{\alpha}'_t = |V_x|\tilde{\alpha}'_c \tag{9.46}$$

from which the variation of the force and moment result:

$$\tilde{F}_y = C_{F\alpha c}\tilde{\alpha}'_c \tag{9.47}$$

$$\tilde{M}_z = -t_{co}\tilde{F}_y - C_{t\alpha c}\tilde{\alpha}'_t F_{yo} + C_{\Delta M}(\tilde{\alpha}'_c - \tilde{\alpha}'_t) \tag{9.48}$$

It is of importance to check the step responses to side slip starting from zero slip angle, especially right after the start of the step change. Initially we have all variables equal to zero while the slip angle input has reached the new value $\alpha_{co}$. The various time derivatives become

$$(t = 0: \quad \alpha_c = \alpha_{co}, \quad \alpha'_c = \alpha'_t = 0, \quad F_y = M_z = 0)$$

$$\frac{d\alpha'_c}{dt} = \frac{1}{\sigma_{co}}|V_x|\alpha_{co}, \quad \frac{d\alpha'_t}{dt} = 0$$
$$\frac{dF_y}{dt} = \frac{|V_x|}{\sigma_{co}}C_{F\alpha co}\alpha_{co}, \quad \frac{dM_z}{dt} = -t_{co}\frac{dF_y}{dt} + C_{\Delta Mo}\frac{d\alpha'_c}{dt} = 0 \tag{9.49}$$

These results are correct. The last equation holds indeed because we have at vanishing slip angle according to Eqn (9.42): $\xi = 1$ and consequently $C_{\Delta M} = t_{co}C_{F\alpha co}$, where $t_{co} = a/3$. Evidently, the extra moment is responsible for the proper start of the course of the aligning torque showing zero slope. This property has been ascertained to occur both in reality and with models, cf. Figures 5.10 and 5.11. Also the response to a lateral wheel displacement $y = \int V_{sy}dt$ at zero forward speed develops correctly. We find with Eqn (7.6) for the force: $F_y = -C_{Fy}\, y$ and for the moment: $M_z = 0$.

The equations may now further be appraised by first introducing tire carcass lateral and torsional compliance, as depicted in Figure 9.5 and in the corresponding block diagram of Figure 9.6, and subsequently comparing the results with analytical solutions obtained by using Eqns (9.25, 9.26).

As a reference, the steady-state characteristics of the model have been presented in Figure 9.7. The diagram contains the curves for both the complete model and for the brush model alone. It is of interest to note the lower cornering

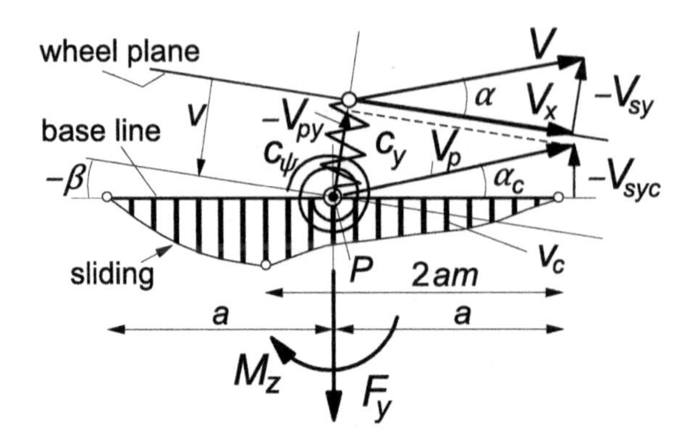


**FIGURE 9.5** The brush model attached to a carcass possessing lateral and torsional compliance.

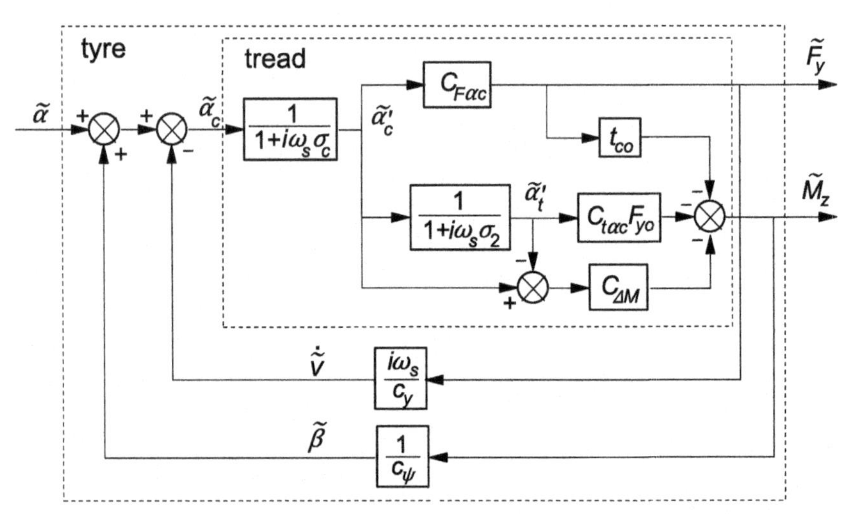


**FIGURE 9.6** Block diagram of the augmented system including carcass lateral and torsional compliance.

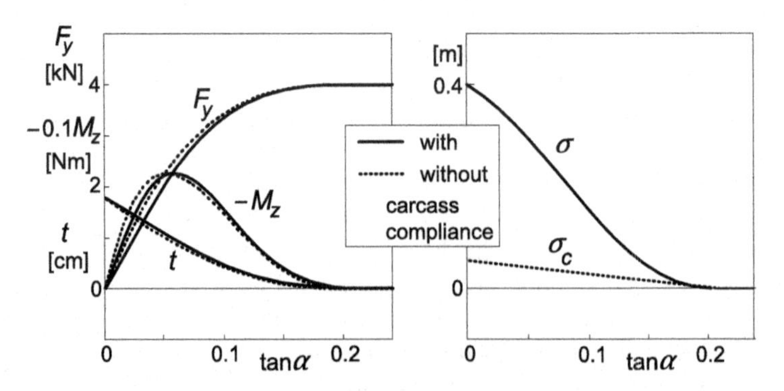


**FIGURE 9.7** Steady-state side slip force and moment characteristics and the relaxation length of the brush model (the contact patch) and of the model including the flexible carcass through which the brush model is attached to the wheel plane.

stiffness due to the introduction of carcass compliance. The expression for the lower side slip stiffness can be found to read

$$C_{F\alpha} = \frac{c_\psi}{c_\psi + C_{M\alpha c}} C_{F\alpha c} \tag{9.50}$$

The relaxation length of the complete model found from the cutoff frequency of the side force response function turns out to become

$$\sigma = \frac{c_\psi}{c_\psi + C_{M\alpha c}} \left( am + \frac{C_{F\alpha c}}{c_y} \right) = \frac{c_\psi}{c_\psi + C_{M\alpha c}} am + \frac{C_{F\alpha}}{c_y} \tag{9.51}$$

The way the relaxation length changes with slip angle is depicted in the right-hand diagram of Figure 9.7. As one might expect, we find that this length multiplied by the increment in wheel slip angle is equal to the increase of the sum of carcass lateral deflection and the average deflection of the adhering tread elements.

At three different levels of side slip, $\alpha_o = 0$, 0.08, and 0.16 rad, the comparison of the simulation model with the analytical model has been conducted in terms of the path frequency response functions. Figure 9.8 shows the results. The general conclusion is that, at least for wavelengths larger than ca. 15 cm, the correspondence can be judged to be very good. The upper pair of diagrams that refer to the side force response shows the sideways 'shift' of the phase lag curves that is caused by the drop in relaxation length with increasing average slip angle. The diagram for the aligning torque response clearly shows the transition from second- to first-order behavior when the average slip angle changes from zero to larger values.

The Eqns (9.45–9.48) can be made applicable to the general case of large slip angle variations by adding to (9.45, 9.46) the steady-state relations and by rewriting the Eqns (9.47, 9.48) in complete nonlinear form so that, when considering a small variation, the linearized equations are recovered. We obtain, for the transient slip quantities,

$$\sigma_c \frac{d\alpha'_c}{dt} + |V_x|\alpha'_c = |V_x|\alpha_c = -V_{syc} \tag{9.52}$$

$$\sigma_2 \frac{d\alpha'_t}{dt} + |V_x|\alpha'_t = |V_x|\alpha'_c \tag{9.53}$$

and, for the force and moment,

$$F_y = F_y(\alpha'_c) \tag{9.54}$$

$$M_z = -t_c(\alpha'_t)F_y + C_{\Delta M}(\alpha'_c - \alpha'_t) \tag{9.55}$$

The last term representing the extra moment is left in linearized form. This term may be replaced by the difference of two equal functions the derivative of which equals $C_{\Delta M}$, one with argument $\alpha'_c$ and the other with $\alpha'_t$. Because of the

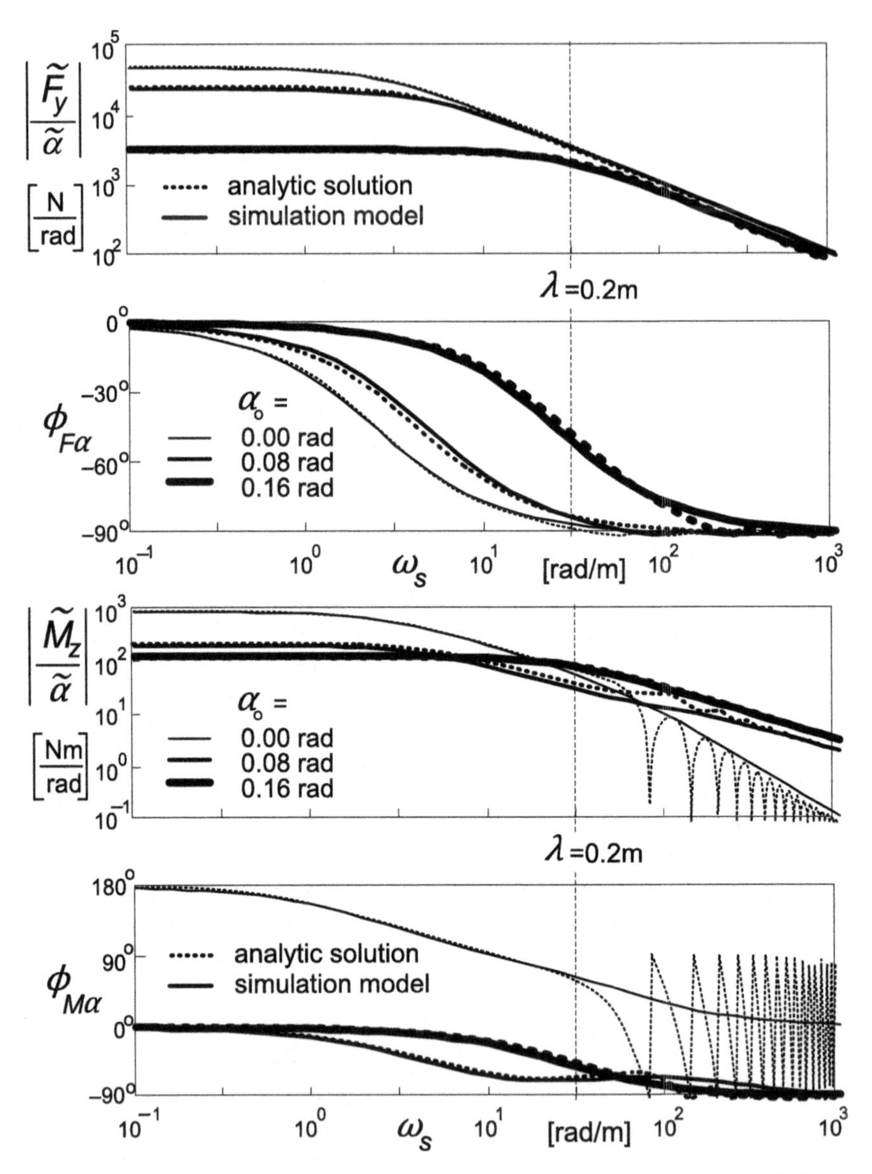


**FIGURE 9.8** Frequency response functions of a linearized system including carcass flexibility at different average slip angle levels according to the analytical solution and to the approximate simulation model. The path frequency at a wavelength of the input wheel plane motion equal to 20 cm has been indicated.

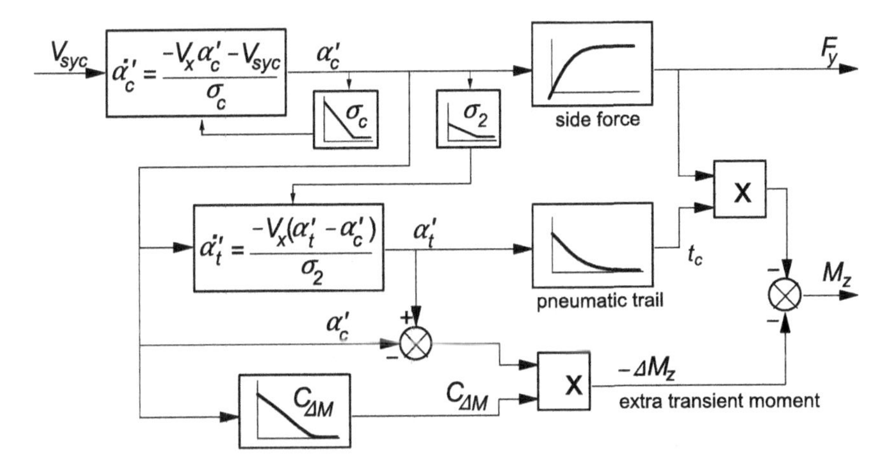


**FIGURE 9.9** Block diagram of the nonlinear model of the contact patch (tread) also valid for larger slip angle variations.

fact that the difference of these two arguments remains small, the linearized version is expected to be sufficiently accurate. The block diagram of the nonlinear system displayed in Figure 9.9 may further clarify the structure of the model.

To ensure that the above equations correctly describe the response to large slip angle variations, the simulation model results have been compared with the response of a physical model. That model features a finite number of tread elements attached to a straight base line. The deflections of carcass and elements are computed at each time step in which the wheel is rolled further over a distance equal to the interval between two successive tread elements and is moved sideways according to the current value of the input slip angle. The actual model employed contains 20 elements. The parameter values used in both models have been listed in Table 9.1.

The slip angle variation is sinusoidal around a given average level. To cover a broad range of operations, the computations have been conducted at three wavelengths: $\lambda = 0.2$, 1, and 5 m, two average slip angle levels: $\alpha_o = 0$ and 0.08 rad, and one amplitude: $\hat{\alpha} = 0.08$ rad.

Figure 9.10 presents the results for the two models. The range of the abscissa has been chosen such that precisely two wavelengths are covered. The distance rolled is large enough to have a situation close to the periodic state. Again, the agreement is quite good and we may have confidence in the model. The bottom diagram presents the variation of the slip angle. The top pair of diagrams shows the responses at the relatively long wavelength of 5 m so that a condition closer to steady state has been reached. From Figure 9.7 it can be seen that at the maximum $\alpha$ of 0.16 rad or almost 9° the aligning torque peak has been surpassed by far. This explains the two dips per wavelength. At a maximum of 0.08 rad the peak has just been surpassed.

**TABLE 9.1 Parameter Values Used for Brush Model with Flexible Carcass**

| | | | |
|---|---|---|---|
| Vertical load | $F_z$ | 4000 | N |
| Friction coefficient | $\mu$ | 1.0 | – |
| Half contact length | $a$ | 0.0535 | m |
| Longitudinal carcass stiffness | $c_x$ | $5.50 \times 10^5$ | N/m |
| Lateral carcass stiffness | $c_y$ | $1.25 \times 10^5$ | N/m |
| Torsional carcass stiffness | $c_\psi$ | 4000 | Nm/rad |
| Tread element stiffness/m | $c_p$ | $10^7$ | N/m$^2$ |
| Composite tire parameter | $\theta$ | 4.77 | – |

The less deep dips occurring in the upper curves of diagrams *b* and *d* are delayed with respect to the slip angle when this has reached its minimum value equal to zero. At steady state, when $\lambda \rightarrow \infty$, the moment (and the force) would have become equal to zero at that instant. The deeper dips belong to the maxima of the slip angle variation. The considerable reduction in amplitude of the response at the shorter wavelengths agrees nicely with the findings of Figure 9.8.

The force responses correspond almost perfectly with the outcome of the physical model. Evidently, a similar correspondence is expected to occur with the response of the longitudinal force to longitudinal slip variations.

### *Turn Slip*

The last item to be studied in the development of the contact model is the response to variations in path curvature while the slip angle remains zero. We will not attempt to develop a background analytical model but will take a more heuristic route. The results will be checked by comparing these with the computed responses of the physical model. The responses derived for the string model with tread elements as presented in Chapter 5, especially the step responses to $\varphi$ as depicted in Figure 5.21 (ex.tr.el.) and Figure 5.10 may be helpful. We observe that the force response is similar to that of the aligning torque to a step change in slip angle. In both cases the slope at the start is zero.

The side force relates to the transient turn slip (if linear) as: $F_y = C_{F\varphi co}\varphi'_c$. The uncorrected further approach of the force to its steady-state level is assumed to occur according to the first-order equation, with $\sigma_c$ as relaxation length. The initial slope of the uncorrected response curve of the normalized transient turn slip of the contact patch $\varphi'_c/\varphi_c$ versus $s$ becomes: $1/\sigma_c$. The zero

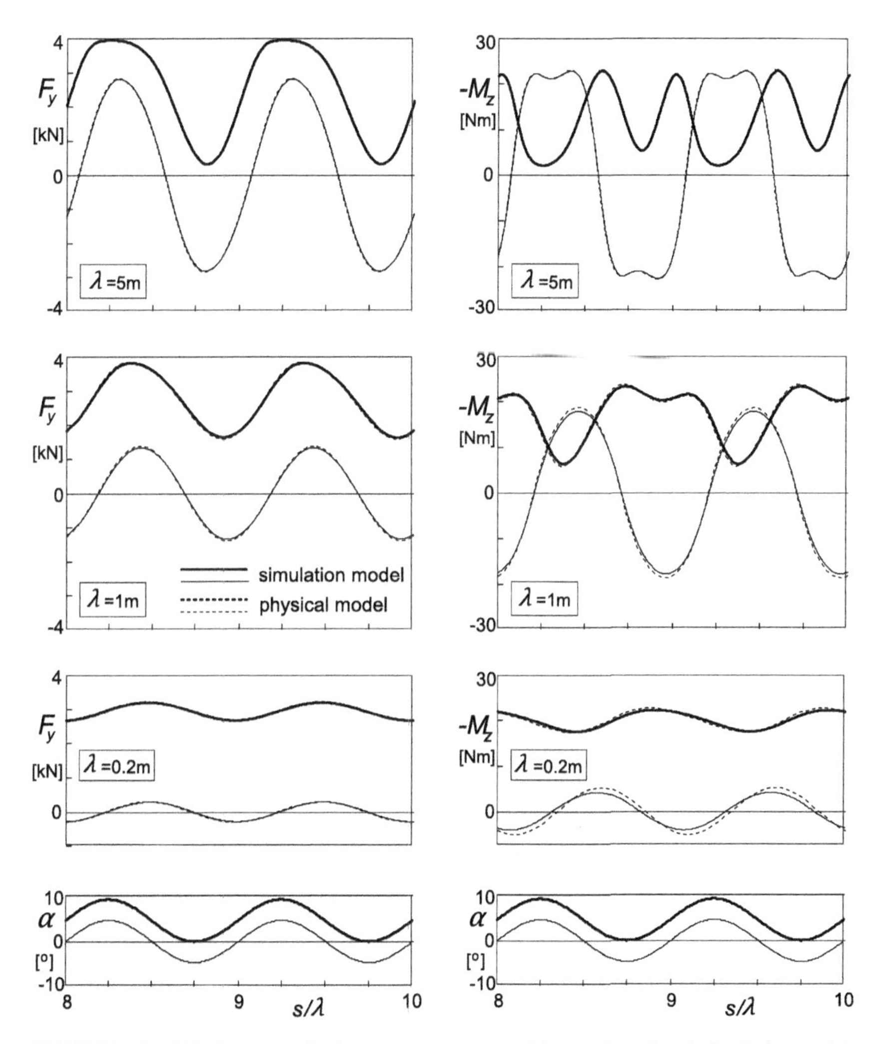


**FIGURE 9.10** Side force and aligning torque responses of the nonlinear brush simulation model with flexible carcass to sinusoidal slip angle input of the wheel plane with a slip angle amplitude of 0.08 rad and average levels of 0 (thin curves) and 0.08 rad (fat curves), compared with results of the physical model (broken curves).

slope at the start may be modeled by subtracting a correction response curve that starts with the same slope but dies out after having reached its peak. Such a short term response may be obtained by taking the difference of two responses, $\varphi'_{F1}$ and $\varphi'_{F2}$, each leading to the same level but starting at different slopes the difference of which should correspond to the initial slope of the uncorrected force response curve. For simplicity we take for one of the two responses ($\varphi'_{F1}$) the uncorrected force response, so that $\sigma_{F1} = \sigma_c$. The relaxation length of the second response ($\varphi'_{F2}$) should then be equal to $\sigma_{F2} = \sigma_c/2$, so that its initial slope becomes: $2/\sigma_c$. Subtracting these two slopes results in the desired initial slope of the correction curve: $1/\sigma_c$. The resulting equations for the force response to the turn slip velocity $\dot{\psi}_c$ read

$$\sigma_c \frac{d\varphi'_c}{dt} + V\varphi'_c = -\dot{\psi}_c \tag{9.56}$$

$$\sigma_{F2} \frac{d\varphi'_{F2}}{dt} + V\varphi'_{F2} = -\dot{\psi}_c \tag{9.57}$$

with $\sigma_{F2} = \sigma_c/2$. The transient turn slip for the force finally becomes

$$\varphi'_F = \{\varphi'_c - (\varphi'_{F2} - \varphi'_{F1})\}\text{sgn } V_x = (2\varphi'_c - \varphi'_{F2})\text{sgn } V_x \tag{9.58}$$

The side force at pure path curvature is obtained from the nonlinear steady-state response function:

$$F_y = F_y(\varphi'_F) \tag{9.59}$$

For the range of turn slip $a|\varphi| < 1/\theta$ the relation remains linear and equals for the brush model:

$$F_y = C_{F\varphi co}\varphi'_F = C_{M\alpha co}\varphi'_F = \frac{2}{3}c_p a^3 \varphi'_F \tag{9.60}$$

The moment response may be divided into the response due to the contact patch length that involves lateral tread deflections, and the response due to tread width giving rise to longitudinal deflections. First, we will address the moment generated by the brush model with zero width. Figures 5.9, 5.10, and 5.21 indicate that we are dealing here with a response that after having reached a peak tends to zero. Again we may model this behavior by subtracting two first-order responses. For this, we introduce two transient turn slip quantities $\varphi'_1$ and $\varphi'_2$ with respective relaxation lengths: $\sigma_{\varphi 1}$ and $\sigma_{\varphi 2}$. The two differential equations become

$$\sigma_{\varphi 1} \frac{d\varphi'_1}{dt} + V\varphi'_1 = -\dot{\psi}_c \tag{9.61}$$

$$\sigma_{\varphi 2} \frac{d\varphi'_2}{dt} + V\varphi'_2 = -\dot{\psi}_c \tag{9.62}$$

At zero speed the response of the difference would become

$$\varphi'_1 - \varphi'_2 = -\left(\frac{1}{\sigma_{\varphi 1}} - \frac{1}{\sigma_{\varphi 2}}\right)\psi_c \tag{9.63}$$

The deflection angle of the tread due to transient spin is defined as

$$\alpha_M = -2a(\varphi'_1 - \varphi'_2) \tag{9.64}$$

which at zero speed becomes

$$\alpha_M = 2a\left(\frac{1}{\sigma_{\varphi 1}} - \frac{1}{\sigma_{\varphi 2}}\right)\psi_c \tag{9.65}$$

The condition to be satisfied is that the deflection angle is equal to the yaw angle: $\alpha_M = \psi_c$. Hence, we have

$$\frac{1}{\sigma_{\varphi 1}} - \frac{1}{\sigma_{\varphi 2}} = \frac{1}{2a} \tag{9.66}$$

In the case of small angles, we may write for the moment

$$M_z = -C_{M\alpha co}\alpha_M \tag{9.67}$$

From Eqns (9.61 and 9.62), it can be assessed that the initial slope of the response of the moment to a step change in spin $\varphi_c = -\dot{\psi}_c/V$ from zero to $\varphi_{c0}$ turns out to be

$$\frac{dM_z}{ds} = 2aC_{M\alpha co}\left(\frac{1}{\sigma_{\varphi 1}} - \frac{1}{\sigma_{\varphi 2}}\right)\varphi_{co} = C_{M\alpha co}\varphi_{co} \tag{9.68}$$

For the second condition to assess the ratios of the $\sigma_\varphi$'s to half the contact length $a$, the best fit of the remaining course of the step response may serve, cf. Eqns (9.114 and 9.115).

When the angle of rotation $\psi$ continues to grow, the state of total sliding will be attained and the moment can be calculated to become

$$M_{z\varphi\infty} = -\frac{3}{8}a\mu F_z \tag{9.69}$$

It has not been tried to derive the functional relationship between moment and increasing steer angle for the standing (nonrolling) brush model. The following nonlinear function to describe the moment response in between the two extremes has been chosen:

$$M_z = -M_{z\varphi\infty}\sin\left\{\arctan\left(\frac{C_{M\alpha co}}{M_{z\varphi\infty}}\varepsilon_\varphi\alpha_M\right)\right\} \tag{9.70}$$

The factor $\varepsilon_\varphi$ has been introduced to have a parameter available to better approach the response shown by the physical model. The value 1.15 appeared to be appropriate. Similar to the relaxation length $\sigma_c$, Eqns (9.29 and 9.24), the lengths $\sigma_{\varphi 1,2}$ are reduced in proportion with the magnitudes of transient turn slip quantities $\varphi'_1$ and $\varphi'_2$.

For the evaluation of the model a comparison with the physical brush model has been executed. First, the flexible carcass is attached to the tread model, cf. Figure 9.5. To simulate this more complex situation, the approach of the enhanced transient model of Section 7.3 has been adopted. The additional dynamic equations for the contact patch with mass $m_c$ and moment of inertia $I_c$ and carcass stiffnesses $c_{y,\psi}$ and damping coefficients $k_{y,\psi}$ read, cf. Figure 9.5:

$$m_c(\dot{V}_{py} + V_x\dot{\psi}_c) + k_y\dot{v} + c_y v = F_y \tag{9.71}$$

$$\dot{v} = V_{py} - V_{sy} \tag{9.72}$$

$$I_c\ddot{\psi}_c + k_\psi\dot{\beta} + c_\psi\beta = M_z \tag{9.73}$$

$$\dot{\beta} = \dot{\psi}_c - \dot{\psi} \tag{9.74}$$

$$V_{syc} = V_{sy} + \dot{v} - V_x\beta \tag{9.75}$$

This extension of the model may, of course, also be used for the previously treated response to pure side slip. In fact, it is to be noted that the model for side slip, Eqns (9.52–9.55), should be added to the spin model, Eqns (9.61, 9.62, 9.64, and 9.70), to correctly account for their interaction in the complete model with carcass compliance included. In the physical model the brush model automatically responds to both the side slip and spin. When the wheel plane is subjected to only side slip, the spin of the base line of the tread model remains very small and may be neglected. On the other hand, when the wheel is being steered with wheel side slip remaining zero (path curvature), the base line does show non-negligible side slip, especially at shorter wavelengths, where the moment becomes considerable and as a result the base line is yawed and thus induces side slip. This effect vanishes at steady-state turning. However, if we would add the effect of tread width, the spin torque also acts at steady state and thereby contributes largely to the side force response to spin of the complete model.

Due to the complexity that arises when adding tread width to the brush model, cf. Chapter 3, it has been decided to consider this aspect when dealing with the ultimate model adapted to the use of the *Magic Formula* in the next section.

The diagrams of Figure 9.11 present the computed responses to varying turn slip. It is seen that the correspondence with the physical model, again with 20 tread elements, is quite good. The deformation of the moment response curve occurring at shorter wavelengths is caused by the extra yaw moment generated through the base line slip angle variation.

As a reference, the steady-state characteristics of the force and moment response to turn slip, as computed for the single row brush model with and without carcass compliance, have been shown in Figure 9.12. Up to $a\varphi = 1/\theta$ the aligning torque remains zero which causes the characteristics for the cases without and with flexible carcass to become identical. The remaining course of the curves for the system including carcass compliance has not been computed as that part lies outside the range of evaluation.

## Combined Slip

To cover the case of combined slip, also including longitudinal slip, the steady-state brush model characteristics are to be adapted as formulated in Chapter 3,

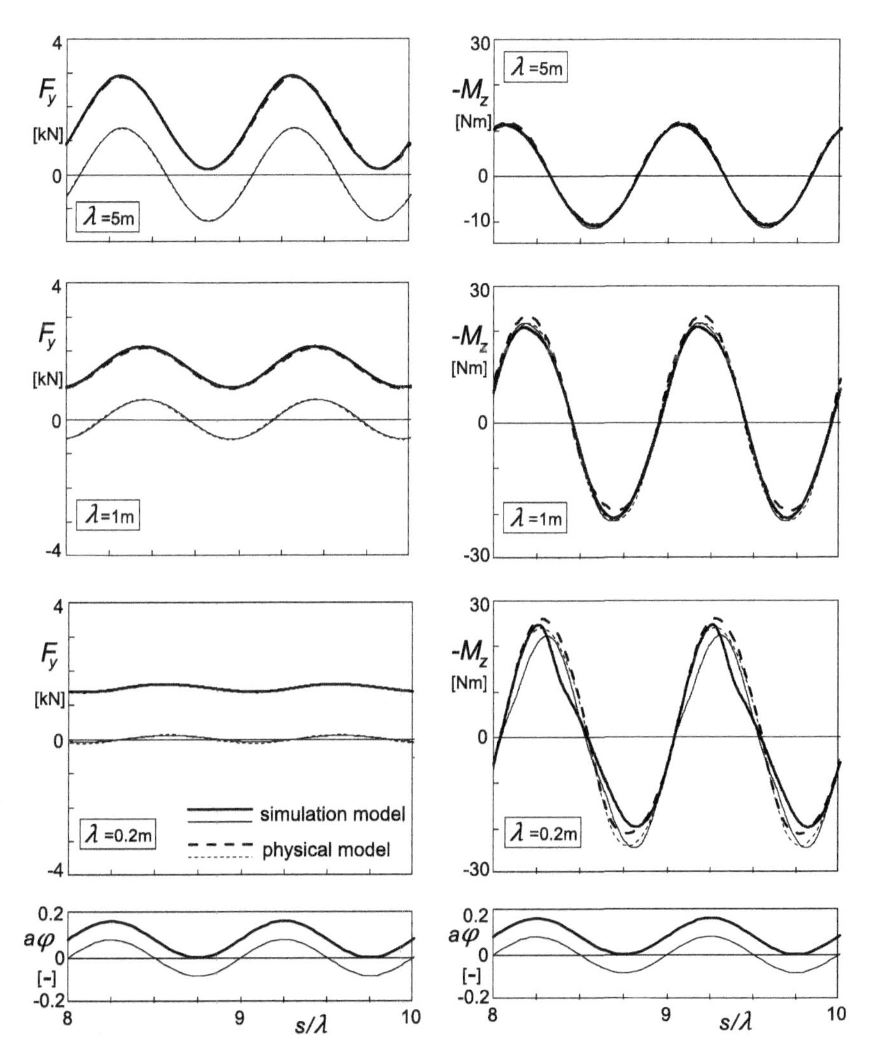


**FIGURE 9.11** Side force and aligning torque responses of the nonlinear brush simulation model with flexible carcass to sinusoidal path curvature variations (turn slip at $\alpha = 0$) of the wheel with an amplitude of $a\varphi$ equal to 0.08 and average levels of 0 (thin curves) and 0.08 (fat curves), compared with results of the physical model (broken curves).

Section 3.2.3. In addition, the factor $m$ that indicates the fraction of the contact length where adhesion occurs and is used to reduce the relaxation length $\sigma_c$ is to be adapted by using the composite magnitude of slip:

$$\zeta'_c = \frac{\sqrt{\alpha'^2_c + \kappa'^2_c}}{1 + \kappa'_c} \tag{9.76}$$

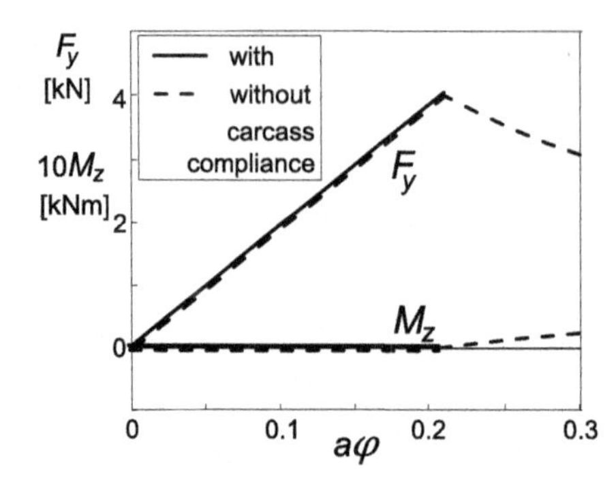


**FIGURE 9.12** Steady-state turn slip force and moment characteristics of the brush model both with and without flexible carcass. The effect of tread width has not been included.

This expression holds because we have assumed that the brush model is isotropic. Using the magnitude of combined slip according to (9.76), the factor $m$ can be assessed:

$$m = 1 - \theta \zeta_c' \quad \text{if} \quad \zeta_c' < 1/\theta \quad \text{else} \quad m = 0 \tag{9.77}$$

Maurice (2000) found excellent agreement between the simulation model and the physical model for the combined slip cases: $\alpha_o = 0.08$ rad and $\hat{\alpha} = 0.08$ rad and $\kappa_o = 0.06$ and $\hat{\kappa} = 0.06$ with a phase difference of 45° and wavelengths of 0.2, 1, and 5 m. For a more precise treatment with the interaction in the sliding range taken into account as well, we refer to the work of Berzeri et al. (1996). Adding turn slip will influence the combined slip response further. The next section approaches this matter in a pragmatic way.

### 9.2.2. The Model Adapted to the Use of the *Magic Formula*

Now that we have treated all ingredients of the force and moment short wavelength responses to longitudinal, lateral, and turn slip and have developed the structure of the contact model, we may carry on and show the performance of the model adapted to the use of the *Magic Formula* the parameters of which may have been assessed through full scale steady-state tire measurements. The model includes the effect of tread width.

To illustrate the matter, we will here consider a simplified set of formulas for the steady-state responses, the complete version of which have been listed in Chapter 4, Sections 4.3.2 and 4.3.3. Only the case of combined side slip and turn slip will be considered. Adding braking or driving will not pose any problems.

A first problem that is encountered is the fact that in the model developed above where the contact patch is represented by the brush model, the steady-state characteristics employed belong to the brush model of the contact patch and not to the total model including the compliant carcass. In Figure 9.7 the calculated total model characteristics can be seen together with those of the contact patch alone.

An obvious solution is to model the contact patch characteristics with the *Magic Formulas*. These, however, will deviate from those assessed for the complete tire because the contact patch 'sees' a slip angle that differs (is smaller) from that of the wheel plane. A set of adapted *MF* parameters may be established off-line for the contact patch or an iteration loop may be included to achieve the correct steady-state behavior of the total model. A practical way has been found which employs a first-order feedback loop with a small time constant. Instead of introducing an additional first-order differential equation, the already present first-order equation for the transient side slip $\alpha'$ has been used. The diagram of Figure 9.13 illustrates the setup. A similar approach is followed in Section 9.3 to account for the camber angle of the belt being different from the camber angle of the wheel plane.

The transient slip first-order differential equations listed below are identical to those derived in the previous section except for the first equation for the lateral transient slip. The factor $\varepsilon_\varphi^*$ in (9.86) accounts for the effect of tread width. Instead of using the axle velocity $V$, we take here the possibly more accurate velocity $V_c$ of the contact center.

*Transient slip equations for side slip and turn slip*:

$$\sigma_c \frac{d\alpha'}{dt} + |V_{cx}|\alpha' = -V_{syc} - |V_{cx}|\beta_{st} \tag{9.78}$$

$$\sigma_2 \frac{d\alpha'_t}{dt} + |V_{cx}|\alpha'_t = |V_{cx}|\alpha' \tag{9.79}$$

$$\sigma_c \frac{d\varphi'_c}{dt} + V_c\varphi'_c = -\dot{\psi}_c \tag{9.80}$$

$$\sigma_{F2} \frac{d\varphi'_{F2}}{dt} + V_c\varphi'_{F2} = -\dot{\psi}_c \tag{9.81}$$

$$\sigma_{\varphi 1} \frac{d\varphi'_1}{dt} + V_c\varphi'_1 = -\dot{\psi}_c \tag{9.82}$$

$$\sigma_{\varphi 2} \frac{d\varphi'_2}{dt} + V_c\varphi'_2 = -\dot{\psi}_c \tag{9.83}$$

To include camber, $-\dot{\psi}_c$ is to be replaced by $V_c\varphi_c$ according to (4.76) with the subscript $c$ (of contact patch) introduced.

In (9.78) the calculated deflection angle has been used:

$$\beta_{st} = M_z/c_\psi \tag{9.84}$$

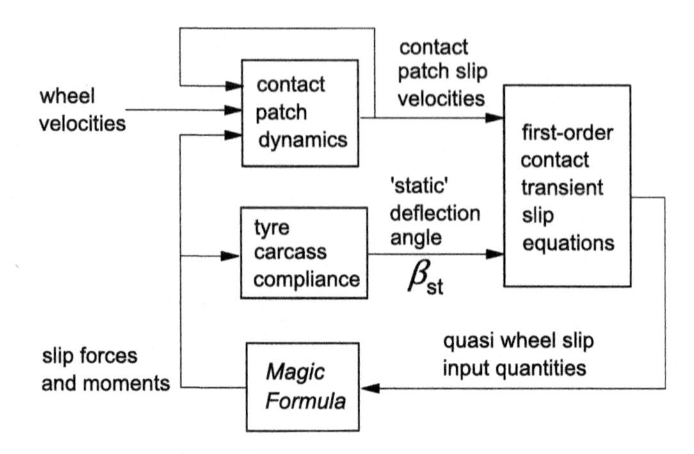


**FIGURE 9.13** Diagram explaining the model structure using the *Magic Formula*.

*Composite transient slip quantities*:

$$\varphi'_F = (2\varphi'_c - \varphi'_{F2})\text{sgn}\, V_{cx} \tag{9.85}$$

$$\varphi'_M = \varepsilon^*_\varphi \varphi'_c + \varepsilon_{\varphi 12}(\varphi'_1 - \varphi'_2) \tag{9.86}$$

*Dynamic contact patch equations*:

$$m_c(\dot{V}_{py} + V_{cx}\dot{\psi}_c) + k_y\dot{v} + c_y v = F_y \tag{9.87}$$

$$\dot{v} = V_{py} - V_{sy} \tag{9.88}$$

$$I_c\ddot{\psi}_c + k_\psi\dot{\beta} + c_\psi\beta = M_z \tag{9.89}$$

$$\dot{\beta} = \dot{\psi}_c - \dot{\psi} \tag{9.90}$$

$$V_{syc} = V_{sy} + \dot{v} - V_x\beta \tag{9.91}$$

*Simplified side force and aligning torque Magic Formulae (MF ed. 2)*:

$$F_y = D_y \sin\{C_y \arctan(B_y\alpha_y)\} \tag{9.92}$$

$$M_z = -tF_y + \Delta M_z + M_{zr} \tag{9.93}$$

$$C_y = p_{Cy1} \tag{9.94}$$

$$D_y = \mu F_z \cdot \zeta_2 \tag{9.95}$$

$$B_y = K_{y\alpha}/(C_y D_y) \tag{9.96}$$

$$K_{y\alpha} = C_{F\alpha o} \cdot \zeta_3 \tag{9.97}$$

$$\alpha_y = \alpha' + S_{Hy} \tag{9.98}$$

$$S_{Hy} = D_{Hy\varphi}\sin\{C_{Hy\varphi}\arctan(B_{Hy\varphi}R_o\varphi'_F)\} \tag{9.99}$$

$$t = D_t\cos\{C_t\arctan(B_t\alpha_t)\} \tag{9.100}$$

$$\alpha_t = \alpha'_t + S_{Ht} \tag{9.101}$$

$$S_{Ht} = 0 \tag{9.102}$$

$$D_t = q_{Dz1}R_o\cdot\zeta_5 \tag{9.103}$$

$$D_{r\varphi} = D_{Dr\varphi}\sin\{C_{Dr\varphi}\arctan(B_{Dr\varphi}R_o\varphi'_M)\} \tag{9.104}$$

$$M_{zr} = M_{z\varphi} \approx D_{r\varphi} \tag{9.105}$$

$$\Delta M_z = C_{\Delta M}(\alpha' - \alpha'_t) \tag{9.106}$$

$$\zeta_2 = \cos\{\arctan(B_{y\varphi}R_o\varphi'_F)\} \tag{9.107}$$

$$\zeta_3 = \cos\{\arctan(p_{Ky\varphi}R_o^2\varphi'^2_F)\} \tag{9.108}$$

$$\zeta_5 = \cos\{\arctan(q_{Dt\varphi1}R_o\varphi'_F)\} \tag{9.109}$$

*Factors reduced with slip*:

$$C_{\Delta M} = C_{F\alpha o}t_o \cdot \max(1 - \theta\zeta', 0) \tag{9.110}$$

$$\sigma_c = a\cdot\max(1 - \theta\zeta', \varepsilon_{\lim}) \tag{9.111}$$

$$\sigma_2 = \frac{t_o}{a}\sigma_c \tag{9.112}$$

$$\sigma_{F2} = b_{F2}\sigma_c = \frac{1}{2}\sigma_c \tag{9.113}$$

$$\sigma_{\varphi1} = b_{\varphi1}\sigma_c \tag{9.114}$$

$$\sigma_{\varphi2} = b_{\varphi2}\sigma_c = \frac{2b_{\varphi1}\sigma_c}{2 - b_{\varphi1}} \tag{9.115}$$

with tire composite parameter

$$\theta = \frac{C_{F\alpha o}}{3\mu F_z} \tag{9.116}$$

and the total magnitude of equivalent side slip

$$\zeta' = \frac{1}{1+\kappa'}\sqrt{\{|\alpha'| + a\varepsilon_{\varphi12}|\varphi'_1 - \varphi'_2|\}^2 + \left(\frac{C_{F\kappa o}}{C_{F\alpha o}}\right)^2\left\{|\kappa'| + \frac{2}{3}b|\varphi'_c|\right\}^2} \tag{9.117}$$

where, in the present application, $\kappa' = 0$.

*Other parameter relations*:

$$
\begin{aligned}
C_{Faco} &= C_{F\kappa co} \\
t_{co} &= q_{Dz1} R_o \\
C_{M\varphi^* co} &= (2/3)^2 b^2 C_{F\kappa co} \\
C_{Maco} &= t_{co} C_{Faco} \\
C_{F\varphi co} &= C_{Maco} \\
C_{M\varphi co} &= C_{M\varphi^* co} - t_{co}\, C_{F\varphi co} \\
t_o &= t_{co} \\
C_{Fao} &= C_{Faco}/(1 + t_o\, C_{Faco}/c_\psi) \\
C_{Mao} &= t_o C_{Fao} \\
C_y &= p_{Cy1} \\
B_t &= q_{Bz1} C_{Fao}/C_{Faco} \\
C_t &= q_{Cz1} \\
C_{F\varphi o} &= C_{F\varphi co} + C_{Faco} C_{M\varphi co}/(c_\psi + C_{Maco}) \\
C_{M\varphi o} &= C_{M\varphi co} c_\psi/(c_\psi + C_{Maco})
\end{aligned}
\tag{9.118}
$$

$$
\begin{aligned}
B_{y\varphi} &= p_{Dy\varphi 1} \\
C_{Hy\varphi} &= p_{Hy\varphi 1} \\
D_{Hy\varphi} &= p_{Hy\varphi 2} \\
B_{Hy\varphi} &= C_{F\varphi o}/(R_o C_{Hy\varphi} D_{Hy\varphi} C_{Fao}) \\
C_{Dr\varphi} &= q_{Dr\varphi 1} \\
D_{Dr\varphi} &= M_{z\varphi\infty}/\sin\,(0.5\pi C_{Dr\varphi}) \\
K_{zR\varphi ro} &= (C_{M\varphi o} + t_o C_{F\varphi o})/R_o \\
B_{Dr\varphi} &= K_{zR\varphi ro}/(C_{Dr\varphi} D_{Dr\varphi}) \\
M_{z\varphi\infty} &= q_{Cr\varphi 1}\; R_o \mu F_z
\end{aligned}
\tag{9.119}
$$

$$
\begin{aligned}
k_{y\mathrm{crit}} &= 2\sqrt{(m_c c_y)} \\
k_{\psi\mathrm{crit}} &= 2\sqrt{(I_c c_\psi)} \\
k_y &= \zeta_k k_{y\mathrm{crit}} \\
k_\psi &= \zeta_k k_{\mathrm{crit}}
\end{aligned}
\tag{9.120}
$$

## Steady-State, Step Response, and Frequency Response Characteristics

To demonstrate the performance of the model, a number of typical characteristics will be presented. The hypothetical steady-state pure side slip and pure turn slip characteristics of the model have been given in Figure 9.14.

The step response graphs of Figure 9.15 show the proper shapes of the various curves, notably the initial horizontal tangent of the response curves of the side force to turn slip and of the moment to side slip, also shown in Figure 5.21 (no tread width). Also, the peak of the moment response curve to spin and the dip of the curve of the moment response to steer angle are exhibited, as expected.

Figure 9.16 presents the frequency response functions of the linearized system at zero average side and turn slip with tread width effect included. The curves may be compared with those of the string model with tread elements, cf. Figure 5.23 (with zero tread width). In Figure 9.17 the Nyquist plots of the moment response to steer angle have been depicted. The upper diagram shows the influence of tread width by changing the parameter $\varepsilon_{\varphi}^{*}$. If equal to zero, the thin tire is represented. The value 1 corresponds with the base line configuration. The lower graph provides insight into the influence of the parameter $\varepsilon_{\varphi 12}$ that governs the magnitude of the effect of the transient yaw deflection angle $\alpha_M$ (9.64, 9.86). The value 4 is used in the base line configuration. When compared with the plots of Figures 5.27 and 5.35, it may be concluded that the model is capable of approaching the responses of more complex infinite order (string) models and of actual tires.

## Large Slip Angle and Turn Slip Response Simulations

The same sinusoidal maneuvers have been simulated as was done before with the brush-based contact model. The complete nonlinear set of Eqns (9.78)–(9.120) has been used with parameters listed in Table 9.2.

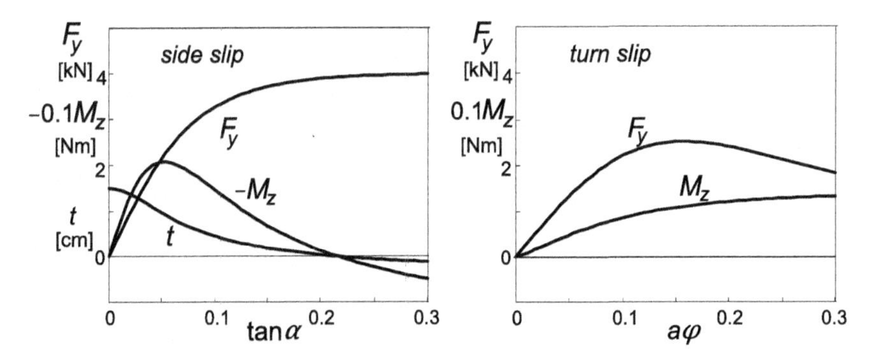


**FIGURE 9.14** Steady-state side slip and turn slip force and moment characteristics of the overall tire model as defined by the *Magic Formula*. The effect of tread width has been included.

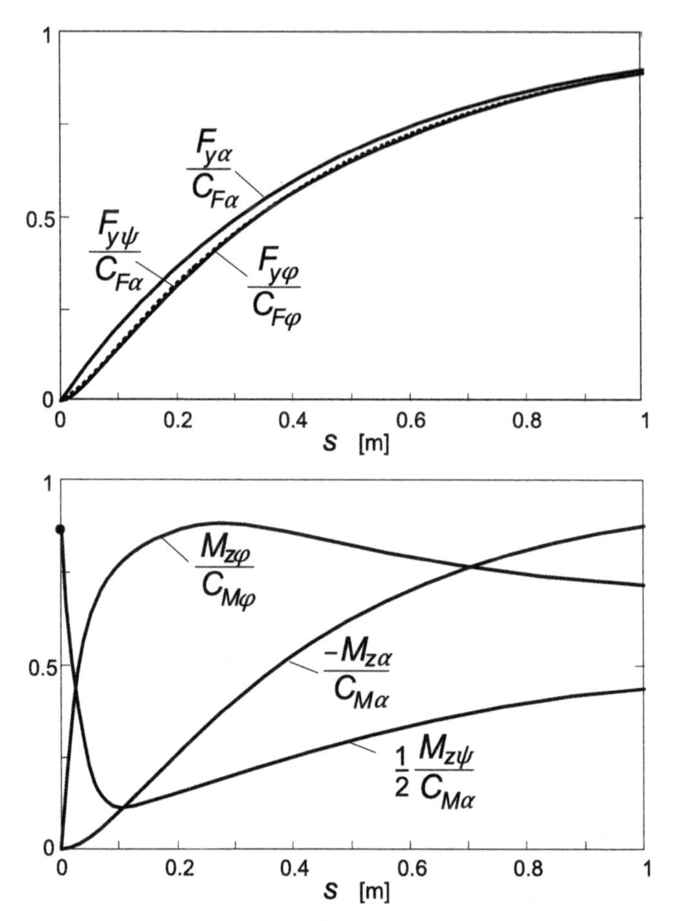


**FIGURE 9.15** Step response curves of the side force and of the aligning torque to side slip $\alpha$, turn slip (path curvature) $\varphi$, and steer angle $\psi$.

The slip angle variation is sinusoidal around average levels $\alpha_o = 0$ and 0.08 rad with one amplitude: $\hat{\alpha} = 0.08$ rad at three wavelengths: $\lambda = 0.2$, 1, and 5 m. Similarly, the turn slip has two average levels: $a\varphi_o = 0$ and 0.08 and one amplitude: $a\hat{\varphi} = 0.08$, also at wavelengths: $\lambda = 0.2$, 1, and 5 m.

The diagrams of Figures 9.18 and 9.19 present the results. We observe that the curves are quite similar to those depicted in the Figures 9.10 and 9.11. Only, as expected, the moment response to turn slip is very much affected by the now introduced effect of the tire tread width.

Maurice has conducted extensive experiments with a 205/60R15 91V tire at 2.2 bar inflation pressure on a 2.5 m drum using the pendulum test rig of Figure 9.38. The diagrams of Figure 9.20a present the computed results compared with experimental data for the case of pure side slip. The experiments have been carried out at very low speed to avoid inertia effects of the moving tire.

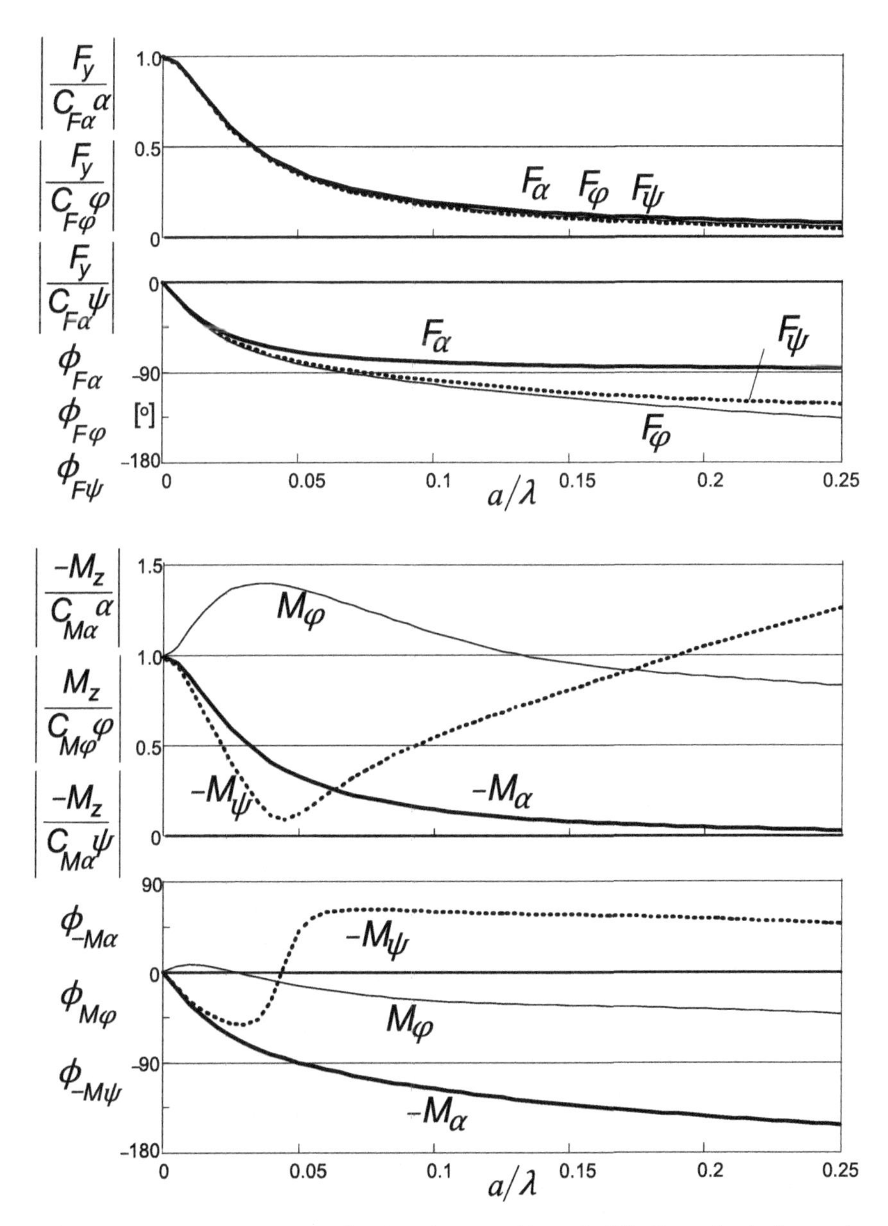


**FIGURE 9.16** Frequency response function of system with tread width. Curves for the force and moment responses to side slip $\alpha$, turn slip (path curvature) $\varphi$, and steer angle $\psi$ have been indicated.

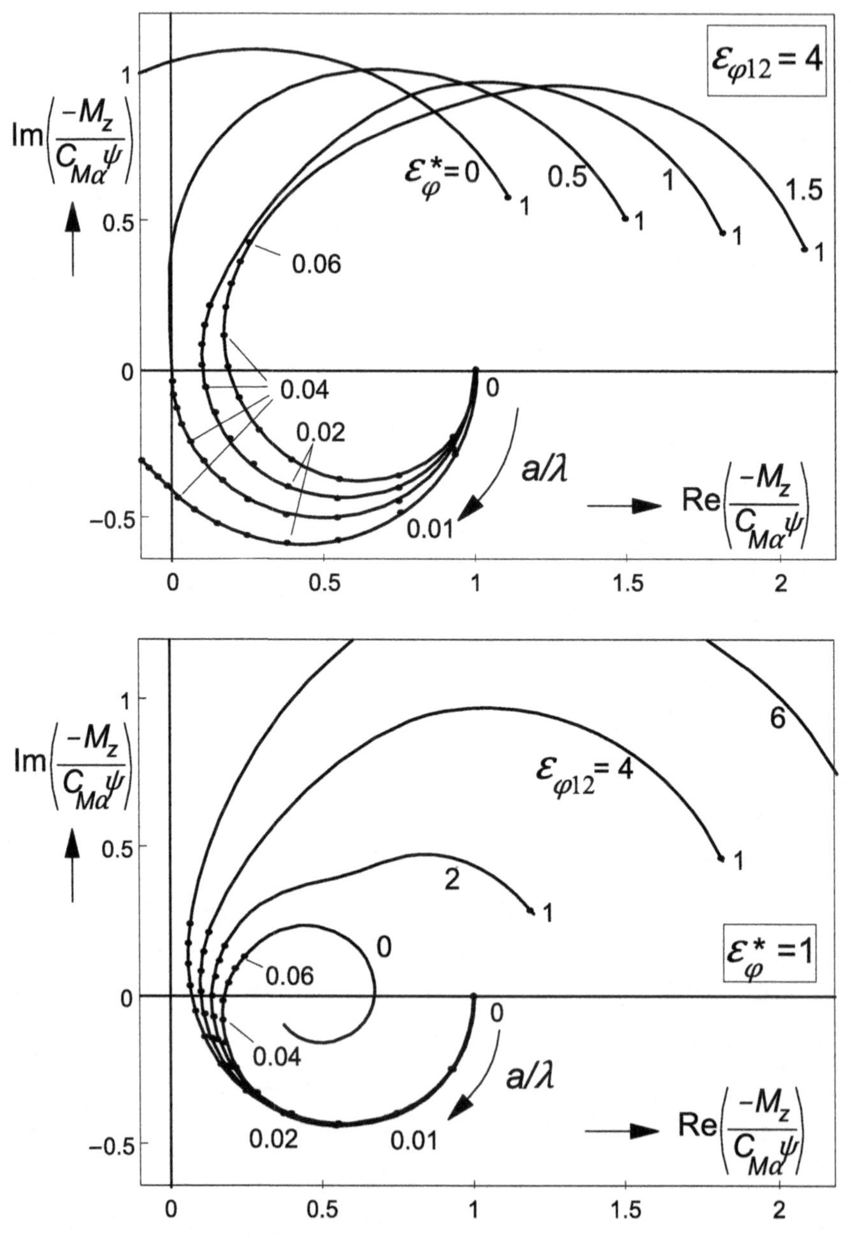


**FIGURE 9.17** Nyquist frequency response plots of the aligning torque to steer angle. Upper diagram: influence of tread width (4 is base line value), lower diagram: influence of the transient yaw deflection response to turn slip (4 is base line value).

**TABLE 9.2** Parameter Values for Tire Model with *Magic Formula* Including Quantities Introduced Later on.

| | | | | | | | |
|---|---|---|---|---|---|---|---|
| $F_z$ | 4000 N | $\mu$ | 1.0 | $p_{Cy1}$ | 1.2 | $p_{K y\varphi 1}$ | 1.0 |
| $c_y$ | $125 \times 10^3$ N/m | $C_{F\kappa c0}$ | 15 $F_z$ | $q_{Bz1}$ | 22 | $p_{Hy\varphi 1}$ | 0.15 |
| $c_\psi$ | $4 \times 10^3$ Nm/rad | $a$ | 0.0535 m | $q_{Cz1}$ | 1.192 | $p_{Hy\varphi 2}$ | 1.0 |
| $m_c$ | 0.5 kg | $b$ | 0.9 $a$ | $q_{Dz1}$ | 0.05 | $q_{Cr\varphi 1}$ | 0.12 |
| $I_c$ | 0.0005 kg m$^2$ | $R_o$ | 0.3 m | $p_{Dy\varphi 1}$ | 0.4 | $q_{Dr\varphi 1}$ | 1.0 |
| $\zeta_k$ | 0.1 | $\varepsilon^*_\varphi$ | 1.0 | $\varepsilon_{\varphi 12}$ | 4.0 | $q_{Dt\varphi 1}$ | 10 |
| $b_{F2}$ | 0.5 | $b_{\varphi 1}$ | 0.5 | $b_{\varphi 2}$ | 1/1.5 | | |
| $\varepsilon_{lim}$ | 0.1 | $V_{low}$ | 1 m/s | $m_{qc}$ | 400 kg | $c_o$ | 2.0 |

The wavelength ranges from 0.3 to 2.4 m. The upper diagram refers to the case of zero average slip angle and 4° amplitude. The lower diagram shows the responses around an average slip angle of 4°, which causes the curves to deviate considerably from the input sinusoidal shape.

The curves clearly show that a shorter wavelength causes the response amplitude to decrease and the phase lag to increase. It can also be observed that the responses occur more quickly at larger levels of slip, which is due to the sharp decrease of the relaxation length with increasing slip. The *Magic Formula* was used to model the steady-state characteristics. The parameters were obtained from separate tests performed on the drum at the much higher speed of 60 km/h. The different conditions may explain the deviation in level of the calculated responses with respect to the measured ones.

Figure 9.20b presents the comparison with the results of a second series of experiments under the same conditions. Here, the vertical load is changed sinusoidally while the slip angle is kept at a low level and at a higher level.

For these experiments the measuring tower of Figure 9.37 (left) has been employed. The results are similar to those discussed in Chapter 8, Figure 8.9. The moment response seems to be improved with the more complex tire model.

The agreement of the computed results with experimental data is quite good. As has been reported by Maurice, also for the moment response a rather good agreement has been established. Since at the maximum slip angle of 8° the peak of the moment characteristic has been surpassed, the result becomes quite sensitive to small differences between actual and model steady-state characteristics of the aligning torque.

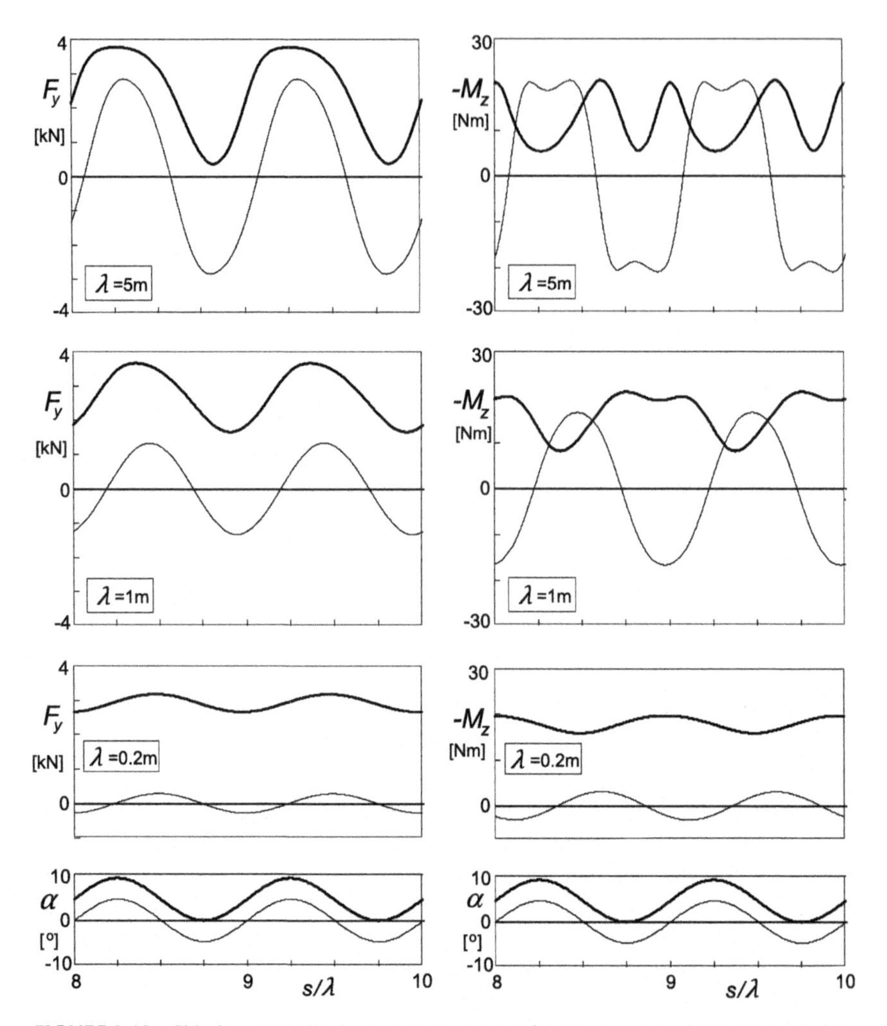


**FIGURE 9.18** Side force and aligning torque responses of the *Magic Formula*-based simulation model with flexible carcass and finite tread width to sinusoidal slip angle variation of the wheel plane with an amplitude of 0.08 rad and average levels of 0 (thin curves) and 0.08 rad (fat curves).

## 9.2.3. Parking Maneuvers

Parking maneuvers take place at very low or zero speed. The torque acting on the tire at such conditions may become very large. The influence of the finite tread width is essential as the response to spin is now predominant. We might employ the equations developed above but then we should take care of the integration of the spin velocity to properly limit the buildup of the yaw transient slip. Similar problems arose when considering the problem of braking to standstill or starting from standstill, cf. Section 8.6, Eqns (8.112, 8.113).

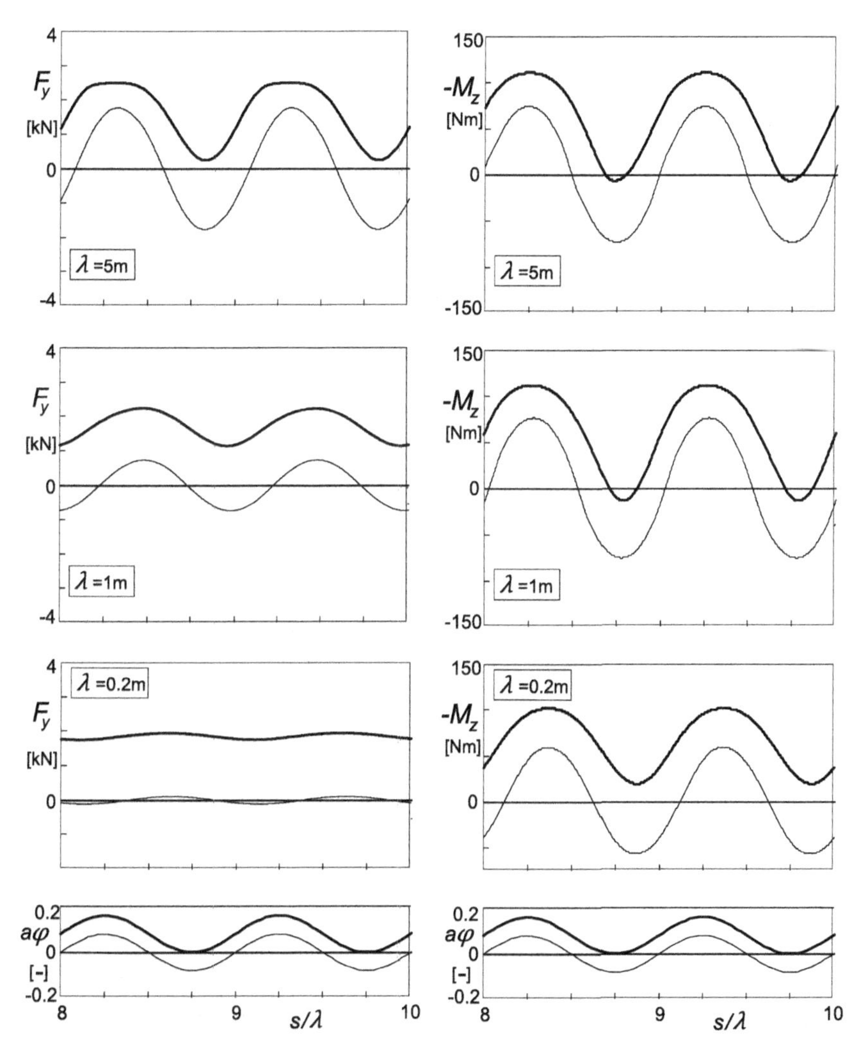


**FIGURE 9.19** Side force and aligning torque responses of the *Magic Formula*-based simulation model with flexible carcass and finite tread width to sinusoidal path curvature variations (turn slip at $\alpha = 0$) of the wheel plane with an amplitude of $a\varphi$ of 0.08 and average levels of 0 (thin curves) and 0.08 (fat curves).

To achieve a much better agreement with experimental evidence, a different approach will be followed in the present application. It may be noted that an important characteristic is actually still missing. For the brush based model, Eqn (9.70) was used. The equation governs the variation of the aligning torque $M_z$ that arises when the nonrolling tire is steered and the steer angle $\psi$ is increased from zero to and beyond the state of full sliding. Ultimately, the torque reaches the magnitude that would also arise when the rolling tire is subjected to a constant rate of turning $d\psi/dt$ (while the slip angle remains zero) at a forward speed $V_x$ that decreases to zero. Then, the radius of turn $R$ reduces to zero and thus the spin approaches infinity. Figure 9.21 illustrates the situation.

The missing characteristic will be modeled by using a for-this-purpose adequate model that has been developed by Van der Jagt (2000). In his dissertation a model study was discussed that is especially aimed at the generation of a proper moment response to steering at very low or zero speed. First, the brush model was used to gain general insight into the phenomena that occur. Qualitatively good results have been obtained using this model, notably when a sinusoidal steer angle variation is imposed and the state of almost full sliding is attained periodically. For practical usage, a special type of model was developed of a nature completely different from the models used so far. Since this model appears to perform very well in the near zero

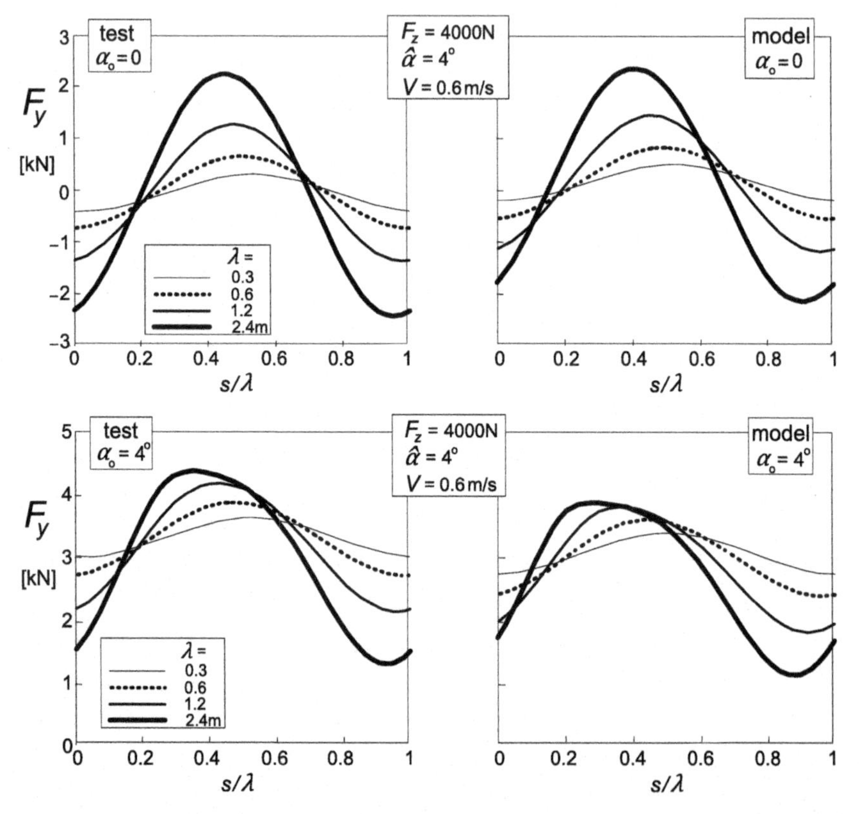


**FIGURE 9.20a** Comparison of theoretical model calculations and experimental results performed at 0.6 m/s on a 2.5 m drum (Maurice). The curves cover one wavelength of the periodic responses to sinusoidal slip angle variations. Force response to slip angle variations at two levels of side slip.

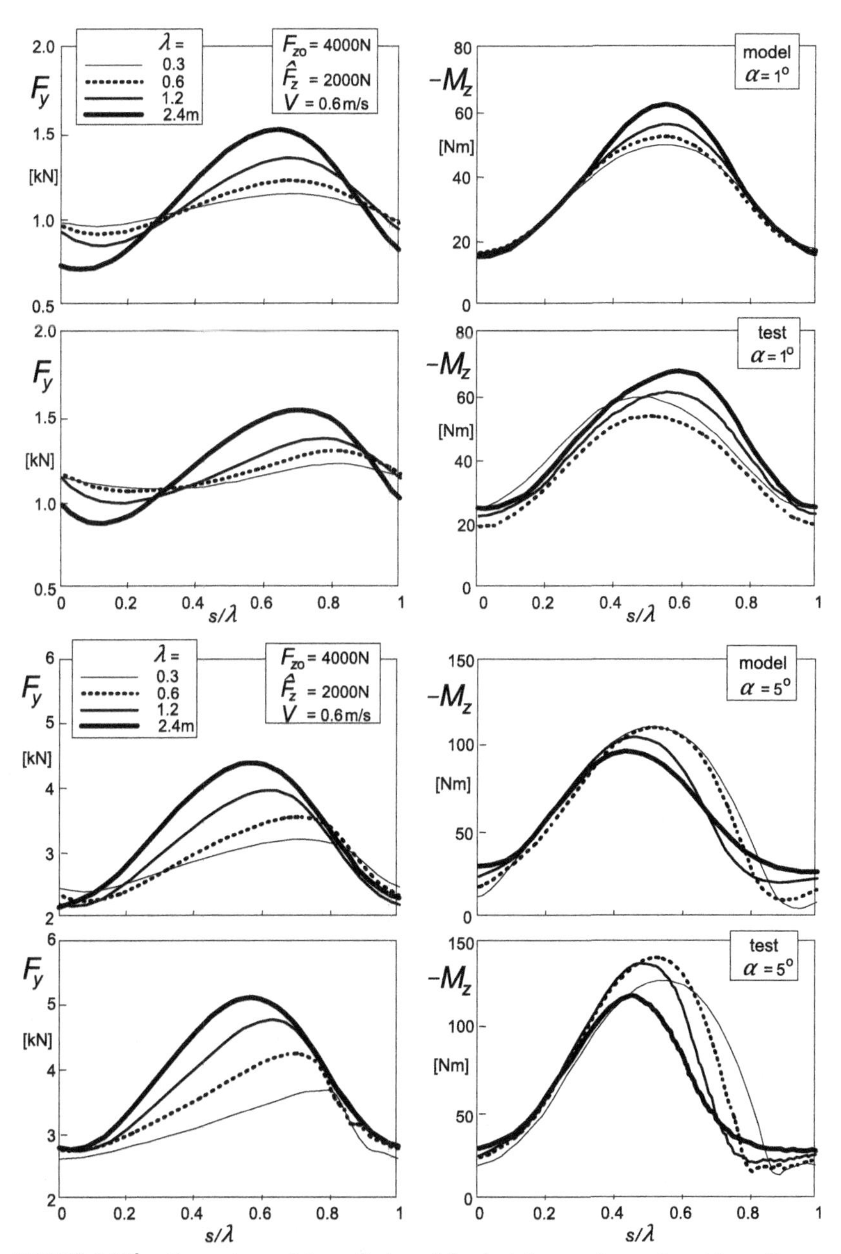


**FIGURE 9.20b** Comparison of theoretical model calculations and experimental results performed at 0.6 m/s on a 2.5 m drum (Maurice). The curves cover one wavelength of the periodic responses to sinusoidal load variations. Force and moment response to vertical load variations at two values of slip angle.

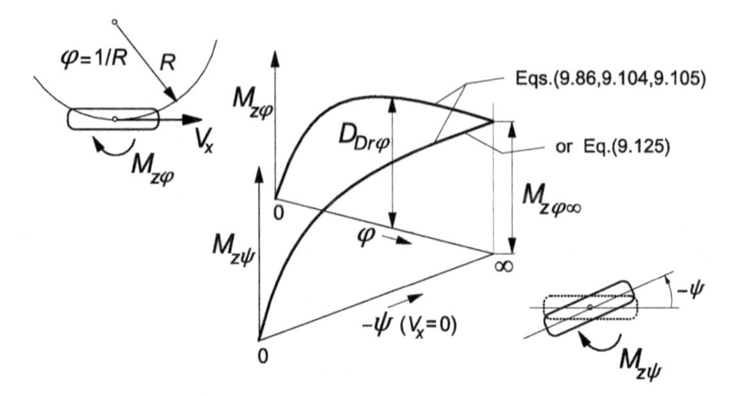


**FIGURE 9.21** Approaching the maximum torque at standstill in two ways: 1. by decreasing the turn radius $R$ to zero and 2. by increasing the steer angle $-\psi$ while standing still.

speed range, we have tried to incorporate Van der Jagt's model in the existing model structure. For a gradual transition from the new type of model to the existing one, when the speed approaches and surpasses a low-speed threshold has been taken care of.

The principle of Van der Jagt's approach is that at a given rate of change of the steer input the growth rate of the tire angular deflection $\beta$ decreases in proportion to a function of the remaining difference between the maximum achievable deflection and the current deflection. The torsional stiffness is assumed to be a constant and the resulting characteristic of the torque becomes similar to a first-order response function. The calculated moment gradually approaches its maximum value. When the direction of rotation of the wheel about the vertical axis is changed, the distance to the new, opposite, peak torque is large and, accordingly, the rate of reduction of the moment is large as well. It is this feature of the model that is attractive since a similar behavior has been found to occur with the actual tire subjected to an alternating left and right sequence of turning. The equations that govern the moment generation at standstill are as follows:

$$\dot{\beta} = -\left(1 - p\left|\frac{M_{z\psi}}{M_{z\varphi\infty}}\right|^{c_o}\right)\dot{\psi} \tag{9.121a}$$

$$M_{z\psi} = C_{M\psi}\beta \tag{9.121b}$$

$$p = 0 \quad \text{if} \quad \operatorname{sgn}\beta \neq -\operatorname{sgn}\dot{\psi} \quad \text{else } p = 1 \tag{9.121c}$$

For the parameter value $c_o = 2$, Figure 9.22 presents the calculated variation of the torque vs the steer angle compared with experimentally obtained results as reported by Van der Jagt. The nonrolling tire (size P205/65R15) is loaded to 4800 N on a flat plate and subsequently steered at a rate of + and −1 deg./s. The correspondence is quite good except perhaps for the initial phase where the wheel starts to be steered from the condition where $M_z = 0$. To improve the model performance Van der Jagt suggests using an exponent $c_o$, the value of which depends on the last extreme of the deflection angle $\beta$. For possible further refinements of the model, we refer to the original work.

When, instead of the new approach, the *Magic Formula* would be used with the integration limitation as suggested according to Eqns (8.112, 8.113), a sharp peak would arise in the curve where the direction of turning is changed. As a result, the moment decreases at a much slower rate than shown by the test result.

The problem is now how to integrate the new model feature in the original model structure. The transient slip quantity $\varphi'_M$, Eqn (9.86), may be recognized to be proportional to the deflection angle. As can be seen from Eqns (9.80, 9.82, 9.83, and 9.86) this quantity is obtained through integration of

$$\dot{\varphi}'_M = \varepsilon^*_\varphi \dot{\varphi}'_c + \varepsilon_{\varphi 12}(\dot{\varphi}'_1 - \dot{\varphi}'_2) \tag{9.122}$$

In the new configuration, the integration is conducted at a gradually decreasing rate while approaching the maximum torque value. We have

$$\dot{\varphi}'_M = \left(1 - w_{V\text{low}} p \left|\frac{M_{z\varphi}}{M_{z\varphi\infty}}\right|^{c_o}\right) \dot{\varphi}'_M \tag{9.123}$$

$$\text{where} \quad p = 0 \quad \text{if} \quad \text{sgn}\, \varphi'_M \neq \text{sgn}\, \dot{\varphi}'_M \quad \text{else} \; p = 1 \tag{9.124}$$

At zero speed $w_{V\text{low}} = 1$. The moment is found with the linear function, cf. (9.104):

$$M_{z\varphi} = D_{Dr\varphi} C_{Dr\varphi} B_{Dr\varphi} R_o \varphi'_M \tag{9.125}$$

For the standing tire with speed $V_x$ equal to zero, the response to alternating steer angle variations will follow a course similar to that of Figure 9.22.

It is now desired to change gradually to the original equations when the tire starts rolling. The transition is accomplished by adding up the following two components. The first one decreases in magnitude with increasing speed until it vanishes at $V_x = V_\text{low}$ while the second part increases from zero to its full value also at $V_x = V_\text{low}$. For the gradual change, the following speed window is used:

$$w_{V\text{low}} = \frac{1}{2}\left\{1 + \cos\left(\pi \frac{V_x}{V_\text{low}}\right)\right\} \quad \text{if} \; |V_x| < V_\text{low} \quad \text{else} \; w_{V\text{low}} = 0 \tag{9.126}$$

With this quantity (already used in (9.123)), the first part that prevails at low speed becomes

$$M_{z\varphi 1} = w_{V\text{low}} \cdot D_{Dr\varphi} C_{Dr\varphi} B_{Dr\varphi} R_o \varphi'_M \tag{9.127a}$$

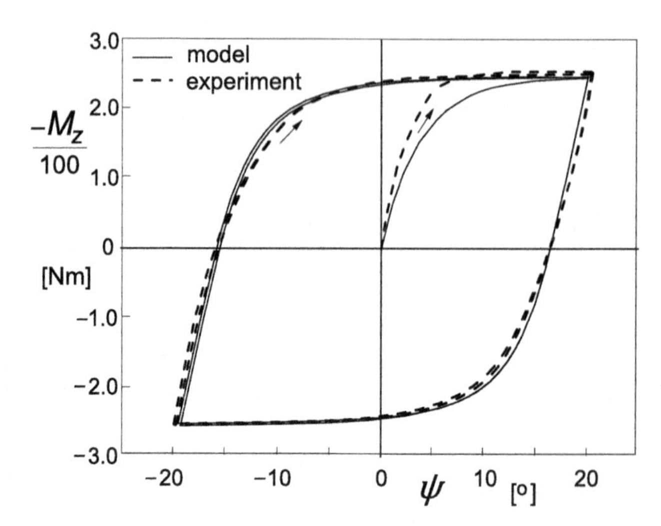


**FIGURE 9.22** Calculated and experimentally assessed variation of the moment vs steer angle for a nonrolling tire pressed against a flat plate at a load $F_z = 4800\,\text{N}$.

and the fraction obtained from the original (here simplified) Eqns (9.105 and 9.104)

$$M_{z\varphi 2} = (1 - w_{V\text{low}}) \cdot D_{Dr\varphi}\sin\{C_{Dr\varphi}\arctan(B_{Dr\varphi}R_o\varphi'_M)\} \tag{9.127b}$$

The resulting expression for the spin moment now reads

$$M_{z\varphi} = M_{z\varphi 1} + M_{z\varphi 2} \tag{9.128}$$

A similar method may be employed to improve the low-speed model for the side force responding to lateral motions of the contact patch (cf. Figure 9.29, point $S$) and for the fore- and-aft force to longitudinal motions of the same point $S$.

The adapted model will now be applied to the simulation of the motion of a rigid quarter car model with mass $m_{qc}$ that, while a sinusoidal steering input is applied, starts moving after 1.6 s with a linearly increasing speed. The lateral acceleration of the quarter car axle results from the action of the side force that begins to develop after the wheel has started to roll:

$$\ddot{y}_{qc} = \frac{F_y}{m_{qc}} \tag{9.129}$$

The lateral wheel slip velocity is now not only a result of the yaw angle at a forward speed of the vehicle $\dot{x}_{qc}$ but also due to the lateral velocity of the wheel axle $\dot{y}_{qc}$. We have

$$V_{sy} = -\dot{x}_{qc}\sin\psi + \dot{y}_{qc}\cos\psi \tag{9.130}$$

which serves as an input into the Eqns (9.88 and 9.91). The additional parameter values have been appended in Table 9.2.

Figure 9.23 shows the courses of variation of various quantities vs time. Simultaneously, in Figure 9.24, the moment is plotted vs steer angle. Several phenomena occur that deserve to be noted. The steer angle has an amplitude that is large enough to attain a level of the moment close to its maximum. The moment starts to decrease in magnitude as soon as the steer angle passes its peak value. The moment changes sign before the steer angle does the same. After 1.6 s, the forward speed increases linearly with time and the side force starts to build up as a result of the slip angle that begins to develop. The car shows a lateral vibration in the low speed range as indicated by the fluctuations of the side force. Evidently, the quarter car vibrates against the lateral tire stiffness. The moment amplitude decreases as the spin diminishes in amplitude due to the increasing speed. The side force amplitude increases because of the

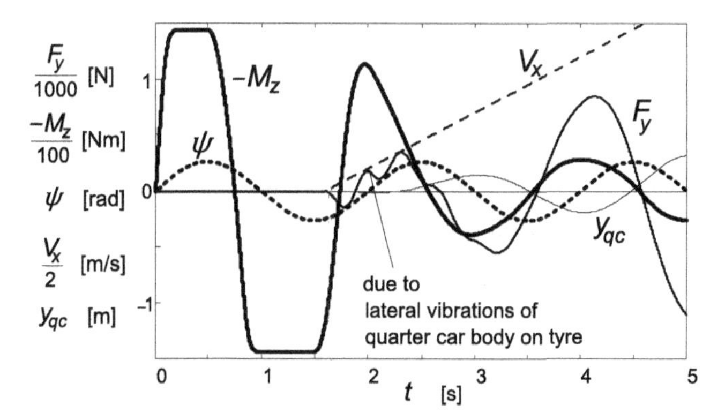


**FIGURE 9.23** Simulation results of a parking maneuver (car leaving the parking lot while steering sinusoidally).

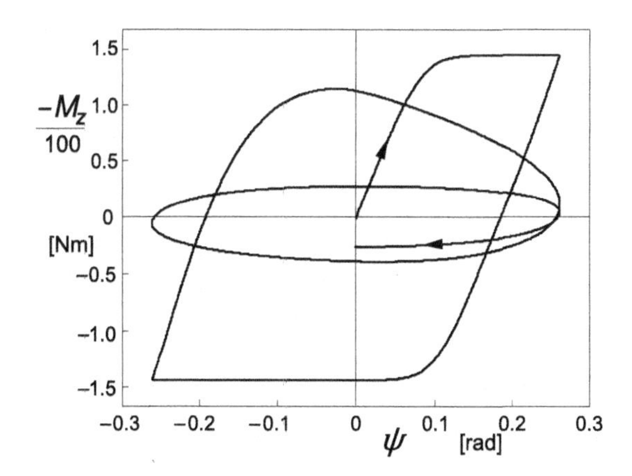


**FIGURE 9.24** The steer torque plotted vs steer angle during the maneuver of Figure 9.23.

larger lateral oscillations of the quarter car mass induced by the increasing speed of travel at the constant steer input pattern with time. The loops shown in Figure 9.24 give a nice impression of the transition from the situation at standstill to the condition at higher speeds. At standstill the moment varies in accordance with the diagram of Figure 9.22.

As mentioned before, to get a more accurate calculation of responses to lateral and circumferential wheel displacements at or near forward speed equal to zero, one might apply, instead of the abrupt integration limitation suggested earlier, Eqns (8.112) and (8.113), the same structure of additional Eqns (9.123) and (9.124), and an adaptation such as achieved in Eqn (9.128).

## 9.3. TIRE DYNAMICS

The contact patch model has been used above in connection with a flexible carcass. In the present section the inertia of the belt will be introduced. Since we restrict the application of the model to frequencies lower than ca. 60 Hz the belt may be approximated as a rigid ring that is attached to the wheel rim through flexible sidewalls. To ensure that the total static tire stiffness remains unchanged, residual springs have been introduced between the contact patch and the belt. In certain cases, a bypass spring directly connecting the rim and the contact patch may be needed to improve model accuracy.

### 9.3.1. Dynamic Equations

As depicted in Figure 9.25, the wheel axle position is defined by the location of the wheel center $(X,Y,Z)$ and orientation $(\gamma,\theta,\psi)$. The wheel speed of

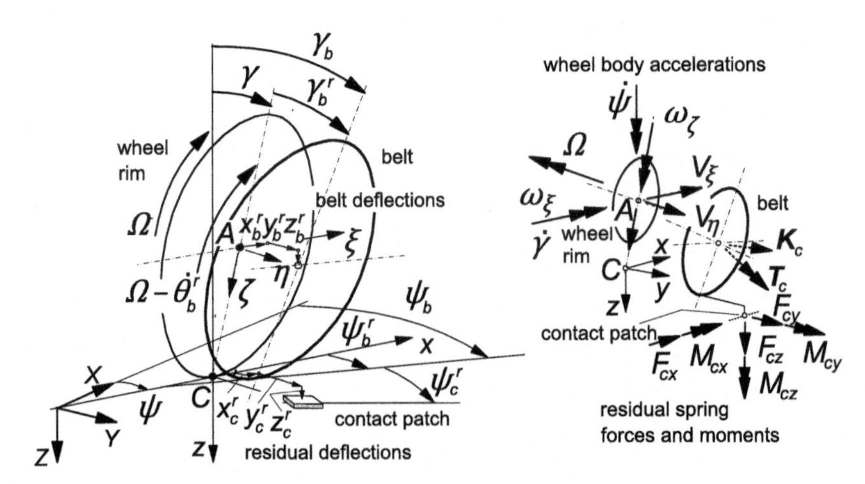


**FIGURE 9.25** Model structure featuring contact patch, residual compliance, rigid belt, carcass compliance, and wheel rim.

revolution is denoted by $\Omega$. A moving axes system ($C$,$x$,$y$,$z$) has been defined of which the $x$ axis points forwards and runs along the line of intersection of the wheel plane and the road plane. The $y$ axis lies in the plane normal to the road plane and passing through the wheel spin axis. The $z$ axis forms the normal to the road surface. The origin of the moving triad is the contact center $C$. Another moving system of axes ($A$,$\xi$,$\eta$,$\zeta$) is introduced of which the origin $A$ is located in the wheel center, the $\xi$ axis is horizontal, and the $\eta$ axis runs along the wheel spindle axis. With respect to the wheel rim the belt shows relative displacements: $(x_b^r, y_b^r, z_b^r)$ and $(\gamma_b^r, \theta_b^r, \psi_b^r)$. The contact patch is displaced horizontally with respect to the belt corresponding to the deflections of the residual springs: $(x_c^r, y_c^r, z_c^r)$ and $\psi_c^r$. The superscript $r$ designates a relative displacement; without the superscript we have the displacement with respect to the inertial system ($X$,$Y$,$Z$). All relative displacements are considered small and the dynamic equations may be linearized.

The wheel motion forms the input to the tire system (possibly together with the road profile). From these, the wheel velocities $V_\xi$, $V_\eta$, and $V_\zeta$ and $\omega_\xi$, $\omega_\eta$, and $\omega_\zeta$ (defined with respect to the axle triad ($A$,$\xi$,$\eta$,$\zeta$)) or alternatively $\dot{\gamma}, \Omega, \dot{\psi}$ (defined with respect to the moving contact triad ($C$,$x$,$y$,$z$)) and the camber angle $\gamma$ and the radial tire deflection $\rho_z$ are available. The camber angle $\gamma$ will be treated here as a small quantity.

The belt considered as a rigid circular body has a mass $m_b$ and moments of inertia $I_{bx,y,z}$. The carcass (sidewalls) possesses stiffnesses $c_{bx,y,z}$ and damping coefficients $k_{bx,y,z}$. In the figure the force and moment vectors $\boldsymbol{K_c}$ and $\boldsymbol{T_c}$ defined to act from contact patch to and about the center of the belt have been indicated. Their components are defined with respect to the ($A$,$\xi$,$\eta$,$\zeta$) triad. The first of the two sets of first-order differential equations for the six degrees of freedom reads:

*Dynamic belt equations*:

$$m_b(\dot{V}_{b\xi} - V_{b\eta}\omega_{b\zeta}) + k_{bx}\dot{x}_b^r + c_{bx}x_b^r + k_{bz}\Omega z_b^r = K_{c\xi} \tag{9.131}$$

$$m_b\dot{V}_{b\zeta} + k_{bz}\dot{z}_b^r + c_{bz}z_b^r - k_{bx}\Omega x_b^r = K_{c\zeta} \tag{9.132}$$

$$I_{by}\dot{\omega}_{b\eta} + k_{b\theta}\dot{\theta}_b^r + c_{b\theta}\theta_b^r = T_{c\eta} \tag{9.133}$$

$$m_b(\dot{V}_{b\eta} + V_{b\xi}\omega_{b\zeta}) + k_{by}\dot{y}_b^r + c_{by}y_b^r = K_{c\eta} \tag{9.134}$$

$$I_{bx}\dot{\omega}_{b\xi} + I_{by}\Omega\omega_{b\zeta} + k_{b\gamma}\dot{\gamma}_b^r + c_{b\gamma}\gamma_b^r + k_{b\psi}\Omega\psi_b^r = T_{c\xi} \tag{9.135}$$

$$I_{bz}\dot{\omega}_{b\zeta} - I_{by}\Omega\omega_{b\xi} + k_{b\psi}\dot{\psi}_b^r + c_{b\psi}\psi_b^r - k_{b\gamma}\Omega\gamma_b^r = T_{c\zeta} \tag{9.136}$$

Several coupling terms show up. These are due to the gyroscopic effect and due to the action of the rotating radial dampers with resulting coefficient $k_{bx} = k_{bz}$ and the lateral dampers with resulting angular damping coefficients

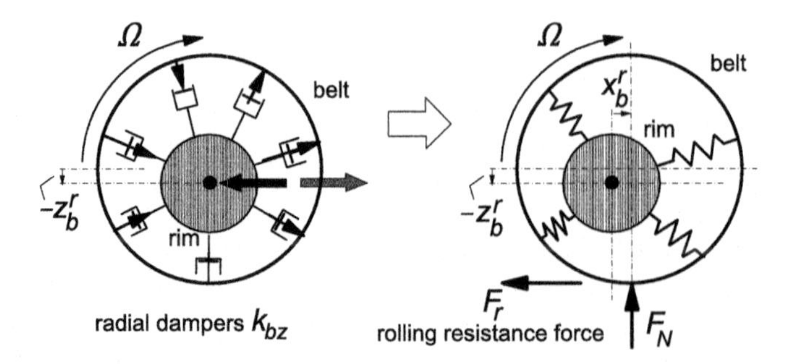


**FIGURE 9.26** The rotating radial dampers of the vertically deflected tire gives rise to a resulting fore-and-aft force acting between the belt and the rim. The resulting longitudinal deflection produces a rolling resistance force $F_r$ through the action of the normal load $F_N$.

$k_{b\psi} = k_{b\gamma}$. Figure 9.26 illustrates the mechanism that gives rise to the interaction terms. The example concerns the term $k_{bz}\Omega z_b^r$ in (9.131). The following relations between the two sets of wheel and axle angular velocities hold:

$$\omega_\xi = \dot{\gamma} \tag{9.137}$$

$$\omega_\eta = -\Omega + \dot{\psi}\sin\gamma \approx -\Omega + \dot{\psi}\gamma \tag{9.138}$$

$$\omega_\zeta = \dot{\psi}\cos\gamma \approx \dot{\psi} \tag{9.139}$$

For the relative displacements between belt and wheel rim we have the second set of six first-order differential equations:

$$\dot{x}_b^r = V_{b\xi} - V_\xi \tag{9.140}$$

$$\dot{y}_b^r = V_{b\eta} - V_\eta \tag{9.141}$$

$$\dot{z}_b^r = V_{b\zeta} - V_\zeta \tag{9.142}$$

and

$$\dot{\gamma}_b^r = \omega_{b\xi} - \dot{\gamma} \tag{9.143}$$

$$\dot{\theta}_b^r = \omega_{b\eta} + \Omega - \dot{\psi}(\gamma + \gamma_b^r) \tag{9.144}$$

$$\dot{\psi}_b^r = \omega_{b\zeta} - \dot{\psi} \tag{9.145}$$

The forces and moments appearing in the right-hand members of Eqns (9.131–136) can be expressed in terms of the forces acting in the residual springs that connect the contact patch with the belt.

The residual spring concept as defined in the model has been addressed in greater detail in Figure 9.27. The left-hand top diagram presents an

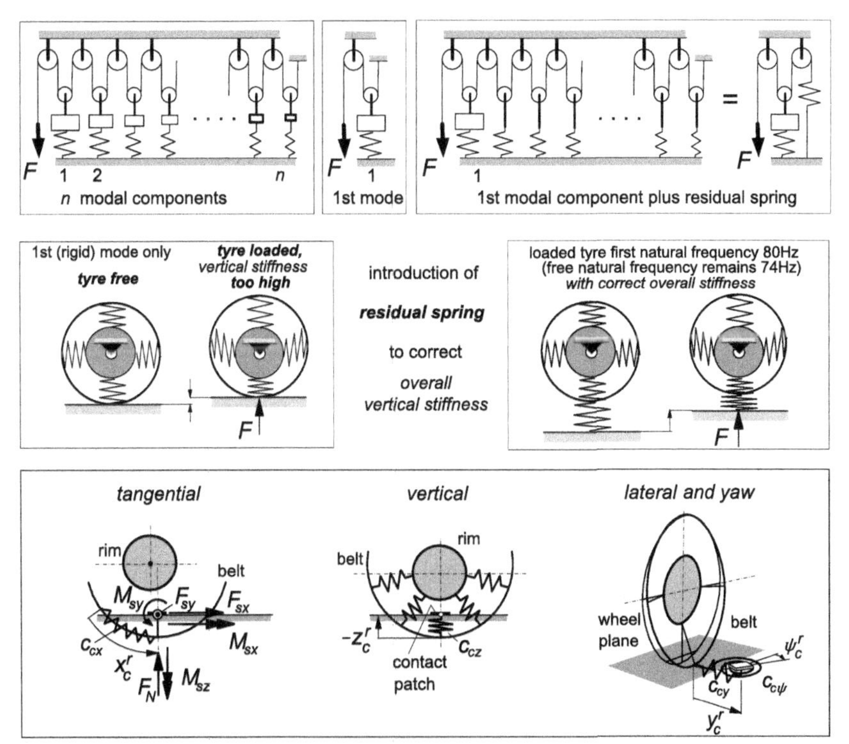


**FIGURE 9.27** *Top*: Analogous system with $n$ modes of vibration; reduction to system with one natural mode plus residual stiffness. *Middle*: Illustration of reduced system without and with residual vertical spring. *Bottom*: Four residual springs introduced in the tire model; deflections of the residual springs attaching the contact patch to the belt.

analogous dynamic system exhibiting $n$ vibrational modes, each belonging to one of the $n$ modal components. The system is formed by a chain of $n$ single mass–spring systems, which are connected by common force joints (realized by rope and pulleys). For a tire, the first mass represents the rigid belt. The remaining modal components with higher natural frequencies show vibrational modes with the belt exhibiting deflections distributed along its circumference. When neglecting the inertias of these higher order modal components (top right diagram), the residual spring remains with compliance equal to the sum of the series of the $n-1$ individual compliances. The middle right diagram shows that attaching the residual spring to the rigid belt results in the correct vertical stiffness and natural frequency of the loaded and the free tire/wheel (cf. Figure 9.40). The lower diagrams of Figure 9.27 indicate the introduced residual springs that connect the belt with the contact patch body in circumferential, vertical, lateral, and yaw directions.

The directions of the residual spring forces and moments are defined to act in parallel to the moving axes system ($C$,$x$,$y$,$z$) with the $z$ axis normal to the road plane. Also, the vertical forces $F_z$ are defined here, in contrast to the definition adopted in the remainder of this chapter and in Chapter 10, according to the consistent SAE convention. For the normal wheel load acting from road to tire we introduce the positive quantity $F_N$. We have in case of a horizontal road plane with products of angles neglected and $r_l$ denoting the loaded radius:

$$K_{c\xi} = F_{cx} \tag{9.146}$$

$$K_{c\zeta} = F_{cz} - \gamma F_{cy} \tag{9.147}$$

$$T_{c\eta} = r_l F_{cx} + M_{cy} \tag{9.148}$$

$$K_{c\eta} = F_{cy} + \gamma F_{cz} \tag{9.149}$$

$$T_{c\xi} = -r_l F_{cy} - (r_l \gamma_b - y_c^r) F_{cz} + M_{cx} \tag{9.150}$$

$$T_{c\zeta} = M_{cz} - y_c^r F_{cx} \tag{9.151}$$

Obviously, a proper axes transformation is to be performed if the road plane is not horizontal. As a result, transverse and forward slopes will affect the terms appearing in the right-hand members of Eqns (9.146–9.151). It is left to the user to introduce these transformations.

The contact patch body is subjected to forces acting in the residual springs (subscript $c$) and external forces acting from road surface to contact patch (subscript $s$). Figure 9.28 illustrates the situation. The contact inertia increases the total order of the system. The introduction of the inertia may be avoided, while maintaining integrational causality, by considering the 'residual' damper not as a resistor but as a conductor. The differential

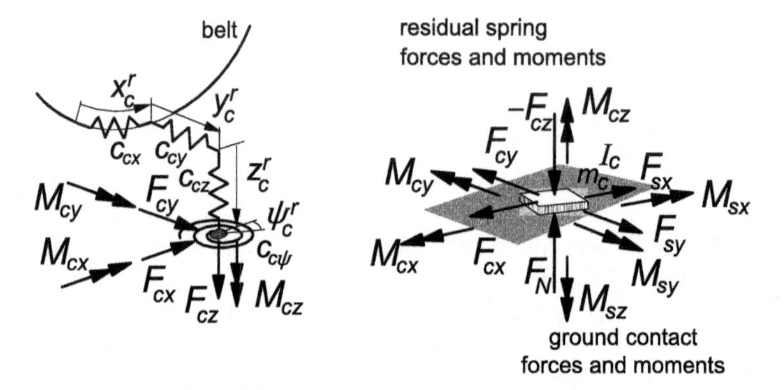


**FIGURE 9.28** Internal and external forces and moments acting on the contact patch body.

equations that govern the horizontal dynamics of the contact patch body read:

*Dynamic contact patch equations*

$$m_c \dot{V}_{px} + k_{cx}\dot{x}_c^r + c_{cx}x_c^r = F_{sx} \quad (9.152)$$

$$m_c(\dot{V}_{py} + V_{cx}\dot{\psi}_c) + k_{cy}\dot{y}_c^r + c_{cy}y_c^r = F_{sy} \quad (9.153)$$

$$I_c\ddot{\psi}_c + k_{c\psi}\dot{\psi}_c^r + c_{c\psi}\psi_c^r = M_{sz} \quad (9.154)$$

and in addition, equations for the residual deflections

$$\dot{x}_c^r = V_{px} - V_{b\xi} + r_e(\Omega - \dot{\theta}_b^r) \quad (9.155)$$

$$\dot{y}_c^r = V_{py} - V_{b\eta} + r_l(\dot{\gamma} + \dot{\gamma}_b^r) \quad (9.156)$$

$$\dot{\psi}_c^r = \dot{\psi}_c - \dot{\psi} - \dot{\psi}_b^r \quad (9.157)$$

Also here, in the right-hand members, a road slope will have an effect. An axes transformation is needed to introduce properly the belt velocities with respect to the location of the contact patch. From the deflections and deflection rates the residual spring and damper forces appearing in Eqns (9.146–9.151) can be determined. We have

$$F_{cx} = k_{cx}\dot{x}_c^r + c_{cx}x_c^r \quad (9.158)$$

$$F_{cy} = k_{cy}\dot{y}_c^r + c_{cy}y_c^r \quad (9.159)$$

$$M_{cz} = k_{c\psi}\dot{\psi}_c^r + c_{c\psi}\psi_c^r \quad (9.160)$$

while

$$M_{cx} = M_{sx} \quad (9.161)$$

$$M_{cy} = M_{sy} \quad (9.162)$$

$$F_{cz} = F_{sz} \quad (9.163)$$

The contact forces and moments result from the contact slip model equations developed in the preceding section. The computed forces and moments have been defined according to the *Magic Formula* model, which at steady state, act with respect to the moving axes system $(C,x,y,z)$. These forces and moments, here provided with subscript $C$, are to be transformed to arrive at the set of forces and moments defined according to the system with lines of action shifted sideways over the calculated 'static' lateral displacement $y_{st}^r$ of the contact patch with respect to the wheel plane. These corrected quantities correspond to the forces and moments provided with subscript $S$ occurring in the Eqns (9.152–9.154) and (9.161–9.163):

$$F_{sx} = F_{x,C} \tag{9.164}$$

$$F_{sy} = F_{y,C} \tag{9.165}$$

$$M_{sz} = M_{z,C} + y_{st}^r \cdot F_{x,C} \tag{9.166}$$

$$M_{sx} = M_{x,C} - y_{st}^r \cdot F_{z,C} \tag{9.167}$$

$$M_{sy} = M_{y,C} \tag{9.168}$$

$$F_{sz} = F_{z,C} \tag{9.169}$$

where the static lateral deflection is computed from the side force and the overall lateral compliance of the standing tire at ground level:

$$y_{st}^r = \frac{F_{sy}}{C_{Fy}} = F_{sy}\left(\frac{1}{c_{cy}} + \frac{1}{c_{by}} + \frac{r_l^2}{c_{b\gamma}}\right) \tag{9.170}$$

With the transient response variables computed with the aid of Eqns (9.78–9.86) plus Eqn (9.20) the contact forces and moments may be found by using these variables as argument in the steady-state equations presented in Sections 4.3.2 and 4.3.3. However, we should properly account for the response to a varying camber angle of the belt plane.

For this purpose, we introduce the tire total spin velocity $\dot{\psi}_\gamma$, cf. Eqn (4.76), which has been corrected for the static belt camber deflection to enable the direct use of the relevant magic formulas (analogous to the use of $\beta_{st}$ in Eqn (9.78)):

$$\dot{\psi}_\gamma = \dot{\psi}_c - (1 - \varepsilon_\gamma)\Omega \sin(\gamma + \gamma_b^r - \gamma_{bst}^r) \tag{9.171}$$

with the 'static' belt deflection angle (neglecting the action of $M_x$)

$$\gamma_{bst}^r = -r_l F_{sy} \frac{1}{c_{b\gamma}} \tag{9.172}$$

Further, we write instead of $\beta_{st}$:

$$\psi_{st}^r = M_{sz}\left(\frac{1}{c_{c\psi}} + \frac{1}{c_{b\psi}}\right) \tag{9.173}$$

The transient slip first-order differential equations are repeated below. They are identical to Eqns (9.78–86) plus Eqn (9.20) except for the now-added effect of the camber angle in the right-hand members. In (9.20) $\kappa_c'$ may be replaced by $\kappa'$ as at steady state, these are equal for the contact patch and overall tire model. The same holds for the spin variables $\varphi_c'$ and $\varphi'$. The input slip velocities have been indicated in Figure 9.29.

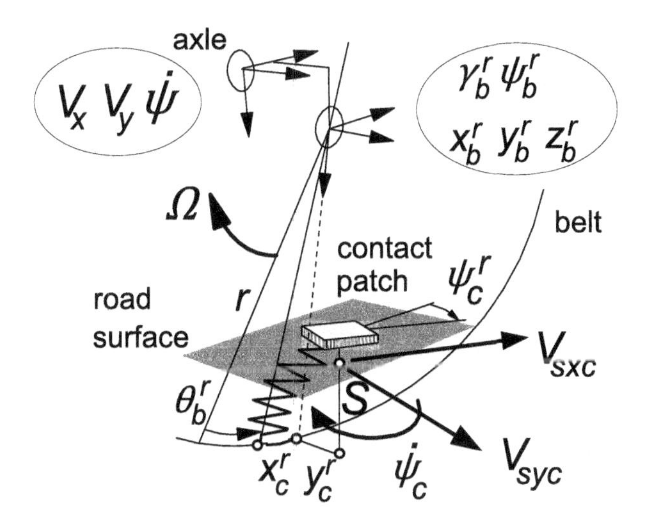


**FIGURE 9.29** Slip velocities of contact patch forming input to transient slip equations.

*Transient slip equations for longitudinal, side slip, turn slip, and camber*:

$$\sigma_c \frac{\mathrm{d}\kappa'}{\mathrm{d}t} + |V_{cx}|\kappa' = -V_{sxc} \tag{9.174}$$

$$\sigma_c \frac{\mathrm{d}\alpha'}{\mathrm{d}t} + |V_{cx}|\alpha' = -V_{syc} - |V_{cx}|\psi^r_{\mathrm{st}} \tag{9.175}$$

$$\sigma_2 \frac{\mathrm{d}\alpha'_t}{\mathrm{d}t} + |V_{cx}|\alpha'_t = |V_{cx}|\alpha' \tag{9.176}$$

$$\sigma_c \frac{\mathrm{d}\varphi'}{\mathrm{d}t} + V_c\varphi' = -\dot{\psi}_\gamma \tag{9.177}$$

$$\sigma_{F2} \frac{\mathrm{d}\varphi'_{F2}}{\mathrm{d}t} + V_c\varphi'_{F2} = -\dot{\psi}_\gamma \tag{9.178}$$

$$\sigma_{\varphi 1} \frac{\mathrm{d}\varphi'_1}{\mathrm{d}t} + V_c\varphi'_1 = -\dot{\psi}_\gamma \tag{9.179}$$

$$\sigma_{\varphi 2} \frac{\mathrm{d}\varphi'_2}{\mathrm{d}t} + V_c\varphi'_2 = -\dot{\psi}_\gamma \tag{9.180}$$

In addition, we need the composite transient slip quantities:

$$\varphi'_F = (2\varphi' - \varphi'_{F2})\mathrm{sgn}\, V_{cx} \tag{9.181}$$

$$\varphi'_M = \varepsilon^*_\varphi \varphi' + \varepsilon_{\varphi 12}(\varphi'_1 - \varphi'_2) \tag{9.182}$$

The slip variables employed in the *Magic Formulas* (*MF*) are replaced by the transient slip variables as indicated in the arguments of the following expressions:

*Output forces and moments*:

$$F_{x,C} = MF_{Fx}(\kappa', \alpha', F_N) \tag{9.183}$$

$$F_{y,C} = MF_{Fy}(\kappa', \alpha', \varphi'_F, F_N) \tag{9.184}$$

$$M_{z,C} = -t_c F'_{y,C} + M_{zr,C} + sF_{x,C} + \Delta M_z \tag{9.185}$$

$$F'_{y,C} = MF_{Fy}(\kappa', \alpha', \varphi' = 0, F_N) \tag{9.185a}$$

$$t_C = MF_t(\kappa', \alpha'_t, F_N) \tag{9.186}$$

$$M_{zr,C} = MF_{Mzr}(\kappa', \alpha', \varphi'_M, F_N) \tag{9.187}$$

$$\Delta M_z = C_{\Delta M}(\alpha' - \alpha'_t) \tag{9.188}$$

$$M_{x,C} = M_x \tag{9.189}$$

$$M_{y,C} = M_y \tag{9.190}$$

$$F_{z,C} = -F_N \tag{9.191}$$

Eqn (9.188) has been added, which is in agreement with the short wavelength transient slip theory, cf. Eqns (9.106 and 9.110). The input to the transient slip Eqns (9.171, 9.174–9.180) is constituted by the velocities of the contact patch, cf. Eqns (9.152–9.154), Figure 9.29:

$$V_{sxc} = V_{px} \tag{9.192}$$

$$V_{syc} = V_{py} - V_{px}(\psi_b^r + \psi_c^r) \tag{9.193}$$

$$\dot{\psi}_c = \dot{\psi}_c \tag{9.194}$$

The overturning couple $M_x$ can be modeled with the function (4.E69) where the $F_y$ part may be replaced by the expression (4.122–4.124) with transient slip angle as argument, if the actual momentary loaded radius $r_l$ (distance between points $A$ and $C$) has been properly accounted for in the (steady-state) measurements and further processing. The wheel load $F_N$ ($=|F_z|$) and the rolling resistance moment $M_y$ depend on the radial deflection and on a number of other variables. The subsequent section provides information on the experimentally assessed functional relationships.

Finally we need to establish the output forces and moments that act from the tire upon and about the wheel center. These quantities are denoted with the symbols $K$ and $T$ and are provided with the subscript $_a$. The components are defined to act along and about the axes of the axle triad ($A$, $\xi$,$\eta$,$\zeta$). We find

$$K_{a\xi} = K_{b\xi} \tag{9.195}$$

$$K_{a\eta} = K_{b\eta} \tag{9.196}$$

$$K_{a\zeta} = K_{b\zeta} \tag{9.197}$$

$$T_{a\xi} = T_{b\xi} + y_b^r K_{b\zeta} - z_b^r K_{b\eta} \tag{9.198}$$

$$T_{a\eta} = T_{b\eta} + z_b^r K_{b\xi} - x_b^r K_{b\zeta} \tag{9.199}$$

$$T_{a\zeta} = T_{b\zeta} + x_b^r K_{b\eta} - y_b^r K_{b\xi} \tag{9.200}$$

where the forces and moments acting from belt center to rim are retrieved from Eqns (9.131–9.136):

$$K_{b\xi} = k_{bx}\dot{x}_b^r + c_{bx}x_b^r + k_{bz}\Omega z_b^r \tag{9.201}$$

$$K_{b\eta} = k_{by}\dot{y}_b^r + c_{by}y_b^r \tag{9.202}$$

$$K_{b\zeta} = k_{bz}\dot{z}_b^r + c_{bz}z_b^r - k_{bx}\Omega x_b^r \tag{9.203}$$

$$T_{b\xi} = k_{b\gamma}\dot{\gamma}_b^r + c_{b\gamma}\gamma_b^r + k_{b\psi}\Omega\psi_b^r \tag{9.204}$$

$$T_{b\eta} = k_{b\theta}\dot{\theta}_b^r + c_{b\theta}\theta_b^r \tag{9.205}$$

$$T_{b\zeta} = k_{b\psi}\dot{\psi}_b^r + c_{b\psi}\psi_b^r - k_{b\gamma}\Omega\gamma_b^r \tag{9.206}$$

Values of inertia parameters normalized with tire mass $m_o$ and reference moment of inertia $m_o r_o^2$, with $r_o$ the unloaded tire radius, have been listed in Appendix 3.

## 9.3.2. Constitutive Relations

In the study of Zegelaar (1998) important observations have been made regarding contact area dimensions, static and dynamic vertical stiffness, and characteristics at different speeds of rolling, static longitudinal stiffness of the standing tire, tire radius growth with speed, rolling resistance, effective rolling radius, and rolling resistance couple. Much of the results will be repeated below.

### *Dimensions of the Contact Area*

Prints of the contact patch may be obtained by using ink or carbon paper. The shape appears to change from an oval shape at very low normal loads to a more rectangular shape at higher values of the load. An effective rectangular contact area may be defined with an area equal to that of the envelope of the actual print. The ratio of the width and length of the rectangle is taken equal to that of the actual contact area. The effective half length and half width are denoted as $a$ and $b$. The dimensions depend on the normal load $F_N$ ($=|F_z|$) and the following formulas have been found to give a good approximation:

$$a = \left(q_{a1}\sqrt{F_N/F_{No}} + q_{a2}F_N/F_{No}\right)r_o \quad (9.207)$$

and

$$b = \left(q_{b1}\sqrt[3]{F_N/F_{No}} + q_{b2}F_N/F_{No}\right)r_o \quad (9.208)$$

with $r_o$ denoting the free tire radius. Figure 9.30 presents the curves compared with the measured effective quantities for the tire pressed on a flat surface and on a curved drum surface with 2.5 m diameter. Obviously, the results are satisfactory. The dimensions of the tire were again: 205/60R15 91V at 2.2 bar inflation pressure. The nondimensional parameter values can be found in Table 9.3. In App. 3(3) an alternative expression for $a$ is presented based on the radial deflection $\rho_z$ instead of on the normal load $F_N$. The resulting value is much less dependent on the possibly changed inflation pressure.

## The Sidewall Stiffnesses and Damping

The rigid ring model of the tire freely rolling and loaded on the road shows three in-plane modes of vibration: the vertical mode and two angular modes. One of these rotational modes vibrates in phase with rim angular vibration while the other moves in anti-phase. The natural frequencies have been estimated with the aid of experiments conducted on the drum test stand where the wheel, at fixed axle position, rolls over a short cleat or is excited by brake torque fluctuations, cf. Section 9.4.2.

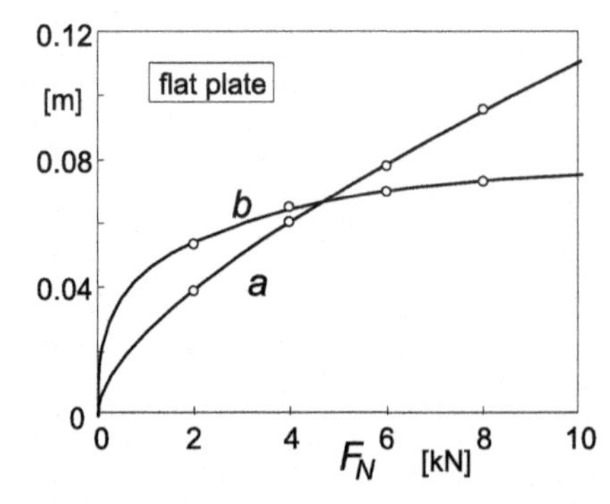


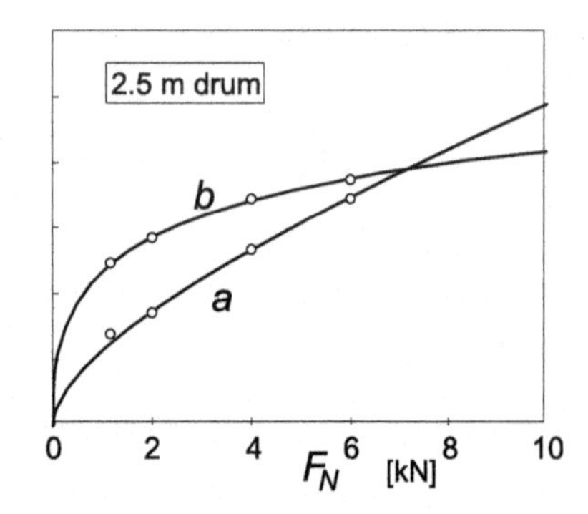


**FIGURE 9.30** Measured and calculated half length $a$ and half width $b$ of the contact patch vs wheel load, for the cases: loading on a flat plate and on a drum surface.

**TABLE 9.3** Parameter Values for Contact Patch Dimensions (205/60R15 91V at 2.2 bar)

| $F_{No}$ (= $\lvert F_{zo}\rvert$) = 4000 N, $r_o$ (= $R_o$) = 0.312 m, drum diameter = 2.5 m | | | |
|---|---|---|---|
| $q_{a1,flat} = 0.140$ | $q_{a2,flat} = 0.053$ | $q_{b1,flat} = 0.240$ | $q_{b2,flat} = -0.037$ |
| $q_{a1,drum} = 0.120$ | $q_{a2,drum} = 0.050$ | $q_{b1,drum} = 0.247$ | $q_{b2,drum} = -0.027$ |

The experiments indicate that the natural frequencies lying in the range of 0–100 Hz decrease with velocity. Other researchers found the same tendency, notably Bruni, Cheli and Resta (1996). Since the sidewall stiffnesses are much larger than the residual stiffnesses it is decided to make the in-plane sidewall stiffnesses dependent on the speed of rolling. As to the out-of-plane vibrations, Maurice did not ascertain the necessity to make the lateral, yaw, and camber stiffnesses speed dependent.

Zegelaar introduces a variable quantity $Q_V$ that is a measure of the time rate of change of the loaded tire deformation due to rolling. We have the nondimensional quantity ($V_o$ representing the reference velocity, cf. Section 4.3.2):

$$Q_V = \frac{|\Omega|}{V_o}\sqrt{(x_b^r)^2 + (z_b^r)^2} \tag{9.209}$$

The following expressions for the sidewall stiffnesses have been found to be appropriate:

$$c_{bx} = c_{bx0}\left(1 - q_{bVx}\sqrt{Q_v}\right) \tag{9.210}$$

$$c_{bz} = c_{bz0}\left(1 - q_{bVz}\sqrt{Q_v}\right) \tag{9.211}$$

$$c_{b\theta} = c_{b\theta 0}\left(1 - q_{bV\theta}\sqrt{Q_v}\right) \tag{9.212}$$

The additional subscript 0 designates the situation of the loaded nonrotating tire. The vertical and longitudinal stiffnesses have been assumed equal to each other. The parameters $q_{bVx,z,\theta}$ govern the speed dependency of the stiffnesses.

The sidewall damping coefficients $k_{bi} = k_{bx,y,z,\gamma,\theta,\psi}$ are considered to be constant quantities. The interaction terms appearing in Eqns (9.131–9.136) containing the coefficients $k_{bi}\Omega$ are omitted since these terms affect the rolling resistance and the aligning torque (also in steady state) and would make these speed dependent. The introduction of material damping being inversely proportional with frequency would be closer to reality. Further on, the rolling resistance will be introduced in an alternative, better controlled way.

For the constant stiffnesses, nondimensional parameters may be introduced. We define with $F_{No}$, $r_o$ (=$R_o$) and $m_o$ (the reference load, free tire radius, and tire mass) the nondimensional parameters $q$:

$$c_{bx0,y,z0} = q_{cbx,y,z} F_{No}/r_o \tag{9.213}$$

$$c_{b\gamma,\theta 0,\psi} = q_{cb\gamma,\theta,\psi} F_{No} r_o \tag{9.214}$$

$$k_{bx,y,z} = 2q_{kbx,y,z}\sqrt{m_o F_{No}/r_o} \tag{9.215}$$

$$k_{b\gamma,\theta,\psi} = 2q_{kb\gamma,\theta,\psi}\sqrt{m_o F_{No} r_o^3} \tag{9.216}$$

To provide more damping when the wheel speed gets close to zero, we may follow the theory of Chapter 7 and introduce $k_{V,\mathrm{low}}$ as demonstrated in Chapter 8, Eqns (8.127, 8.128), where the slip speed $V_{sx}$ may be replaced by $V_{sxc}$. In a similar way the residual stiffness and damping parameters $c_c$ and $k_c$ have been normalized.

## The Normal Force

The spring with residual stiffness $c_{cz}$ indicated in Figure 9.27 hides a structure that is a lot more complex than a spring with constant stiffness. Experiments reveal that the force deflection characteristics are nonlinear: the force develops after contact has been made and increases slightly more than proportionally with the overall normal deflection $\rho_z$. Also, the tire grows with speed due to centrifugal action. Figure 9.31 illustrates both phenomena. Furthermore, it has been found useful to introduce $F_x$ and $F_y$ interaction terms in the vertical stiffness, cf. Reimpell et al. (1986). The following formula is proposed for the

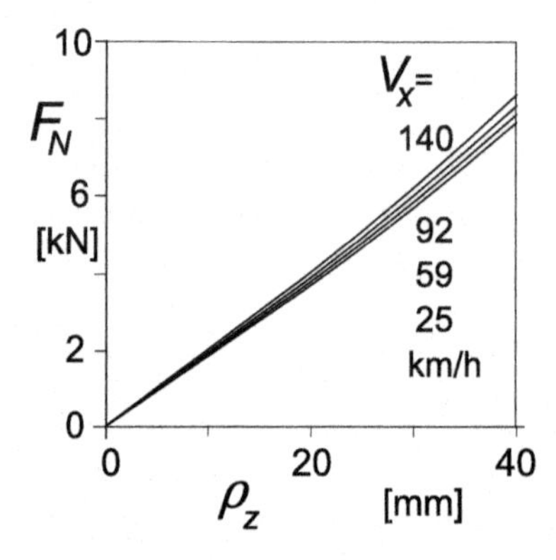


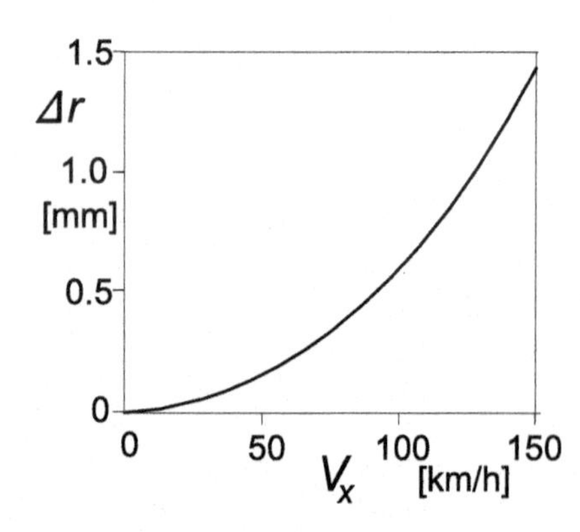


**FIGURE 9.31** Vertical load vs normal deflection characteristics at various forward velocities and tire radius growth with speed.

normal force including interaction and overall stiffness and growth functions (also see Eqn (4.E68) for the influence of inflation pressure):

$$F_N = |F_z| = \left\{1 + q_{V2}|\Omega|\frac{r_o}{V_o} - \left(q_{Fcx1}\frac{F_x}{F_{No}}\right)^2 - \left(q_{Fcy1}\frac{F_y}{F_{No}}\right)^2\right\}\cdot \left(\left(q_{Fz1} + q_{Fz3}\gamma^2\right)\frac{\rho_z}{r_o} + q_{Fz2}\frac{\rho_z^2}{r_o^2}\right)F_{No} \tag{9.217}$$

With radius $r_c$ of assumed circular cross-sectional contour, the deflection becomes at wheel camber angle $\gamma$ (relative to normal to road plane), also cf. Eqn (7.46):

$$\rho_z = \max((r_o - r_l + \Delta r)\cos\gamma + r_c(1 - \cos\gamma), 0) \tag{9.218}$$

The tire radial growth changes quadratically with rotational velocity:

$$\Delta r = q_{V1} r_o (r_o \Omega / V_o)^2 \tag{9.219}$$

Here, $r_o$ denotes the radius of the free nonrotating tire, $r_l$ the loaded radius (distance between wheel center and contact center), and $\Delta r$ the increase in free tire radius due to wheel rotation velocity. The nondimensional parameter $q_{V1}$ governs the influence of tire growth, $q_{V2}$ the stiffness variation with speed, $q_{Fcx,y1}$ the interaction with horizontal forces, and $q_{Fz1,2,3}$ the stiffness and nonlinearity of the force deflection characteristic at zero speed and zero horizontal forces. Appendix 3 presents the parameter values fitted to experimental data. To radically simplify (9.218), we may for small $\gamma$, replace $r_c$ by $r_o - r_l + \Delta r$ which is the calculated tire radial deflection.

From the overall characteristics the properties of the residual spring are to be derived. An exact functional relationship may be established but it can be found that the residual normal spring characteristic can be approximated by the third degree polynomial function:

$$F_N = |F_z| = a_1\rho_{zr} + a_2\rho_{zr}^2 + a_3\rho_{zr}^3 \tag{9.220}$$

with the $F_N$-related normal residual spring deflection (taking into account geometrical interaction terms with horizontal deflections)

$$\rho_{zr} = \rho_z + z_b^r\cos\gamma - q_{Fcx2}\rho_x^2/r_o - q_{Fcy2}\rho_y^2/r_o \tag{9.221}$$

Here, $z_b^r$ is the radial displacement of the center of the belt ring with respect to the wheel center and $\rho_{x,y}$ represent the longitudinal and lateral tire contact deflections. The actual loaded radius $r_l$ results from the calculated deflections (cf. Figure 4.30 for measured evidence) with camber influence included:

$$r_l = r_o + z_b^r + \Delta r + \{r_c(1 - \cos\gamma) - \rho_{zr} - (q_{Fcx2}\rho_x^2 + q_{Fcy2}\rho_y^2)/r_o\}/\cos\gamma \tag{9.222}$$

The coefficients appearing in (9.220) can be expressed in terms of sidewall stiffness $c_{bz}$ and wheel speed of revolution $\Omega$:

$$a_1 = \frac{c_{bz}A_1}{c_{bz} - A_1} \tag{9.223}$$

$$a_2 = \frac{c_{bz}^3 A_2}{(c_{bz} - A_1)^3} \tag{9.224}$$

$$a_3 = 2\frac{c_{bz}^4 A_2^2}{(c_{bz} - A_1)^5} \tag{9.225}$$

where, with Reimpell's terms in (9.217) omitted, we obtain

$$A_1 = (q_{Fz1} + q_{Fz3}\gamma^2)(1 + q_{V2}|\Omega|r_o/V_o)F_{No}/r_o \tag{9.226}$$

$$A_2 = q_{Fz2}A_1/\{(q_{Fz1} + q_{Fz3}\gamma^2)r_o\} \tag{9.227}$$

where the nondimensional parameters $q_{V2}$ and $q_{Fz1,2,3}$ of Eqn (9.217) appear. The horizontal tire deflections at road surface level with respect to the wheel rim are (at small camber):

$$\rho_x = x_b^r + r_o\theta_b^r + x_c^r \tag{9.228}$$

$$\rho_y = y_b^r - r_o\gamma_b^r + y_c^r \tag{9.229}$$

Appendix 3 provides the relevant parameter values for the passenger car tire that has been tested.

### *Free Rolling Resistance*

Experiments show that the rolling resistance force $F_r$ (pointing backward) is proportional to the tire normal load $F_N$. A history on this subject can be found in the publication of Clark (1982). We have

$$F_r = f_r F_N \tag{9.230}$$

The rolling resistance coefficient $f_r$ depends on the forward speed and may be expressed in terms of powers of the speed, cf. Mitschke (1982):

$$f_r = q_{sy1} + q_{sy3}|V_x/V_o| + q_{sy4}(V_x/V_o)^4 \tag{9.231}$$

Parameter $q_{sy1}$ governs the initial level of the rolling resistance force and typically lies in between 1 and 2%. Parameter $q_{sy3}$ controls the slight slope of the resistance with speed. The last parameter $q_{sy4}$ represents the sharp rise of the resistance that occurs after a relatively high critical speed is surpassed. Then, the so-called standing waves show up as a result of instability, cf. Pacejka

(1981), or according to an alternative theory due to resonance, cf. Brockman and Braisted (1994). The formation of standing waves gives rise to large deflection variations and considerable energy loss. The phenomenon may result in failure of the tire and poses an upper limit to the safe range of operation of the tire.

Below, we will see that the rolling resistance will be introduced in the tire model through the rolling resistance moment that is imposed on the tire belt ring as an external torque about the $y$ axis, cf. Eqn (9.236).

### *Effective Rolling Radius, Brake Lever Arm, Rolling Resistance Moment*

In Chapter 8 the notion of the effective rolling radius has been introduced. Figure 8.12 shows the results of experiments of tires running over a drum surface. In Subsection 8.3.1 the theory is restricted to a linearized representation of the variation of the effective rolling radius with radial defection. The complete nonlinear variation versus normal load $F_N$ may be described by the expression:

$$r_e = r_o + \Delta r - \{q_{re1}\rho_z + D_{re}\arctan(B_{re}\rho_z)\} \tag{9.232}$$

with

$$\begin{aligned} D_{re} &= q_{re2}F_{No}/C_{Fzo} \\ B_{re} &= q_{re3}/D_{re} \end{aligned} \tag{9.232a}$$

and the vertical stiffness of the standing tire at nominal load $F_{No}$, as derived from Eqn (9.217):

$$C_{Fzo} = \frac{F_{No}}{r_o}\left(q_{Fz1} + 2q_{Fz2}\frac{\rho_o}{r_o}\right) \tag{9.232b}$$

In App. 3, Eq. (A3.4), an alternative expression has been given. The longitudinal slip velocity is defined with $r_e$ introduced as the slip radius, cf. e.g. Eqn (8.32).

In the present model with the belt ring and contact patch modeled as separate bodies, the longitudinal slip velocity $V_{sxc}$ of the contact patch is used as input in the transient slip differential Eqn (9.174). In Eqn (9.192) with (9.155), the effective rolling radius $r_e$ accomplishes the transmission of the rotational speed of the belt to the residual deflection rate of change. At steady-state condition, the deflection rates vanish and we have the following relation for the longitudinal slip speed:

$$V_{sxc} = V_x - r_e\Omega \tag{9.233}$$

We may consider the power balance of a wheel subjected to a propulsion torque $M_D$ and a drag force $F_D$ acting backward on the wheel in its center. Figure 9.32 depicts the situation. The connected power flow diagram is presented in Figure 9.33. The S represents a power source (the engine) and the

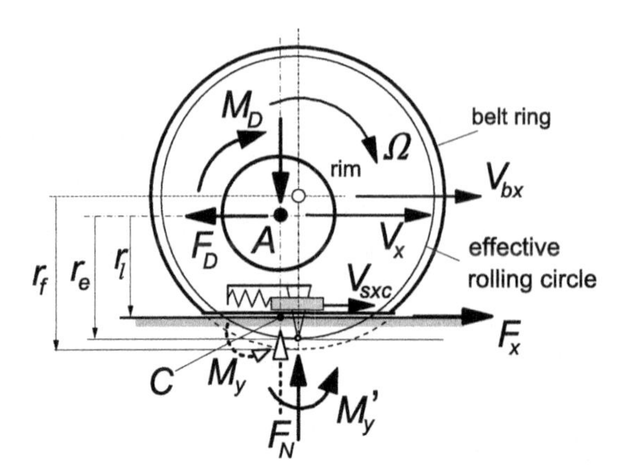


FIGURE 9.32 The driven tire–wheel combination with deflected belt and residual spring.

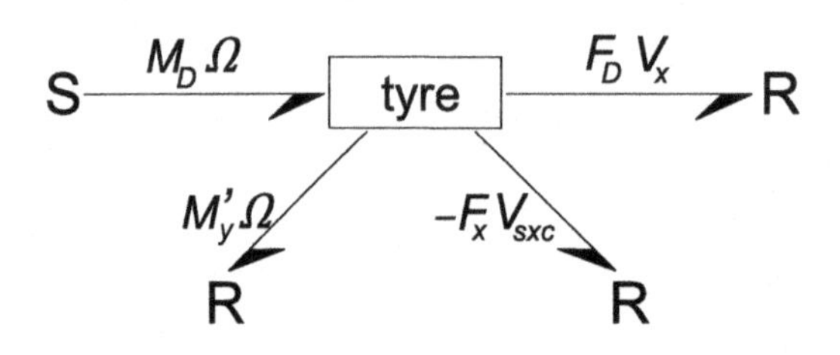


FIGURE 9.33 Power flow diagram (bond graph) of driven tire–wheel combination in steady state.

Rs are resistors where energy is dissipated. The balance of power requires that the equation holds:

$$M_D \Omega = F_D V_x + M_y' \Omega - F_x V_{sxc} \tag{9.234}$$

or with (9.233) and $F_x = F_D$

$$M_D - M_y' = F_x r_e \tag{9.235}$$

where $M_y'$ represents the energy dissipating moment due to rolling. This equation suggests, at least for the model employed, that the moment arm equals the effective rolling radius (defined at zero driving or braking torque: free rolling). Consequently, the block named 'tire' in the diagram of Figure 9.33 represents, when unfolding the bond graph, a junction structure containing a transformer with modulus $r_e$ that transforms the angular speed into (a part of) the slip speed and, in the opposite direction, the slip force into the drive torque.

Experiments have been carried out by Zegelaar on both the flat plank machine and the drum test stand to establish the effective rolling radius and the moment arm. In these tests, a (brake) torque was applied to the wheel. The moment arm may be termed as the brake lever arm. The diagrams of Figure 9.34 have been obtained from tests performed at zero speed (for

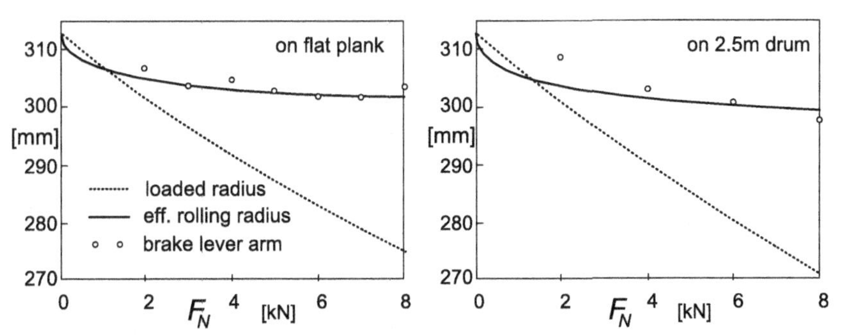


**FIGURE 9.34** Tire radii as function of the normal load measured at zero or very low speed.

assessing the brake lever arm) and at very low speed of travel (for finding the effective rolling radius). Especially in case of the flat surface, an excellent agreement has been found to occur. It is assumed that the growth of the effective rolling radius with speed is equal to that of the free tire radius, cf. Eqn (9.219).

The diagrams of Figure 9.35 present the influence of speed on the two radii. The loaded radius has been kept fixed so that the vertical load rises when the speed is increased. Three different axle heights have been selected corresponding to the indicated initial vertical loads at zero speed $F_{No}$. The left-hand diagram shows the degree of fit for the effective rolling radius. The right-hand diagram shows the correspondence with the brake lever arm. The tests from which the brake lever arm can be assessed have been conducted at low levels of the average and the standard deviation of the brake torque random input (120 and 22 Nm, respectively). The brake lever arm results from the longitudinal force response to the imposed brake torque variation at zero frequency. The influence of the average brake torque on the ratio of the torque amplitude and the force amplitude at zero frequency, $-dM_B/dF_x$, is given in

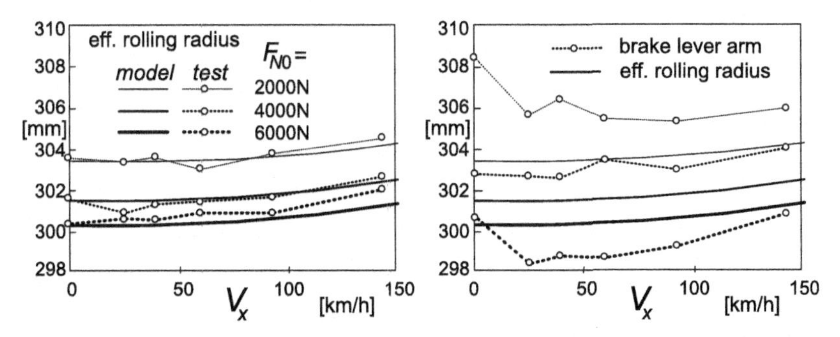


**FIGURE 9.35** Tire radii measured on 2.5 m drum as a function of the forward speed at three axle heights corresponding to the indicated initial loads. Average level and amplitude of brake torque are small.

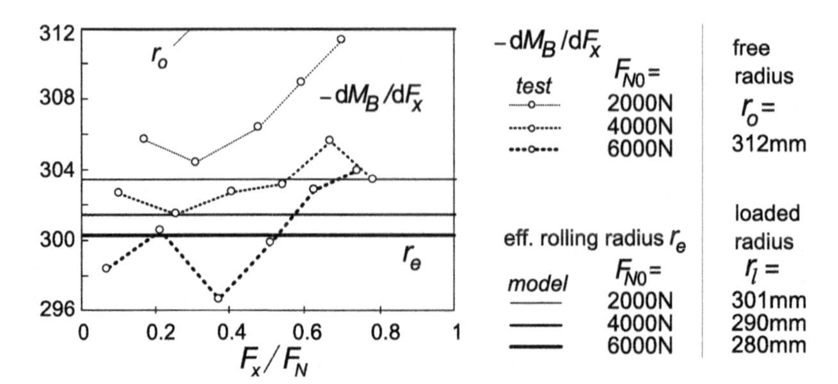


**FIGURE 9.36** Ratio of brake torque increment to brake force increment as a function of level of brake force at a velocity of 25 km/h and three different axle heights (loaded radii).

Figure 9.36. This ratio does not appear to be a constant. Especially at low loads and relatively large braking forces, large deviations arise from the value of the effective rolling radius. Note that, due to its definition, the effective radius is not affected by the magnitude of the brake force.

With the effective radius adopted as the brake or driving torque moment arm, we find from Eqn (9.235) that if $M_D = 0$ the moment $M'_y = -F_x r_e = F_r r_e$.

For the (forward) rolling tire, this moment is assumed to be independent of $F_x$ and $M_D$. Furthermore, we obtain, at $\Omega = 0$ where the moment due to rolling $M'_y = 0$, that $M_D = F_x r_e$. Obviously $r_e$ is the brake lever arm corresponding with the experimental result.

By considering the equilibrium about the wheel axis $A$, the rolling resistance moment $M_y$ (acting about the $C$-$y$ axis) becomes

$$\begin{aligned} M_y &= M_D - F_x r_l = F_x r_e + M'_y - F_x r_l = (F_x + F_r) r_e - F_x r_l \\ &= F_x(r_e - r_l) + F_r r_e \end{aligned} \tag{9.236}$$

with $F_r$ given by Eqn (9.230). Apparently, at free rolling ($M_D = 0$) where $F_x = -F_r$, we correctly have: $M_y = F_r r_l$.

It is clear that theoretical modeling of the mechanism behind the generation of the rolling resistance moment deserves more study. In Chapter 4 the empirical formula Eqn (4.E70) has been presented for $M_y$ that is based on tire test data.

## 9.4. DYNAMIC TIRE MODEL PERFORMANCE

A number of experiments have been conducted at the Delft University of Technology to assess the parameters of the dynamic model and to judge its performance. The steady-state side slip, longitudinal slip, camber force, and moment characteristics have been typically assessed from over-the-road experiments with the Delft TireTest Trailer. For the model performance

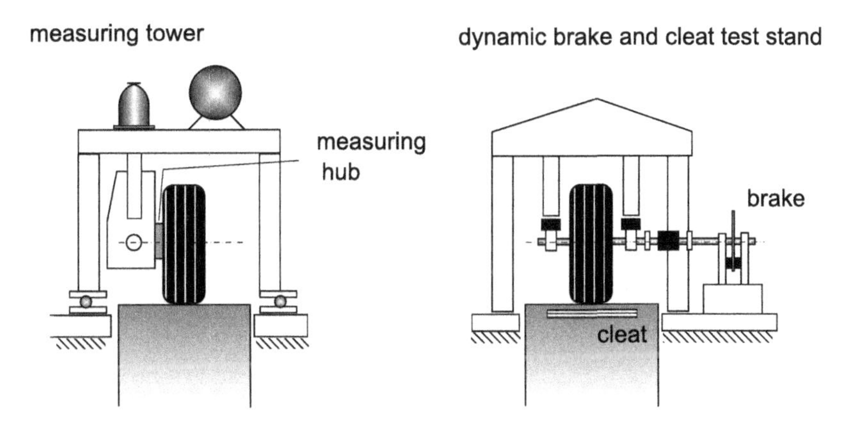


**FIGURE 9.37** Principal setup of the measuring tower and of the dynamic brake and cleat test stand.

evaluation, steady-state characteristics have been used, obtained from tests carried out on the drum with the strain gauge-equipped measuring hub mounted in the measuring tower, cf. Figure 9.37.

In Section 9.4.1 the dedicated dynamic test facilities have been discussed, followed by Section 9.4.2 with the presentation of the model dynamic behavior in comparison with experimental data. In Chapter 10 the model will be extended to include the description of running over road obstacles (cleats). This model extension is followed by the presentation of experimental results compared with model behavior.

The reader is referred to Chapter 12 for a more extensive description of steady-state- and higher frequency test facilities including the dedicated rigs mentioned in the section below. For some practical details regarding parameter assessment for advanced dynamic tire models, Section 10.2 may be of interest. In this context, it is important to note that cleat tests are most useful to determine both the in-plane and the out-of-plane natural frequencies of the model. Also modal testing of the unloaded tire with a fixed wheel axle may provide useful, possibly additional, information.

## 9.4.1. Dedicated Dynamic Test Facilities

### *Dynamic Brake and Cleat Test Rig*

Experiments have been conducted on the 2.5 m steel drum test stand provided with a specially designed rig equipped with a disc brake installation, cf. Figure 9.37 (right-hand diagram), and for more details: Chapter 12, Figure 12.7. Brake torque fluctuation tests (Section 9.4.2) and dynamic cleat tests (Section 10.2) have been carried out. The test facilities with numerous experimental and simulation results have been described in detail by Zegelaar (1998).

**TABLE 9.4** Vertical Load on 2.5 m Drum at Constant Axle Height and Increasing Speed (tire: 205/60R15 91V at 2.2 bar)

| $V_x$ [km/h] | Initial Vertical Deflection $\rho_{z0}$ [mm] | | | |
|---|---|---|---|---|
| 0 | 0 | 11.90 | 22.57 | 32.33 |
| | Vertical load $F_N$ [N] at constant axle height | | | |
| 0 | 0 | 2115 | 4153 | 6133 |
| 25 | 7 | 2166 | 4246 | 6288 |
| 39 | 17 | 2202 | 4307 | 6352 |
| 59 | 40 | 2264 | 4404 | 6483 |
| 92 | 100 | 2388 | 4588 | 6727 |
| 143 | 249 | 2642 | 4939 | 7169 |

The wheel axle height can be adjusted to select the tire initial load. During the tests, the axle position is held fixed causing the wheel load to rise with increasing speed. The wheel axle bearing supports are equipped with piezoelectric load cells. Steady-state- or average force levels cannot be measured very well with these force transducers. To provide an indication of the actual load increase with speed as measured in the measurement tower equipped with a hub provided with strain gauges, Table 9.4 gives for a series of initial deflections ($\rho_{z0}$ at zero speed) the values of the average vertical force derived from measurements at different speeds. The values have been obtained from Eqns (9.217–9.219) after having fitted the parameters involved. The loads shown apply for the cases of nominal loads 2000, 4000, and 6000 N indicated in the graphs presented in the next section.

### *Pendulum and Yaw Oscillation Test Rigs*

To assess the lateral and yaw tire dynamic parameters, two test rigs have been developed. One is the trailing arm 'pendulum' test stand with at one end a vertical hinge and at the other the steering head with a piezoelectric measuring hub. At that point the arm is excited laterally up to ca. 25 Hz through a hydraulic actuator, cf. Figure 9.38, and for more details: Chapter 12, Figure 12.8. The rig is useful to assess the overall relaxation length and the gyroscopic couple coefficient, both needed for the simpler transient models treated in Chapter 7. The idea of the pendulum concept originates from Bandel et al. (1989). They designed and used an actual pendulum rig. The natural frequency of the freely swinging trailing arm with a tire rolling on the drum was

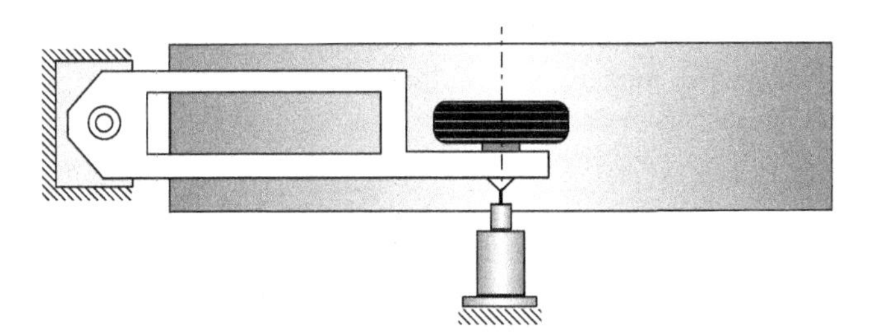

**FIGURE 9.38** Principal sketch of the trailing arm 'pendulum' test rig exciting the tire almost purely laterally. Frequencies up to ca. 25 Hz, adequate for assessing the tire relaxation length and gyroscopic coupling parameter.

taken to establish the relaxation length $\sigma$. This quantity is the parameter of the first-order differential equation such as is used in Chapter 7. Bandel found that $\sigma$ increases with speed. However, when using a model in which a belt ring with mass is used, it turns out that the parameters can be kept constant, cf. Vries and Pacejka (1998b). Consequently, tire inertia, notably the gyroscopic couples, gives rise to the speed dependency of the effective relaxation length.

Another rig was developed to investigate the response of the tire subjected to yaw oscillations at frequencies up to ca. 65 Hz. The structure depicted in Figure 9.39 is light and very stiff; also see Figure 12.8. The two guiding members with flexible hinges intersect in the vertical virtual steering axis that is positioned in the wheel center plane (center point steering). A hydraulic actuator is mounted to generate the yaw vibration. The wheel axle is provided with a piezoelectric measuring hub. The tire is loaded by adjusting the axle height above the drum surface. During the test the loaded radius remains constant.

The measuring tower, cf. Figure 9.37, provided with a hydraulic vertical axle positioning installation is used to conduct pure braking and pure side slip tests as well as combined slip experiments at axle height oscillations and radial dynamic stiffness tests up to ca. 15 Hz.

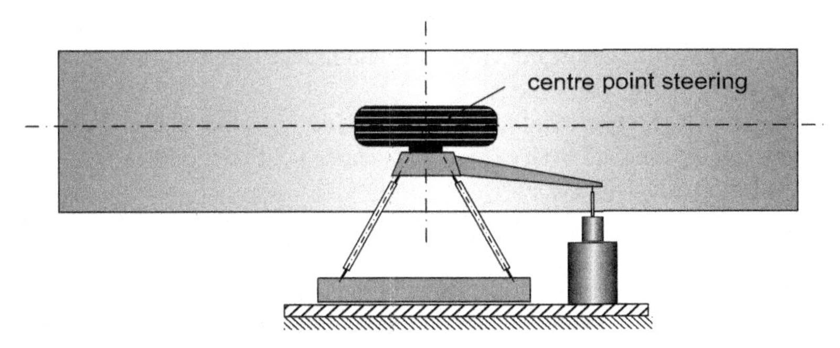


**FIGURE 9.39** Principal sketch of the yaw oscillation test rig featuring center point steering. Frequencies up to ca. 65 Hz enabling the assessment of tire out-of-plane inertia and stiffness parameters including residual stiffnesses and rigid modes.

A detailed description of the various side slip test facilities together with a full account of the numerous experiments conducted and the simulation results of the model have been given by Maurice (2000) and for motorcycle tires by Vries and Pacejka (1998a,b).

### 9.4.2. Dynamic Tire Simulation and Experimental Results

In general, the values of the model parameters can be estimated by minimizing the difference between measured and calculated frequency response functions (both amplitude and phase). In some cases (in particular the in-plane response), special aspects of the response functions may be considered to successfully assist the parameter assessment process. These aspects are: the position and width of resonance peaks (sidewall and residual stiffnesses and damping), the phase relationship in the low frequency range (overall relaxation length which itself is not a parameter!), and the yaw response at zero speed (yaw residual stiffness). The residual damping ratios have been chosen equal to those of the sidewalls. Another help is the establishment of the inertia parameters of the relevant part of the tire by cutting the tire into pieces and considering the parameters of these parts. Rolling over a cleat might be used to estimate some of the parameters but the most accurate way is the identification through frequency response functions. These are obtained with the aid of the random brake test with the measured brake torque (or the measured wheel speed) used as input, the yaw oscillation test, and the vertical axle oscillation test.

#### *Vibrational Modes*

The vibrational modes of the tire may be assessed through modal analysis of the tire wheel system with axle fixed and tire loaded and/or unloaded. When comparing these results with calculated modes using the parameter values assessed by means of the frequency response functions of the rolling tire, it is found that the stiffnesses found from the dynamic rolling experiments are ca. 30% lower than those estimated from experimental modal analysis, cf. Zegelaar. These differences must be due to the different operational conditions and the larger amplitudes of the vibrations and higher temperatures that occur in the realistic rolling experiments.

The calculated vibrational rigid body modes at zero speed using the parameters as established from experiments carried out on the drum have been depicted in Figures 9.40 and 9.41.

We have four in-plane degrees of freedom of the belt ring and the wheel rim (two translational and two rotational) and three out-of-plane degrees of freedom (lateral, yaw, and camber). As a consequence, we can distinguish four in-plane rigid modes and three out-of-plane rigid body modes. The mode shapes change considerably when the tire is making contact with the drum surface. The free rotation (0 Hz) mode changes into a mode with the belt and rim rotating in phase with respect to each other.

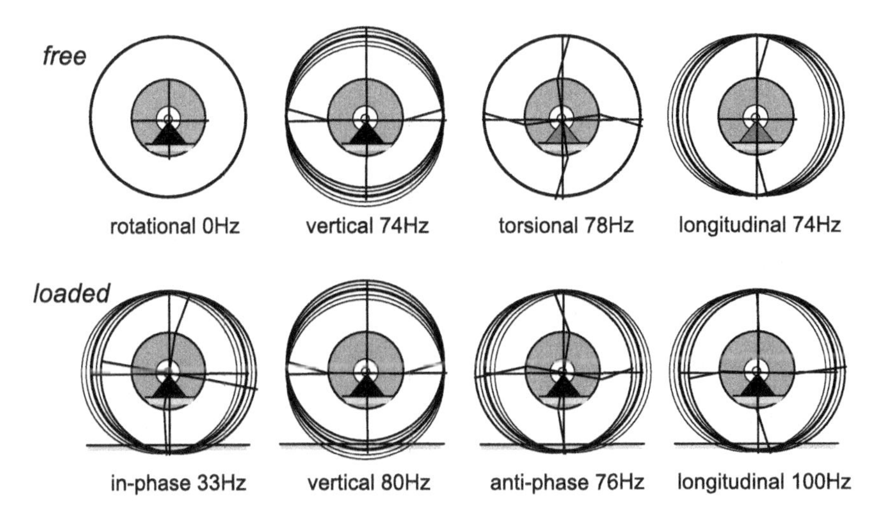


**FIGURE 9.40** Calculated in-plane vibrational modes of tire/wheel system with axle fixed and tire free or loaded on the drum surface with vertical load $F_N = 4000\,\text{N}$ and at zero speed.

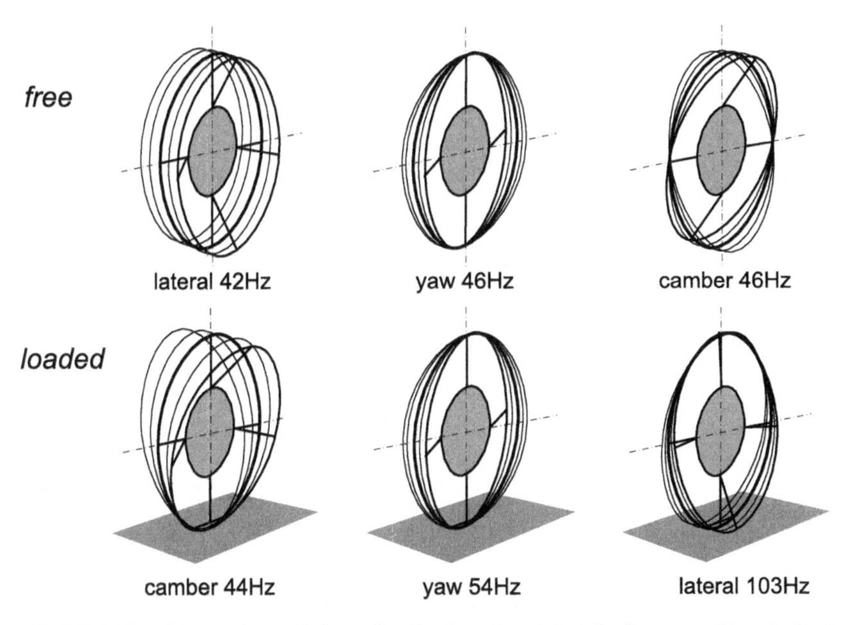


**FIGURE 9.41** Calculated out-of-plane vibrational modes of tire/wheel system with axle fixed and tire free or loaded on the drum surface with vertical load $F_N = 4000\,\text{N}$ and zero speed.

The lateral and camber modes appear to form combinations: one low frequency mode with a low axis of camber oscillations and one high frequency mode with a rotation axis closer to the top of the tire and relatively large lateral deflections in the contact zone. The yaw mode in the loaded case shows a higher

**TABLE 9.5** Natural Frequencies and Damping of Vibrational Modes of Rigid Ring Tire Model Calculated at a Vertical Load $F_N = 4000$ N for Two Values of Forward Speed

| Forward Speed [m/s] | 0 | 30 | 0 | 30 |
|---|---|---|---|---|
| | $n$ [Hz] | | $\zeta$ [%] | |
| ***in-plane modes*** | | | | |
| In-phase | 33 | 30 | 0.10 | 0.25 |
| Vertical | 80 | 75 | 0.05 | 0.05 |
| Anti-phase | 76 | 71 | 0.05 | 0.05 |
| Longitudinal | 100 | 129 | 0.77 | 0.35 |
| ***Out-of-plane modes*** | | | | |
| Camber | 44 | 33 | 0.03 | 0.05 |
| Yaw | 54 | 51 | 0.03 | 0.04 |
| Lateral | 103 | 101 | 0.01 | 0.24 |

calculated natural frequency since the effect of turn slip has been included, which was not the case in Maurice's original model. This means that the yaw stiffness of the contact tread has now been accounted for.

The natural frequencies $n$ and damping ratio $\zeta$ change with the speed of rolling. In Table 9.5 the values have been presented for the loaded tire running at a velocity of 0 and 30 m/s. Especially the out-of-plane modes show considerable changes in frequency and damping. The camber and yaw mode natural frequencies which are identical in the unloaded zero speed case, exhibit a with-speed growing mutual difference with the camber mode frequency becoming smaller and the yaw mode frequency larger.

### *Frequency Response Functions*

A typical example of measured and calculated in-plane frequency response functions has been depicted in Figure 9.42. Coherence functions show that the tests give sensible results up to ca. 80 Hz. Similar response functions have been obtained by Kobiki et al. (1990). The left-hand diagram of the figure represents the response function of the longitudinal force $F_x$ ($=K_{a\xi}$ in Eqn (9.195)) acting on the wheel axle to the imposed brake torque variation considered to be applied in the torque meter. The right-hand diagram shows the response of the force to wheel slip variations. The wheel slip is derived from the measured wheel and drum speeds.

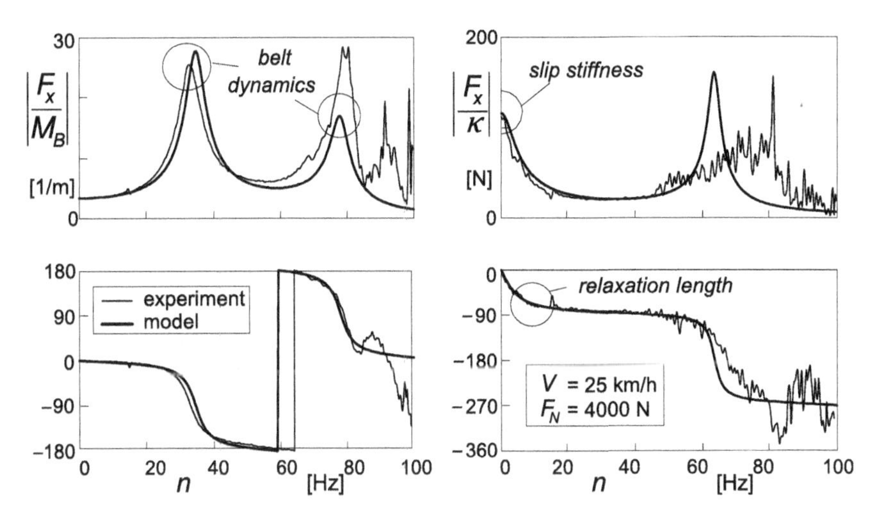


**FIGURE 9.42** Measured and calculated in-plane frequency response functions at an average braking force of 450 N and assessed at a braking force standard deviation of 75 N.

The two peaks occurring in the left-hand diagram belong to the in-phase and the anti-phase modes. The single peak showing up in the right-hand diagram belongs to the mode that would arise if the rim were fixed also in rotation. The natural frequency lies in between the frequencies of the peaks in the left diagram. The natural frequencies contribute to assess the sidewall stiffnesses.

From the right-hand diagram two quantities can be derived: the slip stiffness and the overall relaxation length. Through the latter, additional information is obtained to find the fore-and-aft residual stiffness. From careful interpretation of the frequency response functions assessed at different speeds, the sidewall stiffness dependence on the speed of rolling has been ascertained. Resulting calculated response functions at different loads and brake torque level gave satisfactory agreement with measured behavior. We refer to the original work of Zegelaar (1998) for detailed information.

The out-of-plane frequency response functions of the side force $F_y$ ($=K_{a\eta}$) and the aligning torque $M_z$ ($=T_{a\zeta}$) to yaw oscillations have been presented in Figures 9.43 and 9.44. The parameters have been assessed by minimizing the difference between measured and calculated (complex) response functions. The correspondence achieved between measured and computed curves at different speeds, loads, and side slip level is quite satisfactory; cf. Maurice (2000) for more details. To conduct a proper comparison, the measured data have been corrected for the inertia of the wheel and part of the tire that moves with the wheel.

The expected splitting up of the single peak at low velocity into two peaks, one belonging to the camber mode and the other to the yaw mode, and the growing difference of the two natural frequencies with increasing speed is

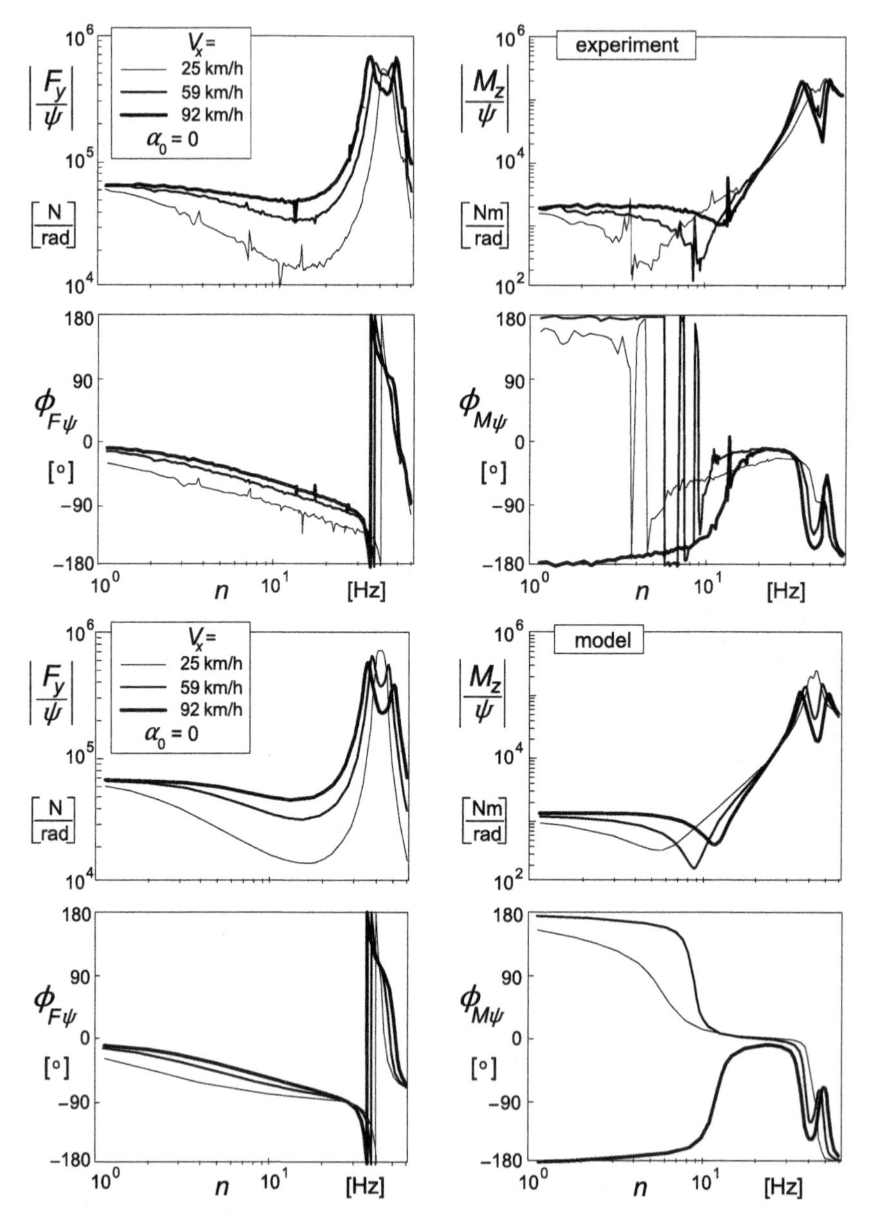


**FIGURE 9.43** Measured and calculated out-of-plane frequency response functions of the side force and aligning torque to steer angle variations at zero average slip angle for three values of forward speed and normal load $F_N = 4000\,\text{N}$.

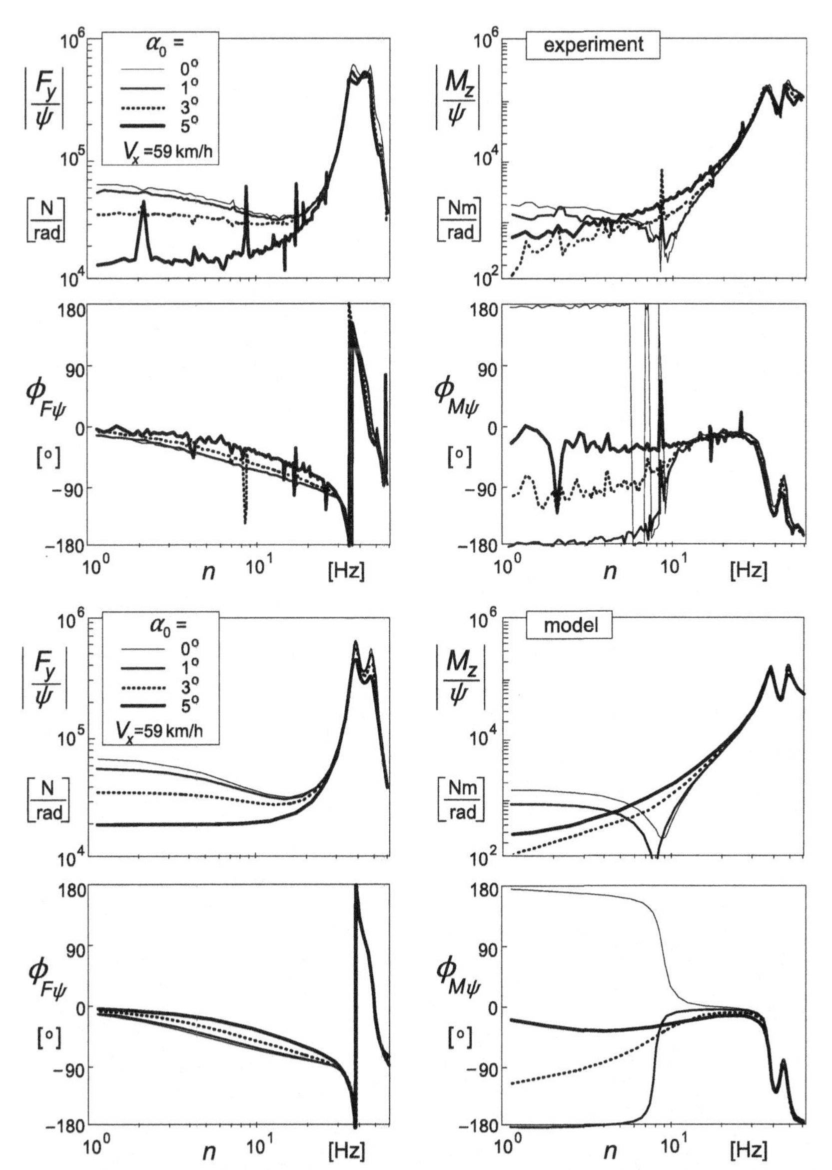


**FIGURE 9.44** Measured and calculated out-of-plane frequency response functions of the side force and aligning torque to steer angle variations at one value of speed and at four levels of average side slip and wheel load $F_N = 4000$ N.

clearly demonstrated. This phenomenon, which is due to gyroscopic action, has already been observed to occur with the stretched string model with inertia included approximately, cf. Figure 5.40. It is noted that the theoretical results of Maurice have been established by using the model that did not include the

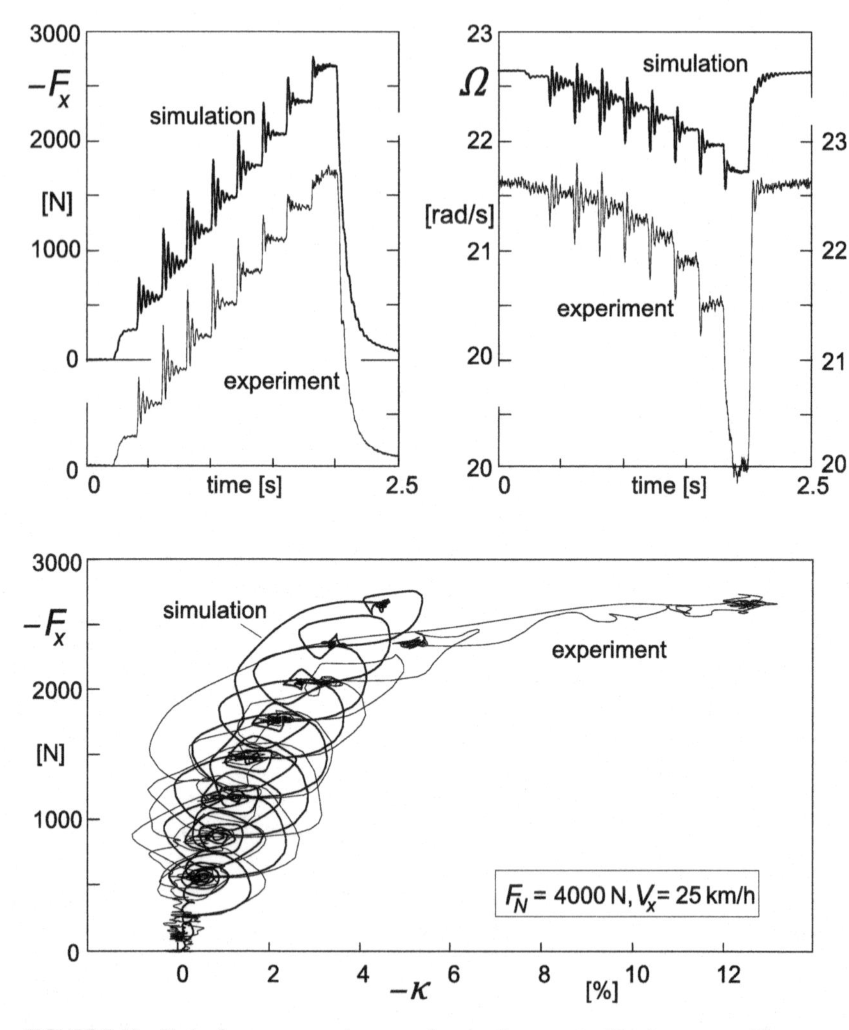


**FIGURE 9.45** Brake force response to successive step increments of brake pressure. The upper diagrams depict the variation of force and wheel speed of revolution with time. The lower figure shows the loops in the force vs wheel slip diagram. Apparently, the actual friction coefficient is lower than assumed in the model. (Zegelaar 1998).

equations for the response to spin (9.177–9.180). Especially the aligning moment is sensitive to turn slip.

The moment response curves to side slip and yaw of the massless tire model as depicted in Figure 9.16 are quite different. The dip in the moment amplitude response curve to yaw oscillations occurring in the curve for $M_\psi$ in Figure 9.16 does not appear in the curve for the response to side slip $M_\alpha$. This dip also appears in the curves of Figure 5.40 where spin is included as well. It is surprising to see that a similar dip occurs in Figures 9.43 and 9.44. The

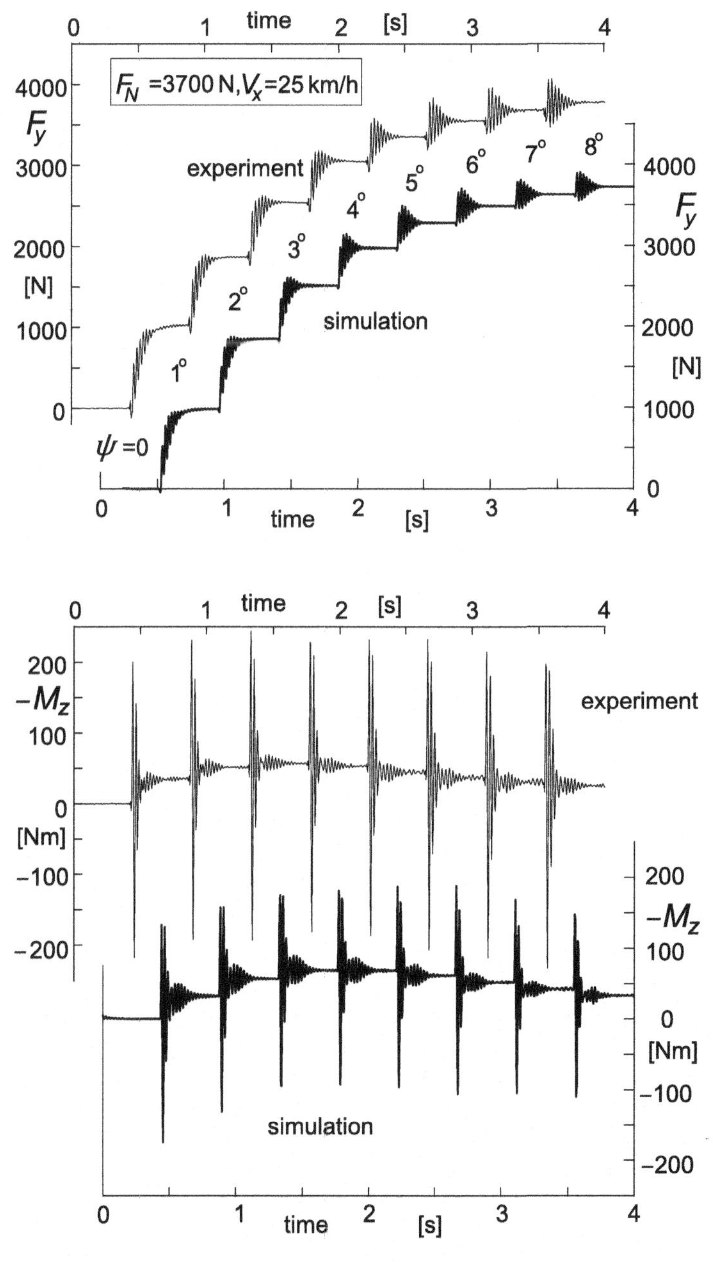


**FIGURE 9.46** Side force and aligning torque response to successive step changes in steer angle. (Maurice 2000).

minimum arises here due to the tendency of the response to side slip to decrease and tend to zero at increasing path frequency while at the same time the moment amplitude increases due to tire inertia when approaching the resonance peak. The dip in the measured curve of the upper right-hand diagram of Figure 9.43 at 25 km/h is deeper than the one shown by the theoretical curve of the lower diagram. This may be due to the additional action of spin in the actual tire. As shown by both the theoretical and measured curves of Figure 9.44, the dip disappears altogether at larger average side slip.

The experiments conducted with the trailing arm 'pendulum' test rig, where the tire is subjected to almost pure side slip and the spin is very small, also show satisfactory correspondence with model behavior (Maurice 2000; also: De Vries and Pacejka, motorcycle tires, 1998b). This also appears to hold for a limited number of conducted combined slip tests and the response to vertical axle motions at side slip and braking carried out with the measuring tower.

Apparently, the rigid ring model provided with the short wavelength transient slip model is very well capable of describing the dynamic tire behavior in the frequency range up to about 60 Hz. Furthermore, it may be concluded that the spin part is only necessary when dealing with short wavelength, especially low-speed phenomena where tire inertia is less important, such as with parking.

### *Time Domain Responses*

To demonstrate the performance of the model, simulations and experiments have been carried out pertaining to successive stepwise increases in brake pressure and steer angle.

The response of the longitudinal force and the associated wheel speed has been presented in Figure 9.45. The lower diagram clearly depicts the oscillatory variation of the force vs slip ratio. The measured response shows a faster decay of the wheel velocity after the highest brake effort has been reached. Apparently, this is due to the friction coefficient being lower in the experiment than assumed in the model. Finally, the brake is released and the wheel spins up again. The oscillations (ca. 28 Hz) correspond to the in-phase vibrational mode of the system, with the brake disc/axle inertia included.

Figure 9.46 shows the responses to successive changes in steer angle for both the side force and the aligning torque at a given load and speed. The responses clearly show vibrations attributed to the yaw/camber mode with natural frequency of ca. 40 Hz (cf. Figure 9.43, 25 km/h). Also, the decrease of the overall relaxation length at larger slip angle can be recognized.

Chapter 10

# Dynamic Tire Response to Short Road Unevennesses

**Chapter Outline**



The actual road surface profile over which the tire rolls may contain spectral components showing relatively short wavelengths. If the wavelength is smaller than two to three times the contact length, a geometric filtering of the profile becomes necessary if the tire model employed is assumed to contact the road in a single point.

## 10.1. MODEL DEVELOPMENT

For the *SWIFT* model a special filter has been developed that takes care of the envelopment properties of the tire and the variation in effective rolling radius

Tire and Vehicle Dynamics. DOI: 10.1016/B978-0-08-097016-5.00010-3

that occurs when the tire rolls over a short obstacle. The envelopment of an obstacle takes place in the contact zone. It is assumed that local dynamic effects can be neglected. The changing conditions that arise for the tire, while quasi-statically traversing an obstacle, are measured and modeled and subsequently used as effective inputs for the tire model also at higher speeds. The belt inertia takes care of the dynamic effects. The central item that is introduced is the effective road plane. Height, slopes, and curvature of the effective surface are used as inputs in the tire model. The orientation of the effective plane is defined such that the resulting force that would act on the assumedly frictionless road surface is directed normal to the effective road plane.

First, much attention will be given to developing the theory concerning rolling over two-dimensional unevennesses. Finally, rolling over three-dimensional unevennesses (such as an oblique cleat) will be addressed. In Section 10.2 model simulations are compared with experimental results.

### 10.1.1. Tire Envelopment Properties

In the literature one finds numerous publications on tire envelopment behavior. We refer to the study of Zegelaar (1998) for an extensive list of references. A number of these will be mentioned here. Important experimental observations have been made by Gough (1963). He indicated that the tire that is slowly rolled at constant velocity and axle height, over a cleat with length much smaller than the contact length, exhibits three distinct responses: (1) variations in the vertical force, (2) variations in the (horizontal) longitudinal force, and (3) variations in the angular velocity of the wheel. Lippmann et al. (1965,1967) studied the responses of both truck and passenger car tires rolling over short sharp unevennesses like cleats and steps of several heights. From the experimental observations it has been concluded that an almost linear relationship exists between tire force variation and step height. The superposition principle may therefore be employed to assess the response to an arbitrarily shaped unevenness by taking the sum of responses to a series of step changes in road surface height. These observations are expected to be approximately true if the obstacle height is not too large.

For three typical road unevennesses, depicted in Figure 10.1, Zegelaar has measured the responses of the tire at three different constant axle heights. To avoid dynamic effects, the velocity of the drum on which the cleat is attached (Figure 9.37) was maintained at the very low level of 0.2 km/h.

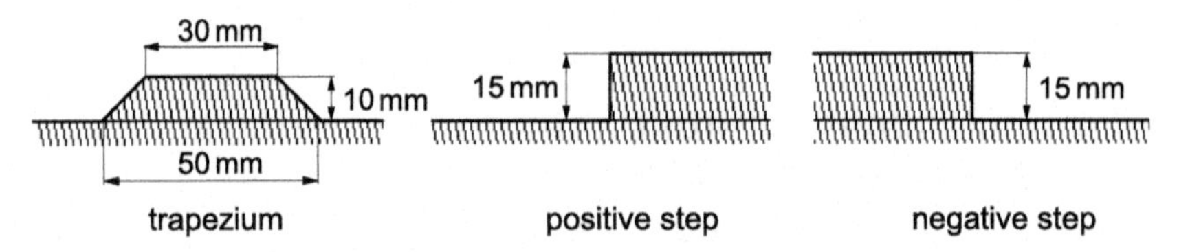


**FIGURE 10.1** Three typical obstacles used in Zegelaar's research.

Figure 10.2 presents the measured vertical load $F_V$, horizontal longitudinal force $F_H$, and the effective rolling radius $r_e$ as derived from the measurements. The latter quantity is obtained from the variation of the wheel rotation rate $d\theta/ds$, which is defined as: the difference of the incremental wheel angular

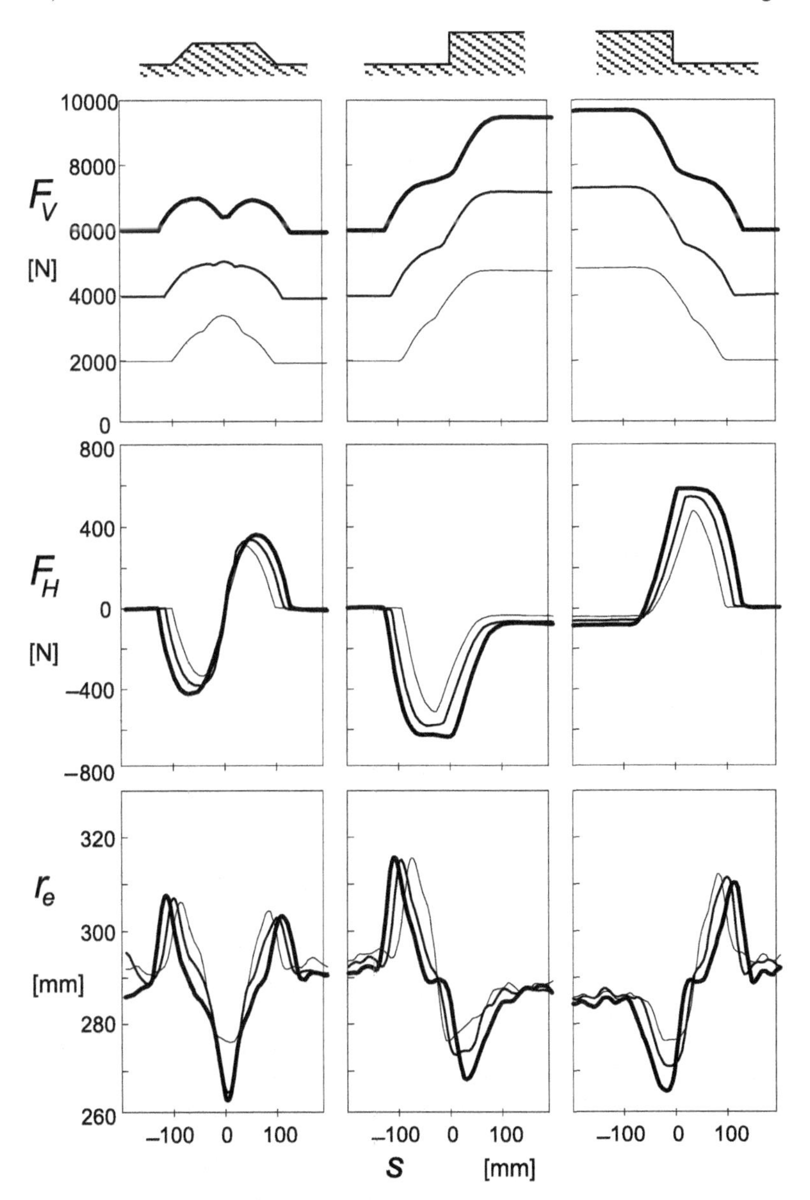


**FIGURE 10.2** Rolling over a trapezium cleat, an upward, and a downward step (Figure 10.1) at very low speed and for three axle heights. Diagrams show measured variations of the vertical force $F_V$ (upward +), the longitudinal horizontal force $F_H$ (forward +), and the effective rolling radius $r_e$, Eqn (10.2). Tire dimensions: 205/60R15.

displacement and the constant (undisturbed) incremental rotation, as a ratio to the increment of the traveled distance of the wheel axle. The following equations apply:

$$r_e = \frac{V_x}{\Omega}, \quad V_x = \frac{ds}{dt}, \quad \Omega = \Omega_o - \frac{d\theta}{dt}, \quad r_{eo} = \frac{V_x}{\Omega_o} \tag{10.1}$$

and hence

$$r_e \approx r_{eo}\left(1 + r_{eo}\frac{d\theta}{dt}\right) \tag{10.2}$$

The peculiar shapes of the various response curves correspond very well with results found in the literature. Several tire models have been used to simulate the experimental observations.

Davis (1974) has developed a model featuring independent radial springs distributed along the circumference. By giving the individual springs a non-linear degressive characteristic the model is able to generate a vertical force response curve with the typical dip that shows up when running at relatively high initial loads. Badalamenti and Doyle (1988) developed a model also consisting of radial springs but now with additional interradial spring elements that connect the end points of adjacent radial springs in the radial direction. When the deflections of neighboring radial springs are not equal to each other, the interradial spring generates a radial 'shear force' that acts on the end points of the radial springs. Mousseau et al. (1994) and Oertel (1997) simulate the tire rolling over a positive step by means of (different types of) finite element models; also cf. Gipser (1987, 1999). Zegelaar uses the flexible ring (belt) model of Gong (1993) that was developed with the aid of the modal expansion method, as a reference model in his research. The addition of tread elements with radial and tangential compliances to Gong's model did enable Zegelaar to employ the model for the study of traversing obstacles. Also this model shows responses very similar to the measured behavior. Schmeitz and Pauwellussen (2001) employ the radial interradial model as a possible basis for the pragmatic model running over an arbitrary road surface.

### 10.1.2. The Effective Road Plane Using Basic Functions

To arrive at a geometrically filtered road profile, Bandel and Monguzzi (1988) follow the idea of Davis and introduce the effective road plane. The effective plane height and slope variation may be established by conducting an experiment where the wheel is rolled at a very low velocity over an uneven road surface at constant axle height (with respect to a horizontal reference plane) and the forces are measured, cf. Figure 10.3. It is argued that the resulting force (with the rolling resistance force omitted) that acts upon the wheel axle is directed perpendicularly to the effective road plane. By dividing the variation of

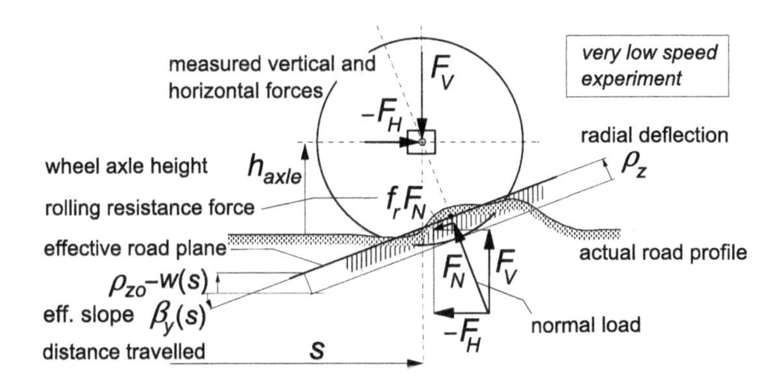


**FIGURE 10.3** The wheel rolled over a road irregularity at constant axle height to establish the effective road plane variation.

the measured vertical force, which is approximately equal to the vertical component of the normal force $F_N$, by the radial stiffness of the tire, the effective height variation, $-w$, is obtained. The effective slope, $\tan \beta_y$, is found by dividing the longitudinal horizontal force (after having subtracted the relatively small rolling resistance force) by the vertical force. Both effective quantities are functions of the longitudinal position $s$ of the wheel center. The following formulas apply for the effective height $w$:

$$-w = \frac{F_V - F_{Vo}}{C_{Fz}} \approx \frac{F_N \cos \beta_y - F_{Vo}}{C_{Fz}} = \rho_z \cos \beta_y - \rho_{zo} \tag{10.3}$$

In Section 10.1.5 a more precise definition and expression for the effective road plane height is given.

For the effective forward slope $\tan \beta_y$ we have

$$\tan \beta_y = -\frac{F_H + f_r F_N / \cos \beta_y}{F_V} \approx -\frac{F_H + f_r F_V}{F_V} \tag{10.4}$$

From Eqn (10.3) the approximate value (effect of small $f_r$ disregarded) of the radial deflection $\rho_z$ can be obtained. If needed, the actual effective road plane height, $w'$ defined below the wheel spin axis, may be assessed, cf. Section 10.1.5. For the description of the effective road input the pragmatic modeling approach initiated by Bandel et al. (1988) and further developed by Zegelaar and extended by Schmeitz is most useful and will be discussed below.

### The Basic Function Technique

Bandel discovered that the function representing the response of the change in vertical force to a short rectangular obstacle, featuring the dip at high load and the nipple at low load, can be decomposed into two identical basic functions, which are each other's mirror image. The basic functions are found to be approximately independent of the initial tire vertical deflection, that is: independent of the axle height. To find the force response curve at a possibly different axle height, the basic force curves are shifted with respect to each other over a distance a bit less than the contact length and then added together. By dividing by the radial stiffness of the tire the basic height functions are found and from these the effective height variations $w$. For the ratio of the measured longitudinal force and vertical load variations, a pair of basic functions can be assessed as well. Again, these are identical but the first must now be subtracted from the second to find the variation of the slope of the effective road plane $\tan\beta_y$, cf. Figure 10.4.

Zegelaar did experiments with the trapezium-shaped cleat, as indicated in Figure 10.1. He found basic functions, which are practically symmetric in shape. Mirror imaging was not necessary and did certainly not apply for non-symmetric unevennesses such as the step. Also, the basic functions assessed for the vertical force appeared to be practically the same as the ones for the longitudinal force. These findings helped a lot to make the principle of the basic function easier and more widely applicable. One basic function established from an experiment with a tire rolling over a given road irregularity at a fixed constant axle position should be sufficient to serve as the source for assessing the equivalent road plane height and slope. These two equivalent

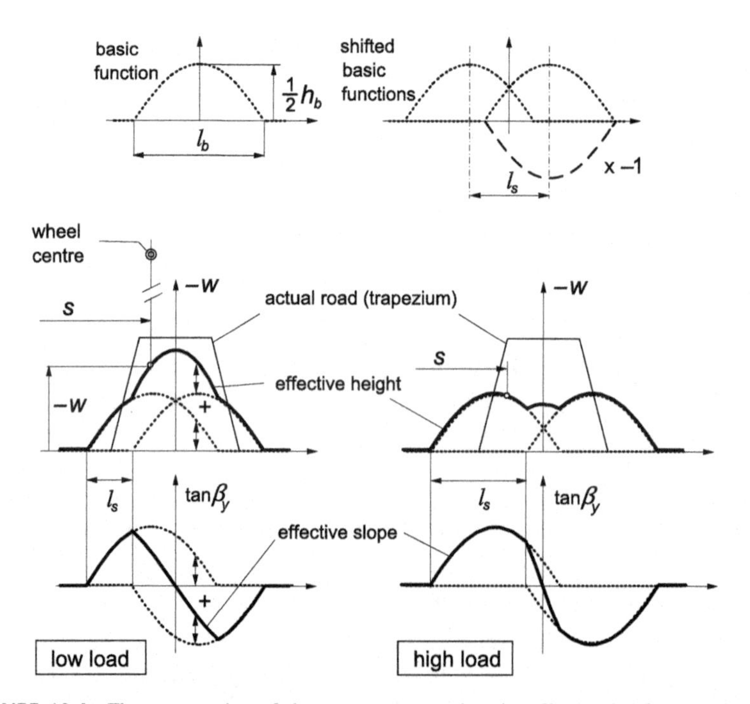


**FIGURE 10.4** The construction of the curves representing the effective height and effective slope from the basic function associated with the trapezium cleat.

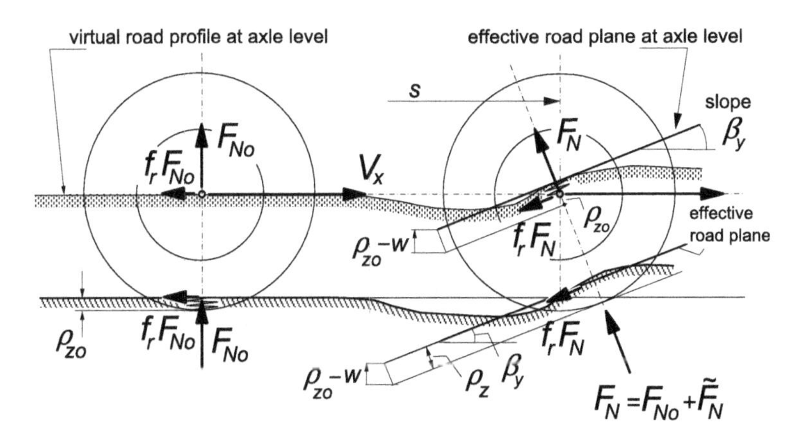


**FIGURE 10.5** The virtual road profile and effective road plane as sensed at wheel axle level.

quantities will later be extended with a third quantity: the effective forward road curvature that may significantly contribute to the variation of the effective rolling radius.

Figure 10.4 demonstrates the use of the basic function for the effective height applicable for the tire rolling over a short trapezoidal cleat at a constant axle height. The vertical scale has been exaggerated. The basic function is approximated by a half sine wave. The base length of the curve is denoted with $l_b$, its height with $0.5h_b$, and the shift with $l_s$. The values of $w$ and $\beta_y$ which vary with the travelled distance $s$ define the local effective road plane, cf. Section 10.1.5. At wheel position $s$ the current effective road plane has been indicated. The virtual road profile defined as the path of the wheel center that would occur at constant normal load has been drawn together with the actual road profile. The distance of the wheel center with respect to the virtual road profile corresponds to the increase of the actual radial tire deflection. The distance of the wheel center to the indicated effective road plane (translated to axle level with $w$ and $\beta_y$ taken into account) is the increase in effective radial deflection that, together with the initial deflection, becomes equal to $\rho_z$.

### The Two-Point Follower Concept

In Figure 10.6 an alternative technique is introduced using a single basic curve with full height $h_b$ and a two-point follower. If the two points are moved along the basic curve, the midpoint describes a curve that represents the characteristic for the effective height. The inclination angle of the follower corresponds to the slope of the effective road plane.

The response to a step change in road level may serve as a building block to compose the response to an arbitrary road unevenness. The corresponding basic function may be termed as the elementary basic function. The elementary basic curve may be represented by a quarter sine wave. For the steps given in

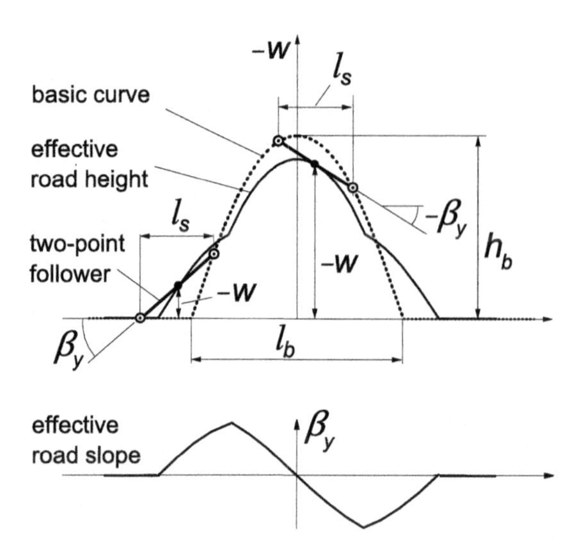


**FIGURE 10.6** Alternative method to determine the effective road height and slope using the basic curve with height $h_b$ and the two-point follower with length equal to the shift $l_s$.

Figure 10.1, the parameters of the basic curve have been determined by fitting the calculated force response to the vertical force variation measured for a series of axle heights (Figure 10.2).

Figure 10.7 shows the elementary basic curve for the upward step and the resulting effective road level and slope characteristics assessed by using the two-point follower technique.

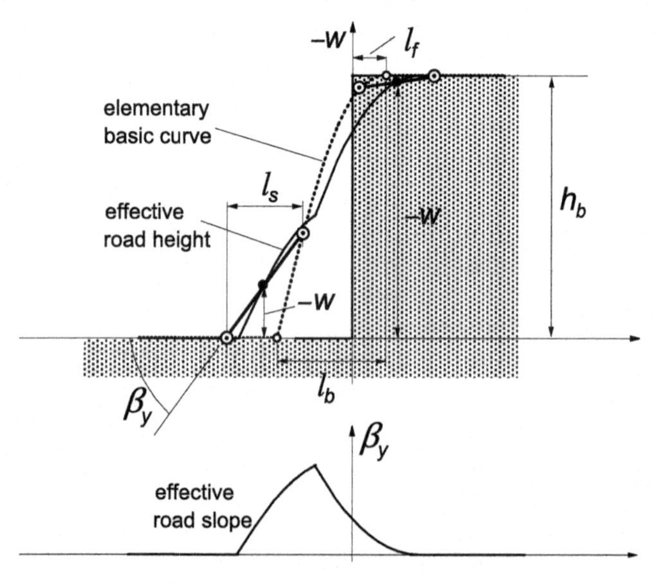


**FIGURE 10.7** The effective road response curves resulting from the (elementary) basic curve associated with the step change in road level by using the two-point follower concept.

For the downward step, the fitted basic function appears to come very close to the mirror image of the one determined for the upward step. The difference may very well be neglected. The parameters that control the size and position of the basic curve are the length $l_b$, the shift $l_s$, and so-called offset $l_f$. The height $h_b$ is equal to the step height. The offset is a new quantity that defines the position of the basic curve with respect to the step, cf. Figure 10.7.

A further important step is taken in the development of the assessment and use of basic functions. It is obvious that during the experiment that is performed at constant axle height, the normal load changes while rolling over an obstacle. The shift has been seen to change with axle height, that is: with a changed vertical load. The shift that corresponds to the length of the two-point follower has been found to be equal to a little less than the contact length. It seems therefore to be practical to adapt the strategy followed so far. We will henceforth define the basic curve to be assessed at constant vertical load. The experiments are to be carried out at constant load and at very low speed of travel. Schmeitz conducted such tests with the flat plank machine, cf. Section 10.2. The effective height variation follows directly from the experiment. It turns out (cf. Section 10.1.5) that $w$ now simply equals the change in axle height, $z_a$. Division of the vertical force by the radial stiffness is not needed anymore, which relieves us from accounting for a possibly non-linear tire compression characteristic, cf. Eqn (9.220). In addition, since the rolling resistance is now assumed to remain constant, it is no longer necessary to take account of the rolling resistance force when determining the effective slope, as was done in Eqn (10.4).

Zegelaar calculated the step response with the flexible ring model provided with tread elements and found good agreement with measured data. By fitting the quarter sine curve representing the basic curve, parameter values have been assessed for a wide range of step heights. The results have been compared with the values calculated for a rigid wheel or zero normal load. Figure 10.8 illustrates the extreme case of the rigid wheel rolling over a step.

In Figure 10.9 Zegelaar's calculated results have been presented for a series of vertical loads $F_V$. The diagrams show that the basic curve length $l_b$ and the offset $l_f$ do change with step height $h_{step}$ but are approximately independent of the vertical load $F_V$. The curve length may be estimated from the circle curve length, Figure 10.8:

$$l_b = \sqrt{r_o^2 - (r_o - h_{step})^2} \tag{10.5}$$

The horizontal shift $l_s$ amounts to approximately 80% of the contact length $2a$. The offset may be approximated by a linear function becoming zero at vanishing step height.

The approach of employing basic curves to assess the effective road height and slope as inputs to the dynamic tire model has been found adequate for the description of the response to single obstacles. Although, in principle, the

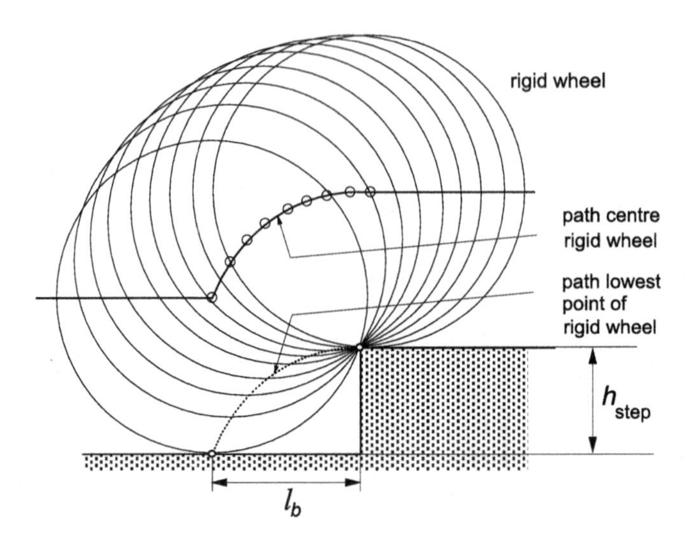


**FIGURE 10.8** The rigid wheel rolling over an upward step change in road level.

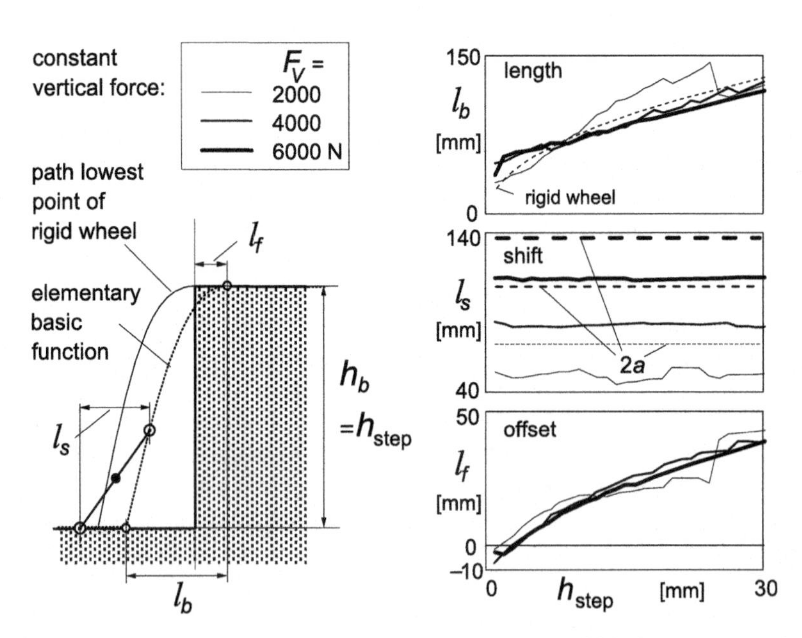


**FIGURE 10.9** Parameter values for the elementary basic curve established at constant vertical loads by computations using a flexible ring model with tread elements, Zegelaar (1998).

method may be used also for a series of obstacles or for an arbitrary road surface profile with the elementary basic curve (that holds for a step unevenness) as a building block, the rules that are to be followed may become rather cumbersome. For such a more general application, the method developed by

Schmeitz (2004), also cf. Schmeitz and Pacejka (2003) based on the so-called tandem cam technique is considered to be the best option.

### 10.1.3. The Effective Road Plane Using the 'Cam' Road Feeler Concept

Instead of using the basic profile and running over that with the two-point follower, we may more closely consider the actual tire shape that moves over the road surface profile. Schmeitz discovered that the principle of the circle moving over the surface (Figure 10.8) may be adopted but with an ellipse instead of the circle. This super ellipse that takes the shape of a standing egg has a height approximately equal to that of the tire but a radius of curvature at the lowest point smaller than that of the free tire. In that way, the 'cam' touches the step later than the circle would. By choosing an optimal shape of the ellipse, the role of the offset $l_f$ (that changes with step height) of the sine-based basic curve (Figure 10.9) can be taken care of automatically. Figure 10.10 depicts the cam moving over a step. The dimensions of the cam are defined by the super ellipse parameters. In terms of the coordinates $x_e$ and $z_e$ the ellipse equation reads

$$\left(\frac{x_e}{a_e}\right)^{c_e} + \left(\frac{z_e}{b_e}\right)^{c_e} = 1 \tag{10.6}$$

The effective road height and slope can be assessed by using two cams following each other at a distance equal to the shift length $l_s$. The change in height of the midpoint of the connecting line and the inclination of this line represent the effective height and effective slope, respectively.

Figure 10.11 illustrates this 'tandem cam' configuration. Fitting the tandem cam parameters follows from assessing the best approximation of low speed responses of a tire running over steps of different heights at a number of

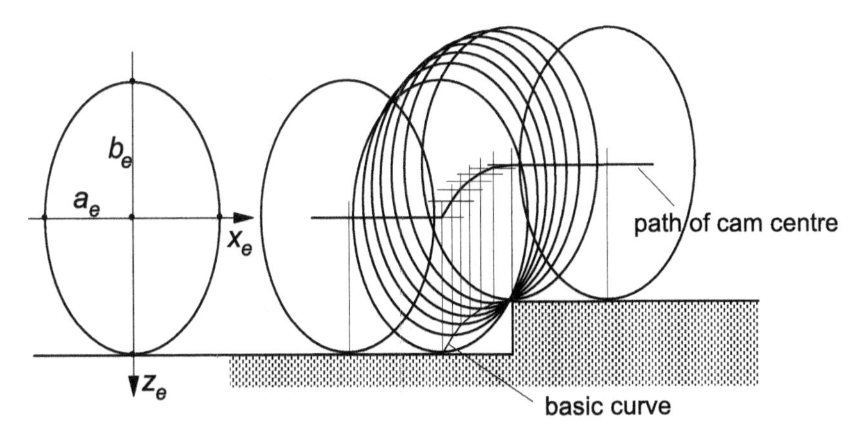


FIGURE 10.10 'Cam' moving over step road profile, producing a basic curve.

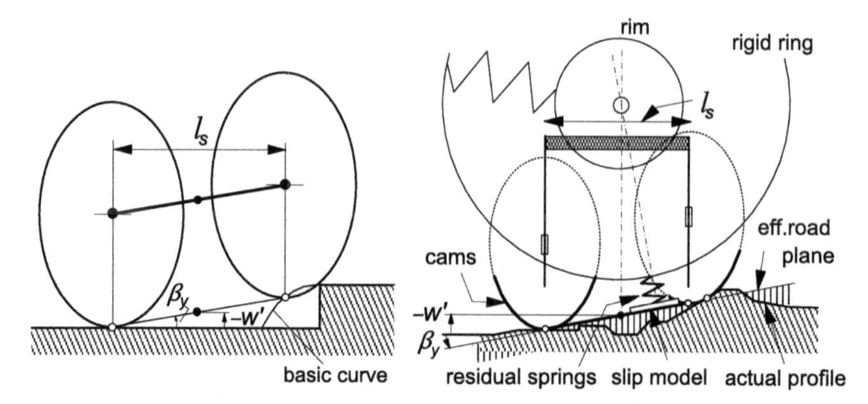


**FIGURE 10.11** The tandem cam configuration that generates the (actual) effective height and slope corresponding to the use of a basic curve and two-point follower. The total model including residual spring and rigid belt ring running over an arbitrary road profile.

constant vertical loads. Once the ellipse parameters have been established, the cam dimensions can be approximately considered to be independent of step height and vertical load. The tandem base length $l_s$, however, does depend on the vertical load. It is interesting that analysis shows that the lower part of the ellipse turns out to be practically identical to the contour of the tire in side view just in front of the contact zone up to the height of the highest step considered in the fitting process (Schmeitz 2004).

In a vehicle simulation, it may be more efficient to assess the basic profile first, that is: before the actual wheel rolls over the road section considered. This is achieved by sending one cam ahead over a given section of the road and having that determine the basic profile. The two-point follower is subsequently moved, concurrently with the actual wheel forward motion, over the basic profile, thereby generating the (actual) effective road inputs, $w'$ and $\beta_y$, which are fed into the tire model. Figure 10.12 illustrates the procedure. When

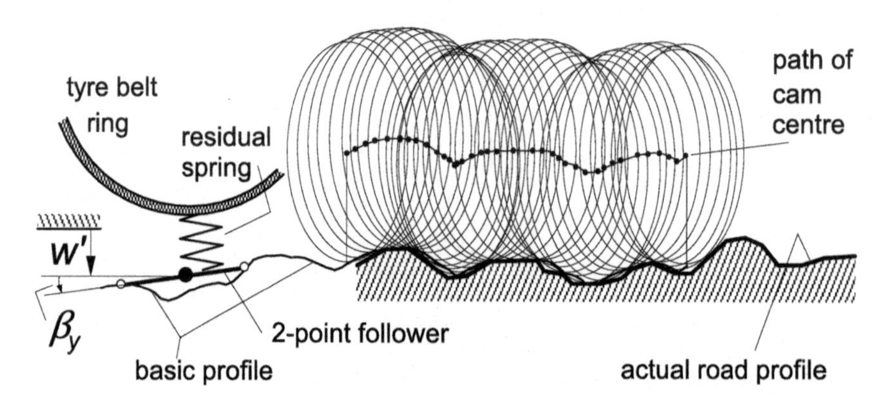


**FIGURE 10.12** The cam generating the basic road profile and two-point follower moving over it.

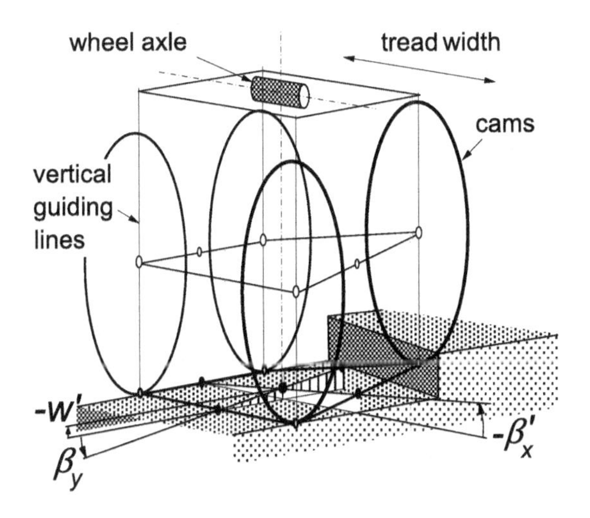


**FIGURE 10.13** The double-track tandem-cam road feeler moving over an oblique step.

traversing a single step it is, of course, more efficient to use an analytic expression for the basic curve based on Eqn (10.6), cf. Figure 10.11 left.

### *The Double-Track Tandem-Cam Road Feeler for Road Camber Variations*

For the assessment of the height and the forward and transverse slopes of the effective road plane, the double-track tandem cam road feeler will be introduced.

In Figure 10.13 the situation is depicted for an upright wheel rolling over an oblique step. The two-dimensional road feeler is used to determine the local effective two-dimensional road plane orientation in addition to the effective road plane height. See Section 10.1.5 for the slight difference between $\beta'_x$ and $\beta_x$. Schmeitz has conducted extensive experimental and model studies for a tire rolling over such types of non-symmetric road unevennesses. For more information, we refer to Schmeitz (2004), and Schmeitz and Pacejka (2003). To achieve more accurate results, additional cams (say three) may be introduced along the four edges when running over non-symmetrical obstacles, such as oblique steps or strips, Figure 10.14, exhibiting transverse slope variations with short wavelength ($\lambda <$ ca. 0.2 m); see Figure 10.15.

## 10.1.4. The Effective Rolling Radius When Rolling Over a Cleat

The third effective input is constituted by the effective road forward curvature that significantly changes the effective rolling radius when a road unevenness is traversed. In Figure 10.2 these variations have been shown in the lower diagrams. The curves are derived from measurements by using the Eqns (10.1) and (10.2).

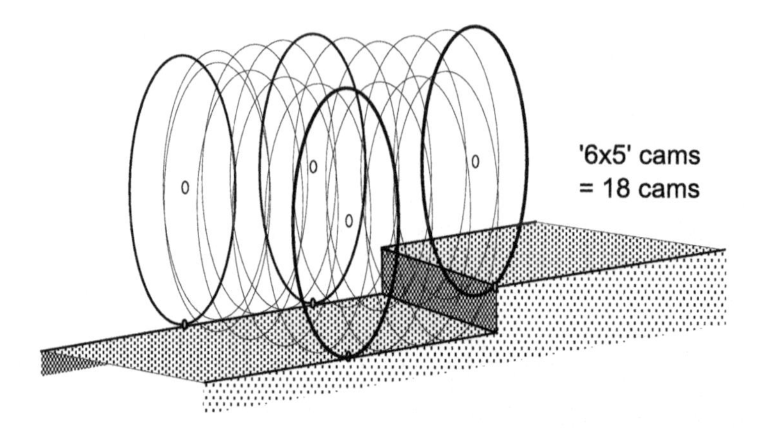


**FIGURE 10.14** Road feeler with more cams on the four edges to improve accuracy.

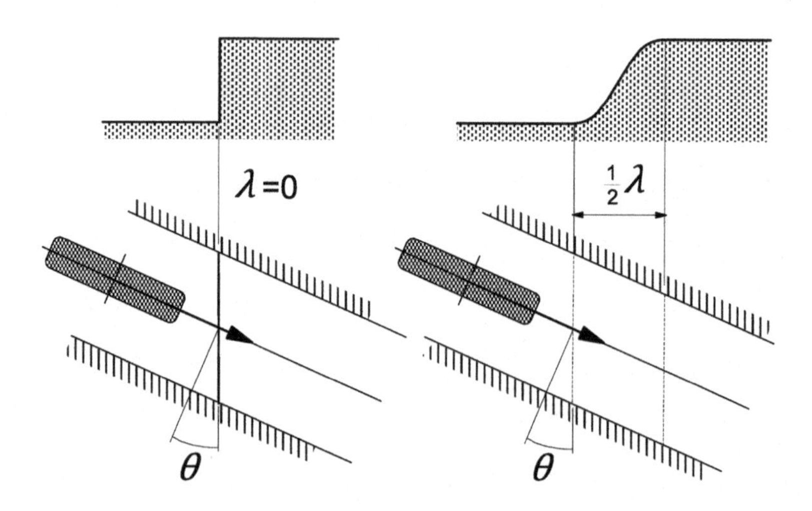


**FIGURE 10.15** Oblique step with zero and larger 'wavelength'.

In the effort to model the aspect of rolling over an obstacle, it is important to realize that we have three elements that contribute to the variation of the rolling radius:

1. Increment in normal load.
2. The local forward slope.
3. The local forward curvature.

Figure 10.16 illustrates the matter. The first item has been dealt with before, cf. Section 8.3.1, Figure 8.12, and Eqn (8.38) and Eqn (9.232). According to the latter equation the effective radius is a function of vertical load and speed of rolling. In Figure 10.17, model considerations and the graph resulting from

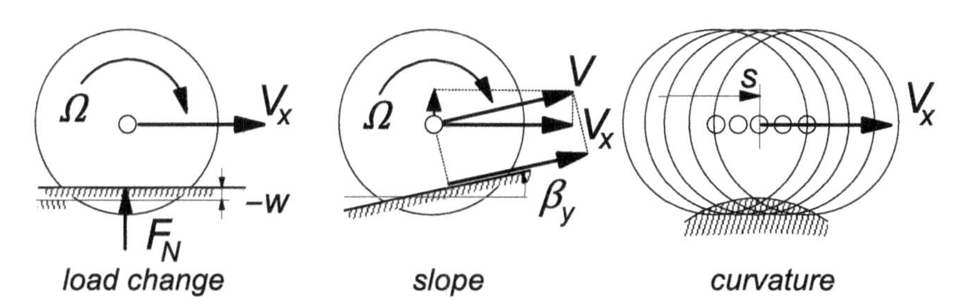


**FIGURE 10.16** Three contributions to the change in apparent effective rolling radius.

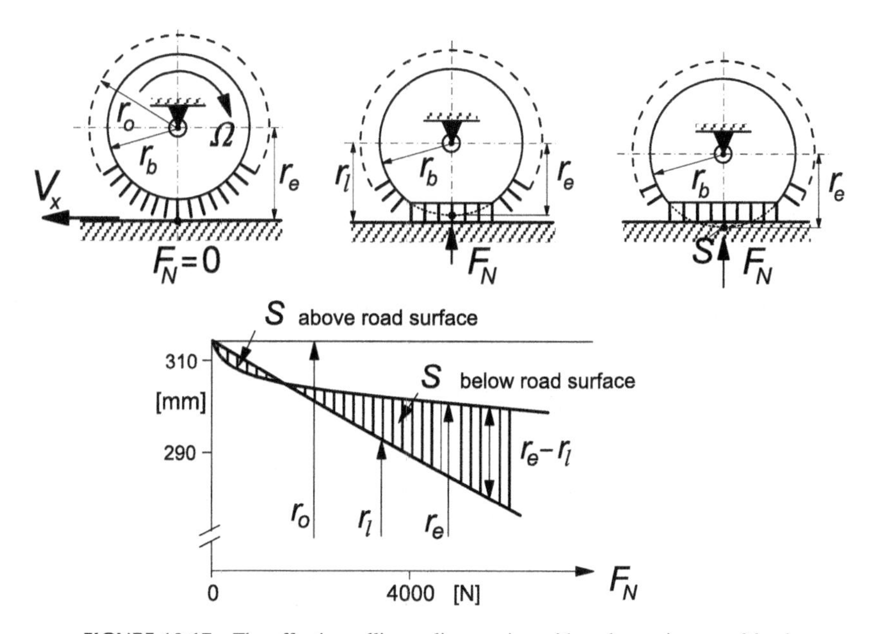


**FIGURE 10.17** The effective rolling radius varying with a change in normal load.

experiments have been repeated. For small variations in radial deflection we may employ the equation

$$\tilde{r}_{e\eta} = -\eta\tilde{\rho}_z \tag{10.7}$$

The second contribution accounts for the fact that at a slope and unchanged normal load, the axle speed parallel to the road surface is larger than the horizontal component $V_x$. We have for the change in the apparent effective rolling radius $r_e = V_x/\Omega$

$$\tilde{r}_{e,\text{slope}} = -r_{eo}(1 - \cos\beta_y) \tag{10.8}$$

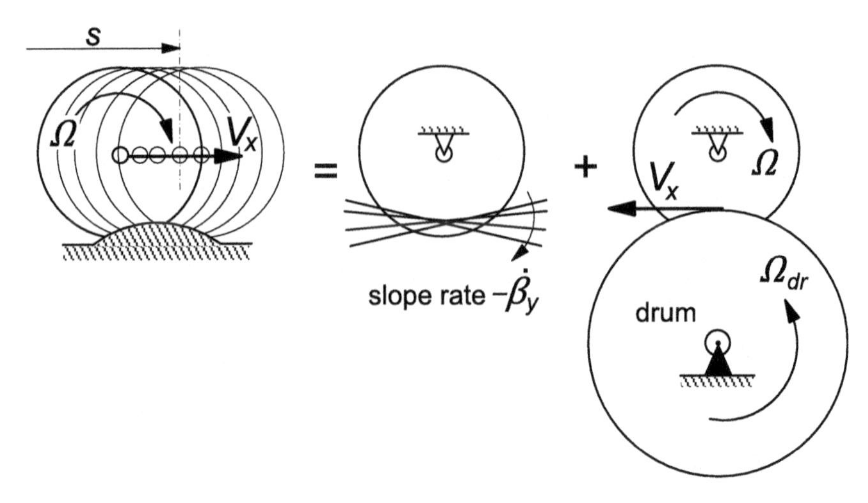


**FIGURE 10.18** Unraveling the process of rolling over a curved obstacle.

The third contribution comes from the road surface curvature. The analysis that attempts to model the relation between curvature and effective rolling radius is more difficult and requires special attention. Figure 10.18 unravels the process of rolling over a curved obstacle and indicates the connection with rolling over a drum surface with same curvature. In contrast to the drum, the obstacle does not rotate. Consequently, to compare the process of rolling over a curved obstacle with that of rolling over a rotating drum surface, we must add the effect of the tire supported by a counter-rotating surface that does not move forward. The left-hand diagram of Figure 10.19 depicts a possible test configuration with a plank that can be tilted about a transverse line in the contact surface. When being tilted, point $S$, that is attached to the wheel rim, must move along with the plank in the longitudinal direction since the wheel

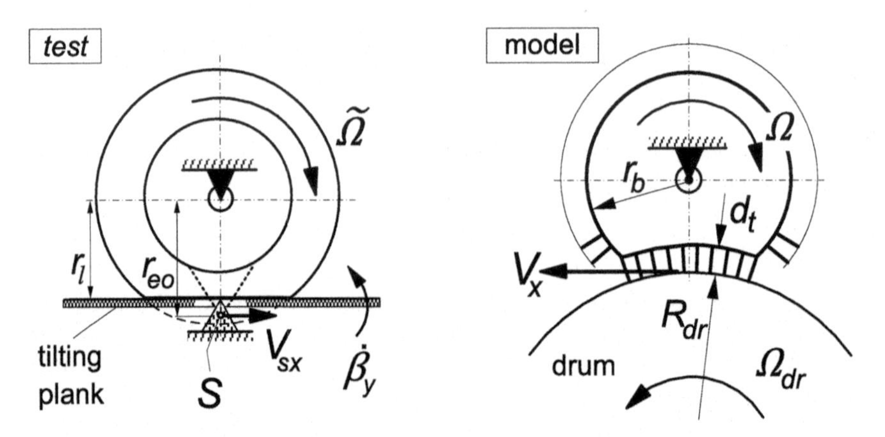


**FIGURE 10.19** Two components of rolling over a curved obstacle.

slip is zero ($V_{sx} = 0$), as brake or drive torque is not applied. Consequently, the wheel rotates slightly and the following equation applies:

$$V_{sx} = -(r_{eo} - r_l)\frac{\mathrm{d}\beta_y}{\mathrm{d}t} - r_{eo}\tilde{\Omega} = 0 \tag{10.9}$$

The apparent variation in effective rolling radius derives from the equation

$$\tilde{r}_e = \frac{V_x}{\Omega} - \frac{V_x}{\Omega_o} \approx -\frac{V_x}{\Omega_o^2}\tilde{\Omega} \tag{10.10}$$

so that we obtain for the contribution from the slope rate

$$\tilde{r}_{e,\text{slope rate}} = \frac{r_{eo} - r_l}{\Omega_o}\frac{\mathrm{d}\beta_y}{\mathrm{d}t} \tag{10.11}$$

The relationship (10.9) has been confirmed to hold through elaborate experiments conducted by Zegelaar (1998) on the tilting plank of the Delft flat plank machine, cf. Figure 12.5.

The other contribution that comes from the drum analog is found by considering the simple model shown in the right-hand diagram of Figure 10.19. The tread elements with length $d_t$ are assumed to stand perpendicularly on the drum surface. The drum has a curvature with radius $R_{dr}$. The belt with radius $r_b$ is considered inextensible. As a consequence, we find the following relation between the wheel and drum velocities:

$$\Omega r_b = \Omega_{dr}(R_{dr} + d_t) = V_x\left(1 + \frac{d_t}{R_{dr}}\right) \tag{10.12}$$

and, for the effective rolling radius for the tire rolling freely over the drum surface,

$$r_{e,\text{drum}} = \frac{V_x}{\Omega} = \frac{r_b}{1 + d_t/R_{dr}} \tag{10.13}$$

For the model, the effective rolling radius of the tire rolling over a flat surface is equal to the radius of the belt:

$$r_{eo} = r_b \tag{10.14}$$

Now, the tread depth is

$$d_t = r_o - r_{eo} \tag{10.15}$$

Hence, the expression for the effective rolling radius on the drum can be rewritten as

$$r_{e,\text{drum}} = \frac{r_{eo}}{1 + (r_o - r_{eo})/R_{dr}} \approx r_{eo}\left(1 - \frac{r_o - r_{eo}}{R_{dr}}\right) \tag{10.16}$$

The variation of the radius becomes

$$\tilde{r}_{e,\text{drum}} = -(r_o - r_{eo})\frac{r_{eo}}{R_{dr}} \tag{10.17}$$

The drum radius is equal to the radius of curvature of the (effective) road surface profile. This curvature corresponds to change in slope $\beta_y$ with traveled distance $s$. Note that for the convex drum surface the $\beta_y$ rate of change is negative. So, we have

$$\frac{1}{R_{dr}} = -\frac{\mathrm{d}\beta_y}{\mathrm{d}s} \tag{10.18}$$

The slope rate of change may be written as

$$\frac{\mathrm{d}\beta_y}{\mathrm{d}s} = \frac{1}{V_x}\frac{\mathrm{d}\beta_y}{\mathrm{d}t} = \frac{1}{\Omega_o r_{eo}}\frac{\mathrm{d}\beta_y}{\mathrm{d}t} \tag{10.19}$$

The drum contribution is now expressed as

$$\tilde{r}_{e,\text{drum}} = \frac{r_o - r_{eo}}{\Omega_o}\frac{\mathrm{d}\beta_y}{\mathrm{d}t} \tag{10.20}$$

Adding up the two contributions (10.11) and (10.20) yields the variation in effective rolling radius due to obstacle forward curvature:

$$\begin{aligned}\tilde{r}_{e,\text{curvature}} &= \tilde{r}_{e,\text{slope rate}} + \tilde{r}_{e,\text{drum}} \\ &= \frac{r_o - r_l}{\Omega_o}\frac{\mathrm{d}\beta_y}{\mathrm{d}t} = \rho_z r_{eo}\frac{\mathrm{d}\beta_y}{\mathrm{d}s}\end{aligned} \tag{10.21}$$

where $\rho_z$ is the radial compression of the tire. By adding up all the contributions, we finally obtain for the variation of the effective rolling radius with respect to the initial condition, where $\beta_y = \mathrm{d}\beta_y/\mathrm{d}s$ and $\rho_z = \rho_{zo}$:

$$\tilde{r}_e = -\eta\tilde{\rho}_z - r_{eo}(1 - \cos\beta_y) + \rho_z r_{eo}\frac{\mathrm{d}\beta_y}{\mathrm{d}s} \tag{10.22}$$

The term with the effective forward curvature $\mathrm{d}\beta_y/\mathrm{d}s$ constitutes by far the most important contribution to the effective rolling radius variation and thus to the wheel rotational acceleration that can only be brought about by a variation in the longitudinal force $F_x$. This force also often outweighs the part of the horizontal longitudinal force $F_H$ that directly results from the slope itself, cf. Eqn (10.4).

### *The Five Effective Road Inputs*

For the simulation of a tire rolling over an irregular road surface it turns out that, all in all, five effective road inputs are required. These are:

- The effective road plane height $w'$
- The effective road plane forward slope $\beta_y$

- The effective road plane transverse slope $\beta_x$
- The effective forward road curvature $d\beta_y/ds$
- The effective road surface warp $d\beta_x/ds$ (Section 10.1.5)

These effective inputs can be established with the use of the two-dimensional (multi) cam road feeler. In Section 10.2 the effective road quantities are used as inputs to the advanced dynamic tire model *SWIFT.*

If we have $n$ cams distributed in the longitudinal direction along the total shift length $L$ and $m$ cams laterally along the tread width $B$, the calculation of the effective inputs may be indicated as follows (subscript $L$: left, $R$: right, 1: front, 2: rear; indices $i$ $(1 \ldots n)$ and $j$ $(1 \ldots m)$ start at rear and left, respectively). The height of lowest point of the individual cam is denoted with $w'_{2L}$, etc.:

$$\begin{aligned} w' &= \sum(w'_{1j} + w'_{2j})/2m \\ \beta_y &= \sum(w'_{2j} - w'_{1j})/mL \\ \beta_x &= \sum(w'_{Ri} - w'_{Li})/nB, \\ d\beta_y/ds &= (\beta_y(s) - \beta_y(s - \Delta s))/\Delta s, \\ d\beta_x/ds &= (\beta_x(s) - \beta_x(s - \Delta s))/\Delta s \end{aligned} \tag{10.22a}$$

## 10.1.5. The Location of the Effective Road Plane

The relationship that exists between the effective height $w$ and the actual height $w'$ of the effective road plane has not been addressed so far. Below, a more precise account will be given of the notion of the effective road plane with a clear definition of the effective road plane height.

In Figures 10.3 and 10.5, the effective height $-w$ is considered as being assessed at a constant axle height above the reference plane that coincides with the initial flat level road surface. In Figure 10.20, the alternative case is illustrated where the effective height is assessed at constant vertical load $F_V$. The following formula covers both cases. The effective height is defined as

$$w = z_a - \rho_{zV} \tag{10.23}$$

where $z_a$ denotes the vertical axle displacement and $\rho_{zV}$ the vertical tire deflection when loaded on a flat level road with load $F_V$. With an assumed linear tire spring characteristic we get

$$\rho_{zV} = \frac{F_V}{C_{Fz}} \tag{10.24}$$

For the case that the effective height is found at a constant vertical load $F_V$, the initial vertical deflection is $z_{ao} = \rho_{zo} = \rho_{zV}$. Consequently by considering (10.24), the variation in effective height equals the change in axle height.

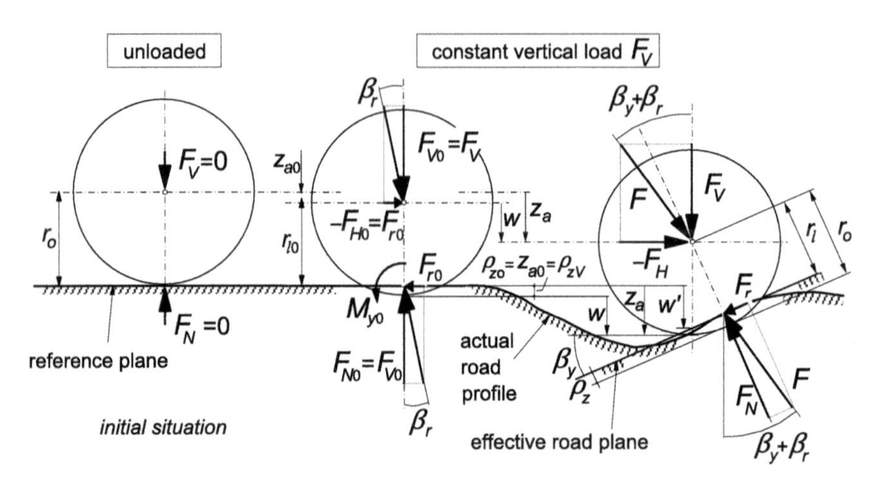


**FIGURE 10.20** The effective road plane assessed at constant vertical load.

If the axle height is kept constant, we have $z_a = z_{ao} = \rho_{zo}$ and the formula becomes: $w = \rho_{zo} - \rho_{zV}$, which corresponds with Eqn (10.3).

In Figure 10.20, the actual position of the effective road plane is defined as the location of the point of intersection of effective road plane and the vertical line through the center of the vertical wheel. Its height below the horizontal reference plane is designated as $w'$, also indicated in Figure 10.11.

With the small effect of the rolling resistance force $F_r = f_r F_N$ first included and then neglected, the normal force becomes in terms of the vertical load

$$F_N = \frac{F_v}{\cos\beta_y - f_r \sin\beta_y} \approx \frac{F_V}{\cos\beta_y} \tag{10.25}$$

Consequently, the normal deflection becomes

$$\rho_z = \frac{F_N}{C_{Fz}} = \frac{\rho_{zV}}{\cos\beta_y - f_r \sin\beta_y} \approx \frac{\rho_{zV}}{\cos\beta_y} \tag{10.26}$$

The expression for the effective road plane height $w'$ follows from Figure 10.20 by inspection. With the vertical distance

$$z_a - w' = \frac{\rho_z}{\cos\beta_y} + r_o \frac{1 - \cos\beta_y}{\cos\beta_y} \tag{10.27}$$

we obtain, using (10.23) and (10.26)

$$w' \approx w - \rho_{zV} \tan^2\beta_y - r_o \frac{1 - \cos\beta_y}{\cos\beta_y} \tag{10.28}$$

For vanishing forward slope, we correctly find

$$w' \to w \quad \text{for} \quad \beta_y \to 0 \tag{10.29}$$

For small values of forward slope, the approximation (10.29) is perfectly acceptable.

## *Road and Wheel Camber*

As indicated in the middle diagram of Figure 10.21, the transverse slope may be detected by a double-track tandem cam 'road feeler'. The four cams are guided along vertical lines that are positioned symmetrically with respect to the two vertical planes, one passing through the wheel spin axis and the other through the line of intersection of the wheel center plane and the horizontal plane that may be approximately defined to pass through the lowest point of the tire undeformed peripheral circle, that is at a distance equal to $r_o$ from the wheel center.

With the effective road plane height and orientation properly defined, we have available the height $w'$ of the point of the effective road plane vertically below the wheel center (or if needed at large camber: defined vertically above the lowest point of the undeformed tire) and the two slopes $\beta_x$ and $\beta_y$. Through these quantities and the wheel axle location and orientation, the normal to the road $\boldsymbol{n}$, the loaded radius $r_l$, and the position vector $\boldsymbol{c}$ of $C$, see Section 2.2, can

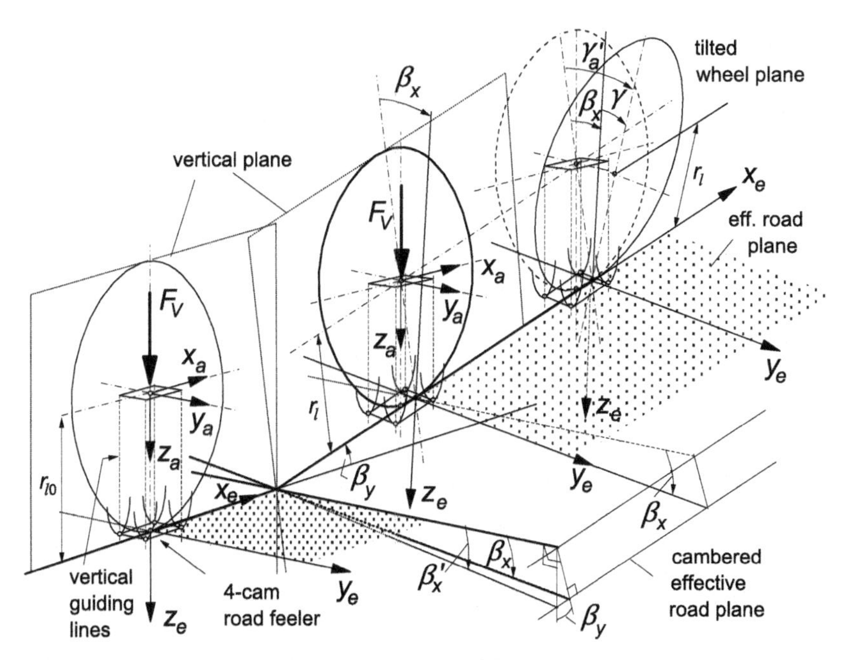


**FIGURE 10.21** The effective road plane showing forward and transverse slope angles $\beta_y$ and $\beta_x$. Wheel plane in the vertical and in the tilted position. Road feeler senses angles $\beta_y$ and $\beta_x'$.

be assessed. With the wheel camber angle $\gamma$ defined with respect to the normal to the local effective road plane and the global inclination angle $\gamma'$ with respect to the vertical plane through the line of intersection (*l*) of road plane and wheel plane, the side slip velocity $V_{sy}$ can be calculated with Eqn (10.30). In case one is interested in the non-lagging part, Eqn (10.31) should be used as well.

The inclusion of the non-lagging part, as discussed in Section 7.2.4, is needed to get good agreement with e.g., low speed flat plank test results. The effect shows up when the effective road plane changes in height and exhibits a varying transverse slope angle $\beta_x$, cf. Figure 10.22. The expressions for the two side slip velocity components parallel to the local effective road plane ($y_e$ axis), cf. Eqns (7.42a,b) with (7.41), now become

$$\begin{aligned} V_{sy} &= V^e_{aye} + \left(V^e_{aze}\sin\gamma - r_l\frac{d\gamma^e}{dt}\right)\frac{1}{\cos\gamma} \\ &= V_{aye} + \left(V_{aze}\sin\gamma - r_l\frac{d\gamma'}{dt}\right)\frac{1}{\cos\gamma} \end{aligned} \tag{10.30}$$

$$\begin{aligned} V_{sy1} &= V^e_{aye} - r_{sy}\frac{d\gamma^e}{dt}\cos\gamma \\ &= V_{aye} - \left\{r_{sy}\frac{d\gamma'}{dt} + (r_l - r_{sy})\frac{d\beta_x}{dt}\right\}\cos\gamma \end{aligned} \tag{10.31}$$

where $d\gamma'/dt$ is the time rate of change of the wheel global tilt angle. The quantities $V_{aye}$ and $V_{aze}$ represent the components of the global axle velocity parallel to the system of current effective road axes ($x_e$, $y_e$, $z_e$). Furthermore, $d\gamma^e/dt = d\gamma'/dt - d\beta_x/dt$ and $V^e_{aye}$ and $V^e_{aze}$ represent the sum of $V_{aye}$ and $V_{aze}$ and the components of the velocity of the virtual axle motion (due to $d\beta_x/dt$) relative to ($x_e$, $y_e$, $z_e$) and parallel to this system of axes. The necessity for including the last term of (10.31) becomes clear when considering the case

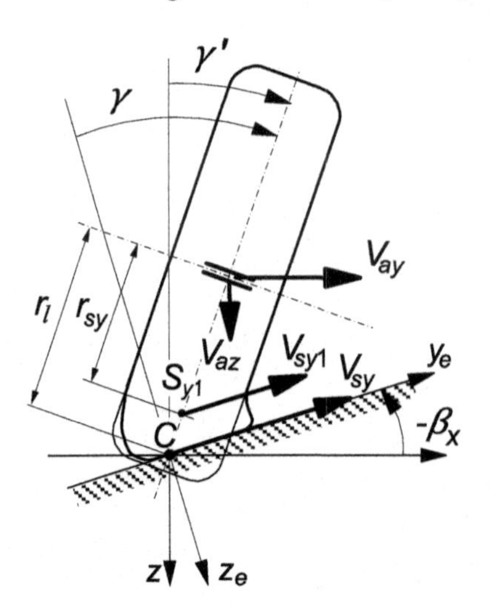


**FIGURE 10.22** Side slip velocity components at road and wheel camber.

that $V_{ay,z}$ and $\mathrm{d}\gamma'/\mathrm{d}t$ remain equal to zero. This situation corresponds with the right-hand case *C* depicted in Figure 7.13. It may be noted that $\mathrm{d}\beta_x/\mathrm{d}t = V_x \mathrm{d}\beta_x/\mathrm{d}s$, with $\mathrm{d}\beta_x/\mathrm{d}s$ representing the effective warp of the road surface along the forward axis.

## 10.2. *SWIFT* ON ROAD UNEVENNESSES (SIMULATION AND EXPERIMENT)

Zegelaar and Schmeitz have performed numerous experiments on the drum test stand (Figure 9.37) and the flat plank machine (Figures 10.23 and 12.6) and used the *SWIFT* model (including enveloping model based on the effective road plane concept) to carry out the simulations.

### 10.2.1. Two-Dimensional Unevennesses

Figure 10.9 shows the obstacle parameter values for the quarter sine basic curve of Zegelaar. Table 10.1 gives the parameters used by Schmeitz, which are based on the ellipse concept (tandem cam technique). The table shows that the height

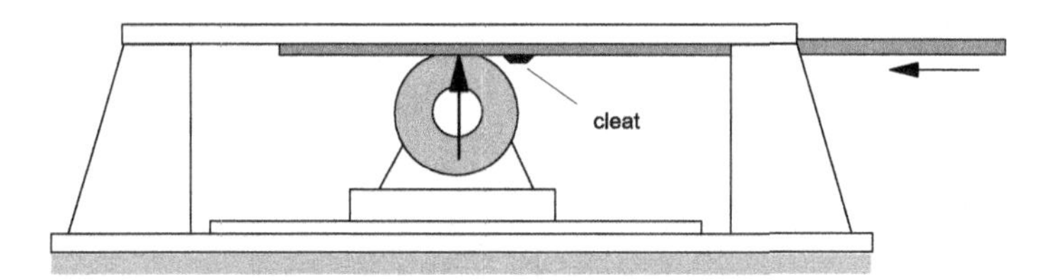


**FIGURE 10.23** Principle of the flat plank machine provided with a trapezium shaped cleat.

**TABLE 10.1 Parameter Values Used for Envelopment Calculations. Tire: 205/60R15, 2.2bar**

| | | |
|---|---|---|
| unloaded radius | $r_o$ | 0.310 m |
| effective rolling radius at $F_N = 4000$ N | $r_{eo}$ | 0.305 m |
| slope effective rolling radius characteristic | $\eta$ | 0.3 |
| vertical tire stiffness at $F_N = 4000$ N | $C_{Fz}$ | 220 N/mm |
| rolling resistance, cf. Eqns (9.230, 9.231), $f_r = q_{sy1}$ | $q_{sy1}$ | 0.01 |
| half contact length, cf. Eqn (9.207) | $a$ | cf. Table 9.3 |
| half ellipse length/unloaded radius $a_e/r_o$ | $p_{ae}$ | 1.0325 |
| half ellipse height/unloaded radius $b_e/r_o$ | $p_{be}$ | 1.0306 |
| ellipse exponent $c_e$ | $p_{ce}$ | 1.8230 |

and length of the ellipse are slightly larger than the dimensions of the free tire. It is the exponent that gives rise to the larger curvature of the ellipse near the ground.

Figure 10.24 presents the measured and calculated variations of the effective height $w$ (equal to the variations of the vertical axle displacement $z_a$ at

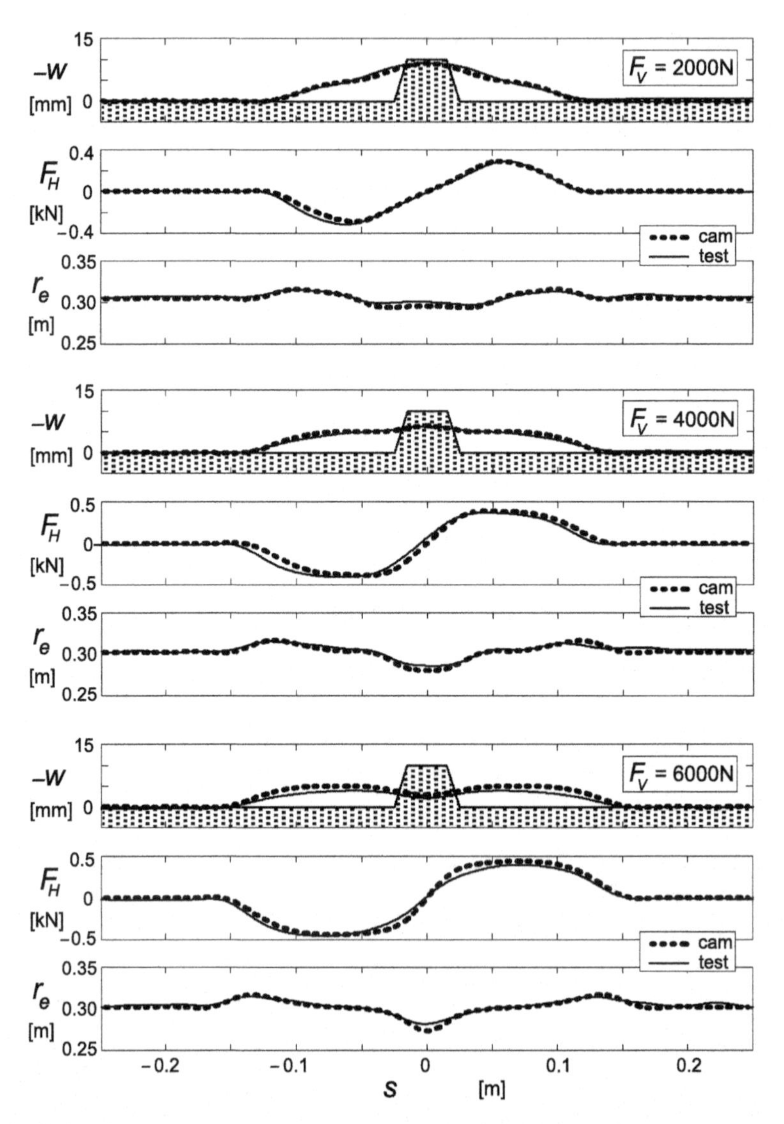


**FIGURE 10.24** Rolling over a trapezium cleat (length: 50 mm, height: 10 mm). Measured and calculated variation of vertical axle displacement ($\Delta z_a = w$), horizontal longitudinal force, and effective rolling radius at three different constant vertical loads. Measurements carried out on the Delft flat plank machine (very low speed) and calculations conducted with the use of the tandem cam technique (Figure 10.11) and Eqn (10.22). Tandem cam parameters according to Table 10.1.

constant vertical load $F_V$), the horizontal fore-and-aft force $F_H$, and the effective rolling radius $r_e$, with the vertical load kept constant while slowly rolling over the trapezium cleat. The effective road plane slope tan $\beta_y$ follows from the ratio of the horizontal force variation and the vertical load. The calculations are based on the two-cam tandem concept of Figure 10.11. The tandem is moved over the original road profile and the effective height and slope are obtained. From the derivative of $\beta_y$, the effective rolling radius is found by using Eqn (10.23), which also contains the two very small additional contributions. It is observed that a good agreement between test and calculation results can be achieved. The use of the quarter sine basic curve function gives very similar results.

Figures 10.25–10.29 present the results of rolling over the same cleat at different speeds while the axle location is kept fixed. The responses of the vertical, fore, and aft forces and the wheel angular speed have been indicated. In addition, the power spectra of these quantities have been shown.

Especially at the higher loads the calculated responses appear to follow the measured characteristics quite well up to frequencies around 50 Hz or higher. Figure 10.30 demonstrates the application to a more general road surface

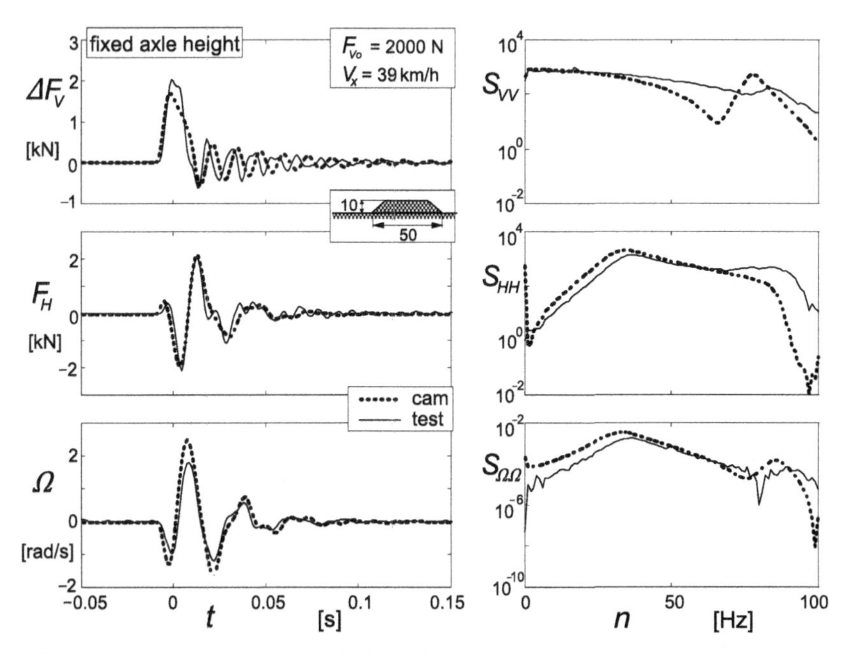


**FIGURE 10.25** Time traces and power spectra of vertical, horizontal, and longitudinal force and angular velocity for wheel running at given speed of 39 km/h and initial vertical load of 2000 N over a trapezium cleat at constant axle height using *SWIFT* tire parameters and obstacle parameters of Table 10.1. (Schmeitz).

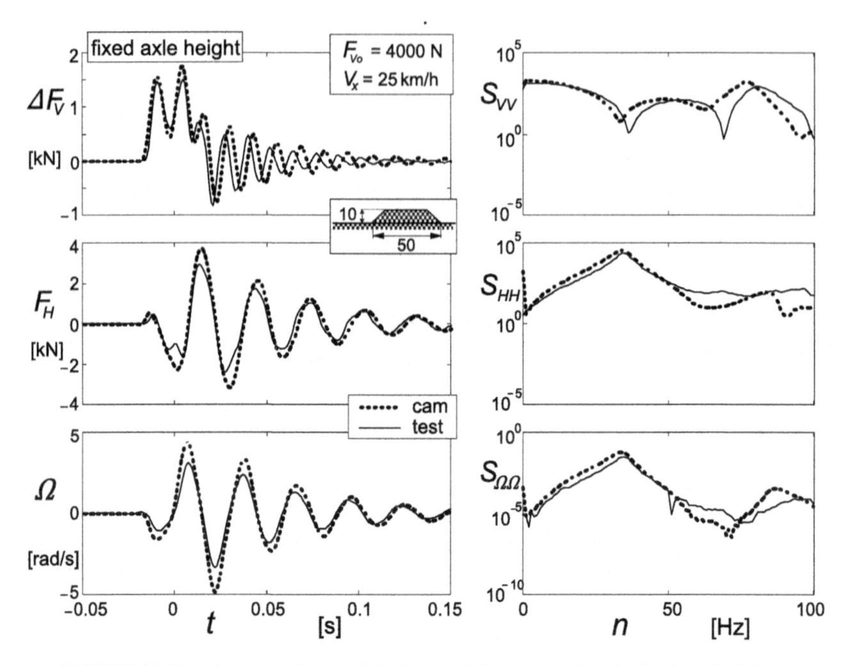


**FIGURE 10.26** Same as Figure 10.25 but at different speed and initial vertical load.

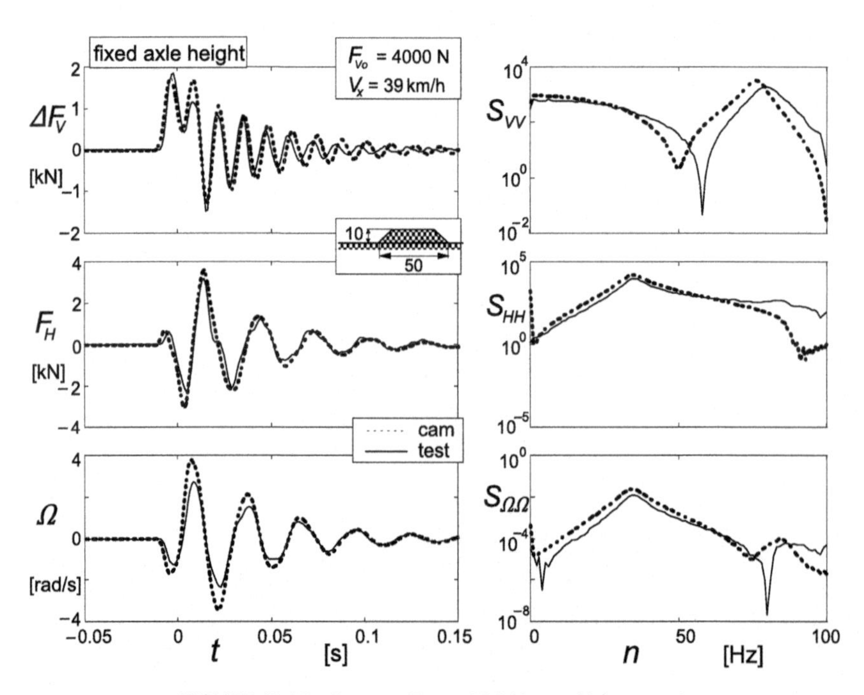


**FIGURE 10.27** Same as Figure 10.26 but at higher speed.

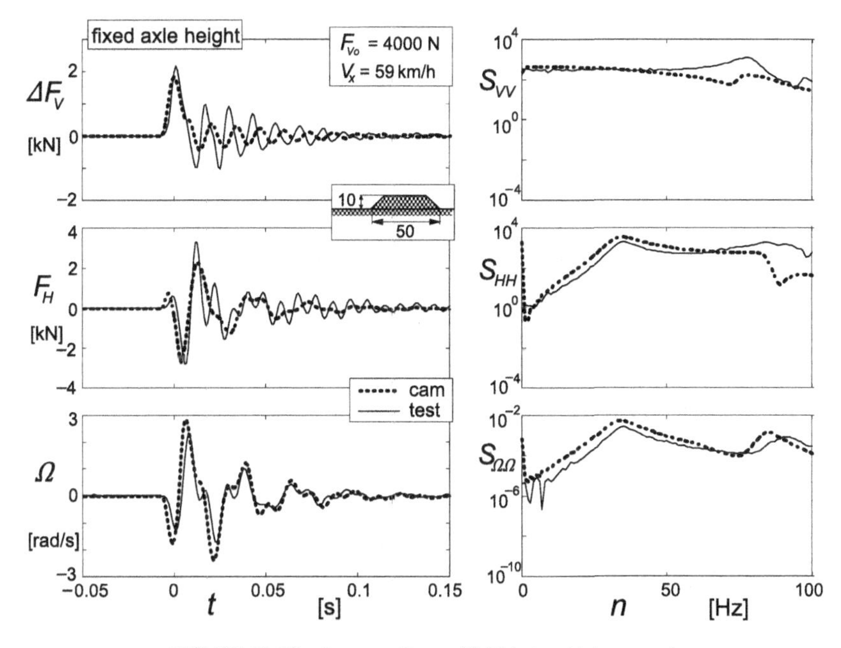


**FIGURE 10.28** Same as Figure 10.27 but at higher speed.

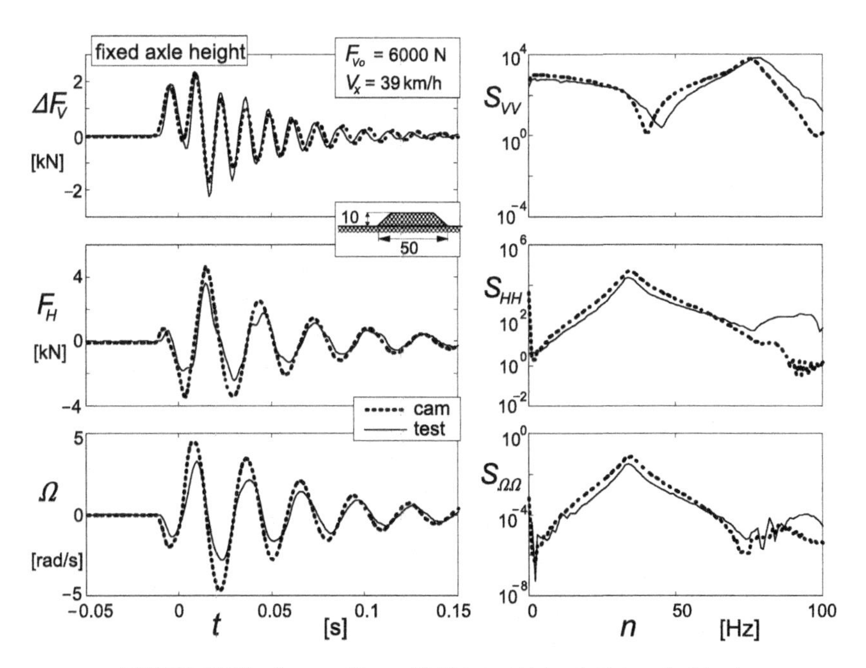


**FIGURE 10.29** Same as Figure 10.27 but at higher initial vertical load.

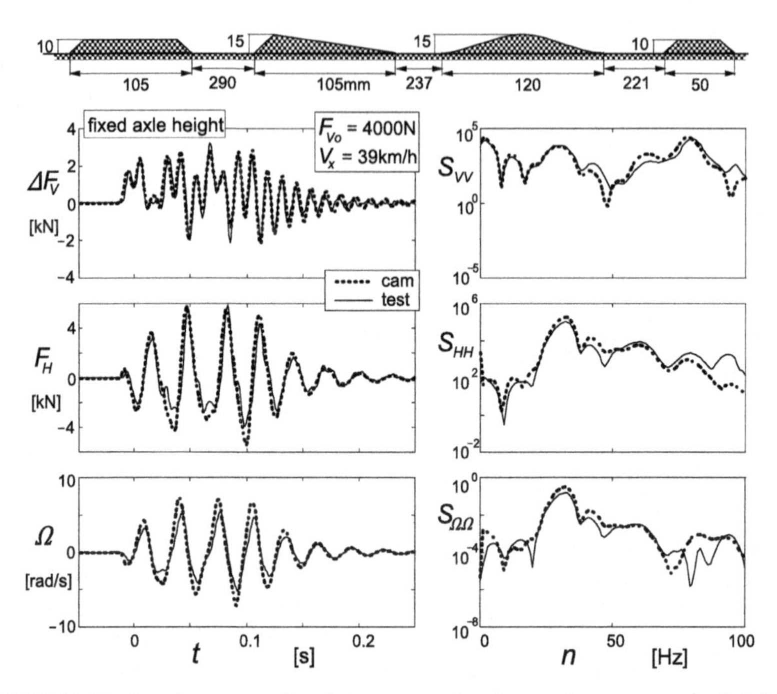


**FIGURE 10.30** Running over a series of cleats mounted on drum surface, at constant axle height.

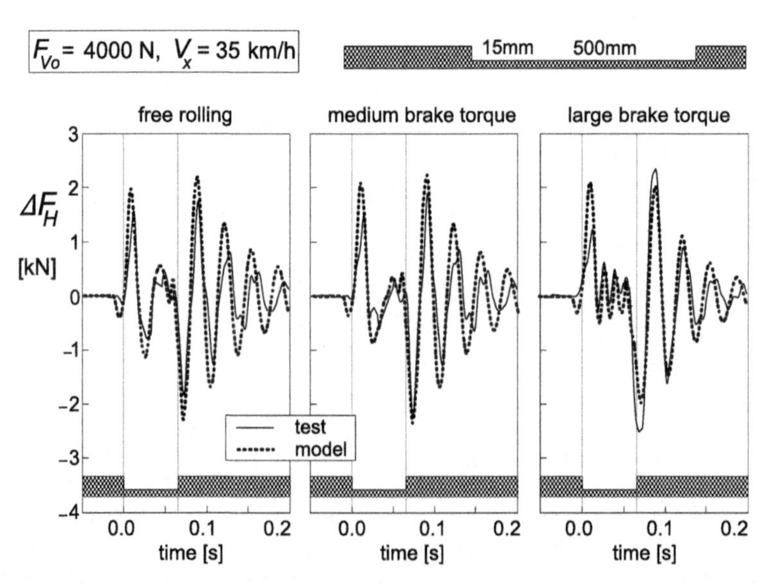


**FIGURE 10.31** The horizontal force variation when traversing a pothole while the wheel is being braked at three different levels of brake torque ($M_B$ = ca. 4, 375, 850 Nm respectively, Zegelaar 1998). Experiments on 2.5 m drum, calculations using parameters according to Figure 10.9.

profile. It shows the responses of the tire when moving over a series of different types of cleats that resembles an uneven stretch of road.

Finally, in Figure 10.31 the test and simulation results conducted by Zegelaar (1998) have been depicted, representing a more complex condition where the tire is subjected to a given brake torque (brake pressure) while the

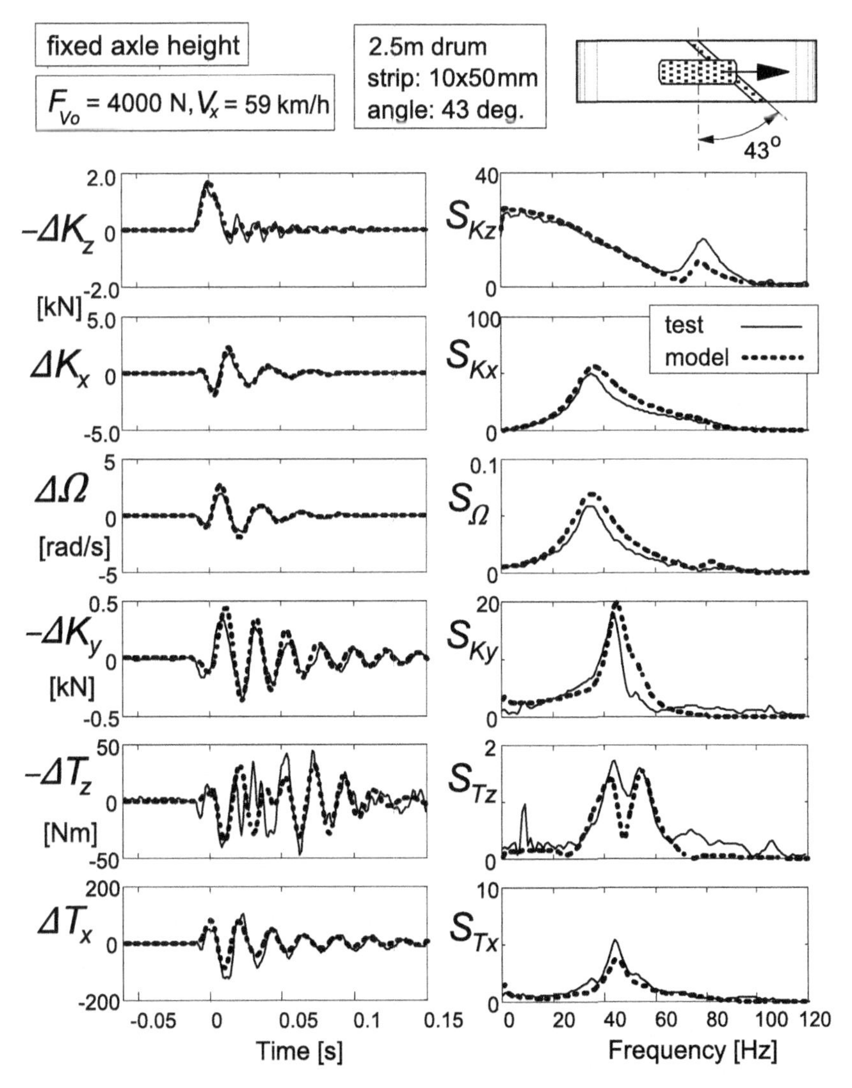


**FIGURE 10.32** Running over an oblique cleat (10 × 50 mm strip) mounted on the drum surface at an angle of 43 degrees, at constant axle height. Measured and computed variations of wheel spindle forces $K$ and moments $T$ and wheel angular speed $\Omega$ versus time and as power spectral densities $S$.

wheel rolls over a pothole at a fixed axle location. The complex longitudinal force response conditions that are brought about by load and slip variations induced by tire modal vibrations and road unevennesses are simulated quite satisfactorily using the *SWIFT* model including obstacle geometric filtering.

### 10.2.2. Three-Dimensional Unevennesses

In Figure 10.32 example results are presented of experimental and model simulation results of the tire moving at constant axle height at a speed of 59 km/h over an oblique strip (height 10 mm, width 50 mm) mounted at an angle of 43 degrees with respect to the drum axis (wheel slip angle and wheel camber angle remain zero). The variations of the wheel spindle forces $K_{x,y,z}$ and moments $T_{x,z}$ and the wheel angular velocity $\Omega$ are presented as functions of time and as power spectral densities versus frequency. Note: The output forces and moments $K_{x,y,z}$ and $T_{x,z}$ are the same as the quantities $K_{a\xi,\eta,\zeta}$ and $T_{a\xi,\zeta}$ expressed by the Eqns (9.195–9.200), also cf. Figure 9.25.

Again, it is seen that a reasonable or good correspondence can be achieved. The various tire–wheel resonance frequencies can be observed to show up: vertical mode at 78 Hz ($K_z$), rotational mode at 35 Hz ($K_x$ and $\Omega$), camber mode at 46 Hz ($K_y$, $T_z$, and $T_x$), and yaw mode at 54 Hz ($T_z$).

Chapter 11

# Motorcycle Dynamics

## Chapter Outline



Tire and Vehicle Dynamics. DOI: 10.1016/B978-0-08-097016-5.00011-5

## 11.1. INTRODUCTION

The single track vehicle is more difficult to study than the double track automobile and poses a challenge to the vehicle dynamicist. Stability of motion is an important issue and it turns out that the stabilizing actions of the human rider are essential to properly handle the vehicle. Steady-state cornering behavior can be analyzed in a straightforward manner together with the examination of the stability of the equilibrium motion. While for an automobile only the lateral and yaw degrees of freedom are minimally needed to perform such an analysis, a single track vehicle requires in addition the inclusion of the roll degree of freedom for the steady-state cornering study and the steer angle as a free motion variable to examine the stability. For better correspondence with reality also the torsion of the front frame with respect to the mainframe about an axis perpendicular to the steering axis is of importance. When the non-linear problem at higher cornering accelerations is investigated, a major difficulty is formed by the fact that the separation of lateral and vertical motions is not possible since due to the roll angle of the motorcycle a strong interaction between in-plane and lateral motions occurs. This is in contrast to the situation of a double track vehicle where the roll angle remains relatively small.

Performance of the vehicle in terms of handling properties is a matter that can be studied theoretically only if a proper model of the stabilization capabilities of the human rider is available. While in an automobile the driver normally uses the steering wheel to control the vehicle direction of motion, the pilot of a motorcycle has two or three quantities to his disposal to steer and stabilize the vehicle. These are: the steer angle or the steer torque and the lean angle (and possibly the lateral shift) of the upper torso.

In the past, a number of researchers studied the performance of the single track vehicle. Noteworthy is the very early theoretical study of Whipple (1899) on the stability of the motion of the bicycle with the tires assumed to be rigid. Sharp (1971) was one of the first to investigate the motorcycle's stability using a proper tire model. Later, the torsional compliance of the frame was introduced (Sharp and Alstead 1980a and Spierings 1981), which appeared to have a marked effect on the stability of the wobble mode (steering oscillations). In 1980 and 1983 Koenen reported an elaborate study on the stability also at large lateral curving accelerations which involve large roll angles and interaction of in-plane and lateral dynamics of the complex system. As the model representing the vehicle becomes more complex and impacts from road obstacles become more demanding to model the tire, e.g. the kick-back phenomenon, multi-body vehicle models and advanced dynamic tire models become indispensable for proper and efficient research. We refer to the following publications: Iffelsberger (1991), Wisselman et al. (1993), Breuer et al. (1998), Sharp et al. (2001a) and Berritta et al. (2000). Significant experimental results relating to the influence of design parameters on the damping of the main oscillatory modes have been given by Bayer (1988), Takahashi et al. (1984) (tire parameters), Hasegawa (1985) and Nishimi et al. (1985). In 1978, 1985, and 2001 Sharp published extensive reviews of the state of the art existing. An extensive more recent interesting review paper on the dynamics of motorcycles and bicycles has been published by Limebeer and Sharp (2006).

Important literature is available on rider behavior both as an active controller and stabilizer and as a passive part of the structure. We refer to the publications: Weir (1972), Nishimi et al. (1985), Katayama et al. (1988,1997), Cossalter et al. (1999), Biral et al. (2001) and Sharp (2010). The first one studies stabilizing feedback control, the second reference deals with passive rider model behavior, the third couple of papers discuss, among other things, maneuvering effort while the latter three papers address the problem of optimal maneuverability and limit maneuvering. Ruijs and Pacejka (1985) uses feed-back control loops to stabilize the unmanned motorcycle with a stabilizing rider-robot. Motorcycle state estimation methods have been investigated by Teerhuis and Jansen (2010) and an evaluation of a motorcycle riding simulator is done by Cossalter, Lot and Rota (2010).

Bicycle dynamics models are addressed by Meijaard et al. (2007) in an extensive detailed review paper. Bicycle dynamics and rider behavior have been studied by Kooijman et al. (2008), Kooijman et al. (2011) and Moore et al. (2011).

In the present chapter we will first establish geometrical relationships of the vehicle also at large roll angles with the steer angle being kept small, then discuss modeling of tire forces, derive the linear equations of motion using the Lagrangean method, and study the motion at relatively small lateral accelerations. For the steady-state cornering motion the understeer coefficient that provides information on the variation of the steer angle with increasing speed of travel at a given radius of turn will be assessed. In addition, the associated steer torque will be determined. For the linear system the stability of motion with its various possibly unstable modes will be investigated. The effects of driving and braking as well as of the aerodynamic drag will be included. In Section 11.5.3 typical changes in vibrational modes that may occur at large roll angles are discussed.

A relatively simple rider model that accomplishes feedback control will be introduced that is able to stabilize vehicle motion. Step response to handling inputs of the rider may then be investigated successfully. Inputs considered are: steer torque and lean torque. The lateral offset of the center of gravity will be treated as a constant small parameter that affects straight running behavior. Similar to the treatment of steady-state cornering behavior of automobiles we will demonstrate the assessment of the handling diagram for the motorcycle also covering large lateral accelerations. From the diagram established, the steer angle required to negotiate a given steady-state cornering maneuver can be assessed. Also, the steer torque is determined. Examples of results will be discussed. The responses to other inputs such as cross wind and transverse slope of the road surface have not been investigated. The introduction of such input quantities into the model may, however, be easily accomplished.

## 11.2. MODEL DESCRIPTION

In Figure 11.1 the motorcycle has been depicted while it moves at a roll angle $\varphi$ of the mainframe and with a steer angle $\delta$ of the handlebar about a steering axis that, in the neutral upright position, shows a steering head (rake) angle $\varepsilon$ with

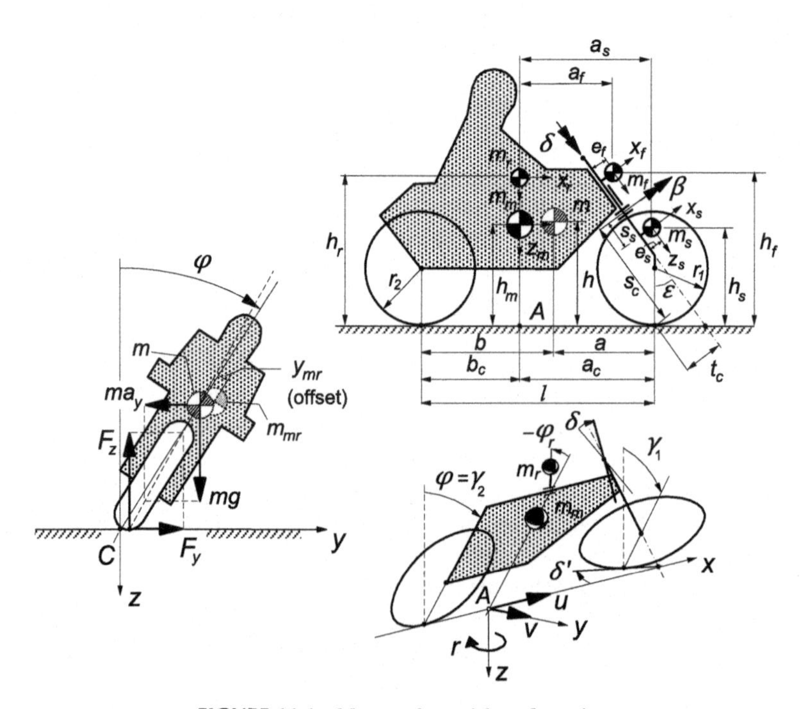


FIGURE 11.1 Motorcycle model configuration.

respect to a vertical line and a caster length $t_c$. The reference point $A$ that lies on the line of intersection of the plane of symmetry of the vehicle and the road plane and is located in the upright position below the center of gravity of the mainframe, moves forward with a velocity $u$ and in lateral direction with a velocity $v$. The line of intersection moves over the road surface and shows a yaw angle $\psi$, the time rate of which is denoted by $r$. The mainframe roll angle $\varphi$ is measured as the angle between the plane of symmetry and the normal to the road surface. As depicted in the figure, an additional degree of freedom may be introduced associated with the torsional flexibility of the front (steerable) frame, possibly including a portion of the mainframe, with respect to the center part of the mainframe. To model this, an axis of rotation is introduced perpendicular to the steering axis about which (a part of) the front frame can rotate with twist angle $\beta$. The rider has a lean degree of freedom (relative angle $\varphi_r$ about a longitudinal axis) of its upper torso (with mass $m_r$) and may exert (internal) moments about the steer and lean axes. Also, a small lateral shift $y_m$ of the c.g. of mainframe and $y_r$ of the rider may be included leading to a joint offset $y_{mr}$. The offsets are of the same order of magnitude as the roll angle and will be used in the steady-state analysis. Air drag is accounted for by the introduction of a longitudinal force $F_d$ acting at a height $h_d$ on the vehicle in the mainframe center plane. Finally, the various feedback control loops have been introduced in the equation for the steer angle to simulate possible rider control.

### 11.2.1. Geometry and Inertia

The geometrical dimensions of the motorcycle and the location of the centers of gravity of the four connected bodies (mainframe including lower part of the rider and rear wheel, upper torso of the rider, front upper frame, and front subframe including front wheel) are defined by quantities given in the figure. The following relations exist between geometrical parameters:

$$\begin{aligned}
a_f &= a_c - \{h_f \sin\varepsilon - (e_f + t_c)\}/\cos\varepsilon \\
a_s &= a_c - \{h_s \sin\varepsilon - (e_s + t_c)\}/\cos\varepsilon \\
s_s &= s_c - \{h_s - (e_s + t_c)\sin\varepsilon\}/\cos\varepsilon \\
h_\beta &= s_c \cos\varepsilon + t_c \sin\varepsilon \\
s_k &= s_c - t_c/\tan\varepsilon \\
h_k &= t_c/\sin\varepsilon
\end{aligned} \tag{11.1}$$

The last three dimensions have not been indicated in the figure.

The masses of the mainframe, the front upper frame, the front subframe, and the upper torso are denoted as $m_m$, $m_f$, $m_s$, and $m_r$ respectively. The magnitude of total mass $m$, the possibly shifted c.g. of $m_{mr} = m_m + m_r$, the wheel base $l$, and the location of the center of the total mass center with respect to the rear and front wheel axles (distances $a$ and $b$) and above the ground (height $h$) become:

$$\begin{aligned}
&m = m_m + m_f + m_s + m_r \\
&m_{mr} = m_m + m_r \\
&l = a_c + b_c \\
&b = \{m_{mr}b_c + m_f(a_f + b_c) + m_s(a_s + b_c)\}/m \\
&a = l - b \\
&h = (h_m m_m + h_f m_f + h_s m_s + h_r m_r)/m
\end{aligned} \tag{11.2}$$

The inertial quantities (provided with subscript 0) that would apply in case of a rigid rider (lean angle remains zero) may be retrieved by using the following equations where $I_{rx}$ is the moment of inertia of the rider's upper torso about its longitudinal axis; $I_{rz}$ is assumed to be incorporated in the mainframe inertia:

$$\begin{aligned}
&m_{m0} = m_{mr0} = m_m + m_r \\
&h_{m0} = (h_m m_m + h_r m_r)/m_{m0} \\
&I_{mx0} = I_{mx} + h_m^2 m_m + I_{rx} + h_r^2 m_r - h_{m0}^2 m_{m0} \\
&I_{mxz0} = I_{mxz} + I_{rxz}
\end{aligned} \tag{11.3}$$

Products of inertia, except $I_{mxz}$ of the mainframe, will be neglected.

The cross-sections of the tires are assumed to have a finite crown radius $r_{c1}$ and $r_{c2}$ (front and rear). These radii are responsible for the creation of the major part of the tire overturning couples $M_{x1,2}$. Moreover, the heights of the c.g.s are affected by these crown radii when running at large roll angles. Figure 11.2 illustrates such a situation.

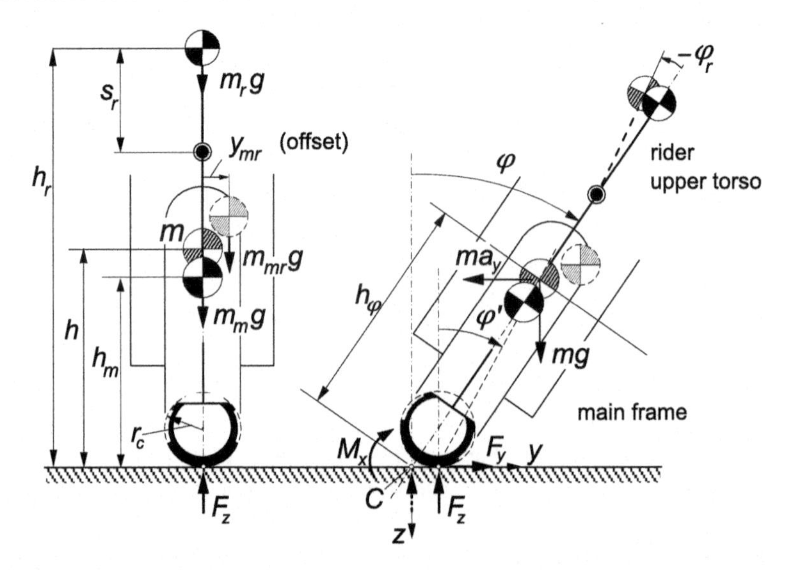


**FIGURE 11.2** Rear view of motorcycle with tire cross-section and possible c.g. offset indicated.

Important is the notion of the so-called contact center or point of intersection $C$ that lies below the wheel spin axis and on the line of intersection of wheel center plane and road surface. We may refer to Figure 4.27 and the related discussion. For the rear wheel the center plane coincides with the plane of symmetry of the (assumedly symmetric) mainframe. Rotation of the mainframe about the line of intersection gives rise to an increase of the normal load of the tire. At constant vertical load, a simultaneous lift of the vehicle must occur. Consequently, the distance of the center of gravity to the line of intersection will increase from $h$ to $h_\varphi$, as indicated in the figure. With a weighted average crown radius

$$r_c = \frac{b}{l} r_{c1} + \frac{a}{l} r_{c2} \tag{11.4}$$

we find:

$$h_\varphi = h + r_c(1 - \cos\varphi)/\cos\varphi \tag{11.5}$$

At large roll angles also the caster length $t_c$ should be adapted. We find approximately with $\delta$ assumed small:

$$t_{c\varphi} = t_c + r_{c1} \sin\varepsilon(1 - \cos\varphi)/\cos\varphi \tag{11.6}$$

In the linear analysis restricted to small angles, these extensions are of no importance.

## 11.2.2. The Steer, Camber, and Slip Angles

To determine the side force $F_y$ and the moments $M_x$ and $M_z$ acting on the front and rear wheels, the respective slip and camber angles are needed as input. For the rear wheel these angles can be obtained in a straightforward way. The front wheel poses a problem because we have an attitude of the wheel plane that is defined by at least three successive rotations. In Figure 11.3 several triads have been introduced which are needed to define the orientation of mainframe and front wheel. The line of intersection of the mainframe center plane and the road plane coincides with the $x$ axis. The origin of the horizontal moving axes system ($x$, $y$, $z$) is the reference point $A$ indicated in Figure 11.1 with forward and lateral velocity components $u$ and $v$. Furthermore, this system rotates about the vertical axis with yaw rate $r = \dot{\psi}$. The mainframe rotates about the $x$ axis giving rise to the roll angle $\varphi$. The rotated system of axes ($x_\varphi$, $y_\varphi$, $z_\varphi$) is attached to the mainframe. In the mainframe center plane the steering axis is positioned at an angle of inclination (the rake angle $\varepsilon$) with respect to the $z_\varphi$ axis. The triad ($x_\varepsilon$, $y_\varepsilon$, $z_\varepsilon$) is also attached to the mainframe but with a $z_\varepsilon$ axis along the inclined steering axis. The system ($x_\delta$, $y_\delta$, $z_\delta$) is attached to the upper part of the front frame that is rotated with steer angle $\delta$ with respect to the ($x_\varepsilon$, $y_\varepsilon$, $z_\varepsilon$) frame. Finally, we may introduce the twist angle $\beta$ (not shown in Figure 11.3, cf.

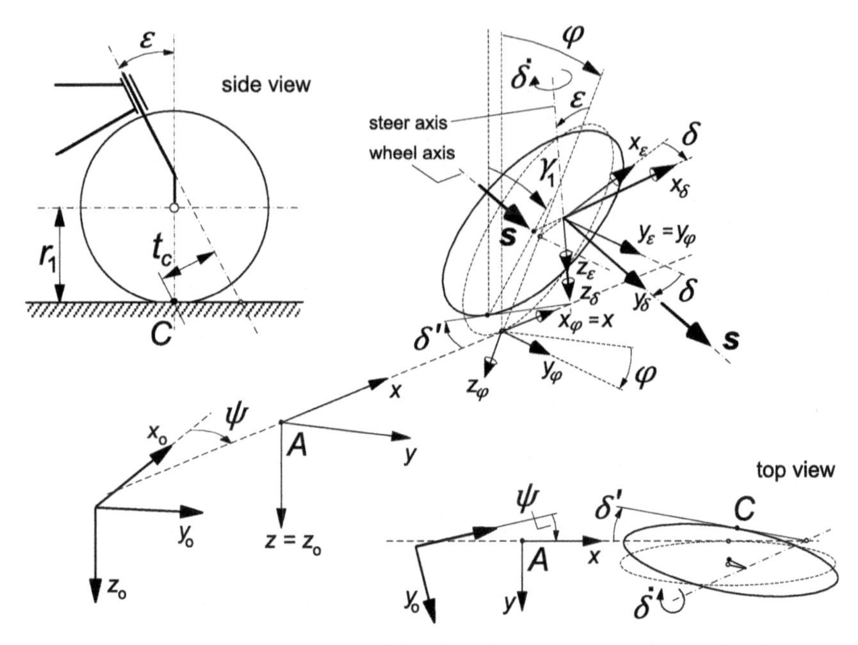


**FIGURE 11.3** View of front wheel assembly with various coordinate system triads to assess the ground steer angle $\delta'$ and camber angle $\gamma_1$ of the front wheel using the unit vector $\boldsymbol{s}$ along the wheel spin axis ($\beta = 0$).

Figure 11.1) giving rise to the system of axes ($x_\beta$, $y_\beta$, $z_\beta$) with the $y_\beta$ axis running along the wheel spin axis. We now introduce a unit vector $\boldsymbol{s}$ directed according to the wheel spin axis, that is, along the $y_\beta$ axis. The components of this vector along the axes of the moving horizontal system ($x$, $y$, $z$) will now be determined by successive rotation transformations. The result of each successive step is indicated by a subscript that denotes the frame with respect to which the unit vector is regarded. We have

$$\boldsymbol{s}_\beta = \begin{pmatrix} 0 \\ 1 \\ 0 \end{pmatrix}, \ \boldsymbol{s}_\delta = \begin{pmatrix} 1 & 0 & 0 \\ 0 & \cos\beta & -\sin\beta \\ 0 & \sin\beta & \cos\beta \end{pmatrix} \boldsymbol{s}_\beta, \ \boldsymbol{s}_\varepsilon = \begin{pmatrix} \cos\delta & -\sin\delta & 0 \\ \sin\delta & \cos\delta & 0 \\ 0 & 0 & 1 \end{pmatrix} \boldsymbol{s}_\delta$$

$$\boldsymbol{s}_\varphi = \begin{pmatrix} \cos\varepsilon & 0 & \sin\varepsilon \\ 0 & 1 & 0 \\ -\sin\varepsilon & 0 & \cos\varepsilon \end{pmatrix} \boldsymbol{s}_\varepsilon, \quad \boldsymbol{s} = \begin{pmatrix} 1 & 0 & 0 \\ 0 & \cos\varphi & -\sin\varphi \\ 0 & \sin\varphi & \cos\varphi \end{pmatrix} \boldsymbol{s}_\varphi \tag{11.7}$$

In the subsequent analysis we will approximate the situation by assuming small steer and twist angles. This is certainly admissible. For the non-linear steady-state cornering problem, the roll angle should be allowed to attain magnitudes larger than 45°. Of course, the rake angle $\varepsilon$ which is a system parameter will be regarded to be large. With $\delta$ and $\beta$ assumed small, the unit vector reduces to:

$$s = \begin{pmatrix} -\delta \cos \varepsilon + \beta \sin \varepsilon \\ \cos \varphi - \sin \varphi(\delta \sin \varepsilon + \beta \cos \varepsilon) \\ \sin \varphi + \cos \varphi(\delta \sin \varepsilon + \beta \cos \varepsilon) \end{pmatrix} \tag{11.8}$$

which in case of a completely linear analysis, with also the roll angle assumed small, reduces to:

$$s = \begin{pmatrix} -\delta \cos \varepsilon + \beta \sin \varepsilon \\ 1 - \varphi(\delta \sin \varepsilon + \beta \cos \varepsilon) \\ \varphi + \delta \sin \varepsilon + \beta \cos \varepsilon \end{pmatrix} \tag{11.9}$$

The ground steer angle $\delta'$ and the camber angle $\gamma_1$ can now easily be determined from the components of the unit vector. We find for the non-linear expressions:

$$\tan \delta' = -\frac{s_x}{s_y} = \frac{\delta \cos \varepsilon - \beta \sin \varepsilon}{\cos \varphi - \sin \varphi(\delta \sin \varepsilon + \beta \cos \varepsilon)} \tag{11.10}$$

$$\sin \gamma_1 = s_z = \sin \varphi + \cos \varphi(\delta \sin \varepsilon + \beta \cos \varepsilon) \tag{11.11}$$

and for the linearized approximations:

$$\delta' = \delta \cos \varepsilon - \beta \sin \varepsilon \tag{11.12}$$

$$\gamma_1 = \varphi + \delta \sin \varepsilon + \beta \cos \varepsilon \tag{11.13}$$

For the non-steered rear wheel we simply have a camber angle

$$\gamma_2 = \varphi \tag{11.14}$$

The slip angles are assumed to remain small and read at steady-state by considering Figure 11.4:

$$\begin{aligned} \alpha_1 &= \delta' - (v + a_c r)/u \\ \alpha_2 &= -(v - b_c r)/u \end{aligned} \tag{11.15}$$

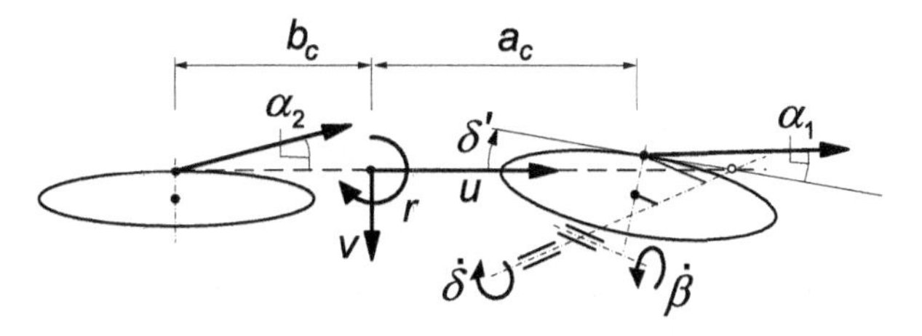


**FIGURE 11.4** Top view of the single track vehicle showing the front and rear slip angles.

For the dynamic non-steady-state situation the linearized model will be employed. The slip angles then become, including the time rate of changes of $\delta$ and $\beta$:

$$\begin{aligned}\alpha_1 &= \delta \cos \varepsilon - \beta \sin \varepsilon - \frac{1}{u}(v + a_c r - t_c \dot{\delta} - s_c \dot{\beta}) \\ \alpha_2 &= -\frac{1}{u}(v - b_c r)\end{aligned} \tag{11.16}$$

With the camber and slip angles now derived, we can formulate the resulting side force and moments. First, however, the normal loads are to be established. These are affected by the fore-and-aft load transfer caused by aerodynamic drag and braking or driving forces.

### 11.2.3. Air Drag, Driving or Braking, and Fore-and-Aft Load Transfer

Apart from the forces and moments acting from road to tires, we will consider the aerodynamic drag force $F_d$ that is assumed to act in the longitudinal backward direction in the pressure center a distance $h_d$ above the road surface (in upright position). We will here define the drag force to depend quadratically on the speed $u$ as follows:

$$F_d = C_{dA} u^2 \tag{11.17}$$

Due to the action of the drag force $F_d$ and the longitudinal tire forces $F_{xi}$, load transfer arises from the front to the rear wheel. The increase of the rear normal load is (by neglecting the overall aerodynamic lift) equal to the decrease of the front normal load.

The sum of the longitudinal tire forces is denoted as $F_{x,\text{tot}}$. The remaining force for the acceleration of the vehicle in longitudinal direction becomes:

$$F_{ax} = F_{x,\text{tot}} - F_d \tag{11.18}$$

which results in the forward acceleration:

$$a_x = \frac{1}{m} F_{ax} \tag{11.19}$$

and from this, the acceleration forces acting on the four individual masses $F_{axm} = a_x m_m$, $F_{axr} = a_x m_r$, $F_{axf} = a_x m_f$ and $F_{axs} = a_x m_s$.

With the moment arms $h_\varphi$ (cf. Figure 11.2) and $h_{d\varphi}$ the amount of load transfer becomes:

$$\Delta F_z = \frac{1}{l}(h_{d\varphi} F_d + h_\varphi F_{ax}) \cos \varphi \tag{11.20}$$

which for small roll angles reduces to:

$$\Delta F_z = \Delta F_{z0} = \frac{1}{l}(h_d F_d + h F_{ax}) \tag{11.21}$$

The resulting vertical wheel loads now become:

$$F_{z1} = F_{z1o} - \Delta F_z, \quad F_{z2} = F_{z2o} + \Delta F_z \tag{11.22}$$

and at small roll angles:

$$F_{z1} = F_{z10} = F_{z1o} - \Delta F_{z0}, \quad F_{z2} = F_{z20} = F_{z2o} + \Delta F_{z0} \tag{11.23}$$

with the initial wheel loads:

$$F_{z1o} = \frac{b}{l} mg, \quad F_{z2o} = \frac{a}{l} mg \tag{11.24}$$

The imposed braking force is assumed to be distributed over the front and rear wheels in proportion to the wheel loads as would occur in straight ahead motion, that is: according to the loads (11.23). We have at braking ($F_{x,\text{tot}} < 0$):

$$F_{x1} = \frac{F_{z10}}{mg} F_{x,\text{tot}}, \quad F_{x2} = \frac{F_{z20}}{mg} F_{x,\text{tot}} \tag{11.25}$$

and at driving:

$$F_{x1} = 0, \quad F_{x2} = F_{x,\text{tot}} \tag{11.26}$$

These forces will act as parameters in the formulae for the tire forces as described in the subsequent subsection.

It may be noted that in the rolled position the longitudinal drag and acceleration forces will also produce a moment about the vertical $z$ axis through reference point $A$. This gives rise to an increase at the rear and a decrease at the front of the side forces to be generated by the tires almost in proportion to the changes in normal load (not exactly because of the effect of the pneumatic trails). This is essentially different from what happens with the automobile.

### 11.2.4. Tire Force and Moment Response

#### *Linear Model*

We will take into account the transient response of the side force $F_y$ and aligning torque $M_z$ to changes in slip angle and camber angle. However, the non-lagging part of the response will be disregarded. On the other hand, the overturning couple $M_x$ is assumed to respond instantaneously to changes in camber. For the transient responses the relaxation length $\sigma$ will be used as parameter in the first-order differential equations. These equations describe the responses of the transient slip or deflection angles $\alpha'$ and $\gamma'$ which in steady-state become equal to the input slip and camber angles $\alpha$ and $\gamma$. We obtain for tire $i$ ($i = 1$ or 2):

$$\frac{1}{u}\sigma_{\alpha i}\dot{\alpha}'_i + \alpha'_i = \alpha_i \tag{11.27}$$

$$\frac{1}{u}\sigma_{\gamma i}\dot{\gamma}'_i + \gamma'_i = \gamma_i \tag{11.28}$$

which gives rise to the side force at small slip and camber angles:

$$F_{yi} = C_{F\alpha i}\alpha'_i + C_{F\gamma i}\gamma'_i \tag{11.29}$$

For the aligning torque stiffness against wheel camber we introduce the effect of the longitudinal force $F_{xi}$ by considering a finite crown radius $r_{ci}$ assuming that the lateral shift of the line of action of the longitudinal force changes instantaneously with the camber angle. We find for the aligning torque:

$$M_{zi} = -C_{M\alpha i}\alpha'_i + C'_{M\gamma i}\gamma'_i - r_{ci}F_{xi}\gamma_i \tag{11.30}$$

Note that we have disregarded here the other influences of the longitudinal force $F_{xi}$ on the side force and the aligning torque, as expressed e.g. by Eqns (4.45–4.48). To $M_z$ one might add the turn slip moment $M_z^*$ (5.82) and $M_{z,gyr}$ (5.178 and 7.49).

Finally, we have the overturning couple indicated in Figure 11.2 assumed here to depend only on the camber angle. In the linearized version we have:

$$M_{xi} = -C_{Mx\gamma i}\gamma_i \tag{11.31}$$

Obviously, we have neglected here the small effect of the lateral distortion due to the side force. The coefficients are assumed to depend on the normal load as follows (omitting subscript $i$):

$$C_{F\alpha} = C_{F\alpha o}/(1 + d_5\gamma^2) \tag{11.32}$$

(in the linear model $d_5 = 0$) with

$$C_{F\alpha o} = d_1F_{zo} + d_2(F_z - F_{zo}) \tag{11.33}$$

$$C_{F\gamma} = d_3F_z \tag{11.34}$$

$$C_{M\alpha} = e_1F_z \tag{11.35}$$

$$C'_{M\gamma} = e_2F_z \tag{11.36}$$

and

$$C_{Mx\gamma} = e_3F_z \quad \text{with } e_3 = r_c \tag{11.37}$$

We introduce the pneumatic trails $t_{\alpha o}$ (>0) of the side force due to side slip and $t_{\gamma o}$ (<0) of the camber force:

$$t_{\alpha o} = \frac{C_{M\alpha}}{C_{F\alpha o}}, \qquad t_{\gamma o} = -\frac{C'_{M\gamma}}{C_{F\gamma}} \tag{11.38}$$

Also the relaxation lengths depend on the normal load. This appears to occur in a way similar to that of the change in cornering stiffness. Experimental evidence shows that the relaxation length for side slip is close to the one for camber. The non-lagging part that (although small) exists in the response to camber changes is disregarded here. We define:

$$\sigma_\alpha = \sigma_\gamma = f_1 F_{zo} + f_2(F_z - F_{zo}) \tag{11.39}$$

## *Non-Linear Model*

For the non-linear force and moment description we will make use of the *Magic Formula* in a simplified version. The values of the parameters involved have been listed in Table 11.1. The similarity method will be employed to incorporate the effect of the imposed fore-and-aft force $F_x$. We will assume here that the cornering stiffness $C_{F\alpha}$ is not affected by $F_x$ and that the vertical shift is small with respect to $D_0$. We have for the side force (again omitting subscript $i$):

$$C = d_8 \tag{11.40}$$

$$K = C_{F\alpha} \tag{11.41}$$

$$D_0 = d_4 F_z/(1 + d_7\gamma^2) \tag{11.42}$$

**TABLE 11.1** Parameters of Hypothetical Front (,1) and Rear (,2) Motorcycle Tire Model.

| | | | | | | | |
|---|---|---|---|---|---|---|---|
| $d_{1,1}$ | 14 | $d_{2,1}$ | 9 | $d_{3,1}$ | 0.8 | $f_{1,1}$ | 0.00015 |
| $d_{1,2}$ | 13 | $d_{2,2}$ | 4 | $d_{3,2}$ | 0.8 | $f_{1,2}$ | 0.00015 |
| $e_{1,1}$ | 0.4 | $e_{2,1}$ | 0.04 | $e_{3,1}$ $(=r_{c1})$ | 0.08 | $f_{2,1}$ | 0.0001 |
| $e_{1,2}$ | 0.4 | $e_{2,2}$ | 0.07 | $e_{3,2}$ $(=r_{c2})$ | 0.1 | $f_{2,2}$ | 0.0001 |
| $d_{4,1}$ | 1.2 | $d_{5,1}$ | 0.15 | $d_{6,1}$ | 0.1 | $d_{7,1}$ | 0.15 |
| $d_{4,2}$ | 1.2 | $d_{5,2}$ | 0.4 | $d_{6,2}$ | 0.1 | $d_{7,2}$ | 0.15 |
| $d_8$ | 1.6 | $e_4$ | 10 | $e_5$ | 2 | $e_6$ | 1.5 |
| $e_7$ | 50 | $e_8$ | 1.1 | $e_9$ | 20 | $e_{10}$ | 1 |

$$D = \sqrt{D_0^2 - F_x^2} \tag{11.43}$$

$$B = K/(CD_0) \tag{11.44}$$

$$S_{Hf} = C_{F\gamma}\gamma'/C_{F\alpha} \tag{11.45}$$

$$S_V = d_6 F_z\, \gamma' D/D_0 \tag{11.46}$$

$$S_H = S_{Hf} - S_V/C_{F\alpha} \tag{11.47}$$

$$\alpha'_{Feq} = (D_0/D)(\alpha' + S_{Hf}) - S_{Hf} \tag{11.48}$$

$$F_y = D \sin[C \arctan\{B(\alpha'_{Feq} + S_H)\}] + S_V \tag{11.49}$$

and for the aligning torque using the pneumatic trail and the side force solely due to side slip:

$$\alpha'_{eq0} = (D_0/D)\alpha' \tag{11.50}$$

$$F_{y\alpha} = D \sin[C \arctan(B\alpha'_{eq0})] \tag{11.51}$$

$$B_t = e_7 \tag{11.52}$$

$$C_t = e_8 \tag{11.53}$$

$$B_r = e_9/(1 + e_4\gamma^2) \tag{11.54}$$

$$C_r = e_{10}/(1 + e_5\gamma^2) \tag{11.55}$$

$$t_\alpha = t_{\alpha o} \cos[C_t \arctan(B_t\alpha'_{eq0})]/(1 + e_5\gamma^2) \tag{11.56}$$

$$M_{zro} = C'_{M\gamma} \arctan(e_6\gamma')/e_6 \tag{11.57}$$

$$M_{zr} = M_{zro} \cos[C_r \arctan(B_r\alpha'_{eq0})] \tag{11.58}$$

$$M_z = -t_\alpha F_{y\alpha} + M_{zr} - r_c F_x \tan\gamma \tag{11.59}$$

in which the term with the product $F_xF_y$ has been disregarded. Finally, we define for the overturning couple using formula (4.126) while neglecting the vertical and lateral tire deflection:

$$M_x = -r_c F_z \tan\gamma \tag{11.60}$$

Since the non-linear analysis will be limited to steady-state conditions, the input slip and camber angles $\alpha$ and $\gamma$ may be used directly instead of the transient angles $\alpha'$ and $\gamma'$.

As an example, in Figures 11.5 and 11.6, the steady-state characteristics for $F_y$ and $M_z$ versus $\alpha$ have been plotted for a number of $\gamma$ values for the front and rear tires for the two cases: free rolling and braking. The characteristics of the freely rolling tire are similar to the experimentally found curves (extending from $-6$ to $+6$ degrees slip angle) reported by De Vries and Pacejka (1998a). The parameter values have been listed in Table 11.1.

The non-linear analysis may be improved by employing the full *Magic Formula* description as given in Chapter 4. In the present analysis these equations may be used for $\kappa = 0$, with the similarity method employed to include the effects of the given longitudinal force $F_x$. If instead of the force $F_x$ the longitudinal slip $\kappa$ is imposed or results from wheel rotational dynamics with imposed braking or driving effort, the complete Magic Formula model with combined slip may be used. For the present study this is a less practical option.

In the diagrams of Figures 11.5 and 11.6 the side force curves show a mainly sideways shift at increasing camber angle. The moment curves are moved upward while their shape is changed. The upward shift is a consequence of the spin torque that results from the wheel inclination angle. At braking

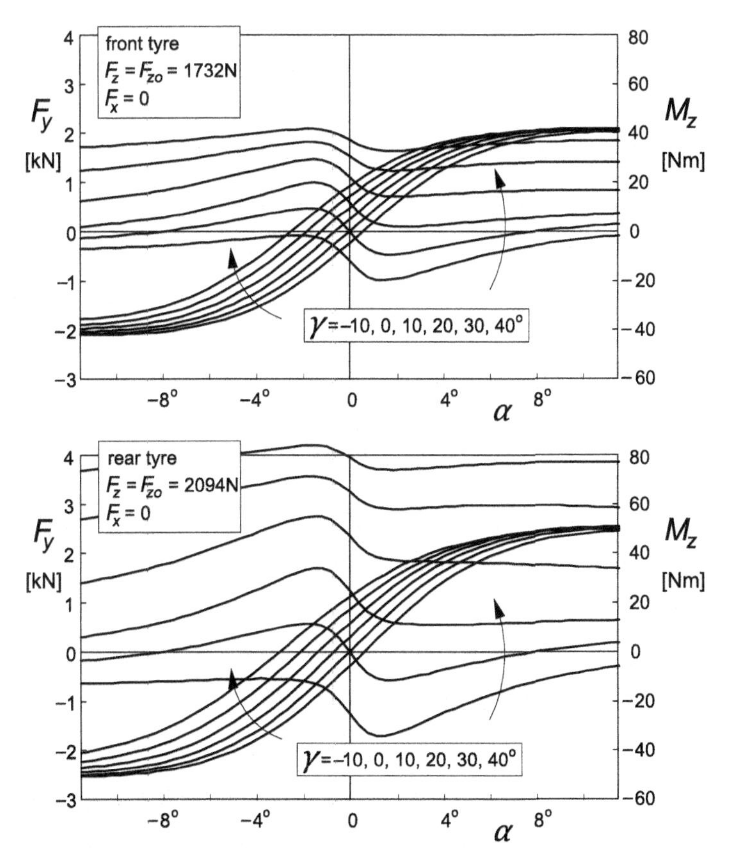


**FIGURE 11.5** Side force and aligning torque characteristics for the freely rolling front and rear motorcycle tire model at a series of camber angles.

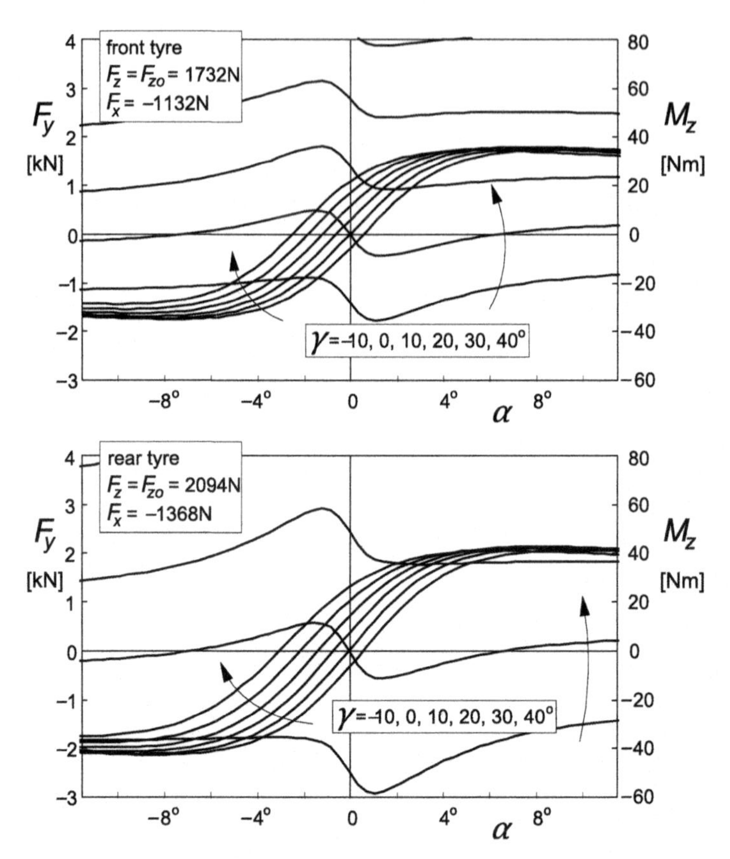


**FIGURE 11.6** Side force and aligning torque characteristics for the braked front and rear motorcycle tire at a series of camber angles.

(Figure 11.6) the aligning torque at camber is considerably increased in magnitude because of the direct contribution of $F_x$ ($<0$), which is represented in (11.59) by the last term. It is observed that due to the longitudinal force, the maximum level of the side force is reduced.

## 11.3. LINEAR EQUATIONS OF MOTION

For the dynamic analysis of the vehicle motion we restrict ourselves to small deviations from the rectilinear path. The differential equations will be kept linear with the forward speed $u$ considered as a constant parameter. We will derive the equations using the modified equations of Lagrange (1.28) derived in Chapter 1. After adding the dissipation function $D$ to include damping in the system, these equations read for the variables $v$ and

$r$ (=d$\psi$/d$t$) and subsequently for the remaining generalized coordinates $q_j$ (here: $j$ = 3, 4, 5, or 6):

$$\frac{\mathrm{d}}{\mathrm{d}t}\frac{\partial T}{\partial v} + r\frac{\partial T}{\partial u} = Q_v$$
$$\frac{\mathrm{d}}{\mathrm{d}t}\frac{\partial T}{\partial r} - v\frac{\partial T}{\partial u} + u\frac{\partial T}{\partial v} = Q_r \tag{11.61}$$
$$\frac{\mathrm{d}}{\mathrm{d}t}\frac{\partial T}{\partial \dot{q}_j} - \frac{\partial T}{\partial q_j} + \frac{\partial U}{\partial q_j} + \frac{\partial D}{\partial \dot{q}_j} = Q_j$$

The generalized forces are found from the virtual work (using $\Delta$ instead of $\delta$):

$$\Delta W = \sum_{j=1}^{6} Q_j \Delta q_j \tag{11.62}$$

with $q_j$ referring to the quasi coordinate $y$ and to the generalized coordinates $\psi$, $\varphi$, $\varphi_r$, $\delta$, and $\beta$. To establish the equations we will assess successively the kinetic and potential energies, the dissipation function, and the virtual work. The terms appearing in the resulting expressions will be restricted to terms of the second order of magnitude. The velocity $u$ and the connected wheel speeds of revolution $\Omega_i = u/r_i$, the vehicle weight $mg$, the aerodynamic drag $F_d$, and longitudinal force $F_{x,\mathrm{tot}}$ are quantities of the zeroth order of magnitude whereas the variables $v$ and the generalized coordinates are of the first order of magnitude. Also, the imposed lateral offset $y_{mr}$ of the combined mainframe and rider mass center is considered as a quantity of the first order of magnitude. Products of first order of magnitude quantities become terms of the second degree. Terms of higher degree will be neglected as these would give rise to second or higher degree terms in the final differential equations, which we intend to keep linear (only first degree terms).

The pitch angle $\theta$ of the mainframe which does not represent a degree of-freedom is of the second order of magnitude and should be accounted for in the expressions for the energies. Expressed in terms of the generalized coordinates we find the second degree constraint equation, cf. Eqn (11.1):

$$l\theta = -h_k(\delta \sin \varepsilon + \beta \cos \varepsilon)\left(\varphi + \frac{1}{2}\,\delta \sin \varepsilon + \frac{1}{2}\,\beta \cos \varepsilon\right)$$
$$-s_k\left(\varphi\beta + \delta\beta \sin \varepsilon + \frac{1}{2}\,\beta^2 \cos \varepsilon\right) \tag{11.63}$$

## 11.3.1. The Kinetic Energy

Translational and angular velocities of the six bodies are to be expressed in terms of $v$ and the generalized coordinates (and tentatively $\theta$) and their time derivatives. We find for the longitudinal, lateral, and vertical velocities of the c.g. of the mainframe plus rear wheel (except polar moment of inertia $I_{wy2}$) with mass $m_m$ (body1):

$$\begin{aligned} u_m &= u - h_m r\varphi - h_m\dot{\theta} - y_m r \\ v_m &= v + h_m\dot{\varphi} \\ w_m &= -b_c\dot{\theta} \end{aligned} \tag{11.64}$$

and for the angular velocities:

$$\begin{aligned} \omega_{mx} &= \dot{\varphi} \\ \omega_{my} &= \dot{\theta} \\ \omega_{mz} &= r \end{aligned} \tag{11.65}$$

The velocities of the rider upper torso with mass $m_r$ (body 2):

$$\begin{aligned} u_r &= u - h_r r\varphi - s_r r\varphi_r - h_r\dot{\theta} - y_r r \\ v_r &= v + h_r\dot{\varphi} + \dot{y}_r + s_r\dot{\varphi}_r \\ w_r &= -b_c\dot{\theta} \end{aligned} \tag{11.66}$$

and for the angular velocities:

$$\begin{aligned} \omega_{rx} &= \dot{\varphi} + \dot{\varphi}_r \\ \omega_{ry} &= \dot{\theta} \\ \omega_{rz} &= r \end{aligned} \tag{11.67}$$

The velocities of the front frame with mass $m_f$ (body 3 with the $z$ axis parallel to the steer axis):

$$\begin{aligned} u_f &= u - h_f r\varphi - e_f r\delta - h_f\dot{\theta} \\ v_f &= v + h_f\dot{\varphi} + e_f\dot{\delta} + a_f r \\ w_f &= -(b_c + a_f)\dot{\theta} \end{aligned} \tag{11.68}$$

and

$$\begin{aligned} \omega_{fx} &= \dot{\varphi}\cos\varepsilon - r\sin\varepsilon \\ \omega_{fy} &= \dot{\theta} \\ \omega_{fz} &= \dot{\varphi}\sin\varepsilon + r\cos\varepsilon + \dot{\delta} \end{aligned} \tag{11.69}$$

For the front subframe plus front wheel (except polar moment of inertia $I_{wy1}$) with mass $m_s$ (body 4 with the $z$ axis parallel to the steer axis):

$$\begin{aligned} u_s &= u - h_s r\varphi - e_s r\delta + s_s r\beta - h_s\dot{\theta} \\ v_s &= v + h_s\dot{\varphi} + e_s\dot{\delta} - s_s\dot{\beta} + a_s r \\ w_s &= -(b_c + a_s)\dot{\theta} \end{aligned} \tag{11.70}$$

and

$$\begin{aligned} \omega_{sx} &= \dot{\varphi}\cos\varepsilon - r\sin\varepsilon + \dot{\beta} \\ \omega_{sy} &= \dot{\theta} \\ \omega_{sz} &= \dot{\varphi}\sin\varepsilon + r\cos\varepsilon + \dot{\delta} \end{aligned} \tag{11.71}$$

For the front wheel angular velocities (bodies 5 and 6 only possessing polar moments of inertia $I_{wy1,2}$ possibly extended with effective moment of inertia $n_g I_{ey}$ of other rotating parts reduced to the wheel rotational speed):

$$\omega_{wy1} = -\Omega_1 + \gamma_1(\dot{\delta}' + r) \tag{11.72}$$

and for the rear wheel:

$$\omega_{wy2} = -\Omega_2 + \varphi r \tag{11.73}$$

where

$$\Omega_1 = \frac{u}{r_1}, \quad \Omega_2 = \frac{u}{r_2} \tag{11.74}$$

with $r_1$ and $r_2$ denoting the front and rear wheel (effective rolling) radii. Due to symmetry of the undisturbed system, the $\Omega$'s do not need to be regarded as variables in our linearized system equations and non-holonomic constraint equations do not occur. The kinetic energy becomes, in general, summed over the six bodies:

$$T = \frac{1}{2}\sum_{k=1}^{6}\Big[m_k(u_k^2 + v_k^2 + w_k^2) + \Big\{I_{xk}\omega_{xk}^2 + I_{yk}\omega_{yk}^2 + I_{zk}\omega_{zk}^2 - 2(I_{xyk}\omega_{xk}\omega_{yk} + I_{yzk}\omega_{yk}\omega_{zk} + I_{zxk}\omega_{zk}\omega_{xk})\Big\}\Big] \tag{11.75}$$

The products of inertia will be neglected except $I_{zx1} = I_{mxz}$ of the main-frame. The velocities $u_k$, etc. appearing in (11.75) correspond to the quantities given by the expressions (11.64–11.74). The time derivative of the pitch angle $\theta$ is obtained from expression (11.63).

## 11.3.2. The Potential Energy and the Dissipation Function

The potential energy is composed of contributions from torsional spring deflections: twist angle $\beta$ and lean angle $\varphi_r$ with stiffnesses $c_\beta$ and $c_{\varphi r}$, respectively, and the heights of the centers of gravity of the various bodies. These heights become expressed in terms of the generalized coordinates and $\theta$:

For body 1 with c.g. lateral offset $y_m$:

$$-z_m = h_m\left(1 - \frac{1}{2}\varphi^2\right) + b_c\theta - y_m\varphi \tag{11.76}$$

For body 2 with c.g. lateral offset $y_r$:

$$-z_r = h_r - \frac{1}{2}(h_r - s_r)\varphi^2 - \frac{1}{2}s_r(\varphi + \varphi_r)^2 + b_c\theta - y_r\varphi \tag{11.77}$$

For body 3:

$$-z_f = h_f - \frac{1}{2}\, h_f\varphi^2 - e_f\delta\varphi - \frac{1}{2}\, e_f\delta^2 \sin\varepsilon + (b_c + a_f)\theta \tag{11.78}$$

For body 4:

$$\begin{aligned} -z_s =& h_s - \frac{1}{2}\, h_s\varphi^2 - e_s\delta\varphi + s_s\beta\varphi - \frac{1}{2}\, e_s\delta^2 \sin\varepsilon + \; s_s\delta\beta \sin\varepsilon \\ &+ \frac{1}{2}\, s_s\beta^2 \cos\varepsilon + (b_c + a_s)\theta \end{aligned} \tag{11.79}$$

Furthermore, we have the torsional deflections $\varphi_r$ and $\beta$. The complete potential energy is now written as follows:

$$U = -g(m_m z_m + m_r z_r + m_f z_f + m_s z_s) + \frac{1}{2}(c_{\varphi r}\varphi_r^2 + c_\beta\beta^2) \tag{11.80}$$

Again, Eqn (11.63) should be employed to express $\theta$, appearing in the formulae (11.76–11.79), in terms of the generalized coordinates.

If we consider viscous damping to be present in the steer bearings and possibly also around the lean and twist axes we obtain for the dissipation function with $k_\delta$, $k_{\varphi r}$, and $k_\beta$ denoting the respective damping coefficients:

$$D = \frac{1}{2}\, k_\delta\dot{\delta}^2 + \frac{1}{2}\, k_{\varphi r}\dot{\varphi}_r^2 + \frac{1}{2}\, k_\beta\dot{\beta}^2 \tag{11.81}$$

### 11.3.3. The Virtual Work

Through the virtual work the generalized forces each associated with a generalized coordinate can be assessed. The forces which act from the environment upon the vehicle are the horizontal ground forces and moments, the aerodynamic forces, and the gravitational force component in the longitudinal direction in case of a forward slope or the dynamic longitudinal d'Alembert forces acting in the mass centers which are in equilibrium with the longitudinal forces generated by the tires. The option of considering a forward slope that at given aerodynamic drag and driving or braking forces is just able to maintain a constant forward velocity would make the analysis correct as in that case the coefficients of the linear equations are time independent (cf. Eqn (3.123) with (3.126) where this is not the case). The longitudinal acceleration $a_x$ appearing in the ensuing equations may be considered equal to the longitudinal component of the acceleration due to gravity that would arise if the vehicle continuously runs over a road with equivalent forward slope.

With the various forces and moments that act upon the vehicle in and around the respective points of application the virtual work becomes:

$$\begin{aligned}\Delta W = & F_{x1}\Delta x_1 + F_{y1}\Delta y_1 + M_{z1}(\Delta\delta' + \Delta\psi) + M_{x1}\Delta\gamma_1 \\ & + F_{x2}\Delta x_2 + F_{y2}\Delta y_2 + M_{z2}\Delta\psi + M_{x2}\Delta\varphi \\ & - F_d\Delta x_d + M_\delta\Delta\delta + M_{\varphi r}\Delta\varphi_r \\ & - a_x(m_m\Delta x_m + m_r\Delta x_r + m_f\Delta x_f + m_s\Delta x_s)\end{aligned} \tag{11.82}$$

The virtual displacements expressed in terms of the generalized coordinates turn out to read:

$$\begin{aligned}\Delta x_1 &= (\delta\cos\varepsilon - \beta\sin\varepsilon)(\Delta y + a_c\Delta\psi) + (t_c\delta + s_c\beta)\Delta\psi + h_\beta\beta\Delta\delta \\ \Delta x_2 &= 0 \\ \Delta y_1 &= \Delta y + a_c\Delta\psi \\ \Delta\delta' &= \cos\varepsilon\Delta\delta - \sin\varepsilon\Delta\beta \\ \Delta\gamma_1 &= \Delta\varphi + \sin\varepsilon\Delta\delta + \cos\varepsilon\Delta\beta \\ \Delta y_2 &= \Delta y - b_c\Delta\psi \\ \Delta x_d &= -h_d\varphi\Delta\psi - h_d\Delta\theta \\ \Delta x_m &= -(h_m\varphi + y_m)\Delta\psi - h_m\Delta\theta \\ \Delta x_r &= -(h_r\varphi + s_r\varphi_r + y_r)\Delta\psi - h_r\Delta\theta \\ \Delta x_f &= -(h_f\varphi + e_f\delta)\Delta\psi - h_f\Delta\theta \\ \Delta x_s &= -(h_s\varphi + e_s\delta - s_s\beta)\Delta\psi - h_s\Delta\theta\end{aligned} \tag{11.83}$$

where $\Delta\theta$ can be expressed in the generalized coordinates by taking the variation of $\theta$ (11.63).

Now, we may compare (11.82), after having substituted herein the expressions (11.83), with the formulation of the virtual work according to Eqn (11.62), which becomes:

$$\begin{aligned}\Delta W &= \sum_{j=1}^{6} Q_j\Delta q_j \\ &= Q_v\Delta y + Q_r\Delta\psi + Q_\varphi\Delta\varphi + Q_{\varphi r}\Delta\varphi_r + Q_\delta\Delta\delta + Q_\beta\Delta\beta\end{aligned} \tag{11.84}$$

As a result, the generalized forces $Q_j$ are obtained which are to be inserted at the right-hand sides of the Lagrangean Eqn (11.61).

### 11.3.4. Complete Set of Linear Differential Equations

All the necessary preparations to set up the equations have been completed. To establish the equations, the operations with the energies as indicated in (11.61) can now be carried out. The resulting set of equations are completed with the four first-order differential equations for the transient slip and camber angles front and rear and the linear equations for the side forces, the aligning torques, and the overturning couples together with expressions of the slip and camber angles all resulting from the analysis of Section 11.2. Note that the tire model coefficients, $C_{F\alpha1}$, $\sigma_{\alpha1}$, etc., depend on the current vertical wheel load $F_{zi} = F_{zio}$, Eqns (11.23, 11.33–11.39). The total order of the system turns out to be 14.

With a given speed of travel $u$ and the total longitudinal tire force $F_{x,\text{tot}}$, the air drag $F_d$, the vertical tire loads $F_{z10}$ and $F_{z20}$ and the individual longitudinal tire forces $F_{x1,2}$ can be assessed using the Eqns (11.17–11.19, 11.21, 11.23–11.26). The initial wheel loads (at stand-still) $F_{z1o}$ and $F_{z2o}$ are directly associated with the vehicle mass distribution, Eqn (11.24). In the equations the imposed handlebar torque $M_\delta$ and lean torque $M_{\varphi r}$ appear in the right-hand members. The tire side forces and moments are tentatively put on the right-hand side as well. In the equation for the steer angle the control terms have been added.

The equations successively for the variables $v$, $r$, $\varphi$, $\varphi_r$, $\delta$, $\beta$, $\alpha'_1$, $\gamma'_1$, $\alpha'_2$, and $\gamma'_2$ become as follows

$v$:

$$\begin{aligned}&(m_m + m_f + m_s + m_r)(\dot{v} + ur) + (m_f a_f + m_s a_s)\dot{r} + (m_m h_m + m_f h_f + m_s h_s \\ &\quad + m_r h_r)\ddot{\varphi} + m_r s_r \ddot{\varphi}_r + (m_f e_f + m_s e_s)\ddot{\delta} - m_s s_s \ddot{\beta} - F_{x1}(\cos\varepsilon\cdot\delta - \sin\varepsilon\cdot\beta) \\ &\quad = F_{y1} + F_{y2}\end{aligned} \tag{11.85}$$

$r$:

$$\begin{aligned}&(m_f a_f + m_s a_s)(\dot{v} + ur) + \{m_f a_f^2 + m_s a_s^2 + I_{mz} + (I_{fx} + I_{sx})\sin^2\varepsilon \\ &\quad + (I_{fz} + I_{sz})\cos^2\varepsilon\}\dot{r} + \{m_f h_f a_f + m_s h_s a_s - I_{mxz} \\ &\quad + (I_{fz} + I_{sz} - I_{fx} - I_{sx})\sin\varepsilon\cos\varepsilon\}\ddot{\varphi} - \{I_{wy1}/r_1 + (I_{wy2} + n_g I_{ey})/r_2\}u\dot{\varphi} \\ &\quad + \{m_f e_f a_f + m_s e_s a_s + (I_{fz} + I_{sz})\cos\varepsilon\}\ddot{\delta} - (I_{wy1}/r_1)u\sin\varepsilon\cdot\dot{\delta} \\ &\quad -(m_s s_s a_s + I_{sx}\sin\varepsilon)\ddot{\beta} - (I_{wy1}/r_1)u\cos\varepsilon\cdot\dot{\beta} - F_d h_d\varphi - F_{x1}\{(t_c + a_c\cos\varepsilon)\delta \\ &\quad + (s_c - a_c\sin\varepsilon)\beta\} - a_x\{mh\varphi + m_r s_r\varphi_r + (m_f e_f + m_s e_s)\delta - m_s s_s\beta\} \\ &\quad = a_c F_{y1} - b_c F_{y2} + M_{z1} + M_{z2} + m_{mr}a_x y_{mr}\end{aligned} \tag{11.86}$$

$\varphi$:

$$\begin{aligned}&(m_m h_m + m_f h_f + m_s h_s + m_r h_r)(\dot{v} + ur) + \{I_{wy1}/r_1 + (I_{wy2} + n_g I_{ey})/r_2\}ur \\ &\quad + \{m_f h_f a_f + m_s h_s a_s - I_{mxz} + (I_{fz} + I_{sz} - I_{fx} - I_{sx})\sin\varepsilon\cos\varepsilon\}\dot{r} \\ &\quad + \{m_f h_f^2 + m_s h_s^2 + m_m h_m^2 + m_r h_r^2 + I_{mx} + I_{rx} + (I_{fx} + I_{sx})\cos^2\varepsilon \\ &\quad + (I_{fz} + I_{sz})\sin^2\varepsilon\}\ddot{\varphi} - (m_m h_m + m_f h_f + m_s h_s + m_r h_r)g\varphi \\ &\quad + (I_{rx} + m_r s_r h_r)\ddot{\varphi}_r - m_r s_r g\varphi_r + \{m_f e_f h_f + m_s e_s h_s + (I_{fz} + I_{sz})\sin\varepsilon\}\ddot{\delta} \\ &\quad + (I_{wy1}/r_1)u\cos\varepsilon\cdot\dot{\delta} - (t_c F_{z1} + m_f e_f g + m_s e_s g)\delta - (m_s s_s h_s - I_{sx}\cos\varepsilon)\ddot{\beta} \\ &\quad - (I_{wy1}/r_1)u\sin\varepsilon\cdot\dot{\beta} - (s_c F_{z1}\cdot m_s s_s g)\beta \\ &\quad = M_{x1} + M_{x2} + m_{mr}g y_{mr}\end{aligned} \tag{11.87}$$

$\varphi_r$:

$$m_r s_r(\dot{v}+ur)+(I_{rx}+m_r s_r h_r)\ddot{\varphi}-m_r s_r g\varphi \\ +(m_r s_r^2+I_{rx})\ddot{\varphi}_r+k_{\varphi r}\dot{\varphi}_r+(c_{\varphi r}-m_r s_r g)\varphi_r=M_r \tag{11.88}$$

$\delta$:

$$(m_f e_f+m_s e_s)(\dot{v}+ur)+(I_{wy1}/r_1)\sin\varepsilon\cdot ur \\ +\{m_f e_f a_f+m_s e_s a_s+(I_{fz}+I_{sz})\cos\varepsilon\}\dot{r} \\ +\{m_f e_f h_f+m_s e_s h_s+(I_{fz}+I_{sz})\sin\varepsilon\}\ddot{\varphi}-(I_{wy1}/r_1)u\cos\varepsilon\cdot\dot{\varphi} \\ -(t_c F_{z1}+m_f e_f g+m_s e_s g)\varphi \\ +(m_f e_f^2+m_s e_s^2+I_{fz}+I_{sz})\ddot{\delta}+k_\delta\dot{\delta}-(t_c F_{z1}+m_f e_f g+m_s e_s g)\sin\varepsilon\cdot\delta \\ -m_s e_s s_s\ddot{\beta}-(I_{wy1}/r_1)u\dot{\beta}-\{(s_c F_{z1}-m_s s_s g)\sin\varepsilon+F_{x1}h_\beta\}\beta \\ +g_v v+g_r r+g_{d\varphi}\dot{\varphi}+g_\varphi\varphi+g_{d\delta}\dot{\delta}+g_\delta\delta \\ =-t_c F_{y1}+M_{z1}\cos\varepsilon+M_{x1}\sin\varepsilon+M_\delta \tag{11.89}$$

$\beta$:

$$-m_s s_s\dot{v}-(m_s s_s a_s+I_{sx}\sin\varepsilon)\dot{r}-\{m_s s_s-(I_{wy1}/r_1)\cos\varepsilon\}ur \\ -(m_s s_s h_s-I_{sx}\cos\varepsilon)\ddot{\varphi}+(I_{wy1}/r_1)u\sin\varepsilon\cdot\dot{\varphi}-(s_c F_{z1}-m_s s_s g)\varphi \\ -m_s e_s s_s\ddot{\delta}+(I_{wy1}/r_1)u\dot{\delta}-(s_c F_{z1}-m_s s_s g)\sin\varepsilon\cdot\delta \\ +(m_s s_s^2+I_{sx})\ddot{\beta}+k_\beta\dot{\beta}+\{c_\beta-(s_c F_{z1}-m_s s_s g)\cos\varepsilon\}\beta \\ =-s_c F_{y1}-M_{z1}\sin\varepsilon+M_{x1}\cos\varepsilon \tag{11.90}$$

transient slip and camber angles:

$$\sigma_{\alpha1}\dot{\alpha}_1'+u\alpha_1'=u\alpha_1 \tag{11.91}$$

$$\sigma_{\gamma1}\dot{\gamma}_1'+u\gamma_1'=u\gamma_1 \tag{11.92}$$

$$\sigma_{\alpha2}\dot{\alpha}_2'+u\alpha_2'=u\alpha_2 \tag{11.93}$$

$$\sigma_{\gamma2}\dot{\gamma}_2'+u\gamma_2'=u\gamma_2 \tag{11.94}$$

tire forces and moments (coefficients depend on wheel load (11.23, 11.33–11.39)):

$$F_{y1}=C_{F\alpha1}\alpha_1'+C_{F\gamma1}\gamma_1' \tag{11.95}$$

$$F_{y2}=C_{F\alpha2}\alpha_2'+C_{F\gamma2}\gamma_2' \tag{11.96}$$

$$M_{z1}=-C_{M\alpha1}\alpha_1'+C_{M\gamma1}'\gamma_1'-r_{c1}F_{x1}\gamma_1 \tag{11.97}$$

$$M_{z2}=-C_{M\alpha2}\alpha_2'+C_{M\gamma2}'\gamma_2'-r_{c2}F_{x2}\gamma_2 \tag{11.98}$$

$$M_{x1}=-C_{Mx\gamma1}\gamma_1 \tag{11.99}$$

$$M_{x2}=-C_{Mx\gamma2}\gamma_2 \tag{11.100}$$

slip and camber angles:

$$\alpha_1 = (-v - a_c r + t_c \dot{\delta} + s_c \dot{\beta})/u + \cos\varepsilon \cdot \delta - \sin\varepsilon \cdot \beta \tag{11.101}$$

$$\gamma_1 = \varphi + \sin\varepsilon \cdot \delta + \cos\varepsilon \cdot \beta \tag{11.102}$$

$$\alpha_2 = (-v + b_c r)/u \tag{11.103}$$

$$\gamma_2 = \varphi \tag{11.104}$$

the rider feedback control gains, occurring in Eqn (11.89), depend on the speed of travel $u$ and are assessed here by trial and error:

$$\begin{aligned} &g_v = g_{vo}/u, \quad g_r = g_{ro}/u, \quad g_{d\varphi} = g_{d\varphi o}(1 - u/u_c)/u, \\ &g_\varphi = g_{\varphi o}, \quad g_{d\delta} = g_{d\delta o}/u, \quad g_\delta = g_{\delta o} \end{aligned} \tag{11.105}$$

The steer torque $M_\delta$ is considered to act in addition to the stabilization steer torque, that results from feedback control, and is represented by the terms containing the control gains in Eqn (11.89).

For the mass center lateral offset we have introduced for abbreviation:

$$m_{mr} y_{mr} = m_m y_m + m_r y_r \quad \text{with} \quad m_{mr} = m_m + m_r \tag{11.106}$$

If the steer and roll angles remain sufficiently small so that it may be assumed that geometric linearity still prevails we may investigate the influence

**TABLE 11.2 Parameters of Vehicle in Baseline Configuration but with Rider Upper Torso Released (in Baseline Configuration $c_{\varphi r} \to \infty$). Rider Control Gains g with Cross-over Velocity $u_c$**

| | | | | | | | |
|---|---|---|---|---|---|---|---|
| $m_m$ | 300 kg | $m_f$ | 15 kg | $m_s$ | 25 kg | $m_r$ | 50 kg |
| $I_{mx}$ | 20 kg m$^2$ | $I_{mz}$ | 20 kg m$^2$ | $I_{mxz}$ | 4 kg m$^2$ | $I_{rx}$ | 4.75 kg m$^2$ |
| $I_{fx}$ | 0.5 kg m$^2$ | $I_{fz}$ | 0.3 kg m$^2$ | $I_{sx}$ | 1.0 kg m$^2$ | $I_{sz}$ | 0.7 kg m$^2$ |
| $I_{wy1}$ | 1.0 kg m$^2$ | $I_{wy2}$ | 1.0 kg m$^2$ | $I_{ey}$ | 0.06 kg m$^2$ | | |
| $a_c$ | 0.9 m | $b_c$ | 0.6 m | $e_f$ | 0.05 m | $e_s$ | 0.05 m |
| $s_c$ | 0.7 m | $t_c$ | 0.1 m | $r_1$ | 0.3 m | $r_2$ | 0.3 m |
| $h_m$ | 0.55 m | $h_f$ | 0.8 m | $h_s$ | 0.4 m | $h_d$ | 0.75 m |
| $s_r$ | 0.4 m | $h_r$ | 0.9 m | | | $k_\delta$ | 0 Nms/rad |
| $k_\beta$ | 50 Nms/rad | $c_\beta$ | 25 k Nm/rad | $k_{\varphi r}$ | 20 Nms/rad | $c_{\varphi r}$ | 350 Nm/rad |
| $g$ | 9.81 m/s$^2$ | $n_g$ | 1.5 | $\varepsilon$ | 0.5 rad | $C_{dA}$ | 0.2 kg/m |
| $u_c$ | 15 m/s | $g_{vo}$ | −500 | $g_{ro}$ | 100 | $g_{d\varphi o}$ | −900 |
| | | $g_{\varphi o}$ | −10 | $g_{d\delta o}$ | 150 | $g_{\delta o}$ | 50 |

of larger wheel slip by using for the tire side forces and moments the non-linear functions of $\alpha'_i$, $\gamma'_i$, $\gamma_i$, wheel load $F_{zi}$, and the braking/driving force $F_{xi}$ according to the Eqns (11.40–11.60).

For the baseline configuration of the linear vehicle model the rider is considered rigid ($c_{\varphi r} \rightarrow \infty$). With the rider upper torso released, the parameter values of Table 11.2 hold. The tire parameter values have been chosen as listed in Table 11.1. The model configuration may be considered as typical for a heavy motorcycle. When a rigidly connected rider upper torso is considered (baseline configuration with $\varphi_r = 0$), the relevant parameters result from Eqns (11.3).

For the linear system defined, the eigenvalues and the steady-state steer angle and steer torque per unit path curvature have been established for a series of values of the speed of travel $u$ ($=V$). In addition, a number of parameters related to both the vehicle construction and the tires have been changed in value to examine their influence.

## 11.4. STABILITY ANALYSIS AND STEP RESPONSES

### 11.4.1. Free Uncontrolled Motion

In a series of diagrams the variation of the eigenvalues as a function of speed has been presented. The diagram of Figure 11.7 represents the situation corresponding to the baseline configuration defined by Tables 11.1 and 11.2 but with the rider regarded as a rigid element of the mainframe body. The upper diagram gives the real part of the eigenvalues of this 12th order system versus speed indicating the degree of instability (if positive) or damping (stable, if negative). The second diagram presents the variation of the imaginary part of the eigenvalues that represent the frequencies of the different modes of vibration (in rad/s). The third diagram depicts the variation of the eigenvalues in the complex plane. Along the curves the speed changes as indicated.

The weave mode is a relatively low-frequency oscillatory motion in which the whole vehicle takes part. This mode exhibits a possibly dangerous unstable yaw and roll motion at higher speeds (in our case above a speed of ca.165 km/h) with a frequency of 3 to 4 Hz. At low speeds the frequency decreases and the mode becomes unstable and constitutes the phenomenon where the uncontrolled vehicle falls over.

At very low speed it is observed that the complex pair of roots with a positive real part changes into two positive real roots constituting the divergent unstable modes associated with capsizing of the whole machine and of the front frame about the inclined steer axis (the latter is observable with the motorcycle on its center stand and with the front wheel clear of the ground). The frequency of the oscillation can be found from the two lower plots. In the high speed unstable range the frequency of the weave mode appears to be around 27 rad/s or ca. 4 Hz.

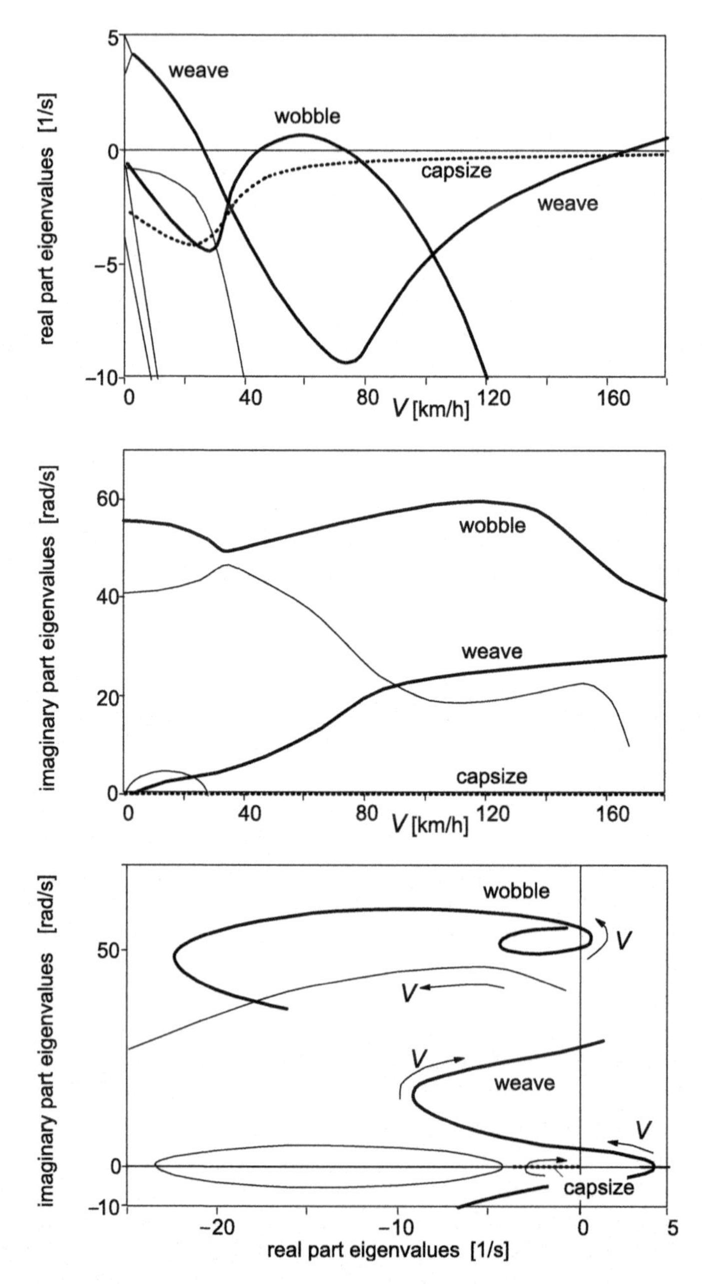


**FIGURE 11.7** Baseline configuration (rigid rider, with air drag, zero forward acceleration, no control). Real and imaginary parts of eigenvalues versus speed of travel $V$.

The capsize mode can in certain cases become (moderately) unstable beyond a relatively low critical velocity. The eigenvalue remains real and, consequently, the motion does not show oscillations. In our case, this mode remains stable. The third mode that is of interest is the wobble mode, which is a steering oscillation that can become unstable in a range of moderate speeds (in our case between ca. 45 and 70 km/h). The frequency in the unstable range appears here to take values around 55 rad/s or ca. 8 Hz as can be seen in the two lower plots with the imaginary part as the ordinate. The frequency is mainly influenced by the front frame inertia, the mechanical trail, and the front tire cornering stiffness. The remaining modes are well damped and consequently of less interest.

Starting out from the baseline configuration (with air drag) the effect of changes in several system parameters has been investigated. To get a clear picture of the role various parameters play in stabilizing or destabilizing this single track vehicle, their values have been changed drastically. Figure 11.8 shows the resulting stability characteristics. For clarity, the weave, capsize, and wobble mode curves have been shown in separate diagrams. From the results we may conclude the following:

1. The model shows that *absence of air drag* (case 1) slightly stabilizes the weave mode and destabilizes the wobble mode.
2. *Torsional rigidity of the mainframe* (case 2, large $c$) appears to be of crucial importance especially for the manner in which the wobble manifests itself. High stiffness or disregarding the torsional compliance gives rise to a (too) high critical speed. A more flexible frame (baseline configuration) causes the wobble to occur in a limited range of speed around 50 km/h, which turns out to be experienced also in reality.
3. Almost vanishing *relaxation lengths* (rapid tire response) drastically suppresses the wobble instability.
4. Making the *tire almost rigid* (with in addition to very small $\sigma$: very large cornering stiffness and very small pneumatic trail, case 4) would drastically suppress the high speed weave but gives rise to violent wobble over a large range of speed. Much larger camber force stiffness cannot be technically realized even with a very stiff solid wheel. It would increase the low speed unstable weave range and stabilize the high speed weave; wobble is almost not affected (not shown).
5. Making only the *aligning torque stiffnesses* very small destabilizes the high speed weave mode and stabilizes the wobble mode. The capsize root is made more negative.
6. Even a considerable decrease of the *camber aligning moment stiffness* has very little effect, at least in the case of zero or very small longitudinal tire forces. The same holds for the *overturning couple stiffness*, that is: a small *crown radius* (not shown). The smaller camber aligning moment stiffness does have an influence on the capsize stability. As indicated in the diagram, capsize instability now shows up and occurs beyond a speed of around 20 km/h.

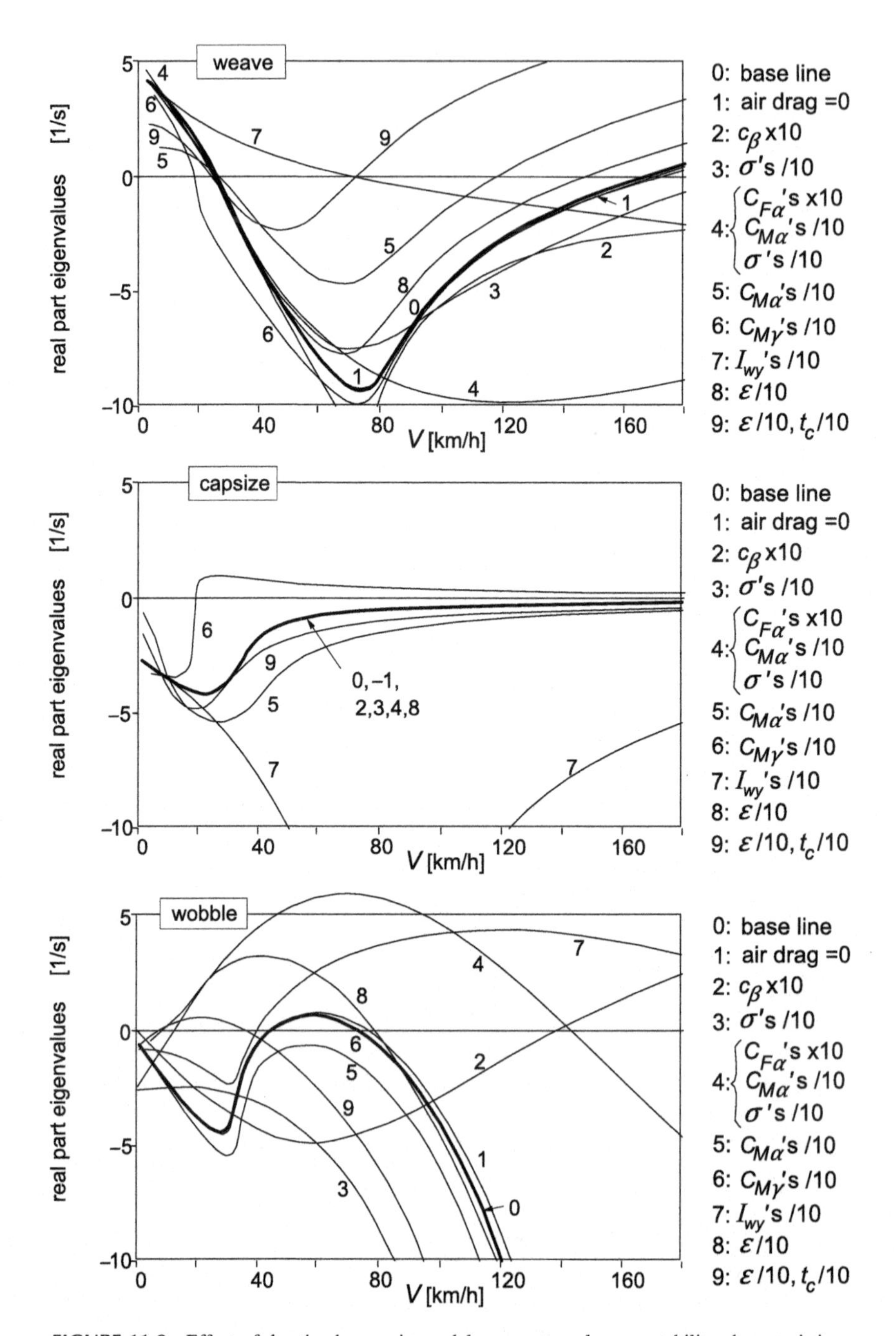


**FIGURE 11.8** Effect of drastic changes in model parameter values on stability characteristics.

7. The fundamental role of the *gyroscopic coupling* terms becomes evident when the polar moments of the wheel are considerably reduced (case 7). First, the low speed unstable weave range is stretched to almost 70 km/h and second, the wobble becomes violently unstable up to very high levels of speed. It may be concluded that the gyroscopic effect of the rotating wheels is largely responsible for the fact that the motorcycle is capable of moving in a stable manner.
8. Reduction of the *rake angle* to an almost vertical steer axis orientation with the caster length kept the same would strongly destabilize the wobble mode.
9. At the same time reducing the *caster length* as well (case 9) would strongly destabilize the weave mode, which unveils the fundamental effect of introducing the steer axis backward inclination.

Effects of more realistic levels of variation of the system parameters have been presented in the stability diagrams of Figure 11.9. Comparison with the curves that represent the baseline configuration (0) reveals that (again, curve numbers correspond to list numbers below):

- A 20% smaller *front relaxation length* stabilizes the wobble mode and has practically no effect on the weave and capsize modes.
- A smaller *rear relaxation length* stabilizes the wobble mode to a lesser degree and stabilizes the weave mode. No effect on the capsize mode.
- Moving the *center of gravity* of the mainframe plus rider 10 cm *forward* turns out to strongly stabilize the weave mode. However, it destabilizes the wobble. Capsize is not affected.
- *Lowering* both *center of gravity* over 10 cm stabilizes both the weave and wobble modes a little. The capsize mode becomes more stable.
- Adding *steer damping* does, of course, stabilize the wobble mode. However, it adversely affects the weave mode stability.

For the free control situation, we finally consider the influence of driving or braking and the addition of a degree of freedom: the rider lean angle measured with respect to the mainframe roll angle.

Figure 11.10 shows the stability diagram of the baseline configuration at zero net acceleration force (11.18) when $F_{ax} = 0$ (rear wheel driving force just withstands the air drag so that speed is constant), at braking (11.25) $F_{ax} = -1500$ N, and at driving (11.26) $F_{ax} = 1500$ N. It is observed that braking destabilizes the wobble mode, which is mainly due to the increased normal load at the front wheel that gives rise to larger relaxation length and cornering and aligning stiffnesses, the effects of which were shown in Figure 11.9. The terms with $F_{x1}\delta$ in Eqn (11.86) (destabilizing) and with $h_\beta\ F_{x1}\beta$ in Eqn (11.89) (stabilizing) have a smaller effect. The terms in Eqns (11.85) and (11.89) with $F_{x1}$ appear to largely neutralize the stabilizing effects of the increased front wheel load.

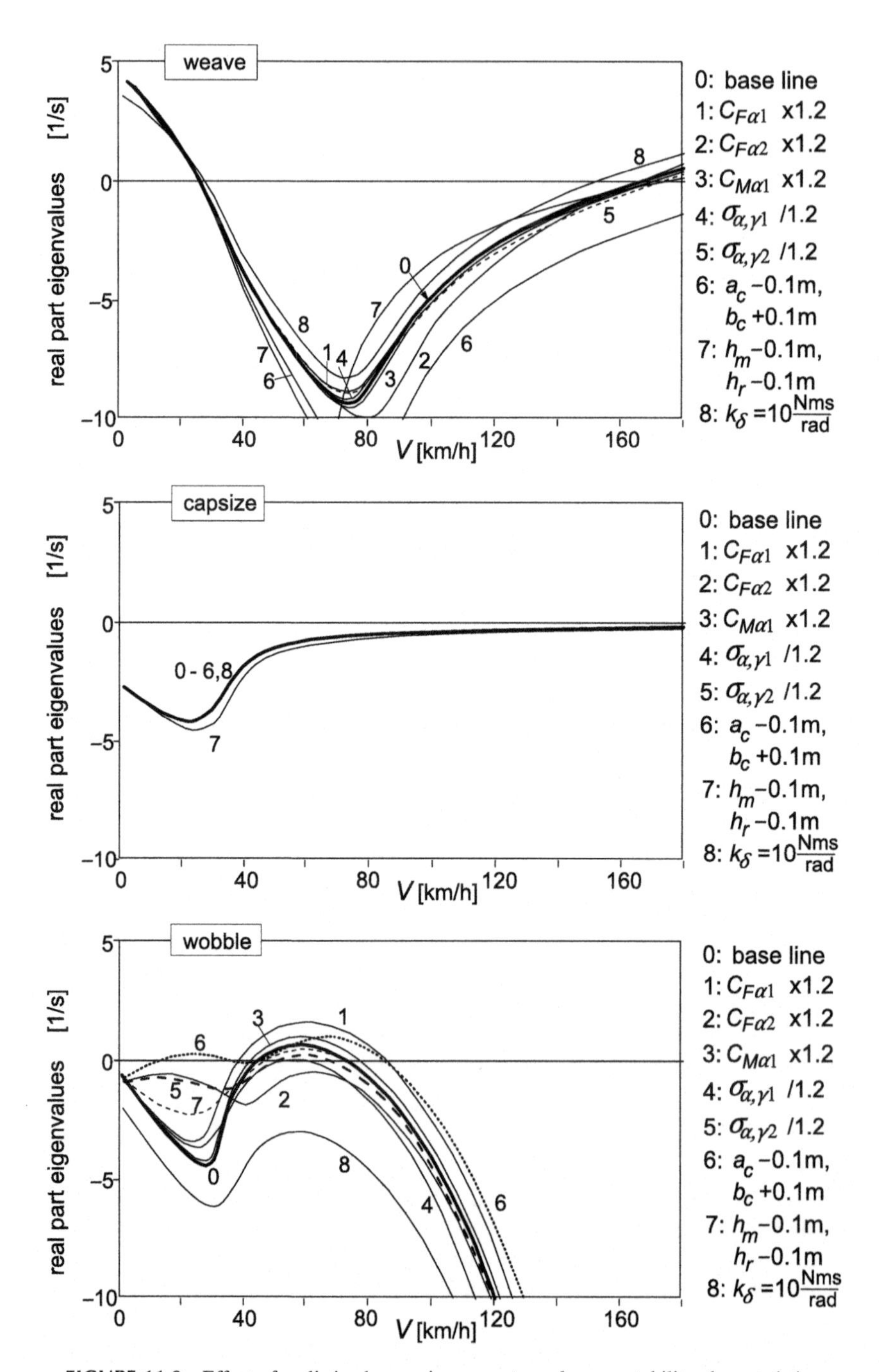


**FIGURE 11.9** Effect of realistic changes in parameter values on stability characteristics.

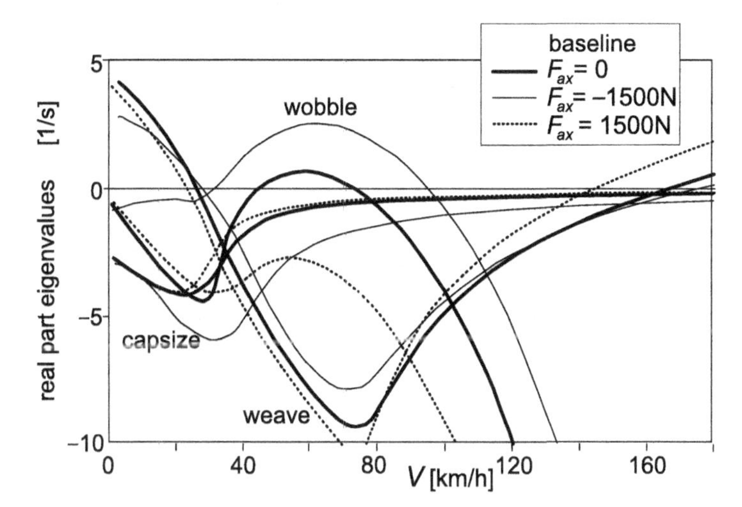


**FIGURE 11.10** Effect of the application of braking ($F_{ax}$, $F_{x1,2} < 0$) and driving forces ($F_{ax}$, $F_{x2} > 0$) on the stability diagram (baseline configuration).

When driving, the longitudinal force is generated at the rear wheel only. Obviously, wobble is completely suppressed. Weave, however, appears to show up already at considerably lower speeds.

As depicted in Figure 11.11, the introduction of the lean angle of the rider as an additional generalized coordinate with torsional stiffness and damping with respect to the mainframe gives rise to a decrease of the critical velocity

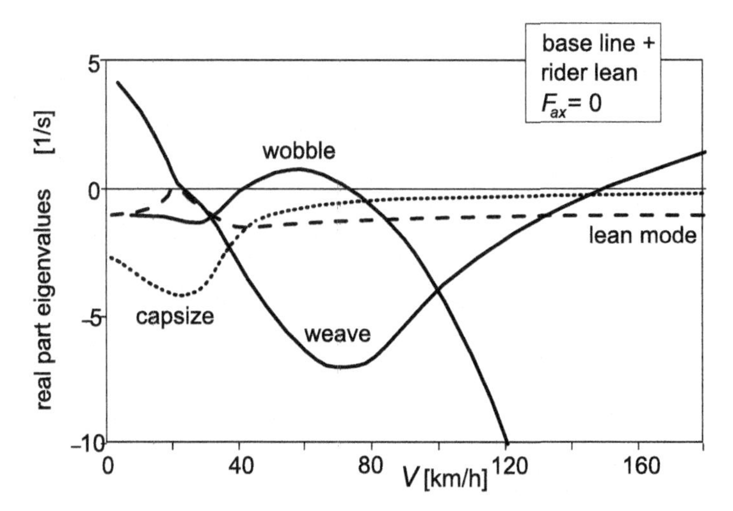


**FIGURE 11.11** The stability diagram with the rider lean degree of freedom introduced.

for weave instability while the wobble mode is hardly affected. The lean angle is introduced to enable the study of the response of the vehicle motion to a step change in lean torque $M_r$, which is defined to act between the upper torso and the mainframe about the longitudinal lean axis.

### 11.4.2. Step Responses of Controlled Motion

We will study the responses of the vehicle motion to a unit step of the imposed steer torque $M_\delta$ and of the lean torque $M_r$. This, as a first attempt to investigate the transient handling quality of the vehicle. First, the motorcycle must be stabilized in the range of speed we want to cover. For this, the various feedback signals have been provided with gains defined by Eqn (11.105) and Table 11.2. In Figure 11.12 the six gains have been presented as a function of the speed of travel. Similar functions have been used by Ruijs (1985) in the feedback control loops to stabilize the unmanned motorcycle with a stabilizing rider-robot. The peculiar change in sign that occurs in the gain from roll rate to steer torque is essential to stabilize both the low- and the high-speed weave.

The resulting stability diagram is presented in Figure 11.13. Comparison with Figure 11.10 shows that adopted feedback gains considerably enlarge the stable velocity range, especially at the low end. An additional control action using the lean torque (and as a consequence moving the rider c.g. laterally) would be necessary to further push back the low speed instability which here appears to correspond to an unstable lean mode. At such low speeds, steering would be almost ineffective to move the contact point laterally and thereby stabilize the motion.

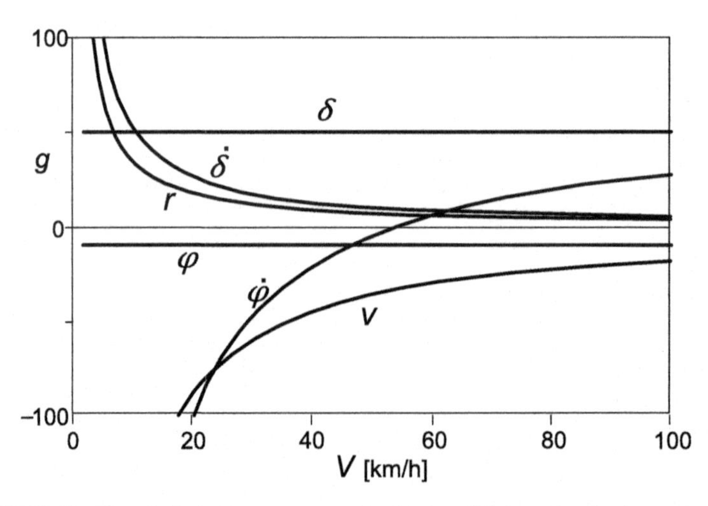


**FIGURE 11.12** Hypothetical rider steer torque feedback stabilizing control gains versus speed.

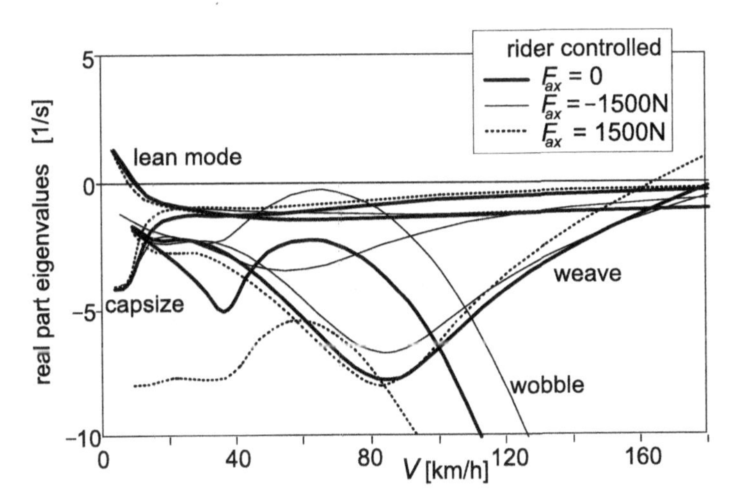


**FIGURE 11.13** The stability diagram with feedback steer torque control of Figure 11.12 activated.

In Figure 11.14 the resulting unit step responses have been shown for the case of zero $F_{ax}$. The figure depicts the variations of the yaw rate $r$, the steer angle $\delta$, the mainframe roll angle $\varphi$, and the upper torso roll angle $\varphi + \varphi_r$. Two values of speed have been considered: 35 and 70 km/h. The results are

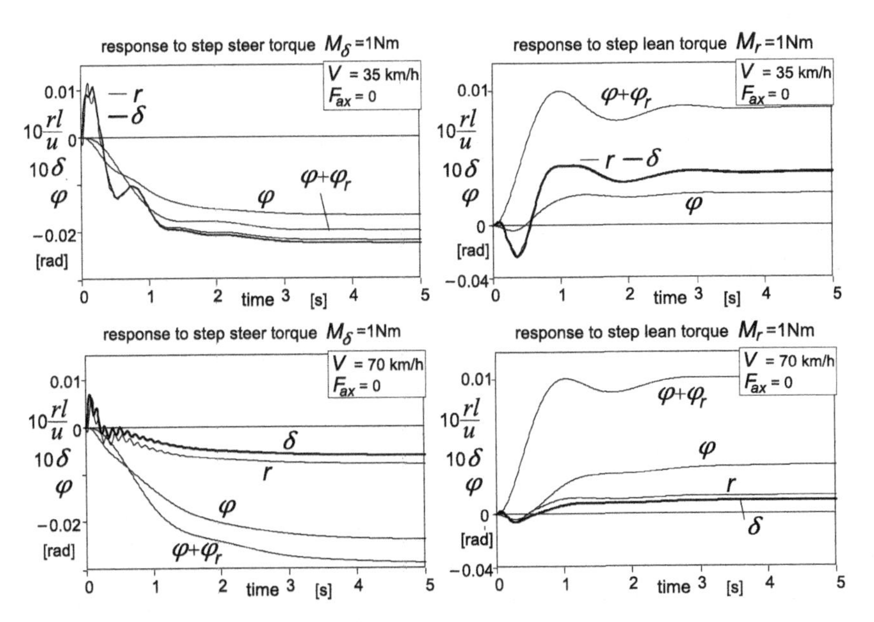


**FIGURE 11.14** The step responses of motion variables to unit steer and rider lean torque at zero forward acceleration, $F_{ax} = 0$ (with rider stabilizing control).

most interesting. We observe that, as a result of a positive change in steer torque (to the right), first a positive steer angle arises with also positive yaw rate (to the right). After a short time, a change in sign occurs and a negative yaw rate is being developed with a negative steer angle. It is seen that during this short transition time the roll angle begins to build up in the correct direction (to the left) that belongs to the final steady-state curving situation. This variation in the motion variables is similar to what would be observed in reality. The uncontrolled vehicle is stable at the speed $V = 35$ km/h (Figure 11.11). If at this speed the stabilization controller would not be activated, the motorcycle exhibits a similar response but with angles and yaw rate becoming larger and with less damped wobble and weave vibrations.

The lean torque response shows initially a negative roll angle, which is the result of counteracting the imposed (internal) lean torque. After the initial phase, the angle becomes positive. Also the steer angle and with that the yaw rate turn out to become negative in the initial phase which causes the contact points to move in the right direction (to the left). The steady-state situation is reached after low-frequency damped lean mode oscillations. If the controller is not activated, a very similar response with now low-frequency damped weave oscillations (Figure 11.11) occurs. Now, it becomes clear that it is mainly the gyroscopic couple (plus camber aligning moment) that steers the front wheel initially in the right direction. Analysis shows that the hardly visible small, positive, steer angle peak right after the lean torque step change is due to the mechanical trail and the aligning stiffness of the front wheel ($t_c$ and $C_{M\alpha 1}$).

At the higher speed of $V = 70$ km/h the responses show the same tendencies but with much smaller steer angles and yaw rates. The weave is more damped and the wobble less. Without the controller the wobble becomes unstable and a violent steering oscillation at ca. 9 Hz is developed.

With the lean degree of freedom disregarded, only the response to steer torque can be considered. The resulting motion responses turn out to get close to those depicted in the left-hand diagrams of Figure 11.14 with the curves for $\varphi + \varphi_r$ deleted. The low-frequency oscillation that is predominant at the lower speed is now attributed to the weave mode. The uncontrolled system (Figure 11.7) is just stable at 35 km/h and behaves similar to the controlled one albeit that a somewhat larger weave and wobble vibrations and a larger steady-state response arise. At 70 km/h the unstable wobble vibration develops again.

Figure 11.15 presents the results for the system while the brakes are applied ($F_{ax} = -1500$ N). The course of the various signals show similar tendencies as occurs in the unbraked situation. The obvious differences are the lowly damped wobble vibration that is seen to occur at 70 km/h (cf. Figure 11.13) and the much smaller steady-state responses (cf. Figure 11.20 of next section). These differences are primarily due to the accompanied load transfer.

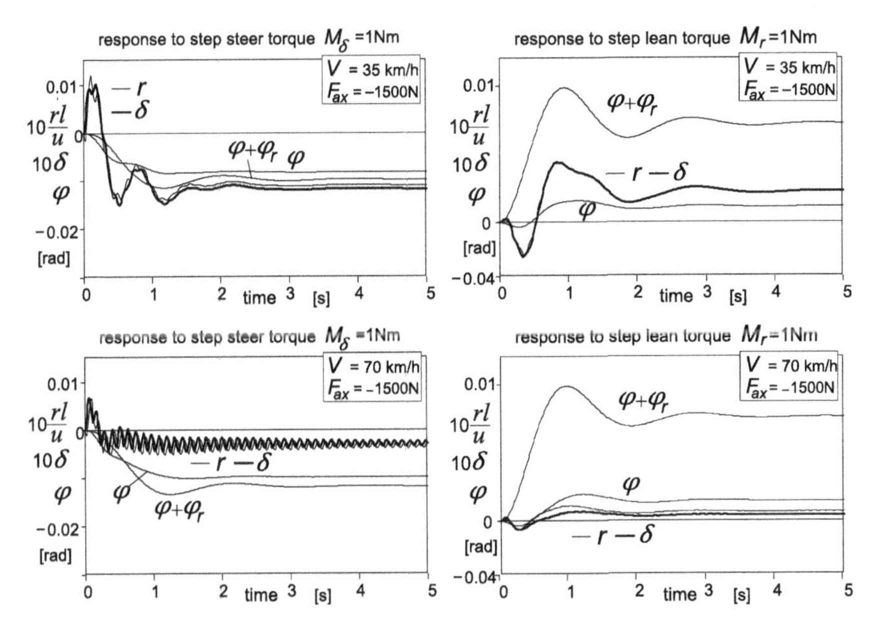


**FIGURE 11.15** The step responses of motion variables to unit steer and rider lean torque while braking, $F_{ax} = -$ 1500N (with rider stabilizing control).

## 11.5. ANALYSIS OF STEADY-STATE CORNERING

When a steady-state cornering condition is reached equilibrium of forces and moments exists. We consider the equilibrium in lateral direction, about the vertical axis, and about the line of intersection of the mainframe center plane and the road plane, this in addition to the static equilibrium of forces in the vertical direction. In the approximate analysis the small c.g. forward offsets of the steerable front and subframes $e_f$ and $e_s$ will be neglected. In addition, the twist angle $\beta$ and the rider lean angle $\varphi_r$ will be disregarded. Instead, we will study the effect of a sideways shift $y_{mr}$ of the center of gravity of the combined mainframe and rider. Once the side forces and the roll angle have been assessed at a given lateral acceleration, the slip angles can be determined, and from their difference with the given non-dimensional path curvature, $l/R$, the steer angle can be estimated. By considering the equilibrium about the steering axis the steer torque needed to maintain the curving motion is derived.

First, an approximate linearized theory is developed. The findings will be compared with the exact steady-state solutions of the linear differential Eqns (11.85)–(11.94) with the $\varphi_r$ degree of freedom deleted. After that, the theory is extended to larger roll angles and the non-linear handling diagram is established for the steadily cornering motorcycle.

### 11.5.1. Linear Steady-State Theory

The theoretical expressions for the coefficients concerning the steady-state turning behavior and the c.g. lateral offset effects will be developed in several successive stages starting with the simplest case in which tire moments, gyroscopic effects, and air drag are disregarded and ending with the ultimate configuration in which all these items are included and driving or braking may be considered. First, the required roll and steer angles will be assessed and, with these results, the expression for the steer torque determined.

#### *Wheel Moments and Air Drag Neglected*

As an introduction, we will first discuss the simple case where the air drag, longitudinal tire forces, and all the tire moments about the vertical axis are disregarded. That means that the aligning moment, the overturning couple, and the spin (camber) moment are set equal to zero. Moreover, we neglect the gyroscopic couples.

For the lateral wheel forces we find as in Chapter 1, Subsection 1.3.2, Eqn (1.61) for the automobile with pneumatic trails neglected (equilibrium in the lateral direction and about the vertical axis)

$$\frac{F_{y1}}{F_{z1o}} = \frac{F_{y2}}{F_{z2o}} = \frac{a_y}{g} \quad \left(= \frac{ur}{g}\right) \tag{11.107}$$

The roll angle $\varphi_y$ needed to maintain equilibrium about the longitudinal $x$ axis when a lateral offset $y_{mr}$ of the center of gravity of the combined main-frame and rider exists (cf. Figures 11.1 and 11.2) equals

$$\varphi_y = -\frac{m_{mr} y_{mr}}{mh} \tag{11.108}$$

The additional roll angle of the motorcycle that arises while cornering becomes

$$\varphi - \varphi_y = \frac{a_y}{g} \quad \left(= \frac{ur}{g}\right) \tag{11.109}$$

The front and rear camber angles read in terms of the roll angle and the ground steer angle, cf. Eqns (11.12)–(11.14):

$$\gamma_1 = \varphi + \delta' \tan \varepsilon \tag{11.110}$$

$$\gamma_2 = \varphi \tag{11.111}$$

The slip angles result from Eqn (11.29), primes omitted, with (11.107–11.111). We find:

$$C_{F\alpha 1}\alpha_1 = (F_{z1o} - C_{F\gamma 1})\varphi + C_{F\gamma 1}\delta' \tan \varepsilon - F_{z1o}\varphi_y \tag{11.112}$$

$$C_{F\alpha 2}\alpha_2 = (F_{z2o} - C_{F\gamma 2})\varphi - F_{z2o}\varphi_y \tag{11.113}$$

These equations show the importance of the difference between camber stiffness (N/rad) and initial tire load (N) in view of the sign of the slip angles. Obviously, in general, side slip is needed to accomplish lateral/yaw equilibrium. It may be realized that at higher speeds the steer angle is much smaller than the roll angle, $|\delta'| \ll |\varphi|$.

With the wheel base $l = a_c + b_c$ and the Eqn (11.15) we find the relationship between the ground steer angle $\delta' (= \delta \cos \varepsilon)$ on the one hand and the non-dimensional path curvature $l/R = lr/u$ and the difference of slip angles on the other:

$$\delta' = \frac{l}{R} + \alpha_1 - \alpha_2 \tag{11.114}$$

With the use of (11.112 and 11.113) the ground steer angle may now be written in terms of the non-dimensional path curvature $l/R$, the lateral acceleration $a_y$, and the c.g. offset $y_{mr}$:

$$\zeta\delta' = \frac{l}{R} + \eta\frac{a_y}{g} + \eta_y\frac{m_{mr}y_{mr}}{mh} \tag{11.115}$$

where we have introduced the steer angle coefficient $\zeta$ which is a bit larger than unity:

$$\zeta = \zeta_{oo} = 1 + \frac{C_{F\gamma 1}}{C_{F\alpha 1}}\tan \varepsilon \tag{11.116}$$

the understeer coefficient $\eta$ of the simplified system,

$$\eta = \eta_{oo} = \frac{F_{z1o}}{C_{F\alpha 1}} - \frac{F_{z2o}}{C_{F\alpha 2}} - \left(\frac{C_{F\gamma 1}}{C_{F\alpha 1}} - \frac{C_{F\gamma 2}}{C_{F\alpha 2}}\right) \tag{11.117}$$

and the c.g. offset coefficient $\eta_y$,

$$\eta_y = \eta_{yoo} = \frac{C_{F\gamma 1}}{C_{F\alpha 1}} - \frac{C_{F\gamma 2}}{C_{F\alpha 2}} \tag{11.118}$$

Since we have proportions occurring in the right-hand member of (11.117) which are all of the same order of magnitude it is evident that the understeer coefficient of the single track vehicle is quite different from its counterpart that is found for the automobile where only the first two terms appear in the simplified analysis while $\zeta$ is assumed equal to one. The ratio of camber and cornering stiffness front relative to rear is of decisive importance for the sign and magnitude of the understeer coefficient and consequently also for the difference in slip angle front and rear which follows from (11.114 and 11.115):

$$\zeta(\alpha_1 - \alpha_2) = (1 - \zeta)\frac{l}{R} + \eta\frac{a_y}{g} + \eta_y\frac{m_{mr}y_{mr}}{mh} \tag{11.119}$$

where the first term in the right-hand member vanishes if the rake angle equals zero making $\zeta = 1$.

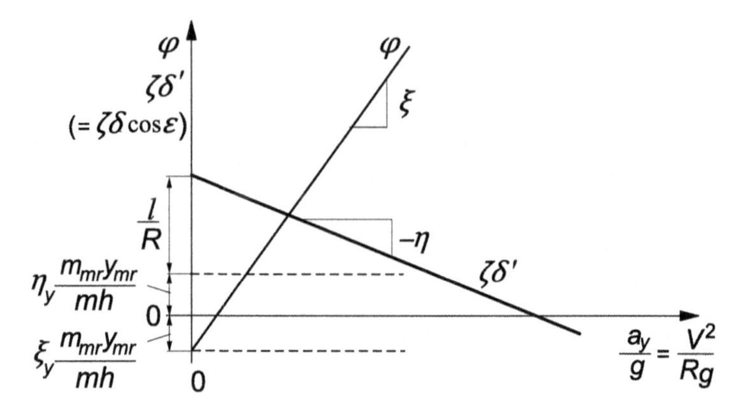


**FIGURE 11.16** The ground steer angle $\delta'$ and the roll angle $\varphi$ due to path curvature $1/R$ and c.g. lateral offset $y_{mr}$ and the variation with lateral acceleration $a_y$.

When the motorcycle moves straight ahead ($l/R = a_y = 0$) Eqns (11.115 and 11.119) show that a steer angle and difference in slip angles would arise solely as a result of the c.g. offset and depend on the difference of the ratios of camber and side slip stiffnesses front and rear.

The general course of steer angle and roll angle versus speed of travel at a given path curvature and c.g. lateral offset has been illustrated in Figure 11.16. The coefficients $\xi$ and $\xi_y$ will be defined below.

The actual steer angle $\delta$ of the handlebar about the inclined steer axis with rake angle $\varepsilon$ is obtained from the ground steer angle $\delta'$ by dividing this angle by cos $\varepsilon$ as has been formulated by Eqn (11.12) with twist angle $\beta$ set equal to zero.

If fore-and-aft load transfer due to aerodynamic drag is considered, the lines of Figure 11.16 are not quite straight anymore. The understeer coefficient $\eta$ corresponds to the slope of the line in the diagram if air drag is neglected. This may be compared with the theory for the automobile where $\zeta$ is taken equal to unity and air drag is disregarded (Figure 1.10). If, at high speed, air drag is not negligible, $\eta$ is found as the slope of the characteristic that arises if $\zeta\delta' - l/R$ is plotted against $a_y/g$ with the speed $V$ held constant while the curvature increases with growing lateral acceleration. If we would plot $\zeta\delta'$ versus $a_y/g$, again, in this linear theory with constant speed, a straight line arises which now has a slope equal to $lg/V^2 + \eta$. For the sake of convenience the above definition for the degree of understeer has been adopted although a more proper definition might be $\eta' = (\partial\delta / \partial(a_y/g))_R = (\eta/\zeta)/\cos\varepsilon$.

Discussion of numerical results (Table 11.3) follows after more adequate approximations have been developed. For the simple theory developed so far the coefficients $\xi$ and $\xi_y$ appearing in the diagram are equal to unity. Below, we will see the effect of tire width and gyroscopic couples on these roll angle coefficients.

**TABLE 11.3 Handling Coefficients with Tire Aligning Moments Neglected. $F_{z1o} = 1732$ N, $F_{z2o} = 2094$ N, cos $\varepsilon = 0.878$; Eqns (11.116–11.118 and 11.123–11.129)**

| $V$ km/h | $F_d$ N | $F_{z1}$ N | $F_{z2}$ N | $\xi$ – | $\xi_y$ – | $\zeta_o$ – | $\eta_{oo}$ – | $\eta_o$ – | $\eta_{yoo}$ – | $\eta_{yo}$ – |
|---|---|---|---|---|---|---|---|---|---|---|
| 0 | 0 | 1732 | 2094 | 1.214 | 1.18 | 1.031 | 0 | 0 | 0 | 0 |
| 160 | 0 | 1732 | 2094 | 1.214 | 1.18 | 1.031 | 0 | 0 | 0 | 0 |
| 160 | 395 | 1534 | 2291 | 1.217 | 1.183 | 1.03 | 0.013 | 0.016 | −0.01 | −0.01 |

## *Tire Overturning and Gyroscopic Couples Included*

Closer consideration of the roll equilibrium around the line of intersection reveals that instead of (11.109) we actually have with also the caster length $t_c$ taken into account (cf. Eqn (11.87) with (11.2)):

$$mgh\varphi - mha_y - \left(\frac{I_{wy1}}{r_1} + \frac{I_{wy2}}{r_2}\right)ur + F_{z1}t_c\delta - M_{x1} - M_{x2} - m_{mr}gy_{mr} = 0 \tag{11.120}$$

The third term represents the sum of gyroscopic couples. With Eqns (11.13), (11.14) and (11.31), we obtain for the roll angle:

$$\varphi = \frac{(mh + I_{wy1}/r_1 + I_{wy2}/r_2)a_y + (C_{Mx\gamma1}\sin\varepsilon - F_{z1}t_c)\delta - m_{mr}gy_{mr}}{mgh - C_{Mx\gamma1} - C_{Mx\gamma2}} \tag{11.121}$$

Inspection of the values of parameters as listed in Tables 11.1 and 11.2 leads to the conclusion that the steer angle $\delta$ has a negligible effect on the relationship between roll angle $\varphi$ and $y_{mr}$ and also between $\varphi$ and lateral acceleration $a_y = ur$ if $\delta \ll \sim 30(a_y/g)$. With $\delta$ expressed in radians it is expected that at not too low speed, this condition will soon be satisfied in the more interesting range of operation.

We will henceforth use the approximate expression for the roll angle with the $\delta$ term in (11.121) neglected:

$$\varphi = \xi\frac{a_y}{g} - \xi_y\frac{m_{mr}y_{mr}}{mh} \tag{11.122}$$

where the tilt coefficients $\xi$ and $\xi_y$ read:

$$\xi = g\,\frac{mh + I_{wy1}/r_1 + I_{wy2}/r_2}{mgh - C_{Mx\gamma1} - C_{Mx\gamma2}} \tag{11.123}$$

$$\xi_y = \frac{mgh}{mgh - C_{Mx\gamma1} - C_{Mx\gamma2}} \tag{11.124}$$

The quantities appear to be a little larger than unity due to the gyroscopic action and the width of the tires (finite crown radii). The tilt coefficients become in the baseline configuration (Tables 11.1 and 11.2):

$$\xi = 1.21, \quad \xi_y = 1.18 \tag{11.125}$$

An effective tilt angle $\varphi'$ (the angle of the dashed line, through tire contact point and mass center, indicated in Figure 11.1 with respect to the vertical) may be defined. We have:

$$\varphi' = \frac{\varphi}{\xi_y} \tag{11.126}$$

With the additional moments about the $x$ axis now introduced, we find for the understeer coefficient:

$$\eta = \eta_o = \frac{F_{z1o}}{C_{F\alpha1}} - \frac{F_{z2o}}{C_{F\alpha2}} - \xi\left(\frac{C_{F\gamma1}}{C_{F\alpha1}} - \frac{C_{F\gamma2}}{C_{F\alpha2}}\right) \tag{11.127}$$

and the c.g. offset coefficient:

$$\eta_y = \eta_{yo} = \xi_y\left(\frac{C_{F\gamma1}}{C_{F\alpha1}} - \frac{C_{F\gamma2}}{C_{F\alpha2}}\right) \tag{11.128}$$

while the steer angle coefficient $\zeta$ remains unchanged:

$$\zeta = \zeta_o = 1 + \frac{C_{F\gamma1}}{C_{F\alpha1}} \tan \varepsilon \tag{11.129}$$

The numerical values listed in Table 11.3 will be discussed later on.

### *Tire Yaw Moments Included*

Because of the obvious sensitivity to slight variations in tire parameters we should examine the influence of the remaining (yaw) moments due to side slip and camber acting on the tire. When taking into account the tire aligning torques and the moment applied by the air drag force (assumed to act parallel to the $x$ axis) we find for the tire side forces from the equilibrium conditions:

$$F_{y1} = C_{F\alpha1}\alpha_1 + C_{F\gamma1}\gamma_1 = \frac{1}{l}(bma_y - h_d F_d \varphi - M_{z1} - M_{z2}) \tag{11.130}$$

$$F_{y2} = C_{F\alpha2}\alpha_2 + C_{F\gamma2}\gamma_2 = \frac{1}{l}(ama_y + h_d F_d \varphi + M_{z1} + M_{z2}) \tag{11.131}$$

in which the lateral acceleration $a_y$ can, with (11.122), be expressed in terms of the roll angle $\varphi$ and the c.g. offset $y_{mr}$. The aligning torques are

$$M_{zi} = -C_{F\alpha i}t_{\alpha i}\alpha_i + C_{M\gamma i}\gamma_i \tag{11.132}$$

with camber aligning torque stiffness that without the action of the longitudinal tire force is defined by (11.36). The camber angles can be expressed in terms of the roll angle and the slip angles by using the relations (11.109, 11.110, and 11.113). We have:

$$\gamma_1 = \varphi + \delta' \tan\varepsilon = \varphi + \left(\frac{l}{R} + \alpha_1 - \alpha_2\right)\tan\varepsilon \tag{11.133}$$

$$\gamma_2 = \varphi \tag{11.134}$$

From the thus created two equations the two unknown slip angles $\alpha_i$ can be solved and expressed in terms of the known path curvature $l/R$, roll angle $\varphi$, and c.g. offset $y_{mr}$. By substituting the expression (11.114):

$$\delta' = \frac{l}{R} + \alpha_1 - \alpha_2$$

in the relationship (11.115):

$$\zeta\delta' = \frac{l}{R} + \eta\frac{a_y}{g} + \eta_y\frac{m_{mr}y_{mr}}{mh}$$

and using again (11.122) the steer angle, understeer, and c.g. offset coefficients $\zeta$, $\eta$, and $\eta_y$ can finally be assessed. The resulting expressions read:

$$\zeta = 1 + \left\{\frac{C_{F\gamma 1}}{C_{F\alpha 1}} + \left(\frac{1}{C_{F\alpha 1}} + \frac{1}{C_{F\alpha 2}}\right)\frac{t_{\alpha 1}C_{F\gamma 1} + C_{M\gamma 1}}{l^*}\right\}\tan\varepsilon \tag{11.135}$$

with the effective wheel base

$$l^* = l - t_{\alpha 1} + t_{\alpha 2} \tag{11.136}$$

and

$$\eta = \lambda_1\frac{F_{z1o} - \xi C_{F\gamma 1}}{C_{F\alpha 1}} - \lambda_2\frac{F_{z2o} - \xi C_{F\gamma 2}}{C_{F\alpha 2}} - \frac{\xi}{l^*}\left(\frac{1}{C_{F\alpha 1}} + \frac{1}{C_{F\alpha 2}}\right)(C_{M\gamma 1} + C_{M\gamma 2} + F_d h_d) \tag{11.137}$$

and

$$\eta_y = \xi_y\left\{\lambda_1\frac{C_{F\gamma 1}}{C_{F\alpha 1}} - \lambda_2\frac{C_{F\gamma 2}}{C_{F\alpha 2}} + \frac{1}{l^*}\left(\frac{1}{C_{F\alpha 1}} + \frac{1}{C_{F\alpha 2}}\right)(C_{M\gamma 1} + C_{M\gamma 2} + F_d h_d)\right\} \tag{11.138}$$

**TABLE 11.4** Handling Coefficients with Aligning Moments and Driving and Braking Included. $F_{z1o} = 1732$ N, $F_{z2o} = 2094$ N, $\cos\varepsilon = 0.878$; Eqns (11.123), (11.124), (11.135)–(11.138), (11.144)–(11.146)

| $V$ km/h | $F_d$ N | $F_{ax}$ N | $F_{x1}$ N | $F_{x2}$ N | $F_{z1}$ N | $F_{z2}$ N | $\xi$ – | $\xi_y$ – | $\zeta$ – | $\eta$ – | $\eta_y$ – |
|---|---|---|---|---|---|---|---|---|---|---|---|
| 1 | 0 | 0 | 0 | 0 | 1732 | 2094 | 1.214 | 1.18 | 1.034 | 0 | 0.014 |
| 160 | 0 | 0 | 0 | 0 | 1732 | 2094 | 1.214 | 1.18 | 1.034 | 0 | 0.014 |
| 160 | 395 | 0 | 0 | 395 | 1534 | 2292 | 1.217 | 1.183 | 1.033 | 0 | 0.023 |
| 1 | 0 | 1500 | 0 | 1500 | 1137 | 2689 | 1.222 | 1.188 | 1.028 | 0 | 0 |
| 1 | 0 | −1500 | −912 | −588 | 2327 | 1499 | 1.207 | 1.173 | 1.008 | 0 | 0.034 |
| 160 | 395 | 1500 | 0 | 1895 | 940 | 2886 | 1.225 | 1.19 | 1.026 | 0 | 0 |
| 160 | 395 | −1500 | −615 | −490 | 2129 | 1697 | 1.209 | 1.176 | 1.016 | 0 | 0.042 |

in which the coefficients $\lambda_1$ and $\lambda_2$ have been introduced:

$$\lambda_1 = 1 + \frac{1}{l^*}\frac{C_{F\alpha1} + C_{F\alpha2}}{C_{F\alpha2}} t_{\alpha1} \tag{11.139}$$

$$\lambda_2 = 1 - \frac{1}{l^*}\frac{C_{F\alpha1} + C_{F\alpha2}}{C_{F\alpha1}} t_{\alpha2} \tag{11.140}$$

which are close to unity (around 1.03 and 0.95, respectively, in the baseline configuration changing a little depending on load transfer).

In Table 11.4, the values that $\eta$ and $\eta_y$ take in the baseline configuration (Tables 11.1 and 11.2) for the cases without and with air drag (in the latter case at speeds 1 and 160 km/h) have been listed together with the current wheel loads and air drag force.

### *Driving and Braking Forces Included*

Applying longitudinal forces to the tires through braking or driving have three effects on the handling behavior of the vehicle occurring through the tires. First, we have the direct effect (only at braking) due to the sideways component of the front wheel longitudinal tire force that arises as a result of the steer angle. Then, we have two indirect effects brought about by changes in tire behavior due to the longitudinal force (notably in the camber aligning torque coefficient $C_{M\gamma}$), and due to the induced load transfer. In addition, the longitudinal acceleration inertia forces acting at the mass centers exert a moment about the vertical $z$ axis if we have a roll angle $\varphi$ and possibly a c.g. lateral offset $y_{mr}$. The front and rear lateral tire forces now become, instead of the right-most members of Eqns (11.130) and (11.131),

$$F_{y1} = \frac{1}{l}\{bma_y - lF_{x1}\delta' - (h_dF_d\varphi + M_{z1} + M_{z2} + ma_xh\varphi + m_{mr}a_xy_{mr})\} \tag{11.141}$$

$$F_{y2} = \frac{1}{l}\{ama_y + (h_dF_d\varphi + M_{z1} + M_{z2} + ma_xh\varphi + m_{mr}a_xy_{mr})\} \tag{11.142}$$

with the tire aligning torques according to (11.132) but with the total camber aligning torque stiffness now expressed as (according to (11.30)):

$$C_{M\gamma i} = C'_{M\gamma i} - r_{ci}F_{xi} \tag{11.143}$$

The same procedure is followed as before and we find the coefficients $\zeta$, $\eta$, and $\eta_y$ augmented for the presence of vehicle acceleration $a_x$ and longitudinal tire forces $F_{x1}$. The terms that are to be added to the expressions (11.135, 11.137, and 11.138) (for zero acceleration, $F_{ax} = 0$, cf. (11.19)) read respectively:

$$\Delta\zeta_{Fx} = \lambda_1\frac{F_{x1}}{C_{F\alpha 1}} \tag{11.144}$$

$$\Delta\eta_{Fx} = -\frac{\xi}{l^*}\left(\frac{1}{C_{F\alpha 1}} + \frac{1}{C_{F\alpha 2}}\right)mha_x \tag{11.145}$$

$$\Delta\eta_{y,Fx} = -\frac{\xi_y}{l^*}\left(\frac{1}{C_{F\alpha 1}} + \frac{1}{C_{F\alpha 2}}\right)(C_{Mx\gamma 1} + C_{Mx\gamma 2})\frac{a_x}{g} \tag{11.146}$$

### *Numerical Results*

For the parameter values of the baseline configuration the various handling coefficients have been presented in Tables 11.3 and 11.4 together with the calculated air drag, wheel loads, and possibly braking or driving forces (for a given acceleration force $F_{ax}$).

Table 11.3 shows the values obtained when using the simpler expressions for the coefficients which result when the gyroscopic and overturning couples and/or the aligning torques are neglected, according to the theories covered by the Eqns (11.107–118) and (11.120–11.129). The influence of air drag results here from the induced load transfer. Table 11.4 presents the results when all the moments are accounted for and also braking and driving may be considered. The equations concerned are (11.130–140) and (11.141–146). First of all, when we compare the different stages of approximation, it is observed that the influence of the various wheel and tire moments is essential in generating more correct values of understeer and c.g. offset coefficients, $\eta$ and $\eta_y$.

The steer angle coefficient $\zeta$ is much less affected by the tire moments while the roll angle coefficients $\xi$ and $\xi_y$ are influenced considerably by the gyroscopic and/or overturning couple coefficients; note that these coefficients take the value one in the simplest approximation according to Eqns (11.108 and 11.109).

The last two columns of Table 11.4 indicate that, for the conditions considered, the understeer coefficient shows large changes but keeps the same negative sign, which means: steering less to the right for a right-hand turn if speed is increased. In automobile terms one would speak of an oversteered vehicle. However, the steer angle does not serve as an input quantity and the speed where the steer angle changes sign is not a critical speed beyond which instability occurs. It is the steer torque that acts as the input variable and when at a given path radius a change in sign of the steer torque would arise at a certain velocity, the system becomes unstable because in that situation the last term of the characteristic equation of the system becomes negative. We then have divergent instability corresponding to the capsize mode that becomes unstable beyond a critical speed as illustrated in the diagram of Figure 11.8, case 6, with very small camber aligning stiffness.

The sign of the c.g. offset coefficient appears to change from positive to negative in the case of accelerating at low speed. When the rider hangs to right-hand side ($y_{mr} > 0$ making $\varphi < 0$) while moving straight ahead, steering to the right is normally required to compensate for the camber forces pointing to the left. This appears to be true except when the rear wheel driving force produces sufficient positive yaw moment through the finite crown radius.

As was already clear from the relevant expressions, the influence of speed on the coefficients occurs if air drag is included in the model. At the high speed of 160 km/h the effect of air drag becomes quite noticeable. This would be much less at a speed of say 100 km/h because of the quadratic speed influence.

As was to be expected, the longitudinal tire forces have a large effect on the two coefficients. When driving with $F_{ax} = 1500$ N, coefficient $\eta$ changes from $-0.0177$ to $-0.0204$. At the high speed, much more driving force at the rear wheel is needed to overcome the air drag and realize the aimed acceleration. Oversteer has increased a lot with respect to the situation at zero acceleration (first and third row). The driving force and the opposite inertia force form a couple around the vertical axis (because of the roll angle) that tries to turn the vehicle more into the curve. At braking, more understeer arises caused by the forward inertia force exerting an outward couple about the $z$ axis and the sideways component of the front wheel braking force that does the same, trying to straighten the vehicle's path.

As an example, we might further analyze the case represented by the last row of Table 11.4. If the motorcycle would be negotiating a circular path with a radius of 200 times the wheel base, $l/R = 0.005$, and $R = 300$ m, at $V =$ 160 km/h (=44.4 m/s) the lateral acceleration becomes $a_y = V^2/R = 6.58$ m/s$^2$ $= 0.67\,g$. As a consequence, the roll angle becomes $\varphi = \xi\, a_y/g = 1.209 \times 0.67$ $= 0.81$rad $= 46°$. The ground steer angle takes the value: $\delta' = (l/R + \eta a_y/g)/\zeta$

$= -0.0027/1.016 = -0.00266$ rad $= -0.15°$ and the handlebar steer angle $\delta = \delta'/\cos\varepsilon = -0.17°$. It is obvious that with this high lateral acceleration the assumption of linearity is not valid. A smaller path curvature would have been a better choice. If the center of gravity of the mainframe plus rider is located a distance $y_{mr} = 1$ cm to the right of the vehicle center plane, a roll angle $\varphi = -\xi_y m_{mr} y_{mr}/mh = -1.176 \times 350 \times 0.01/(390 \times 0.59) = -0.018$ rad $= -1.03°$ results at straight line running. The ground steer angle is predicted to amount to $\delta' = \eta_y m_{mr} y_{mr}/mh = 0.0423 \times 350 \times 0.01/(390 \times 0.59) = 6.38 \times 10^{-4}$ rad $= 0.036°$and the handlebar steer angle $\delta = 0.041°$. In Figures 11.17 and 11.18 this case where the brakes are applied is further examined for speeds ranging from 1 to 100 km/h, also showing the variation of the slip angles.

In Figure 11.17 the left-hand diagram is presented for the non-accelerating vehicle with air drag included (rows 1 and 3 of Table 11.4) showing the variation of the roll angle and the ground steer angle vs speed squared. In addition, the variation of the front and rear slip angles has been depicted. The almost straight lines result from computations with the steady-state version of the differential Eqns (11.85–11.94) with the rider lean degree of freedom deleted. An additional (dotted) line shows the variation of the ground steer angle according to the approximate analytical results corresponding to those listed in Table 11.4. Only very slight differences appear to occur between the approximate ($\beta$ disregarded and the $\delta$ term in Eqn (11.121) neglected) and exact results for the ground steer angle, slip angles, and roll angle. According to the exact computations, the twist angle $\beta$ amounts to ca. 0.5% of the roll angle $\varphi$. Note that the difference in slip angles $\alpha_2 - \alpha_1$ practically equals $l/R$ minus the ground steer angle $\delta'$ ($\zeta$ being very close to unity). We may calculate the slip angles following the analytical approach through an explicit expression by

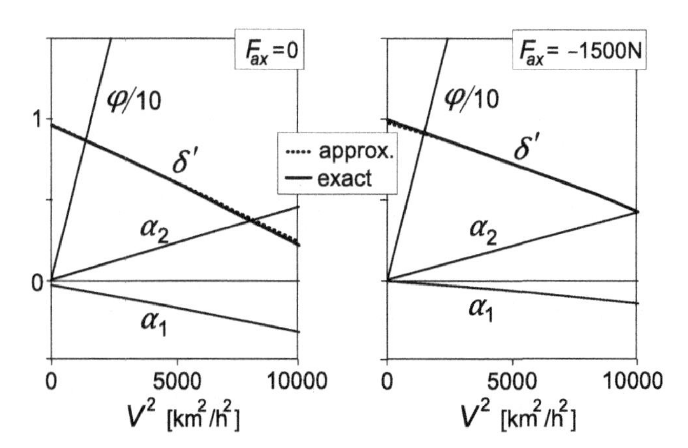


**FIGURE 11.17** Steer, roll, and slip angles per unit non-dimensional path curvature $l/R$ versus speed squared $V^2 = a_y R$ (left: non-accelerating, right: braking) according to the exact and approximate theories.

using the Eqns (11.148) and (11.149) for the lateral tire force components caused by side slip, given later on.

The right-hand diagram of Figure 11.17 refers to the situation that arises when the brakes are applied (fifth and last row of Table 11.4). Again, an excellent correspondence between the approximate and exact solutions occurs. Obviously, the vehicle performs now in a less oversteered manner as was already concluded from the less negative value of $\eta$ in Table 11.4. The front slip angle is less negative to make the side force more positive to counteract the sideways component of the front wheel braking force. At the same time, the rear slip angle is made slightly larger to help compensate for the positive yaw moment that arises from the braking forces, their points of application being shifted to the right because of the finite crown radii. Under this condition of braking, $\beta$ appears to become somewhat smaller.

Finally, Figure 11.18 shows the variation of the various angles per unit lateral c.g. offset. It is seen that the influence of speed is very small and that the agreement between approximate and exact results is good. Both slip angles are positive thereby neutralizing the camber forces which point to the left because of the negative roll angle that arises to compensate for the c.g. location lateral offset. At braking, the center of gravity remains above the line connecting the contact points of the tires. The forward load transfer reduces the rear camber force, which allows a decrease of the rear slip angle. As $\alpha_1$ increases only a little to balance both the increased camber force and the sideways component of the front wheel braking force, the steer angle needs to become larger to keep $\alpha_1 - \delta'$ equal to $\alpha_2$. In both cases, $\beta$ is equal to ca. 5% of $\varphi$ and will have a relatively large effect on the actual value of steer angle of the handlebar $\delta$ (cf. Eqn (11.12)).

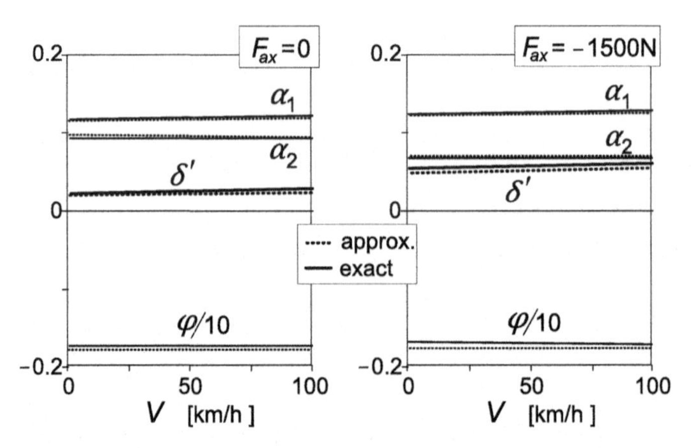


**FIGURE 11.18** Steer, roll, and slip angles per m lateral c.g. offset $y_{mr}$ that occur at straight line motion (according to the exact and approximate theories).

### The Steer Torque

By considering Eqn (11.89) with $\beta$ neglected the following steady-state expression for the steering couple is obtained:

$$\begin{aligned} M_\delta = & \ t_c F_{y1} \\ & -t_c F_{z1}\gamma_1 \\ & +(1/r_1) I_{wy1} a_y \sin\varepsilon \\ & -M_{x1}\sin\varepsilon \\ & -M_{z1}\cos\varepsilon \\ & +(m_f e_f + m_s e_s)(a_y - g\gamma_1) \end{aligned} \qquad (11.147)$$

As we did with the steer angle, we wish to develop an expression for the steer torque in terms of path curvature $l/R$, lateral acceleration $a_y$, and c.g. offset $y_{mr}$. For this, first the relationship with $a_y$, $\varphi$, $\gamma_1$, and $\delta'$ will be established. With the aid of relationships assessed before, the desired expression can be derived.

In the first term of expression (11.147) the side force of the front wheel appears, which is determined using Eqn (11.141). For this, and also for the fifth term of (11.147), the sum of aligning torques is needed. The torque due to side slip follows by multiplying the side force due to side slip $F_{y\alpha i}$ with the pneumatic trail $t_{\alpha i}$. This part of the side force is found by subtracting from the total side force the part due to camber. The also appearing side force due to side slip of the rear wheel can be eliminated by employing the expression for the sum of side slip forces obtained from (11.130, 11.131, 11.141, and 11.142):

$$F_{y\alpha 1} + F_{y\alpha 2} = ma_y - lF_{x1}\delta' - C_{F\gamma 1}\gamma_1 - C_{F\gamma 2}\varphi \qquad (11.148)$$

The following relation for the front wheel side slip force is finally obtained:

$$\begin{aligned} F_{y\alpha 1}(l - t_{\alpha 1} + t_{\alpha 2}) = & \ bma_y - (l + t_{\alpha 2})F_{x1}\delta' - (h_d F_d + mha_x)\varphi \\ & - m_{mr} y_{mr} a_x + t_{\alpha 2}(ma_y - C_{F\gamma 1}\gamma_1 - C_{F\gamma 2}\varphi) - C_{M\gamma 1}\gamma_1 - C_{M\gamma 2}\varphi - lC_{F\gamma 1}\gamma_1 \end{aligned} \qquad (11.149)$$

The expression for $F_{y\alpha 1}$ can be used to determine the slip angle $\alpha_1$ ($= F_{y\alpha 1}/C_{F\alpha 1}$) or directly the aligning torque due to the side slip $M_{z\alpha 1}$ by multiplying $-F_{y\alpha 1}$ with the pneumatic trail $t_{\alpha 1}$. Similarly, we obtain $M_{z\alpha 2}$. The remaining part of the aligning torque due to the wheel camber $M_{z\gamma 1}$ is obviously found by multiplying the camber aligning torque stiffness $C_{M\gamma i}$ (11.143) with the camber angle $\gamma_i$. The resulting expression for the steer torque turns out to read:

$$\begin{aligned} M_\delta = & \ \frac{t_c + t_{\alpha 1}\cos\varepsilon}{l^*}\{(b + t_{\alpha 2})ma_y - (l + t_{\alpha 1})C_{F\gamma 1}\gamma_1 - t_{\alpha 2}C_{F\gamma 2}\varphi \\ & - C_{M\gamma 1}\gamma_1 - C_{M\gamma 2}\varphi - lF_{x1}\delta' - (h_d F_d + mha_x)\varphi - m_{mr} y_{mr} a_x\} \\ & - \{t_c(F_{z1} - C_{F\gamma 1}) - C_{Mx\gamma 1}\sin\varepsilon + C_{M\gamma 1}\cos\varepsilon\}\gamma_1 \\ & + (I_{wy1}/r_1)\sin\varepsilon - a_y + (m_f e_f + m_s e_s)(a_y - g\gamma_1) \end{aligned} \qquad (11.150)$$

The front wheel camber angle $\gamma_1$ may be written in terms of $\varphi$ and $\delta'$ with Eqn (11.133). In their turn, $\varphi$ and $\delta'$ can be expressed in $l/R$, $a_y$, and $y_{mr}$ with the use of Eqns (11.122) and (11.115). At this stage we may improve the result for $M_\delta$ by approximately accounting for the term with $\delta$ in (11.121), which we neglected. Due to a steer angle, the contact point shifts sideways and a roll angle is needed to keep the motorcycle in balance. The approximate correction for $\varphi$ becomes

$$\Delta\varphi = -\frac{bt_c}{lh\cos\varepsilon}\delta' \tag{11.151}$$

with the steer angle equation

$$\zeta\delta' = \frac{l}{R} + \eta\frac{a_y}{g} + \eta_y\frac{m_{mr}y_{mr}}{mh} \tag{11.115}$$

and the roll angle Eqn (11.122), the corrected roll angle can be assessed:

$$\varphi = \xi\frac{a_y}{g} - \xi_y\frac{m_{mr}y_{mr}}{mh} - \frac{bt_c}{lh\cos\varepsilon}\delta' \tag{11.152}$$

and also the front wheel camber angle:

$$\gamma_1 = \varphi + \delta'\tan\varepsilon$$

Especially at low speeds and when the front brake is applied, the improvement can become considerable. Obviously, this is due to the camber aligning torque in which the longitudinal force may play a predominant role, cf. Eqn (11.143).

It is possible now to write Eqn (11.150) in the following non-dimensional form:

$$\frac{M_\delta}{F_{z1o}l} = \mu_R\frac{l}{R} + \mu_a\frac{a_y}{g} + \mu_y\frac{m_{mr}y_{mr}}{mgh} \tag{11.153}$$

The expressions for the steer torque coefficients are too elaborate to be reproduced here. Numerically, however, their values can be easily assessed directly from the original Eqn (11.150) together with Eqns (11.122), (11.152), and (11.133).

In Figure 11.19 the general course of the non-dimensional steer torque (11.152) has been depicted. The initial value at speed nearly zero is governed by the coefficients $\mu_R$ and $\mu_y$ while the slope equals $\mu_a$.

### *Numerical Results*

For the baseline configuration, and the various conditions examined in Table 11.4, the values for the three steer torque coefficients have been listed in Table 11.5.

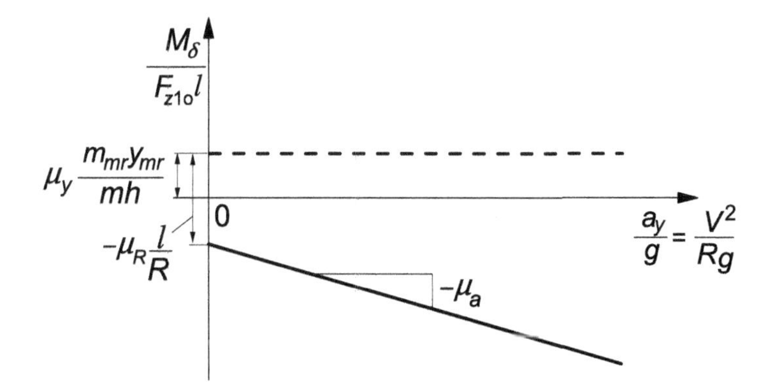


**FIGURE 11.19** The non-dimensional steer torque due to path curvature $1/R$ and c.g. lateral offset $y_{mr}$ and the variation with lateral acceleration $a_y$ or speed $V$.

**TABLE 11.5 Steer Torque Coefficients; Eqn (11.152)**

| $V$ km/h | $F_d$ N | $F_{ax}$ N | $F_{x1}$ N | $F_{x2}$ N | $F_{z1}$ N | $F_{z2}$ N | $\mu_R$ – | $\mu_a$ – | $\mu_y$ – |
|---|---|---|---|---|---|---|---|---|---|
| 1 | 0 | 0 | 0 | 0 | 1732 | 2094 | −0.05 | −0.01 | 0.1118 |
| 160 | 0 | 0 | 0 | 0 | 1732 | 2094 | −0.05 | −0.01 | 0.1118 |
| 160 | 395 | 0 | 0 | 395 | 1534 | 2292 | −0.04 | −0.01 | 0.1104 |
| 1 | 0 | 1500 | 0 | 1500 | 1137 | 2689 | −0.03 | −0.01 | 0.080 |
| 1 | 0 | −1500 | −912 | −588 | 2327 | 1499 | −0.02 | −0.05 | 0.1729 |
| 160 | 395 | 1500 | 0 | 1895 | 940 | 2886 | −0.03 | −0.01 | 0.079 |
| 160 | 395 | −1500 | −615 | −490 | 2129 | 1697 | −0.03 | −0.04 | 0.1624 |

Figure 11.20 depicts the variation of the steer moment with the speed squared computed according to the approximate analytical Eqn (11.150) (dotted line) and with the exact Eqns (11.85–11.94) (including $\beta$ but with $\varphi_r = 0$) together with contributing components. These components correspond with the terms of expression (11.147) and with the relevant terms of Eqn (11.89). In the contribution from the front wheel aligning torque (11.59) we distinguish the part directly attributed to $F_{x1}$ and the remaining part indicated with $M_z''$. The component indicated in the diagram by: $F_{x1}$ contains the former part plus the term with $F_{x1}$ already appearing in the $\beta$ coefficient in Eqn (11.89). In both cases examined (without and with braking) the approximate total moment closely follows the course of the exact solution. The difference is largely due to the fact that in the approximate theory the twist angle has been neglected.

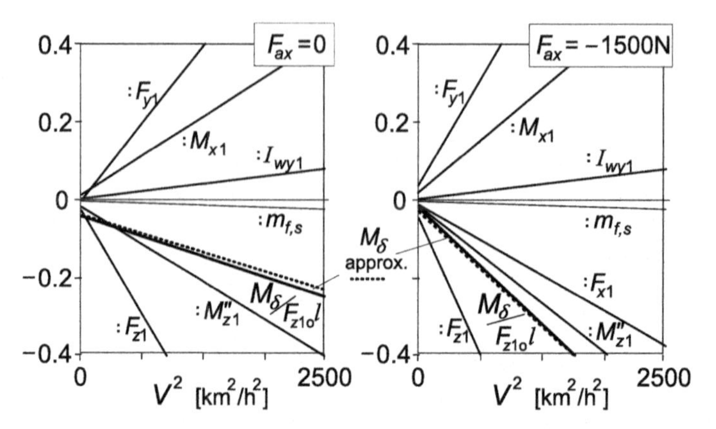


**FIGURE 11.20** The non-dimensional steer torque $(M_\delta)/(F_{z10l})$ and its components (cf. Eqn (11.147)) per unit non-dimensional path curvature $l/R$. Component $M_z''$ denotes the aligning torque, Eqn (11.59), with $F_x$ term deleted.

In all cases, the initial torque ($V \rightarrow 0$) at cornering is negative ($\mu_R < 0$), which means that when turning to the right an anti-clockwise steer torque is required. This appears to be mainly due to the direct action of the front normal load the axial component of which ($F_{z1}\gamma_1$) forms with the arm $t_c$ a moment about the steering axis. Also the tire aligning torque with the negative slip angle of the front wheel (Figure 11.17) helps to make the steer torque negative. In the braking case, the front wheel brake force considerably increases the negative steer torque. The weights of the steerable front and subframes have an almost negligible effect, while the gyroscopic action and the front wheel side force have a tendency to make the steer torque positive.

Finally, Figure 11.21 gives the moment response to lateral offset of the center of mass of the combined rider and mainframe. The approximation is less

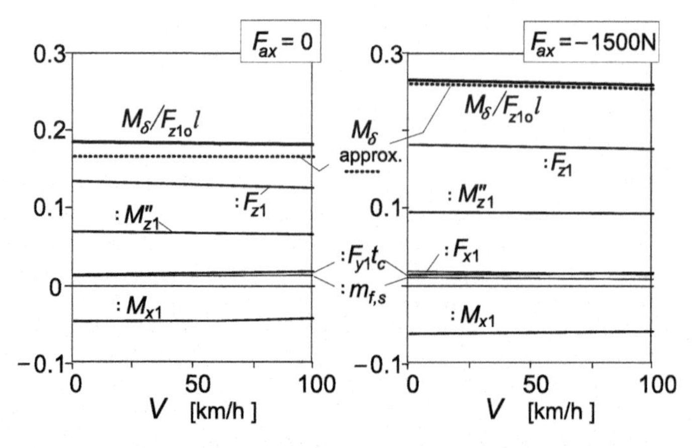


**FIGURE 11.21** The non-dimensional steer torque $(M_\delta)/(F_{z10l})$ and its components (cf. Eqn (11.147)) per m lateral offset of the center of gravity $y_{mr}$.

good than we have seen so far. The deviation is largely due to the omission of the twist compliance. Because of the small side forces in this straight ahead motion we see only a very small direct effect of $F_{y1}$. At braking, the tire torques become larger in magnitude due to the increased front slip angle and steer angle (camber), cf. Figure 11.18.

As an illustration, we consider again the case that when the motorcycle runs in a curve with a radius of 200 times the wheel base, $l/R = 0.005$, $R = 300$ m, at $V = 160$ km/h ($= 44.4$ m/s), the lateral acceleration becomes $a_y = V^2/R = 6.58$ m/s$^2$ $= 0.67$ g. The vehicle is being braked and the last row of Table 11.5 applies. As a consequence, the non-dimensional torque becomes: $M_\delta/(F_{z1o}l) = \mu_R\, l/R + \mu_a\, a_y/g = -0.029 \times 0.005 - 0.0363 \times 0.67 = -0.0245$ or in dimensional form: $M_\delta = -0.0245 \times 1732 \times 1.5 = -63.7$ Nm. At 50 km/h we would obtain: $M_\delta/(F_{z1o}l) = -0.0245 \times 0.005 - 0.0451 \times 0.0655 = -0.63 \times 0.005 = -0.0031$ and $M_\delta = -8.05$ Nm. The value $-0.63$ corresponds to the value found in Figure 11.20 at $V = 50$ km/h.

For a c.g. offset of 1 cm one would find a steer torque at straight ahead running at 160 km/h: $M_\delta/(F_{z1o}l) = \mu_y m_{mr} y_{mr}/mh = 0.1624 \times 350 \times 0.01/(390 \times 0.59) = 0.0025$ and $M_\delta = 0.0025 \times 1732 \times 1.5 = 6.5$ Nm. The value $100 \times 0.0025 = 0.25$ may be predicted from Figure 11.21, where $y_{mr} = 1$ m, at 160 km/h.

## 11.5.2. Non-Linear Analysis of Steady-State Cornering

In this section the so-called handling diagram will be established for the motor cycle. In that diagram the variation of the (ground) steer angle can be found as a function of the lateral acceleration for a given speed of travel or for a selected value of path curvature. As additional information, the variation of the associated steer torque will be given. The diagram is similar to the handling diagrams established for the motor car in Chapter 1. The relatively large roll angle calls for special attention to assess the equilibrium condition for the front and for the rear tire.

### *A Simple Approximation*

To introduce the problem we will again first neglect all the tire moments and the gyroscopic couples as well as the air drag. We then have the equilibrium conditions stated before:

$$\frac{F_{y1}}{F_{z1}} = \frac{F_{y2}}{F_{z2}} = \frac{a_y}{g} \qquad \left(= \frac{ur}{g}\right) \tag{11.154}$$

while the roll angle and the front and rear camber angles become, with $\delta$ neglected with respect to the large $\varphi$:

$$\tan\varphi = \frac{a_y}{g} \tag{11.155}$$

and

$$\gamma_1 = \varphi \tag{11.156}$$

$$\gamma_2 = \varphi \tag{11.157}$$

Consequently, for a given roll angle $\varphi$ we know the lateral acceleration, both side forces and the camber angles. From the non-linear tire side force characteristic that holds for the camber angle at hand and the vertical load (here equal to the static load) it must now be possible to assess the slip angle that produces, together with the camber angle, the desired side force. When we do this for both the front and rear wheels the difference in slip angles can be found and the handling curve can be drawn. At a given path curvature the ground steer angle is then easily found by using the approximate relationship:

$$\delta' = \frac{l}{R} + \alpha_1 - \alpha_2 \tag{11.158}$$

Figure 11.22 shows the normalized side force characteristics according to the baseline configuration together with the equilibrium curves where the camber angle equals the roll angle belonging to the side force. In Figure 11.23 the resulting handling curve with speed lines forming the handling diagram has been presented. The speed lines indicate the relationship between lateral acceleration and path curvature for given values of speed:

$$\frac{l}{R} = \frac{gl}{V^2}\frac{a_y}{g} \tag{11.159}$$

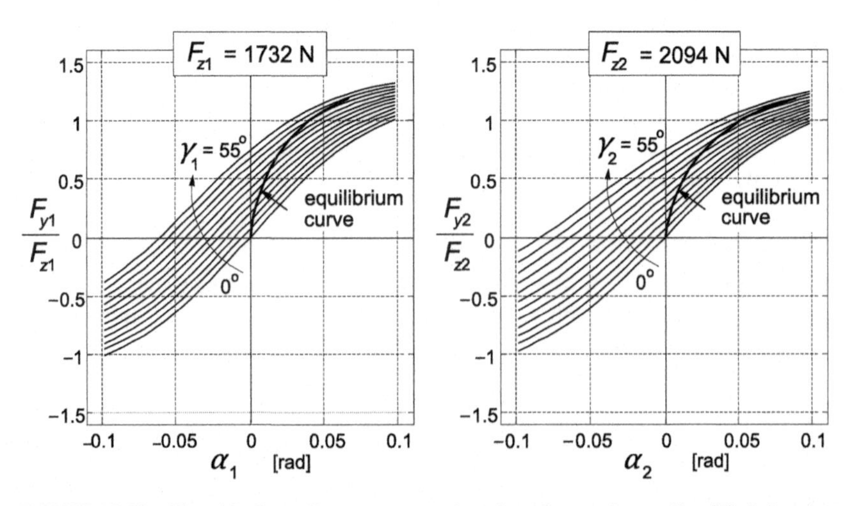


**FIGURE 11.22** Tire side force diagrams at a series of camber angles $\gamma$. Simplified situation where wheel loads remain unchanged (no air drag) and tire moments and gyroscopic couples are left out of consideration. Equilibrium curves have been shown where the tire side force agrees with the roll angle ($\approx \gamma_{1,2}$).

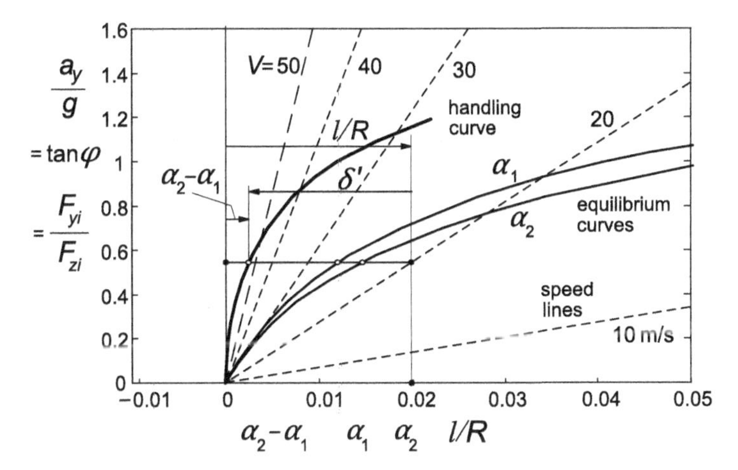


**FIGURE 11.23** The handling diagram for the simplified case where tire moments and gyroscopic couples are deleted and air drag is disregarded. The normalized side force versus slip angle curves (in this simplified case equal to the equilibrium curves) have been depicted as well.

In the figure, the equilibrium curves of Figure 11.22 have been reproduced. By horizontal subtraction of these curves, the handling curve is obtained. The horizontal distance between selected speed line and the handling curve equals the ground steer angle. The slope of the handling curve with respect to the vertical axis at zero lateral acceleration corresponds to the understeer coefficient in the linear theory $\eta/\zeta = \partial\delta'/\partial(a_y/g)$ at constant *l/R*. By using Eqn (11.10) with twist angle neglected and steer angle assumed small with respect to the roll angle we find the actual steer angle:

$$\delta = \frac{\delta'}{\cos\varepsilon}\cos\varphi \tag{11.160}$$

The theory given above turns out to be too crude to give a reasonably accurate result. To improve the analysis, the tire moments and the gyroscopic couples should be accounted for. We may also include the effect of air drag and braking or driving forces.

To set up the equations for the steady-state cornering situation, we need to consider the conditions for equilibrium in the directions of the $x$, $y$, and $z$ axes and about these three axes and for the steer torque: about the steering axis. We will do this for the general case with air drag introduced and with braking or driving forces acting on the front and rear wheels and possibly with a given sideways shift of the center of gravity. The twist angle $\beta$ is neglected and the lean angle $\varphi_r$ is disregarded since, at steady-state turning, its effect is similar to that of the c.g. lateral offset. The steer angle $\delta$ is considered small with respect to the roll angle. Also the wheel slip angles are assumed to be relatively small (smaller than ca. 10 degrees).

### Equilibrium Conditions for the Complex Configuration

The condition of equilibrium in the longitudinal direction can be met by making the driving force equal to the air drag force. This means that $F_{x,\text{acc}} = 0$ or $F_{x2} = F_d$. When a different driving force or when braking forces are applied, equilibrium can only be assured ($a_x = 0$) when the vehicle runs on an upward or downward slope. If this is not the case, we may speak of a quasi-equilibrium situation in the longitudinal direction. We have the equation (where the small longitudinal component of the side force $F_{y1}\delta'$ has been neglected):

$$ma_x = F_{x,\text{acc}} = F_{x1} + F_{x2} - F_d \tag{11.161}$$

In the lateral direction equilibrium occurs if

$$ma_y = F_{y1} + F_{y2} + F_{x1}\delta' \tag{11.162}$$

and in the vertical direction if

$$mg = F_{z1} + F_{z2} \tag{11.163}$$

About the $x$ axis we have the condition:

$$\begin{aligned} &-ma_y h_\varphi \cos\varphi + mgh_\varphi \sin\varphi + m_{mr} g y_{mr} \cos\varphi \\ &\quad -(I_{wy1}\Omega_1 + I_{wy2}\Omega_2) r \cos\varphi + F_{z1} t_{c\varphi} \delta \cos\varphi + M_{x1} + M_{x2} = 0 \end{aligned} \tag{11.164}$$

about the $y$ axis:

$$ma_x h_\varphi \cos\varphi - m_{mr} a_x y_{mr} \sin\varphi + aF_{z1} - bF_{z2} + h_{d\varphi} F_d \cos\varphi = 0 \tag{11.165}$$

and about the $z$ axis:

$$\begin{aligned} &ma_x h_\varphi \sin\varphi + m_{mr} a_x y_{mr} \cos\varphi + aF_{y1} - bF_{y2} + aF_{x1}\delta' \\ &\quad + M_{z1} + M_{z2} + h_{d\varphi} F_d \sin\varphi = 0 \end{aligned} \tag{11.166}$$

The wheel normal loads $F_{z1,2}$ are found from the Eqns (11.163) and (11.165). The longitudinal tire forces $F_{x1,2}$ are distributed according to Eqn (11.25) and as before, the acceleration force $F_{x,\text{acc}}$ is used as parameter in the handling diagrams to be developed.

With $M_{ya}$ introduced for abbreviation:

$$M_{ya} = ma_x h_\varphi \cos\varphi - m_{mr} a_x y_{mr} \sin\varphi + h_{d\varphi} F_d \cos\varphi \tag{11.167}$$

where $h_\varphi$ and $h_{d\varphi}$ are defined by Eqn (11.5) with if relevant $h$ replaced by $h_d$, we obtain the loads:

$$\begin{aligned} F_{z1} &= \frac{1}{l}(bmg - M_{ya}) \\ F_{z2} &= \frac{1}{l}(amg + M_{ya}) \end{aligned} \tag{11.168}$$

The two side forces are found from the Eqns (11.162) and (11.166). If we introduce for abbreviation:

$$M_{za} = ma_x h_\varphi \sin\varphi + m_{mr} a_x y_{mr} \cos\varphi + M_{z1} + M_{z2} + h_{d\varphi} F_d \sin\varphi \qquad (11.169)$$

We obtain:

$$F_{y1} = \frac{1}{l}(bma_y - lF_{x1}\delta' - M_{za})$$

$$F_{y2} = \frac{1}{l}(ama_y + M_{za}) \qquad (11.170)$$

with (from Eqn (11.10) with $\beta = 0$)

$$\delta' = \delta \frac{\cos\varepsilon}{\cos\varphi} \qquad (11.171)$$

From Eqn (11.164) an expression for the lateral acceleration can be obtained. After using Eqn (11.60) for $M_{xi}$ and the corrected effective rolling radii $r_{i\varphi}$:

$$r_{i\varphi} = r_i - r_{ci} + r_{ci} \cos\varphi \qquad (11.172)$$

and neglecting the very small contribution of $\delta$ to the moment about the $x$-axis we find by using the factor $\beta_x$ defined as

$$\beta_x = \frac{mh_\varphi}{mh_\varphi + \dfrac{I_{wy1}}{r_{1\varphi}} + \dfrac{I_{wy2}}{r_{2\varphi}}} \qquad (11.173)$$

for the lateral acceleration $a_y = ur$:

$$\frac{a_y}{g} = \beta_x \left[ \left( 1 - \frac{r_{c1}F_{z1} + r_{c2}F_{z2}}{mgh_\varphi \cos\varphi} \right) \tan\varphi + \frac{m_{mr} y_{mr}}{mh_\varphi} \right] \qquad (11.174)$$

### *The Handling Diagram*

For a given $\tan\varphi$ and $y_{mr}$ the corresponding lateral acceleration $a_y$ can be computed by using (11.174). If we disregard, as a first step, the influence of $\delta'$ and the aligning torques $M_{z1,2}$ in Eqns (11.170) and (11.169), we can calculate the front and rear side forces $F_{y1,2}$ belonging to $a_y$. With the approximation (11.156 and 11.157) the camber angles are taken equal to the roll angle of the mainframe. It is then possible through iteration to assess the values for the slip angles $\alpha_{1,2}$ belonging to the calculated $F_y$s and the known camber angles $\gamma_{1,2}$ with the loads obtained from (11.168 and 11.167). From the slip angles and the camber angles established, the aligning torques $M_{z1,2}$ can be estimated for the next computation step (for a given next incremented roll angle $\varphi$) by extrapolation. Also the ground steer angle $\delta'$ is estimated for the next step. This is done by using the Eqn (11.158) after having selected a value for the

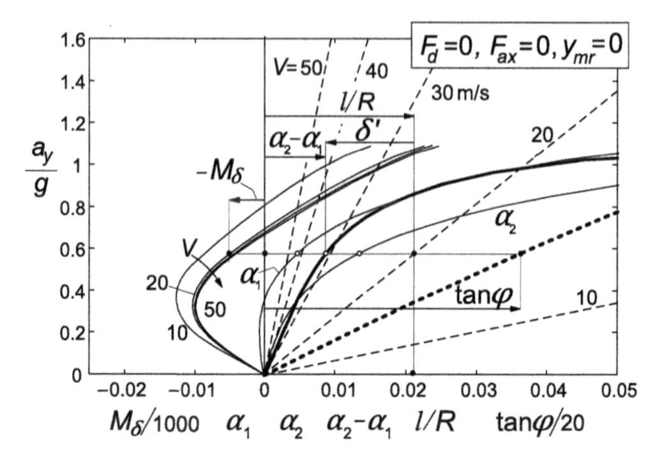


**FIGURE 11.24** The handling diagram for complex model with all tire moments and gyroscopic couples included. Air drag = 0 which makes the diagram valid for all speeds *V*. Curves for roll angle, front and rear slip angles, and steer torque (different for each speed).

forward speed $V = u$. Now, a more accurate value for the side forces that belong to the equilibrium state can be found from Eqns (11.170), with (11.169). For a series of successive values of $\varphi$ the values of $\alpha_1$ and $\alpha_2$ may be computed in this way, resulting, when plotted against $a_y/g$, in the equilibrium curves for the front and rear wheels (equivalent to the effective axle characteristics of Chapter 1). From the difference $\alpha_1 - \alpha_2$ the handling curve is obtained. If the air drag $F_d = C_{dA}\, u^2$ and/or a front wheel longitudinal force $F_{x1}$ are considered, the handling curve changes with speed and the handling diagram holds only for the selected value of speed $u$. If air drag and front wheel braking are disregarded, a generally valid single handling curve with a series of speed lines arises as depicted in Figure 11.24. The slope at $a_y = 0$ is $(\partial\delta'/\partial(a_y/g))_R = \eta/\zeta$.

In these diagrams also the slip angles and the actual roll angle have been plotted (in abscissa direction). As an additional information, the diagram contains curve(s) for the steer torque $M_\delta$ possibly changing with speed of travel $V(\approx u)$. How this quantity is assessed will be discussed in the subsection below.

### The Steer Torque

The steer torque that is applied to the handlebar by the rider in a steady turn can be found by considering the equilibrium about the steering axis. The resulting expression is similar to the one given by Eqn (11.147) but now extended to cover large roll angles. We find with the caster length $t_{c\varphi}$ according to Eqn (11.6):

$$\begin{aligned} M_\delta = \; & t_{c\varphi} F_{y1} \cos\gamma_1 - t_{c\varphi} F_{z1} \sin\gamma_1 \\ & + (I_{wy1}/r_{1\varphi}) ur \sin\varepsilon \cos\varphi \\ & - M_{x1} \sin\varepsilon - M_{z1} \cos\varepsilon \cos\gamma_1 \\ & + (m_f e_f + m_s e_s)(ur \cos\gamma_1 - g \sin\gamma_1) \end{aligned} \tag{11.175}$$

with the aligning torque according to Eqn (11.59):

$$M_{z1} = -t_{\alpha 1} F_{y\alpha 1} + M_{zr1} - r_{c1} F_{x1} \tan \gamma_1 \tag{11.176}$$

In the computations, we can use the correct expression for the camber angle $\gamma_1$ according to Eqn (11.13) with $\beta = 0$ because the steer angle $\delta$ is now available. This $\delta$ dependency causes the moment to change with path curvature and thus produces different curves for each speed value even if air drag is disregarded (Figure 11.24). The slope of the steer torque curve with respect to the vertical at $a_y = 0$ corresponds to $\mu_a \, F_{z1o} \, l$, cf. Table 11.5.

### Results and Discussion of the Non-Linear Handling Analysis

For a number of situations the handling diagram has been established for the baseline configuration and presented in the diagrams of the Figures 11.24–11.28. The following cases have been examined:

- No air drag, no braking or driving forces, and no offset of c.g. (Figure 11.24).
- With air drag and a rear wheel driving force necessary to withstand the air drag so that the forward acceleration equals zero ($F_{ax} = 0$: neutral situation) ($V = 30$ m/s $= 108$ km/h) (Figure 11.25).
- Braking: total deceleration force $-F_{ax} = 1500$ N ($V = 108$ km/h) (Figure 11.26).
- Hard braking: total deceleration force $-F_{ax} = 2500$ N ($V = 108$ km/h) (Figure 11.27).
- Neutral but now with lateral shift $y_{mr} = 5$ cm of the combined mass of main-frame and rider ($V = 108$ km/h) (Figure 11.28).

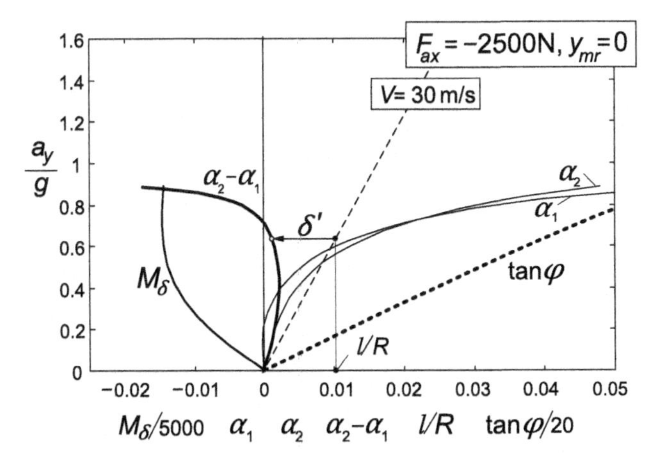


**FIGURE 11.25** The handling diagram with air drag considered and zero forward acceleration ($V = 108$ km/h).

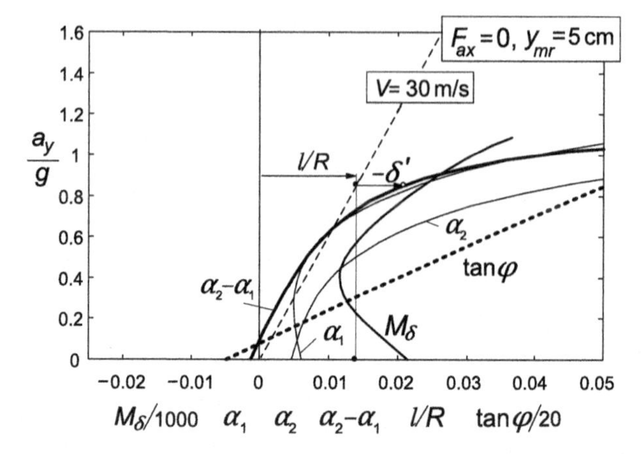


**FIGURE 11.26** The handling diagram at braking ($V = 108$ km/h).

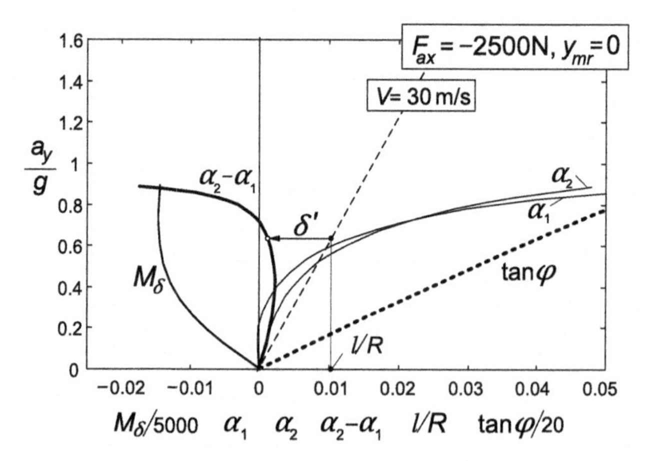


**FIGURE 11.27** The handling diagram at hard braking ($V = 108$ km/h).

The following interesting observations may be made:

- The tire/wheel moments have a considerable influence on the resulting handling characteristic, Figure 11.24. Much more oversteer appears to occur when compared with Figure 11.23. This was already concluded from the linear theory.
- Air drag does not change handling behavior very much if the speed is not much higher than 100 km/h. The steer torque at the speed of 108 km/h appears to reach a peak, decreases, and changes in sign in the higher lateral acceleration range (Figure 11.25), where $F_{y1}$ and $M_{x1}$ in (11.175) become dominant.

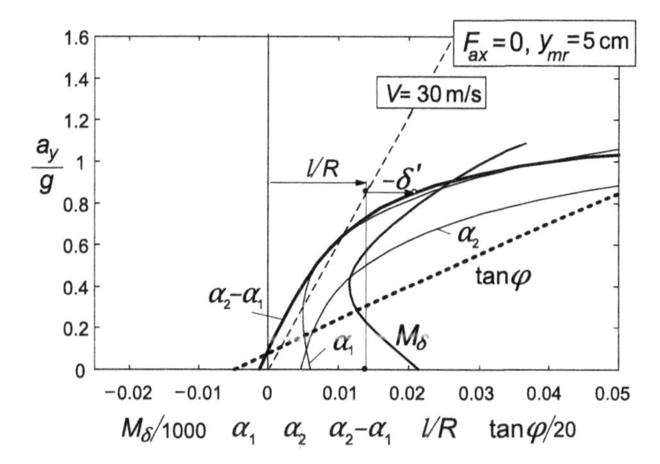


**FIGURE 11.28** The handling diagram at zero forward acceleration with 5 cm lateral shift of the center of gravity of combined rider and mainframe mass ($V = 108$ km/h).

- Braking causes the ground steer angle to become larger while the steer torque changes toward the negative direction especially at higher lateral accelerations (Figure 11.26, note the change in scale for the steer torque).
- Hard braking (Figure 11.27) tends to change the slope of the handling curve into understeer at large lateral accelerations. Also, the maximum lateral acceleration that can be reached is reduced. Larger steer angles arise in the higher $a_y$ range of operation. Also the steer torque increases considerably in magnitude.
- The lateral c.g. shift of 5 cm gives rise to large changes in the diagram as indicated in Figure 11.28. It shows that the changes are largely due to horizontal shifts of the curves. We may compare the graph with that of Figure 11.25. Initially, at zero lateral acceleration (corresponding to Figures 11.18 and 11.21) a negative roll angle occurs while the ground steer angle $\delta'$ and the steer torque $M_\delta$ are positive (which corresponds to $(\eta_y/\zeta)m_{mr}y_{mr}/mh$ and $(\mu_y m_{mr}y_{mr}/mh)F_{z1o}l$, respectively). At a certain level of lateral acceleration the steer torque reaches a minimum.

The path curvature where these minima of the steer moment or maxima of $-M_\delta$ occur constitutes the boundary of monotonous instability. At larger curvature the capsize mode becomes unstable. This is analogous to what was found for the automobile where the input is the steer angle and its maximum at a given speed of travel corresponds to the boundary of divergent (monotonous) instability (cf. Figure 1.20).

## 11.5.3. Modes of Vibration at Large Lateral Accelerations

The free vibrations that occur after a slight disturbance in a high lateral acceleration cornering maneuver are considerably complex in nature. Koenen

(1980, 1983) studied the eigenvalues and eigenvectors of the linear homogeneous differential equations that result after linearization of the non-linear set of equations at given steady-state levels of operation. The original non-coupled in-plane and lateral out-of-plane modes of vibration appear to strongly interact with each other at larger roll angles. The weave mode occurring at straight ahead running intertwines with the bounce and pitch modes and forms three different modes of vibration in which both the in-plane and lateral degrees of freedom play a role. As a result, unstable cornering weave oscillations may arise. Similarly, the wobble mode may interact with the wheel hop mode of the front wheel, which manifests itself by a violent combined steering and vertical wheel oscillation. As an example, we present in Figure 11.29 the set of root loci in the complex plane (the eigenvalues) for both the case of straight ahead motion where we have uncoupled in-plane and out-of plane vibrations and the case where the motorcycle moves along a circular path at an average roll angle of one radian (from Koenen's work). The bounce, pitch, and hop modes appear to become almost speed independent at vanishing roll angle. In a curve, all the modes do depend on the speed of travel. In the sharp, high, lateral acceleration curving motion considered, the cornering weave root loci appear to be shifted completely to the right-hand side of the imaginary axis. The wobble mode is destabilized as well. The higher frequency twist mode (flexible frame torsion mode) keeps an individual course.

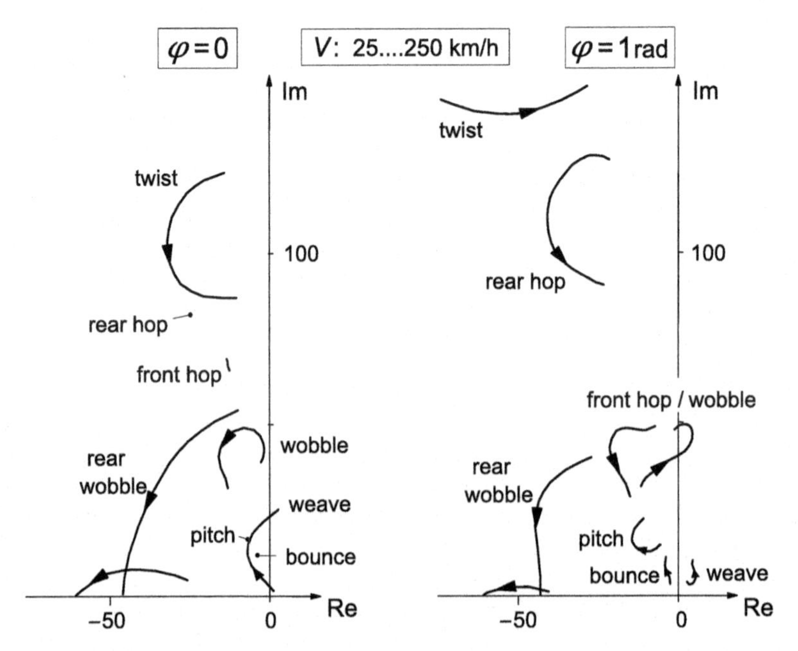


**FIGURE 11.29** Eigenvalues of Koenen's motorcycle model at straight ahead running ($\varphi = 0$) and at a roll angle of one radian. The speed varies from 25 to 250 km/h.

## 11.6. THE MAGIC FORMULA TIRE MODEL

For single track vehicles that may corner at possibly very large roll angles, a modified set of equations has been developed. To properly cover large camber angles De Vries and Pacejka (1998a) have adapted the original *Magic Formulae*. Since the equations listed in Chapter 4 have been revised and made more versatile and suitable to cover large camber angles, the listing of the special modified model equations has not been repeated in this new edition of the book. We may refer to the 2nd edition of this book and to the above reference for details of this special MC-MF Tire model.

The MC-MF formulae contain special features, parts of which have been incorporated in the revised set of equations presented in Section 4.3.2. Here by, these equations became suited to be employed for motorcycle tires as well.

As an example, the characteristics of a motorcycle tire, as tested by TNO with camber angle up to 45°, have been collected in the diagrams of Figure 4.33 and 4.34 in Chapter 4, similar to the presentation of the car and the truck tire characteristics.

Chapter 12

# Tire Steady-State and Dynamic Test Facilities

At various automotive and tire industrial companies and at a number of universities and institutes, test facilities are available for performing full-scale tire measurements to assess the tire force and moment generation properties. The test installation may be built on a truck or trailer that is equipped with a special wheel suspension and guidance system to which a measuring hub is attached. Typically, with such an over the road testing equipment, moderate speeds up to ca. 120 km/h can be reached. The frequencies with which the steer angle can be varied are relatively low. The camber angle is changed mechanically or through a hydraulic cylinder. The vertical load may be set at a desired average level; the load variations caused by road unevennesses may be filtered out. Commonly, the longitudinal slip results from the controlled application of the brake pressure. In very few cases, the wheel angular velocity is controlled through a hydraulic motor that acts relative to the vehicle's road wheel speed of revolution. In such a way, the test wheel drive and brake slip can be varied in a controlled manner. In some devices, water can be sprayed in front of the test wheel to create wet road conditions. The measuring hub contains strain gauge or piezo-electric force measuring elements. With these test facilities, usually measurements are conducted at quasi-steady-state conditions. Typically, a side-slip or brake-slip sweep may be imposed at a low rate. After processing, correcting and averaging the signals, the steady-state force and moment slip characteristics are obtained.

The large indoor test stands usually operate along similar lines. These rigs are based on an imitated road surface provided by the surface of a drum or by that of a flat track (endless belt). Somewhat higher frequencies of changing vertical axle position and yaw angle can usually be achieved (up to ca. 2–8 Hz). Drum test stands have been built with diameters ranging from 2 to ~4–5 m. With the larger drums, the tire usually runs or can also be run on the inner surface. This configuration makes it possible to mount realistic road surface segments and to maintain a layer of water on the inner surface, thereby enabling testing at wet or icy (and even snowy) conditions. External drums of 2.5–3 m diameter are more often encountered. Also, flat bed and flat plank test rigs are

**Tire and Vehicle Dynamics. DOI: 10.1016/B978-0-08-097016-5.00012-7**

used. Typically, but not always, these machines operate at very low speed of travel of either the plank or the wheel axle. Also, turn table or swing arm devices exist which are constructed to measure turn slip properties.

On the drum test stand, a measuring tower or a special wheel guidance system with a force measurement platform or hub can be positioned. For dynamic higher frequency tire tests, very stiff rigs are required (or very soft seismic systems). Special equipment is used that is often limited to a specific application to allow the system to become light and sufficiently rigid, that is: high first natural frequency. Devices exist dedicated to specific tasks such as systems with: axle fixed (but adjustable in height) to assess tire non-uniformity, cleat response, or response to brake pressure variations; axle forced to perform only vertical axle oscillations to assess the vertical dynamic stiffness and the response of the longitudinal force; and axle to perform only yaw angle variations to assess the tire dynamic steer response.

The original TU-Delft Tire Test Trailer, cf. Eldik Thieme (1960), later owned and operated by TNO-Automotive, Helmond, the Netherlands, has been replaced by a newly constructed semi-trailer (cf. Figure 12.1), and is equipped with the two original measuring stations. One for passenger car size tires and e.g., F1 tires limited to not so large camber angles and on the other side a new device specially designed for motorcycle tires that can handle also large camber angles, cf. Figure 12.2.

The car wheel can be subjected to a fixed or sweeping steer angle ($-18$ to $+18$ deg.) and the camber angle can be mechanically set at given values

**FIGURE 12.1** The new TNO Tire Test Semi-Trailer.

FIGURE 12.2 The large camber measuring device of TNO here mounted in the old test trailer.

from −5 up to 30 deg., cf. Sec.4.3.6 for example test results. At the motorcycle test side, the wheel camber angle can be swept from −20 to 70 deg. or set at a fixed angle, cf. Sec.4.3.6. The wheels can be braked up to wheel lock. The three forces and two moments (that is: except the brake torque) are measured with measuring hubs: a large strain gauge based and a compact piezo-electric based system respectively. Water can be sprayed at a controlled rate in front of the test tire.

Figure 12.3 depicts the flat track machine (TIRF, 1973) constructed and operated by Calspan, Buffalo. The upper structure can be steered about a vertical axis. With respect to this part, the wheel axle can be tilted about a line that forms the line of intersection of the wheel center plane and the belt that imitates the road surface. Through this configuration, a pure camber or wheel inclination angle can be created also at nonzero steer angle. The belt is supported by two drums and underneath the tire by a flat air bearing surface. The lateral stabilization of the belt is accomplished by letting one of the drums tilt against a stiff spring. The test wheel can be driven and braked in a slip-controlled manner. The facility can also handle larger truck tires. For some test results, cf. Sec.4.3.6, Figures 4.31 and 4.32. MTF Systems Corporation has developed and produces a similar device but with the steer axis inclining with camber variations, cf. Figure 4.30 for a test result.

Figures 12.4 and 12.5 show the layout of two internal drum test stands, cf. Bröder, Haardt, Paul (1973), and Krempel (1967). The drum of Figure 12.4 can also be used on its outer surface with a maximum speed of 250 km/h.

These large rigs are operated at the Karlsruhe University of Technology. The arrangement of the wheel loading, tilting, and steering system enables

**FIGURE 12.3** The flat track machine (TIRF) of Calspan.

center point steering and tilting about a line that touches the inner drum surface and lies in the wheel center plane. The order of changing the wheel attitude angles – first steering about the always vertical axis and then tilting about the new (still horizontal) $x$-axis – ensures that each of the angles, defined about and with respect to the vertical, remains unchanged when the other is varied. Of course, this principle also holds for the system of Figure 12.1 but not for the flat track machine of Figure 12.3. The configuration of Figure 12.5 exhibits

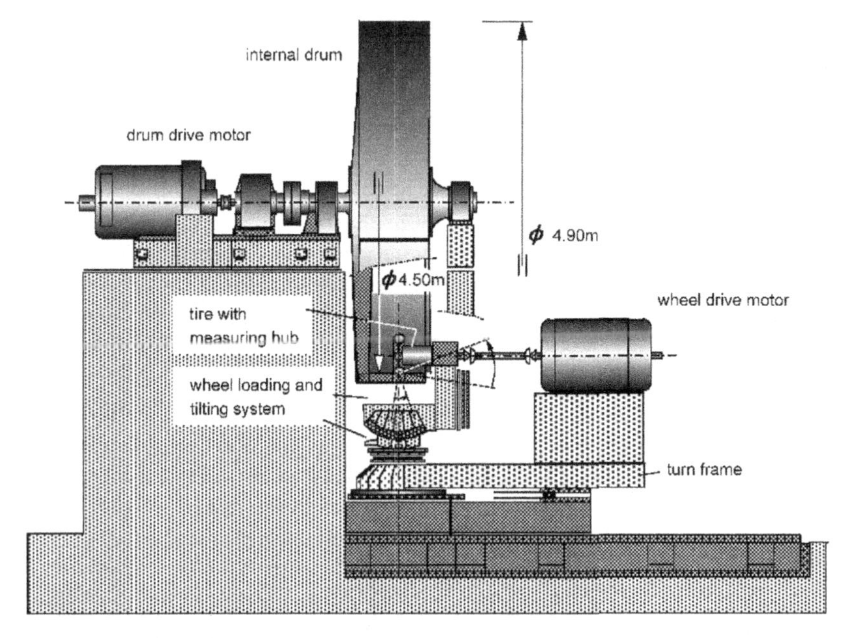


**FIGURE 12.4** Internal drum test stand originally possessed and operated by Porsche.

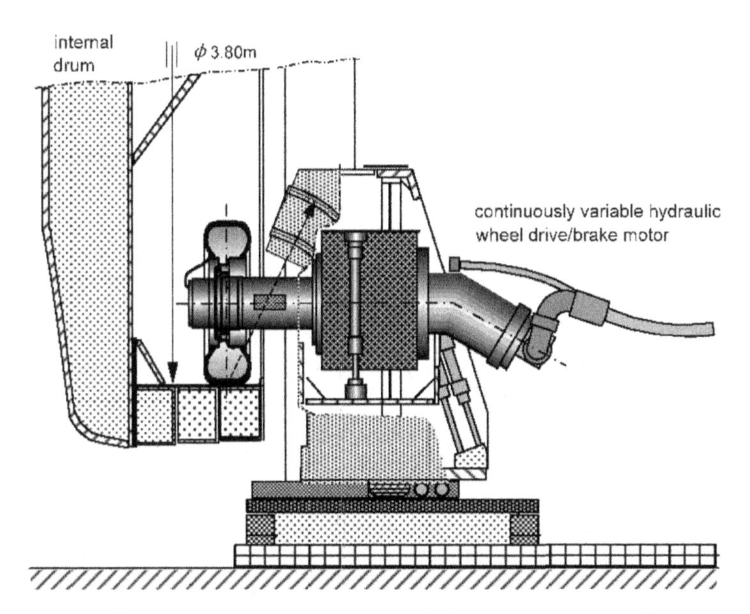


**FIGURE 12.5** The Karlsruhe University internal drum test stand.

a hydraulic wheel drive motor with which the wheel slip ratio can be controlled. Here, the force-measuring unit rotates together with the wheel. This avoids the otherwise necessary constructional measure to suppress parasitic forces and cross-talk such as the brake torque interaction with aligning torque that arises due to slight misalignment.

With a stationary measuring hub, a double Cardan coupling drive shaft (including a length change compensation element) is an example of such a measure which ensures that practically only the drive/brake torque is transmitted to the wheel. Instead of a pair of these couplings, a set of two membranes (thin flexible discs) or other devices (see Figure 12.7) may be used. Accurate alignment remains necessary.

With the UMTRI configuration (University of Michigan), cross-talk has been prevented by positioning the brake system between the nonrotating, stationary measuring system and the wheel. One of the drawbacks of rotating measuring systems is the fact that the sensitivity about the vertical axis (for the small aligning torque) cannot be chosen larger than the sensitivity about the horizontal longitudinal axis about which the moment may become relatively large.

Figure 12.6 shows the Delft flat plank machine, cf. Eldik Thieme (1960) and Higuchi (1997), now operated by the Eindhoven University of Technology. With this rig, accurate measurements can be conducted at a low speed of 2.3 cm/s. The plank has a maximum stroke of 7.5 m and can be tilted about the longitudinal center line on the test surface. Hereby, pure camber step responses can be established. The wheel axle, equipped with a measuring hub, can be steered, cambered, and the wheel can be braked. The vertical axle position or the vertical tire load can be adjusted. Cleats can be mounted on the plank surface. Typically, tire static stiffness tests (vertical, longitudinal, lateral, yaw, and camber), transient (step) side slip and camber tests (relaxation lengths and nonlagging part), impulse turning, and low speed cleat tests are performed on this machine, cf. Figure 5.12, Figures 7.5–7.10 and Figure 10.24 and Pacejka (2004).

The indoor drum test facility of the Delft University of Technology is based on two coupled drums with a diameter of 2.5 m that can run up to a maximum speed of 300 km/h. On top of one of the drums, a measuring tower (for low or moderate frequency yaw and brake experiments, cf. Figure 9.37) can be installed (not shown). This rig, cf. Eldik Thieme (1960) and Maurice (2000), can be turned about a vertical axis that passes through the tire contact center and through the top of the drum surface.

On the other drum, a rig can be mounted that is designed for measuring in-plane tire dynamics, Figure 12.7, cf. Zegelaar (1998). Much care has been taken to make the rig sufficiently rigid. This resulted in a lowest natural frequency of just over 100 Hz which allows the use of test data up to ca. 70 Hz. In addition, to avoid force and moment cross-talk, the brake shaft is connected with the wheel shaft through an intermediate shaft with two flexible couplings. These

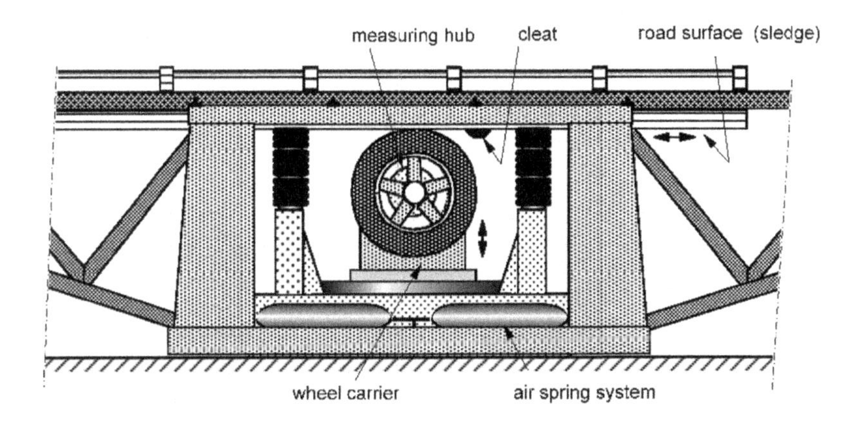


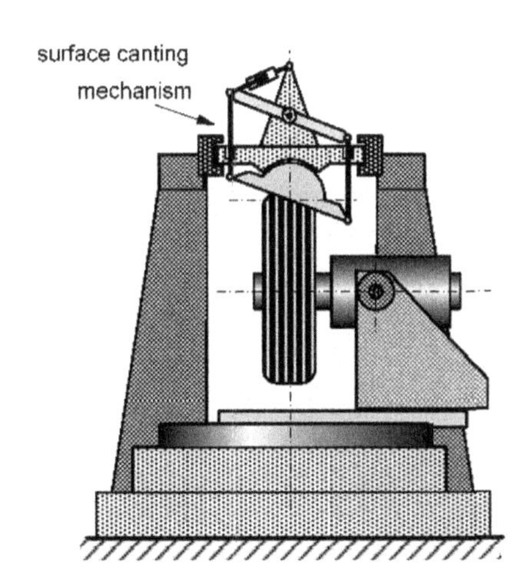


**FIGURE 12.6** The flat plank machine with road canting system.

couplings represent an alternative solution for the double Cardan coupling of Figure 12.4. They are flexible in all directions except about the axis of rotation. If properly aligned, this ensures that only the brake torque is transmitted to the wheel and that other parasitic forces and moments are largely suppressed. The brake torque is measured with strain gauges attached to the intermediate coupling shaft. A hydraulic servo system is used to control the brake pressure fluctuations (bandwidth up to ca. 60 Hz). Piezo-electric load cells placed on top of the bearings of the wheel shaft provide the signals from which the forces and moments that act on the wheel (except the brake torque) can be derived. With this machine, dynamic brake tests (Figures 9.42 and 9.45) and cleat tests (Figures 10.25–10.32) have been conducted. In a different setup, the machine may be used for response measurements to vertical axle oscillations ($<$20 Hz), cf. Figure 8.6.

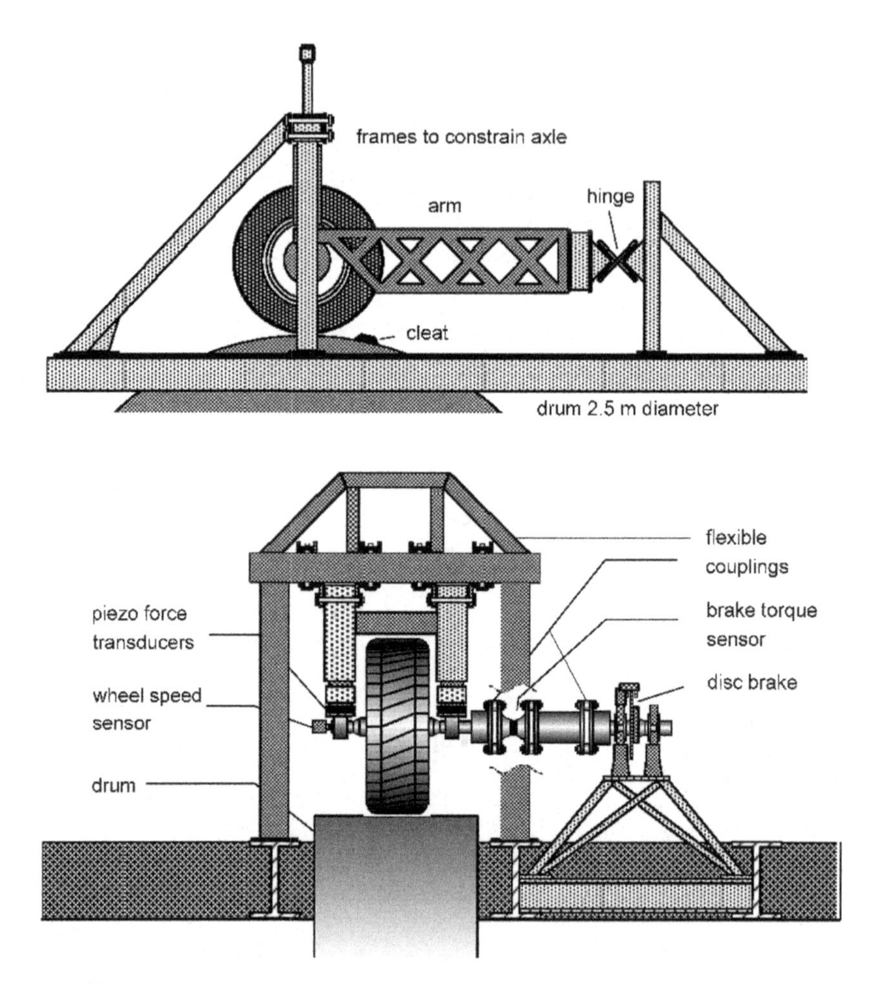


**FIGURE 12.7** Dynamic brake and cleat test facility in side and front view. Tests may be conducted up to ca. 65 Hz enabling the assessment of tire/wheel in-plane inertia and stiffness parameters including residual stiffnesses and rigid modes of vibration.

On the other drum, rigs may be mounted for out-of-plane tire dynamic experiments. One is the trailing arm 'pendulum' test stand, cf. Vries et al. (1998a) and Maurice (2000), with at one end a vertical hinge and at the other the steering head (for adjusting the average slip angle) with piezo-electric measuring hub (cf. Figure 12.8). At that point, the arm is excited laterally up to ca. 25 Hz by means of a hydraulic actuator. The wheel load is adjusted by tilting the vertical hinge slightly forward. The arm length is 1.65 m and the tire is subjected to an almost purely lateral slip variation. The rig is useful to assess the overall relaxation length and the gyroscopic couple parameter. The idea of the pendulum concept originates from

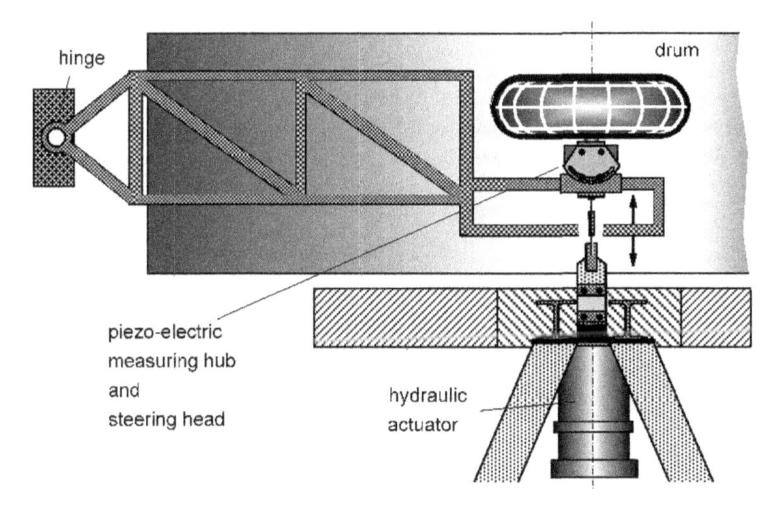


**FIGURE 12.8** The trailing arm 'pendulum' test rig exciting the tire almost purely laterally. Frequencies up to ca. 25 Hz adequate for assessing the tire relaxation length and gyroscopic couple parameter.

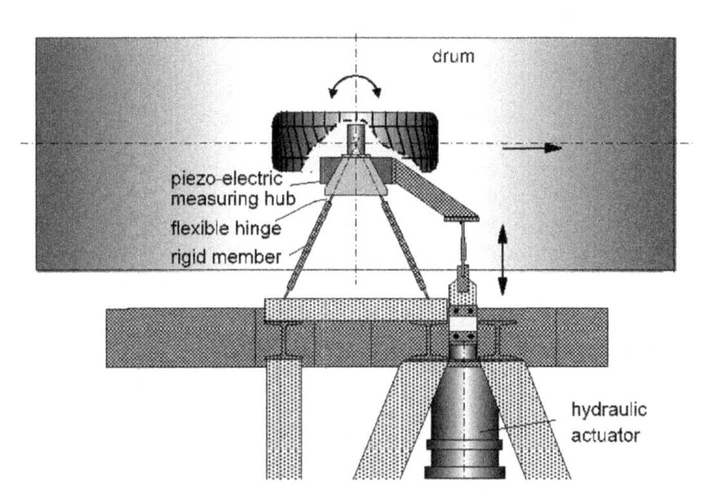


**FIGURE 12.9** The yaw oscillation test rig featuring center point steering. Frequencies up to ca. 65 Hz enable the assessment of tire out-of-plane inertia and stiffness parameters including residual stiffnesses and rigid modes.

Bandel et al. (1989). They designed and used an actual freely swinging pendulum rig.

In Figure 12.9, the so-called yaw oscillation test stand, Maurice (2000), is depicted that can be used for tests around an average steer angle that can be set at a value between −5 and +5 degrees. The structure is light and very stiff.

The two guiding members with flexible hinges intersect in the vertical virtual steering axis that is positioned in the wheel center plane (center point steering). A hydraulic actuator is mounted to generate the yaw vibration (typically random with a bandwidth of 65 Hz). The wheel axle is provided with a piezo-electric measuring hub. The tire is loaded by mechanically adjusting the axle height above the drum surface. Test results have been presented in Figures 9.43, 9.44 and Figure 9.46.

Chapter 13

# Outlines of Three Advanced Dynamic Tire Models

## Chapter Outline



## INTRODUCTION

Three relatively recent and advanced dynamic tire models have been discussed at a CCG Seminar held at the Technical University in Vienna, September 2010. These commercially available models are: *FTire*, *RMOD-K*, and *SWIFT*. Over the years, the models have been further developed and the applicability has been widened. The present chapter outlines the present state of these models. The contents of the ensuing sections have been contributed by the original developers. For further study, we refer to the indicated literature.

The three models manifest different ways of approach, different levels of complexity, and as a result differences in computational effort. Agreement with experimental data may be significantly different depending on the type of application. The three models all aim at similar motion input ranges and types of application. These include: steady-state (combined) slip, transients, and

**Tire and Vehicle Dynamics. DOI: 10.1016/B978-0-08-097016-5.00013-9**

higher frequency responses, covering at least the rigid body modes of vibration of the tire belt. The models are designed to roll over three-dimensional road unevennesses, typically exhibiting the enveloping properties of the tire.

The program packages offer simplified and/or more refined versions that may be chosen depending on less or more demanding types of application. Of course, considerable differences in required computation time are involved.

### Remarks

Obviously, the complex physically based models (*RMOD-K* and *FTire*) are better suited to examine the effects of the change of a physical parameter such as material stiffness and details of tire structure. The more empirically oriented models are better equipped to investigate the effect of changing performance parameters such as cornering and vertical stiffnesses without affecting remaining properties.

## 13.1. THE *RMOD-K* TIRE MODEL (Christian Oertel)

### 13.1.1. The Nonlinear FEM Model

Within the *RMOD-K* tire model family, the latest model is *RMOD-K FEM*, a single purpose nonlinear finite element code including HEX8 and HEX20 elements. It is able – by use of the Green–Lagrange strain tensor – to handle large rotations which occur during rolling and includes embedded continuous rebar layers for carcase, belt, and ply-caps as well as a kind of micro-buckling feature in the fiber material model.

The contact algorithm in *RMOD-K FEM* is divided into the normal contact problem based on the total Lagrangian formulation of structure deformation and the tangential contact problem in Eulerian formulation, including adhesion and sliding area (both formulations combined in the framework of the ALE formulation in steady-state rolling case). An interface transfers information between Lagrangian and Eulerian meshes, which normally differ in density. Pressure distribution and contact velocities are transferred toward the Eulerian mesh and tangential contact forces are transferred back to the Lagrangian mesh. In Figure 13.1, the variable mesh density and the penalty contact sensors are shown. Due to the single purpose orientated software design, the code has performance as well as robustness advantages in relation to multipurpose codes. The model produces results such as global static stiffness, steady-state force and moments, modal analysis, transfer functions from contact patch to spindle, and dynamic transient analysis with arbitrary uneven roads. Most of the model parameters are obtained from the geometry of the tire cross section including the layer positions via a pre-processor. The remaining values such as material parameters – if unknown or unavailable – are determined automatically by optimization based on measurements of the vertical stiffness

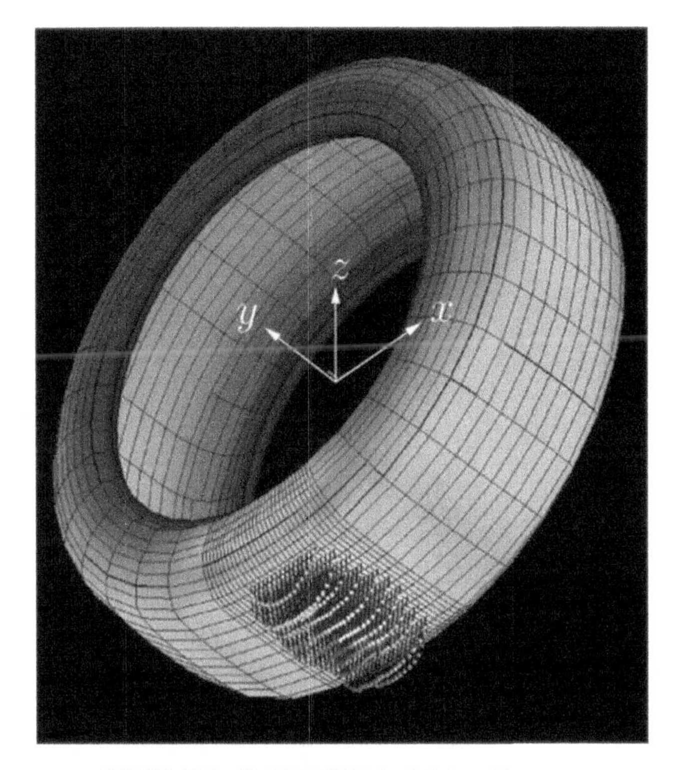


**FIGURE 13.1** *RMOD-K FEM*, loaded tire with contact.

and few of the inflated tire modes. Once the model parameters are known, the model can also serve as parametrization tool for simplified models. This can be done directly by comparing the element stiffness and mass matrixes of the simplified model with those of *RMOD-K FEM* or in the conventional way by comparing model results.

## 13.1.2. The Flexible Belt Model

The model *RMOD-K FB (flexible belt),* cf. Oertel (1997) and Oertel and Fandre (1999), gives also a detailed finite element description of the actual tire structure, but uses a number of simplifications in order to obtain a massive reduction of computational effort. The model uses the original load path of tires like *RMOD-K FEM* – load transfer via the pre-stressed sidewall without radial spokes. It features a flexible belt modeled by discrete rebars and the simplification of combing all different layers in one position relative to the cross section by analytically pre-processed QUAD 4 elements (cf. see Figure 13.2). Additional bending stiffnesses represent the rubber matrix. The belt is connected to the rim with a simplified sidewall model with pressurized

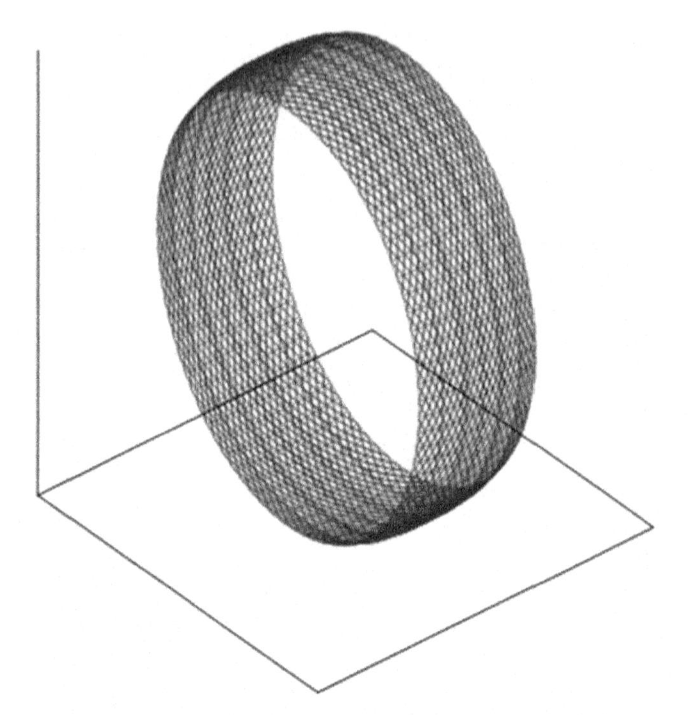

**FIGURE 13.2** *RMOD-K FB*, belt model.

air. Road contact is realized through an additional tread or sensor layer, a simplified HEX8 element. In a grid of sensor points at the outer tire surface, the normal and frictional forces are calculated. The contact area shape – with possible gaps – and contact pressure distribution result from computing the normal contact problem with the rolling and the structural deformation of the tire. Three-dimensional uneven surfaces can be generated and dealt with very well by this sophisticated model. A friction function included allows the generation of both adhesion and sliding areas with friction levels that depend on temperature (*WLF* transformation) and contact pressure. The complexity of the model may be controlled, depending on the type of application by changing the mesh density in the belt or sidewall area. The computational effort, however, is much smaller than that of *RMOD-K FEM* and the latest version of the simplified model offers parallel computing and therefore reduces the computational effort with respect to the hardware. Interfaces to the main MBS codes as well as to some FE codes are available. An additional feature is the misuse module, where a second load path between the inner surface of the tire and the rim can establish and very high wheel load can be handled up to plastic deformation of the rim, which is included in the misuse module, cf. Oertel and Fandre (2009).

### 13.1.3. Comparison of Various *RMOD-K* Models

Comparing the results obtained from *RMOD-K FEM* with those of the reduced finite element approach *RMOD-K FB*, it can be seen that both models lead to similar global results like vertical stiffness or steady-state force and moments while internal measures like Cauchy fiber stress may differ due to the simplifications. In Figure 13.3, the radial displacements for some circumferential node paths of both models are shown. Despite the slightly different meshes, the results are nearly identical. The two models address different fields of applications. *RMOD-K FEM* is used to work on tire design and durability by virtual

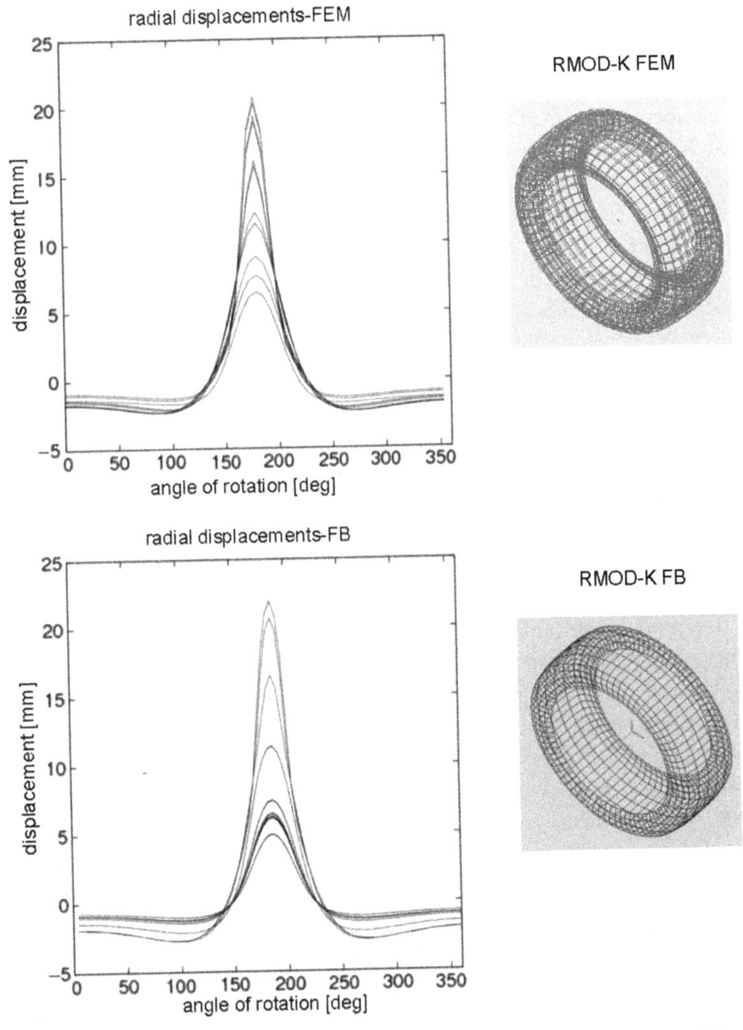


**FIGURE 13.3** Radial structure deformation, *RMOD-K FEM and RMOD-K FB.*

prototypes of the tire, typical questions dealt with in tire industry. *RMOD-K FB* is mainly used in the field of full vehicle dynamics by OEM with multi-body or hybrid models to investigate ride comfort, to compute load collectives and to calculate vehicle misuse loads. A much simpler model is available, which is based on a rigid belt representation and a separate model for the calculation of the footprint dimensions and the pressure distribution, *RMOD-K RB* (rigid belt).

Contact dynamics is analytically treated in the areas of adhesion. Applications are limited to smooth roads and ca. 100 Hz. The field of application is handling and controller design, so this type of model may be called *mechatronics* tire and used together with *simulink* vehicle and controller models.

Finally, a model for steady-state tire behavior – the *RMOD-K formula* – completes the tire model family. It is based on physical modeling, including adhesion and sliding area and contains a discrete contact area in order to handle larger camber angles. *RMOD-K formula* is much faster than real time and can be used in microcontroller applications. It becomes available as freeware in 2011. For these models also handling and fitting, aids are available.

One of the subjects of ongoing research and development is to build up a unique parametrization and a process chain across the entire *RMOD-K* model family, which will result in a reduction of cost and time with respect to the amount of needed measurements and which will simplify the process of parametrization.

The reader is referred to Oertel and Fandre (2009) for more details. Furthermore, we mention the website for up-to-date information: www.rmod-k.com.

## 13.2. THE *FTIRE* TIRE MODEL (Michael Gipser)

### 13.2.1. Introduction

*FTire* (Flexible Ring Tire Model) belongs to the class of strictly mechanics-based tire models, suitable for use in general vehicle dynamics simulations. *FTire* development started in 1998, using certain ideas and numerical concepts of the 'coarse-mesh' FE model *DNS-Tire* (Dynamic Nonlinear Spatial Tire Model), cf. Gipser (1996, 1998), as well as the nonlinear 'rigid-ring' model *BRIT* (Brush and Ring Tire Model), cf. Gipser (1997, 1998).

*FTire*'s complexity is below that of detailed FE models, but far above classical 'point contact' models. Consequent use of mechanically consistent, highly nonlinear structure, and friction models allows 'safe' extrapolation into operating conditions not covered by respective laboratory experiments. *FTire* returns plausible dynamic tire forces even at multiple high-frequent excitation, caused by road height profile and deformation, friction variation, suspension vibrations, drive and brake torque, tire nonuniformity and imbalance, temperature and pressure variation, and misuse events. This is *not* achieved by overly compromising computing time. Depending on activation of subsystems and on timely and spatial resolution, FTire simulation only takes about 1–50 times real time. Due to *FTire*'s multicore support, all tires of a vehicle can be simulated in parallel at the same computation speed.

*FTire* constitutes a full tire simulation environment. More than just a single model, it provides a scalable tire model kit, ranging from parallelized, real time capable versions for hardware-in-the-loop application, up to high-resolution realizations, connected to explicit FEA solvers. Upon demand, *FTire* provides a tread pattern, a tread temperature distribution, and a tread wear model, as well as visco-elastic rim and road models. Assisting tools are available for editing the model data file, for parametrization and data fit, for static, steady-state, and modal analysis, for visualization, for linearization, for DOE studies, for model export, and more. Parametrization may be based on laboratory measurements, on tire design data, on similarity considerations, or on combinations of these.

Using numerically robust co-simulation, *FTire* is made available as tire model plug-in for most of the relevant commercial simulation environments, covering MBS, FEA, specialized vehicle dynamics, and system simulation approaches. It is been used for nearly all rubber-tired types of vehicles, including motorcycles, aircrafts, and all-terrain vehicles.

*FTire* applications comprise primary and secondary ride, handling on flat and uneven road surfaces, tire forces influenced by suspension control systems, NVH, mobility, tire-imperfection-induced suspension and steering vibrations, misuse, and road load prediction for durability.

### 13.2.2. Structure Model

Like in many other detailed tire models, the kernel of *FTire* consists of two main components. The first one, the *structure* model, describes the tire's structural stiffness, damping, and inertia properties. The second one, the *tread* model, comprises evaluation of height, compliance, and friction coefficient of the road surface, as well as the computation of resulting ground pressure and shear stresses in the contact patch. Basis of the structure model is a set of flexible bodies (called *belt segments*, typically using nearly 50–500 of them), being *FTire*'s image of the tire's belt layer structure. Each belt segment has $4+x$ dynamic degrees of freedom (Figure 13.4):

- longitudinal, lateral, and vertical *displacement* of the center point,
- *rotation* angle about the circumferential axis, and
- flexible *bending* in lateral direction, described by $x$-independent shape functions.

These segments, their internal degrees of freedom, and the rim (which itself is either assumed to be stiff, elastic, or visco-elastic) are interconnected by several nonlinear, inflation pressure-dependent force elements. These force elements comprise nonlinear translational and rotational stiffnesses, bending stiffnesses along all three axes, respective damping elements (most of them assumed to be

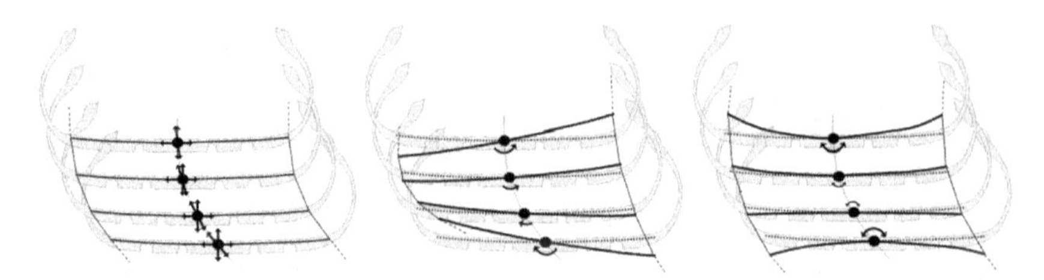

**FIGURE 13.4** Belt segments' degrees of freedom: translation, torsion, lateral bending.

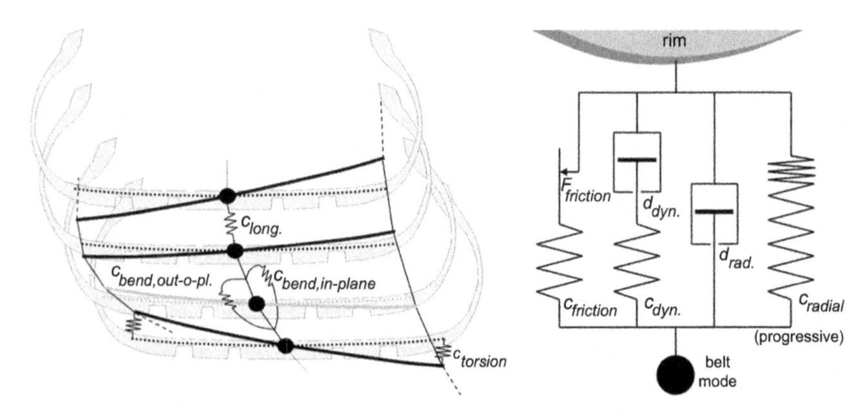


**FIGURE 13.5** Some structural force elements.

linearly viscous), as well as certain Maxwell and hysteresis elements (Figure 13.5). In addition, the segments are subject to inflation pressure forces in radial direction, and to forces and moments generated by the tread model.

As an example, the right image of Figure 13.5 shows the combined *radial force element* of one single belt segment. Circumferential and transversal elements are placed analogously. Maxwell elements are used to describe the dynamic tire stiffening at higher rolling speeds, whereas several dry friction elements (only one of them being shown in the image) accurately approximate rubber hysteresis.

All stiffness values depend on inflation pressure. The *cold* inflation pressure is treated as an *operating condition* and might arbitrarily be modified during a running simulation. The *actual* inflation pressure depends on cold inflation pressure and tire temperature, which is the output of the thermal model, if activated.

Several types of optional tire imperfections complete the structure model, comprising static and dynamic imbalance, radial and tangential nonuniformity, ply-steer, conicity, and detailed geometrical run-out.

## 13.2.3. Tread Model

Between each two adjacent belt segments, several mass-less contact and friction elements (Figure 13.6 *top*) are placed. Their number per segment, typically between 10 and 200, can be chosen by the user, depending on the desired road height resolution. These contact elements constitute the *tread* model.

The contact elements are placed along approximately parallel lines (cf. Figure 13.6 *top*). The actual global positions of the belt points to which the contact elements are attached to are determined by smooth interpolation, using the coordinates of the four nearest-by belt segments.

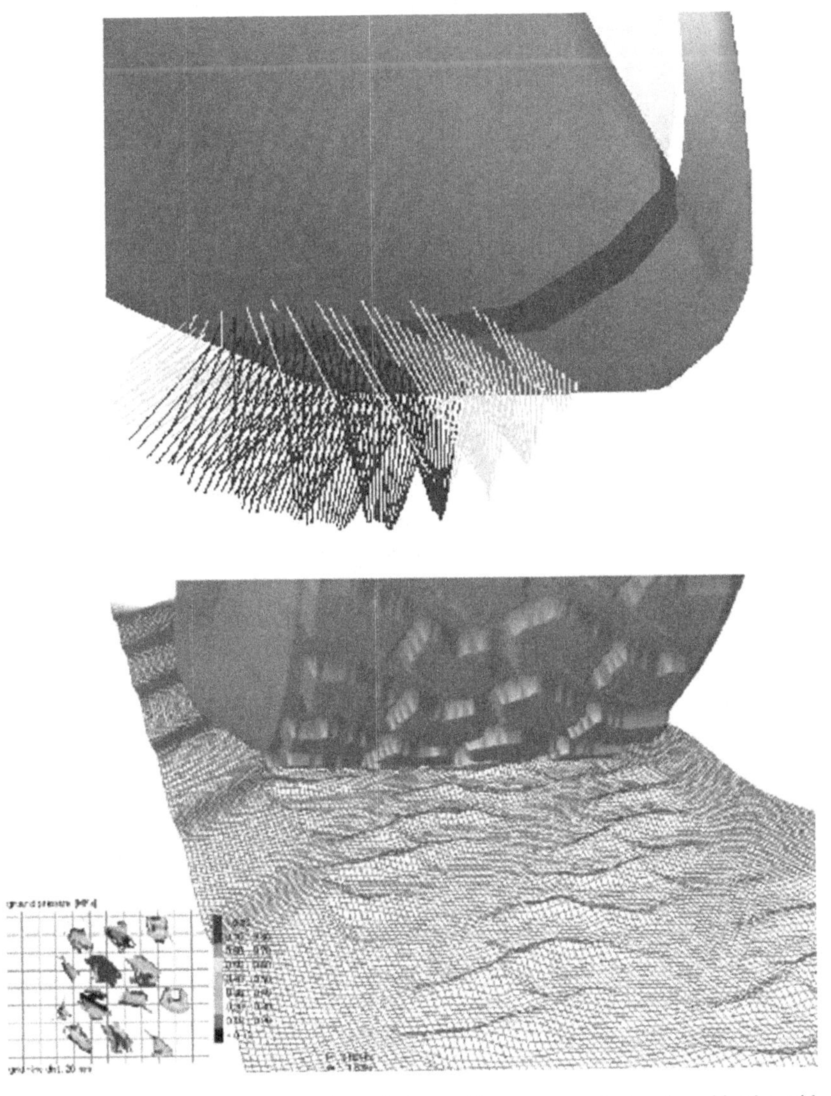

**FIGURE 13.6** Tread model: contact forces during parking maneuver (*top*) and combination with visco-elastic road surface (*bottom*).

If the tread pattern is given in terms of a black-and-white or gray-scale bitmap file, the contact elements' lengths (that is, the local tread rubber thickness) is set by *FTire*, according to this pattern. In Figure 13.6 *bottom*, the computed contact pressure of such a model is shown, when rolling over a visco-elastic road surface. Clearly, a larger number of contact elements is required to exactly resolve more detailed patterns, like those of passenger-car tires. *FTire*'s computing time increases less than linearly with the total number of contact elements.

The road tangential plane is computed individually for each contact element, by evaluating the road height in three different locations near the contact element. This is necessary to resolve obstacle sizes far below the contact patch length. Road surface may depend on time, like it does for four-post test-rigs and rotating drums. In that case, both normal and tangential surface velocities are taken into account as well.

Each contact element's normal force is a nonlinear function of radial deflection and deflection velocity, describing tread rubber compression stiffness and damping. The normal force, the tangential sliding velocity, and the local friction coefficient are used to compute the tangential friction forces, and by this the shear stress. The local friction coefficient is a function of position, ground pressure, sliding velocity, and tread temperature.

### 13.2.4. Model Data and Parametrization

Ease of parametrization had been an important objective during development of *FTire*. A clear distinction is made between data used in the model equations (*pre-processed* data) and data to be supplied by the user (*basic* data). The idea is to define only basic data which can be obtained by standard laboratory measurements in a cheap, repeatable, and reliable way, and which at the same time yield enough information to completely and unambiguously determine all internal pre-processed tire data.

*FTire* can be parametrized by using a standardized measurement procedure, recently defined by a working group of major German car manufacturers. Given the respective measurement files, the software *FTire/fit* can nearly automatically derive from this an *FTire* data file, together with a respective validation report.

Detailed information on *FTire* and its tools can be found on www.cosin.eu, *FTire*'s homepage. For further reading, we refer to the references: Gipser (2000), (2006), (2007).

## 13.3. THE *MF-Swift* TIRE MODEL (Igo Besselink)

### 13.3.1. Introduction

Already after the first publications in 1987 and 1989, the *Magic Formula* has quickly become a very popular tire model to describe the steady-state forces and moment occurring under various slip conditions. Its typical field of application is vehicle-handling studies. With the advent of vehicle control systems such as anti-lock brakes, possibly on rough roads, the need developed for a tire model also capable of describing the transient and dynamic tire behavior accurately. This marked the starting point of the development of the *SWIFT* tire model, the acronym indicating *Short Wavelength Intermediate Frequency Tire model*. The model should be able to handle frequencies up to 60–80 Hz and wavelengths of 0.1–0.2 m. An important feature is that the *Magic Formula* has been maintained, as this has proved to be an accurate and well-established model. Research on the *SWIFT* model was done by three PhD students at the Delft University of Technology under the supervision of Professor Pacejka. Peter Zegelaar focused on the in-plane dynamics and Jan-Pieter Maurice looked at the out-of-plane dynamics, notably in connection with lateral slip variations. Later, Hans Pacejka developed the turn slip and camber aspect (spin) that covers the final input slip component of the dynamic model. After initial work of Zegelaar, Antoine Schmeitz developed the obstacle enveloping model that enables the assessment of the flat so-called effective road plane. Recent developments, like the inclusion of tire inflation pressure in the model, have been developed in cooperation with the Eindhoven University of Technology.

### 13.3.2. Model Overview

Various aspects of the *SWIFT* model have already been explained in detail in Chapters 9 and 10 of this book; the discussion will be limited to an overview here. For an impression of the model configuration, see Figure 13.7. The four main elements comprising the *SWIFT* tire model are:

- *Magic Formula*: describing the steady-state forces and moments of a tire rolling at various slip conditions. The full set of equations can be found in Section 4.3.2.
- *Contact patch slip model*: the tire forces and moments do not respond instantaneously to variations in slip of the contact patch body; the differential equations describing this behavior have been given in Section 9.3.1.
- *Rigid ring*: at higher frequencies the wheel (rim and tire combination) cannot be considered as a single body anymore. Therefore, the wheel has been split into two rigid bodies, which are interconnected via springs and dampers. The ring is connected with the contact patch through so-called residual springs with dampers (Figure 9.27).
- *Obstacle enveloping model*: to cope with short wavelength road undulations, a method using elliptical cams has been developed (Figure 10.11).

The structure of the tire model and software implementation allows the user to select the level of complexity: for handling studies, it may be sufficient to use only the *Magic Formula*; for the analysis of wheel shimmy on a flat road

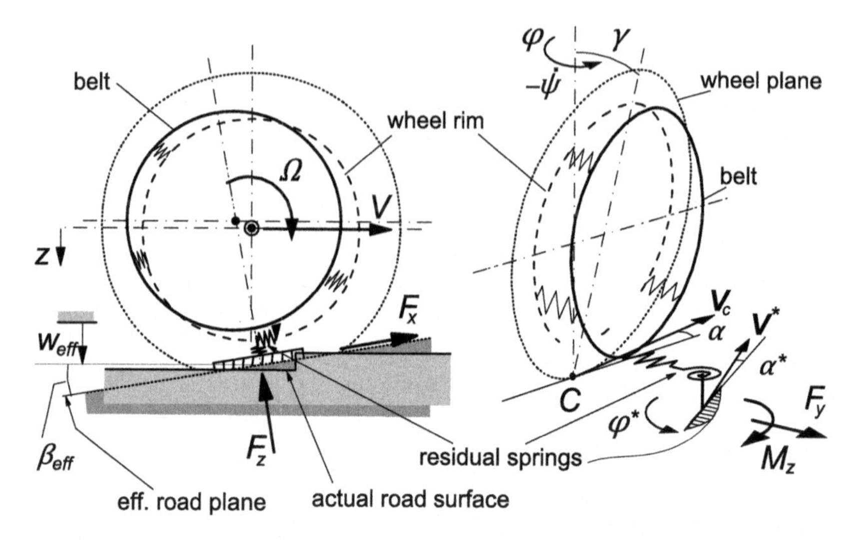


**FIGURE 13.7** General configuration of the *SWIFT* model featuring rigid belt ring, residual stiffnesses, contact patch slip model, and obstacle enveloping model that produces the effective road inputs.

surface, contact patch and rigid ring dynamics may be switched on and for calculation of suspension forces on a 3D uneven road surface the enveloping model and all further features are switched on. This is advantageous for calculation time, as an appropriate representation without overhead can be selected for the task at hand.

### 13.3.3. MF-Tire/MF-Swift

The Dutch organization for applied research TNO also contributed to the development of the tire model and has turned them into commercial software, which is sold under the name "Delft-Tire". The first software product was the *Magic Formula Tire Model* implementation in various multi-body software packages, known as *MF-Tire*. In particular, *MF-Tire* version 5.2 achieved the status of an industry standard for modeling passenger-car tires in the late 1990's. The software program *MF-Tool* is used to process measurements and determine the parameters of the tire model. For some years, various separate tire models were available (e.g., motorcycle tires: *MF-MCTire*, early versions of *SWIFT*), but more recently all functionality has been combined into a single tire model known as *MF-Tire/MF-Swift*. Figure 13.8 gives an impression of the TNO *Delft-Tire* tool chain, which supports all steps from tire measurements, parameter identification to multi-body simulations.

For more detailed information, we suggest to contact TNO. The website address reads: www.delft-tire.com.

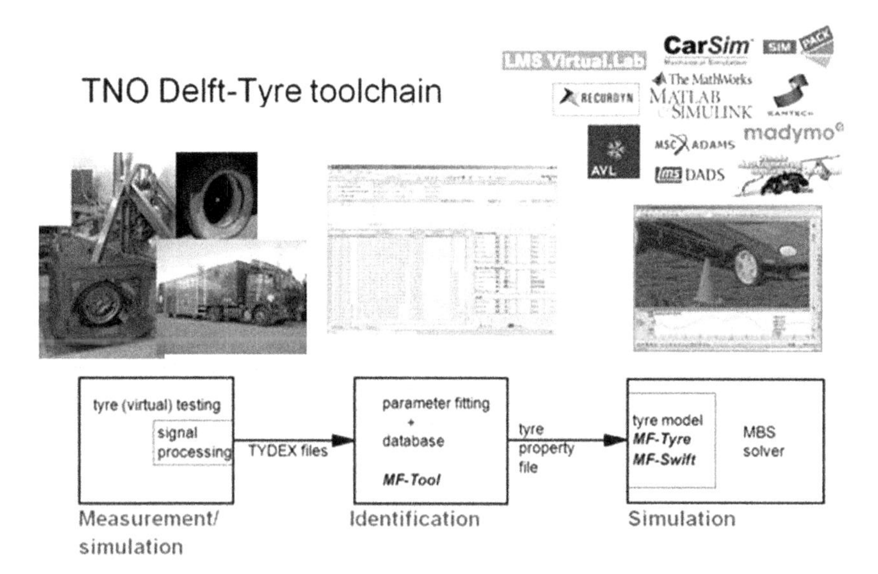


**FIGURE 13.8** TNO Delft-Tire tool chain.

## 13.3.4. Parameter Identification

The parameter identification process should not be underestimated, as it may prove to be difficult to find a parameter set, which matches the various tests best. Typically, numerical optimization techniques are used to obtain the "best" parameters by minimizing the difference between measurements and model output. The empirical *SWIFT* tire model has advantages here, as the model consists of a number of relatively independent elements. In a physical tire model, all characteristics are interlinked: changing e.g., a material stiffness could affect everything from vertical stiffness, to cornering stiffness and relaxation length. This means that all tire tests have to be analyzed, to evaluate the effect of a parameter change. For the *SWIFT* model, the identification process is split up in a number of consecutive, small optimization problems. With *MF-Tool* a *Magic Formula* fit is made within 10 minutes and a full *SWIFT* data set (including *Magic Formula*) can be fitted in less than an hour, which is comparatively fast. The structure of the TNO *MF-Tire/MF-Swift* tire property file is explained in Appendix 3. This appendix also gives an introduction to the tests required and ways to estimate tire model parameters. It is important to note that the tests required for the identification of tire model parameters have been standardized, so the measurement data requirements for *MF-Swift*, *FTire*, and *RMOD-K* are almost the same.

## 13.3.5. Test and Model Comparison

In Chapters 4, 9, and 10, various graphs have been given, comparing test and tire model results, see e.g., Sections 4.3.6, 9.4.2 and 10.2. These cases concern the reproduction of tire behavior measured on a test bench. Obviously, the tire model combined with an accurate vehicle model should be in agreement with measurements on an instrumented vehicle. A good example is given by Schmeitz, Versteden and Eguchi (2011). This study focuses on the calculation

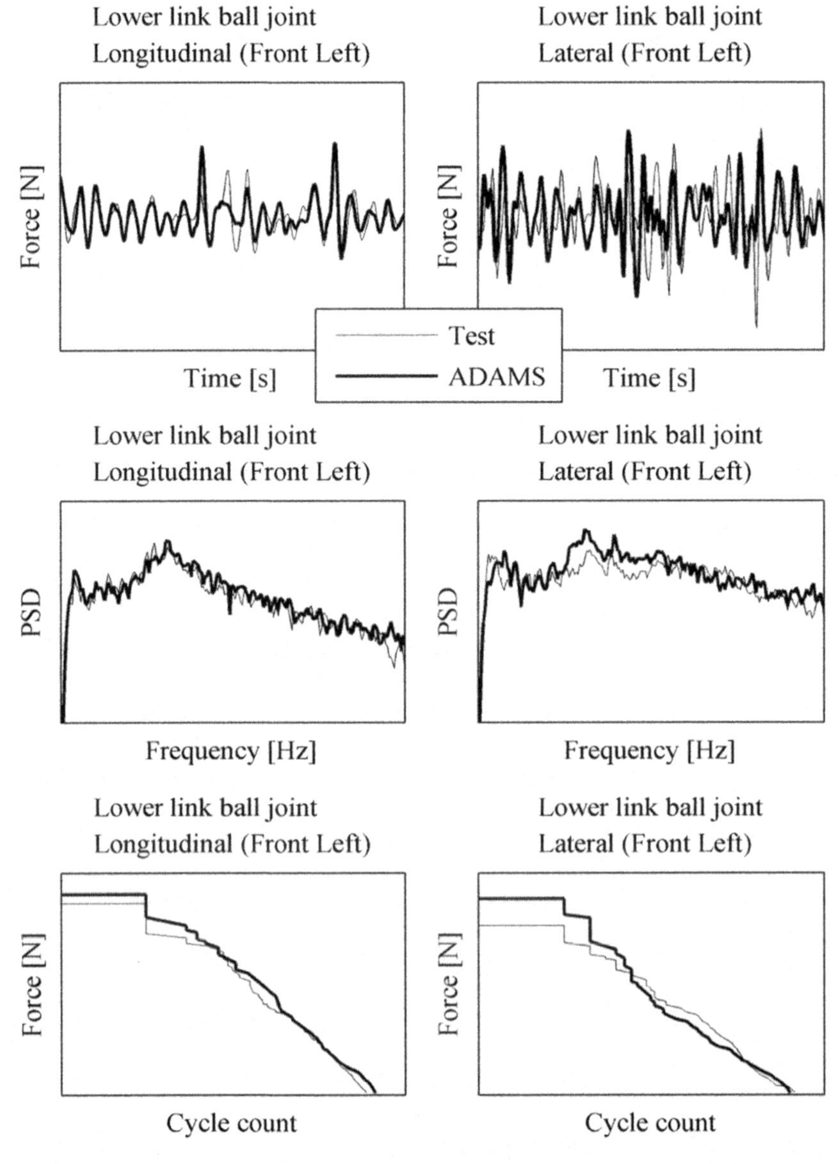


**FIGURE 13.9** Measured and simulated lower link ball joint forces in the time domain (detail shown), frequency domain, and cycle count for the base vehicle (*ADAMS*: full vehicle model including *MF-Swift*).

of suspension loads while driving over a digitized 3D road surface. It appears that the calculated forces from the full vehicle simulation model agree quite well with measurements. Also, the effects of changes to tire and suspension parameters are captured fairly accurately. Figure 13.9 gives an impression of the results obtained.

# References

Abe, M. (2009). *Vehicle handling dynamics*. Oxford: Butterworth-Heinemann.
Abramowitz, M., & Stegun, I. A. (1965). *Handbook of mathematical functions*. New York: Dover.
TNO Automotive. (2010). *MF-TYRE/MF-SWIFT 6.1.2*. Equation manual. Cf. www.delft-tyre.nl.
Badalamenti, J. M., & Doyle, G. R. (1988). Radial-interradial spring tire models. *Journal of Vibration, Acoustic, Stress and Reliability in Design, 110*(1).
Bakker, E., Nyborg, L., & Pacejka, H. B. (1987). *Tire modeling for use in vehicle dynamics studies*. SAE Paper No. 870421.
Bakker, E., Pacejka, H. B., & Lidner, L. (1989). *A new tire model with an application in vehicle dynamics studies*. SAE Paper No. 890087.
Bandel, P., & Bernardo, C. di (1989). A test for measuring transient characteristics of tires. *Tire Science and Technology, 17*(2).
Bandel, P., & Monguzzi, C. (1988). Simulation model of the dynamic behavior of a tire running over an obstacle. *Tire Science and Technology, TSTCA, 16*(2).
Bayer, B. (1988). Flattern und pendeln bei krafträdern. *Automobil Industrie, 2*.
Bayle, P., Forissier, J. F., & Lafon, S. (1993). A new tyre model for vehicle dynamics simulation. *Automotive Technology International*.
Becker, G., Fromm, H., & Maruhn, H. (1931). *Schwingungen in Automobil-lenkungen* ('Shimmy'). Berlin.
Bergman, W. (1965). Theoretical prediction of the effect of traction on cornering force. *SAE Transactions, 614*.
Bernard, J. E., Segel, L., & Wild, R. E. (1977). *Tire shear force generation during combined steering and braking maneuvers*. SAE Paper 770852.
Berritta, R., Biral, F., & Garbin, S. (2000). Evaluation of motorcycle handling and multibody modelling and simulation. In: *Proceedings of 6th int. conference on high tech. engines and cars*, Modena.
Berzeri, M., & Maurice, J. P. (1996). *A mathematical model for studying the out-of-plane behaviour of a pneumatic tyre under several kinematic conditions*. Prague: FISITA Youth Congress.
Besselink, I. J. M., Schmeitz, A. J. C., & Pacejka, H. B. (2009). An improved magic formula/SWIFT tyre model that can handle inflation pressure changes. In M. Berg & A. Stensson (Eds.), *Selected and extended papers of 21st IAVSD symposium*. Trigell, Stockholm, 2009. Suppl. Vehicle System Dynamics, 48.
Besselink, I. J. M. (2000). *Shimmy of aircraft main landing gears*. Dissertation, TU Delft.
Biral, F., & Da Lio, M. (2001). Modelling drivers with the optimal maneuver method. In: *Proceedings of ATA 2001, the role of experiments in the automotive product development process*, Florence.
Böhm, F. (1963). Der rollvorgang des automobil-rades. *ZAMM, 43*, T56–T60.
Borgmann, W. (1963). *Theoretische und experimentelle Untersuchungen an Luftreifen bei Schräglauf*. Dissertation, Braunschweig.
Breuer, T., & Pruckner, A. (1998). Advanced dynamic motorbike analysis and driver simulation. In: *13th European ADAMS users' conference*, Paris.

Brockman, R. A., & Braisted, W. R. (1994). Critical speed estimation of aircraft tires. *Tire Science and Technology, 22*(2).

Bröder, K., Haardt, H., & Paul, U. (1973). Reifenprüfstand mit innerer und äusserer fahrbahn. *ATZ, 75*(2).

Bruni, S., Cheli, F., & Resta, F. (1996). On the identification in time domain of the parameters of a tyre model for the study of in-plane dynamics. In F. Böhm & H. P. Willumeit (Eds.), *Proceedings of 2nd colloquium on tyre models for vehicle analysis.* Berlin 1997, Suppl. Vehicle System Dynamics, 27.

CCG. (2004). *Tyre models for vehicle dynamics simulation.* Seminar TV 4.08 lecture notes, Coord. P. Lugner, Vienna, Sept. 1–2. 82234 Oberpfaffenhofen, Germany: Carl Cranz Gesellschaft.

Chiesa, A., & Rinonapoli, L. (1967). *Vehicle stability studied with a nonlinear seven degree of freedom model.* SAE Paper 670476.

Clark, S. K. (1982). A brief history of tire rolling resistance. In: *Proceedings of the rubber division symposia, 1*, Chicago.

Cossalter, V., Da Lio, M., Lot, R., & Fabbri, L. (1999). A general method for the evaluation of vehicle maneuverability with special emphasis on motorcycles. *Vehicle System Dynamics, 31.*

Cossalter, V., Lot, R., & Rota, S. (2010). Objective and subjective evaluation of an advanced motorcycle riding simulator. In: *Proceedings bicycle and motorcycle dynamics 2010.* Cf. Schwab.

Davis, D. C. (1974). A radial-spring terrain-enveloping tire model. *Vehicle System Dynamics, 3.*

Dijks, A. (1974). *A multifactor examination of wet skid resistance of car tires.* SAE Paper 741106.

Dugoff, H., Fancher, P. S., & Segel, L. (1970). An analysis of tire traction properties and their influence on vehicle dynamics performance. In: *Proceedings FISITA Int. Auto. safety conference.* SAE Paper 700377.

Fiala, E. (1954). Seitenkräfte am rollenden luftreifen. *VDI Zeitschrift, 96.*

Frank, F. (1965a). Grundlagen zur berechnung der seitenführungskennlinien von reifen. *Kautchuk und Gummi, 18*(8).

Frank, F. (1965b). *Theorie des reifenschräglaufs.* Dissertation, Braunschweig.

Freudenstein, G. (1961). Luftreifen bei Schräg- und Kurvenlauf. *Deutsche Kraftfahr-zeugforschung und Str. Verk. techn., 152.*

Fritz, W. (1977). *Federhärte von reifen und frequenzgang der reifenkräfte bei periodischer vertikalbewegung der felge.* Dissertation, Karlsruhe.

Fromm, H. (1941). *Kurzer bericht über die geschichte der theorie des radflatterns.* Bericht 140 der Lilienthal Gesellschaft, 1941; NACA TM 1365.

Gillespie, T. D. (1992). *Fundamentals of vehicle dynamics.* SAE.

Gipser, M., Hofer, R., & Lugner, P. (1997). Dynamical tire forces response to road unevenness. In: *Proc. 2nd int. coll. on tyre models for vehicle dynamic analysis.* Suppl. to Vehicle System Dynamics, 27.

Gipser, M. (1996). *DNS-tire 3.0-die weiterentwicklung eines bewährten strukturmechanischen reifenmodells.* In: *Darmstädter reifenkolloquium (pp. 52–62). VDI Berichte, 512.* Düsseldorf: VDI Verlag.

Gipser, M. (1998). Reifenmodelle für komfort- und schlechtwegsimulationen. In: *Proc. 7. aachener kolloquium fahrzeug- und motorentechnik.* Aachen: IKA, RWTH. VDI Verlag.

Gipser, M. (2000). *ADAMS/FTire – A tire model for ride & durability simulations.* Tokyo: ADAMS user's conference.

Gipser, M. (2006). FTire software: advances in modelization and data supply. In: *Proc. Tire Society Meeting*. Akron.

Gipser, M. (2007). FTire – the tire simulation model for all applications related to vehicle dynamics. *Vehicle System Dynamics, 45*(1), 217–225.

Goncharenko, V. I., Lobas, L. S., & Nikitina, N. V. (1981). Wobble in guide wheels. *Soviet Applied Mechanics, 17*(8).

Gong, S., Savkoor, A. R., & Pacejka, H. B. (1993). *The influence of boundary conditions on the vibration transmission properties of tires*. SAE Paper 931280.

Gong, S. (1993). *A study of in-plane dynamics of tires*. Dissertation, TU Delft.

Gough, V. E. (1963). Tyres and air suspension. In G. H. Tidbury (Ed.), *Advances in automobile engineering*. Oxford: Pergamon Press.

Guan, D. H., Shang, J., & Yam, L. H. (1999). *Modelling of tire cornering properties with experimental modal parameters*. SAE 1999-01-0784.

Guntur, R. R. (1975). *Adaptive brake control systems*. Dissertation, TU Delft.

Guo, K. H., & Liu, Q. (1997). Modelling and simulation of non-steady state cornering properties and identification of structure parameters of tyres. In F. Böhm & H. P. Willumeit (Eds.), *Proceedings of 2nd colloquium on tyre models for vehicle analysis*, Berlin Suppl. Vehicle System Dynamics, *Vol. 27.*

Guo, K. H. (1994). *The effect of longitudinal force and vertical load distribution on tire slip properties*. SAE Paper 945087.

Hadekel, R. (1952). *The mechanical characteristics of pneumatic tyres*. S & T Memo 10/52. British Ministry of Supply. TPA 3/TIB.

Hartog, J. P. den (1940). *Mechanical vibrations*, New York.

Hasegawa, A. (1985). Analysis of controllability and stability of motorcycles – analysis of stability of high speed driving. In: *Proceedings of 10th int. tech. conf. on experimental safety vehicles.*

Henker, E. (1968). *Dynamische Kennlinien von PKW Reifen. Wissenschaftlich-Technische Veröffentlichungen aus dem Automobilbau (IFA-DDR)* Heft 3.

Higuchi, A., & Pacejka, H. B. (1997). The relaxation length concept at large wheel slip and camber. In F. Böhm & H. P. Willumeit (Eds.), *Proceedings of 2nd colloquium on tyre models for vehicle analysis*, Berlin, Suppl. Vehicle System Dynamics, 27.

Higuchi, A. (1997): *Transient response of tyres at large wheel slip and camber.* Dissertation, TU Delft.

Ho, F. H., & Hall, M. F. (1973). *An experimental study of the pure-yaw frequency response of the 18x5.5 type VII aircraft tires. AFFDL-TR-73–79.*

Iffelsberger, L. (1991). Application of vehicle dynamic simulation in motorcycle development. *Safety Environment Future, Forschungshefte Zweiradsicherheit, 7.*

Jagt, P. van der (2000). *The road to virtual vehicle prototyping*. Dissertation. TU Eindhoven.

Jenkinson, D. (1958). *The racing driver.* London: Batsford Ltd.

Jianmin, G., Gall, R., & Zuomin, W. (2001). Dynamic damping and stiffness characteristics of the rolling tire. *Tire Science and Technology, 29*(4).

Katayama, T., Aoki, A., & Nishimi, T. (1988). Control behaviour of motorcycle riders. *Vehicle System Dynamics, 17.*

Katayama, T., Nishimi, T., Okoyama, T., & Aoki, A. (1997). A simulation model for motorcycle rider's behaviours. In: *Proceedings of SETC'97.* Yokohama, SAE of Japan.

Keldysh, M.V. (1945). *Shimmy of the front wheel of a three-wheeled landing gear.* Tr. Tsentr. Aerogidrodinamicheskogo Inst., 564.

Klotter, K. (1960). *Technische Schwingungslehre II.* Berlin.

Kluiters, M. A. M. (1969). An investigation into F-28 main gear vibrations. *Fokker Report X.* 28–430.

Kobiki, Y., Kinoshita, A., & Yamada, H. (1990). Analysis of interior booming noise caused by tire and power train-suspension system vibration. *International Journal of Vehicle Design, 11*(3).

Koenen, C., & Pacejka, H. B. (1980). *Vibrational modes of motorcycles in curves.* In: *Proceedings of the int. motorcycle safety conference*, Vol. II. Wash. D.C.: Motorcycle Safety Foundation.

Koenen, C. (1983). *The dynamic behaviour of a motorcycle when running straight ahead and when cornering*. Dissertation, TU Delft.

Koiter., W. T., & Pacejka, H. B. (1969). On the skidding of vehicles due to locked wheels. In: *Proceedings of the symposium on handling of vehicles under emergency conditions. Inst. of Mech. Engnrs.* 1968–69, 183 Pt 3H, 19.

Kooijman, J. D. G., Schwab, A. L., & Meijaard, J. P. (2008). Experimental validation of a model of an uncontrolled bicycle. *Multibody System Dynamics, 19*(1–2).

Kooijman, J. D. G., Meijaard, J. P., Papadopoulos, J. M., Ruina, A., & Schwab, A. L. (2011). A bicycle can be self-stable without gyroscopic or caster effects. *Science, 15*. 332(6027).

Kortüm, W., & Lugner, P. (1994). *Systemdynamik und regelung von fahrzeugen.* Berlin: Springer Verlag.

Krempel, G. (1967). Untersuchungen an kraftfahrzeugreifen. *ATZ, 69*(1), 8, (Cf. also dissertation Karlsruhe University, 1965).

Krylov, N., & Bogoljubov, N. (1947). *Introduction to non-linear mechanics*. Princeton.

Laerman, F. J. (1986). *Seitenführungsverhalten von Kraftfahrzeugreifen bei schnellen Radlaständerungen*. Dissertation, Braunschweig: VDI-Fortschritt Berichte, 12, 73.

Lee, Jung-Hwan (2000). Analysis of tire effect on the simulation of vehicle straight line motion. *Vehicle System Dynamics, 33.*

Leipholz, H. (1987). *Stability theory, an introduction to the stability of dynamic systems and rigid bodies*. Stuttgart: John Wiley & Sons, B.G.Teubner.

Limebeer, D. J. N., & Sharp, R. S. (2006). *Bicycles, motorcycles and models.* IEEE Control Systems Magazine.

Lippmann, S. A., & Nanny, J. D. (1967). *A quantitative analysis of the enveloping forces of passenger car tires*. SAE Paper 670174.

Lippmann, S. A., Piccin, W. A., & Baker, T. P. (1965, 1967). *Enveloping characteristics of truck tires – a laboratory evaluation.* SAE Paper 650184.

Lugner, P., Pacejka, H. B., & Plöchl, M. (2005). *Recent advances in tyre models and testing procedures.* State of the art paper of 19th IAVSD symposium on the dynamics of vehicles on roads and tracks, Milano, Veh. System Dynamics, *43*.

Magnus, K. (1955). Ueber die verfahren zur untersuchung nicht-linearer schwingungs- und regelungs-systeme. *VDI-Forschungsheft, 451 B*, 21.

Mastinu, G. (1997). A semi-analytical tyre model for steady and transient state simulations. In F. Böhm & H. P. Willumeit, (Eds.), *Proceedings of 2nd colloquium on tyre models for vehicle analysis*, Berlin, Suppl. Vehicle System Dynamics, 27.

Maurice, J. P., Berzeri, M., & Pacejka, H. B. (1999). Pragmatic tyre model for short wavelength side slip variations. *Vehicle System Dynamics, 31*(2).

Maurice, J.P. (2000). *Short wavelength and dynamic tyre behaviour under lateral and combined slip conditions*. Dissertation, TU Delft.

Meijaard, J. P., Papadopoulos, J. M., Ruina, A., & Schwab, A. L. (2007). Linearized dynamics equations for the balance and steer of a bicycle: a benchmark and review. In: *Proceedings of the Royal Society A*. 463, (pp. 1955–1982) (with Elec.Supp.Mat.).

Metcalf, W. H. (1963). *Effect of a time-varying load on side force generated by a tire operating at constant slip angle*. SAE Paper 713c.

Milliken, W. F., & Milliken, D. L. (1995). *Race car vehicle dynamics*. SAE.

Milliken, W. F., et al. (1956). *Research in automobile stability and control and tire performance*. London: The Inst. of Mech. Engnrs.

Milliken, W. F. (2006). *Equations of motion, an engineering autobiography*. Bentley Publishers.

Mitschke, M. (1982). *Dynamik der kraftfahrzeuge, band A, antrieb und bremsung*. Berlin: Springer Verlag.

Mitschke, M. (1990). *Dynamik der kraftfahrzeuge, Band C, fahrverhalten*. Berlin: Springer Verlag.

Moore, J. K., Kooijman, J. D. G., Schwab, A. L., & Hubbard, M. (2011). Rider motion identification during normal bicycling by means of principal component analysis. *Multibody System Dynamics, 25*(2).

Moreland, W. J. (1954). The story of shimmy. *Journal of the Aeronautical Sciences*.

Mousseau, C. W., & Clark, S. K. (1994). An analytical and experimental study of a tire rolling over a stepped obstacle at low velocity. *Tire Science and Technology, TSTCA, 16*(2).

Nishimi, T., Aoki, A., & Katayama, T. (1985). Analysis of straight-running stability of motorcycles. In: *Proceedings of 10th Int. Tech. Conf. on Experimental Safety Vehicles*.

Nordeen, D. L., & Cortese, A. D. (1963). *Force and moment characteristics of rolling tires*. SAE Paper 713A, 1963; SAE Transactions 325.

Oertel, Ch., & Fandre, A. (1999). *Ride comfort simulations and steps towards life time calculations: RMOD-K and ADAMS*. International ADAMS User Conference, Berlin.

Oertel, Ch., & Fandre, A. (2009). *Tire Model RMOD-K 7 and misuse load cases*. SAE international, SAE-Paper 2009-01-0582.

Oertel, Ch. (1997). On modelling contact and friction – calculation of tyre response on uneven roads. In F. Böhm & H. P. Willumeit (Eds.), *Proceedings of 2nd colloquium on tyre models for vehicle analysis*, Berlin 1997, Suppl. Vehicle System Dynamics, 27.

Olley, M. (1947). Road manners of the modern car. *J. Inst. Auto. Engnrs., 15*.

Oosten, J. J. M. van, & Bakker, E. (1993). Determination of magic tyre model parameters. In H. B. Pacejka (Ed.), *Proceedings of 1st colloquium on tyre models for vehicle analysis*. Delft 1991, Suppl. Vehicle System Dynamics, 21.

Pacejka, H. B., & Bakker, E. (1993). The magic formula tyre model. In H. B. Pacejka (Ed.), *Proceedings of 1st colloquium on tyre models for vehicle analysis, delft 1991*. Suppl. Vehicle System Dynamics, 21.

Pacejka, H. B., & Besselink, I. J. M. (1997). Magic formula tyre model with transient properties. In F. Böhm & H. P. Willumeit (Eds.), *Proceedings of 2nd colloquium on tyre models for vehicle analysis*, Berlin, Suppl. Vehicle System Dynamics, 27.

Pacejka, H. B., & Fancher, P. S. (1972a). Hybrid simulation of shear force development of a tire experiencing longitudinal and lateral slip. In: *Proceedings of XIV FISITA Int. Auto. Tech. Congress*, London.

Pacejka, H. B., & Sharp, R. S. (1991). Shear force development by pneumatic tyres in steady state conditions: a review of modelling aspects. *Vehicle System Dynamics, 20*.

Pacejka, H. B., & Takahashi, T. (1992). Pure slip characteristics of tyres on flat and on undulated road surfaces. In: *Proceedings of AVEC'92*. Yokohama, SAE of Japan.

Pacejka, H. B., Van der Berg, J., & Jillesma, P. J. (1977). Front wheel vibrations. In: A. Slibar & H. Springer (Eds.), *Proceedings of 5th VSD-2nd IUTAM Symposium*, Vienna 1977, Swets and Zeitlinger, Lisse.

Pacejka, H. B. (1958). Study of the lateral behaviour of an automobile moving upon a flat level road. Cornell aeronautical laboratory report YC-857-F-23, 1958. Pacejka, H. B. (1965): Analysis of the shimmy phenomenon. In: *Proceedings of the Auto. Division of I. Mech. E., 180, Part 2A, Inst. of Mech. Engnrs.*

Pacejka, H. B. (1966). The wheel shimmy phenomenon. Dissertation, TU Delft.

Pacejka, H. B. (1971). *The Tyre as a vehicle component. Chapter 7 of: Mechanics of pneumatic tires*, Ed. S. K. Clark, N. B. S. Monograph 122, Washington D.C., (new edition 1981).

Pacejka, H. B. (1972). Analysis of the dynamic response of a string-type tire model to lateral wheel-plane vibrations. *Vehicle System Dynamics, 1*(1).

Pacejka, H. B. (1973a). Approximate dynamic shimmy response of pneumatic tyres. *Vehicle System Dynamics, 2.*

Pacejka, H. B. (1973b). Simplified analysis of steady-state turning behaviour of motor vehicles. *Vehicle System Dynamics, 2.* p.161,173,185.

Pacejka, H. B. (1974). Some recent investigations into dynamics and frictional behavior of pneumatic tires. In D. F. Hays & A. L. Browne (Eds.), *Proceedings of G.M. symposium physics of tire traction*. New York: Plenum Press.

Pacejka, H. B. (1981). Analysis of tire properties. In S. K. Clark (Ed.), *Chapter 9, Mechanics of pneumatic tires*. DOT HS-805 952.

Pacejka, H. B. (1981a). In-plane and out-of-plane dynamics of pneumatic tyres. *Vehicle System Dynamics, 10.*

Pacejka, H. B. (1986). Non-linearities in road vehicle dynamics. *Vehicle System Dynamics, 15*(5).

Pacejka, H. B. (1996). The tyre as a vehicle component. In M. Apetaur (Ed.), *Proceedings of XXVI FISITA Congress*, Prague.

Pacejka, H. B. (2004). Spin: camber and turning. In P. Lugner (Ed.), *Proceedings of 3rd colloquium on tyre models for vehicle analysis*. Vienna 2004, Suppl. Vehicle System Dynamics.

Pevsner, Ja. M. (1947). *Theory of the stability of automobile motions* (In Russian). Leningrad: Masjgiz.

Radt, H. S., & Milliken, W. F. (1983). *Non-dimensionalizing tyre data for vehicle simulation*. Road vehicle handling. Inst. of Mech. Engnrs. (C133/83).

Radt, H. S., & Pacejka, H. B. (1963). Analysis of the steady-state turning behavior of an automobile. In: *Proceedings of the symposium on control of vehicles. Inst. of Mech. Engnrs.* London.

Reimpell, J., and Sponagel, P. (1986): Fahrwerktechnik: Reifen und Räder. Vogel Buchverlag, Würzburg.

Oosten, J. J. M. van, Unrau, J. H., Riedel, G., and Bakker, E. (1996). TYDEX Workshop: Standardisation of data exchange in tyre testing and tyre modelling. In F. Böhm & H. P. Willumeit (Eds.), *Proceedings of 2nd colloquium on tyre models for vehicle analysis*, Berlin, Suppl. Vehicle System Dynamics, 27.

Riekert, P., & Schunck, T. E. (1940). Zur fahrmechanik des gummi-bereiften kraftfahrzeugs. *Ingenieur Archiv, 11*(210).

Rocard, Y. (1949). *Dynamique général des vibrations*. Paris.

Rogers, L. C., & Brewer, H. K. (1971). Synthesis of tire equations for use in shimmy and other dynamic studies. *Journal of Aircraft, 8*(9).

Rogers, L. C. (1972). Theoretical tire equations for shimmy and other dynamic studies. *AIAA J. of Aircraft.*

Ruijs, P. A. J., & Pacejka, H. B. (1985). Research in lateral dynamics of motorcycles. In O. Nordström (Ed.), *Proceedings of 9th IAVSD symposium on the dynamics of vehicles on roads and tracks*. Linköping. Suppl. Vehicle system dynamics, 15.

SAE J670e. (1976). *Vehicle dynamics terminology*. Warrendale, PA: SAE J670e, Society of Automotive Engineers, Inc.

Saito, Y. (1962). A study of the dynamic steering properties of tyres. In: *Proceedings IX FISITA congress*, London.

Sakai, H. (1981). Theoretical and experimental studies on the dynamic cornering properties of tyres. *International Journal of Vehicle Design, 2*(1–4).

Sakai, H. (1989). Study on cornering properties for tire and vehicle. *The 8th annual meeting of the tire society*, Akron.

Sakai, H. (1990). Study on cornering properties of tire and vehicle. *Tire Science and Technology, TCTCA, 18*(3).

Savkoor, A. R. (1970). The lateral flexibility of a pneumatic tyre and its application to the lateral contact problem. In *Proceedings FISITA Int. Auto. safety conference*, SAE Paper 700378.

Jagt, P. van der, Pacejka, H. B., & Savkoor, A. R. (1989). Influence of tyre and suspension dynamics on the braking performance of an anti-lock system on uneven roads. In: *Proceedings of EAEC conference*, Strassbourg, C382/047 IMechE.

Schlippe, B. von, & Dietrich, R. (1941). Das flattern eines bepneuten rades. Bericht 140 der Lilienthal Gesellschaft, 1941: NACA TM 1365.

Schlippe, B. von, & Dietrich, R. (1942). Zur mechanik des luftreifens. Zentrale für wissenschaftliches Berichtwesen, Berlin-Adlershof.

Schlippe, B. von, & Dietrich, R. (1943). Das flattern eines mit luftreifen versehenen rades. In *Jahrbuch der deutsche Luftfahrtforschung*.

Schmeitz, A. J. C., & Pacejka, H. B. (2003). A semi-empirical three-dimensional tyre model for rolling over arbitrary road unevennesses. In M. Abe (Ed.), *Proceedings of the 18th IAVSD Symposium on the dynamics of vehicles on roads and tracks*, Kanagawa, Japan, Swets and Zeitlinger, Suppl. of Vehicle System Dynamics.

Schmeitz, A. J. C., & Pauwelussen, J. P. (2001). An efficient dynamic ride and handling tyre model for arbitrary road unevennesses. VDI-Berichte.

Schmeitz, A. J. C. (2004). *A semi-empirical three-dimensional model of the pneumatic tyre rolling over arbitrarily Uneven Road Surfaces*. Dissertation, TU Delft.

Schwab, A. L. & Meijaard, J. P. (2010) (Eds.); *Proceedings bicycle and motorcycle dynamics 2010. Symposium on the dynamics and control of single track vehicles*. TU-Delft, Oct. 2010, The Netherlands. http://bicycle.tudelft.nl/bmd2010/?page_id=483.

Segel, L., & Ervin, R. D. (1981). The influence of tire factors on the stability of trucks and tracktor trailers. *Vehicle System Dynamics, 10*(1).

Segel, L., & Wilson, R. (1976): Requirements on describing the mechanics of tires used on single-track vehicles. In H. B. Pacejka (Ed.), *Proceedings of IUTAM Symposium on the dynamics of vehicles*, TU Delft 1975, Swets and Zeitlinger, Lisse.

Segel, L. (1956). Theoretical prediction and experimental substantiation of the response of the automobile to steering control. In: *Proceedings of auto. Division of I.Mech.E.* 7.

Segel, L. (1966). Force and moment response of pneumatic tires to lateral motion inputs. *Transactions ASME, Journal of Engineering for Industry, 88B*.

Sekula, P. J., et al. (1976). Dynamic indoor tyre testing and Fourier transform analysis. *Tire Science and Technology, 4*(2).

Shang, J., Guan, D., & Yam, L. H. (2002). Study on tyre dynamic cornering properties using experimental modal parameters. *Vehicle System Dynamics, 37*(2).

Sharp, R. S., & Alstead, C. J. (1980a). The influence of structural flexibilities on the straight-running stability of motorcycles. *Vehicle System Dynamics, 9*.

Sharp, R. S., & El-Nashar, M. A. (1986). A generally applicable digital computer based mathematical model for the generation of shear forces by pneumatic tyres. *Vehicle System Dynamics, 15*.

Sharp, R. S., & Jones, C. J. (1980). A comparison of tyre representations in a simple wheel shimmy problem. *Vehicle System Dynamics, 9*(1).

Sharp, R. S., & Limebeer, D. J. N. (2001a). A motorcycle model for stability and control analysis. *Multibody System Dynamics, 6*(2).

Sharp, R. S. (1971). The Stability and control of motorcycles. *J. of Mech. Engng. Sci., 13*. 5, I.Mech.E.

Sharp, R. S. (1978). *A review of motorcycle steering behaviour and straight line stability characteristics*. SAE Paper 780303.

Sharp, R. S. (1985). The lateral dynamics of motorcycles and bicycles. *Vehicle System Dynamics, 14*, 4–6.

Sharp, R. S. (2001). Stability, control and steering responses of motorcycles. *Vehicle System Dynamics, 35*, 4–5.

Sharp, R. S. (2010). Rider Control of a motorcycle near to its cornering limits. In: *Proceedings bicycle and motorcycle dynamics 2010*. Cf. Schwab.

Smiley, R. F. (1957). Correlation and extension of linearized theories for tire motion and wheel shimmy. NACA Report 1299.

Smiley, R. F. (1958). Correlation, evaluation and extension of linearized theories for tire motion and wheel shimmy. NACA (NASA) Tech. Note 4110.

Sperling, E. (1977): *Zur Kinematik und Kinetik elastischer Räder aus der Sicht verschiedener Theorien*. Dissertation, TU Munich.

Spierings, P. T. J. (1981). The effects of lateral front fork flexibility on the vibrational modes of straight running single track vehicles. *Vehicle System Dynamics, 10*.

Stepan, G. (1997). Delay, nonlinear oscillations and shimmying wheels. In: *Proceedings of symposium CHAOS'97*, Ithaca, N.Y., Kluwer Ac. Publ., Dordrecht.

Stoker, J. J. (1950): Non-linear vibrations. New York.

Strackerjan, B. (1976): Die querdynamik von kraftfahrzeugreifen. In: *Proceedings of VDI-Schwingungstagung*.

Takahashi, T., & Hoshino, M. (1996). The tyre cornering model on uneven roads for vehicle dynamics studies. In H. Wallentowitz (Ed.), *Proceedings of AVEC'96, Int. Symp. on advanced vehicle control.*, Aachen.

Takahashi, T., & Pacejka, H. B. (1987). Cornering on uneven roads. In M. Apetaur (Ed.), *Proceedings of 10th IAVSD symposium on the dynamics of vehicles on roads and tracks, Prague 1987, Suppl. Vehicle System Dynamics, 17*.

Takahashi, T., Yamada, T., & Nakamura, T. (1984). *Experimental and theoretical study of the influence of tires on straight-running motorcycle weave response*. SAE Paper 840248.

Tanguy, G. (1986). *Antiskid systems and vehicle suspension*. SAE Paper 865134.

Teerhuis, A. P., & Jansen, S. T. H. (2010). Motorcycle state estimation for lateral dynamics. In: *Proceedings bicycle and motorcycle dynamics 2010*. Cf. Schwab.

Eldik Thieme, H. C. A. van (1960). Experimental and theoretical research on mass-spring systems. In: *Proceedings of FISITA congress*, The Hague.

Troger, H., & Zeman, K. (1984). A non-linear analysis of the generic types of loss of stability of the steady-state motion of a tractor-semi-trailer. *Vehicle System Dynamics, 13.*

Vågstedt, N. G. (1995). *On the cornering characteristics of ground vehicle axles.* Dissertation, KTH Stockholm.

Valk, R. van der, & Pacejka, H. B. (1993). An analysis of a civil aircraft main landing gear shimmy failure. *Vehicle System Dynamics, 22.*

Vries, E. J. H. de, & Pacejka, H. B. (1998a). Motorcycle tyre measurements and models. In L. Palkovics (Ed.), *Proceedings of 15th IAVSD symposium on the dynamics of vehicles on roads and tracks*, Budapest, Suppl. Vehicle system dynamics 28.

Vries, E. J. H. de, & Pacejka, H. B. (1998b). The effect of tire modelling on the stability analysis of a motorcycle. In: *Proceedings AVEC'98*, Nagoya, SAE of Japan.

Weir, D. H. (1972). Motorcycle handling dynamics and rider control and the effect of design configuration on response and performance. Dissertation, UCLA.

Whipple, F. J. W. (1899). The stability of the motion of a bicycle. *Quarterly Journal of Pure and Applied Mathematics, 30.*

Whitcomb, D. W., & Milliken, W. F. (1956). Design implications of a general theory of automobile stability and control. In: *Proceedings of auto. Division of I.Mech.E.* 7.

Willumeit, H. P. (1969). *Theoretisch Untersuchungen an einem Modell des Luftreifens.* Dissertation, Berlin.

Winkler, C. B. (1998). Simplified analysis of the steady-state turning of complex vehicles. *Vehicle System Dynamics, 29*(3).

Wisselman, D., Iffelsberger, D., & Brandlhuber, B. (1993). Einsatz eines fahrdynamik-simulationsmodells in der motorradentwicklung bei BMW. *ATZ, 95*(2).

Zegelaar, P. W. A., & Pacejka, H. B. (1995). The in-plane dynamics of tyres on uneven roads. In L. Segel (Ed.), *Proceedings of 14th IAVSD symposium on the dynamics of vehicles on roads and tracks*, Ann Arbor, Suppl. Vehicle System Dynamics, 25.

Zegelaar, P. W. A., & Pacejka, H. B. (1997). Dynamic tyre responses to brake torque variations. In F. Böhm & H. P. Willumeit (Eds.), *Proceedings of 2nd colloquium on tyre models for vehicle analysis*, Berlin, Suppl. Vehicle System Dynamics, 27.

Zegelaar, P. W. A., Gong, S., & Pacejka, H. B. (1993). Tyre models for the study of in-plane dynamics. In Z. Shen (Ed.), *Proceedings of 13th IAVSD symposium on the dynamics of vehicles on roads and tracks*, Chengdu 1993, Suppl. Vehicle System Dynamics, 23.

Zegelaar, P. W. A. (1998). *The dynamic response of tyres to brake torque variations and road unevennesses.* Dissertation, TU Delft.

# List of Symbols

| | |
|---|---|
| $a$ | distance front axle to c.g.; half of contact length |
| $a_x$ | longitudinal acceleration |
| $a_y$ | lateral acceleration |
| $a_\mu$ | slip velocity dependency coefficient for friction |
| $A_r$ | rolling resistance coefficient |
| $b$ | distance rear axle to c.g.; half contact width |
| $B$ | stiffness factor in *'Magic Formula'* |
| $B_1$ | brake force of rolling wheel |
| $c$ | stiffness; factor |
| $c_c$ | lateral carcass stiffness per unit length |
| $c_{\rm gyr}$ | non-dimensional gyroscopic coefficient |
| $c_{px,y}$ | tread element stiffness per unit length of circumference |
| $c'_{px}$ | tread element longitudinal stiffness per unit area |
| $C$ | cornering stiffness ; sum front and rear |
| $C_i$ | cornering stiffness, sum left and right |
| $C$ | contact centre (point of intersection) |
| $C$ | shape factor in *'Magic Formula'* |
| $C_{dA}$ | air drag coefficient |
| $C_{Fx}$ | longitudinal stiffness of standing tyre |
| $C_{Fy}$ | lateral stiffness of standing tyre |
| $C_{Fz}$ | stiffness of tyre normal to the road |
| $C_{F\alpha}$ | cornering stiffness |
| $C_{F\kappa}$ | longitudinal slip stiffness |
| $C_{F\gamma}$ | camber stiffness for side force |
| $C_{F\varphi}$ | spin stiffness for side force |
| $C_{\rm gyr}$ | tyre gyroscopic coefficient |
| $C_{M\alpha}$ | aligning torque stiffness |
| $C_{M\gamma}$ | camber stiffness for aligning torque |
| $C_{M\varphi}$ | spin stiffness for aligning torque |
| $C_{M\psi}$ | torsional yaw stiffness of standing tyre |
| $C_{Mx\gamma}$ | overturning couple stiffness against camber |
| $C_{cx,y}$ | carcass horizontal stiffness of standing tyre |
| $C_{\rm gyr}$ | gyroscopic coefficient |
| $df_z$ | normalised change in normal load, Eq.(4.E2a) |
| $dp_i$ | normalised change in inflation pressure, Eq.(4.E2b) |

| | |
|---|---|
| $d_t$ | tread depth |
| $D$ | peak factor in *'Magic Formula'*; dissipation function |
| $E$ | curvature factor in 'Magic Formula' |
| $e$ | caster length; tread element deflection |
| $f$ | trail of c.g.; frequency [Hz] |
| $f_r$ | rolling resistance coefficient |
| $F_{ax}$ | force for forward acceleration |
| $F_d$ | air drag force |
| $F_{x,\mathrm{tot}}$ | sum of longitudinal tyre forces |
| $F_x$ | longitudinal tyre force |
| $F_y$ | lateral tyre force |
| $F_z$ | vertical (normal) tyre force (load) (>0), in Chap.9,10: $F_z <0$ |
| $F_r$ | rolling resistance force (>0) |
| $F_N$ | tyre normal force (>0) |
| $F_{No}$ | reference vertical load, nominal load ($= \|F_{zo}\|$) |
| $F_V$ | tyre vertical force |
| $F_H$ | tyre longitudinal horizontal force |
| $g$ | acceleration due to gravity; feedback rider control gain |
| $G$ | weighting factor |
| $h$ | height |
| $H$ | height; sharpness factor in 'Magic Formula' |
| $H$ | transform; Hurwitz determinant |
| $i$ | $\sqrt{-1}$ |
| $i_z$ | radius of inertia |
| $I$ | moment of inertia |
| $I_w$ | wheel polar moment of inertia |
| $I_p$ | wheel polar moment of inertia |
| $j$ | $\sqrt{-1}$ |
| $k$ | radius of inertia; viscous damping coefficient |
| $K$ | centrifugal force; force acting on belt, wheel centre |
| $l$ | wheel base |
| $l_s$ | shift; two-point follower length |
| $l_b$ | length of basic curve |
| $l_f$ | offset |
| $\boldsymbol{l}$ | unit vector along line of intersection |
| $m$ | mass; fraction of contact length $2a$ where adhesion occurs |
| $m_c$ | contact patch mass |
| $m_t$ | tyre mass |
| $m_m$ | mass of mainframe (including lower part of rider) |
| $m_{mr}$ | mass of mainframe plus rider |
| $m_r$ | mass of upper torso |
| $M_{B,D}$ | brake, drive torque |

$M_x$ overturning couple
$M_y$ rolling resistance moment
$M_z$ (self) aligning torque
$M'_z$ (self) aligning torque due to lateral deflections
$M^*_z$ aligning torque due to longitudinal deflections
$M_{z,\text{gyr}}$ gyroscopic couple
$M_\delta$ steer torque
$n$ number of elements; frequency [Hz]
$\boldsymbol{n}$ unit vector normal to the road $=(0,0,-1)^T$
$n_{st}$ steer system ratio
$p$ Laplace variable [1/m]
$p_i$ inflation pressure
$q$ average vehicle yaw resistance arm; generalised coordinate
$q$ contact force per unit length of circumference, vector
$Q$ generalised force
$r$ yaw rate; tyre (loaded) radius
$r_c$ radius of carcass (belt), unloaded; cross section crown radius
$r_{yo}$ free tyre radius varying along cross section contour, $r_{yo} = r_{yo}(y_{co})$
$r_e$ effective rolling radius of freely rolling wheel
$r_f$ free unloaded tyre radius
$r_l$ loaded radius
$r_o$ free unloaded tyre radius ($= R_o$)
$R$ radius of curvature
$R_o$ free unloaded tyre radius ($= r_o$)
$s$ forward position of neutral steer point; half track width
$s$ Laplace variable; travelled distance
$s_{sx}$ $\kappa$ (practical longitudinal slip component)
$s_{sy}$ $\tan\alpha$ (practical lateral slip component)
$\boldsymbol{s}$ unit vector along wheel spin axis
$S$ wheel slip point; impulse; string tension force
$S_{V,H}$ vertical, horizontal shift
$t$ pneumatic trail; time
$t_c$ caster length
$t_r$ rise time
$\boldsymbol{t}$ unit vector in road plane perpendicular to line of intersection $\boldsymbol{l}$
$T$ kinetic energy; moment acting on belt, wheel centre
$u$ forward velocity of c.g.; longitudinal deflection
$U$ potential energy
$v$ lateral velocity of c.g.; lateral deflection
$V$ speed of travel of c.g. (with $x$, $y$ components)
$V$ speed of travel of wheel centre (with $x$, $y$ components)
$V_c$ speed of contact centre $C$ (with $x$, $y$ components)

| | |
|---|---|
| $V_g$ | speed of sliding (with $x$, $y$ components) |
| $V_o$ | reference velocity $=\sqrt{(gR_o)}$ |
| $V_r$ | wheel linear speed of rolling ($= V_{cx} - V_{sx}$) |
| $V_s$ | wheel slip velocity of slip point $S$ (with $x$, $y$ components) |
| $V_x$ | longitudinal speed component of wheel centre |
| $V_s^*$ | velocity of contact patch mass (with $x$, $y$ components) |
| $w$ | vertical road (effective) profile (positive downwards) |
| $W$ | work |
| $x,y,z$ | longitudinal, lateral, vertical displacement |
| $x,y,z$ | coordinates with respect to moving axes system, $z$ axis vertical |
| $x^o,y^o,z^o$ | global coordinates |
| $\overline{x}, \overline{y}, \overline{z}$ | global coordinates |
| $X$ | longitudinal horizontal tyre force |
| $X,Y,Z$ | global coordinates |
| $y_{co}$ | distance from wheel centre plane |
| $y_{mr}$ | lateral offset of $m_{mr}$ c.g. |
| | |
| $\alpha$ | wheel (side) slip angle; axle (side) slip angle |
| $\alpha$ | road transverse slope angle |
| $\alpha'$ | transient tyre slip angle |
| $\alpha_a$ | virtual axle slip angle |
| $\beta$ | vehicle side slip angle; tyre yaw torsion angle |
| $\beta_{x,y}$ | road transverse, forward (effective) slope angle |
| $\beta_{\mathrm{gyr}}$ | gyroscopic wheel coupling coefficient, Eq.(6.35) |
| $\gamma$ | camber (wheel inclination) angle |
| $\gamma'$ | transient tyre camber angle |
| $\Gamma$ | unit step response function |
| $\delta$ | steer angle of front wheels |
| $\delta_o$ | $\approx l/R$, steer angle at $V\rightarrow 0$ |
| $\Delta$ | increment |
| $\varepsilon$ | roll steer coefficient; rake angle of steering axis |
| $\varepsilon$ | string length ratio, Eq.(5.153); eff. roll. radius gradient $-\partial r_e/\partial d_t$ |
| $\varepsilon$ | small quantity to avoid singularity |
| $\varepsilon_\gamma$ | camber stiffness reduction factor |
| $\varepsilon_{NL}$ | non-lagging part |
| $\zeta$ | damping ratio; spin factor (=1 if spin influence is disregarded) |
| $\zeta_h$ | height ratio, Eq.(6.36) |
| $\zeta_\alpha$ | cornering stiffness load transfer coefficient |
| $\zeta_\gamma$ | camber stiffness load transfer coefficient |
| $\eta$ | understeer coefficient; effective rolling radius gradient $-\partial r_e/\partial \rho_z$ |
| $\eta_y$ | c.g. offset steer coefficient |
| $\theta$ | tyre model parameter, Eqs.(3.6,3.24,3.46) |

$\theta$ angular displacement about $\eta$ axis; pitch angle
$\theta_c$ string model composite parameter, Eq.(5.160)
$\kappa$ longitudinal wheel slip
$\kappa'$ transient longitudinal tyre slip
$\kappa^*$ damping coefficient due to tread width
$\lambda$ wavelength; root characteristic equation
$\lambda$ fraction of $2a$ where adhesion occurs; user scaling factor
$\mu$ coefficient of friction
$\rho$ tyre radial (vertical) deflection
$\rho_{x,y,z}$ tyre longitudinal, lateral, normal deflection
$\sigma$ relaxation length; load transfer coefficient
$\sigma$ theoretical slip, vector, Eq.(3.34)
$\sigma^*$ intersection length in string model with tread elements
$\sigma_c$ string model length parameter, Eq.(5.153)
$\sigma_c$ contact patch relaxation length
$\tau$ roll camber coefficient
$\varphi$ body roll angle; spin slip
$\varphi'$ transient spin slip
$\varphi_t$ turn slip
$\phi$ phase angle
$\psi$ yaw angle; steer angle
$\psi_{c1}$ compliance steer angle
$\psi_{io}$ toe angle
$\omega$ frequency [rad/s]
$\omega_o$ undamped natural frequency
$\omega_{1,2}$ natural frequencies
$\omega_n$ damped natural frequency
$\omega_s$ path frequency [rad/m]
$\Omega$ wheel speed of revolution
$\xi,\eta,\zeta$ moving axes system, $\eta$ axis along spin axis, $\xi$ horizontal

## SUBSCRIPTS AND SUPERSCRIPTS

$a$ axle; from belt to wheel rim centre
$b$ belt; from belt centre to rim
$c$ compliance (steer)
$c$ contact patch; from contact patch centre to belt; crown; contour
$D$ drag
$e$ effective
$eff$ effective (cornering stiffness)

| | |
|---|---|
| $eq$ | equivalent |
| $f$ | free, unloaded; of front frame |
| $g$ | global |
| $i$ | 1: front, 2: rear |
| $L,R$ | left, right |
| $m$ | of mainframe |
| $mr$ | of mainframe plus rider |
| $NL$ | non-lagging |
| $o$ | original; initial; average; unloaded; nominal; at vanishing speed; natural |
| $r$ | roll; rolling; rolling resistance; of residual spring; of rider |
| $s$ | slip; from road surface to contact patch; of front sub-frame |
| $sl$ | at verge of total sliding |
| $sf$ | side force (steer) |
| $ss$ | steady state |
| $st$ | static |
| $stw$ | steering wheel |
| $t$ | transition from adhesion to sliding |
| $w$ | wheel |
| $x,y,z$ | forward (longitudinal), lateral (to the right), downward |
| $zr$ | residual (torque) |
| $\xi,\eta,\zeta$ | along, around $\xi,\eta,\zeta$ axes |
| 0 | at zero condition |
| 1,2 | front, rear; leading, trailing edge |

# Appendix 1

# Sign Conventions for Force and Moment and Wheel Slip

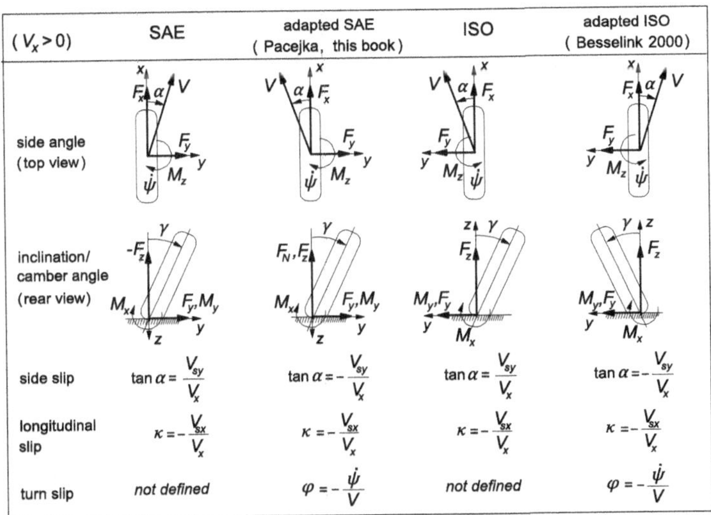


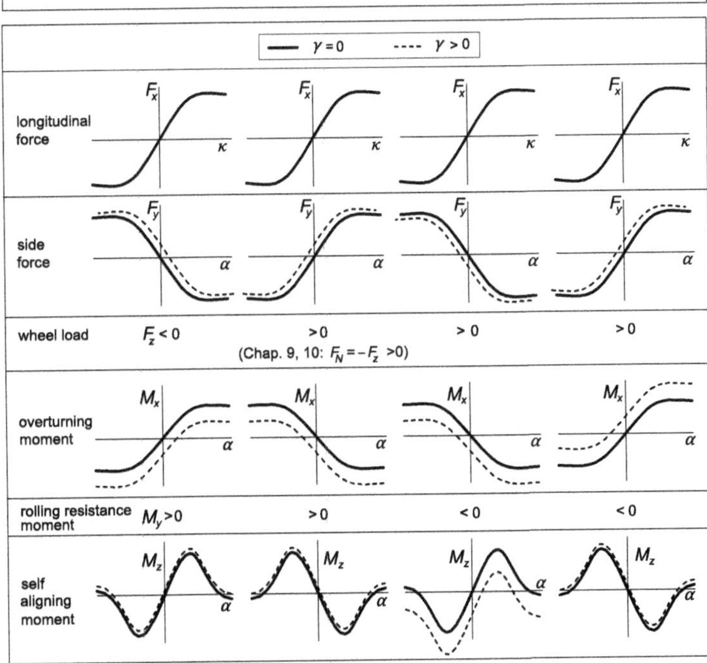

Appendix 2

# Online Information

Igo Besselink

In addition to this book, a number of simulation models can be found on the companion website:

http://www.elsevierdirect.com/companion.jsp?ISBN=9780080970165.
The software has been developed with *MATLAB* R2007b, but also works with newer *MATLAB* versions. A *MATLAB* installation including *Simulink* is needed to use the models, but no additional toolboxes are required.

The models provided consist of:

**TreadSim**
*TreadSim* is the brush tire model as discussed in Section 3.3 of this book. It is capable of returning the steady state force and moment characteristics for various slip conditions, including turn slip. Only a limited number of physical parameters are needed to obtain these results.

**Single track vehicle model**
The theory of the linear single-track vehicle model with two degrees of freedom is explained in Section 1.3.2. A comparison of this model with experimental data is made using different vehicle tests, e.g. steady state cornering, J-turn, double lane change, and random steer. Extensions to include tire relaxation behavior and saturation of the tire forces are also made available.

**Double track roll axis vehicle model**
Based on the equations provided in Section 1.3.1 a vehicle model with four tires is developed. This model includes the longitudinal and body roll degree of freedom. It allows us to investigate, e.g., braking in a turn and the effect of suspension compliance on vehicle handling behavior. It can be extended easily to include four-wheel steering and torque vectoring.

Additional models and exercises may be provided by the companion website, but they are still to be defined at the time of writing.

Appendix 3

# MF-Tire/MF-Swift Parameters and Estimation Methods

Igo Besselink

This appendix starts with a complete listing of the tire model parameters of a passenger car tire, which is closely linked to the TNO *MF-Tire/MF-Swift* 6.1 model. The tests required to obtain these parameters are discussed briefly. Furthermore, parameter estimation methods are presented in case no or limited measurement data are available. This is important as, for some classes of tires (e.g. aircraft and truck tires), dedicated measurement equipment may either simply not be available or not suited to handle very large tires.

Table A3.1 gives a listing of the various tire model parameters. Typically, this data is stored in a so-called 'tire property file', which normally has the filename extension '.tir' or '.tpf'. Please note that the ISO sign convention is used.

**TABLE A3.1** TNO MF-Tire/MF-Swift 6.1 Tire Model Parameters

| Tire Designation: 205/60R15 91 V |
|---|
| [model]<br>$V_o = 16.7$ m/s, $V_{x,\mathrm{low}} = 1$ m/s |
| [dimension]<br>$r_o = 0.3135$ m, $w = 0.205$ m, $r_{\mathrm{rim}} = 0.1905$ m |
| [operating_conditions]<br>$p_{io} = 220{,}000$ Pa |
| [inertia]<br>$m_{\mathrm{tire}} = 9.3$ kg, $I_{xx,\mathrm{tire}} = 0.391$ kg m$^2$, $I_{yy,\mathrm{tire}} = 0.736$ kg m$^2$<br>$m_{\mathrm{belt}} = 7.247$ kg, $I_{xx,\mathrm{belt}} = 0.3519$ kg m$^2$, $I_{yy,\mathrm{belt}} = 0.5698$ kg m$^2$ |

(*Continued*)

**TABLE A3.1** TNO MF-Tire/MF-Swift 6.1 Tire Model Parameters—Cont'd

**Tire Designation: 205/60R15 91 V**

[vertical]
$F_{zo} = 4000$ N
$c_{zo} = 209651$ N/m, $k_z = 50$ Ns/m, $q_{V2} = 0.04667$, $q_{Fz2} = 15.4$, $q_{Fcx} = 0$, $q_{Fcy} = 0$, $p_{Fz1} = 0.7098$
$B_{reff} = 8.386$, $D_{reff} = 0.25826$, $F_{reff} = 0.07394$, $q_{reo} = 0.9974$, $q_{V1} = 7.742 \cdot 10^{-4}$

[structural]
$c_{xo} = 358066$ N/m, $p_{cfx1} = 0.17504$, $p_{cfx2} = 0$, $p_{cfx3} = 0$
$c_{yo} = 102673$ N/m, $p_{cfy1} = 0.16365$, $p_{cfy2} = 0$, $p_{cfy3} = 0.24993$
$c_{\psi} = 4795$ Nm/rad, $p_{cmz1} = 0$
$f_{long} = 77.17$ Hz, $f_{lat} = 42.41$ Hz, $f_{yaw} = 53.49$ Hz, $f_{windup} = 58.95$ Hz
$\zeta_{long} = 0.056$, $\zeta_{lat} = 0.037$, $\zeta_{yaw} = 0.0070$, $\zeta_{windup} = 0.050$
$q_{bvx} = 0.364$, $q_{bv\theta} = 0.065$

[contact_patch]
$q_{ra1} = 0.671$, $q_{ra2} = 0.733$, $q_{rb1} = 1.059$, $q_{rb2} = -1.1878$
$p_{ls} = 0.8335$, $p_{ae} = 1.471$, $p_{be} = 0.9622$, $c_e = 1.5174$

[longitudinal_coefficients]
$p_{Cx1} = 1.579$, $p_{Dx1} = 1.0422$, $p_{Dx2} = -0.08285$, $p_{Dx3} = 0$
$p_{Ex1} = 0.11113$, $p_{Ex2} = 0.3143$, $p_{Ex3} = 0$, $p_{Ex4} = 0.001719$
$p_{Kx1} = 21.687$, $p_{Kx2} = 13.728$, $p_{Kx3} = -0.4098$
$p_{Hx1} = 2.1615e\text{-}4$, $p_{Hx2} = 0.0011598$, $p_{Vx1} = 2.0283e\text{-}5$, $p_{Vx2} = 1.0568e\text{-}4$
$p_{px1} = -0.3485$, $p_{px2} = 0.37824$, $p_{px3} = -0.09603$, $p_{px4} = 0.06518$
$r_{Bx1} = 13.046$, $r_{Bx2} = 9.718$, $r_{Bx3} = 0$, $r_{Cx1} = 0.9995$, $r_{Ex1} = -0.4403$, $r_{Ex2} = -0.4663$, $r_{Hx1} = -9.968e\text{-}5$

[overturning_coefficients]
$q_{sx1} = -0.007764$, $q_{sx2} = 1.1915$, $q_{sx3} = 0.013948$, $q_{sx4} = 4.912$, $q_{sx5} = 1.02$
$q_{sx6} = 22.83$, $q_{sx7} = 0.7104$, $q_{sx8} = -0.023393$, $q_{sx9} = 0.6581$, $q_{sx10} = 0.2824$
$q_{sx11} = 5.349$, $q_{sx12} = 0$, $q_{sx13} = 0$, $q_{sx14} = 0$, $p_{pmx1} = 0$

[lateral_coefficients]
$p_{Cy1} = 1.338$, $p_{Dy1} = 0.8785$, $p_{Dy2} = -0.06452$, $p_{Dy3} = 0$
$p_{Ey1} = -0.8057$, $p_{Ey2} = -0.6046$, $p_{Ey3} = 0.09854$, $p_{Ey4} = -6.697$, $p_{Ey5} = 0$
$p_{Ky1} = -15.324$, $p_{Ky2} = 1.715$, $p_{Ky3} = 0.3695$, $p_{Ky4} = 2.0005$, $p_{Ky5} = 0$, $p_{Ky6} = -0.8987$, $p_{Ky7} = -0.23303$
$p_{Hy1} = -0.001806$, $p_{Hy2} = 0.00352$, $p_{Vy1} = -0.00661$, $p_{Vy2} = 0.03592$, $p_{Vy3} = -0.162$, $p_{Vy4} = -0.4864$
$p_{py1} = -0.6255$, $p_{py2} = -0.06523$, $p_{py3} = -0.16666$, $p_{py4} = 0.2811$, $p_{py5} = 0$
$r_{By1} = 10.622$, $r_{By2} = 7.82$, $r_{By3} = 0.002037$, $r_{By4} = 0$, $r_{Cy1} = 1.0587$
$r_{Ey1} = 0.3148$, $r_{Ey2} = 0.004867$, $r_{Hy1} = 0.009472$, $r_{Hy2} = 0.009754$
$r_{Vy1} = 0.05187$, $r_{Vy2} = 4.853e\text{-}4$, $r_{Vy3} = 0$, $r_{Vy4} = 94.63$, $r_{Vy5} = 1.8914$, $r_{Vy6} = 23.8$

**TABLE A3.1** TNO MF-Tire/MF-Swift 6.1 Tire Model Parameters—Cont'd

| Tire Designation: 205/60R15 91 V |
|---|
| [rolling_coefficients]<br>$q_{sy1} = 0.00702$, $q_{sy2} = 0$, $q_{sy3} = 0.001515$, $q_{sy4} = 8.514\text{e-}5$, $q_{sy5} = 0$<br>$q_{sy6} = 0$, $q_{sy7} = 0.9008$, $q_{sy8} = -0.4089$ |
| [aligning_coefficients]<br>$q_{Bz1} = 12.035$, $q_{Bz2} = -1.33$, $q_{Bz3} = 0$, $q_{Bz4} = 0.176$, $q_{Bz5} = -0.14853$, $q_{Bz9} = 34.5$,<br>$q_{Bz10} = 0$<br>$q_{Cz1} = 1.2923$, $q_{Dz1} = 0.09068$, $q_{Dz2} = -0.00565$, $q_{Dz3} = 0.3778$, $q_{Dz4} = 0$,<br>$q_{Dz6} = 0.0017015$<br>$q_{Dz7} = -0.002091$, $q_{Dz8} = -0.1428$, $q_{Dz9} = 0.00915$, $q_{Dz10} = 0$, $q_{Dz11} = 0$<br>$q_{Ez1} = -1.7924$, $q_{Ez2} = 0.8975$, $q_{Ez3} = 0$, $q_{Ez4} = 0.2895$, $q_{Ez5} = -0.6786$<br>$q_{Hz1} = 0.0014333$, $q_{Hz2} = 0.0024087$, $q_{Hz3} = 0.24973$, $q_{Hz4} = -0.21205$<br>$p_{pz1} = -0.4408$, $p_{pz2} = 0$<br>$s_{sz1} = 0.00918$, $s_{sz2} = 0.03869$, $s_{sz3} = 0$, $s_{sz4} = 0$ |
| [turnslip_coefficients]<br>$p_{Dx\varphi1} = 0.4$, $p_{Dx\varphi2} = 0$, $p_{Dx\varphi3} = 0$<br>$p_{Ky\varphi1} = 1$, $p_{Dy\varphi1} = 0.4$, $p_{Dy\varphi2} = 0$, $p_{Dy\varphi3} = 0$, $p_{Dy\varphi4} = 0$<br>$p_{Hy\varphi1} = 1$, $p_{Hy\varphi2} = 0.15$, $p_{Hy\varphi3} = 0$, $p_{Hy\varphi4} = -4$<br>$p_{E\gamma\varphi1} = 0.5$, $p_{E\gamma\varphi2} = 0$<br>$q_{Dt\varphi1} = 10$, $q_{Cr\varphi1} = 0.2$, $q_{Cr\varphi2} = 0.1$, $q_{Br\varphi1} = 0.1$, $q_{Dr\varphi1} = 1$ |

## 1. Constants

As can be seen from Table A3.1, many parameters of the tire model are non-dimensional. This is achieved by introducing four reference parameters: $r_o$, $F_{zo}$, $V_o$, and $p_{io}$. Parameter $r_o$ represents the unloaded tire radius of the non-rolling tire. Parameter $F_{zo}$ denotes the nominal load on the tire. It does not exactly have to match the actual load of the tire, but it should be in the same range. This ensures that no extreme values for the *Magic Formula* parameters will be required, which makes parameter identification faster and more robust. A typical value would be 4000 N for a passenger car tire. Parameter $V_o$ represents a reference velocity; it is used to make parameters governing velocity-dependent effects dimensionless, e.g. rolling resistance, tire centrifugal growth, and velocity dependent changes in stiffnesses. Normally, $V_o$ is set to the velocity at which the Force and Moment testing is done – so typically 60 km/h or 16.7 m/s. The nominal inflation pressure $p_{i0}$ is required in the description of inflation pressure-dependent effects. If the tire is evaluated for a single tire pressure, $p_{io}$ should be set to this value. If a more elaborate testing program is done at different tire pressures, e.g. $p_{io}$–0.5 bar, $p_{io}$, and $p_{io}$ + 0.5 bar, then the tire pressure $p_i$ in the simulation model can take any value within this range; otherwise $p_i = p_{io}$.

Please note that $r_o$, $F_{zo}$, $V_o$, and $p_{io}$ should be selected with care before starting the parameter identification process and should absolutely be kept unchanged afterward. Changing them would require completely redoing the identification process!

## 2. Force and moment testing

The *Magic Formula* parameters are identified using Force and Moment tests. These tests can be executed on the road with a tire test trailer or in a lab using a flat-track tire tester, (cf. Chapter 12). In Table A3.1, the *Magic Formula* parameters start with [longitudinal_coefficients] up to [aligning_coefficients]. The parameters regarding turn slip, identified with [turnslip_coefficients], also belong to the *Magic Formula*, but work is still ongoing to define a suitable test and identification procedure.

***Some points of attention***:

*Velocity dependency*

Note that the contribution of the forward velocity $V_x$ is present in the rolling resistance formula (9.231):

$$f_r = q_{sy1} + q_{sy3}|V_x/V_o| + q_{sy4}(V_x/V_o)^4 \tag{9.231}$$

that is used in connection with Eqns (9.230, 9.236). Of course, in cases such as moving on wet roads, the friction coefficient may be formulated as functions of the speed, cf. Eqn (4.E23).

*Symmetric tire behavior*

The measured tire characteristics may be not entirely symmetric, for example $F_y(\alpha,\gamma) \neq -F_y(-\alpha,-\gamma)$. This can be caused the tire characteristics conicity and ply steer or inaccuracies in the measurements. In some cases, it is preferable to eliminate these offsets and asymmetry and have a completely symmetric tire in the simulation environment. In that case the following parameters should be set zero and kept zero in the identification process: $r_{Hx1}$, $q_{sx1}$, $p_{Ey3}$, $p_{Hy1}$, $p_{Hy2}$, $p_{Vy1}$, $p_{Vy2}$, $r_{By3}$, $r_{Vy1}$, $r_{Vy2}$, $q_{Bz4}$, $q_{Dz6}$, $q_{Dz7}$, $q_{Ez4}$, $q_{Hz1}$,$q_{Hz2}$, $s_{sz1}$, and $q_{Dz3}$.

*Incomplete measurement set*

The Force and Moment testing can be divided into three different types of tests:

- Kappa sweep: variation of the longitudinal slip $\kappa$, while keeping the side slip angle $\alpha$ equal to zero.
- Alpha sweep: variation of the slip angle $\alpha$ for a freely rolling tire, ($\kappa = 0$).
- Combined slip: variation of the longitudinal slip $\kappa$ for non-zero values of the side slip angle $\alpha$.

The structure of the *Magic Formula* is such that different parts of the formulas are responsible for modeling different aspects of the tire behavior.

All parameters in the [longitudinal_coefficients] section starting with a $p$ are determined by kappa sweeps, combined slip is modeled by coefficients starting with an $r$ and $s$, etc.

A situation, which may occur in practice, is that only alpha sweeps are available but no measurements including longitudinal slip $\kappa$ variations. Then it is still possible to combine the parameters of an existing tire with the newly identified *Magic Formula* parameters for the alpha sweeps to obtain a new tire parameter set. This set then combines the longitudinal and combined slip parameters of the existing tire with the lateral behavior based on the latest measurements. Practical experience has shown that the parameters of the combined slip weighting functions are relatively constant for various tires.

It is clear that the approach described above is not ideal, but may be necessary due to a restricted number of measurements being available. Once combining different tire parameter sets, it is in any case important to visually check the resulting tire characteristics. Another consideration is the application of the tire model: e.g. if only straight line simulations are done, the accuracy of the lateral tire behavior may be less important.

*No measurement data at all*

In some cases no Force and Moment data are available at all, e.g. when considering the tires of a fork lift truck, agricultural vehicle, etc. Still, wanting to simulate such a vehicle, an estimate for some *Magic Formula* parameters can be made and it can be improved when measurement data become available.

To start, all *Magic Formula* parameters are set to zero, with the following exceptions: $p_{Ky2} = 2$, $q_{sy7} = 1$, and $q_{sy8} = 1$. For the combined slip weighting functions, the following values are suggested: $r_{bx1} = 8.3$, $r_{Bx2} = 5$, $r_{Cx1} = 0.9$, $r_{By1} = 4.9$, $r_{By2} = 2.2$, and $r_{Cy1} = 1$.

The friction coefficient in longitudinal and lateral directions at the nominal vertical tire force is controlled by $p_{Dx1}$ and $p_{Dy1}$, respectively. Generally, the friction coefficient for highly loaded tires is below 1, possibly 0.8 on a dry surface, whereas for high performance tires used on motorcycles or on racing cars the friction coefficient may become 1.5 or even 2 in extreme cases. If desired, load dependency of the friction coefficient can be introduced by adapting $p_{Dx2}$ and $p_{Dy2}$. Note that generally $p_{Dy2} < p_{Dx2}$ and that they normally are below zero (typically between $-0.1$ and 0). If different values are used for $p_{Dx2}$ and $p_{Dy2}$, it is good to ensure that, at a vertical load tending to zero, the longitudinal and lateral friction coefficients become equal to each other.

The longitudinal slip stiffness is for many tires (almost) linearly dependent on the vertical force. This is controlled by parameter $p_{Kx1}$; typically, the value of this parameter is in the range of 14 to 18, although it may be much higher for racing tires. The dependency of the cornering stiffness on the vertical tire force is less linear. The shape is controlled by $p_{Ky1}$ and $p_{Ky2}$ as shown in the left (picture) of Figure 4.12. The maximum cornering stiffness normally does not occur at the nominal vertical force but is generally higher, so $p_{Ky2}$ may be in the range 1.5 to 3. The maximum value of the cornering stiffness is controlled by $p_{Ky1}$ for which no narrow range can be given: expect values between 10 and 20 (or higher for racing tires). In the TNO MF-Tire model, the ISO sign convention is used and then $p_{Ky1}$ should be negative.

Finally, the shape of the longitudinal slip curve is determined by parameters $p_{Cx1}$ and $p_{Ex1}$ and the same applies to the lateral direction: parameters $p_{Cy1}$ and $p_{Ey1}$ control the shape of the lateral force as a function of the side slip angle. Taking $p_{Cx1} = 1.6$ and $p_{Cy1} = 1.3$ and leaving $p_{Ex1}$ and $p_{Ey1}$ equal to zero should be a fair starting point. The effect of modifications to $p_{Ex1}$ and $p_{Ey1}$ is illustrated by Figure 4.10: an increasing negative $E$ value will make the resulting characteristic more 'peaky'. In any case, $p_{Cx1} \geq 1, p_{Cy1} \geq 1, p_{Ex1} \leq 1$ and $p_{Ey1} \leq 1$.

## 3. Loaded radius/effective rolling radius

Figure A3.1 illustrates the various tire radii: the free tire radius of the rotating tire $r_\Omega$, the loaded tire radius $r_l$ and effective rolling radius $r_e$.

First, centrifugal growth of the free tire radius $r_\Omega$ is calculated using the following formula:

$$r_\Omega = r_o\left(q_{reo} + q_{V1}\left(\frac{r_o\Omega}{V_o}\right)^2\right) \tag{A3.1}$$

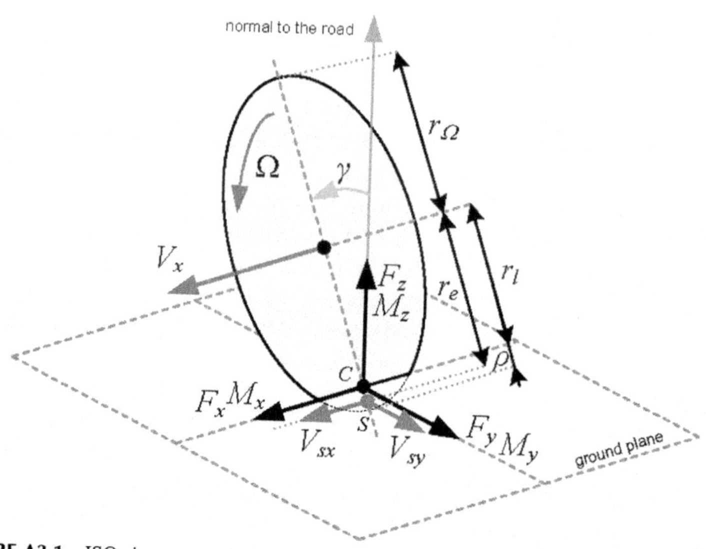


**FIGURE A3.1** ISO sign convention for the tire forces and moments and definition of tire radii.

The parameter $q_{re0}$ is introduced to adapt the free unloaded radius to the measurements. The tire normal deflection $\rho$ is the difference between the free tire radius of the rotating tire $r_\Omega$ and the loaded tire radius $r_l$ (cf. Eqn (9.218)):

$$\rho = \max(r_\Omega - r_l, 0) \quad (A3.2)$$

The vertical tire force $F_z$ is then calculated using the following formula:

$$\begin{aligned} F_z &= \left\{1 + q_{V2}|\Omega|\frac{r_o}{V_o} - \left(q_{Fcx1}\frac{F_x}{F_{zo}}\right)^2 - \left(q_{Fcy1}\frac{F_y}{F_{zo}}\right)^2\right\} \\ &\cdot F_{zo}\left(q_{Fz1}\frac{\rho}{r_o} + q_{Fz2}\frac{\rho^2}{r_o^2}\right)(1 + p_{Fz1}\mathrm{d}p_i) \end{aligned} \quad (A3.3)$$

Various effects are included in this expression: a stiffness increase with velocity ($q_{v2}$), vertical sinking due to longitudinal and lateral forces ($q_{Fcx1}$ and $q_{Fcy1}$), a quadratic force deflection characteristic ($q_{Fz1}$ *and* $q_{Fz2}$) and the influence of the tire inflation pressure ($p_{Fz1}$). The vertical stiffness $c_{zo}$ at the nominal vertical load, nominal inflation pressure, no tangential forces, and zero forward velocity can be calculated as

$$c_{zo} = \frac{F_{zo}}{r_o}\left(q_{Fz1} + 2q_{Fz2}\frac{\rho_o}{r_o}\right) = \frac{F_{zo}}{r_o}\sqrt{q_{Fz1}^2 + 4q_{Fz2}} \quad (A3.4a)$$

The derivation of the last expression is given in section 10 at the end of this Appendix. In the expressions for the effective rolling radius and contact patch dimensions, the vertical stiffness adapted for tire inflation pressure is used:

$$c_z = c_{zo}(1 + p_{Fz1}\mathrm{d}p_i) \quad (A3.5)$$

The effective rolling radius, similar to (9.232), is calculated as

$$r_e = r_\Omega - \frac{F_{zo}}{c_z}\left\{F_{\text{reff}}\frac{F_z}{F_{zo}} + D_{\text{reff}}\arctan\left(B_{\text{reff}}\frac{F_z}{F_{zo}}\right)\right\} \quad (A3.6)$$

The parameters $D_{\text{reff}}$, etc., can be found in Table A3.1.

The measurement data for assessing the loaded and the effective rolling radius typically consist of values for $r_l$, $r_e$, $V$, $F_x$, $F_y$, and $F_z$ carried out for a number of different vertical loads, forward velocities, and possibly longitudinal and/or lateral forces. Based on the relation $V = \Omega r_e$, it is sufficient to specify two out of three variables: forward velocity $V$, angular velocity $\Omega$, and effective rolling radius $r_e$. The data for $F_x$ and $F_y$ can be considered optional for passenger car tires, but generally should be included for racing tires. Fitting of the parameters is generally done in two steps. In order to obtain maximum accuracy for the loaded radius, the Eqns (A3.1) and (A3.3) are fitted first. In a second step, the coefficients for the effective rolling radius, as given by Eqn (A3.6), are determined.

When only a tire radius $r_o$ and vertical stiffness $c_{zo}$ are available, many parameters can be set to zero: $q_{V1}$, $q_{V2}$, $q_{Fcx1}$, $q_{Fcy1}$, $q_{Fz2}$, and $p_{Fz1}$ while $q_{reo}$ is set to one. In this case, the parameter $q_{Fz1}$ can be calculated easily using eqn (A3.4) since $q_{Fz2} = 0$. The parameters in the expression for the effective rolling radius ($B_{reff}$, $D_{reff}$, and $F_{reff}$) may be set to zero, but for a radial tire the following values are suggested: $B_{reff} = 8$, $D_{reff} = 0.24$, and $F_{reff} = 0.01$. For a bias ply tire: $B_{reff} = 0$, $D_{reff} = 0$, and $F_{reff} = 1/3$. To include the effect of a changing tire inflation pressure on the tire vertical stiffness, the parameter $p_{Fz1}$ needs to be non-zero; the suggested value would be in the range of 0.7 to 0.9.

## 4. Contact patch dimensions

The following formulas are used for the contact patch dimensions. The length $a$ represents half of the contact length and $b$ half of the width of the contact patch:

$$a = r_o\left(q_{ra2}\frac{F_z}{c_z r_o} + q_{ra1}\sqrt{\frac{F_z}{c_z r_o}}\right) \approx r_0\left(q_{ra2}\frac{\rho}{r_o} + q_{ra1}\sqrt{\frac{\rho}{r_o}}\right) \tag{A3.7}$$

$$b = w\left(q_{rb2}\frac{F_z}{c_z r_o} + q_{rb1}\left(\frac{F_z}{c_z r_o}\right)^{\frac{1}{3}}\right) \approx w\left(q_{rb2}\frac{\rho}{r_o} + q_{rb1}\left(\frac{\rho}{r_o}\right)^{\frac{1}{3}}\right) \tag{A3.8}$$

Consequently, measurement data should provide half the contact length and half the contact width as a function of the vertical load $F_z$. Actually, the contact length is not so much directly dependent on the vertical tire force, but is more a geometrical property and dependent on the vertical tire deflection $\rho$, as indicated in the right-hand side part of Eqns (A3.7) and (A3.8). If no measurement data are available, based on the work of Besselink (2000) the following values are suggested: $q_{ra2} = 0.35$ and $q_{ra1} = 0.79$.

## 5. Overall longitudinal and lateral stiffness, relaxation lengths

Relaxation behavior of the tire can be modeled using a first order differential equation and an explicit expression for the relaxation length as a function of the vertical load. When rigid ring dynamics are included on the other hand, the relaxation lengths become a function of the longitudinal and lateral stiffness of the tire, respectively. As the TNO MF-Tire/MF-Swift 6.1 tire model is able to handle both conditions, the relaxation length is included in the model using explicit expressions for the non-rolling longitudinal and lateral stiffness:

$$\sigma_x = \frac{C_{F\kappa}}{c_x} \tag{A3.9}$$

$$\sigma_y = \frac{C_{F\alpha}}{c_y}$$

where $\sigma_x$ is the longitudinal relaxation length, $C_{F\kappa}$ the longitudinal slip stiffness of the tire, $c_x$ the nonrolling longitudinal stiffness at ground level, $\sigma_y$ the lateral relaxation length, $C_{F\alpha}$ the cornering stiffness of the tire, $c_y$ the nonrolling lateral stiffness at ground level.

The stiffnesses are made dependent on both the vertical force $df_z$ and the inflation pressure increment $dp_i$:

$$c_x = c_{xo}(1 + p_{cfx1}df_z + p_{cfx1}df_z^2)(1 + p_{cfx3}dp_i) \quad \text{(A3.10)}$$

$$c_y = c_{yo}(1 + p_{cfy1}df_z + p_{cfy1}df_z^2)(1 + p_{cfy3}dp_i) \quad \text{(A3.11)}$$

where $c_{xo}$ and $c_{yo}$ are the longitudinal and lateral stiffness of the tire at the nominal vertical force and inflation pressure.

In principle, the stiffness $c_{xo}$ and $c_{yo}$ may be measured on a non-rolling tire, but in general the relaxation length based on the non-rolling lateral stiffness is too short compared to the measurements on a rolling tire. The preferred approach therefore is to measure the cornering stiffness and lateral relaxation length in a transient test for a number of different inflation pressures and vertical loads. Subsequently, the overall lateral stiffness equation can be fitted to these measurement points accordingly.

In case no transient tests are available, the non-rolling stiffnesses can still be used to model relaxation behavior, albeit with a reduced accuracy. If no non-rolling static stiffness tests are available, a first, crude rule of thumb is that the lateral stiffness $c_{yo}$ equals half of the vertical stiffness $c_{zo}$ and the longitudinal stiffness $c_{xo}$ is twice the vertical stiffness. To include the effect of the inflation pressure on these stiffnesses, the parameters $p_{cfx3}$ and $p_{cfy3}$ can be set to 0.2 and 0.5, respectively, as a first estimate.

## 6. Tire inertia

When rigid ring dynamics are employed, the tire is not considered as a single rigid body any more. A part of the tire near the rim is assumed to be rigid and fixed to the rim. The other part that represents the belt is considered to move as a rigid ring. Consequently, the total tire mass has to be divided and a distribution has to be made over the inertia of the ring and the inertia of the part rigidly attached to the rim. The latter part is further regarded as a part of the wheel body. An estimate of this division can be made based on past experience or on a detailed weight breakdown provided by the tire manufacturer. The following rough initial estimate is suggested:

75% of the tire mass is assigned to the tire belt
85% of the tire moments of inertia is assigned to the tire belt

Tentatively, the contact patch body is considered as an additional small mass.

## 7. Carcass compliances

The rigid belt ring is elastically suspended with respect to the rim in all directions. So, various stiffnesses are associated with these motions: $c_{bx}$, $c_{by}$, $c_{b\theta}$, and $c_{b\gamma}$ as is illustrated in Figure A3.2.

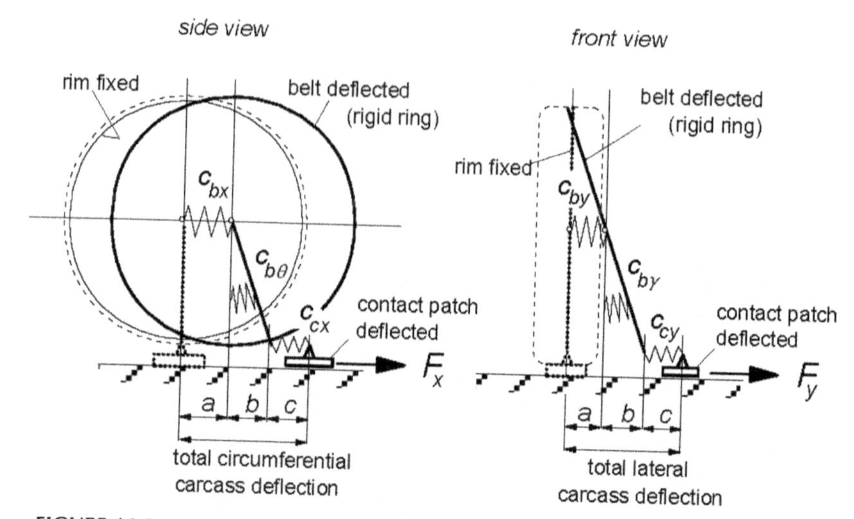


**FIGURE A3.2** Springs connecting rigid ring to the rim and contact patch to the rigid ring.

In the tire property file, the choice has been made to allow the user to specify the eigenfrequencies $f$ [Hz] of the belt for a tire not in contact with the ground and the rim fixed. The corresponding stiffnesses then become

$$c_{bx} = c_{bz} = 4\pi^2 m_{\text{belt}} f_{\text{long}}^2 \tag{A3.12}$$

$$c_{by} = 4\pi^2 m_{\text{belt}} f_{\text{lat}}^2 \tag{A3.13}$$

$$c_{b\theta} = 4\pi^2 I_{yy,\text{belt}} f_{\text{windup}}^2 \tag{A3.14}$$

$$c_{b\gamma} = c_{b\psi} = 4\pi^2 I_{zz,\text{belt}} f_{\text{yaw}}^2 \tag{A3.15}$$

where $f_{\text{long}}$ equals the eigenfrequency (in Hz) of the longitudinal/vertical motion of the tire belt, $f_{\text{lat}}$ equals the eigenfrequency of the lateral motion, $f_{\text{windup}}$ the eigenfrequency of the in-plane torsional mode, and $f_{\text{yaw}}$ the eigenfrequency of the yaw (and roll) mode of the belt. Obviously, $m_{\text{belt}}$ refers to the mass of the belt and $I_{yy,\text{belt}}$ and $I_{zz,\text{belt}}$ to the belt moment of inertia about the $y$- and $z$-axes. The natural frequencies are assessed for a free unloaded tire mounted on a rim fixed with respect to space.

As can be seen from Figure A3.2, the deflection of the contact patch is the result of the deflection of a number of successive springs. The expressions for the overall longitudinal and lateral stiffness have already been given in step (5). Therefore, the stiffness of the residual springs $c_{cx}$ and $c_{cy}$ should be chosen such that the overall stiffnesses become correct. The expression for the overall longitudinal stiffness reads

$$\frac{1}{c_x} = \frac{1}{c_{bx}} + \frac{r_l^2}{c_{b\theta}} + \frac{1}{c_{cx}} + \frac{a}{C_{F\kappa}} \tag{A3.16}$$

In the lateral direction, the overall stiffness becomes

$$\frac{1}{c_y} = \frac{1}{c_{by}} + \frac{r_l^2}{c_{b\gamma}} + \frac{1}{c_{cy}} + \frac{a}{C_{F\alpha}} \tag{A3.17}$$

Making sure that Eqn (A3.16) and (A3.17) hold for all vertical loads and inflation pressures is not trivial. In any case, it is clear that from a physical point of view the residual springs never should get a negative stiffness.

If no measurement data are available to identify the eigenfrequencies of the tire belt, another approach can be followed to determine the stiffness of the various springs. Based on past measurement results, the contribution of the various springs on the overall carcass deflection at ground level has been identified, see Table A3.2. Here it is assumed that a longitudinal or lateral force is applied at ground level and that the rim is held fixed, cf. Figure A3.2.

Table A3.2 shows the relative contributions *a*, *b*, and *c* that have been assessed for three different tires. Based on these data, it may be concluded that although the compliance values themselves are different, the relative contributions to the overall carcass deflection are fairly constant. Since in step (5) the overall longitudinal and lateral carcass stiffnesses have been determined, it is now possible to assess approximate individual stiffness values using the suggested 'rule of thumb'.

Additional measurement data (e.g. yaw stiffness or *FEM* model results) may also be used to enhance the estimates or to gain additional confidence in the stiffness/deflection distribution.

**TABLE A3.2 Distribution of Longitudinal and Lateral Carcass Compliance Components**

| **Longitudinal at Ground Level** | **Tire 1** | **Tire 2** | **Tire 3** | **Rule of Thumb** |
|---|---|---|---|---|
| Rigid ring translation (a) | 27% | 28% | 31% | 30% |
| Rigid ring rotation (b) | 63% | 62% | 60% | 60% |
| Contact patch translation (c) | 10% | 10% | 9% | 10% |
| **Lateral at Ground Level** | **Tire 1** | **Tire 2** | **Tire 3** | **Rule of Thumb** |
| Rigid ring translation (a) | 34% | 25% | 27% | 25% |
| Rigid ring rotation (b) | 62% | 56% | 55% | 55% |
| Contact patch translation (c) | 4% | 19% | 18% | 20% |

## 8. Carcass damping

Generally, the exact amount of damping of the tire is very difficult to assess experimentally. For instance, it is observed that a large difference in vertical damping exists between a tire that stands still and a tire that rolls: under rolling conditions the apparent damping may be a factor 10 smaller compared to the damping of a tire standing still, also cf. Jianmin et al. (2001). A simple model with finite contact length and provided with radial dry friction dampers (Pacejka 1981a) may explain this phenomenon.

Based on experience, the following guideline may be provided: when the tire is not in contact with the ground (and the rim is fixed) the damping will be relatively low. Typical values of the damping coefficients lie in the range of 1 to 6% of the corresponding critical damping coefficients. Generally, the modes that contain a large translational component are more heavily damped than the modes with a large rotational component.

In the tire property file, the dimensionless damping coefficient for the four vibration modes is specified: $\zeta_{\text{long}}$, $\zeta_{\text{lat}}$, $\zeta_{\text{yaw}}$, and $\zeta_{\text{windup}}$. Given the observations above, their values should normally be in the range of 0.01 to 0.06.

## 9. Enveloping

The enveloping model uses elliptical cams that approximate the tire contour at the leading and trailing parts of the contact point. The shape of the ellipse is described by

$$\left(\frac{x_e}{a_e}\right)^{c_e} + \left(\frac{z_e}{b_e}\right)^{c_e} = 1 \tag{A3.18}$$

with $a_e = p_{ae} r_o$ and $b = p_{be} r_o$. The parameter $p_{ae}$ defines the length of the ellipse and $p_{be}$ the height of the ellipse. As both parameters are dimensionless, the size of the ellipse scales linearly with the tire radius $r_o$, which is plausible from a physical point of view. When no measurements are available, the following parameters could be used as a starting point: $p_{ae} = 1$, $p_{be} = 1$, and $c_e = 1.8$.

The distance between the ellipses is given by $l_s$. This distance depends on the contact length $2a$. The following relation is used:

$$l_s = 2 p_{ls} a \tag{A3.19}$$

If no measurement data are available, the suggested value for $p_{ls}$ is 0.8.

## 10. Derivation of expression for vertical stiffness

Starting with (A3.3) for the vertical force and eliminating all velocity, force, and inflation pressure dependent effects, results in

$$F_z = F_{zo}\left(q_{Fz1}\frac{\rho}{r_o} + q_{Fz2}\frac{\rho^2}{r_o^2}\right) \tag{A3.20}$$

When the vertical force is equal to the nominal force, ($F_z = F_{zo}$) the next equation holds for the vertical tire deflection:

$$q_{Fz2}\frac{\rho^2}{r_o^2} + q_{Fz1}\frac{\rho}{r_o} - 1 = 0 \tag{A3.21}$$

The possible solutions to this equation are

$$\frac{\rho}{r_o} = \frac{-q_{Fz1} \pm \sqrt{q_{Fz1}^2 + 4q_{Fz2}}}{2q_{Fz2}} \tag{A3.22}$$

The vertical stiffness derived from (A3.20) is given by

$$\frac{\partial F_z}{\partial \rho} = F_{zo}\left(\frac{q_{Fz1}}{r_o} + \frac{2q_{Fz2}}{r_o}\left(\frac{\rho}{r_o}\right)\right) \tag{A3.23}$$

The vertical stiffness at the nominal vertical force $F_{zo}$ can be obtained by substituting (A3.22) in (A3.23):

$$c_{zo} = F_{zo}\left(\frac{q_{Fz1}}{r_o} + \frac{2q_{Fz2}}{r_o}\left(\frac{-q_{Fz1} \pm \sqrt{q_{Fz1}^2 + 4q_{Fz2}}}{2q_{Fz2}}\right)\right) \tag{A3.24}$$

After simplification, this results in the second expression of (A3.4b), as a positive stiffness is the only physical solution:

$$c_{zo} = \frac{F_{zo}}{r_o}\sqrt{q_{Fz1}^2 + 4q_{Fz2}} \tag{A3.4b}$$

This expression is of interest because at given $c_{zo}$ and $q_{Fz2}$ (cf. the tire property file shown in Table A3.1) $q_{Fz1}$ and with this the vertical force (A3.3) can be calculated.

# Index

*Note*: Page numbers followed by *f* indicate figures, *t* indicate tables and *b* indicate boxes.

# T

## U



## V



## W



## Y



## Z

Printed in the United States
By Bookmasters